Texts in Applied Mathematics

Volume 82

The mathematization of all sciences, the fading of traditional scientific boundaries, the impact of computer technology, the growing importance of computer modelling and the necessity of scientific planning all create the need both in education and research for books that are introductory to and abreast of these developments. The aim of this series is to provide such textbooks in applied mathematics for the student scientist. Books should be well illustrated and have clear exposition and sound pedagogy. Large number of examples and exercises at varying levels are recommended. TAM publishes textbooks suitable for advanced undergraduate and beginning graduate courses, and complements the Applied Mathematical Sciences (AMS) series, which focuses on advanced textbooks and research-level monographs.

Qinian Jin

Lectures on Nonsmooth Optimization

Springer

Qinian Jin
Mathematical Sciences Inst
Australian National Univ
Canberra, ACT, Australia

ISSN 0939-2475 ISSN 2196-9949 (electronic)
Texts in Applied Mathematics
ISBN 978-3-031-91416-4 ISBN 978-3-031-91417-1 (eBook)
https://doi.org/10.1007/978-3-031-91417-1

Mathematics Subject Classification: 65Y20, 90C25, 90C26, 90C06, 65K05

This Springer imprint is published by the registered company Springer Nature Switzerland AG
The registered company address is: Gewerbestrasse 11, 6330 Cham, Switzerland

In memory of my mother Xianlan Zhou
(20/04/1951–13/02/2024)

Preface

This book provides an in-depth exploration of nonsmooth optimization, covering foundational algorithms, theoretical insights, and a wide range of applications. Nonsmooth optimization, characterized by nondifferentiable objective functions or constraints, plays a crucial role across various fields, including machine learning, imaging, inverse problems, statistics, optimal control, and engineering. Its scope and relevance continue to expand, as many real-world problems are inherently nonsmooth or benefit significantly from nonsmooth regularization techniques.

Nonsmooth optimization problems frequently arise in real-world applications. For instance, in inverse problems, nonsmooth regularization techniques stabilize solutions and enhance interpretability, particularly in imaging, where data is often noisy or incomplete. In fields such as medical imaging, nonsmooth methods help in reconstructing clear images from incomplete or corrupted data, thereby enabling better diagnostic tools. In machine learning, nonsmooth functions that promote sparsity or robustness are essential for developing effective models that can generalize well to unseen data. Similarly, in statistics, nonsmooth optimization plays a critical role in tasks like parameter estimation and model selection, where the need to balance accuracy and computational efficiency is paramount. These challenges demand innovative algorithmic designs and a deep understanding of the underlying theory. This book serves as a comprehensive guide to nonsmooth optimization, spanning theoretical foundations, algorithmic development, and diverse applications, offering both depth and practical insights.

Organized with a systematic, progressive structure, this book leads readers from fundamental concepts in nonsmooth optimization to advanced methods for tackling complex problems. Having had the privilege of teaching nonsmooth optimization to students at the Australian National University (ANU), I designed this book to maintain the clarity and accessibility that I found effective in the classroom. Lecturing on this subject has highlighted the importance of striking a balance between rigorous theory and practical examples, and I have endeavored to make this balance central to the book. As a result, each chapter is crafted to be as self-contained as possible, allowing readers to selectively engage with specific topics or follow a structured progression from foundational topics to specialized applications. To make learning more accessible, concepts in convex analysis are introduced and developed gradually, integrated within relevant chapters rather than consolidated into a single section. This allows readers to encounter these tools within context, easing comprehension and application to real-world problems. I hope that this structure will provide a more seamless and accessible learning experience, especially for students approaching this area for the first time.

The book is organized into 11 chapters, progressing from basic principles to more advanced methods and recent developments.

This book offers several distinctive features compared to other texts on optimization. First, it emphasizes recent developments in both convex and non-convex optimization, with a dedicated chapter focusing on methods based on the Kurdyka-Lojasiewicz (KL) conditions. These conditions provide a versatile framework for analyzing the convergence of non-convex algorithms, enabling us to extend the reach of optimization techniques into more complex problem settings. The book dedicates considerable attention to understanding these conditions, how they can be applied to algorithm design, and why they matter for practical problem-solving. By exploring convergence under these conditions, the book brings readers to the frontier of current research in non-convex optimization. Additionally, this book addresses the rapid evolution of stochastic algorithms for large-scale optimization, which are indispensable in applications ranging from data science to machine learning. These algorithms play a central role in handling the immense volumes of data generated in modern applications, and we explore how recent stochastic approaches are designed to maintain both efficiency and convergence. Another unique aspect of this book is the integration of recent advancements in acceleration methods, including Nesterov's acceleration and primal-dual hybrid gradient methods, which have proven highly effective for nonsmooth optimization tasks.

My research interest primarily centers on the computation of ill-posed inverse problems through regularization methods, where nonsmooth optimization techniques have proven invaluable. This experience has significantly influenced the content of this book, which is intentionally selective, reflecting my personal insights and perspectives developed over many years of teaching and research. Rather than attempting to provide exhaustive coverage of the field, this book offers a carefully curated selection of topics and references, predominantly from recent publications, to showcase both the breadth and depth of nonsmooth optimization. While the reference list is extensive, it is by no means exhaustive; instead, it serves as a curated guide rather than a comprehensive mapping of the field. My aim is for readers to find value in the synthesis presented, which integrates recent results and insights from a variety of publications, thereby contributing to an ongoing dialogue in optimization theory and practice. In a landscape where many excellent books on optimization are available, this text seeks to complement existing resources by highlighting recent developments that represent some of the most exciting and impactful directions in nonsmooth optimization.

This book is written for a wide-ranging audience, including senior undergraduates, graduate students, researchers, and practitioners who are interested in gaining a comprehensive understanding of nonsmooth optimization. For students and newcomers to the field, this book offers an accessible yet rigorous introduction to nonsmooth optimization. The foundational concepts are presented in a clear, systematic manner, with intuitive explanations and step-by-step development of topics like subgradients, convex analysis, and the most commonly used nonsmooth optimization algorithms. Students will find helpful examples and applications that will aid in solidifying their understanding and preparing them to tackle more advanced topics. For researchers and academics, the book explores the theoretical foundations and recent advancements in nonsmooth optimization. It offers in-depth discussions of advanced topics and cutting-edge techniques specifically tailored to solving complex, large-scale problems. For practitioners, particularly those in areas such as imaging, machine learning, engineering, operations research, and data science, the book presents a variety of state-of-the-art algorithms and methods that are directly applicable to real-world optimization problems. These techniques provide practical solutions to some of the most demanding problems encountered in modern industries.

This book can serve as both an in-depth textbook for students new to nonsmooth optimization and a reference for researchers and practitioners. Whether used for learning the basics, exploring advanced theories, or seeking practical techniques for complex optimization challenges, we hope this book will serve as a valuable resource for a wide range of readers dedicated to the study and application of nonsmooth optimization.

This book is built upon the lectures I have delivered at the Australian National University (ANU) over the past decade, refined through interactions with students and ongoing research in the field. I would like to thank my students, whose curiosity and probing questions have continually shaped and enriched my understanding of nonsmooth optimization. Their engagement has been a constant source of inspiration and has significantly influenced the direction of this work.

Finally, I would like to express my deepest gratitude to my family for their unwavering support. Their patience and encouragement have been instrumental in the completion of this book.

Canberra, ACT, Australia
October 2024

Qinian Jin

Contents

Chapter 1
Introduction

This book focuses on optimization problems and the design of algorithms in finite-dimensional real Hilbert spaces. However, most of the results can be extended to infinite dimensional Hilbert spaces either directly or with minor modifications. For clarity, we work within the Euclidean space $\mathbb{R}^n$ of dimension n.

Let us start with some notation conventions and necessary terminologies. Any point $x \in \mathbb{R}^n$ will be written as a column vector, i.e.,

$$x = \begin{pmatrix} x_1 \\ \vdots \\ x_n \end{pmatrix} = (x_1, \cdots, x_n)^T.$$

Recall that $\mathbb{R}^n$ is a vector space under the addition and the scalar multiplication

$$x + y = (x_1 + y_1, \cdots, x_n + y_n)^T, \qquad \lambda x = (\lambda x_1, \cdots, \lambda x_n)^T$$

for any $x = (x_1, \cdots, x_n)^T$ and $y = (y_1, \cdots, y_n)^T$ in $\mathbb{R}^n$ and $\lambda \in \mathbb{R}$.

A function $\|\cdot\| : \mathbb{R}^n \to \mathbb{R}$ is called a *norm* on $\mathbb{R}^n$ if it satisfies the following properties:

(i) $\|x\| \geq 0$ for all $x \in \mathbb{R}^n$; moreover $\|x\| = 0$ if and only if $x = 0$.
(ii) $\|\lambda x\| = |\lambda|\|x\|$ for any $\lambda \in \mathbb{R}$ and $x \in \mathbb{R}^n$.
(iii) $\|x + y\| \leq \|x\| + \|y\|$ for any $x, y \in \mathbb{R}^n$.

The Euclidean space $\mathbb{R}^n$ endowed with a norm is a *normed vector space*. Given a norm $\|\cdot\|$ on $\mathbb{R}^n$, we can define

$$d(x, y) := \|x - y\|, \quad x, y \in \mathbb{R}^n$$

Q. Jin, *Lectures on Nonsmooth Optimization*, Texts in Applied Mathematics 82,
https://doi.org/10.1007/978-3-031-91417-1_1

which is called the *distance* induced by the given norm. For any $\bar{x} \in \mathbb{R}^n$ and $r > 0$, the set

$$B_r(\bar{x}) := \{x \in \mathbb{R}^n : \|x - \bar{x}\| < r\}$$

is called the *open ball* of radius r with center at $\bar{x}$.

Various norms can be defined on the Euclidean space $\mathbb{R}^n$. Given $1 \leq p \leq \infty$, the function

$$\|x\|_p := \begin{cases} \left(\sum_{i=1}^{n} |x_i|^p\right)^{1/p} & \text{if } 1 \leq p < \infty \\ \max_{1 \leq i \leq n} |x_i| & \text{if } p = \infty \end{cases}$$

for any $x = (x_1, \cdots, x_n)^T$ defines a norm on $\mathbb{R}^n$, which is called the *p-norm*. The 2-norm is also called the *Euclidean norm*, and it can be induced from the inner product

$$\langle x, y\rangle := x^T y, \quad \forall x, y \in \mathbb{R}^n$$

by the formula $\|x\|_2^2 = \langle x, x\rangle$. Under this inner product, two vectors $x, y \in \mathbb{R}^n$ are called *orthogonal/perpendicular* if $\langle x, y\rangle = 0$. There holds the *polarization identity*

$$\|x + y\|_2^2 = \|x\|_2^2 + \|y\|_2^2 + 2\langle x, y\rangle$$

for any $x, y \in \mathbb{R}^n$. If $\langle x, y\rangle = 0$, then

$$\|x + y\|_2^2 = \|x\|_2^2 + \|y\|_2^2$$

which is called the *Pythagorean identity*. Moreover, there holds the Cauchy-Schwarz inequality:

$$|\langle x, y\rangle| \leq \|x\|_2 \|y\|_2, \quad x, y \in \mathbb{R}^n.$$

In this book, unless otherwise stated, the notation $\|\cdot\|$ always means the Euclidean norm $\|\cdot\|_2$.

Optimization concerns the minimization or maximization of a given function $f(x)$ defined on a set $\Omega \subset \mathbb{R}^n$. Mathematically, an optimization problem can be formulated as

$$\min_{x \in \Omega} f(x), \tag{1.1}$$

where f is called the *objective function* and Ω is called the *constraint set*. In case $\Omega = \mathbb{R}^n$, (1.1) is called an *unconstrained* optimization problem. Otherwise, (1.1) is called a *constrained* optimization problem.

Definition 1.1 Let f be a real valued function defined on a subset Ω of $\mathbb{R}^n$. A point x^* in Ω is called:

(i) A *global minimizer* of f on Ω if $f(x^*) \le f(x)$ for all $x \in \Omega$
(ii) A *strictly global minimizer* of f on Ω if $f(x^*) < f(x)$ for all $x \in \Omega$ with $x \neq x^*$
(iii) A *local minimizer of f over Ω* if there is a ball $B_r(x^*)$ such that

$$f(x^*) \le f(x) \quad \text{for all } x \in B_r(x^*) \cap \Omega$$

(iv) A *strictly local minimizer of f over Ω* if there is a ball $B_r(x^*)$ such that

$$f(x^*) < f(x) \quad \text{for all } x \in B_r(x^*) \cap \Omega \text{ with } x \neq x^*$$

where $B_r(x^*)$ denotes the open ball of radius r with center at x^*.

The main task in optimization includes (i) determining if the minimum is achieved, (ii) determining the conditions on the minimizers, and (iii) finding the minimizers via efficient numerical algorithms. In this monograph, we will focus on the situation that Ω is a closed subset in $\mathbb{R}^n$ and f is an objective function that is not necessarily differentiable everywhere. We will develop theories and algorithms to deal with such optimization problems.

1.1 Examples

Let us give some examples to illustrate that nonsmooth optimization problems can arise naturally in many situations.

Example 1.2 (Finding Common Points) Let C_1 and C_2 be two closed convex subsets in $\mathbb{R}^n$ with $C_1 \cap C_2 \neq \emptyset$. How to find a common point in C_1 and C_2? If the dimension n is small, this is an easy problem. However, if n is large, it turns out to be a nontrivial problem. To solve this problem in general, we introduce the functions

$$\delta_{C_1}(x) = \begin{cases} 0 & \text{if } x \in C_1, \\ +\infty & \text{otherwise;} \end{cases} \qquad \delta_{C_2}(x) = \begin{cases} 0 & \text{if } x \in C_2, \\ +\infty & \text{otherwise} \end{cases}$$

which are called the *indicator functions* of C_1 and C_2, respectively. Note that

$$\delta_{C_1}(x) + \delta_{C_2}(x) = \begin{cases} 0 & \text{if } x \in C_1 \cap C_2, \\ +\infty & \text{otherwise.} \end{cases}$$

Therefore, our problem can be formulated as the minimization problem:

$$\min_{x\in\mathbb{R}^n} \left\{\delta_{C_1}(x) + \delta_{C_2}(x)\right\}.$$

This is a nonsmooth optimization problem because $\delta_{C_1} + \delta_{C_2}$ is a non-differentiable function.

Example 1.3 (Compressed Sensing) There is an image $f \in \mathbb{R}^n$ to be sent out. When n is large, the storage of f is huge. Choose a basis $\boldsymbol{\varphi_1}, \cdots, \boldsymbol{\varphi_n}$ of $\mathbb{R}^n$ and write

$$f = x_1\varphi_1 + \cdots + x_n\varphi_n.$$

A wise choice of (φ_i) can make all but a few of x_i to be zero. Form $x = (x_1, \cdots, x_n)^T \in \mathbb{R}^n$ and define the counting function

$$\|x\|_0 := \#\{i : x_i \neq 0\}$$

which is the number of nonzero components in x. When $\|x\|_0$ is small, we call x *sparse*.

For security reasons, we encode x before sending out. We choose an $m \times n$ matrix A with $m \ll n$ and form $b = Ax$. Then, $b \in \mathbb{R}^m$ has smaller size and is easy to send. The question is to how to reconstruct f from b. Equivalently, we need to reconstruct x from b by solving the linear system:

$$Ax = b.$$

This system is under-determined, and it has infinitely many solutions. One can reconstruct x by solving the optimization problem:

$$\min \|x\|_0 \quad \text{subject to } Ax = b.$$

Since $\|x\|_0$ is non-convex and non-differentiable, this problem turns out to be very hard to solve. Instead, we may replace $\|x\|_0$ by

$$\|x\|_p^p := |x_1|^p + \cdots + |x_n|^p, \quad p > 0.$$

It is easy to check that:

- $\|x\|_p^p \to \|x\|_0$ as $p \searrow 0$ for any $x \in \mathbb{R}^n$
- $x \to \|x\|_p^p$ is convex for $p \geq 1$ and non-convex for $0 \leq p < 1$

Replacing $\|x\|_0$ by $\|x\|_1$ leads to the nonsmooth convex optimization problem:

$$\min \|x\|_1 \quad \text{subject to } Ax = b.$$

which is called the *basis pursuit* in *compressed sensing*. □

Example 1.4 (De-noising) In signal processing, when a signal is contaminated by noise, we will get a noisy signal y. It is often desirable to get a clean signal from y by removing the noise. Noisy signals are usually oscillatory. The oscillation of a signal $x \in \mathbb{R}^n$ can be reflected by its (discrete) gradient:

$$\begin{pmatrix} x_2 - x_1 \\ x_3 - x_2 \\ \vdots \\ x_n - x_{n-1} \end{pmatrix} = Dx, \text{ where } D = \begin{pmatrix} -1 & 1 & & & \\ & -1 & 1 & & \\ & & \ddots & \ddots & \\ & & & -1 & 1 \end{pmatrix},$$

A clean signal obtained from y should be not far from y with less oscillation. This motivates one of the classical denoising methods that defines the clean signal $\hat{x}$ as the minimizer of the minimization problem:

$$\hat{x} := \arg\min_{x \in \mathbb{R}^n} \left\{ \frac{1}{2}\|x - y\|_2^2 + \frac{\lambda}{2}\|Dx\|_2^2 \right\} \tag{1.2}$$

which is called H^1-denoising, where $\lambda > 0$ is a parameter that is crucial for the accuracy of reconstruction. Equation (1.2) is a quadratic optimization problem whose solution has the following explicit formula:

$$\hat{x} = \left(I + \lambda D^T D\right)^{-1} y.$$

Figure 1.1 presents the original signal of length 4096, the noisy signal, and the reconstruction result by (1.2) with $\lambda = 400$.

When the original signal is piecewise constant, the H^1-denoising may not produce good result. It was suggested by Rudin et al. in [195] to consider (1.2) with $\frac{1}{2}\|Dx\|_2^2$ replaced by

$$\|Dx\|_1 := \sum_{i=1}^{n-1} |x_{i+1} - x_i|$$

which is called the *total variation* of x. This leads to define a clean signal by solving

$$\hat{x} = \arg\min_{x \in \mathbb{R}^n} \left\{ \frac{1}{2}\|x - y\|_2^2 + \lambda\|Dx\|_1 \right\} \tag{1.3}$$

which is called the *total variation (TV) denoising* or the *ROF model* for denoising.

Figure 1.2 presents an original signal of length 4096, the noisy signal, the reconstruction result by (1.2) with $\lambda = 200$, and the reconstruction result by TV-denoising with $\lambda = 20$. TV-denoising clearly gives a better result.

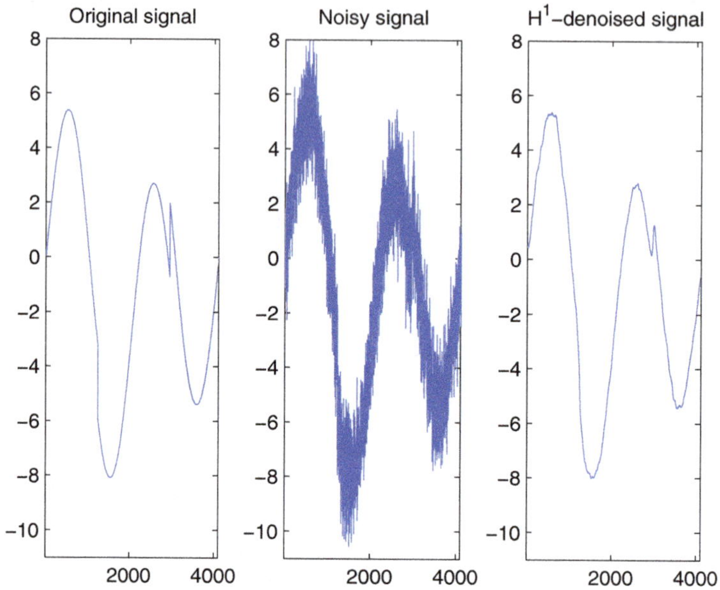

Fig. 1.1 Reconstruction of a smooth signal from a noisy version by the H^1-denoising model (1.2)

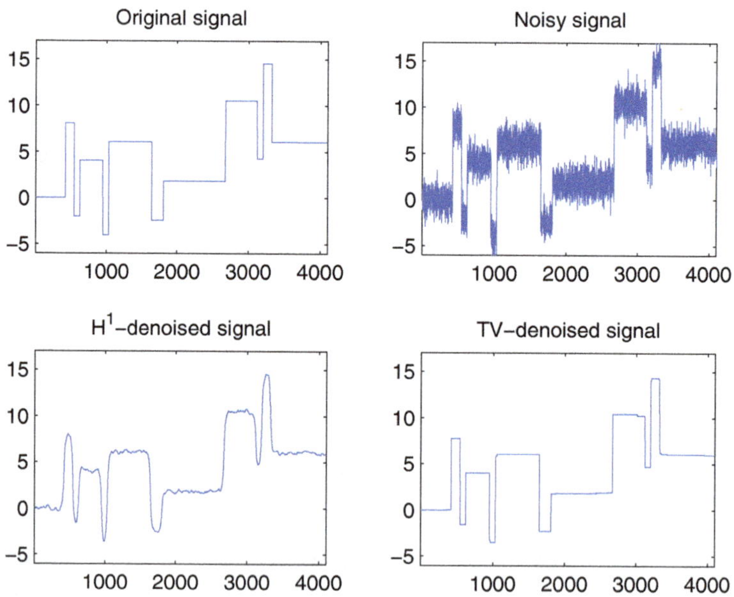

Fig. 1.2 Reconstruction of a piecewise constant signal from a noisy version by the H^1-denoising model (1.2) and the total variation denoising model (1.3)

Since the total variation $x \to \|Dx\|_1$ is non-differentiable, (1.3) is a nonsmooth convex optimization problem. The numerical implementation of (1.3) will be discussed in later chapters.

Example 1.5 (Data Fitting) Given a set of data points $(t_1, y_1), \cdots, (t_m, y_m)$, we want to find a function $y = f(t)$ to approximately fit the data. One common choice of f is a polynomial of the form

$$f(t) = x_1 + x_2 t + \cdots + x_n t^{n-1}$$

for some given n with coefficients $x_1, \cdots, x_n$ to be determined. For each $x = (x_1, \cdots, x_n)^T$, we form the vector of residuals:

$$r = (f(t_1) - y_1, \cdots, f(t_n) - y_n)^T.$$

It is easily seen that $r = Ax - y$, where $y = (y_1, \cdots, y_m)^T$ and $A = (a_{ij})$ with $a_{ij} = t_i^{j-1}, i = 1, \cdots, m, j = 1, \cdots, n$. We may determine f by minimizing $\|r\|$ over x with respect to a suitable norm $\|\cdot\|$. This leads to the optimization problem:

$$\min_{x \in \mathbb{R}^n} \|Ax - y\|.$$

The most common choice of $\|\cdot\|$ is the 2-norm $\|\cdot\|_2$, which leads to the *least squares approximation* problem:

$$\min_{x \in \mathbb{R}^n} \|Ax - y\|_2^2$$

whose solution is given by the equation $A^T Ax = A^T y$.

Another important choice of $\|\cdot\|$ is the 1-norm, which leads to the *ℓ^1-norm approximation* problem:

$$\min_{x \in \mathbb{R}^n} \|Ax - y\|_1.$$

which is a nonsmooth convex optimization problem.

Note that $|x| \gg x^2$ for small x and $x^2 \gg |x|$ for large x. Thus, the ℓ^1-norm approximation puts more emphasis on small residuals and less emphasis on large residuals, compared to the least squares approximation. Thus, the ℓ^1-norm approximation will tend to have more zero and very small residuals and is less sensitive to outliers that occur due to flawed measurements. Therefore, ℓ^1-norm approximation is more suitable when outliers occur. For instance, assume we have the data-set

$$(0, 3), (1, 4), (2, 5), (3, 20), (4, 7), (5, 9), (6, 8), (7, 10).$$

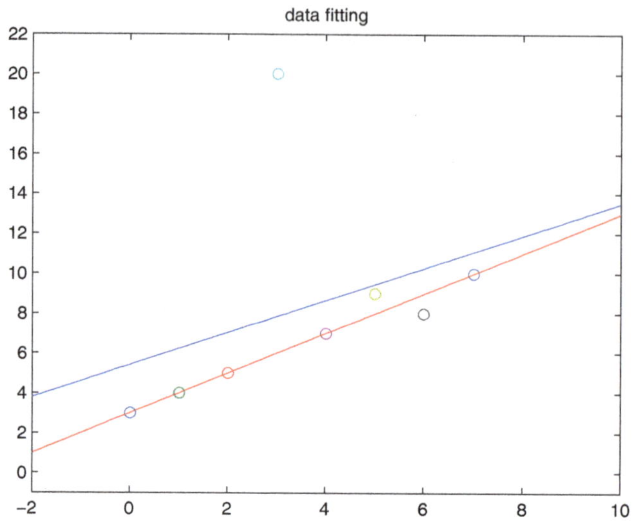

Fig. 1.3 Data fitting: The blue line is the result of least squares approximation and the red line is the result of ℓ^1-norm approximation

We want to find a linear function $f(t) = \alpha_0 + \alpha_1 t$ to approximately fit the data. The least squares approximation gives $\alpha_0 = 5.4167$ and $\alpha_1 = 0.8095$, while the ℓ^1-norm approximation gives $\alpha_0 = 3$ and $\alpha_1 = 1$. See Fig. 1.3; the ℓ^1-norm approximation gives more robust result.

1.2 Existence of Optimizers

A basic question in optimization is to determine the existence of a global minimizer for a given function f. In the following, we will give some sufficient conditions on f to guarantee the existence.

Definition 1.6 A subset Ω in $\mathbb{R}^n$ is called:

(i) *Bounded* if there is a constant $R > 0$ such that $\|x\| \leq R$ for all $x \in \Omega$.
(ii) *Closed* if the limit point of any convergent sequence in Ω lies in Ω.
(iii) *Compact* if any sequence in Ω contains a subsequence that converges to a point in Ω.

The following result gives a characterization of compact sets in $\mathbb{R}^n$.

Lemma 1.7 (Bolzano-Weierstrass Theorem) *A subset of $\mathbb{R}^n$ is compact if and only if it is bounded and closed. Consequently, every bounded sequence in $\mathbb{R}^n$ has a convergent subsequence.*

Proof See any standard textbook on analysis. □

Theorem 1.8 *If f is a continuous function on a compact set Ω in $\mathbb{R}^n$, then f has a global minimizer x^* on Ω, i.e., $f(x^*) \leq f(x)$ for all $x \in \Omega$.*

Proof Let $m := \inf\{f(x) : x \in \Omega\}$. Pick a sequence $\{x_k\} \subset \Omega$ such that $f(x_k) \to m$ as $k \to \infty$. This is possible by the definition of m. Since Ω is compact, we may extract a convergent subsequence $\{x_{k_j}\}$ from $\{x_k\}$. Let $x^* \in \Omega$ denote the limit point of $\{x_{k_j}\}$. Since f is continuous, $f(x^*) = \lim_{j\to\infty} f(x_{k_j}) = m$. Thus, m is finite and x^* is a global minimizer of f on Ω. □

We next give an existence result on minimizers for functions that are not necessarily continuous. The functions we will consider do not take $-\infty$ value; however, they are allowed to take $+\infty$ value. For a function $f : \mathbb{R}^n \to (-\infty, \infty]$, the set

$$\text{dom}(f) := \{x \in \mathbb{R}^n : f(x) < +\infty\}$$

is called its *domain*. A function $f : \mathbb{R}^n \to (-\infty, \infty]$ is called *proper* if $\text{dom}(f) \neq \emptyset$. As a replacement of continuity, we introduce the following definition.

Definition 1.9 A function $f : \mathbb{R}^n \to [-\infty, \infty]$ is called *lower semi-continuous* at a point $x_0 \in \mathbb{R}^n$ if

$$f(x_0) \leq \liminf_{x\to x_0} f(x).$$

f is called lower semi-continuous on $\mathbb{R}^n$ if f is lower semi-continuous at every point.

Clearly, continuous functions are always lower semi-continuous. However, lower semi-continuous functions are not necessarily continuous. For instance, consider the function

$$f(x) = \begin{cases} 0, & x \leq 0, \\ 1, & x > 0 \end{cases}$$

which is not continuous at $x_0 = 0$. However, since $f(x) \geq 0$ for all x and $f(0) = 0$, we have

$$f(0) = 0 \leq \liminf_{x\to 0} f(x),$$

i.e., f is lower semi-continuous at $x_0 = 0$.

Definition 1.10 (Sublevel Set and Epigraph) For a function $f : \mathbb{R}^n \to [-\infty, \infty]$, its *epigraph* is defined to be

$$\text{epi}(f) := \{(x, t) \in \mathbb{R}^n \times \mathbb{R} : t \geq f(x)\};$$

for a given number $\alpha \in \mathbb{R}$, the set

$$L_f^-(\alpha) := \{x \in \mathbb{R}^n : f(x) \leq \alpha\}$$

is called a sublevel set of f with level α.

We can give some characterizations of lower semi-continuity of a function $f : \mathbb{R}^n \to [-\infty, \infty]$ in terms of its sublevel sets and epigraph.

Proposition 1.11 *For a function $f : \mathbb{R}^n \to [-\infty, \infty]$, the following are equivalent:*

(i) *f is lower semi-continuous.*
(ii) *$L_f^-(\alpha) := \{x \in \mathbb{R}^n : f(x) \leq \alpha\}$ is closed in $\mathbb{R}^n$ for any $\alpha \in \mathbb{R}$.*
(iii) *epi(f) is closed in $\mathbb{R}^n \times \mathbb{R}$.*

Proof (i) $\Longrightarrow$ (ii). Let $\{x_k\}$ be a sequence in $L_f^-(\alpha)$ with $x_k \to x$ as $k \to \infty$. Since $f(x_k) \leq \alpha$, by the lower semi-continuity of f, we have $f(x) \leq \liminf_{k\to\infty} f(x_k) \leq \alpha$. Thus, $x \in L_f^-(\alpha)$. So $L_f^-(\alpha)$ is closed.

(ii) $\Longrightarrow$ (iii). Let $\{(x_k, t_k)\} \subset \text{epi}(f)$ be such that $(x_k, t_k) \to (x, t)$ as $k \to \infty$. Then, for any $\alpha > t$, we have $f(x_k) \leq t_k < \alpha$ for large k. So $x_k \in L_f^-(\alpha)$ for large k. By the closedness of $L_f^-(\alpha)$, we have $x \in L_f^-(\alpha)$, i.e., $f(x) \leq \alpha$ for all $\alpha > t$. Thus, $f(x) \leq t$, i.e., $(x, t) \in \text{epi}(f)$.

(iii) $\Longrightarrow$ (i). Let $\{x_k\} \subset \mathbb{R}^n$ be such that $x_k \to x$ as $k \to \infty$. For any $t > \liminf_{k\to\infty} f(x_k)$, there exists a subsequence $\{x_{k_j}\}$ such that $f(x_{k_j}) < t$. Thus, $(x_{k_j}, t) \in \text{epi}(f)$. By the closedness of epi(f), we have $(x, t) \in \text{epi}(f)$, i.e., $f(x) \leq t$ for any such t. Therefore, $f(x) \leq \liminf_{k\to\infty} f(x_k)$. This shows that f is lower semi-continuous. □

Now we are ready to prove the following existence result.

Theorem 1.12 *Let $f : \mathbb{R}^n \to (-\infty, \infty]$ be lower semi-continuous. If there is an $\alpha \in \mathbb{R}$ such that the sublevel set $L_f^-(\alpha) := \{x \in \mathbb{R}^n : f(x) \leq \alpha\}$ is nonempty and compact, then f achieves a global minimizer on $\mathbb{R}^n$.*

Proof Let $m = \inf\{f(x) : x \in \mathbb{R}^n\}$. We may assume $\alpha > m$. Let $\{x_k\}$ be a minimizing sequence for f, i.e., $f(x_k) \to m$ as $k \to \infty$. Then, there is an N such that $f(x_k) \leq \alpha$ for all $k \geq N$, that is, $x_k \in L_f^-(\alpha)$ for all $k \geq N$. Since $L_f^-(\alpha)$ is compact, $\{x_k\}$ has a convergent subsequence $\{x_{k_j}\}$ with $x_{k_j} \to x^* \in L_f^-(\alpha)$ as $j \to \infty$. In view of the lower semi-continuity of f, we have $f(x^*) \leq \lim_{j\to\infty} f(x_{k_j}) = m$. By the definition of m, we must have $f(x^*) = m$. Therefore, m is finite and f achieves its minimum on $\mathbb{R}^n$ at x^*. □

For a proper function $f : \mathbb{R}^n \to (-\infty, \infty]$, in order to guarantee the existence of nonempty compact sublevel set, we need to impose certain conditions on f at infinity.

Definition 1.13 A function $f : \mathbb{R}^n \to (-\infty, \infty]$ is called *coercive* if $f(x) \to +\infty$ as $\|x\|_2 \to \infty$.

Example 1.14

(a) Let $\|\cdot\|$ be any norm on $\mathbb{R}^n$. Then, the function $f(x) := \|x\|$ is coercive. Indeed, let $e_i \in \mathbb{R}^n$ denote the vector whose entries are 0 except the ith spot where the entry is 1. By the triangle inequality and the Cauchy-Schwarz inequality, we have

$$\begin{aligned}|f(x) - f(y)| \le \|x - y\| = \left\|\sum_{i=1}^{n}(x_i - y_i)e_i\right\| &\le \sum_{i=1}^{n}|x_i - y_i|\|e_i\| \\ &\le \left(\sum_{i=1}^{n}\|e_i\|^2\right)^{1/2}\left(\sum_{i=1}^{n}|x_i - y_i|^2\right)^{1/2} \\ &= \left(\sum_{i=1}^{n}\|e_i\|^2\right)^{1/2}\|x - y\|_2.\end{aligned}$$

This shows that f is continuous on $\mathbb{R}^n$. Thus, by Theorem 1.8, f achieves its minimum on the compact set $\Omega := \{x \in \mathbb{R}^n : \|x\|_2 = 1\}$, i.e., there exists $\bar{x} \in \Omega$ such that $f(x) \ge f(\bar{x}) = \|\bar{x}\|$ for all $x \in \Omega$. Since $\bar{x} \ne 0$, there must exists a constant $c > 0$ such that $f(x) \ge c$ for all $x \in \Omega$. Consequently, for any $x \in \mathbb{R}^n$, we have

$$f(x) = \|x\|_2 f\left(\frac{x}{\|x\|_2}\right) \ge c\|x\|_2$$

which implies that $f(x) \to +\infty$ as $\|x\|_2 \to +\infty$, i.e., f is coercive.

(b) Given a vector $a \in \mathbb{R}^n$ and an $n \times n$ symmetric matrix A, i.e., $A^T = A$, consider the function

$$f(x) = x^T Ax + a^T x, \qquad x \in \mathbb{R}^n.$$

We can show that f is coercive if A is *positive definite* in the sense that

$$x^T Ax > 0 \qquad \text{for all } x \in \mathbb{R}^n \text{ with } x \ne 0.$$

To see this, let $m = \min\{x^T Ax : \|x\|_2 = 1\}$. Then $m > 0$ by Theorem 1.8. Consequently, $x^T Ax \ge m\|x\|_2^2$ for all $x \in \mathbb{R}^n$. By using the Cauchy-Schwartz inequality, we thus obtain

$$f(x) \ge x^T Ax - |a^T x| \ge m\|x\|_2^2 - \|a\|_2\|x\|_2 = m\|x\|_2\left(\|x\|_2 - \frac{\|a\|_2}{m}\right).$$

This demonstrates that $f(x) \to +\infty$ as $\|x\|_2 \to \infty$.

Corollary 1.15 *Let $f : \mathbb{R}^n \to (-\infty, \infty]$ be a proper, lower semi-continuous, coercive function. Then, f achieves a global minimizer on $\mathbb{R}^n$.*

Proof We first claim that $L_f^-(\alpha)$ must be bounded for every $\alpha \in \mathbb{R}$. If not, then there exists $\{x_k\} \subset L_f^-(\alpha)$ such that $\|x_k\|_2 \to \infty$. The coercivity of f then implies $f(x_k) \to +\infty$ as $k \to \infty$, which is a contradiction because $f(x_k) \leq \alpha$ for all k. Thus, by using the lower semi-continuity and coercivity of f, we can conclude that $L_f^-(\alpha)$ is closed and bounded and therefore compact for every $\alpha \in \mathbb{R}$. Furthermore, by the properness of f, we have $L_f^-(\alpha) \neq \emptyset$ for sufficiently large $\alpha \in \mathbb{R}$. Therefore, we may use Theorem 1.12 to conclude the existence of a minimizer. □

Problems

1.1 For a subset $\Omega \subset \mathbb{R}^n$, let δ_Ω denote its indicator function, i.e.,

$$\delta_\Omega(x) = \begin{cases} 0 & \text{if } x \in \Omega, \\ +\infty & \text{otherwise.} \end{cases}$$

Show that δ_Ω is lower semi-continuous if and only if Ω is closed.

1.2 Let $f_\alpha : \mathbb{R}^n \to (-\infty, \infty]$, with $\alpha \in I$, be a family of lower semi-continuous functions. Define

$$f(x) := \sup_{\alpha \in I} f_\alpha(x), \quad x \in \mathbb{R}^n.$$

Show that f is also lower semi-continuous.

1.3 Let $f : \mathbb{R}^n \to (-\infty, \infty]$ be a proper function bounded from below. Show that f is lower semi-continuous if and only if there is an increasing sequence of Lipschitz continuous functions $\{f_k\}$ on $\mathbb{R}^n$ such that $\lim_{k\to\infty} f_k(x) = f(x)$ for all $x \in \mathbb{R}^n$.

1.4 Determine the coercivity of the following functions and explain the reason:

(i) $f(x_1, x_2) = x_1^2 + x_2^2 - 4x_1x_2$.
(ii) $f(x_1, x_2) = x_1^2 + x_2^2 - x_1x_2 - x_1$.

1.5 Show that any norm $|||\cdot|||$ on $\mathbb{R}^n$ is equivalent to the Euclidean norm $\|\cdot\|_2$, i.e., there exist constants $0 < A \leq B < \infty$ such that

$$A\|x\|_2 \leq |||x||| \leq B\|x\|_2 \quad \forall x \in \mathbb{R}^n.$$

1.6 The fundamental theorem of algebra says that any nonconstant polynomial has a complex root. This problem is designed to give an optimization proof of this theorem.

Let $p(z)$ be a polynomial of degree n. Recall that any complex number z can be written as $z = x + \mathtt{i}y$ with $x, y \in \mathbb{R}$, where $\mathtt{i}$ denote the imaginary unit. Thus, $|p(z)|$ is a continuous function from $\mathbb{R}^2$ to $\mathbb{R}$.

(i) Show that $|p(z)|$ is a coercive function on $\mathbb{R}^2$ and conclude that $|p(z)|$ has a global minimizer z^*. Therefore, it suffices to show that $p(z^*) = 0$.

(ii) By introducing $q(z) := p(z + z^*)$, we only need to show $q(0) = 0$. Write

$$q(z) = a_0 + a_1 z + \cdots + a_n z^n.$$

We will use a contradiction argument to show $a_0 = 0$. Assume that $a_0 \neq 0$. Please justify that

$$q(z) = a_0 + a_k z^k + z^{k+1} h(z),$$

where $a_k \neq 0$ is the first nonzero coefficient after a_0 and $h(z)$ is a polynomial of degree $n - k - 1$.

(iii) Let ω be a kth root of $-a_0/a_k$, i.e., $\omega^k = -a_0/a_k$. Show that for any $t > 0$, there holds

$$q(t\omega) = (1 - t^k)a_0 + t\left[t^k \omega^{k+1} h(t\omega)\right].$$

(iv) Show that there is $t_0 > 0$ such that for all $0 < t \leq t_0$, there holds

$$|q(t\omega)| < |a_0|.$$

(v) This is a contradiction. Why?

Chapter 2
Convex Sets and Convex Functions

In this chapter, we begin by recalling the definition of convex sets and introducing the concept of the convex hull, along with its key characterizations. We then explore the separation properties of convex sets using hyperplanes. Next, we examine convex functions and their fundamental properties. In particular, we provide equivalent characterizations of convexity for continuously differentiable functions and establish the co-coercivity result for continuously differentiable convex functions with Lipschitz continuous gradients. These results form the foundation for many developments in this book and will be used throughout.

2.1 Convex Sets

We begin with the definition of convex sets. Given any two points $x_0, x_1 \in \mathbb{R}^n$, the set of points of the form $tx_1 + (1-t)x_0$ with $0 \le t \le 1$ represents the line segment between x_0 and x_1.

Definition 2.1 (Convex Set) A set $\Omega \subset \mathbb{R}^n$ is called *convex* if the line segment between any two points in Ω always lies in Ω, i.e., for any $x_0, x_1 \in \Omega$ and $0 \le t \le 1$, we always have $tx_1 + (1-t)x_0 \in \Omega$.

Example 2.2

(a) Let $\|\cdot\|$ be any norm on $\mathbb{R}^n$. Then, for any $x \in \mathbb{R}^n$ and $r > 0$, the open ball $B_r(x)$ and the closed ball $\overline{B}_r(x)$ are convex sets.
(b) Given $\bar{x} \in \mathbb{R}^n$ and an $n \times n$ symmetric positive definite matrix A, the set

$$\{x \in \mathbb{R}^n : (x-\bar{x})^T A(x-\bar{x}) \le 1\}$$

is called an *ellipsoid* with center $\bar{x}$. Ellipsoids are convex sets.

Q. Jin, *Lectures on Nonsmooth Optimization*, Texts in Applied Mathematics 82,
https://doi.org/10.1007/978-3-031-91417-1_2

(c) Given $\xi \in \mathbb{R}^n$ with $\xi \neq 0$ and $b \in \mathbb{R}$, the set

$$\{x \in \mathbb{R}^n : \langle \xi, x \rangle = b\}$$

is called a *hyperplane* in $\mathbb{R}^n$. A hyperplane divides $\mathbb{R}^n$ into two *half-spaces*:

$$\{x \in \mathbb{R}^n : \langle \xi, x \rangle \leq b\} \quad \text{and} \quad \{x \in \mathbb{R}^n : \langle \xi, x \rangle \geq b\}.$$

All hyperplanes and half-spaces are convex sets.

The following lists some properties of convex sets that can be proved easily:

(i) The intersection of convex sets is also convex.
(ii) Let Ω_1 and Ω_2 be two convex sets in $\mathbb{R}^n$. For any numbers $\alpha, \beta \in \mathbb{R}$, the set $\alpha\Omega_1 + \beta\Omega_2$ defined by

$$\alpha\Omega_1 + \beta\Omega_2 = \{\alpha x_1 + \beta x_2 : x_1 \in \Omega_1 \text{ and } x_2 \in \Omega_2\}$$

is also convex.
(iii) Given a set $\Omega \subset \mathbb{R}^n$. A point $\bar{x} \in \mathbb{R}^n$ is called an *interior point* of Ω if there is an open ball $B_r(\bar{x})$ such that $B_r(\bar{x}) \subset \Omega$ and the *interior* of Ω, denoted by $\text{int}(\Omega)$, is the set consisting of all interior points of Ω. The interior of Ω is the largest open set contained in Ω. If Ω is convex, so is $\text{int}(\Omega)$.
(iv) Given a set $\Omega \subset \mathbb{R}^n$, its *closure* in $\mathbb{R}^n$, denoted by $\overline{\Omega}$, is the smallest closed set containing Ω. It is known that $\bar{x} \in \overline{\Omega}$ if and only if for any $r > 0$, the ball $B_r(\bar{x})$ always contains a point in Ω. We can show that if Ω is convex, then so is $\overline{\Omega}$.
(v) By induction, it is easy to show that a set $\Omega \subset \mathbb{R}^n$ is convex if and only if $t_1x_1 + \cdots + t_kx_k \in \Omega$ for any integer $k \geq 1$, any $x_1, \cdots, x_k \in \Omega$, and any nonnegative $t_1, \cdots, t_k$ satisfying $t_1 + \cdots + t_k = 1$.

A point of the form $t_1x_1 + \cdots + t_kx_k$, where $t_1 + \cdots + t_k = 1$ and $t_i \geq 0$, $i = 1, \cdots, k$, is called a *convex combination* of the points $x_1, \cdots, x_k$. Thus, a set is convex if and only if it contains every convex combination of its points.

Definition 2.3 (Convex Hull) Given a set $\Omega \subset \mathbb{R}^n$, the smallest convex set containing Ω is called the *convex hull* of Ω and is denoted by $\text{conv}(\Omega)$.

The following result gives a characterization of the convex hull of a given set in terms of the convex combinations of its points.

Lemma 2.4 *For any set* $\Omega \subset \mathbb{R}^n$, *we have*

$$\text{conv}(\Omega) = \left\{ t_1x_1 + \cdots + t_kx_k : \begin{array}{l} k \geq 1,\ x_1, \cdots, x_k \in \Omega, \\ t_1, \cdots, t_k \geq 0, \\ t_1 + \cdots + t_k = 1 \end{array} \right\},$$

i.e., $\text{conv}(\Omega)$ *consists of all convex combinations of points from* Ω.

Proof Let K denote the set on the right side. Clearly, $\Omega \subset K$. We first show that K is a convex set. Take any two points $x, y \in K$; we can write

$$x = t_1 x_1 + \cdots + t_k x_k, \quad t_1 + \cdots + t_k = 1, \quad t_1, \cdots, t_k \geq 0;$$
$$y = s_1 y_1 + \cdots + s_l y_l, \quad s_1 + \cdots + s_l = 1, \quad s_1, \cdots, s_l \geq 0.$$

Then, for any $0 \leq \lambda \leq 1$, we have

$$\lambda x + (1-\lambda)y = \lambda t_1 x_1 + \cdots + \lambda t_k x_k + (1-\lambda)s_1 y_1 + \cdots + (1-\lambda)s_l y_l.$$

Note that

$$\lambda t_1 + \cdots + \lambda t_k + (1-\lambda)s_1 + \cdots + (1-\lambda)s_l = 1$$

Therefore, $\lambda x + (1-\lambda)y \in K$.

Next, if C is a convex set containing Ω, then C contains every convex combination of points from Ω and hence $K \subset C$. Thus, K is the smallest convex set containing Ω. By definition, $K = \text{conv}(\Omega)$. □

Theorem 2.5 (Caratheodory's Theorem) *Let Ω be a subset of $\mathbb{R}^n$. Then, every element in* $\text{conv}(\Omega)$ *can be written as a convex combination of at most $n+1$ elements in Ω.*

Proof Let $x \in \text{conv}(\Omega)$ be any element. It follows from Lemma 2.4 that

$$x = \sum_{i=1}^{k} t_i x_i$$

for some integer $k \geq 1$, where $x_i \in \Omega$ and $t_i > 0$ for all $i = 1, \cdots, k$ and $\sum_{i=1}^{k} t_i = 1$. If $k \leq n+1$, there is nothing to prove. So we may assume $k > n+1$. As $x_2 - x_1, \cdots, x_k - x_1$ are $k-1$ elements in $\mathbb{R}^n$ with $k-1 > n$, they are linearly dependent. Thus, there exist real numbers $\mu_2, \cdots, \mu_k$ not all zero such that

$$\sum_{j=2}^{k} \mu_j (x_j - x_1) = 0.$$

Let $\mu_1 := -\sum_{j=2}^{k} \mu_j$. Then,

$$\sum_{i=1}^{k} \mu_i x_i = 0 \quad \text{and} \quad \sum_{i=1}^{k} \mu_i = 0.$$

Consequently,

$$x = \sum_{i=1}^{k} t_i x_i - \alpha \sum_{i=1}^{k} \mu_i x_i = \sum_{i=1}^{k} (t_i - \alpha\mu_i) x_i \tag{2.1}$$

for any $\alpha \in \mathbb{R}$. Since $\mu_1, \cdots, \mu_k$ are not all zero and $\sum_{i=1}^{k} \mu_i = 0$, at least one of them must be positive. We choose α as

$$\alpha := \min \left\{ \frac{t_i}{\mu_i} : \mu_i > 0 \right\}$$

Then, $t_1 - \alpha\mu_1, \cdots, t_k - \alpha\mu_k$ are all nonnegative and at least one of them is zero. Note that

$$\sum_{i=1}^{k} (t_i - \alpha\mu_i) = \sum_{i=1}^{k} t_i - \alpha \sum_{i=1}^{k} \mu_i = 1.$$

It then follows from (2.1) that x is a convex combination of at most $k - 1$ elements in Ω. Repeating the above procedure, we can eventually represent x as a convex combination of at most $n + 1$ elements in Ω. □

Corollary 2.6 *If* $\Omega \subset \mathbb{R}^n$ *is a compact set, then* $\mathrm{conv}(\Omega)$ *is also compact.*

Proof Let $\{x^k\}$ be any sequence in $\mathrm{conv}(\Omega)$. By Caratheodory's theorem, each x^k can be written as

$$x^k = \sum_{j=1}^{n+1} t_j^k x_j^k, \quad x_j^k \in \Omega, \quad t_j^k \geq 0 \text{ and } \sum_{j=1}^{n+1} t_j^k = 1.$$

By the compactness of Ω and the properties of $\{t_j^k\}$, we can find a subsequence $\{k_l\}$ of positive integers such that $x_j^{k_l} \to x_j$ and $t_j^{k_l} \to t_j$ as $l \to \infty$ for each $j = 1, \cdots, n+1$ with $x_j \in \Omega$, $t_j \geq 0$ and $\sum_{j=1}^{n+1} t_j = 1$. Consequently, $x^{k_l} \to \sum_{j=1}^{n+1} t_j x_j \in \mathrm{conv}(\Omega)$. This shows that $\mathrm{conv}(\Omega)$ is compact. □

2.2 Projection onto Convex Sets

Given a set $\Omega \subset \mathbb{R}^n$ and a point $\bar{x} \in \mathbb{R}^n$, we define the distance from $\bar{x}$ to Ω as

$$d(\bar{x}, \Omega) := \inf_{x \in \Omega} \|\bar{x} - x\|.$$

We adopt the convention that $d(x, \Omega) = +\infty$ if $\Omega = \emptyset$. Sometimes, we also use the notation $d_\Omega(\bar{x})$ to denote $d(\bar{x}, \Omega)$.

When $\Omega \subset \mathbb{R}^n$ is a nonempty closed convex set, we will show that there exists a unique point that realizes the distance from $\bar{x}$ to Ω. The existence of such a point is equivalent to finding a minimizer of the optimization problem:

$$\min_{x\in\Omega} \|x - \bar{x}\|^2$$

which can be equivalently rewritten as

$$\min_{x\in\mathbb{R}^n} \left\{ f(x) := \|x - \bar{x}\|^2 + \delta_\Omega(x) \right\}.$$

Since Ω is closed, the function f is proper, lower semi-continuous, and coercive. By Corollary 1.15, it follows that the minimization problem has a global minimizer, ensuring the existence of a point in Ω closest to $\bar{x}$. The closest point on Ω to $\bar{x}$ must be unique. Indeed, if x_1 and x_2 are two points on Ω that are closest to $\bar{x}$, then, by the convexity of Ω, we have $(x_1 + x_2)/2 \in \Omega$ and thus

$$\left\| \frac{1}{2}(x_1 + x_2) - \bar{x} \right\| \geq \|x_1 - \bar{x}\| = \|x_2 - \bar{x}\|.$$

By using the parallelogram law $\|a + b\|^2 + \|a - b\|^2 = 2(\|a\|^2 + \|b\|^2)$ with $a = \frac{1}{2}(x_1 - \bar{x})$ and $b = \frac{1}{2}(x_2 - \bar{x})$, we have

$$\begin{aligned}
\left\| \frac{1}{2}(x_1 + x_2) - \bar{x} \right\|^2 + \left\| \frac{1}{2}(x_1 - x_2) \right\|^2 &= 2\left(\left\| \frac{1}{2}(x_1 - \bar{x}) \right\|^2 + \left\| \frac{1}{2}(x_2 - \bar{x}) \right\|^2 \right) \\
&= \frac{1}{2}\left(\|x_1 - \bar{x}\|^2 + \|x_2 - \bar{x}\|^2 \right) \\
&\leq \left\| \frac{1}{2}(x_1 + x_2) - \bar{x} \right\|^2
\end{aligned}$$

which implies that $\|x_1 - x_2\|^2 \leq 0$ and hence $x_1 = x_2$.

The unique closest point on Ω to $\bar{x}$ is denoted by $P_\Omega(\bar{x})$ and is called the *Euclidean projection* of $\bar{x}$ onto Ω. The mapping $P_\Omega : \mathbb{R}^n \to \Omega$ is called the *Euclidean projection* of $\mathbb{R}^n$ onto Ω. One can check

$$\langle \bar{x} - P_\Omega(\bar{x}), x - P_\Omega(\bar{x}) \rangle \leq 0 \tag{2.2}$$

for $x \in \Omega$ and $\bar{x} \in \mathbb{R}^n$. Indeed, since the function

$$\varphi(t) := \|\bar{x} - (tx + (1 - t)P_\Omega(\bar{x}))\|^2, \qquad 0 \leq t \leq 1,$$

achieves its minimum at $t = 0$, we have

$$0 \leq \varphi'(0) = 2\langle \bar{x} - P_\Omega(\bar{x}), P_\Omega(\bar{x}) - x \rangle$$

which implies (2.2).

Lemma 2.7 *Let $\Omega \subset \mathbb{R}^n$ be a closed convex set. Then, P_Ω is firmly nonexpansive, i.e.,*

$$\|P_\Omega(x) - P_\Omega(y)\|^2 + \|(x - P_\Omega(x)) - (y - P_\Omega(y))\|^2 \leq \|x - y\|^2$$

for all $x, y \in \mathbb{R}^n$.

Proof Applying (2.2) to $x, y \in \mathbb{R}^n$, it follows that

$$\langle x - P_\Omega(x), P_\Omega(y) - P_\Omega(x) \rangle \leq 0, \quad \langle y - P_\Omega(y), P_\Omega(x) - P_\Omega(y) \rangle \leq 0.$$

Adding these two equations gives

$$\langle x - y - (P_\Omega(x) - P_\Omega(y)), P_\Omega(y) - P_\Omega(x) \rangle \leq 0$$

which implies

$$\|P_\Omega(x) - P_\Omega(y)\|^2 \leq \langle x - y, P_\Omega(x) - P_\Omega(y) \rangle.$$

This is equivalent to the result to be proved. □

2.3 Separation Theorems

Definition 2.8 Let C and D be two sets in $\mathbb{R}^n$ and let $L := \{x \in \mathbb{R}^n : \langle \xi, x \rangle = b\}$ be a hyperplane.

(i) L is said to *separate* C and D if $\langle \xi, x \rangle \leq b$ for all $x \in C$ and $\langle \xi, x \rangle \geq b$ for all $x \in D$.

(ii) L is said to *strongly separate* C and D if

$$\sup_{x \in C} \langle \xi, x \rangle < b < \inf_{y \in D} \langle \xi, y \rangle.$$

In the following, we will give some results of using hyperplanes to separate convex sets. We start with some useful lemmas.

Lemma 2.9 *Let $\Omega \subset \mathbb{R}^n$ be a convex set. If $x_0 \in \overline{\Omega}$ and $x_1 \in \operatorname{int}(\Omega)$, then*

$$x_t := tx_1 + (1-t)x_0 \in \operatorname{int}(\Omega)$$

for every $t \in (0, 1]$.

Proof Since $x_1 \in \operatorname{int}(\Omega)$, we can find $r > 0$ such that $B_r(x_1) \subset \Omega$. We will show for each $t \in (0, 1]$ that $B_\rho(x_t) \subset \Omega$ for $\rho < rt$. Since $x_0 \in \overline{\Omega}$, for $\varepsilon := rt - \rho > 0$, we can find $y_0 \in \Omega$ such that $\|x_0 - y_0\| < \varepsilon$. For any $y \in B_\rho(x_t)$, define

$$y_1 := \frac{1}{t}\left(y - (1-t)y_0\right).$$

Then,

$$y_1 - x_1 = \frac{1}{t}\left(y - x_t - (1-t)(y_0 - x_0)\right).$$

Therefore,

$$\|y_1 - x_1\| \le \frac{1}{t}\|y - x_t\| + \frac{1-t}{t}\|x_0 - y_0\| < \frac{\rho}{t} + \frac{1-t}{t}\varepsilon \le \frac{1}{t}(\rho + \varepsilon) = r$$

which shows that $y_1 \in B_r(x_1) \subset \Omega$. Note that $y = ty_1 + (1-t)y_0$. By the convexity of Ω, we have $y \in \Omega$. Since $y \in B_\rho(x_t)$ is arbitrary, we have $B_\rho(x_t) \subset \Omega$ and hence $x_t \in \operatorname{int}(\Omega)$. □

Lemma 2.10 *Let $\Omega \subset \mathbb{R}^n$ be a convex set with $\operatorname{int}(\Omega) \neq \emptyset$. Then, $\overline{\Omega} = \overline{\operatorname{int}(\Omega)}$.*

Proof $\overline{\operatorname{int}(\Omega)} \subset \overline{\Omega}$ is obvious. To show $\overline{\Omega} \subset \overline{\operatorname{int}(\Omega)}$, let $x \in \overline{\Omega}$ be any element. Pick $\tilde{x} \in \operatorname{int}(\Omega)$. It then follows from Lemma 2.9 that $x_t := t\tilde{x} + (1-t)x \in \operatorname{int}(\Omega)$ for all $t \in (0, 1]$. Since $x_t \to x$ as $t \to 0$, we have $x \in \overline{\operatorname{int}(\Omega)}$. □

Lemma 2.11 *Let $\Omega \subset \mathbb{R}^n$ be a convex set. Then,* $\operatorname{int}(\Omega) = \operatorname{int}(\overline{\Omega})$.

Proof $\operatorname{int}(\Omega) \subset \operatorname{int}(\overline{\Omega})$ is obvious. In the following, we will show the converse. We first consider the case $\operatorname{int}(\Omega) \neq \emptyset$. Pick $\tilde{x} \in \operatorname{int}(\Omega)$. Let $x \in \operatorname{int}(\overline{\Omega})$ be any element. Then, there exists $r > 0$ such that $B_r(x) \subset \overline{\Omega}$. For any $t > 0$, consider $x_t := x + t(x - \tilde{x})$. We have $\|x_t - x\| = t\|x - \tilde{x}\| < r$ and hence $x_t \in B_r(x) \subset \overline{\Omega}$ for sufficiently small $t > 0$. Note that

$$x = \frac{1}{1+t}x_t + \frac{t}{1+t}\tilde{x}$$

and $\tilde{x} \in \operatorname{int}(\Omega)$. It follows from Lemma 2.9 that $x \in \operatorname{int}(\Omega)$. Therefore, $\operatorname{int}(\overline{\Omega}) \subset \operatorname{int}(\Omega)$.

We next consider the case $\operatorname{int}(\Omega) = \emptyset$. Assume $\operatorname{int}(\overline{\Omega}) \neq \emptyset$ and let $x \in \operatorname{int}(\overline{\Omega})$. Then, there exists $r > 0$ such that $B_r(x) \subset \overline{\Omega}$. As $x + re_i, x - re_i \in \overline{\Omega}$ for

$i = 1, \cdots, n$, we can find $x_i, y_i \in \Omega$ for $i = 1, \cdots, n$ such that

$$\|x_i - (x + re_i)\|, \quad \|y_i - (x - re_i)\|, \quad i = 1, \cdots, n$$

to be sufficiently small so that the convex set $\operatorname{conv}(x_1, y_1, \cdots, x_n, y_n)$, which is contained in Ω, contains x as an interior point. Consequently, x is also an interior point of Ω, which is a contradiction. □

Theorem 2.12 *Let $\Omega \subset \mathbb{R}^n$ be a nonempty convex set and let $\bar{x} \in \mathbb{R}^n$ be such that $\bar{x} \notin \operatorname{int}(\Omega)$. Then, $\bar{x}$ and Ω can be separated by a hyperplane, i.e., there is $0 \neq \xi \in \mathbb{R}^n$ such that $\langle \xi, x\rangle \leq \langle \xi, \bar{x}\rangle$ for all $x \in \Omega$.*

Proof Recall that $\operatorname{int}(\Omega) = \operatorname{int}(\overline{\Omega})$ from Lemma 2.11. We have $\bar{x} \notin \operatorname{int}(\Omega)$ if and only if $\bar{x} \notin \operatorname{int}(\overline{\Omega})$. Thus, by replacing Ω by $\overline{\Omega}$, we may assume Ω is closed. Since $\bar{x} \notin \operatorname{int}(\Omega)$, there is a sequence $\{x_k\}$ such that $x_k \notin \Omega$ and $x_k \to \bar{x}$ as $k \to \infty$. Let $P_\Omega(x_k)$ be the Euclidean projection of x_k onto Ω. Then, $x_k \neq P_\Omega(x_k)$. We define

$$\xi_k := \frac{x_k - P_\Omega(x_k)}{\|x_k - P_\Omega(x_k)\|}, \qquad k = 1, 2, \cdots.$$

Note that $\|\xi_k\| = 1$ for all k. Therefore, by taking a subsequence if necessary, we can find $\xi \in \mathbb{R}^n$ such that $\xi_k \to \xi$ as $k \to \infty$. By using the property (2.2) of P_Ω, we have

$$\langle x_k - P_\Omega(x_k), x - P_\Omega(x_k)\rangle \leq 0, \quad \forall x \in \Omega$$

which implies that

$$\langle x_k - P_\Omega(x_k), x - x_k\rangle \leq -\|x_k - P_\Omega(x_k)\|^2 < 0, \quad \forall x \in \Omega.$$

Consequently, using the definition of ξ_k gives

$$\langle \xi_k, x\rangle \leq \langle \xi_k, x_k\rangle, \quad \forall x \in \Omega.$$

By taking $k \to \infty$, we then obtain $\langle \xi, x\rangle \leq \langle \xi, \bar{x}\rangle$ for all $x \in \Omega$. □

Let Ω be a set in $\mathbb{R}^n$. A point $\bar{x} \in \mathbb{R}^n$ is called a *boundary point* of Ω if every open ball $B_r(\bar{x})$ contains points in Ω and points not in Ω.

Definition 2.13 (Supporting Hyperplane) Let $\Omega \subset \mathbb{R}^n$ and let $\bar{x}$ be a boundary point of Ω. A hyperplane $\{x \in \mathbb{R}^n : \langle \xi, x\rangle = b\}$ is called a *supporting hyperplane* of Ω at $\bar{x}$ if $\langle \xi, \bar{x}\rangle = b$ and $\langle \xi, x\rangle \leq b$ for all $x \in \Omega$.

As an immediate consequence of Theorem 2.12, we have the following.

Theorem 2.14 (Supporting Hyperplane Theorem) *For any nonempty convex set $\Omega \subset \mathbb{R}^n$ and any boundary point $\bar{x}$ of Ω, there exists a supporting hyperplane to Ω at $\bar{x}$.*

Now we are ready to prove some hyperplane separation theorems for general convex sets.

Theorem 2.15 (Separation Theorem) *Let C and D be two nonempty convex sets in $\mathbb{R}^n$ with $C \cap D = \emptyset$. Then, C and D can be separated by a hyperplane, i.e., there is $0 \neq \xi \in \mathbb{R}^n$ and $b \in \mathbb{R}$ such that*

$$\langle \xi, x \rangle \leq b \leq \langle \xi, y \rangle, \qquad \forall x \in C \text{ and } y \in D.$$

Proof Let $\Omega := C - D$. Since C and D are nonempty convex sets with $C \cap D = \emptyset$, Ω is a nonempty convex set with $0 \notin \Omega$. According to Theorem 2.12, there is $0 \neq \xi \in \mathbb{R}^n$ such that

$$\langle \xi, z \rangle \leq 0, \quad \forall z \in \Omega.$$

Therefore,

$$\langle \xi, x \rangle \leq \langle \xi, y \rangle, \quad \forall x \in C \text{ and } y \in D$$

which implies that

$$\sup_{x \in C} \langle \xi, x \rangle \leq \inf_{y \in D} \langle \xi, y \rangle.$$

Consequently, there is a number b such that $\langle \xi, x \rangle \leq b \leq \langle \xi, y \rangle$ for all $x \in C$ and $y \in D$. □

Corollary 2.16 *Let C and D be nonempty convex sets in $\mathbb{R}^n$ with $\text{int}(D) \neq \emptyset$. If $C \cap \text{int}(D) = \emptyset$, then C and D can be separated by a hyperplane.*

Proof Applying Theorem 2.15 to C and $\text{int}(D)$, we can find $0 \neq \xi \in \mathbb{R}^n$ and $b \in \mathbb{R}$ such that

$$\langle \xi, x \rangle \leq b \leq \langle \xi, y \rangle, \qquad \forall x \in C \text{ and } y \in \text{int}(D).$$

Recall that $\overline{D} = \overline{\text{int}(D)}$ from Lemma 2.10. Thus, every point in $D \subset \overline{D}$ is a limit point of a sequence from $\text{int}(D)$. By taking the limit in the above equation, we can obtain

$$\langle \xi, x \rangle \leq b \leq \langle \xi, y \rangle, \qquad \forall x \in C \text{ and } y \in D$$

which completes the proof. □

Theorem 2.17 (Strong Separation Theorem) *Let C and D be nonempty convex sets in $\mathbb{R}^n$ such that $C \cap D = \emptyset$ and $C - D$ is closed. Then, C and D can be strongly* *separated by hyperplanes, i.e., there exist* $0 \neq \xi \in X$ *and* $\alpha, \beta \in \mathbb{R}$ *such that*

$$\langle \xi, x \rangle \leq \alpha < \beta \leq \langle \xi, y \rangle$$

for all $x \in C$ *and* $y \in D$.

Proof Since $0 \notin C - D$ and $\Omega := C - D$ is closed and convex, we have $P_\Omega(0) \neq 0$. Let $\xi = -P_\Omega(0)$. Then, since

$$\langle 0 - P_\Omega(0), z - P_\Omega(0) \rangle \leq 0, \quad \forall z \in \Omega,$$

we have $\langle \xi, z \rangle \leq -\|\xi\|^2$ for all $z \in \Omega$. Consequently,

$$\langle \xi, x \rangle \leq -\|\xi\|^2 + \langle \xi, y \rangle, \quad \forall x \in C \text{ and } y \in D$$

which implies that

$$\sup_{x \in C} \langle \xi, x \rangle \leq -\|\xi\|^2 + \inf_{y \in D} \langle \xi, y \rangle < \inf_{y \in D} \langle \xi, y \rangle.$$

The proof is complete. □

Corollary 2.18 *Let C and D be two nonempty convex sets in* $\mathbb{R}^n$. *If C is closed, D is compact, and* $C \cap D = \emptyset$, *then they can be strongly separated by hyperplanes.*

Proof According to Theorem 2.17, it suffices to show that $C - D$ is closed. To see this, let $\{z_k\}$ be any sequence in $C - D$ with $z_k \to z$ as $k \to \infty$. We can write $z_k = x_k - y_k$ with $x_k \in C$ and $y_k \in D$. Since D is compact, by taking a subsequence if necessary, we can find $y \in D$ such that $y_k \to y$ as $k \to \infty$. Then, $x_k = z_k + y_k \to x := z + y$ as $k \to \infty$. Since C is closed, we have $x \in C$. As $z = x - y$, we have $z \in C - D$. Therefore, $C - D$ is closed. □

2.4 Convex Functions

Definition 2.19 (Convex Function) A function $f : \mathbb{R}^n \to [-\infty, \infty]$ is called *convex* if

$$f(tx_1 + (1-t)x_0) \leq tf(x_1) + (1-t)f(x_0)$$

for all $0 \leq t \leq 1$ and any $x_0, x_1 \in \mathrm{dom}(f)$.

Example 2.20 Let $\{f_\alpha\}_{\alpha \in \Lambda}$ be a family of convex functions from $\mathbb{R}^n$ to $[-\infty, \infty]$. Then, the function $f : \mathbb{R}^n \to [-\infty, \infty]$ defined by

$$f(x) := \sup_{a \in \Lambda} f_\alpha(x), \quad x \in \mathbb{R}^n$$

is also convex. Indeed, for any $0 \le t \le 1$ and $x, y \in \text{dom}(f)$, we have

$$\begin{aligned} f(tx + (1-t)y) &= \sup_{\alpha \in \Lambda} f_\alpha(tx + (1-t)y) \le \sup_{\alpha \in \Lambda} \{t f_\alpha(x) + (1-t) f_\alpha(y)\} \\ &\le t \sup_{\alpha \in \Lambda} f_\alpha(x) + (1-t) \sup_{\alpha \in \Lambda} f_\alpha(y) \\ &= t f(x) + (1-t) f(y). \end{aligned}$$

Example 2.21 Let $\Phi : \mathbb{R}^m \times \mathbb{R}^n \to [-\infty, \infty]$ be a convex function. Then, the function $f : \mathbb{R}^m \to [-\infty, \infty]$ defined by

$$f(x) := \inf_{y \in \mathbb{R}^n} \Phi(x, y), \quad x \in \mathbb{R}^m$$

is also convex. To see this, let $x_1, x_2 \in \text{dom}(f)$ be any two points. By definition, for any $a > f(x_1)$ and $b > f(x_2)$, we can find $y_1, y_2 \in \mathbb{R}^n$ such that $\Phi(x_1, y_1) < a$ and $\Phi(x_2, y_2) < b$. Thus, for any $0 \le t \le 1$, we have from the convexity of Φ that

$$\begin{aligned} f(tx_1 + (1-t)x_2) &= \inf_{y \in \mathbb{R}^n} \Phi(tx_1 + (1-t)x_2, y) \\ &\le \Phi(tx_1 + (1-t)x_2, ty_1 + (1-t)y_2) \\ &\le t\Phi(x_1, y_1) + (1-t)\Phi(x_2, y_2) \\ &\le ta + (1-t)b. \end{aligned}$$

Letting $a \to f(x_1)$ and $b \to f(x_2)$ gives $f(tx_1+(1-t)x_2) \le tf(x_1)+(1-t)f(x_2)$, which shows the convexity of f. □

Example 2.22 Let C be a nonempty closed convex set in $\mathbb{R}^n$. For each $x \in \mathbb{R}^n$, let $d_C(x)$ denote the Euclidean distance of x to C, i.e.,

$$d_C(x) := \min_{z \in C} \|x - z\| = \|x - P_C(x)\|,$$

where P_C denotes the Euclidean projection onto C. Note that for any $x_1, x_2 \in \mathbb{R}^n$ and $0 \le t \le 1$, we have $tP_C(x_1) + (1-t)P_C(x_2) \in C$ and hence

$$\begin{aligned} d_C(tx_1 + (1-t)x_2) &\le \|tx_1 + (1-t)x_2 - (tP_C(x_1) + (1-t)P_C(x_2))\| \\ &= \|t(x_1 - P_C(x_1)) + (1-t)(x_2 - P_C(x_2))\| \\ &\le t\|x_1 - P_C(x_1)\| + (1-t)\|x_2 - P_C(x_2)\| \\ &= td_C(x_1) + (1-t)d_C(x_2). \end{aligned}$$

This shows that $d_C(x)$ is a convex function on $\mathbb{R}^n$. □

In the definition of convex function, we allow f to take both the values $-\infty$ and $+\infty$. The following result shows that if f takes a finite value at some point where f is lower semi-continuous, then f does not take $-\infty$ value.

Lemma 2.23 *Let $f : \mathbb{R}^n \to [-\infty, \infty]$ be a convex function. If there is an $\bar{x} \in \mathbb{R}^n$ such that $f(\bar{x})$ is finite and f is lower semi-continuous at $\bar{x}$, then there exists $(\xi, \gamma) \in \mathbb{R}^n \times \mathbb{R}$ such that*

$$f(x) \geq \langle \xi, x \rangle + \gamma, \qquad \forall x \in \mathbb{R}^n.$$

Consequently, $f(x) > -\infty$ for all $x \in \mathbb{R}^n$.

Proof Take a number $\bar{t} < f(\bar{x})$. We claim that $(\bar{x}, \bar{t}) \notin \overline{\text{epi}(f)}$, where $\overline{\text{epi}(f)}$ denotes the closure of $\text{epi}(f)$. If not, then there is a sequence $\{(x_k, t_k)\} \subset \text{epi}(f)$ such that $(x_k, t_k) \to (\bar{x}, \bar{t})$ as $k \to \infty$. This together with the lower semi-continuity of f at $\bar{x}$ shows that

$$f(\bar{x}) \leq \liminf_{k\to\infty} f(x_k) \leq \liminf_{k\to\infty} t_k = \bar{t} < f(\bar{x})$$

which is a contradiction. Since $\overline{\text{epi}(f)}$ is closed, convex, and nonempty, by Corollary 2.18, there exists $(\eta, \alpha) \in \mathbb{R}^n \times \mathbb{R}$ such that

$$\langle \eta, x \rangle + \alpha t > \langle \eta, \bar{x} \rangle + \alpha \bar{t}, \qquad \forall (x, t) \in \text{epi}(f). \tag{2.3}$$

Since $(\bar{x}, f(\bar{x})) \in \text{epi}(f)$, we can obtain, using $x = \bar{x}$ and $t = f(\bar{x})$, that $\alpha f(\bar{x}) > \alpha \bar{t}$, which implies $\alpha > 0$. By dividing both sides of (2.3) by α, we obtain

$$t > \langle \xi, x \rangle + \gamma, \quad \forall t > f(x) \text{ and } x \in \text{dom}(f),$$

where $\xi := -\eta/\alpha$ and $\gamma := \langle \eta/\alpha, \bar{x} \rangle + \bar{t}$. This implies $f(x) \geq \langle \xi, x \rangle + \gamma$ for all $x \in \mathbb{R}^n$. The proof is complete. □

Because of Lemma 2.23 and because the functions we encounter in applications are usually lower semi-continuous, therefore, throughout the book, we always assume that convex functions do not take $-\infty$ value.

Proposition 2.24 (Jensen's Inequality) *A function $f : \mathbb{R}^n \to (-\infty, \infty]$ is convex if and only if for any $x_1, \cdots, x_k \in \text{dom}(f)$ and $t_1, \cdots, t_k \geq 0$ with $t_1 + \cdots + t_k = 1$, there holds*

$$f(t_1 x_1 + \cdots + t_k x_k) \leq t_1 f(x_1) + \cdots + t_k f(x_k).$$

Proof It can be proved by an induction argument. □

Theorem 2.25 *Any local minimizer of a convex function is a global minimizer.*

Proof Let x^* be a local minimizer of a convex function $f : \mathbb{R}^n \to (-\infty, \infty]$. Then, there exists a ball $B_r(x^*)$ such that $f(x^*) \leq f(x)$ for all $x \in B_r(x^*)$. For any $x \in \mathbb{R}^n$, we have $x^* + t(x - x^*) \in B_r(x^*)$ for small $t > 0$. By convexity, we have

$$f(x^*) \leq f(x^* + t(x - x^*)) \leq tf(x) + (1 - t)f(x^*)$$

which implies that $f(x^*) \leq f(x)$ for all $x \in \mathbb{R}^n$. Therefore, x^* is a global minimizer of f. □

Definition 2.26 (Strictly Convex Function) A function $f : \mathbb{R}^n \to (-\infty, \infty]$ is called *strictly convex* if

$$f(tx_1 + (1 - t)x_0) < tf(x_1) + (1 - t)f(x_0)$$

for any points $x_0, x_1 \in \text{dom}(f)$ with $x_0 \neq x_1$ and $0 < t < 1$.

Note that strictly convex functions are convex; however, convex functions are not necessarily strictly convex.

For a proper convex function $f : \mathbb{R}^n \to (-\infty, \infty]$, a constant μ is called the *modulus of convexity* of f if it is the largest number such that

$$f(tx_1 + (1 - t)x_2) + \mu t(1 - t)\|x_1 - x_2\|^2 \leq tf(x_1) + (1 - t)f(x_2)$$

holds for all $x_1, x_2 \in \text{dom}(f)$ and $0 \leq t \leq 1$. Clearly, $\mu \geq 0$. If $\mu > 0$, then f is called *strongly convex*. Strongly convex functions must be strictly convex.

Example 2.27

(a) Any norm on $\mathbb{R}^n$ is a convex function. Thus, $\|x\|_1$ is a convex function on $\mathbb{R}^n$. However, $\|x\|_1$ is not strictly convex because for $x = (1, 0, \cdots, 0)^T$ and $y = (2, 0, \cdots, 0)^T$, we have for any $0 < t < 1$ that

$$\|tx + (1 - t)y\|_1 = t\|x\|_1 + (1 - t)\|y\|_1.$$

(b) For the function $f(x) = \|x\|_2^2$ on $\mathbb{R}^n$, by the polarization identity, we have for $x, y \in \mathbb{R}^n$ and $0 \leq t \leq 1$ that

$$\begin{aligned}\|tx + (1 - t)y\|_2^2 &= t^2\|x\|_2^2 + (1 - t)^2\|y\|_2^2 + 2t(1 - t)\langle x, y\rangle \\ &= t\|x\|_2^2 + (1 - t)\|y\|_2^2 - t(1 - t)\left(\|x\|_2^2 + \|y\|_2^2 - 2\langle x, y\rangle\right) \\ &= t\|x\|_2^2 + (1 - t)\|y\|_2^2 - t(1 - t)\|x - y\|_2^2.\end{aligned}$$

Thus, $f(x) = \|x\|_2^2$ is strongly convex with modulus of convexity $\mu = 1$ and hence it is strictly convex.

Proposition 2.28 *If $f : \mathbb{R}^n \to (-\infty, \infty]$ is a proper, strict convex function, then it has at most one minimizer.*

Proof Assume that f has two distinct minimizers x_0 and x_1. By the strict convexity of f, we have for $0 < t < 1$ that

$$f(x_0) = f(x_1) \le f(tx_0 + (1-t)x_1) < tf(x_0) + (1-t)f(x_1) = f(x_0)$$

which is a contradiction. □

Proposition 2.29 *Let $f : \mathbb{R}^n \to (-\infty, \infty]$ be convex. Then, f is locally Lipschitz continuous in a neighborhood of each point in the interior of* dom(f)*, i.e., for any $x_0 \in$* int(dom(f))*, there exist $L \ge 0$ and $r > 0$ such that*

$$|f(x) - f(y)| \le L\|x - y\|$$

for all $x, y \in B_r(x_0)$.

Proof Given $x_0 \in \text{int}(\text{dom}(f))$, we can find a cubic K centered at x_0 that is contained in dom(f). Let $v_1, \cdots, v_m$ be the vertices of the cubic, where $m = 2^n$. Then, each $x \in K$ is a convex combination of $v_1, \cdots, v_m$, that is,

$$x = \sum_{i=1}^{m} t_i v_i, \quad t_i \ge 0 \text{ for } i = 1, \cdots, m \text{ and } \sum_{i=1}^{m} t_i = 1.$$

By the convexity of f, we then obtain

$$f(x) \le \sum_{i=1}^{m} t_i f(x_i) \le \max\{f(v_i) : i = 1, \cdots, m\} =: M_0.$$

Furthermore, for any $x \in K$, we have by symmetry that $y := 2x_0 - x \in K$. As $x_0 = (x + y)/2$, from the convexity of f we can obtain

$$f(x_0) \le \frac{1}{2}(f(x) + f(y))$$

which implies $f(x) \ge 2f(x_0) - f(y) \ge 2f(x_0) - M_0$. Therefore, f is bounded on K.

The above argument shows that we can find $M \ge 0$ and $r > 0$ such that $|f(x)| \le M$ for all $x \in B_{2r}(x_0)$. For any two distinct points $x, y \in B_r(x_0)$, let $a = \|x - y\|$ and set $z = x + \frac{r}{a}(x - y)$. Then, we have $z \in B_{2r}(x_0)$ and

$$x = \frac{r}{a+r}y + \frac{a}{a+r}z.$$

Thus, it follows from the convexity of f that

$$f(x) \le \frac{r}{a+r} f(y) + \frac{a}{a+r} f(z)$$

which implies that

$$f(x) - f(y) \le \frac{a}{a+r}(f(z) - f(y)) \le \frac{a}{r}|f(z) - f(y)| \le \frac{2Ma}{r} = \frac{2M}{r}\|x - y\|.$$

Since $x, y \in B_r(x_0)$ are arbitrary, we may interchange x and y in the above inequality to obtain

$$|f(x) - f(y)| \le \frac{2M}{r}\|x - y\|$$

which shows the locally Lipschitz continuity of f around x_0. □

We next consider differentiable convex functions. Recall that a function $f : \mathbb{R}^n \to \mathbb{R}$ is called differentiable at a point x if there is a vector $u \in \mathbb{R}^n$ such that

$$|f(y) - f(x) - \langle u, y - x\rangle| = o(\|y - x\|)$$

as $y \to x$. The vector u is called the gradient of f at x and is denoted as $\nabla f(x)$. If f is differentiable at x, by the chain rule, we have

$$\frac{d}{dt} f(x + tv)\bigg|_{t=0} = \langle \nabla f(x), v\rangle, \quad \forall v \in \mathbb{R}^n.$$

Furthermore, if f differentiable on $\mathbb{R}^n$, there holds the mean value theorem, i.e., for any $x_0, x_1 \in \mathbb{R}^n$, there exists $0 < t < 1$ such that

$$f(x_1) - f(x_0) = \langle \nabla f(x_0 + t(x_1 - x_0)), x_1 - x_0\rangle.$$

A differentiable function $f : \mathbb{R}^n \to \mathbb{R}$ is called L-smooth for some constant $L \ge 0$ if its gradient ∇f is L-Lipschitz continuous, i.e.,

$$\|\nabla f(x) - \nabla f(y)\|_2 \le L\|x - y\|_2 \quad \text{for all } x, y \in \mathbb{R}^n.$$

We have the following simple but useful result.

Lemma 2.30 (Descent Lemma) *If a differentiable function $f : \mathbb{R}^n \to \mathbb{R}$ is L-smooth for some constant $L \ge 0$, then*

$$f(y) \le f(x) + \langle \nabla f(x), y - x\rangle + \frac{L}{2}\|y - x\|^2$$

for all $x, y \in \mathbb{R}^n$.

Proof By the fundamental theorem of calculus, we have

$$f(x) - f(z) = \int_0^1 \frac{d}{ds}[f(z + s(x - z))]\,ds = \int_0^1 \langle \nabla f(z + s(x - z)), x - z\rangle ds.$$

Therefore,

$$f(x) - f(z) - \langle \nabla f(z), x - z\rangle = \int_0^1 \langle \nabla f(z + s(x - z)) - \nabla f(z), x - z\rangle ds.$$

By the Cauchy-Schwarz inequality and the Lipschitz continuity of ∇f, we then obtain

$$f(x) - f(z) - \langle \nabla f(z), x - z\rangle \leq L \int_0^1 s\|x - z\|^2 ds = \frac{L}{2}\|x - z\|^2.$$

The proof is complete. □

We have the following characterization result on the convexity of differentiable functions.

Proposition 2.31 *Let $f : \mathbb{R}^n \to \mathbb{R}$ be differentiable. The following are equivalent:*

(i) *f is convex in $\mathbb{R}^n$.*
(ii) *$f(x_1) \geq f(x_0) + \langle \nabla f(x_0), x_1 - x_0\rangle$ for all $x_0, x_1 \in \mathbb{R}^n$.*
(iii) *$\langle \nabla f(x_1) - \nabla f(x_0), x_1 - x_0\rangle \geq 0$ for all $x_0, x_1 \in \mathbb{R}^n$.*

Proof (i) $\Longrightarrow$ (ii). By differentiability and convexity of f, we have

$$\begin{aligned}\langle \nabla f(x_0), x_1 - x_0\rangle &= \lim_{t\searrow 0} \frac{f(x_0 + t(x_1 - x_0)) - f(x_0)}{t}\\ &\leq \lim_{t\searrow 0} \frac{tf(x_1) + (1 - t)f(x_0) - f(x_0)}{t}\\ &= f(x_1) - f(x_0).\end{aligned}$$

(ii) $\Longrightarrow$ (iii). For any $x_0, x_1 \in \mathbb{R}^n$, we have

$$\begin{aligned}f(x_1) &\geq f(x_0) + \langle \nabla f(x_0), x_1 - x_0\rangle,\\ f(x_0) &\geq f(x_1) + \langle \nabla f(x_1), x_0 - x_1\rangle.\end{aligned}$$

Adding this two inequalities gives $0 \geq \langle \nabla f(x_0) - \nabla f(x_1), x_1 - x_0\rangle$.

(iii) $\Longrightarrow$ (ii). Let $x_0, x_1 \in \mathbb{R}^n$. Then, there is $0 < \theta < 1$ such that

$$f(x_1) - f(x_0) = \langle \nabla f(x_\theta), x_1 - x_0\rangle,$$

where $x_\theta = x_0 + \theta(x_1 - x_0)$. Since $x_\theta - x_0 = \theta(x_1 - x_0)$, we have

$$\begin{aligned} f(x_1) - f(x_0) &= \langle \nabla f(x_0), x_1 - x_0 \rangle + \frac{1}{\theta}\langle \nabla f(x_\theta) - \nabla f(x_0), x_\theta - x_0 \rangle \\ &\geq \langle \nabla f(x_0), x_1 - x_0 \rangle. \end{aligned}$$

(ii) $\Longrightarrow$ (i). For $x_0, x_1 \in \mathbb{R}^n$ and $0 \leq t \leq 1$, let $x_t = tx_1 + (1-t)x_0$. Then,

$$\begin{aligned} f(x_0) &\geq f(x_t) + \langle \nabla f(x_t), x_0 - x_t \rangle, \\ f(x_1) &\geq f(x_t) + \langle \nabla f(x_t), x_1 - x_t \rangle. \end{aligned}$$

Since $x_0 - x_t = t(x_0 - x_1)$ and $x_1 - x_t = (1-t)(x_1 - x_0)$, we have

$$\begin{aligned} f(x_0) &\geq f(x_t) + t\langle \nabla f(x_t), x_0 - x_1) \rangle, \\ f(x_1) &\geq f(x_t) + (1-t)\langle \nabla f(x_t), x_1 - x_0 \rangle. \end{aligned}$$

Multiplying the first inequality by $1-t$ and the second by t and adding together gives $(1-t)f(x_0) + tf(x_1) \geq f(x_t)$. □

Definition 2.32 For a differentiable function $f : \mathbb{R}^n \to \mathbb{R}$, any point x^* satisfying $\nabla f(x^*) = 0$ is called a *critical point* of f.

Corollary 2.33 *Let $f : \mathbb{R}^n \to \mathbb{R}$ be a convex differentiable function. Then, any critical point of f is a global minimizer of f.*

Proof Let x^* be a critical point of f, i.e., $\nabla f(x^*) = 0$. Then, from Proposition 2.31, it follows that

$$f(x) \geq f(x^*) + \langle \nabla f(x^*), x - x^* \rangle = f(x^*)$$

for any $x \in \mathbb{R}^n$. Thus, x^* is a global minimizer of f. □

Finally, we conclude this chapter by proving the following important co-coercivity property on differentiable convex functions, which is known as Baillon-Haddad theorem [11].

Proposition 2.34 *Let $f : \mathbb{R}^n \to \mathbb{R}$ be a differentiable convex function whose gradient is L-Lipschitz continuous for some constant $L > 0$. Then,*

$$\frac{1}{2L}\|\nabla f(y) - \nabla f(x)\|^2 \leq f(y) - f(x) - \langle \nabla f(x), y - x \rangle \tag{2.4}$$

and

$$\langle \nabla f(y) - \nabla f(x), y - x \rangle \geq \frac{1}{L}\|\nabla f(y) - \nabla f(x)\|^2 \tag{2.5}$$

for all $x, y \in \mathbb{R}^n$.

Proof For any fixed $x \in \mathbb{R}^n$, we consider the function

$$\phi(y) := f(y) - f(x) - \langle \nabla f(x), y - x \rangle, \quad y \in \mathbb{R}^n.$$

According to Proposition 2.31, we have $\phi(y) \geq 0$ for all y. Note that $\nabla\phi(y) = \nabla f(y) - \nabla f(x)$, and we have

$$\|\nabla\phi(y) - \nabla\phi(z)\| = \|\nabla f(y) - \nabla f(z)\| \leq L\|y - z\|$$

for any $y, z \in \mathbb{R}^n$. Thus, $\nabla\phi$ is Lipschitz continuous with Lipschitz constant L. By Lemma 2.30, we thus have

$$0 \leq \phi(z) \leq \phi(y) + \langle \nabla\phi(y), z - y \rangle + \frac{L}{2}\|z - y\|^2 \tag{2.6}$$

for all y and z. Observe that the right-hand side of (2.6) is a quadratic function in z, and it achieves its minimum at $\bar{z}$ satisfying $\nabla\phi(y) + L(\bar{z} - y) = 0$, i.e., $\bar{z} = y - \frac{1}{L}\nabla\phi(y)$. Thus, it follows from (2.6) that

$$0 \leq \phi(y) + \langle \nabla\phi(y), \bar{z} - y \rangle + \frac{L}{2}\|\bar{z} - y\|^2 = \phi(y) - \frac{1}{2L}\|\nabla\phi(y)\|^2$$

which implies $\frac{1}{2L}\|\nabla\phi(y)\|^2 \leq \phi(y)$, that is,

$$\frac{1}{2L}\|\nabla f(y) - \nabla f(x)\|^2 \leq f(y) - f(x) - \langle \nabla f(x), y - x \rangle$$

which shows (2.4). Since this inequality is true for any x and y, we may switch x and y to obtain

$$\frac{1}{2L}\|\nabla f(x) - \nabla f(y)\|^2 \leq f(x) - f(y) - \langle \nabla f(y), x - y \rangle.$$

Adding the above two inequalities gives (2.5). □

Problems

2.1 Let $\Omega \subset \mathbb{R}^n$ be a convex set. Show that both $\mathrm{int}(\Omega)$ and $\overline{\Omega}$ are convex sets.

2.2 Show that a set $\Omega \subset \mathbb{R}^n$ is convex if and only if $t_1x_1 + \cdots + t_kx_k \in \Omega$ for any $x_1, \cdots, x_k \in \Omega$ and $t_1, \cdots, t_k \geq 0$ with $t_1 + \cdots + t_k = 1$.

2.3 Show that a function $f : \mathbb{R}^n \to (-\infty, \infty]$ is convex if and only if

$$f(t_1x_1 + \cdots + t_kx_k) \leq t_1 f(x_1) + \cdots + t_k f(x_k)$$

for any $x_1, \cdots, x_k \in \mathrm{dom}(f)$ and $t_1, \cdots, t_k \geq 0$ with $t_1 + \cdots + t_k = 1$.

2.4 Let $a \in \mathbb{R}^n$ be a nonzero vector and $b \in \mathbb{R}$. Consider the hyperplane

$$H := \{x \in \mathbb{R}^n : \langle a, x\rangle = b\}$$

in $\mathbb{R}^n$. Show that the projection of any $x \in \mathbb{R}^n$ onto H is given by

$$P_H(x) = x + \frac{b - \langle a, x\rangle}{\|a\|^2} a.$$

2.5 Let $\Omega \subset \mathbb{R}^n$ be a nonempty closed convex set and let $x \in \mathbb{R}^n$. Show that $\hat{x} = P_\Omega(x)$ if and only if $\hat{x} \in \Omega$ and $\langle x - \hat{x}, y - \hat{x}\rangle \leq 0$ for all $y \in \Omega$.

2.6 Let $A, B \subset \mathbb{R}^n$ be nonempty convex sets. If B is closed and

$$\sup_{x \in A} \langle \xi, x\rangle \leq \sup_{x \in B} \langle \xi, x\rangle, \quad \forall \xi \in \mathbb{R}^n,$$

show that $A \subset B$.

2.7 Let $\Omega \subset \mathbb{R}^n$ be a nonempty closed convex set. Show that the function $f(x) := \|x\|^2 - d_\Omega^2(x)$ is convex.

2.8 Let $\|\cdot\|$ be any norm on $\mathbb{R}^n$. Show that for any $x, y \in \mathbb{R}^n$, the function

$$f(t) := \frac{1}{t}\left(\|x + ty\| - \|x\|\right)$$

is an increasing function on $(0, \infty)$, i.e., $f(s) \leq f(t)$ for all $0 < s \leq t < \infty$.

2.9 A function $f : \mathbb{R}^n \to (-\infty, \infty]$ is called quasi-convex if

$$f(tx + (1-t)y) \leq \max\{f(x), f(y)\}$$

for all $x, y \in \mathbb{R}^n$ and $0 \leq t \leq 1$.

(i) Show that every convex function is quasi-convex.

(ii) Consider the function

$$f(x_1, x_2) = \begin{cases} -x_1 x_2, & x_1 \geq 0, x_2 \geq 0, \\ +\infty, & \text{otherwise} \end{cases}$$

on $\mathbb{R}^2$. Show that f is not convex but is quasi-convex.

(iii) Show that $f : \mathbb{R}^n \to (-\infty, \infty]$ is quasi-convex if and only if the sublevel sets

$$L_f^-(\alpha) = \{x \in \mathbb{R}^n : f(x) \leq \alpha\}$$

are convex for all $\alpha \in \mathbb{R}$.

(iv) Let $f : \mathbb{R}^n \to \mathbb{R}$ be a differentiable function. Show that f is quasi-convex if and only if

$$\langle \nabla f(x), y - x \rangle \leq 0$$

for all $x, y \in \mathbb{R}^n$ satisfying $f(y) \leq f(x)$.

2.10 Let $f : \mathbb{R}^n \to \mathbb{R}$ be a continuously differentiable function. Show that the following statements are equivalent:

(i) f is strictly convex on $\mathbb{R}^n$.
(ii) $f(x_1) > f(x_0) + \langle \nabla f(x_0), x_1 - x_0 \rangle$ for all $x_0, x_1 \in \mathbb{R}^n$ with $x_0 \neq x_1$.
(iii) $\langle \nabla f(x_1) - \nabla f(x_0), x_1 - x_0 \rangle > 0$ for all $x_0, x_1 \in \mathbb{R}^n$ with $x_0 \neq x_1$.

2.11 Let $f : \mathbb{R}^n \to \mathbb{R}$ be a continuously differentiable function with L-Lipschitz continuous gradient ∇f. Let $\alpha > 0$ be a number. Show that the function

$$\varphi_\alpha(x) := \frac{1}{2}\alpha \|x\|^2 - f(x)$$

is convex if $\alpha \geq L$ and is strictly convex if $\alpha > L$.

2.12 Let $f : \mathbb{R}^n \to \mathbb{R}$ be a continuously differentiable convex function and let L be a positive constant. Show that the following statements are equivalent:

(i) ∇f is L-Lipschitz continuous on $\mathbb{R}^n$, i.e.,

$$\|\nabla f(x) - \nabla f(y)\| \leq L\|x - y\|, \quad \forall x, y \in \mathbb{R}^n.$$

(ii) The function $g(x) := \frac{L}{2}\|x\|^2 - f(x)$ is convex on $\mathbb{R}^n$.
(iii) For all $x, y \in \mathbb{R}^n$, there holds

$$f(y) \leq f(x) + \langle \nabla f(x), y - x \rangle + \frac{L}{2}\|x - y\|^2.$$

(iv) For all $x, y \in \mathbb{R}^n$, there holds

$$\langle \nabla f(x) - \nabla f(y), x - y \rangle \leq L\|x - y\|^2.$$

(v) For all $x, y \in \mathbb{R}^n$ and $t \in [0, 1]$, there holds

$$f(tx + (1 - t)y) \geq tf(x) + (1 - t)f(y) - \frac{1}{2}t(1 - t)L\|x - y\|^2.$$

(vi) For all $x, y \in \mathbb{R}^n$, there holds

$$\frac{1}{2L}\|\nabla f(y) - \nabla f(x)\|^2 \leq f(y) - f(x) - \langle \nabla f(x), y - x \rangle.$$

(vii) For all $x, y \in \mathbb{R}^n$, there holds

$$\langle \nabla f(y) - \nabla f(x), y - x \rangle \geq \frac{1}{L} \|\nabla f(y) - \nabla f(x)\|^2.$$

(viii) For all $x, y \in \mathbb{R}^n$ and $t \in [0, 1]$, there holds

$$f(tx + (1-t)y) \leq tf(x) + (1-t)f(y) - \frac{1}{2L} t(1-t)\|\nabla f(x) - \nabla f(y)\|^2.$$

2.13 Let $f : \mathbb{R}^n \to \mathbb{R}$ be a continuously differentiable function bounded from below with L-Lipschitz continuous gradient ∇f. Show that

$$\inf_{x \in \mathbb{R}^n} \left\{ f(x) - \frac{1}{2L} \|\nabla f(x)\|^2 \right\} > -\infty.$$

Chapter 3
Subgradient and Mirror Descent Methods

Consider the minimization problem

$$\min_{x \in \mathbb{R}^n} f(x), \tag{3.1}$$

where $f : \mathbb{R}^n \to (-\infty, \infty]$ is a proper convex function. In case f is continuous differentiable everywhere, the gradient method solves (3.1) iteratively by using the gradient of f to produce a sequence toward a minimizer. Assuming $x^0 \in \mathbb{R}^n$ is an initial guess, the iterative scheme takes the form

$$x^{k+1} = x^k - t_k \nabla f(x^k), \quad k = 0, 1, \cdots$$

where $t_k > 0$ is a step size and $\nabla f(x^k)$ denotes the gradient of f at x^k. Note that, when $\nabla f(x^k) \neq 0$, there holds

$$\frac{d}{dt}\Big|_{t=0} f(x^k - t\nabla f(x^k)) = -\langle \nabla f(x^k), \nabla f(x^k) \rangle < 0.$$

which implies that

$$f(x^k - t\nabla f(x^k)) < f(x^k) \quad \text{for small } t > 0.$$

This means that, by a careful choice of the step size t_k to produce x^{k+1}, we can guarantee $f(x^{k+1}) < f(x^k)$, and thus the objective function value decreases along the iterative sequence. In this sense, the gradient method is a descent method and the negative gradient direction is a descent direction. The convergence property of the gradient method has been well-studied. In particular, it has been shown that if f has a minimizer over $\mathbb{R}^n$ and if f is convex and differentiable with Lipschitz continuous gradient, then the sequence $\{x^k\}$ produced by the gradient method converges to a minimizer of f whenever the step sizes are chosen carefully by

Q. Jin, *Lectures on Nonsmooth Optimization*, Texts in Applied Mathematics 82,
https://doi.org/10.1007/978-3-031-91417-1_3

various strategies, including the constant step size, the exact line search, and inexact line search strategies. One may refer to [165, 169], for instance, for various results on convergence and convergence rates of the gradient method.

In case f is not differentiable everywhere, we may not be able to use the gradient method to solve (3.1) since the gradient might not be available. In this chapter, we will introduce the notion of subgradient as a replacement for gradient and consider various subgradient methods to solve (3.1).

3.1 Subdifferential

We will consider convex optimization problems with non-differentiable objective functions. As a substitute for gradient, we will introduce the notion of subgradient. Recall that for a convex function f on $\mathbb{R}^n$ that is differentiable at a point x, there holds

$$f(y) \geq f(x) + \langle \nabla f(x), y - x \rangle, \qquad \forall y \in \mathbb{R}^n;$$

see Proposition 2.31 and its proof. This motivates us to introduce the subgradient for convex functions in the following manner.

Definition 3.1 Let $f : \mathbb{R}^n \to (-\infty, \infty]$ be a proper convex function and let $x \in \mathbb{R}^n$.

(i) A vector $\xi \in \mathbb{R}^n$ is called a *subgradient* of f at x if

$$f(y) \geq f(x) + \langle \xi, y - x \rangle, \quad \forall y \in \mathbb{R}^n.$$

(ii) The set $\partial f(x)$ consisting of all subgradients of f at x is called the *subdifferential* of f at x.
(iii) If $\partial f(x) \neq \emptyset$, then f is said to be *subdifferentiable* at x. We set $\text{dom}(\partial f) := \{x \in \mathbb{R}^n : \partial f(x) \neq \emptyset\}$.

By definition, $\partial f(x) = \emptyset$ if $f(x) = +\infty$. Thus, $\partial f(x) \neq \emptyset$ implies $f(x)$ is finite. Furthermore, it is straightforward to show that $\partial f(x)$ is a closed convex subset of $\mathbb{R}^n$.

Proposition 3.2 *If $f : \mathbb{R}^n \to (-\infty, \infty]$ is a convex function that is differentiable at x, then $\partial f(x) = \{\nabla f(x)\}$.*

Proof Since f is convex and differentiable at x, by the argument in the proof of Proposition 2.31, we have

$$f(y) \geq f(x) + \langle \nabla f(x), y - x \rangle, \qquad \forall y \in \mathbb{R}^n.$$

Thus, $\nabla f(x) \in \partial f(x)$. Conversely, if $\xi \in \partial f(x)$, then by definition, we have

$$f(y) \geq f(x) + \langle \xi, y - x \rangle, \qquad \forall y \in \mathbb{R}^n.$$

By taking $y = x + tv$ with $v \in \mathbb{R}^n$ and $t > 0$, we obtain

$$\frac{f(x + tv) - f(x)}{t} \geq \langle \xi, v \rangle, \qquad \forall v \in \mathbb{R}^n \text{ and } t > 0.$$

Letting $t \to 0$ gives $\langle \nabla f(x), v \rangle \geq \langle \xi, v \rangle$ for all $v \in \mathbb{R}^n$, i.e.,

$$\langle \nabla f(x) - \xi, v \rangle \geq 0, \qquad \forall v \in \mathbb{R}^n.$$

Taking $v = \xi - \nabla f(x)$ gives $-\|\nabla f(x) - \xi\|^2 \geq 0$, which implies $\xi = \nabla f(x)$. □

Example 3.3 Let $\|\cdot\|$ be any norm on $\mathbb{R}^n$. Its *dual norm* is defined by

$$\|\xi\|^* := \max_{\|x\| \leq 1} \langle \xi, x \rangle, \quad \forall \xi \in \mathbb{R}^n.$$

It is easy to check that $\|\cdot\|^*$ indeed defines a norm on $\mathbb{R}^n$ with

$$\langle \xi, x \rangle \leq \|\xi\|^* \|x\|, \quad \forall \xi, x \in \mathbb{R}^n.$$

We now show that

$$\partial(\|\cdot\|)(x) = \{\xi \in \mathbb{R}^n : \|\xi\|^* \leq 1 \text{ and } \langle \xi, x \rangle = \|x\|\}, \quad \forall x \in \mathbb{R}^n.$$

In fact, if ξ is such that $\|\xi\|^* \leq 1$ and $\langle \xi, x \rangle = \|x\|$, then for any $y \in \mathbb{R}^n$, we have

$$\|y\| - \|x\| - \langle \xi, y - x \rangle = \|y\| - \langle \xi, y \rangle \geq \|y\| - \|\xi\|^* \|y\| \geq 0$$

which shows that $\xi \in \partial(\|\cdot\|)(x)$. Conversely, let $\xi \in \partial(\|\cdot\|)(x)$, and we have

$$\|y\| \geq \|x\| + \langle \xi, y - x \rangle, \qquad \forall y \in \mathbb{R}^n.$$

Taking $y = 0$ and $y = 2x$ respectively give $\langle \xi, x \rangle \geq \|x\|$ and $\langle \xi, x \rangle \leq \|x\|$. Thus, $\langle \xi, x \rangle = \|x\|$ and consequently $\|y\| \geq \langle \xi, y \rangle$ for all $y \in \mathbb{R}^n$. By definition, we then obtain $\|\xi\|^* = \max_{\|y\| \leq 1} \langle \xi, y \rangle \leq 1$. □

Example 3.4 For any $1 \leq p \leq \infty$, one can show that the dual norm of $\|\cdot\|_p$ on $\mathbb{R}^n$ is $\|\cdot\|_{p^*}$, where p^* is the number conjugate to p, i.e., $1/p + 1/p^* = 1$. Thus, it follows from Example 3.3 that

$$\partial(\|\cdot\|_p)(x) = \{\xi \in \mathbb{R}^n : \|\xi\|_{p^*} \leq 1 \text{ and } \langle \xi, x \rangle = \|x\|_p\}, \quad \forall x \in \mathbb{R}^n.$$

This in particular implies that $\operatorname{sign}(x) \in \partial(\|\cdot\|_1)(x)$ for all $x \in \mathbb{R}^n$. □

Example 3.5 Let A be an $m \times n$ matrix. We use $\mathcal{N}(A)$ and $\mathcal{R}(A)$ to denote its null space and range space, respectively. From linear algebra, we have $\mathcal{N}(A)^{\perp} = \mathcal{R}(A^T)$. Given $b \in \mathcal{R}(A)$, we define

$$L := \{x \in \mathbb{R}^n : Ax = b\}.$$

Then, for the indicator function δ_L of L, we have for $x \in L$ that

$$\begin{aligned} \xi \in \partial\delta_L(x) &\Longleftrightarrow \delta_L(y) \geq \delta_L(x) + \langle \xi, y - x \rangle, \quad \forall y \in \mathbb{R}^n \\ &\Longleftrightarrow 0 \geq \langle \xi, y - x \rangle, \quad \forall y \in L \\ &\Longleftrightarrow 0 \geq \langle \xi, v \rangle, \quad \forall v \in \mathcal{N}(A) \\ &\Longleftrightarrow \xi \in \mathcal{N}(A)^{\perp} = \mathcal{R}(A^T). \end{aligned}$$

Therefore, $\partial\delta_L(x) = \mathcal{R}(A^T)$ for all $x \in L$. □

Theorem 3.6 (Existence of Subgradients) *Let $f : \mathbb{R}^n \to (-\infty, \infty]$ be proper and convex. If f is continuous at $\bar{x} \in \mathbb{R}^n$, then $\partial f(\bar{x}) \neq \emptyset$.*

Proof Since f is proper and convex, $\text{epi}(f)$ is a nonempty convex set in $\mathbb{R}^n \times \mathbb{R}$. Note that $(\bar{x}, f(\bar{x}))$ is not an interior point of $\text{epi}(f)$. By Theorem 2.12, there is nonzero $(\xi, \alpha) \in \mathbb{R}^n \times \mathbb{R}$ such that

$$\langle \xi, x \rangle + \alpha t \leq \langle \xi, \bar{x} \rangle + \alpha f(\bar{x}), \quad \forall (x, t) \in \text{epi}(f). \tag{3.2}$$

Taking $x = \bar{x}$ and letting $t \to \infty$ shows that $\alpha \leq 0$. If $\alpha = 0$, then

$$\langle \xi, x \rangle \leq \langle \xi, \bar{x} \rangle, \qquad \forall x \in \text{dom}(f).$$

Since f is continuous at $\bar{x}$, there is an open ball $B_r(\bar{x})$ contained in $\text{dom}(f)$. Consequently,

$$\langle \xi, x - \bar{x} \rangle \leq 0, \quad \forall x \in B_r(\bar{x}).$$

This implies $\xi = 0$ and thus $(\xi, \alpha) = (0, 0)$, which is a contradiction. Hence $\alpha < 0$. It then follows from (3.2) that

$$t \geq f(\bar{x}) + \left\langle -\frac{\xi}{\alpha}, x - \bar{x} \right\rangle, \qquad \forall (x, t) \in \text{epi}(f).$$

This in particular implies

$$f(x) \geq f(\bar{x}) + \left\langle -\frac{\xi}{\alpha}, x - \bar{x} \right\rangle, \qquad \forall x \in \mathbb{R}^n.$$

Therefore, $-\xi/\alpha \in \partial f(\bar{x})$. □

Example 3.7 Let $\|\cdot\|$ be any norm on $\mathbb{R}^n$ and denote by $\|\cdot\|^{**}$ the dual of the dual norm $\|\cdot\|^*$, i.e., $\|\cdot\|^{**} = (\|\cdot\|^*)^*$. Then,

$$\|\cdot\|^{**} = \|\cdot\|. \tag{3.3}$$

To see this, we first note that

$$\|x\|^{**} = \max_{\|\xi\|^* \le 1} \langle \xi, x\rangle \le \max_{\|\xi\|^* \le 1} \|\xi\|^* \|x\| = \|x\|, \quad \forall x \in \mathbb{R}^n.$$

On the other hand, since $\|x\|$ is a continuous convex function on $\mathbb{R}^n$, it follows from Theorem 3.6 that $\partial(\|\cdot\|)(x) \neq \emptyset$ for any $x \in \mathbb{R}^n$. Thus, we may use Example 3.3 to conclude that for any $x \in \mathbb{R}^n$, there is $\xi_x \in \mathbb{R}^n$ with $\|\xi_x\|^* \le 1$ and $\langle \xi_x, x\rangle = \|x\|$. Consequently,

$$\|x\|^{**} = \max_{\|\xi\|^* \le 1} \langle \xi, x\rangle \ge \langle \xi_x, x\rangle = \|x\|$$

Therefore, $\|x\|^{**} = \|x\|$ for all $x \in \mathbb{R}^n$.

Based on (3.3), we may use Example 3.3 to obtain

$$\partial(\|\cdot\|^*)(\xi) = \left\{x \in \mathbb{R}^n : \|x\| \le 1 \text{ and } \langle \xi, x\rangle = \|\xi\|^*\right\}$$

for any $\xi \in \mathbb{R}^n$. □

Example 3.8 Let C be a nonempty closed convex set in $\mathbb{R}^n$ and consider $d_C(x)$, which is the distance of x to C. According to Example 2.22, d_C is a convex function. Note that for any $x, z \in \mathbb{R}^n$, we have

$$d_C(x) - d_C(z) \le \|x - P_C(z)\| - \|z - P_C(z)\| \le \|x - z\|$$

and hence

$$|d_C(x) - d_C(z)| \le \|x - z\|, \quad \forall x, z \in \mathbb{R}^n.$$

Thus, d_C is Lipschitz continuous on $\mathbb{R}^n$ with Lipschitz constant 1. According to Theorem 3.6, $\partial d_C(x) \neq \emptyset$ for all x. Let us determine an element in $\partial d_C(x)$ for each $x \in \mathbb{R}^n$. One can check $0 \in \partial d_C(x)$ if $d_C(x) = 0$ and

$$\xi := \frac{x - P_C(x)}{d_C(x)} \in \partial d_C(x) \qquad \text{if } d_C(x) > 0.$$

Indeed, for any $y \in \mathbb{R}^n$, we have

$$\begin{aligned} d_C(y) - d_C(x) - \langle \xi, y - x \rangle &= d_C(y) - d_C(x) - \frac{1}{d_C(x)}\langle x - P_C(x), y - x \rangle \\ &= d_C(y) - \frac{1}{d_C(x)}\langle x - P_C(x), y - P_C(x) \rangle \\ &= d_C(y) - \frac{1}{d_C(x)}\langle x - P_C(x), y - P_C(y) \rangle \\ &\quad - \frac{1}{d_C(x)}\langle x - P_C(x), P_C(y) - P_C(x) \rangle. \end{aligned}$$

Using (2.2) and the Cauchy-Schwartz inequality, it follows that

$$\begin{aligned} d_C(y) - d_C(x) - \langle \xi, y - x \rangle &\geq d_C(y) - \frac{1}{d_C(x)}\langle x - P_C(x), y - P_C(y) \rangle \\ &\geq d_C(y) - \frac{1}{d_C(x)}\|x - P_C(x)\| \|y - P_C(y)\| \\ &= 0 \end{aligned}$$

which shows the claim. □

Theorem 3.9 (Optimality) *Let $f : \mathbb{R}^n \to (-\infty, \infty]$ be a proper convex function. Then,*

$$x^* \in \arg\min_{x \in \mathbb{R}^n} f(x) \Longleftrightarrow 0 \in \partial f(x^*).$$

Proof We have

$$\begin{aligned} x^* \in \arg\min_{x \in \mathbb{R}^n} f(x) &\Longleftrightarrow f(x) \geq f(x^*) \text{ for all } x \in \mathbb{R}^n \\ &\Longleftrightarrow f(x) \geq f(x^*) + \langle 0, x - x^* \rangle \text{ for all } x \in \mathbb{R}^n \\ &\Longleftrightarrow 0 \in \partial f(x^*). \end{aligned}$$

The proof is complete. □

Definition 3.10 (Bregman Distance) Let $f : \mathbb{R}^n \to (-\infty, \infty]$ be a proper convex function and let $x \in \mathbb{R}^n$ be such that $\partial f(x) \neq \emptyset$. For $\xi \in \partial f(x)$, we define

$$D_f^\xi(y, x) := f(y) - f(x) - \langle \xi, y - x \rangle, \quad \forall y \in \mathbb{R}^n$$

which is called *the Bregman distance* induced by f at x in the direction ξ.

The Bregman distance was first introduced in [42] and now it becomes a useful tool for studying optimization problems. By the convexity of f, we always have $D_f^{\xi}(y, x) \geq 0$. However, the Bregman distance is not a distance on $\mathbb{R}^n$ in general because it does not necessarily satisfy the symmetry and the triangle inequality. Nevertheless, there holds the following nice identity.

Lemma 3.11 *Let $f : \mathbb{R}^n \to (-\infty, \infty]$ be a proper convex function. Let $x, y \in \operatorname{dom}(\partial f)$ with any $\xi \in \partial f(x)$ and $\eta \in \partial f(y)$, there holds*

$$D_f^{\xi}(z, x) - D_f^{\eta}(z, y) + D_f^{\eta}(x, y) = \langle \xi - \eta, x - z \rangle$$

for any $z \in \mathbb{R}^n$.

Proof This can be verified directly by the definition of the Bregman distance. □

Proposition 3.12 (Monotonicity) *Let $f : \mathbb{R}^n \to (-\infty, \infty]$ be a proper convex function with modulus of convexity $\mu \geq 0$.*

(i) *For $x \in \operatorname{dom}(\partial f)$ and $\xi \in \partial f(x)$, there holds*

$$D_f^{\xi}(y, x) \geq \mu \|y - x\|^2, \quad \forall y \in \mathbb{R}^n.$$

(ii) *For any $x, y \in \operatorname{dom}(\partial f)$ and any $\xi \in \partial f(x)$, $\eta \in \partial f(y)$, there holds*

$$\langle \xi - \eta, x - y \rangle \geq 2\mu \|x - y\|^2.$$

Proof By the definition of modulus of convexity and $\xi \in \partial f(x)$, we have for any $y \in \mathbb{R}^n$ and $0 < t < 1$ that

$$\begin{aligned} tf(y) + (1-t)f(x) &\geq f(ty + (1-t)x) + \mu t(1-t)\|y - x\|^2 \\ &\geq f(x) + t\langle \xi, y - x \rangle + \mu t(1-t)\|y - x\|^2. \end{aligned}$$

This implies that

$$f(y) - f(x) \geq \langle \xi, y - x \rangle + \mu(1-t)\|y - x\|$$

for all $0 < t < 1$. Taking $t \to 0$ shows (i). Next, by using (i), we have

$$2\mu \|x - y\|^2 \leq D_f^{\xi}(y, x) + D_f^{\eta}(x, y) = \langle \xi - \eta, x - y \rangle$$

which shows (ii). The proof is thus complete. □

Finally, we give some calculus rules on subdifferentials. Let $f : \mathbb{R}^n \to (-\infty, \infty]$ be a convex function. By definition, it is straightforward to show that $\partial(\lambda f)(x) = \lambda \partial f(x)$ for $x \in \mathbb{R}^n$ and $\lambda > 0$.

Theorem 3.13 (Composition with Linear Mapping) *Let A be an $m \times n$ matrix and let $f : \mathbb{R}^m \to (-\infty, \infty]$ be a convex function. Then,*

$$A^T \partial f(Ax) \subset \partial(f \circ A)(x), \quad \forall x \in \mathbb{R}^n.$$

If there is $\bar{x} \in \mathbb{R}^n$ such that f is continuous at $A\bar{x}$, then

$$\partial(f \circ A)(x) = A^T \partial f(Ax), \quad \forall x \in \mathbb{R}^n.$$

Proof We first show $A^T \partial f(Ax) \subset \partial(f \circ A)(x)$. Let $\xi \in \partial f(Ax)$. Then,

$$f(u) \geq f(Ax) + \langle \xi, u - Ax \rangle, \qquad \forall u \in \mathbb{R}^m.$$

Taking $u = Ay$ with $y \in \mathbb{R}^n$ gives

$$\begin{aligned}(f \circ A)(y) = f(Ay) &\geq f(Ax) + \langle \xi, A(y - x) \rangle \\ &= (f \circ A)(x) + \langle A^T \xi, y - x \rangle\end{aligned}$$

for all $y \in \mathbb{R}^n$, which implies $A^T \xi \in \partial(f \circ A)(x)$.

Next, we show the converse under the additional given conditions. Let $\eta \in \partial(f \circ A)(x)$. We have

$$f(Ay) \geq f(Ax) + \langle \eta, y - x \rangle, \qquad \forall y \in \mathbb{R}^n.$$

Consider the convex set

$$\mathscr{L} := \{(Ay, f(Ax) + \langle \eta, y - x \rangle) : y \in \mathbb{R}^n\} \subset \mathbb{R}^m \times \mathbb{R}.$$

Then, $\mathscr{L}$ and $\operatorname{epi}(f)$ can only have boundary point in common. Since f is continuous at $A\bar{x}$, we may find $r > 0$ such that

$$B_r(A\bar{x}) \times [f(A\bar{x}) + 1, \infty) \subset \operatorname{epi}(f).$$

In particular, $\operatorname{epi}(f)$ has nonempty interior. Thus, by Corollary 2.16, there exist nonzero $(\xi, \alpha) \in \mathbb{R}^m \times \mathbb{R}$ and $b \in \mathbb{R}$ such that

$$\langle \xi, Ay \rangle + \alpha \left(f(Ax) + \langle \eta, y - x \rangle \right) \geq b, \quad \forall y \in \mathbb{R}^n, \tag{3.4}$$

$$\langle \xi, u \rangle + \alpha t \leq b, \qquad \forall (u, t) \in \operatorname{epi}(f). \tag{3.5}$$

Since $(A\bar{x}, t) \in \operatorname{epi}(f)$ for $t \geq f(A\bar{x})$, we may deduce from (3.5) that $\alpha \leq 0$. If $\alpha = 0$, then (3.4) and (3.5) become

$$\begin{aligned}\langle \xi, Ay \rangle &\geq b, \quad \forall y \in \mathbb{R}^n, \\ \langle \xi, u \rangle &\leq b, \qquad \forall u \in B_r(A\bar{x}).\end{aligned}$$

Taking $y = \bar{x}$, we deduce that $\langle \xi, u - A\bar{x} \rangle \le 0$ for all $u \in B_r(A\bar{x})$. This implies $\xi = 0$ and thus $(\xi, \alpha) = (0, 0)$, a contradiction.

Thus, $\alpha < 0$. By rescaling ξ and b, we may assume $\alpha = -1$. Then, (3.4) and (3.5) become

$$\langle \xi, Ay \rangle - (f(Ax) + \langle \eta, y - x \rangle) \ge b, \quad \forall y \in \mathbb{R}^n, \tag{3.6}$$

$$\langle \xi, u \rangle - t \le b, \qquad \forall (u, t) \in \text{epi}(f). \tag{3.7}$$

Equation (3.6) can be written as

$$\langle A^T \xi - \eta, y \rangle \ge b + f(Ax) - \langle \eta, x \rangle, \quad \forall y \in \mathbb{R}^n.$$

Since it is true for all y, we must have $\eta = A^T \xi$. Consequently, (3.6) becomes

$$- f(Ax) + \langle \xi, Ax \rangle \ge b.$$

Combining this with (3.7) gives

$$\langle \xi, u \rangle - t \le \langle \xi, Ax \rangle - f(Ax), \qquad \forall (u, t) \in \text{epi}(f).$$

Thus,

$$\langle \xi, u \rangle - f(u) \le \langle \xi, Ax \rangle - f(Ax), \qquad \forall u \in \text{dom}(f).$$

This implies $\xi \in \partial f(Ax)$ and hence $\eta \in A^T \partial f(Ax)$. □

Theorem 3.14 (Moreau-Rockafellar) *Let $f, g : \mathbb{R}^n \to (-\infty, \infty]$ be two convex functions. Then, for any $x \in \mathbb{R}$, there holds*

$$\partial f(x) + \partial g(x) \subset \partial (f + g)(x). \tag{3.8}$$

If there is $x_0 \in \text{dom}(f) \cap \text{dom}(g)$ such that f is continuous at x_0, then for any $x \in \mathbb{R}^n$, there holds

$$\partial (f + g)(x) = \partial f(x) + \partial g(x). \tag{3.9}$$

Proof We first prove (3.8). For any $\xi \in \partial f(x)$ and $\eta \in \partial g(x)$, we have

$$f(y) \ge f(x) + \langle \xi, y - x \rangle \quad \text{and} \quad g(y) \ge g(x) + \langle \eta, y - x \rangle$$

for all $y \in \mathbb{R}^n$. Thus,

$$(f + g)(y) \ge (f + g)(x) + \langle \xi + \eta, y - x \rangle, \quad \forall y \in \mathbb{R}^n.$$

This shows that $\xi + \eta \in \partial (f + g)(x)$.

Next, we prove (3.9) under the additional conditions. We may assume that both $f(x)$ and $g(x)$ are finite. Let $\xi \in \partial(f+g)(x)$, and then

$$f(y)+g(y) \geq f(x)+g(x)+\langle \xi, y-x\rangle, \quad \forall y \in \mathbb{R}^n.$$

Consider the convex sets in $\mathbb{R}^n \times \mathbb{R}$:

$$\begin{aligned}\Lambda_f &:= \{(y,t) : f(y) - \langle \xi, y-x\rangle - f(x) < t\} \\ \Lambda_g &:= \{(y,t) : t \leq g(x) - g(y)\}.\end{aligned}$$

Then, $\Lambda_f \neq \emptyset$, $\Lambda_g \neq \emptyset$, and $\Lambda_f \cap \Lambda_g = \emptyset$. It follows from Theorem 2.15 that there exist nonzero $(\xi_0, \alpha) \in \mathbb{R}^n \times \mathbb{R}$ and $\gamma \in \mathbb{R}$ such that

$$\begin{aligned}\langle \xi_0, y\rangle + \alpha t &\leq \gamma, \quad \forall (y,t) \in \Lambda_f, \\ \langle \xi_0, y\rangle + \alpha t &\geq \gamma, \quad \forall (y,t) \in \Lambda_g.\end{aligned} \tag{3.10}$$

Since $(x,0) \in \Lambda_g$, we obtain $\gamma \leq \langle \xi_0, x\rangle$. Since $(x,t) \in \Lambda_f$ for any $t > 0$, we have $\langle \xi_0, x\rangle + \alpha t \leq \gamma$ for $t > 0$. Letting $t \searrow 0$ gives $\gamma \geq \langle \xi_0, x\rangle$. Therefore, $\gamma = \langle \xi_0, x\rangle$ and $\alpha \leq 0$.

If $\alpha = 0$, then we have from (3.10) and $\gamma = \langle \xi_0, x\rangle$ that

$$\begin{aligned}\langle \xi_0, y-x\rangle &\leq 0, \qquad \forall y \in \mathrm{dom}(f), \\ \langle \xi_0, y-x\rangle &\geq 0, \qquad \forall y \in \mathrm{dom}(g).\end{aligned}$$

Since $x_0 \in \mathrm{dom}(f) \cap \mathrm{dom}(g)$, this implies $\langle \xi_0, x_0 - x\rangle = 0$. Since f is continuous at x_0, there is $r > 0$ such that $B_r(x_0) \subset \mathrm{dom}(f)$. So for any $z \in B_r(0)$, we have $x_0 + z \in B_r(x_0)$ and hence

$$\langle \xi_0, z\rangle = \langle \xi_0, (x_0+z) - x\rangle \leq 0, \qquad \forall z \in B_r(0).$$

This implies $\xi_0 = 0$. Thus, $(\xi_0, \alpha) = (0,0)$, which is a contradiction.

Thus, $\alpha < 0$. It follows from (3.10) and $\gamma = \langle \xi_0, x\rangle$ that

$$\begin{aligned}\langle -\xi_0/\alpha, y-x\rangle &\leq t, \qquad \forall (y,t) \in \Lambda_f, \\ \langle -\xi_0/\alpha, y-x\rangle &\geq t, \qquad \forall (y,t) \in \Lambda_g.\end{aligned}$$

The second equation gives $\langle -\xi_0/\alpha, y-x\rangle \geq g(x) - g(y)$ for all $y \in \mathrm{dom}(g)$. Thus, $\xi_0/\alpha \in \partial g(x)$. From the first equation, we have

$$\langle -\xi_0/\alpha, y-x\rangle \leq f(y) - f(x) - \langle \xi, y-x\rangle, \quad \forall y \in \mathrm{dom}(f).$$

Therefore, $f(y) \geq f(x) + \langle \xi - \xi_0/\alpha, y - x \rangle$ for all $y \in \mathbb{R}^n$. This implies $\xi - \xi_0/\alpha \in \partial f(x)$ and thus

$$\xi = (\xi - \xi_0/\alpha) + \xi_0/\alpha \in \partial f(x) + \partial g(x).$$

The proof is complete. □

3.2 Projected Subgradient Method

Let $f : \mathbb{R}^n \to (-\infty, \infty]$ be a convex function that is not necessarily differentiable everywhere. Consider the minimization problem:

$$\min_{x \in \mathbb{R}^n} f(x). \tag{3.11}$$

Due to the lack of gradients, the standard gradient method cannot be applied directly to solve this problem. However, we may replace $\nabla f(x^k)$ in the gradient method by a subgradient $\xi^k \in \partial f(x^k)$ to perform the iteration. This leads to the following subgradient method for solving (3.11).

Algorithm 3.15 (Subgradient Method) *Choose an initial guess x^0 and repeat*

$$x^{k+1} = x^k - t_k \xi^k, \qquad k = 0, 1, \cdots,$$

where ξ_k is any subgradient of f at x^k and $t_k > 0$ denotes a step size.

Subgradient method in general is not necessarily a descent method because, for a given $\xi^k \in \partial f(x^k)$, $-\xi^k$ may not be a descent direction of f at x^k. For instance, consider

$$f(x) = |x_1| + 2|x_2| \qquad \text{on } \mathbb{R}^2.$$

Then, $\xi = (1, 2)^T$ is a subgradient of f at $x = (1, 0)^T$. However,

$$\lim_{t \searrow 0} \frac{f(x - t\xi) - f(x)}{t} = 1 > 0.$$

Therefore, the step size in subgradient method cannot be chosen by the exact line search or the backtracking line search. Step sizes in subgradient methods are usually pre-specified.

The subgradient method can be extended to solve constrained convex optimization problems of the form

$$\min_{x \in C} f(x), \tag{3.12}$$

where $f : \mathbb{R}^n \to (-\infty, \infty]$ is a convex function and $C \subset \mathbb{R}^n$ is a closed convex set. The corresponding algorithm takes the following form.

Algorithm 3.16 (Projected Subgradient Method) *Take an initial guess $x^0 \in C$ and iteratively define*

$$x^{k+1} = P_C\left(x^k - t_k\xi^k\right), \quad k = 0, 1, \cdots,$$

where ξ^k is any element in $\partial f(x^k)$, $t_k > 0$ is a step size, and P_C is the projection onto C.

For the convergence analysis of the (projected) subgradient method, we make the following assumptions:

Assumption 3.17

(i) *$f : \mathbb{R}^n \to (-\infty, \infty]$ is convex, C is closed convex, and f has a minimizer over C.*
(ii) *$\partial f(x) \neq \emptyset$ for each $x \in C$, and there is a constant $L > 0$ such that $\|\xi\| \leq L$ for all $\xi \in \partial f(x)$ and $x \in C$.*

The above assumption has the following consequence, which implies the Lipschitz continuity of f on C.

Lemma 3.18 *Let Assumption 3.17 hold. Then, f is Lipschitz continuous on C, i.e.,*

$$\|f(x) - f(y)\| \leq L\|x - y\|, \quad \forall x, y \in C.$$

Proof For $x, y \in C$, take $\xi \in \partial f(x)$ and $\eta \in \partial f(y)$. Then,

$$\langle \xi, y - x\rangle \leq f(y) - f(x) \leq \langle \eta, y - x\rangle.$$

Using $\|\xi\| \leq L$, $\|\eta\| \leq L$, and the Cauchy-Schwarz inequality, we have

$$-L\|x - y\| \leq f(y) - f(x) \leq L\|x - y\|$$

which completes the proof. □

The next result gives a sufficient condition to guarantee the validity of (ii) in Assumption 3.17.

Lemma 3.19 *Let $f : \mathbb{R}^n \to (-\infty, \infty]$ be convex and $C \subset \mathbb{R}^n$ be closed convex. If there is an open set Ω containing C such that f is Lipschitz continuous on Ω, i.e.,*

$$|f(x) - f(y)| \leq L\|x - y\|, \quad \forall x, y \in \Omega,$$

then $\|\xi\| \leq L$ for any $\xi \in \partial f(x)$ and $x \in C$.

Proof According to Theorem 3.6, $\partial f(x) \neq \emptyset$ for all $x \in C$. For any $\xi \in \partial f(x)$, we have $x + t\xi \in \Omega$ for small $t > 0$. Thus,

$$t\|\xi\|^2 = \langle \xi, t\xi \rangle \leq f(x + t\xi) - f(x) \leq Lt\|\xi\|$$

which implies $\|\xi\| \leq L$. □

For the convergence analysis of the projected subgradient method, we will use the following elementary fact.

Lemma 3.20 *If a sequence $\{a_k\} \subset \mathbb{R}$ is bounded from below and satisfies*

$$a_{k+1} - a_k \leq \beta_k, \quad k = 0, 1, \cdots,$$

where $\{\beta_k\}$ is a sequence of nonnegative numbers with $\sum_{k=0}^{\infty} \beta_k < \infty$, then $\lim_{k\to\infty} a_k$ must exist.

Proof Observing $a_k \leq a_0 + \sum_{l=0}^{\infty} \beta_l$ shows that $\{a_k\}$ is bounded and hence $\limsup_{k\to\infty} a_k$ is finite. Observing also that $a_k \leq a_j + \sum_{l=j}^{k-1} \beta_l$ for all $k > j$ shows that $\limsup_{k\to\infty} a_k \leq a_j + \sum_{l=j}^{\infty} \beta_l$. Consequently

$$\liminf_{j\to\infty} a_j \geq \limsup_{k\to\infty} a_k - \lim_{j\to\infty} \sum_{l=j}^{\infty} \beta_l = \limsup_{k\to\infty} a_k.$$

Therefore, $\liminf_{k\to\infty} a_k = \limsup_{k\to\infty} a_k$, i.e., $\lim_{k\to\infty} a_k$ exists. □

Now, we are ready to prove the following convergence result for the projected subgradient method.

Theorem 3.21 *Let Assumption 3.17 hold. If $\{t_k\}$ is chosen such that*

$$\sum_{k=0}^{\infty} t_k = \infty \quad \text{and} \quad \sum_{k=0}^{\infty} t_k^2 < \infty,$$

then, for the projected subgradient method given in Algorithm 3.16, there hold the following:

(i) *$\{x^k\}$ converges to a minimizer x^* of f over C.*
(ii) *Let $f_* = \min_{x\in C} f(x)$. Then,*

$$\min_{0\leq i\leq k} f(x^i) - f_* \leq \frac{\|x^0 - x^*\|^2 + L^2 \sum_{i=0}^{k} t_i^2}{2\sum_{i=0}^{k} t_i}.$$

Proof Let $\bar{x}$ be any minimizer of f over C. We have

$$\|x^k - t_k\xi^k - \bar{x}\|^2 = \|x^k - \bar{x}\|^2 + t_k^2\|\xi^k\|^2 + 2t_k\langle \xi^k, \bar{x} - x^k\rangle.$$

Since $\xi^k \in \partial f(x^k)$, we have $\langle \xi^k, \bar{x} - x^k\rangle \le f(\bar{x}) - f(x^k)$. Thus,

$$\|x^k - t_k\xi^k - \bar{x}\|^2 \le \|x^k - \bar{x}\|^2 + t_k^2\|\xi^k\|^2 + 2t_k(f_* - f(x^k)).$$

Using the nonexpansiveness of P_C and $P_C(\bar{x}) = \bar{x}$, we have

$$\|x^{k+1} - \bar{x}\| = \|P_C(x^k - t_k\xi^k) - P_C(\bar{x})\| \le \|x^k - t_k\xi^k - \bar{x}\|.$$

Therefore,

$$\|x^{k+1} - \bar{x}\|^2 \le \|x^k - \bar{x}\|^2 + t_k^2\|\xi^k\|^2 + 2t_k(f_* - f(x^k)). \tag{3.13}$$

Using $\|\xi^k\| \le L$ and $f_* - f(x^k) \le 0$, it follows

$$\|x^{k+1} - \bar{x}\|^2 \le \|x^k - \bar{x}\|^2 + L^2t_k^2.$$

Since $\sum_k t_k^2 < \infty$, it follows from Lemma 3.20 that $\{\|x^k - \bar{x}\|^2\}$ must be convergent, which in particular implies that $\{x^k\}$ is bounded.

We next show that $\{x^k\}$ has a subsequence $\{x^{k_\ell}\}$ such that $f(x^{k_\ell})$ converges to f_*. If not, then there is $\varepsilon > 0$ such that

$$f(x^k) \ge f_* + \varepsilon \qquad \text{for all large } k.$$

It then follows from (3.13) and $\|\xi^k\| \le L$ that

$$\|x^{k+1} - \bar{x}\|^2 \le \|x^k - \bar{x}\|^2 + t_k(L^2t_k - 2\varepsilon).$$

Since $t_k \to 0$ as $k \to \infty$, there is k_0 such that

$$\|x^{k+1} - \bar{x}\|^2 \le \|x^k - \bar{x}\|^2 - \varepsilon t_k, \qquad \forall k \ge k_0.$$

This implies $\varepsilon\sum_{k=k_0}^{\infty} t_k \le \|x^{k_0} - \bar{x}\|^2 < \infty$, a contradiction.

According to the above analysis, now we can extract a subsequence $\{x^{k_\ell}\}$ from $\{x^k\}$ such that

$$x^{k_\ell} \to x^* \quad \text{and} \quad f(x^{k_\ell}) \to f_* \qquad \text{as } \ell \to \infty.$$

The closedness of C and the continuity of f imply $x^* \in C$ and $f(x^*) = f_*$, i.e., x^* is a minimizer of f over C. Consequently, $\{\|x^k - x^*\|^2\}$ is convergent. Since $\|x^{k_\ell} - x^*\| \to 0$, we obtain $\|x^k - x^*\| \to 0$ as $k \to \infty$.

Finally, it follows from (3.13) with $\bar{x} = x^*$ that

$$2t_i \left(f(x^i) - f_* \right) \leq \|x^i - x^*\|^2 - \|x^{i+1} - x^*\|^2 + L^2 t_i^2.$$

Summing over i from 0 to k gives

$$2\sum_{i=0}^{k} t_i \left(f(x^i) - f_* \right) \leq \|x^0 - x^*\|^2 - \|x^{k+1} - x^*\|^2 + L^2 \sum_{i=0}^{k} t_i^2.$$

Consequently,

$$2\left(\sum_{i=0}^{k} t_i\right)\left(\min_{0\leq i\leq k} f(x^i) - f_*\right) \leq \|x^0 - x^*\|^2 + L^2 \sum_{i=0}^{k} t_i^2.$$

The proof is complete. □

The projected subgradient method can also be guaranteed to converge if the step size $\{t_k\}$ is chosen as $t_k = \alpha_k / \|\xi^k\|$ with $\{\alpha_k\}$ satisfying suitable summability conditions. Furthermore, with such choices of $\{t_k\}$, the condition (ii) in Assumption 3.17 can be relaxed.

Theorem 3.22 *Let Assumption 3.17 (i) hold and assume that $\partial f(x) \neq \emptyset$ for each $x \in C$ and for any $r > 0$, there is a constant $L > 0$ such that $\|\xi\| \leq L$ for all $\xi \in \partial f(x)$ and $x \in C \cap B_r(x^0)$. If $\{t_k\}$ is chosen as $t_k = \alpha_k / \|\xi^k\|$ with $\{\alpha_k\}$ satisfying*

$$\alpha_k > 0, \quad \sum_{k=0}^{\infty} \alpha_k = \infty \quad \textit{and} \quad \sum_{k=0}^{\infty} \alpha_k^2 < \infty,$$

then, for the projected subgradient method given in Algorithm 3.16, there hold the following:

(i) *$\{x^k\}$ converges to a minimizer x^* of f over C.*
(ii) *Let $f_* = \min_{x\in C} f(x)$. Then, there is a constant L' such that*

$$\min_{0\leq i\leq k} f(x^i) - f_* \leq \frac{L'(\|x^0 - x^*\|^2 + \sum_{i=0}^{k} \alpha_i^2)}{2\sum_{i=0}^{k} \alpha_i}.$$

Proof The result can be proved by essentially following the proof of Theorem 3.21 with minor modifications. □

In case f_* is known, we may take t_k to be the so-called Polyak step size [182]. This kind of step size is obtained by minimizing the right-hand side of (3.13) with respect to $t_k \in [0, \infty)$. Direct calculation shows that the right-hand side of (3.13) achieves its minimum at

$$t_k = \frac{f(x^k) - f_*}{\|\xi^k\|^2} \tag{3.14}$$

which is the *Polyak step size*. With this step size t_k, we obtain from (3.13) that

$$\|x^{k+1} - \bar{x}\|^2 \le \|x^k - \bar{x}\|^2 - \frac{(f(x^k) - f_*)^2}{\|\xi^k\|^2}. \tag{3.15}$$

Theorem 3.23 *Let all the conditions in Theorem 3.22 hold and assume that $f_* := \min_{x \in C} f(x)$ is known. Consider the projected subgradient method given in Algorithm 3.16 with $\{t_k\}$ chosen as the Polyak step size (3.14). Then, $\{x^k\}$ converges to a minimizer x^* of f over C, and there is a constant L such that*

$$\min_{0 \le i \le k} f(x^i) - f_* \le \frac{L\|x^0 - x^*\|}{\sqrt{k+1}}, \quad k = 0, 1, \cdots.$$

Proof Let $\bar{x}$ be any minimizer of f over C. From (3.15), it follows immediately that $\{\|x^k - \bar{x}\|\}$ is monotonically decreasing and thus is convergent. Furthermore, by the given conditions, we can find a constant L such that $\|\xi^k\| \le L$ for all k. Thus,

$$(f(x^k) - f_*)^2 \le L^2 \left(\|x^k - \bar{x}\|^2 - \|x^{k+1} - \bar{x}\|^2\right)$$

which implies that

$$\sum_{i=0}^{k} \left(f(x^i) - f_*\right)^2 \le L^2\|x^0 - \bar{x}\|^2 \tag{3.16}$$

for all k which in particular shows that $\sum_{k=0}^{\infty}(f(x^k) - f_*)^2 < \infty$ and hence $f(x^k) - f_* \to 0$ as $k \to \infty$. Now, we can follow the argument in the proof of Theorem 3.21 to conclude that $\{x^k\}$ converges to a minimizer x^* of f over C. By using (3.16) with $\bar{x} = x^*$, we also have

$$(k+1) \min_{0 \le i \le k} (f(x^i) - f_*)^2 \le L^2\|x^0 - x^*\|^2.$$

The proof is therefore complete. □

Finally, we conclude this section by presenting examples to demonstrate the applicability of the (projected) subgradient methods in various applications.

Example 3.24 (ℓ^1-norm Approximation) Given an $m \times n$ matrix A and a vector $b \in \mathbb{R}^m$, consider the convex minimization problem:

$$\min_{x \in \mathbb{R}^n} \|Ax - b\|_1;$$

see Example 1.5. For the objective function $f(x) = \|Ax - b\|_1$, we know from Theorem 3.13 and Example 3.4 that

$$A^T \mathrm{sign}(Ax - b) \in \partial f(x), \qquad \forall x \in \mathbb{R}^n.$$

Thus, using the subgradient method to solve this problem gives the iteration scheme:

$$x^{k+1} = x^k - t_k A^T \mathrm{sign}(Ax^k - b), \quad k = 0, 1, \cdots$$

with an initial guess x^0 and suitable step sizes $t_k > 0$. □

Example 3.25 (Lasso) Consider the linear system.

$$Ax = b,$$

where $b \in \mathbb{R}^m$ and A has size $m \times n$ with $m \ll n$, In general, it may have infinite many solutions. Assume the linear system has a sparse solution. In applications, b is usually corrupted by noise. Thus, instead of b, what we have is a noisy data $\tilde{b}$. How to use $\tilde{b}$ to reconstruct an approximate sparse solution? An approximate solution should be close to $\tilde{b}$ and should have small 1-norm. To keep this balance, we may define the approximate solution as the minimizer of

$$\min_{x \in \mathbb{R}^n} \left\{ \frac{1}{2} \|Ax - \tilde{b}\|_2^2 + \lambda \|x\|_1 \right\}$$

where the balance parameter $\lambda > 0$ should be carefully chosen. This is the so-called *ℓ^1-norm regularization*; in statistics, it is called *lasso* (least absolute shrinkage and selection operator); see [211].

The objective function in lasso is $f(x) = \frac{1}{2}\|Ax - \tilde{b}\|_2^2 + \lambda \|x\|_1$. By Theorem 3.14 and Example 3.4, we know

$$A^T(Ax - \tilde{b}) + \lambda \mathrm{sign}(x) \in \partial f(x), \qquad \forall x \in \mathbb{R}^n.$$

Thus, the subgradient method for solving lasso becomes

$$x^{k+1} = x^k - t_k \left(A^T(Ax^k - \tilde{b}) + \lambda \mathrm{sign}(x^k) \right), \quad k = 0, 1, \cdots.$$

starting from an initial guess x^0 with suitably chosen step sizes $t_k > 0$. It is easy to see that the conditions in Theorem 3.22 are satisfied. □

Example 3.26 (Convex Feasibility Problem) Given N closed convex sets $C_1, \cdots, C_N$ in $\mathbb{R}^n$ with nonempty intersection $C := C_1 \cap \cdots \cap C_N$, we consider the problem of finding a point in C, which is called the *convex feasibility problem* [13]. This problem amounts to solving the minimization problem:

$$\min_{x \in \mathbb{R}^n} \{f(x) := \max\{d_1(x), \cdots, d_N(x)\}\},$$

where $d_j(x) = d_{C_j}(x)$ for $j = 1, \cdots, N$. Let f_* denote the optimal value. Then, $x^* \in C_1 \cap \cdots \cap C_N$ if and only if $f(x^*) = f_*$ and $f_* = 0$. According to Example 2.22, each d_j is convex and hence f is also convex. Furthermore, for any $x, z \in \mathbb{R}^n$, let $j \in \{1, \cdots, N\}$ be such that $f(x) = d_j(x)$. Then,

$$f(x) - f(z) = d_j(x) - \max\{d_1(z), \cdots, d_N(z)\} \leq d_j(x) - d_j(z) \leq \|x - z\|.$$

Interchanging the order of x and z then gives $|f(x) - f(z)| \leq \|x - z\|$ for all $x, z \in \mathbb{R}^n$, which shows that f is Lipschitz continuous on $\mathbb{R}^n$ with Lipschitz constant 1.

To apply the subgradient method, we need to find an element in $\partial f(x)$ for $x \in \mathbb{R}^n$. This can be done as follows: choose j such that $d_j(x) = f(x)$, and then $\partial d_j(x) \subset \partial f(x)$ because for any $\xi \in \partial d_j(x)$, we have

$$f(y) \geq d_j(y) \geq d_j(x) + \langle \xi, y - x \rangle = f(x) + \langle \xi, y - x \rangle.$$

Thus, when using the subgradient method, at k-th iteration, we pick $i_k \in \{1, \cdots, N\}$ such that $d_{i_k}(x_k) = f(x_k)$ and, according to Example 3.8, we may take

$$\xi^k = \frac{x^k - P_{C_{i_k}}(x^k)}{d_{i_k}(x^k)} \in \partial f(x^k).$$

Using the Polyak step size $t_k = \frac{f(x^k) - f_*}{\|\xi^k\|} = d_{i_k}(x^k)$ gives

$$x^{k+1} = x^k - t_k \xi^k = P_{C_{i_k}}(x^k)$$

which leads to the following algorithm.

Algorithm 3.27 *Take an initial x^0. For $k = 0, 1, \cdots$, define*

$$x^{k+1} = P_{C_{i_k}}(x^k),$$

where $i_k \in \{1, \cdots, N\}$ is an index such that C_{i_k} is farthest from x^k.

As an application, we consider the linear system $Ax = b$, where A is a matrix of size $m \times n$ and $b \in \mathbb{R}^m$ is a column vector. Let a_i denote the i-th column of A^T and let b_i be the i-th component of b. We define the hyperplanes:

$$H_i := \{x \in \mathbb{R}^n : \langle a_i, x \rangle = b_i\}, \quad i = 1, \cdots, m.$$

Clearly, x is a solution of $Ax = b$ if and only if $x \in H_1 \cap H_2 \cap \cdots \cap H_m$. For any $x \in \mathbb{R}^n$, let $P_{H_i}(x)$ be the projection of x onto H_i. Then,

$$P_{H_i}(x) = x + \frac{b_i - \langle a_i, x \rangle}{\|a_i\|_2^2} a_i.$$

Thus, the Euclidean distance of x to H_i is given by

$$d_{H_i}(x) = \frac{|b_i - \langle a_i, x \rangle|}{\|a_i\|_2}.$$

Using Algorithm 3.27, we may derive the following algorithm for computing a solution of $Ax = b$.

Algorithm 3.28 (Greedy Kaczmarz Algorithm) *Take an initial guess x^0. For $k = 0, 1, \cdots$, define*

$$x^{k+1} = x^k + \frac{b_{i(k)} - \langle a_{i_k}, x^k \rangle}{\|a_{i_k}\|_2^2} a_{i_k}$$

where $i_k \in \{1, \cdots, m\}$ is an index j such that $|b_j - \langle a_j, x^k \rangle| / \|a_j\|_2$ is the largest.

Kaczmarz algorithm was first proposed in [124] to solve a linear system $Ax = b$ by projecting onto each hyperplane H_i according to a designated order in a cyclical manner. Algorithm 3.28 is a greedy version of the Kaczmarz algorithm in the sense that at each iteration step, x^k is projected onto the farthest hyperplane among $H_1, \cdots, H_m$. Kaczmarz algorithm has been extended to deal with inverse problems and imaging reconstruction; see [67, 96, 115, 120, 148]. In computed tomography, Kaczmarz algorithm is also called the algebraic reconstruction technique [103, 159]. In [206], a randomized Kaczmarz algorithm has been proposed in which at each iteration, x^k is projected onto a hyperplane randomly among $H_1, \cdots, H_m$ according to a certain probability distribution, and the method enjoys a linear convergence in expectation. □

Example 3.29 (Image Deblurring) Let x^* be a true image, which is corrupted by salt and pepper noise. Instead of x^*, we only have the contaminated image b. In applications, it is important to consider how to recover the true image from a contaminated one. Many models have been proposed to solve this problem. Assume images are grayscale and can be represented by $m \times n$ matrices. One successful model for reconstructing the original image from b is to consider the convex minimization problem:

$$\min_{x\in\mathbb{R}^{m\times n}} \left\{\alpha\|x-b\|_{1,1}+|x|_{TV,a}\right\}, \tag{3.17}$$

where $\alpha > 0$ is the regularization parameter,

$$\|x\|_{1,1}=\sum_{i=1}^{m}\sum_{j=1}^{n}|x_{i,j}|,$$

and $|x|_{TV,a}$ is the anisotropic total variation of x defined by

$$|x|_{TV,a}:=\sum_{i=1}^{m-1}\sum_{j=1}^{n}|x_{i+1,j}-x_{i,j}|+\sum_{i=1}^{m}\sum_{j=1}^{n-1}|x_{i,j+1}-x_{i,j}|.$$

Let

$$f(u,v):=\sum_{i=1}^{m}\sum_{j=1}^{n}(|u_{i,j}|+|v_{i,j}|),\qquad u,v\in\mathbb{R}^{m\times n}$$

and introduce the linear operator $D:\mathbb{R}^{m\times n}\to\mathbb{R}^{m\times n}\times\mathbb{R}^{m\times n}$ by

$$Dx=(D_1x,D_2x),\qquad \forall x\in\mathbb{R}^{m\times n},$$

where $D_1, D_2 : \mathbb{R}^{m\times n}\to\mathbb{R}^{m\times n}$ are linear operators such that for any $x\in\mathbb{R}^{m\times n}$, the matrices $p := D_1x$ and $q := D_2x$ are defined by

$$p_{i,j}=\begin{cases}x_{i+1,j}-x_{i,j}, & i=1,\cdots,m-1;\ j=1,\cdots,n,\\ 0 & i=m;\ j=1,\cdots,n,\end{cases}$$

$$q_{i,j}=\begin{cases}x_{i,j+1}-x_{i,j}, & i=1,\cdots,m;\ j=1,\cdots,n-1,\\ 0 & i=1,\cdots,m;\ j=n.\end{cases}$$

Then, $|x|_{TV,a} = f(Dx)$. Recall from Example 3.4 that $(\text{sign}(u),\text{sign}(v))\in\partial f(u,v)$. We have

$$D^T(\text{sign}(D_1x),\text{sign}(D_2x))\in\partial|x|_{TV,a},\qquad \forall x\in\mathbb{R}^{m\times n}$$

where $D^T:\mathbb{R}^{m\times n}\times\mathbb{R}^{m\times n}\to\mathbb{R}^{m\times n}$ is defined by

$$\langle D^T(p,q),x\rangle_F:=\langle(p,q),Dx\rangle_F=\langle p,D_1x\rangle_F+\langle q,D_2x\rangle_F$$

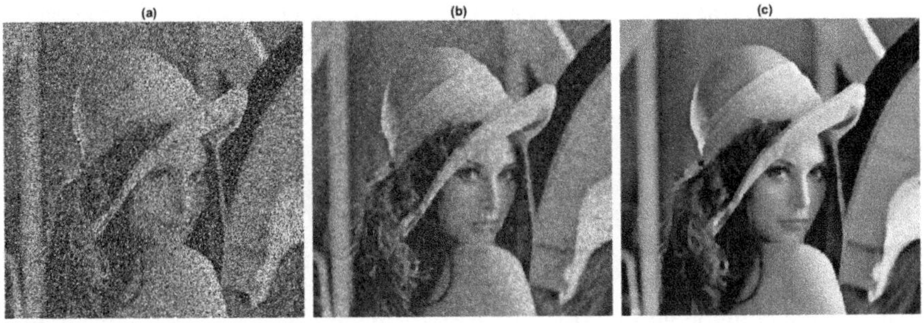

Fig. 3.1 Reconstruction images from blurred images by the subgradient method (3.18). (**a**) Blurred image. (**b**) Reconstruction after 20 iterations. (**c**) Reconstruction after 100 iterations

for all $(p, q) \in \mathbb{R}^{m\times n} \times \mathbb{R}^{m\times n}$ and $x \in \mathbb{R}^{m\times n}$. Direct calculation shows that $z := D^T(p, q)$ is given by

$$z_{i,j} = p_{i-1,j} - p_{i,j} + q_{i,j-1} - q_{i,j},$$

where we set $p_{0,j} = p_{m+1,j} = q_{i,0} = q_{i,n+1} = 0$ for $i = 1, \cdots, m$ and $j = 1, \cdots, n$. Note also that

$$\text{sign}(x - b) \in \partial\|x - b\|_{1,1}, \quad \forall x \in \mathbb{R}^{m\times n}.$$

Thus, the subgradient method for solving (3.17) takes the form

$$x^{k+1} = x^k - t_k \left(\alpha \text{sign}(x^k - b) + D^T(\text{sign}(D_1x^k), \text{sign}(D_2x^k))\right), \tag{3.18}$$

where $t_k > 0$ is the step size.

As an illustration on the performance of the subgradient method (3.18), we blur an image by adding salt and pepper noise with density 0.4 and plot the blurred image in Fig. 3.1a. We next use the method (3.18) with $x^0 = b$, $\alpha = 1$, and $t_k = 100/(k+1)$ to do the reconstruction. The computational results after 20 and 100 iterations are plotted in Fig. 3.1b and c. □

3.3 Incremental Subgradient Method

This section concerns the minimization problem of the form

$$\min_{x\in C} \left\{ f(x) := \sum_{i=1}^{N} f_i(x) \right\}, \tag{3.19}$$

where the constraint set $C \subset \mathbb{R}^n$ is a nonempty closed convex set and the objective function f is the finite sum of a large number of component functions $f_i : \mathbb{R}^n \to \mathbb{R}$ that are convex but not necessarily differentiable everywhere. If we apply the projected subgradient method given in Algorithm 3.16 to solve (3.19), at each iteration step, we need to determine a subgradient of f, which requires to consider all the component functions f_i. As N is large, this requires expensive computation and a huge amount of storage. To reduce the computational complexity per iteration, we now consider an incremental subgradient method [29, 125, 160, 161] in which at each step, a subgradient iteration is performed for a single component function f_i and all the component functions are used cyclically. The precise statement of the algorithm is as follows.

Algorithm 3.30 (Incremental Subgradient Method) *Pick an initial guess $x^0 \in C$. For $k = 0, 1, \cdots$, let $i_k = k \ (mod\ N) + 1$, take $\xi^k \in \partial f_{i_k}(x^k)$, and define*

$$x^{k+1} = P_C(x^k - t_k \xi^k)$$

with a suitably chosen step size $t_k \geq 0$, where $P_C : \mathbb{R}^n \to C$ denotes the Euclidean projection.

The convergence property of Algorithm 3.30 depends crucially on the choices of the step sizes. We first provide a convergence result when $\{t_k\}$ is chosen as a diminishing sequence. We need the following assumption.

Assumption 3.31 *$C \subset \mathbb{R}^n$ is a closed convex set. Each $f_i : \mathbb{R}^n \to \mathbb{R}$ is convex, and there is a constant L_i such that $\|\xi\| \leq L_i$ for all $\xi \in \partial f_i(x)$ with $x \in C$. Moreover, $S := \arg\min_{x \in C} f(x) \neq \emptyset$.*

Theorem 3.32 *Let Assumption 3.31 hold and consider Algorithm 3.30. If $\{t_k\}$ is chosen such that $t_{Nd} = t_{Nd+1} = \cdots = t_{Nd+N-1}$ for $d = 0, 1, \cdots$ and*

$$t_k > 0, \quad \sum_{k=0}^{\infty} t_k = \infty \quad \text{and} \quad \sum_{k=0}^{\infty} t_k^2 < \infty,$$

then $\{x^k\}$ converges to an optimal solution of (3.19).

Proof Since $x^{k+1} = P_C(x^k - t_k \xi^k)$ and $\xi^k \in \partial f_{i_k}(x^k)$, one may use the same argument in the proof of Theorem 3.21 to derive for any $\bar{x} \in S$ that

$$\|x^{k+1} - \bar{x}\|^2 \leq \|x^k - \bar{x}\|^2 - 2t_k \left(f_{i_k}(x^k) - f_{i_k}(\bar{x}) \right) + t_k^2 \|\xi^k\|^2. \tag{3.20}$$

Let $d \geq 0$ be any integer. Summing the above inequality over k from Nd to $N(d+1)-1$ and using the condition on $\{t_k\}$, we have

$$\begin{aligned}
\|x^{N(d+1)} - \bar{x}\|^2 &\leq \|x^{Nd} - \bar{x}\|^2 - \sum_{k=Nd}^{N(d+1)-1} \left\{2t_k \left(f_{i_k}(x^k) - f_{i_k}(\bar{x})\right) + t_k^2 \|\xi^k\|^2\right\} \\
&= \|x^{Nd} - \bar{x}\|^2 - 2t_{Nd} \left(\sum_{k=Nd}^{N(d+1)-1} f_{i_k}(x^k) - f(\bar{x}) \right) \\
&\quad + t_{Nd}^2 \sum_{k=Nd}^{N(d+1)-1} \|\xi^k\|^2 \\
&= \|x^{Nd} - \bar{x}\|^2 - 2t_{Nd} \left(f(x^{Nd}) - f(\bar{x})\right) + t_{Nd}^2 \sum_{k=Nd}^{N(d+1)-1} \|\xi^k\|^2 \\
&\quad - 2t_{Nd} \sum_{k=Nd}^{N(d+1)-1} \left(f_{i_k}(x^k) - f_{i_k}(x^{Nd})\right).
\end{aligned}$$

By using Assumption 3.31, we then obtain

$$\begin{aligned}
\|x^{N(d+1)} - \bar{x}\|^2 &\leq \|x^{Nd} - \bar{x}\|^2 - 2t_{Nd} \left(f(x^{Nd}) - f(\bar{x})\right) + t_{Nd}^2 \sum_{k=Nd}^{N(d+1)-1} L_{i_k}^2 \\
&\quad + 2t_{Nd} \sum_{k=Nd}^{N(d+1)-1} L_{i_k} \|x^k - x^{Nd}\|.
\end{aligned}$$

Note that for $Nd \leq k \leq Nd + N - 1$, we have

$$\begin{aligned}
\|x^k - x^{Nd}\| &\leq \sum_{l=Nd}^{k-1} \|x^{l+1} - x^l\| = \sum_{l=Nd}^{k-1} \|P_C(x^l - t_l \xi^l) - P_C(x^l)\| \\
&\leq \sum_{l=Nd}^{k-1} t_l \|\xi^l\| \leq t_{Nd} \sum_{l=Nd}^{k-1} L_{i_l}.
\end{aligned} \tag{3.21}$$

Therefore,

$$\|x^{N(d+1)} - \bar{x}\|^2 \leq \|x^{Nd} - \bar{x}\|^2 - 2t_{Nd} \left(f(x^{Nd}) - f(\bar{x})\right) + t_{Nd}^2 \sum_{k=Nd}^{N(d+1)-1} L_{i_k}^2$$

$$+ 2t_{Nd}^2 \sum_{k=Nd}^{N(d+1)-1} L_{i_k} \left(\sum_{l=Nd}^{k-1} L_{i_l} \right)$$

$$= \|x^{Nd} - \bar{x}\|^2 - 2t_{Nd} \left(f(x^{Nd}) - f(\bar{x}) \right) + t_{Nd}^2 L^2,$$

where

$$L := \sum_{k=Nd}^{N(d+1)-1} L_{i_k} = \sum_{i=1}^{N} L_i.$$

Since the conditions on $\{t_k\}$ imply that $\sum_{d=0}^{\infty} t_{Nd} = \infty$ and $\sum_{d=0}^{\infty} t_{Nd}^2 < \infty$, we may use the same argument in the proof of Theorem 3.21 to conclude that $\{x^{Nd}\}$ converges to an optimal solution x^* of (3.19). For any integer $k \geq 0$, let $d \geq 0$ be the largest integer such that $Nd \leq k$, and then $k \to \infty$ implies $d \to \infty$. Thus, we may use (3.21) and $t_{Nd} \to 0$ as $d \to \infty$ to obtain $\|x^k - x^*\| \leq \|x^{Nd} - x^*\| + \|x^k - x^{Nd}\| \to 0$ as $k \to \infty$. □

Next, by generalizing Polyak's strategy, we consider an adaptive rule to choose the step sizes for the incremental subgradient method under the assumption that the minimum of f over C is zero and every component function f_i is nonnegative.

Theorem 3.33 *Consider Algorithm 3.30 for solving (3.19) where each $f_i : \mathbb{R}^n \to \mathbb{R}$ is nonnegative and convex. Assume that $S := \arg\min_{x \in C} f(x) \neq \emptyset$ with $f_* := \min_{x \in C} f(x) = 0$. If $\{t_k\}$ is chosen such that*

$$t_k = \begin{cases} \frac{f_{i_k}(x^k) - f_*}{\|\xi^k\|^2} & \text{if } \xi^k \neq 0, \\ 0 & \text{otherwise,} \end{cases}$$

then $\{x^k\}$ converges to an optimal solution of (3.19).

Proof Let $\bar{x} \in S$ be any element. By the given condition, we have $f_i(\bar{x}) = f_* = 0$ for all $i = 1, \cdots, N$. According to (3.20), we have

$$\|x^{k+1} - \bar{x}\|^2 \leq \|x^k - \bar{x}\|^2 - 2t_k \left(f_{i_k}(x^k) - f_* \right) + t_k^2 \|\xi^k\|^2.$$

By the choice of t_k, we have $t_k^2 \|\xi^k\|^2 = t_k (f_{i_k}(x^k) - f_*)$. Therefore,

$$\|x^{k+1} - \bar{x}\|^2 \leq \|x^k - \bar{x}\|^2 - t_k \left(f_{i_k}(x^k) - f_* \right), \tag{3.22}$$

which implies that $\{\|x^k - \bar{x}\|\}$ is monotonically decreasing. Consequently, $\{x^k\}$ is bounded. Since each $f_i : \mathbb{R}^n \to \mathbb{R}$ is convex, by Proposition 2.29, it is locally

Lipschitz, and thus there is a constant L such that

$$\|\xi^k\| \leq L \quad \text{and} \quad |f_i(x^k) - f_i(x^l)| \leq L\|x^k - x^l\|, \quad i = 1, \cdots, N \tag{3.23}$$

for all integers $k, l = 0, 1, \cdots$. From (3.22), the monotonicity of $\{\|x^k - \bar{x}\|\}$, and $\|\xi^k\| \leq L$, we also have

$$t_k(f_{i_k}(x^k) - f_*) \to 0 \quad \text{and} \quad f_{i_k}(x^k) - f_* \to 0 \quad \text{as } k \to \infty. \tag{3.24}$$

We claim that

$$\lim_{k\to\infty} f_i(x^k) = f_*, \quad i = 1, \cdots, N. \tag{3.25}$$

To see this, recall $i_k = k \pmod N + 1$ and let $d := (k - i_k + 1)/N$. By using (3.23), we have

$$\begin{aligned}
&f_i(x^k) - f_* \\
&= f_i(x^{Nd+i_k-1}) - f_* = \left(f_i(x^{Nd+i-1}) - f_*\right) + \left(f_i(x^{Nd+i_k-1}) - f_i(x^{Nd+i-1})\right) \\
&\leq \left(f_i(x^{Nd+i-1}) - f_*\right) + L\|x^{Nd+i_k-1} - x^{Nd+i-1}\|.
\end{aligned}$$

Let us assume $i_k \leq i$; the case $i_k > i$ can be considered similarly. Since $x^k \in C$ for all k, by the triangle inequality and the definition of the algorithm, we have

$$\begin{aligned}
\|x^{Nd+i_k-1} - x^{Nd+i-1}\| &\leq \sum_{l=Nd+i_k-1}^{Nd+i-2} \|x^l - x^{l+1}\| \\
&= \sum_{l=Nd+i_k-1}^{Nd+i-2} \|P_C(x^l) - P_C(x^l - t_l\xi^l)\| \\
&\leq \sum_{l=Nd+i_k-1}^{Nd+i-2} t_l\|\xi^l\| = \sum_{l=Nd+i_k-1}^{Nd+i-2} \sqrt{t_l(f_{i_l}(x^l) - f_*)} \\
&\leq \sum_{l=Nd}^{Nd+N-1} \sqrt{t_l(f_{i_l}(x^l) - f_*)}.
\end{aligned}$$

Therefore,

$$f_i(x^k) - f_* \leq \left(f_i(x^{Nd+i-1}) - f_*\right) + L\sum_{l=Nd}^{Nd+N-1} \sqrt{t_l(f_{i_l}(x^l) - f_*)}. \tag{3.26}$$

Since $k \to \infty$ implies $d \to \infty$, we can obtain (3.25) immediately from the above inequality and (3.24).

Since $\{x^k\}$ is bounded, it has a convergent subsequence, say $\{x^{k_l}\}$, with $x^{k_l} \to x^*$ for some $x^* \in \mathbb{R}^n$. Since $x^{k_l} \in C$ and C is closed, we have $x^* \in C$. Moreover, by using (3.25) and the continuity of f_i, we also have $f_i(x^*) = f_*$ for all $i = 1, \cdots, N$. Thus, x^* is an optimal solution of (3.19). Consequently, $\{\|x^k - x^*\|\}$ is monotonically decreasing with $\|x^{k_l} - x^*\| \to 0$ as $l \to \infty$ and hence $\|x^k - x^*\| \to 0$ as $k \to \infty$. □

Example 3.34 As an important application of Theorem 3.33, we consider again the convex feasibility problem; see Example 3.26. It can be reformulated as the equivalent convex minimization problem:

$$\min_{x \in \mathbb{R}^n} \left\{ f(x) := \sum_{i=1}^{N} d_i(x) \right\},$$

where $d_i(x) := d_{C_i}(x)$ for $i = 1, \cdots, N$. Let f_* denote the minimum of f over $\mathbb{R}^n$. Clearly, $x^* \in \bigcap_{i=1}^N C_i$ if and only if $f(x^*) = f_*$ and $f_* = 0$. When applying Algorithm 3.30, at the k-th iteration, we take

$$\xi^k = \frac{x^k - P_{C_{i_k}}(x^k)}{d_{i_k}(x^k)} \in \partial d_{i_k}(x^k)$$

and use the adaptive step size $t_k = d_{i_k}(x^k)$ as suggested by Theorem 3.33. Then, the updating formula for x^{k+1} becomes

$$x^{k+1} = x^k - t_k \xi^k = P_{C_{i_k}}(x^k).$$

This leads to the following cyclic projection algorithm for determining a point in the intersection $\bigcap_{i=1}^N C_i$.

Algorithm 3.35 (Cyclic Projection Algorithm) *Given N closed convex subsets $C_1, \cdots, C_N$ in $\mathbb{R}^n$ with $\bigcap_{i=1}^N C_i \neq \emptyset$. Take any initial guess $x^0 \in \mathbb{R}^n$. For $k = 0, 1, \cdots$, define*

$$x^{k+1} = P_{C_{i_k}}(x^k),$$

where $i_k = k \pmod N + 1$.

Comparing with Algorithm 3.27 in which at each iteration step a farthest convex set from the current iterate is determined and then the projection is performed, Algorithm 3.35 performs the projections cyclically among $C_1, \cdots, C_N$ via a preassigned order. According to Theorem 3.33, the sequence $\{x^k\}$ produced by Algorithm 3.35 converges to a point in $\bigcap_{i=1}^N C_i$ for any initial guess x^0. In

Theorem 5.21, we will investigate the convergence rate of Algorithm 3.35 under further conditions on the sets $C_1, \cdots, C_N$.

3.4 Distributed Projected Subgradient Method

Let $(\mathcal{V}, \mathcal{E})$ be an undirected network of N agents, where $\mathcal{V} = \{1, \cdots, N\}$ is the node set representing the collection of the N agents, and $\mathcal{E} \subset \mathcal{V} \times \mathcal{V}$ is the set of undirected edges that are unordered pairs of nodes. Note that $(i, j) \in \mathcal{E}$ if and only if agents i and j can exchange information with each other. For each $i \in \mathcal{V}$, we use O_i to denote the collection of all agents that can exchange information with agent i, including itself.

In this section, we will consider the convex constrained minimization problems of the form

$$\min_{x \in C} \left\{ f(x) := \sum_{i=1}^{N} f_i(x) \right\} \tag{3.27}$$

which needs to be solved cooperatively by the N agents, where $C \subset \mathbb{R}^n$ is a closed convex set that is known to every agent. However, due to privacy and security reasons or due to the communication cost, each function $f_i : \mathbb{R}^n \to \mathbb{R}$ is privately known only to agent i, and each agent can only communicate its computational results with its neighbors. In order to solve the optimization problem (3.27) cooperatively, each agent needs to make full use of the information received from its neighborhoods in the network. without explicitly revealing their functions f_i.

We will assume that problem (3.27) satisfies Assumption 3.31, which in particular assumes that each f_i is a convex function that is not necessarily differentiable everywhere. We will use a distributed projected subgradient method [162, 186] to solve (3.27). By running this algorithm, every agent will produce an iterative sequence eventually. During the computation, at the k-th iteration step, each agent $j \in O_i$ will send its computed result x_j^k to agent i, and agent i will use these information to form an average of the form

$$v_i^k := \sum_{j \in O_i} w_{ij} x_j^k,$$

where $w_{ij} > 0$ are suitable weights with $\sum_{j \in O_i} w_{ij} = 1$. With this v_i^k at hand, agent i will compute x_i^{k+1} by applying one step of the projected subgradient method, i.e., taking $\xi_i^k \in \partial f_i(x_i^k)$ and updating

$$x_i^{k+1} = P_C(v_i^k - t_k \xi_i^k),$$

where $t_k > 0$ is a step size. By defining $w_{ij} = 0$ for $j \notin O_i$ and $i = 1, \cdots, N$, we may obtain a matrix $W = (w_{ij}) \in \mathbb{R}^{N\times N}$, which is called a mixing matrix. Then, the above procedure can be formulated as the following algorithm.

Algorithm 3.36 (Distributed Projected Subgradient Method) *Pick an initial guess $x_i^0 \in C$ for each $i = 1, \cdots, N$ and a mixing matrix $W = (w_{ij}) \in \mathbb{R}^{N\times N}$. For each $k = 0, 1, \cdots$, take a step size t_k and calculate*

$$v_i^k = \sum_{j=1}^{N} w_{ij} x_j^k \quad \text{and} \quad x_i^{k+1} = P_C(v_i^k - t_k \xi_i^k) \tag{3.28}$$

for $i = 1, \cdots, N$, where $\xi_i^k \in \partial f_i(x_i^k)$.

In the following, we will analyze Algorithm 3.36. We expect that for any $i, j \in \mathcal{V}$, the disagreement estimate $\|x_i^k - x_j^k\|$ can be as small as possible when k is sufficiently large and every $\{x_i^k\}$ converges to a common optimal solution of (3.27) as $k \to \infty$. The convergence behavior of Algorithm 3.36 depends crucially on the topological structure of the network $(\mathcal{V}, \mathcal{E})$ and the property of the mixing matrix W. We make the following assumption.

Assumption 3.37

(i) *The undirected network $(\mathcal{V}, \mathcal{E})$ is connected, i.e., for any $i, j \in \mathcal{V}$, there exist indices $i_1, \cdots, i_p \in \mathcal{V}$ such that $i = i_1$, $j = i_p$, and $i_l \in O_{i_{l+1}}$ for $l = 1, \cdots, p-1$.*

(ii) *The mixing matrix $W = (w_{ij}) \in \mathbb{R}^{N\times N}$ is symmetric doubly stochastic, i.e.,*

$$w_{ij} = w_{ji} \geq 0 \quad \text{and} \quad \sum_{j=1}^{N} w_{ij} = 1$$

for $i, j = 1, \cdots, N$. Moreover, $w_{ij} > 0$ if and only if $j \in O_i$.

We first provide some properties on mixing matrices satisfying Assumption 3.37. Let e denote the column vector of size N with every entry equal to 1, i.e., $e = (1, 1, \cdots, 1)^T$. We define the $N \times N$ matrix

$$J = \frac{1}{N} ee^T = \frac{1}{N} \begin{pmatrix} 1 & 1 & \cdots & 1 \\ 1 & 1 & \cdots & 1 \\ \vdots & \vdots & \ddots & \vdots \\ 1 & 1 & \cdots & 1 \end{pmatrix}$$

which is clearly a symmetric doubly stochastic matrix. For any doubly stochastic matrix W, by definition, we have $e^T W = e^T$ and $We = e$. Consequently, $JW = WJ = J$. Therefore, $(W - J)^k = W^k - J$ for any integer $k \geq 1$.

Lemma 3.38 *Let Assumption 3.37 hold. Then,* $\|W\|_2 = 1$ *and* $\beta := \|J - W\|_2 < 1$.

Proof Since W is symmetric, all its eigenvalues are real. Because W is doubly stochastic, for $e = (1, 1, \cdots, 1)^T$, we have $We = e$, which means $\lambda = 1$ is an eigenvalue of W. Now let λ be any eigenvalue of W and let $x = (x_j)$ be a corresponding eigenvector. We may choose an index i such that $|x_i| = \|x\|_\infty$. Since $\lambda x = Wx$, we have

$$\lambda x_i = \sum_{j=1}^{N} w_{ij} x_j = w_{ii} x_i + \sum_{j \neq i} w_{ij} x_j$$

which implies

$$|(\lambda - w_{ii}) x_i| = \left| \sum_{j \neq i} w_{ij} x_j \right| \leq \sum_{j \neq i} w_{ij} |x_j| \leq \sum_{j \neq i} w_{ij} |x_i| = (1 - w_{ii}) |x_i|.$$

Since $|x_i| > 0$, we can obtain $|\lambda - w_{ii}| \leq 1 - w_{ii}$, which shows that $-1 + 2w_{ii} \leq \lambda \leq 1$. Consequently, $\|W\|_2 = 1$ and as $w_{ii} > 0$, we also have $\lambda > -1$ for any eigenvalue λ of W.

Let $1 = \lambda_1 \geq \lambda_2 \geq \cdots \geq \lambda_N > -1$ be the eigenvalues of W, counting by the multiplicities. We now show that

$$\beta := \|J - W\|_2 = \max\{|\lambda_2|, |\lambda_N|\}. \tag{3.29}$$

To see this, let $u_1, \cdots, u_N$ be the eigenvectors of W corresponding to $\lambda_1, \cdots, \lambda_N$, respectively. We may take $u_1 = e/\|e\|$ and make $\{u_1, \cdots, u_N\}$ to be orthonormal. For any $x \in \mathbb{R}^N$, we may write $x = \sum_{i=1}^N \gamma_i u_i$ for some scalars γ_i. Then, as $e^T u_i = 0$ for $i = 2, \cdots, N$, we have

$$(J - W)x = \sum_{i=1}^{N} \gamma_i \left(\frac{1}{N} e e^T u_i - W u_i \right) = -\sum_{i=2}^{N} \gamma_i \lambda_i u_i.$$

Thus,

$$\begin{aligned} \|(J - W)x\|^2 &= \left\| \sum_{i=2}^{N} \gamma_i \lambda_i u_i \right\|^2 = \sum_{i=2}^{N} |\gamma_i|^2 |\lambda_i|^2 \leq \max\{|\lambda_2|^2, |\lambda_N|^2\} \sum_{i=2}^{N} |\gamma_i|^2 \\ &\leq \max\{|\lambda_2|^2, |\lambda_N|^2\} \|x\|^2 \end{aligned}$$

which implies that $\|J - W\|_2 \leq \max\{|\lambda_2|, |\lambda_N|\}$. Noting that $(J - W)u_i = -\lambda_i u_i$ for $i = 2, \cdots, N$, we must have the equality hold.

Finally, we show $\beta < 1$. According to (3.29), we need only to show $\lambda_2 < 1$, which amounts to showing that the geometric multiplicity of the eigenvalue $\lambda_1 = 1$

is one. To see this, let x be any eigenvector of W corresponding to the eigenvalue $\lambda_1 = 1$. Let i be an index such that $|x_i| = \|x\|_\infty$, and we may assume that $x_i > 0$. Because $x = Wx$, there holds

$$x_i = \sum_{j=1}^{N} w_{ij} x_j = \sum_{j \in O_i} w_{ij} x_j.$$

Since $x_j \le x_i$, we have $w_{ij}x_j \le w_{ij}x_i$ for all j. If there is a $j \in O_i$ such that $x_j < x_i$, then $w_{ij}x_j < w_{ij}x_i$ as $w_{ij} > 0$. Thus, $x_i < \sum_{j \in O_i} w_{ij} x_i = x_i$, which is a contradiction. Therefore, $x_j = x_i$ for all $j \in O_i$. Since the network $(\mathcal{V}, \mathcal{E})$ is connected, any $j \in \{1, \cdots, N\}$ is connected to i by a path consisting of links from $\mathcal{E}$. Thus, $x_j = x_i$ for all j. This shows that $x = x_i e \in \text{span}\{e\}$ and hence the geometric multiplicity of λ_1 is one. □

We next consider the disagreement estimates between different agents. Instead of estimate $\|x_i^k - x_j^k\|$ directly, we will use the mean

$$\bar{x}^k := \frac{1}{N} \sum_{i=1}^{N} x_i^k$$

of $x_1^k, \cdots, x_N^k$ as a reference and estimate $\|x_i^k - \bar{x}^k\|$ for each agent i. To this end, we set

$$y_i^k := x_i^{k+1} - v_i^k = x_i^{k+1} - \sum_{j=1}^{N} w_{ij} x_j^k$$

for $k = 0, 1, \cdots$ and $i = 1, \cdots, N$ and introduce the following matrices:

$$X_k := \begin{pmatrix} (x_1^k)^T \\ (x_2^k)^T \\ \vdots \\ (x_N^k)^T \end{pmatrix}, \quad V_k := \begin{pmatrix} (v_1^k)^T \\ (v_2^k)^T \\ \vdots \\ (v_N^k)^T \end{pmatrix}, \quad \bar{X}_k := e(\bar{x}^k)^T \quad \text{and} \quad Y_k := \begin{pmatrix} (y_1^k)^T \\ (y_2^k)^T \\ \vdots \\ (y_N^k)^T \end{pmatrix}$$

Then, $V_k = WX_k$ and $Y_k = X_{k+1} - V_k$ and hence

$$X_{k+1} = Y_k + WX_k. \tag{3.30}$$

By recursively using (3.30), we can obtain

$$X_k = W^k X_0 + \sum_{l=0}^{k-1} W^{k-1-l} Y_l. \tag{3.31}$$

According to the definition of $\bar{X}_k$, we can see that $\bar{X}_k = JX_k$. Then, from (3.31) and $JW = J$, it follows that

$$\bar{X}_k = J\left(W^k X_0 + \sum_{l=0}^{k-1} W^{k-1-l} Y_l\right) = JX_0 + \sum_{l=0}^{k-1} JY_l.$$

Therefore, by using $W^k - J = (W - J)^k$ for all $k \geq 1$, we have

$$\begin{aligned} X_k - \bar{X}_k &= \left(W^k - J\right) X_0 + \sum_{l=0}^{k-1} \left(W^{k-1-l} - J\right) Y_l \\ &= (W - J)^k X_0 + (I - J)Y_{k-1} + \sum_{l=0}^{k-2} (W - J)^{k-1-l} Y_l. \end{aligned} \tag{3.32}$$

In order to proceed further, we need the following result concerning the Frobenius and spectral norms for matrices.

Lemma 3.39 *For any two matrices A and B such that AB is well-defined, there holds*

$$\|AB\|_F \leq \|A\|_2 \|B\|_F.$$

Proof Let b_j be the j-th column of B and then Ab_j be the j-th column of AB. Thus,

$$\|AB\|_F^2 = \sum_j \|Ab_j\|^2 \leq \sum_j \|A\|_2^2 \|b_j\|^2 = \|A\|_2^2 \sum_j \|b_j\|^2 = \|A\|_2^2 \|B\|_F^2$$

which shows the inequality. □

Based on (3.32) and Lemma 3.39, we can derive the following consensus estimate.

Proposition 3.40 *Let Assumptions 3.31 and 3.37 hold. Then, for Algorithm 3.36, there holds the estimate*

$$\|X_k - \bar{X}_k\|_F \leq \beta^k \|X_0\|_F + L \sum_{l=0}^{k-1} \beta^{k-1-l} t_l, \tag{3.33}$$

where $L = (\sum_{i=1}^N L_i^2)^{1/2}$ *and* $\beta := \|J - W\|_2 < 1$. *If in addition the step size* $\{t_k\}$ *is chosen such that* $0 < t_k \to 0$, *then for each* $i = 1, \cdots, N$, *there holds* $\|x_i^k - \bar{x}^k\| \to 0$ *as* $k \to \infty$.

Proof From (3.32), Lemma 3.39, and $\|I - J\|_2 \le 1$, it follows that

$$\begin{aligned}\|X_k - \bar{X}_k\|_F &\le \|(W-J)^k X_0\|_F + \|(I-J)Y_{k-1}\|_F + \sum_{l=0}^{k-2} \|(W-J)^{k-1-l} Y_l\|_F \\ &\le \|(W-J)^k\|_2 \|X_0\|_F + \|I-J\|_2 \|Y_{k-1}\|_F \\ &\quad + \sum_{l=0}^{k-2} \|(W-J)^{k-1-l}\|_2 \|Y_l\|_F \\ &\le \|W-J\|_2^k \|X_0\|_F + \|Y_{k-1}\|_F + \sum_{l=0}^{k-2} \|W-J\|_2^{k-1-l} \|Y_l\|_F \\ &= \beta^k \|X_0\|_F + \sum_{l=0}^{k-1} \beta^{k-1-l} \|Y_l\|_F. \end{aligned} \tag{3.34}$$

According to the definition of Y_l and noting that $v_i^k \in C$, we can see that

$$\|Y_l\|_F^2 = \sum_{i=1}^N \|y_i^l\|^2 = \sum_{i=1}^N \|x_i^{l+1} - v_i^l\|^2 = \sum_{i=1}^N \|P_C(v_i^l - t_l \xi_i^l) - P_C(v_i^l)\|^2.$$

By using the nonexpansiveness of projection and Assumption 3.31, we can obtain

$$\|Y_l\|_F^2 \le \sum_{i=1}^N t_l^2 \|\xi_i^l\|^2 \le L^2 t_l^2.$$

Combining this with (3.34) shows (3.33).

Next, we show $\|x_i^k - \bar{x}^k\| \to 0$ as $k \to \infty$ for each $i = 1, \cdots, N$, which is equivalent to showing $\|X^k - \bar{X}^k\|_F \to 0$ as $k \to \infty$. According to (3.33) and the fact $\beta < 1$, it remains only to show that

$$\sum_{l=0}^{k-1} \beta^{k-1-l} t_l \to 0 \quad \text{as } k \to \infty. \tag{3.35}$$

Since $t_k \to 0$ as $k \to \infty$, we may find a constant M such that $0 \le t_k \le M$ for all k, and furthermore, for any $\varepsilon > 0$, we can find an integer k_0 such that $0 \le t_k < \varepsilon$ for all $k \ge k_0$. Thus, for $k \ge k_0$, we have

$$0 \le \sum_{l=0}^{k-1} \beta^{k-1-l} t_l \le M \sum_{l=0}^{k_0-1} \beta^{k-1-l} + \varepsilon \sum_{l=k_0}^{k-1} \beta^{k-1-l} \le \frac{M\beta^{k-k_0}}{1-\beta} + \frac{\varepsilon}{1-\beta}.$$

Since $0 \le \beta < 1$, this implies that

$$0 \le \liminf_{k\to\infty} \sum_{l=0}^{k-1} \beta^{k-1-l} t_l \le \limsup_{k\to\infty} \sum_{l=0}^{k-1} \beta^{k-1-l} t_l \le \frac{\varepsilon}{1-\beta}.$$

Since $\varepsilon > 0$ is arbitrary, this shows (3.35) and the proof is complete. □

Finally, we will show that x_i^k converges to a common optimal solution as $k \to \infty$ for each i under further conditions on the step size $\{t_k\}$. For any optimal solution z of (3.27), let $Z := ez^T$. Then, by the definition of the algorithm and the nonexpansiveness of the Euclidean projection, we have

$$\begin{aligned}
\|X_{k+1} - Z\|_F^2 &= \sum_{i=1}^N \|x_i^{k+1} - z\|^2 = \sum_{i=1}^N \|P_C(v_i^k - t_k\xi_i^k) - P_C(z)\|^2 \\
&\le \sum_{i=1}^N \|v_i^k - t_k\xi_i^k - z\|^2 \\
&= \sum_{i=1}^N \left(\|v_i^k - z\|^2 + 2t_k\langle \xi_i^k, z - v_i^k\rangle + t_k^2\|\xi_i^k\|^2\right) \\
&\le \|V_k - Z\|_F^2 + L^2 t_k^2 + 2t_k \sum_{i=1}^N \langle \xi_i^k, z - v_i^k\rangle.
\end{aligned}$$

Note that $V_k = WX_k$ and $Z = WZ$ and recall that $\|W\|_2 = 1$; see Lemma 3.38. We may use Lemma 3.39 to obtain

$$\|V_k - Z\|_F = \|W(X_k - Z)\|_F \le \|W\|_2\|X_k - Z\|_F = \|X_k - Z\|_F.$$

Therefore,

$$\|X_{k+1} - Z\|_F^2 \le \|X_k - Z\|_F^2 + L^2 t_k^2 + 2t_k \sum_{i=1}^N \langle \xi_i^k, z - v_i^k\rangle.$$

Since $\xi_i^k \in \partial f_i(x_i^k)$, by using Assumption 3.31 and the Cauchy-Schwarz inequality, we have

$$\begin{aligned}
\sum_{i=1}^{N} \langle \xi_i^k, z - v_i^k \rangle &= \sum_{i=1}^{N} \left\{ \langle \xi_i^k, z - x_i^k \rangle + \langle \xi_i^k, x_i^k - v_i^k \rangle \right\} \\
&\le \sum_{i=1}^{N} \left\{ (f_i(z) - f_i(x_i^k)) + L_i \|x_i^k - v_i^k\| \right\} \\
&\le f(z) - f(\bar{x}^k) + \sum_{i=1}^{N} \left(f_i(\bar{x}^k) - f_i(x_i^k) \right) + L\|X_k - V_k\|_F \\
&\le f(z) - f(\bar{x}^k) + \sum_{i=1}^{N} L_i \|\bar{x}^k - x_i^k\| + L\|X_k - V_k\|_F \\
&\le f(z) - f(\bar{x}^k) + L\|X_k - \bar{X}_k\|_F + L\|X_k - V_k\|_F .
\end{aligned}$$

Consequently,

$$\begin{aligned}
\|X_{k+1} - Z\|_F^2 \le \|X_k - Z\|_F^2 - 2t_k(f(\bar{x}^k) - f(z)) + 2Lt_k\|X_k - \bar{X}_k\|_F \\
+ 2Lt_k\|X_k - V_k\|_F + L^2 t_k^2 .
\end{aligned}$$

Since $\bar{X}_k = W\bar{X}_k$ and $\|W\|_2 = 1$, we have

$$\begin{aligned}
\|X_k - V_k\|_F &\le \|X_k - \bar{X}_k\|_F + \|V_k - \bar{X}_k\|_F \\
&= \|X_k - \bar{X}_k\|_F + \|W(X_k - \bar{X}_k)\|_F \\
&\le 2\|X_k - \bar{X}_k\|_F .
\end{aligned}$$

We thus obtain the following result.

Lemma 3.41 *Let Assumptions 3.31 and 3.37 hold. For any optimal solution z of (3.27), let $Z = ez^T$. Then, for Algorithm 3.36, there holds*

$$\|X_{k+1} - Z\|_F^2 \le \|X_k - Z\|_F^2 - 2t_k(f(\bar{x}^k) - f(z)) + 6Lt_k\|X_k - \bar{X}_k\|_F + L^2 t_k^2$$

for all $k = 0, 1, \cdots$.

Now, we are ready to prove the main convergence result on the distributed projected subgradient method, i.e., Algorithm 3.36.

Theorem 3.42 *Let Assumptions 3.31 and 3.37 hold. Consider Algorithm 3.36. If the sequence of the step sizes $\{t_k\}$ is chosen such that*

$$t_k > 0, \quad \sum_{k=0}^{\infty} t_k = \infty \quad \textit{and} \quad \sum_{k=0}^{\infty} t_k^2 < \infty,$$

then there is an optimal solution x^* of (3.27) such that $\|x_i^k - x^*\| \to 0$ as $k \to \infty$ for all $i = 1, \cdots, N$.

Proof Let z be any optimal solution of (3.27) and set $Z := ez^T$. Then, $f(z) = f_* := \arg\min_{x\in C} f(x)$. It follows from Lemma 3.41 that

$$\|X_{k+1} - Z\|_F^2 \le \|X_k - Z\|_F^2 - 2t_k(f(\bar{x}^k) - f_*) + 6Lt_k\|X_k - \bar{X}_k\|_F + L^2 t_k^2.$$

Based on this, we will use Lemma 3.20 to conclude the proof. To this end, we need to show $\sum_{k=0}^{\infty} t_k\|X_k - \bar{X}_k\|_F < \infty$. According to the estimate (3.33) in Proposition 3.40, it suffices to show

$$\sum_{k=0}^{\infty} \beta^k t_k < \infty \quad \text{and} \quad \sum_{k=1}^{\infty} t_k \sum_{l=0}^{k-1} \beta^{k-1-l} t_l < \infty. \tag{3.36}$$

Due to the given conditions, $\{t_k\}$ is bounded, and thus the first one in (3.36) follows from the fact $\beta < 1$ immediately. For the second one in (3.36), we have

$$\begin{aligned}
\sum_{k=1}^{\infty} t_k \sum_{l=0}^{k-1} \beta^{k-1-l} t_l &\le \frac{1}{2}\sum_{k=1}^{\infty}\sum_{l=0}^{k-1} \beta^{k-1-l}(t_k^2 + t_l^2) \\
&= \frac{1}{2}\sum_{k=1}^{\infty}\left(\sum_{l=0}^{k-1} \beta^{k-1-l}\right) t_k^2 + \frac{1}{2}\sum_{l=0}^{\infty}\left(\sum_{k=l+1}^{\infty} \beta^{k-1-l}\right) t_l^2 \\
&\le \frac{1}{2(1-\beta)}\sum_{k=1}^{\infty} t_k^2 + \frac{1}{2(1-\beta)}\sum_{l=0}^{\infty} t_l^2 \\
&\le \frac{1}{1-\beta}\sum_{k=0}^{\infty} t_k^2 < \infty.
\end{aligned}$$

Since $f(\bar{x}^k) - f_* \ge 0$, we may use Lemma 3.20 to conclude that

$$\sum_{k=0}^{\infty} t_k(f(\bar{x}^k) - f_*) < \infty \tag{3.37}$$

and

$$\lim_{k\to\infty} \|X_k - Z\|_F^2 \text{ exists.} \tag{3.38}$$

In Proposition 3.40, we have shown that $\|X_k - \bar{X}_k\|_F \to 0$ as $k \to \infty$. Since

$$\|\bar{X}_k - Z\|_F^2 = \|X_k - Z\|_F^2 + \|X_k - \bar{X}_k\|_F^2 + 2\langle X_k - Z, \bar{X}_k - X_k\rangle,$$

we can use (3.38) to conclude that $\|\bar{X}_k - Z\|_F$ converges as $k \to \infty$, which in particular implies that $\{\bar{X}_k\}$ and hence $\{\bar{x}^k\}$ is bounded.

From (3.37) and $\sum_k t_k = \infty$, we can conclude that there is a subsequence $\{k_l\}$ of positive integers such that $f(\bar{x}^{k_l}) \to f_*$ as $l \to \infty$. Since $\{\bar{x}^{k_l}\}$ is bounded, by taking a subsequence if necessary, we may assume that $\bar{x}^{k_l} \to x^*$ as $l \to \infty$. Then, by the continuity of f, we have $f(x^*) = \lim_{l\to\infty} f(\bar{x}^{k_l}) = f_*$. Thus, x^* is an optimal solution of (3.27), and consequently, by setting $X^* = e(x^*)^T$, we know that $\|\bar{X}_k - X^*\|_F$ is convergent. Because $\|\bar{X}_{k_l} - X^*\|_F \to 0$ as $l \to \infty$, we must have $\|\bar{X}_k - X^*\|_F \to 0$ as $k \to \infty$. Finally, by using Proposition 3.40, we have

$$\|X_k - X^*\|_F \le \|X_k - \bar{X}_k\|_F + \|\bar{X}_k - X^*\|_F \to 0$$

as $k \to \infty$. Therefore, $\|x_i^k - x^*\| \to 0$ as $k \to \infty$ for each $i = 1, \cdots, N$. □

3.5 Mirror Descent Methods

Let $f : \mathbb{R}^n \to (-\infty, \infty]$ be a proper convex function. The subgradient method for solving the convex minimization problem

$$\min_{x\in\mathbb{R}^n} f(x) \tag{3.39}$$

takes the form $x^{k+1} = x^k - t_k\xi^k$, where $\xi^k \in \partial f(x^k)$ and $t_k > 0$ is the step size. The definition of x^{k+1} can be equivalently stated as

$$\begin{aligned} x^{k+1} &= \arg\min_{x\in\mathbb{R}^n} \left\{\frac{1}{2}\|x - (x^k - t_k\xi^k)\|^2\right\} \\ &= \arg\min_{x\in\mathbb{R}^n} \left\{\frac{1}{2}\|x - x^k\|^2 + t_k\langle\xi^k, x\rangle\right\}. \end{aligned}$$

Note that in the above formula, we used the Euclidean distance to measure the discrepancy between two points. In many applications, the Euclidean distance may be replaced by other distance-like function to better reflect the geometry of the underlying problems. The Bregman distance induced by a convex function is usually used for this purpose. Let $h : \mathbb{R}^n \to (-\infty, \infty]$ be a proper convex function. If $\partial h(x^k) \neq \emptyset$, we may take $\theta^k \in \partial h(x^k)$ suitably and then use the Bregman distance

$$D_h^{\theta^k}(x, x^k) := h(x) - h(x^k) - \langle \theta^k, x - x^k \rangle$$

to replace $\frac{1}{2}\|x - x^k\|^2$ in the above equation. This leads to the new updating formula of x^{k+1} as follows:

$$x^{k+1} \in \arg\min_{x \in \mathbb{R}^n} \left\{ D_h^{\theta^k}(x, x^k) + t_k \langle \xi^k, x \rangle \right\}.$$

By using the definition of $D_h^{\theta^k}(x, x^k)$, we have

$$x^{k+1} \in \arg\min_{x \in \mathbb{R}^n} \left\{ h(x) - \langle \theta^k - t_k \xi^k, x \rangle \right\}.$$

Let $\theta^{k+1} := \theta^k - t_k \xi^k$. Then,

$$x^{k+1} \in \arg\min_{x \in \mathbb{R}^n} \left\{ h(x) - \langle \theta^{k+1}, x \rangle \right\}.$$

If x^{k+1} is well-defined, by the first-order optimality condition, we have $\theta^{k+1} \in \partial h(x^{k+1})$. Therefore, we can repeat the above procedure, leading to the following algorithm for solving (3.39).

Algorithm 3.43 (Mirror Descent Method) *Take an initial guess $\theta^0 \in \mathbb{R}^n$ and iteratively define*

$$\begin{cases} x^k \in \arg\min_{x \in \mathbb{R}^n} \left\{ h(x) - \langle \theta^k, x \rangle \right\}, \\ \theta^{k+1} = \theta^k - t_k \xi^k \end{cases}$$

for $k = 0, 1, \cdots$, where $\xi^k \in \partial f(x^k)$ and $t_k > 0$ is a step size.

Algorithm 3.43 is known as the mirror descent method in the literature [17, 163]. In order to give a convergence analysis, we make the following assumption on h and f.

Assumption 3.44

(i) $h : \mathbb{R}^n \to (-\infty, \infty]$ *is proper, lower semi-continuous, and strongly convex with modulus of convexity* $\mu_h > 0$.
(ii) $f : \mathbb{R}^n \to (-\infty, \infty]$ *is proper convex and has a minimizer over* $\mathrm{dom}(h)$. *Moreover,* $\partial f(x) \neq \emptyset$ *for each* $x \in \mathrm{dom}(h)$, *and there is a constant* $M > 0$ *such that* $\|\xi\| \leq M$ *for all* $\xi \in \partial f(x)$ *and* $x \in \mathrm{dom}(h)$.

Under Assumption 3.44, we can show that Algorithm 3.43 is well-defined. It suffices to show that each x^k is uniquely determined. The uniqueness follows from the strong convexity of h; see Proposition 2.28. It remains only to show the existence. According to Lemma 2.23, we have

$$h(x) \geq \langle \xi, x \rangle + \gamma, \quad \forall x \in \mathbb{R}^n$$

for some $(\xi, \gamma) \in \mathbb{R}^n \times \mathbb{R}$. Take $x_0 \in \mathrm{dom}(h)$. Then, by the strong convexity of h, we have for any $x \in \mathbb{R}^n$ that

$$\langle \xi, (x + x_0)/2 \rangle + \gamma \leq h((x + x_0)/2) \leq \frac{1}{2}h(x) + \frac{1}{2}h(x_0) - \frac{1}{4}\mu_h \|x - x_0\|^2$$

and hence

$$h(x) \geq \frac{1}{2}\mu_h \|x - x_0\|^2 + \langle \xi, x \rangle + \langle \xi, x_0 \rangle + 2\gamma - h(x_0)$$

which implies that $h(x)/\|x\| \to \infty$ as $\|x\| \to \infty$. Thus, the function $x \to h(x) - \langle \theta^k, x \rangle$ is proper, lower semi-continuous, and coercive, and hence it has a global minimizer over $\mathbb{R}^n$ by Corollary 1.15.

For Algorithm 3.43, we have the following convergence result under suitable choices of the step sizes t_k.

Theorem 3.45 *Let Assumption 3.44 hold. Consider Algorithm 3.43 where* $\{t_k\}$ *is chosen such that*

$$t_k > 0, \quad \sum_{k=0}^{\infty} t_k = \infty \quad \text{and} \quad \sum_{k=0}^{\infty} t_k^2 < \infty.$$

Then, $\{x^k\}$ *is a bounded sequence, and each cluster point of* $\{x^k\}$ *is a minimizer of* f *over* dom(h). *Moreover, letting* $f_* := \min_{x \in \mathrm{dom}(h)} f(x)$, *then*

$$\min_{0 \leq i \leq k} f(x^i) - f_* \leq \frac{4\mu_h D_h^{\theta^0}(\bar{x}, x^0) + M^2 \sum_{i=0}^k t_i^2}{4\mu_h \sum_{i=0}^k t_i} \tag{3.40}$$

for all k, where $\bar{x}$ *denotes any minimizer of* f *over* dom(h).

Proof By the optimality condition on x^k, we have $\theta^k \in \partial h(x^k)$ for $k = 0, 1, \cdots$. Thus, it follows from Lemma 3.11 and $\theta^{k+1} = \theta^k - t_k \xi^k$ that

$$\begin{aligned} D_h^{\theta^{k+1}}(\bar{x}, x^{k+1}) - D_h^{\theta^k}(\bar{x}, x^k) &= -D_h^{\theta^k}(x^{k+1}, x^k) - \langle \theta^{k+1} - \theta^k, \bar{x} - x^{k+1} \rangle \\ &= -D_h^{\theta^k}(x^{k+1}, x^k) + t_k \langle \xi^k, \bar{x} - x^{k+1} \rangle \end{aligned}$$

By the strong convexity of h, we then obtain

$$D_h^{\theta^{k+1}}(\bar{x}, x^{k+1}) - D_h^{\theta^k}(\bar{x}, x^k) \le -\mu_h \|x^{k+1} - x^k\|^2 + t_k \langle \xi^k, x^k - x^{k+1} \rangle + t_k \langle \xi^k, \bar{x} - x^k \rangle.$$

Since $\xi^k \in \partial f(x^k)$, we have $\langle \xi^k, \bar{x} - x^k \rangle \le f(\bar{x}) - f(x^k)$. By the Cauchy-Schwarz inequality, we also have

$$t_k \langle \xi^k, x^k - x^{k+1} \rangle \le t_k \|\xi^k\| \|x^k - x^{k+1}\| \le \mu_h \|x^k - x^{k+1}\|^2 + \frac{1}{4\mu_h} t_k^2 \|\xi^k\|^2.$$

Therefore,

$$D_h^{\theta^{k+1}}(\bar{x}, x^{k+1}) - D_h^{\theta^k}(\bar{x}, x^k) \le \frac{1}{4\mu_h} t_k^2 \|\xi^k\|^2 + t_k \left(f(\bar{x}) - f(x^k) \right). \tag{3.41}$$

Based on this inequality, we may follow the proof of Theorem 3.21 to conclude that $\{D_h^{\theta^k}(\bar{x}, x^k)\}$ is convergent and hence $\{\|\bar{x} - x^k\|\}$ is bounded by the strong convexity of h; moreover, $\{x^k\}$ has a convergent subsequence $\{x^{k_l}\}$ such that $x^{k_l} \to x^*$ and $f(x^{k_l}) \to f(\bar{x}) = f_*$ as $l \to \infty$. Since $\{x^k\} \subset \text{dom}(h)$ and the lower semi-continuity of h implies that $\text{dom}(h)$ is closed, we have $x^* \in \text{dom}(h)$. By the given condition, f is Lipschitz continuous on $\text{dom}(h)$. Thus, $f(x^*) = \lim_{l\to\infty} f(x^{k_l}) = f_*$, which shows that x^* is a minimizer of f over $\text{dom}(h)$. Finally, the estimate (3.40) can be derived from (3.41). □

As an application, we consider the linear system $Ax = b$, where A is an $m \times n$ matrix and $b \in \text{Ran}(A)$. This system may have many solutions. We will use the mirror descent method to determine a solution with desired feature. To this end, we first reformulate the linear system $Ax = b$ as the equivalent quadratic minimization problem

$$\min_{x \in \mathbb{R}^n} \{\|Ax - b\|^2\}$$

and then choose a suitable convex function $h : \mathbb{R}^n \to (-\infty, \infty]$ with the purpose of capturing the feature of the sought solution. Applying Algorithm 3.43 to this situation leads immediately to the following algorithm.

Algorithm 3.46 *Take $\theta^0 = 0$ and iteratively define*

$$\begin{cases} x^k \in \arg\min_{x \in \mathbb{R}^n} \left\{ h(x) - \langle \theta^k, x \rangle \right\}, \\ \theta^{k+1} = \theta^k - t_k A^T (Ax^k - b) \end{cases}$$

for $k = 0, 1, \cdots$, where $t_k > 0$ is a step size.

We will show that Algorithm 3.46 enjoys nicer convergence result than Algorithm 3.43, and it allows much more flexible choices of the step sizes. We will make the following assumption.

Assumption 3.47 $h : \mathbb{R}^n \to (-\infty, \infty]$ *is proper, lower semi-continuous, and strongly convex with modulus of convexity* $\mu_h > 0$. *Moreover,* $\text{dom}(h) \cap \{x \in \mathbb{R}^n : Ax = b\} \neq \emptyset$.

In the following, we will show, under Assumption 3.47, that the sequence $\{x^k\}$ defined by Algorithm 3.46 converges to the unique solution of the convex minimization problem

$$\min_{x \in \mathbb{R}^n} \{h(x) : Ax = b\} \tag{3.42}$$

if the step size t_k is chosen properly. To see this, let $x \in \text{dom}(h)$ be any point. It follows from Lemma 3.11 and $\theta^{k+1} = \theta^k - t_k A^T(Ax^k - b)$ that

$$\begin{aligned}
D_h^{\theta^{k+1}}(x, x^{k+1}) - D_h^{\theta^k}(x, x^k) &= -D_h^{\theta^k}(x^{k+1}, x^k) - \langle \theta^{k+1} - \theta^k, x - x^{k+1} \rangle \\
&= -D_h^{\theta^k}(x^{k+1}, x^k) + t_k \langle Ax^k - b, A(x - x^{k+1}) \rangle \\
&= -D_h^{\theta^k}(x^{k+1}, x^k) + t_k \langle Ax^k - b, A(x^k - x^{k+1}) \rangle \\
&\quad + t_k \langle Ax^k - b, A(x - x^k) \rangle.
\end{aligned}$$

By using the polarization identity, we then obtain

$$\begin{aligned}
&D_h^{\theta^{k+1}}(x, x^{k+1}) - D_h^{\theta^k}(x, x^k) \\
&= -D_h^{\theta^k}(x^{k+1}, x^k) - \frac{t_k}{2}\left(\|Ax^{k+1} - b\|^2 - \|Ax^k - b\|^2 - \|A(x^{k+1} - x^k)\|^2\right) \\
&\quad + \frac{t_k}{2}\left(\|Ax - b\|^2 - \|Ax^k - b\|^2 - \|A(x - x^k)\|^2\right) \\
&= -D_h^{\theta^k}(x^{k+1}, x^k) - \frac{t_k}{2}\left(\|Ax^{k+1} - b\|^2 - \|Ax - b\|^2\right) \\
&\quad + \frac{t_k}{2}\|A(x^{k+1} - x^k)\|^2 - \frac{t_k}{2}\|A(x - x^k)\|^2.
\end{aligned}$$

Since h is strongly convex, it follows from Proposition 3.12 that

$$D_h^{\theta^k}(x^{k+1}, x^k) \geq \mu_h \|x^{k+1} - x^k\|^2.$$

Therefore,

$$\begin{aligned} D_h^{\theta^{k+1}}(x, x^{k+1}) - D_h^{\theta^k}(x, x^k) \le & -\left(\mu_h - \frac{t_k}{2}\|A\|^2\right)\|x^{k+1} - x^k\|^2 \\ & - \frac{t_k}{2}\left(\|Ax^{k+1} - b\|^2 - \|Ax - b\|^2\right) \\ & - \frac{t_k}{2}\|A(x - x^k)\|^2. \end{aligned} \tag{3.43}$$

Based on this inequality, we can obtain the following lemma.

Lemma 3.48 *Let Assumption 3.47 hold and let* $0 < t_k \le 2\mu_h/\|A\|^2$. *Let* $\bar{x}$ *be any solution of* $Ax = b$ *in* dom(h). *Then, for the sequence* $\{x^k\}$ *defined by Algorithm 3.46, there hold the following:*

(i) $\{D_h^{\theta^k}(\bar{x}, x^k)\}$ *is monotonically decreasing.*
(ii) $\{\|Ax^k - b\|\}$ *is monotonically decreasing and*

$$t_k\|Ax^k - b\|^2 \le 2\left(D_h^{\theta^k}(\bar{x}, x^k) - D_h^{\theta^{k+1}}(\bar{x}, x^{k+1})\right), \quad k = 0, 1, \cdots.$$

Proof By taking $x = x^k$ in (3.43) and using $0 < t_k \le \mu_h/\|A\|^2$, we can obtain

$$0 \le D_h^{\theta^{k+1}}(x^k, x^{k+1}) \le \frac{t_k}{2}\left(\|Ax^k - b\|^2 - \|Ax^{k+1} - b\|^2\right)$$

which shows the monotonicity of $\{\|Ax^k - b\|\}$. Next, we take $x = \bar{x}$ in (3.43) and use $A\bar{x} = b$ to obtain

$$\begin{aligned} D_h^{\theta^{k+1}}(\bar{x}, x^{k+1}) - D_h^{\theta^k}(\bar{x}, x^k) & \le -\frac{t_k}{2}\|Ax^{k+1} - b\|^2 - \frac{t_k}{2}\|Ax^k - b\|^2 \\ & \le -\frac{t_k}{2}\|Ax^k - b\|^2. \end{aligned}$$

This in particular implies the monotonicity of $\{D_h^{\theta^k}(\bar{x}, x^k)\}$ and the proof is complete. □

Now, we are ready to prove the following convergence result for Algorithm 3.46.

Theorem 3.49 *Let Assumption 3.47 hold and let* $\{t_k\}$ *be chosen such that* $0 < t_k \le 2\mu_h/\|A\|^2$ *and* $\sum_{k=0}^{\infty} t_k = \infty$. *Let* x^* *denote the unique solution of (3.42). Then, for the sequence* $\{x^k\}$ *defined by Algorithm 3.46, there hold*

$$D_h^{\theta^k}(x^*, x^k) \to 0 \quad \textit{and} \quad h(x^k) \to h(x^*)$$

as $k \to \infty$. *Consequently,* $\|x^k - x^*\| \to 0$ *as* $k \to \infty$.

Proof We first show that $\{x^k\}$ converges to a solution of $Ax = b$. To this end, we show that $\{x^k\}$ is a Cauchy sequence. Let $\bar{x}$ be any solution of $Ax = b$ in $\mathrm{dom}(h)$. For $0 \le l < k < \infty$, we have from Lemma 3.11 that

$$D_h^{\theta^l}(x^k, x^l) = D_h^{\theta^l}(\bar{x}, x^l) - D_h^{\theta^k}(\bar{x}, x^k) + \langle \theta^k - \theta^l, x^k - \bar{x} \rangle.$$

By the definition of $\{\theta^k\}$, we have

$$\begin{aligned} |\langle \theta^k - \theta^l, x^k - \bar{x} \rangle| &= \left| \sum_{j=l}^{k-1} \langle \theta^{j+1} - \theta^j, x^k - \bar{x} \rangle \right| = \left| \sum_{j=l}^{k-1} t_j \langle Ax^j - b, Ax^k - b \rangle \right| \\ &\le \sum_{j=l}^{k-1} t_j \|Ax^j - b\| \|Ax^k - b\|. \end{aligned}$$

By using monotonicity of $\{\|Ax^j - b\|\}$ shown in Lemma 3.48, we then obtain

$$|\langle \theta^k - \theta^l, x^k - \bar{x} \rangle| \le \sum_{j=l}^{k-1} t_j \|Ax^j - b\|^2.$$

By using (ii) in Lemma 3.48, we further have

$$|\langle \theta^k - \theta^l, x^k - \bar{x} \rangle| \le 2 \left(D_h^{\theta^l}(\bar{x}, x^l) - D_h^{\theta^k}(\bar{x}, x^k) \right). \tag{3.44}$$

Consequently,

$$D_h^{\theta^l}(x^k, x^l) \le 3 \left(D_h^{\theta^l}(\bar{x}, x^l) - D_h^{\theta^k}(\bar{x}, x^k) \right).$$

Since $\{D_h^{\theta^k}(\bar{x}, x^k)\}$ is monotonically decreasing, we obtain $D_h^{\theta^l}(x^k, x^l) \to 0$ as $k, l \to \infty$. In view of the strong convexity of h, we can conclude that $\{x^k\}$ is a Cauchy sequence. Thus, $x^k \to x^*$ for some $x^* \in \mathbb{R}^n$ as $k \to \infty$. In view of (ii) in Lemma 3.48, we have

$$\|Ax^k - b\|^2 \sum_{j=0}^{k} t_j \le \sum_{j=0}^{k} t_j \|Ax^j - b\|^2 \le 2 D_h^{\theta^0}(\bar{x}, x^0) = 2(h(\bar{x}) - h(x^0)).$$

Since $\sum_{k=0}^{\infty} t_k = \infty$, we have $\|Ax^k - b\| \to 0$ as $k \to \infty$ and hence $Ax^* = b$.

Next, we show that

$$x^* \in \mathrm{dom}(h), \quad \lim_{k \to \infty} h(x^k) = h(x^*) \quad \text{and} \quad \lim_{k \to \infty} D_h^{\theta^k}(x^*, x^k) = 0.$$

From the convexity of h and $\theta^k \in \partial h(x^k)$, it follows that

$$h(x^k) \le h(\bar{x}) + \langle \theta^k, x^k - \bar{x} \rangle. \tag{3.45}$$

In view of (3.44) and $\theta^0 = 0$, we have

$$h(x^k) \le h(\bar{x}) + 2D_h^{\theta^0}(\bar{x}, x^0) = h(\bar{x}) + 2(h(\bar{x}) - h(x^0)).$$

Since $x^k \to x^*$ as $k \to \infty$, by using the lower semi-continuity of h, we obtain

$$h(x^*) \le \liminf_{k\to\infty} h(x^k) \le h(\bar{x}) + 2(h(\bar{x}) - h(x^0)) < \infty. \tag{3.46}$$

This implies that $x^* \in \mathrm{dom}(h)$. We next use (3.44) to derive for $l < k$ that

$$|\langle \theta^k, x^k - x^* \rangle| \le 2\left(D_h^{\theta^l}(x^*, x^l) - D_h^{\theta^k}(x^*, x^k)\right) + |\langle \theta^l, x^k - x^* \rangle|.$$

By taking $k \to \infty$ and using $x^k \to x^*$, we can derive that

$$\limsup_{k\to\infty} |\langle \theta^k, x^k - x^* \rangle| \le 2\left(D_h^{\theta^l}(x^*, x^l) - \varepsilon_0\right),$$

where $\varepsilon_0 := \lim_{k\to\infty} D_h^{\theta^k}(x^*, x^n)$ whose existence is guaranteed by the monotonicity of $\{D_h^{\theta^k}(x^*, x^k)\}$. Since the above inequality holds for all l, by taking $l \to \infty$, we obtain

$$\limsup_{k\to\infty} |\langle \theta^k, x^k - x^* \rangle| \le 2(\varepsilon_0 - \varepsilon_0) = 0. \tag{3.47}$$

Using (3.45) with $\bar{x}$ replaced by x^*, we thus obtain $\limsup_{k\to\infty} h(x^k) \le h(x^*)$. Combining this with (3.46), we therefore obtain $\lim_{k\to\infty} h(x^k) = h(x^*)$. This together with (3.47) then implies that $\lim_{k\to\infty} D_h^{\theta^k}(x^*, x^k) = 0$.

Finally, we show that x^* is the unique solution of (3.42). To see this, by using (3.45) and (3.44), we have for any integers $0 \le l < k$ that

$$\begin{aligned} h(x^k) &\le h(\bar{x}) + \langle \theta^k - \theta^l, x^k - \bar{x} \rangle + \langle \theta^l, x^k - \bar{x} \rangle \\ &\le h(\bar{x}) + 2\left(D_h^{\theta^l}(\bar{x}, x^l) - D_h^{\theta^k}(\bar{x}, x^k)\right) + \langle \theta^l, x^k - \bar{x} \rangle. \end{aligned}$$

Since $\theta^0 = 0$, by the definition of θ^l, we have $\xi^l = A^T v_l$ for some $v_l \in \mathbb{R}^m$. Consequently,

$$\left|\langle \theta^l, x^k - \bar{x} \rangle\right| = \left|\langle v_l, Ax^k - b \rangle\right| \le \|v_l\| \|Ax^k - b\|.$$

Therefore,

$$h(x^k) \leq h(\bar{x}) + 2\left(D_h^{\theta^l}(\bar{x}, x^l) - D_h^{\theta^k}(\bar{x}, x^k)\right) + \|v_l\|\|Ax^k - b\|.$$

Recall that $\|Ax^k - b\| \to 0$ as $k \to \infty$. In view of the lower semi-continuity of h and the fact $x^k \to x^*$, we thus obtain

$$h(x^*) \leq \liminf_{k\to\infty} h(x^k) \leq h(\bar{x}) + 2\left(D_h^{\theta^l}(\bar{x}, x^l) - \varepsilon_0\right).$$

Letting $l \to \infty$ then gives $h(x^*) \leq h(\bar{x})$ for any $\bar{x} \in \text{dom}(h)$ satisfying $A\bar{x} = b$. □

Remark 3.50

(i) The proof of Theorem 3.49 is taken from [120] where the argument is performed for a more general method in arbitrary Banach spaces.
(ii) Algorithm 3.46 can be interpreted as a dual gradient method that enables us to derive the rate of convergence; see Sect. 5.8, where we may further consider its acceleration.

Example 3.51 (Matrix Completion) Let $X^* \in \mathbb{R}^{m\times n}$ be an unknown matrix for which only sampled entries $\{X_{ij} : (i, j) \in \Omega\}$ are known, where Ω is a subset of $\{(i, j) : i = 1, \cdots, m;\ j = 1, \cdots, n\}$. We consider the problem of recovering X^* from this sampling of its entries such that the rank of X^* is as low as possible. Let $Y := P_\Omega(X^*)$, where $P_\Omega : \mathbb{R}^{m\times n} \to \mathbb{R}^{m\times n}$ denotes the orthogonal projection such that for any $X \in \mathbb{R}^{m\times n}$, the entries of $P_\Omega(X)$ at the spot (i, j) are equal to X_{ij} if $(i, j) \in \Omega$ and zero otherwise. It is easy to see that P_Ω is a linear self-adjoint mapping with $P_\Omega^2 = P_\Omega$ and $\|P_\Omega\| = 1$. It is now natural to recover X^* by solving the optimization problem:

$$\min\{\text{rank}(X) : X \in \mathbb{R}^{m\times n} \text{ and } P_\Omega(X) = Y\}$$

Since rank(X) is nonsmooth and non-convex, solving this optimization problem is extremely challenging. Instead, we may consider its convex relaxation:

$$\min\{\|X\|_* : X \in \mathbb{R}^{m\times n} \text{ and } P_\Omega(X) = Y\}, \tag{3.48}$$

where $\|X\|_*$ denotes the nuclear norm of X, which is defined as the sum of all singular values of X. We can confirm the convexity of the nuclear norm by showing that it is the dual norm of the spectral norm, i.e.,

$$\|X\|_* = \|X\|_2^* := \sup_{Y\in\mathbb{R}^{m\times n}, \|Y\|_2\leq 1} \text{tr}(X^T Y), \quad \forall X \in \mathbb{R}^{m\times n}, \tag{3.49}$$

where $\|Y\|_2$ denotes the spectral norm of Y, which equals the largest singular value of Y. To see this, let $r = \text{rank}(X)$ and let $X = U\Sigma V^T$ be a singular value decomposition of X, where $U \in \mathbb{R}^{m\times r}$ and $V \in \mathbb{R}^{n\times r}$ are such that $U^T U = V^T V = I_r$, the $r \times r$ identity matrix, and $\Sigma \in \mathbb{R}^{r\times r}$ is a diagonal matrix with positive singular values of X on the main diagonal. By taking $Y = UV^T$ and noting that $\|Y\|_2 = 1$, we have

$$\|X\|_2^* \geq \text{tr}(X^T Y) = \text{tr}(V\Sigma U^T UV^T) = \text{tr}(V\Sigma V^T)=\text{tr}(V^T V\Sigma)=\text{tr}(\Sigma)=\|X\|_*.$$

On the other hand, letting $U = (u_1, \cdots, u_r)$ and $V = (v_1, \cdots, v_r)$, we have $\|u_i\| = \|v_i\| = 1$, and hence for any $Y \in \mathbb{R}^{m\times n}$, there holds

$$\begin{aligned}\text{tr}(X^T Y) &= \text{tr}(V\Sigma U^T Y) = \text{tr}(\Sigma U^T YV) = \sum_{i=1}^r \sigma_i (U^T YV)_{ii} = \sum_{i=1}^r \sigma_i u_i^T Y v_i \\ &\leq \sum_{i=1}^r \sigma_i \|u_i\|\|Y\|_2\|v_i\| = \sum_{i=1}^r \sigma_i \|Y\|_2 = \|X\|_*\|Y\|_2\end{aligned}$$

which implies that $\|X\|_2^* \leq \|X\|_*$. Therefore, $\|X\|_* = \|X\|_2^*$ and hence $\|X\|_*$ is indeed a norm. Consequently, (3.48) is indeed a convex minimization problem.

In order to apply Algorithm 3.46 and its convergence result, we modify (3.48) by adding a small positive multiple of $\|X\|_F^2$ to $\|X\|_*$ and consider the strongly convex minimization problem:

$$\min\left\{\|X\|_* + \frac{1}{2\tau}\|X\|_F^2 : X \in \mathbb{R}^{m\times n} \text{ and } P_\Omega(X) = Y\right\}$$

or equivalently

$$\min\left\{\tau\|X\|_* + \frac{1}{2}\|X\|_F^2 : X \in \mathbb{R}^{m\times n} \text{ and } P_\Omega(X) = Y\right\}, \tag{3.50}$$

where $\tau > 0$ is a large number. Applying Algorithm 3.46 leads to the method

$$\begin{aligned}X^k &= \arg\min_{X\in\mathbb{R}^{m\times n}} \left\{\tau\|X\|_* + \frac{1}{2}\|X\|_F^2 - \langle X, \Lambda^k\rangle_F\right\}, \\ \Lambda^{k+1} &= \Lambda^k - \gamma P_\Omega(P_\Omega(X^k) - Y) = \Lambda^k - \gamma(P_\Omega(X^k) - Y)\end{aligned} \tag{3.51}$$

We need to determine an explicit formula for calculating X^k. Let $\Lambda^k = U^k\Sigma^k(V^k)^T$ be a singular value decomposition. We claim that

$$X^k = \tilde{X}^k := U^k\mathcal{D}_\tau(\Sigma^k)(V^k)^T, \tag{3.52}$$

where $\mathcal{D}_\tau(\Sigma^k) := \max\{\Sigma^k - \tau I, 0\}$ with I being the identity matrix of the same size as Σ^k. To see this, according to Theorem 3.9, it suffices to show

$$(\Lambda^k - \tilde{X}^k)/\tau \in \partial(\|\cdot\|_*)(\tilde{X}^k).$$

By virtue of (3.49) and Example 3.7, we only need to check

$$\|(\Lambda^k - \tilde{X}^k)/\tau\|_2 \le 1 \quad \text{and} \quad \langle(\Lambda^k - \tilde{X}^k)/\tau, \tilde{X}^k\rangle_F = \|\tilde{X}^k\|_*.$$

Note that $\Lambda^k - \tilde{X}^k = U^k(\Sigma^k - \mathcal{D}_\tau(\Sigma^k))(V^k)^T$. The singular values of $\Lambda^k - \tilde{X}^k$ are $(\sigma_i - (\sigma_i - \tau)_+)/\tau$, which belong to $[0, 1]$, where σ_i are the singular values of Λ^k. Therefore, $\|(\Lambda^k - \tilde{X}^k)/\tau\|_2 \le 1$. We also have

$$\begin{aligned}
\langle(\Lambda^k - \tilde{X}^k)/\tau, \tilde{X}^k\rangle_F &= \frac{1}{\tau}\mathrm{tr}(V^k(\Sigma^k - \mathcal{D}_\tau(\Sigma^k))(U^k)^T U^k \mathcal{D}_\tau(\Sigma^k)(V^k)^T) \\
&= \frac{1}{\tau}\mathrm{tr}(V^k(\Sigma^k - \mathcal{D}_\tau(\Sigma^k))\mathcal{D}_\tau(\Sigma^k)(V^k)^T) \\
&= \frac{1}{\tau}\mathrm{tr}((\Sigma^k - \mathcal{D}_\tau(\Sigma^k))\mathcal{D}_\tau(\Sigma^k)(V^k)^T V^k) \\
&= \frac{1}{\tau}\mathrm{tr}((\Sigma^k - \mathcal{D}_\tau(\Sigma^k))\mathcal{D}_\tau(\Sigma^k)) \\
&= \sum_i \frac{1}{\tau}(\sigma_i - (\sigma_i - \tau)_+)(\sigma_i - \tau)_+ \\
&= \sum_i (\sigma_i - \tau)_+ = \|\tilde{X}^k\|_*.
\end{aligned}$$

We thus obtain (3.52). Combining (3.51) and (3.52) gives the following algorithm for solving (3.50).

Algorithm 3.52 (Singular Value Thresholding) *Input $Y \in \mathbb{R}^{m\times n}$ and a subset Ω of $\{(i, j) : i = 1, \cdots, m;\ j = 1, \cdots, n\}$. Take a number $\tau > 0$ and set $\Lambda^0 = 0$, the zero matrix of $m \times n$. For $k = 0, 1, \cdots$, do*

(i) *Find a singular value decomposition of Λ^k: $\Lambda^k = U^k\Sigma^k(V^k)^T$.*
(ii) *Set $D_\tau(\Sigma^k) := (\Sigma^k - \tau I)_+$ and calculate $X^k = U^k D_\tau(\Sigma^k)(V^k)^T$.*
(iii) *Set $\Lambda^{k+1} = \Lambda^k - \gamma(P_\Omega(X^k) - Y)$.*

Algorithm 3.52 is known as the singular value thresholding algorithm for matrix completion proposed in [46]. The main computational load per iteration is the determination of a singular value decomposition of Λ^k. Determining a singular value decomposition is expensive in general. However, since $\Lambda^0 = 0$, one can easily see that each Λ^k is supported on Ω, i.e., Λ^k vanishes outside of Ω and is thus sparse. Therefore, the sparse matrix techniques can be applied to reduce computational costs significantly. According to Theorem 3.49, if $0 < \gamma \leq 1$, the sequence $\{X^k\}$ defined by Algorithm 3.52 converges to the unique solution X^* of (3.50) as $k \to \infty$. Actually, we have the convergence rate $\|X^k - X^*\| = O(1/\sqrt{k})$ as $k \to \infty$; see Sect. 5.8 where the strategy can even be used to accelerate Algorithm 3.52.

Problems

3.1 For the function $f(x_1, x_2) = |x_1| + 2|x_2|$ on $\mathbb{R}^2$, find $\partial f(0, 0)$.

3.2 Consider the function $f(x) = \|x\|_2$ on $\mathbb{R}^n$. Show that

$$\partial f(x) = \begin{cases} \{x/\|x\|_2\} & \text{if } x \neq 0, \\ D_1(0) & \text{if } x = 0, \end{cases}$$

where $D_1(0) = \{\xi \in \mathbb{R}^n : \|\xi\|_2 \leq 1\}$.

3.3 Let $f : \mathbb{R}^n \to \mathbb{R}$ be a convex function and let $\Omega \subset \mathbb{R}^n$ be a nonempty convex set. Show that the following are equivalent:

(i) $x^* \in \arg\min_{x \in \Omega} f(x)$.
(ii) There exists $\xi \in \partial f(x^*)$ such that $\langle \xi, x - x^* \rangle \geq 0$ for all $x \in \Omega$.

3.4 Consider the minimization problem:

$$\begin{aligned} &\text{minimize} \quad f(x) + g(z) \\ &\text{subject to} \quad x \in \mathbb{R}^n, \ z \in \mathbb{R}^p, \ Ax + Bz = c, \end{aligned}$$

where $f : \mathbb{R}^n \to \mathbb{R}$ and $g : \mathbb{R}^p \to \mathbb{R}$ are convex functions; A and B are matrices of sizes $m \times n$ and $m \times p$, respectively; and $c \in \mathbb{R}^m$. Show that the following are equivalent:

(i) (x^*, z^*) is an optimal solution.
(ii) $Ax^* + Bz^* = c$ and there is $\lambda^* \in \mathbb{R}^m$ such that $-A^T\lambda^* \in \partial f(x^*)$ and $-B^T\lambda^* \in \partial g(z^*)$.

3.5 Consider the convex minimization problem:

$$\min_{x \in \mathbb{R}^n} \{f(x) + g(x)\},$$

where $g : \mathbb{R}^n \to (-\infty, \infty]$ is a proper convex function, and

$$f(x) = h(Ax) + \langle c, x \rangle$$

with $c \in \mathbb{R}^n$, $A \in \mathbb{R}^{m\times n}$, and $h : \mathbb{R}^m \to \mathbb{R}$ being a continuous differentiable strictly convex function. Let S denote the set of optimal solutions and assume $S \neq \emptyset$. Show that there exists $\bar{y} \in \mathbb{R}^m$ such that

$$S = \left\{x \in \mathbb{R}^n : Ax = \bar{y} \text{ and } -A^T\nabla h(\bar{y}) - c \in \partial g(x)\right\}.$$

3.6 Let $\ell : \mathbb{R}\times\mathbb{R} \to \mathbb{R}$ be a function that is convex with respect to the first variable and let $\lambda > 0$. Consider the minimization problem:

$$\min_{x\in\mathbb{R}^n} \left\{\sum_{i=1}^{N} \ell(u_i^T x, b_i) + \frac{\lambda}{2}\|x\|^2\right\}$$

which may arise from a supervised machine learning model, where $(u_i, b_i) \in \mathbb{R}^n \times \mathbb{R}$, $i = 1, \cdots, N$, are training data. Show that the optimal solution x^* of this minimization problem, if exists, must take the form

$$x^* = \sum_{i=1}^{N} \alpha_i u_i$$

for some real numbers α_i.

3.7 Complete the proof of Theorem 3.22.

3.8 Given a noisy data $b \in \mathbb{R}^n$, consider the total variation denoising problem:

$$x^* = \arg\min_{x\in\mathbb{R}^n} \left\{\frac{1}{2}\|x - b\|_2^2 + \lambda \sum_{i=1}^{n-1} |x_{i+1} - x_i|\right\},$$

where $\lambda > 0$ is a regularization parameter. It can be written as

$$x^* = \arg\min_{x\in\mathbb{R}^n} \left\{\frac{1}{2}\|x - b\|_2^2 + \lambda\|Dx\|_1\right\}$$

by introducing the $(n-1)\times n$ matrix:

$$D = \begin{pmatrix} -1 & 1 & & \\ & \ddots & \ddots & \\ & & -1 & 1 \end{pmatrix}.$$

(i) Show that the total variation denoising problem has a unique minimizer.
(ii) When the subgradient method is used to solve it, what is the iteration formula?

(iii) Write a code to implement the algorithm obtained in (ii) and test your code in the following way: create a piecewise constant signal x^* of length $n = 4000$ with at least 10 discontinuous points, and obtain noisy data $b = x^* + \texttt{randn}(n, 1)$ by adding to x^* a Gaussian white noise with standard deviation 1. Use total variation denoising with $\lambda = 10, 20, 40$ to reconstruct x^*. The total variation denoising problem is solved by the subgradient method with step size $t_k = 1/k$.

(*Hint*: In your code, you may need to calculate Dx and $D^T x$; in MATLAB they can be implemented by defining

```
D = @(x)diff(x);     and     DT = @(x)[-x(1); -diff(x); x(end)];
```

and set $Dx = \texttt{D}(x)$ and $D^T x = \texttt{DT}(x)$. This requires no explicit formula of D and can make the algorithm fast)

3.9 Consider the constrained convex minimization problem (3.12) for which Assumption 3.17 holds. Assume $f_* := \min_{x \in C} f(x)$ is known. Consider the projected subgradient method given in Algorithm 3.16 with $\{t_k\}$ chosen as the Polyak step size (3.14). It is known from Theorem 3.23 that $\{x^k\}$ converges to a solution x^* of (3.12).

(i) If f is strongly convex over $\mathbb{R}^n$ with the modulus of convexity $\mu > 0$, show that

$$\|x^k - x^*\| \leq \frac{L\|x^0 - x^*\|}{\sqrt{L^2 + 3\mu^2\|x^0 - x^*\|^2 k}}, \qquad k = 0, 1, \cdots .$$

(ii) If $C = \mathbb{R}^n$ and $f : \mathbb{R}^n \to \mathbb{R}$ is continuous differentiable with L-Lipschitz continuous gradient and is strongly convex with the modulus of convexity $\mu > 0$, show that

$$\|x^k - x^*\| \leq \left(1 - \frac{\mu}{L}\right)^{k/2} \|x^0 - x^*\|, \quad k = 0, 1, \cdots .$$

3.10

(i) Consider on $\mathbb{R}^n$ the negative entropy function:

$$\varphi(x) := \begin{cases} \sum_{i=1}^{n} x_i \log x_i, & \text{if } x = (x_1, \cdots, x_n) \geq 0, \\ +\infty, & \text{otherwise.} \end{cases}$$

Show that h is a convex function.

(ii) Show that $\partial h(x) \neq \emptyset$ if and only if $x > 0$. Moreover,

$$\partial\varphi(x) = \{\nabla\varphi(x)\} = \{(1 + \log x_1, \cdots, 1 + \log x_n)\}, \quad \forall x > 0.$$

(iii) Given $x \in \mathbb{R}^n$ with $x > 0$, show that the Bregman distance induced by φ at x is given by

$$D_{KL}(z, x) := \sum_{i=1}^{n} \left(z_i \log \frac{z_i}{x_i} - z_i + x_i \right), \quad \forall z \geq 0$$

which is called the Kullback-Leibler divergence.

(iv) Let Δ_n denote the unit simplex in $\mathbb{R}^n$, i.e.,

$$\Delta_n := \left\{ x \in \mathbb{R}^n : x \geq 0 \text{ and } \sum_{i=1}^{n} x_i = 1 \right\}$$

and let $\Delta_n^+ := \{x \in \Delta_n : x > 0\}$. Show that

$$D_{KL}(z, x) \geq \frac{1}{2}\|x - z\|_1^2, \quad \forall x \in \Delta_n^+, \ z \in \Delta_n$$

which is called Pinsker's inequality.

(v) Let $f : \mathbb{R}^n \to \mathbb{R}$ be a continuously differentiable convex function. We may determine approximately a solution of the minimization problem:

$$\min_{x \in \Delta_n} f(x) \tag{3.53}$$

by the mirror descent method given in Algorithm 3.43 with $h := \varphi + \delta_{\Delta_n}$. Show that the corresponding method takes the form

$$x_i^k = e^{\theta_i^k} \Big/ \sum_{j=1}^{n} e^{\theta_j^k}, \quad i = 1, \cdots, n; \quad \theta^{k+1} = \theta^k - t_k \nabla f(x^k)$$

which is called the exponential gradient descent method.

(vi) Assume (3.53) has a solution and $\|\nabla f(x)\|_\infty \leq M$ for all $x \in \Delta_n$. Mimic the proof of Theorem 3.45 to show that if $\{t_k\}$ is chosen to satisfy

$$t_k > 0, \quad \sum_{k=1}^{\infty} t_k = \infty \quad \text{and} \quad \sum_{k=1}^{\infty} t_k^2 < \infty,$$

then

$$\min_{0 \leq i \leq k} f(x^i) - f_* \leq \frac{2D_{KL}(\bar{x}, x^0) + M^2 \sum_{i=0}^{k} t_i^2}{2\sum_{i=0}^{k} t_i}$$

for all k, where $\bar{x}$ denotes any solution of (3.53) and $f_* := \min_{x \in \Delta_n} f(x)$.

3.11 Let A be an $m \times n$ real matrix and $b \in \mathbb{R}^n$. Let $C \subset \mathbb{R}^n$ be a nonempty closed convex set. Assume that the equation $Ax = b$ has a solution in C. The unique solution $x^\dagger$ of $Ax = b$ in C with the shortest distance to the origin can be obtained by solving the minimization problem:

$$\min \left\{ \frac{1}{2} \|x\|^2 + \delta_C(x) : Ax = b \right\},$$

where δ_C denotes the indicator function of C. Show that an application of Algorithm 3.46 yields the iteration scheme:

$$x^k = P_C(\theta^k), \quad \theta^{k+1} = \theta^k - t_k A^T(Ax^k - b)$$

with $\theta^0 = 0$. Show further that if $0 < t_k \leq 1/\|A\|^2$ and $\sum_{k=0}^{\infty} t_k = \infty$, then $\|x^k - x^\dagger\| \to 0$ as $k \to \infty$.

Chapter 4
Proximal Algorithms

In this chapter, we explore various proximal algorithms for solving nonsmooth convex minimization problems. To this end, we begin by introducing the Legendre-Fenchel conjugate of convex functions and developing the corresponding subdifferential calculus. Next, we define the proximal mapping and investigate its key properties. Using the proximal mapping, we reformulate the underlying convex minimization problem as a fixed-point equation. This perspective naturally leads to a range of proximal algorithms, including the proximal point algorithm, the proximal gradient method and its acceleration, and the Douglas-Rachford splitting method. In particular, we provide a detailed study of the accelerated proximal gradient method, presenting state-of-the-art convergence results. These include the convergence of iterates and linear convergence under strong convexity. For a comprehensive survey on proximal algorithms, we refer the reader to [175].

4.1 Legendre-Fenchel Conjugate

Let $f : \mathbb{R}^n \to (-\infty, \infty]$ be a given function. Given $\xi \in \mathbb{R}^n$, we may consider the linear function $x \to \langle \xi, x\rangle - \alpha$ for any $\alpha \in \mathbb{R}$. It is natural to ask for what value α the hyperplane determined by this linear function is below the graph of f. Clearly, α should be chosen to satisfy

$$\langle \xi, x\rangle - \alpha \le f(x), \quad \forall x \in \mathbb{R}^n,$$

i.e.,

$$\alpha \ge \langle \xi, x\rangle - f(x), \quad \forall x \in \mathbb{R}^n.$$

This leads us to define the following function.

Q. Jin, *Lectures on Nonsmooth Optimization*, Texts in Applied Mathematics 82,
https://doi.org/10.1007/978-3-031-91417-1_4

Definition 4.1 Let $f : \mathbb{R}^n \to (-\infty, \infty]$ be a given function. For any $\xi \in \mathbb{R}^n$, define

$$f^*(\xi) := \sup_{x\in\mathbb{R}^n} \{\langle \xi, x\rangle - f(x)\}$$

which is called the *Legendre-Fenchel conjugate* of f.

By definition, there holds

$$f^*(\xi) + f(x) \geq \langle \xi, x\rangle, \qquad \forall \xi, x \in \mathbb{R}^n$$

which is called the *Fenchel-Young inequality*. If f is proper, then there is $\bar{x}$ such that $f(\bar{x})$ is finite and thus

$$f^*(\xi) \geq \langle \xi, \bar{x}\rangle - f(\bar{x}) > -\infty, \quad \forall \xi \in \mathbb{R}^n.$$

Moreover, $f^* : \mathbb{R}^n \to (-\infty, \infty]$ is a lower semi-continuous convex function because it is the supremum of a family of linear functions. The following result shows that f^* is also proper if f is lower semi-continuous at some point in $\mathrm{dom}(f)$.

Proposition 4.2 *Let $f : \mathbb{R}^n \to (-\infty, \infty]$ be a proper convex function. If f is lower semi-continuous at some point $\bar{x} \in \mathrm{dom}(f)$, then $f^* : \mathbb{R}^n \to (-\infty, \infty]$ is proper, lower semi-continuous, and convex.*

Proof It is already known that f^* is convex, lower semi-continuous, and $f^*(\xi) > -\infty$ for all $\xi \in \mathbb{R}^n$. It remains only to show $\mathrm{dom}(f^*) \neq \emptyset$. In view of Lemma 2.23, there is $(\xi, \gamma) \in \mathbb{R}^n \times \mathbb{R}$ such that $f(x) \geq \langle \xi, x\rangle + \gamma$ for all $x \in \mathbb{R}^n$. Therefore,

$$f^*(\xi) = \sup_{x\in\mathbb{R}^n} \{\langle \xi, x\rangle - f(x)\} \leq -\gamma < \infty$$

which means that $\xi \in \mathrm{dom}(f^*)$ and the proof is complete. □

Example 4.3 Given $1 < p < \infty$, consider $f(x) = \frac{1}{p}|x|^p$ on $\mathbb{R}$. We calculate

$$f^*(\xi) = \sup_{x\in\mathbb{R}} \left\{ \xi x - \frac{1}{p}|x|^p \right\}.$$

For each ξ, $x \to \xi x - \frac{1}{p}|x|^p$ attains maximum at $x = |\xi|^{\frac{1}{p-1}}\mathrm{sign}(\xi)$. Thus,

$$f^*(\xi) = \xi|\xi|^{\frac{1}{p-1}}\mathrm{sign}(\xi) - \frac{1}{p}|\xi|^{\frac{p}{p-1}} = \frac{p-1}{p}|\xi|^{\frac{p}{p-1}}.$$

Let p^* be the number conjugate to p, i.e., $1/p + 1/p^* = 1$. Then, $f^*(\xi) = \frac{1}{p^*}|\xi|^{p^*}$ for all $\xi \in \mathbb{R}$. Using the result on f^* and the Fenchel-Young inequality $f(x) + f^*(\xi) \geq \langle \xi, x \rangle$, we can obtain *Young's inequality:*

$$\frac{1}{p}|x|^p + \frac{1}{p^*}|\xi|^{p^*} \geq \xi x$$

for all $x, \xi \in \mathbb{R}$. □

Example 4.4 Given a convex set $C \subset \mathbb{R}^n$ and considering its indicator function $\delta_C(x)$, we have

$$\delta_C^*(\xi) = \sup_{x \in \mathbb{R}^n} \{\langle \xi, x \rangle - \delta_C(x)\} = \sup_{x \in C} \langle \xi, x \rangle$$

which is called the *support function* of C. If $C = \{x \in \mathbb{R}^n : \|x\|_\infty \leq 1\}$, then $\delta_C^*(\xi) = \|\xi\|_1$. Indeed, since $\langle \xi, x \rangle \leq \|\xi\|_1 \|x\|_\infty$, we have $\delta_C^*(\xi) \leq \|\xi\|_1$; on the other hand, when $\xi \neq 0$, by taking $x = \text{sign}(\xi)$, we have $\|x\|_\infty = 1$ and $\langle \xi, x \rangle = \|\xi\|_1$. Thus, $\delta_C^*(\xi) \geq \|\xi\|_1$. □

Example 4.5 Consider the function $f(x) = \|x\|_1$ on $\mathbb{R}^n$, and we have

$$f^*(\xi) = \begin{cases} 0 & \text{if } \|\xi\|_\infty \leq 1, \\ +\infty & \text{if } \|\xi\|_\infty > 1. \end{cases}$$

Indeed, if $\|\xi\|_\infty \leq 1$, then $\langle \xi, x \rangle \leq \|x\|_1$ and "=" holds if $x = 0$. Thus, $f^*(\xi) = 0$. If $\|\xi\|_\infty > 1$, take $x = t\xi_j e_j$, where j is an index such that $|\xi_j| = \|\xi\|_\infty$. Then,

$$\langle \xi, x \rangle - \|x\|_1 = t\|\xi\|_\infty^2 - t\|\xi\|_\infty = t\|\xi\|_\infty(\|\xi\|_\infty - 1) \to \infty$$

as $t \to \infty$. Thus, $f^*(\xi) = +\infty$. □

In the following, we will provide some subdifferential calculus rules related to the Legendre-Fenchel conjugate.

Proposition 4.6 *Let $f : \mathbb{R}^n \to (-\infty, \infty]$ be a proper convex function. Then,*

$$\xi \in \partial f(x) \Longleftrightarrow f(x) + f^*(\xi) \leq \langle \xi, x \rangle \Longleftrightarrow f(x) + f^*(\xi) = \langle \xi, x \rangle.$$

Proof The first equivalence can be derived as follows:

$$\begin{aligned} \xi \in \partial f(x) &\Longleftrightarrow f(y) \geq f(x) + \langle \xi, y - x \rangle \quad \text{for all } y \in \mathbb{R}^n \\ &\Longleftrightarrow \langle \xi, x \rangle \geq f(x) + \langle \xi, y \rangle - f(y) \quad \text{for all } y \in \mathbb{R}^n \\ &\Longleftrightarrow \langle \xi, x \rangle \geq f(x) + f^*(\xi). \end{aligned}$$

Since the inequality $f^*(\xi) + f(x) \geq \langle \xi, x \rangle$ is always true, the second equivalence is obvious. □

Proposition 4.7 *Let $f : \mathbb{R}^n \to (-\infty, \infty]$ be proper and convex. Then,*

$$\xi \in \partial f(x) \Longrightarrow x \in \partial f^*(\xi).$$

Proof Since $f(x) \geq \langle \eta, x \rangle - f^*(\eta)$ for any $\eta \in \mathbb{R}^n$, it follows that

$$\begin{aligned} \xi \in \partial f(x) &\Longrightarrow f(y) \geq f(x) + \langle \xi, y - x \rangle, \quad \forall y \in \mathbb{R}^n \\ &\Longrightarrow f(y) \geq \langle \eta, x \rangle - f^*(\eta) + \langle \xi, y - x \rangle, \quad \forall y, \eta \in \mathbb{R}^n \\ &\Longrightarrow f^*(\eta) \geq \langle \eta - \xi, x \rangle + \langle \xi, y \rangle - f(y), \quad \forall y, \eta \in \mathbb{R}^n \\ &\Longrightarrow f^*(\eta) \geq \langle \eta - \xi, x \rangle + f^*(\xi), \quad \forall \eta \in \mathbb{R}^n \\ &\Longrightarrow x \in \partial f^*(\xi) \end{aligned}$$

which shows the result. □

When f is lower semi-continuous, the converse of Proposition 4.7 is also true. To see this, we introduce the *biconjugate* of f:

$$f^{**}(x) := \sup_{\xi \in \mathbb{R}^n} \left\{ \langle \xi, x \rangle - f^*(\xi) \right\}, \qquad x \in \mathbb{R}^n.$$

For $x \in \mathbb{R}^n$, we have $f(x) + f^*(\xi) \geq \langle \xi, x \rangle$ for all $\xi \in \mathbb{R}^n$. Thus,

$$f(x) \geq \sup_{\xi \in \mathbb{R}^n} \left\{ \langle \xi, x \rangle - f^*(\xi) \right\} = f^{**}(x).$$

Theorem 4.8 *Let $f : \mathbb{R}^n \to (-\infty, \infty]$ be a proper convex function. If f is lower semi-continuous at $\bar{x} \in \mathbb{R}^n$, then $f^{**}(\bar{x}) = f(\bar{x})$.*

Proof Assume that $f^{**}(\bar{x}) < f(\bar{x})$. Then, as in the proof of Lemma 2.23, we have $(\bar{x}, f^{**}(\bar{x})) \notin \overline{\text{epi}(f)}$. Since $\overline{\text{epi}(f)}$ is closed, convex, and nonempty, by Corollary 2.18, there is $(\eta, \alpha) \in \mathbb{R}^n \times \mathbb{R}$ and $\gamma \in \mathbb{R}$ such that

$$\langle \eta, z \rangle + \alpha t < \gamma < \langle \eta, \bar{x} \rangle + \alpha f^{**}(\bar{x}), \quad \forall (z, t) \in \text{epi}(f).$$

Letting $c := \gamma - \langle \eta, \bar{x} \rangle - \alpha f^{**}(\bar{x})$, we have

$$\langle \eta, z - \bar{x} \rangle + \alpha \left(t - f^{**}(\bar{x}) \right) < c < 0, \qquad \forall (z, t) \in \text{epi}(f). \tag{4.1}$$

By taking $z \in \text{dom}(f)$ and letting $t \to \infty$ in (4.1), it follows that $\alpha \leq 0$.

If $\alpha < 0$, we may divide (4.1) by $-\alpha$ and set $\xi = -\frac{\eta}{\alpha}$, $b = -\frac{c}{\alpha}$ to obtain

$$\langle \xi, z - \bar{x} \rangle - t + f^{**}(\bar{x}) < b < 0, \qquad \forall (z, t) \in \text{epi}(f).$$

Thus,

$$\sup_{z \in \mathbb{R}^n} \{\langle \xi, z \rangle - f(z)\} - \langle \xi, \bar{x} \rangle + f^{**}(\bar{x}) \le b < 0,$$

that is, $f^*(\xi) - \langle \xi, \bar{x} \rangle + f^{**}(\bar{x}) \le b < 0$, contradicting the Fenchel-Young inequality.

If $\alpha = 0$, then (4.1) becomes

$$\langle \eta, z - \bar{x} \rangle < c < 0, \qquad \forall z \in \mathrm{dom}(f).$$

Take $\upsilon \in \mathrm{dom}(f^*)$, which is possible by Proposition 4.2. We may take a small $\varepsilon > 0$ such that

$$\begin{aligned}
\langle \eta + \varepsilon \upsilon, z - \bar{x} \rangle - \varepsilon \left(t - f^{**}(\bar{x})\right) &< c + \varepsilon \left(\langle \upsilon, z \rangle - t - \langle \upsilon, \bar{x} \rangle + f^{**}(\bar{x})\right) \\
&\le c + \varepsilon \left(\sup_{z \in \mathbb{R}^n} \{\langle \upsilon, z \rangle - f(z)\} - \langle \upsilon, \bar{x} \rangle + f^{**}(\bar{x}) \right) \\
&= c + \varepsilon \left(f^*(\upsilon) - \langle \upsilon, \bar{x} \rangle + f^{**}(\bar{x})\right) \\
&\le c/2 < 0
\end{aligned}$$

for all $(z, t) \in \mathrm{epi}(f)$. Now, we can apply the argument for the case $\alpha < 0$ to derive a contradiction. □

Theorem 4.9 *Let $f : \mathbb{R}^n \to (-\infty, \infty]$ be proper, lower semi-continuous, and convex. Then,*

$$\xi \in \partial f(x) \iff x \in \partial f^*(\xi).$$

Proof We have from Theorem 4.8 and Proposition 4.6 that

$$\begin{aligned}
x \in \partial f^*(\xi) &\iff f^*(\xi) + f^{**}(x) = \langle \xi, x \rangle \\
&\iff f^*(\xi) + f(x) = \langle \xi, x \rangle \\
&\iff \xi \in \partial f(x).
\end{aligned}$$

This completes the proof. □

4.2 Proximal Mapping of Moreau

Let $f : \mathbb{R}^n \to (-\infty, \infty]$ be a proper convex function. From Theorem 3.9, we know that

$$x^* \in \arg\min_{x \in \mathbb{R}^n} f(x) \iff 0 \in \partial f(x^*).$$

We want to interpret x^* as a fixed point of a mapping from $\mathbb{R}^n$ to itself. Multiplying $0 \in \partial f(x^*)$ by a number $t > 0$ and adding x^*, we obtain

$$x^* \in x^* + t\partial f(x^*) = (I + t\partial f)(x^*),$$

where I denotes the identity map on $\mathbb{R}^n$. Note that $I + t\partial f$ is a multivalued mapping. However, it has the following nice property.

Lemma 4.10 *Let $f : \mathbb{R}^n \to (-\infty, \infty]$ be a proper, lower semi-continuous, convex function, and let $t > 0$ be a given number. Then, for any $v \in \mathbb{R}^n$, there is a unique $z \in \mathbb{R}^n$ such that*

$$v \in (I + t\partial f)(z).$$

Proof We first prove the uniqueness. Let there be two $z_1, z_2 \in \mathbb{R}^n$ such that

$$v \in (I + t\partial f)(z_i), \quad i = 1, 2.$$

Then,

$$\frac{v - z_i}{t} \in \partial f(z_i), \quad i = 1, 2.$$

Recall that ∂f is monotone; see Proposition 3.12. Therefore,

$$0 \leq \left\langle \frac{v - z_1}{t} - \frac{v - z_2}{t}, z_1 - z_2 \right\rangle = -\frac{1}{t}\|z_1 - z_2\|^2$$

which implies $z_1 = z_2$.

Next, we show the existence by considering the minimization problem:

$$\min_{x \in \mathbb{R}^n} \left\{ f(x) + \frac{1}{2t}\|x - v\|^2 \right\}. \tag{4.2}$$

The objective function is lower semi-continuous and strictly convex. By Lemma 2.23, we can find $(\xi, \gamma) \in \mathbb{R}^n \times \mathbb{R}$ such that $f(x) \geq \langle \xi, x \rangle + \gamma$ for all $x \in \mathbb{R}^n$. Therefore, the objective function in (4.2) is coercive. By Theorem 1.15 and Proposition 2.28, (4.2) has a unique minimizer, denoted by z. In view of Theorem 3.9 and Theorem 3.14, we have

$$0 \in \partial f(z) + \frac{1}{t}(z - v).$$

Consequently, $v \in z + t\partial f(z) = (I + t\partial f)(z)$. □

Lemma 4.10 motivates the following definition ([157, 158]).

Definition 4.11 (Resolvent) Given $v \in \mathbb{R}^n$, the unique $z \in \mathbb{R}^n$ such that $v \in (I + t\partial f)(z)$ is denoted as $z = (I + t\partial f)^{-1}(v)$. The mapping

$$(I + t\partial f)^{-1} : \mathbb{R}^n \to \mathbb{R}^n$$

is called the *resolvent of* ∂f *with parameter* $t > 0$.

We often write $(I + t\partial f)^{-1}$ as prox_{tf} and call it the *proximal mapping of* f *with parameter* t. From the proof of Lemma 4.10, we have

$$\text{prox}_{tf}(v) = (I + t\partial f)^{-1}(v) = \arg\min_{x \in \mathbb{R}^n} \left\{ f(x) + \frac{1}{2t}\|x - v\|^2 \right\}$$

for all $v \in \mathbb{R}^n$, whenever $f : \mathbb{R}^n \to (-\infty, \infty]$ is proper, lower semi-continuous, and convex. Thus, $\text{prox}_{tf}(v)$ is a point that compromises between minimizing f and being near to v. For this reason, $\text{prox}_{tf}(v)$ is called a *proximal point* of v with respect to f.

Lemma 4.12 *Let* $f : \mathbb{R}^n \to (-\infty, \infty]$ *be a proper, lower semi-continuous, convex function. Then,*

$$x^* \in \arg\min_{x \in \mathbb{R}^n} f(x) \Longleftrightarrow x^* = (I + t\partial f)^{-1}(x^*) \Longleftrightarrow x^* = \text{prox}_{tf}(x^*)$$

for any $t > 0$.

Proof Let $t > 0$ be any number. It follows from Theorem 3.9 and Lemma 4.10 that

$$\begin{aligned} x^* \in \arg\min_{x \in \mathbb{R}^n} f(x) &\Longleftrightarrow 0 \in \partial f(x^*) \Longleftrightarrow x^* \in (I + t\partial f)(x^*) \\ &\Longleftrightarrow x^* \in (I + t\partial f)^{-1}(x^*) \\ &\Longleftrightarrow x^* = \text{prox}_{tf}(x^*). \end{aligned}$$

The proof is thus complete. □

Therefore, finding a minimizer of f is equivalent to finding a fixed point of the proximal mapping prox_{tf}. Based on this fact, we will develop the proximal point algorithm in the next section. We now give examples on evaluating $\text{prox}_{tf}(v)$.

Example 4.13 For the indicator function δ_C of a closed convex set C in $\mathbb{R}^n$, we have

$$\text{prox}_{\delta_C}(v) = \arg\min_{x \in C} \|x - v\|^2 = P_C(v),$$

where P_C denotes the Euclidean projection onto C. Therefore, the proximal mapping can be viewed as a generalization of the Euclidean projection.

Example 4.14 Consider $f(x) = |x|$ on $\mathbb{R}$. Given $t > 0$ and $\tau \in \mathbb{R}$, let

$$\hat{\tau} = \text{prox}_{tf}(\tau) = \arg\min_{x\in\mathbb{R}} \left\{ |x| + \frac{1}{2t}(x-\tau)^2 \right\}.$$

Let $\phi(x) = |x| + \frac{1}{2t}(x-\tau)^2$, which is a convex function. Then,

$$\begin{aligned}\hat{\tau} > 0 \text{ is a minimizer of } \phi &\Longleftrightarrow \phi'(\hat{\tau}) = 0 \text{ and } \hat{\tau} > 0 \\ &\Longleftrightarrow 1 + \frac{1}{t}(\hat{\tau} - \tau) = 0 \text{ and } \hat{\tau} > 0 \\ &\Longleftrightarrow \hat{\tau} = \tau - t \text{ and } \tau > t\end{aligned}$$

and

$$\begin{aligned}\hat{\tau} < 0 \text{ is a minimizer of } \phi &\Longleftrightarrow \phi'(\hat{\tau}) = 0 \text{ and } \hat{\tau} < 0 \\ &\Longleftrightarrow -1 + \frac{1}{t}(\hat{\tau} - \tau) = 0 \text{ and } \hat{\tau} < 0 \\ &\Longleftrightarrow \hat{\tau} = \tau + t \text{ and } \tau < -t\end{aligned}$$

From these two observations, we can see that if $|\tau| \le t$, then as a minimizer, there must hold $\hat{\tau} = 0$. Combining the above analysis, we therefore obtain

$$\text{prox}_{tf}(\tau) = s_t(\tau) := \begin{cases} \tau - t & \text{if } \tau > t, \\ 0 & \text{if } |\tau| \le t, \\ \tau + t & \text{if } \tau < -t \end{cases}$$

which is called the one-dimensional *soft thresholding* function. It is easy to see that

$$s_t(\tau) = \text{sign}(\tau)\max\{|\tau| - t, 0\} = \tau - P_{[-t,t]}(\tau),$$

where $P_{[-t,t]}$ denotes the projection of $\mathbb{R}$ onto the interval $[-t, t]$.

Example 4.15 Consider $f(x) = \|x\|_1$ on $\mathbb{R}^n$. Then, for $v = (v_1, \cdots, v_n)^T \in \mathbb{R}^n$ and $t > 0$, we have

$$\begin{aligned}\text{prox}_{tf}(v) &= \arg\min_{x\in\mathbb{R}^n} \left\{ \|x\|_1 + \frac{1}{2t}\|x - v\|^2 \right\} \\ &= \arg\min_{x\in\mathbb{R}^n} \left\{ \sum_{i=1}^{n} \left(|x_i| + \frac{1}{2t}(x_i - v_i)^2 \right) \right\}.\end{aligned}$$

By using Example 4.14, we obtain

$$\operatorname{prox}_{tf}(v) = S_t(v) := (s_t(v_1), \cdots, s_t(v_n))$$

which is called the *soft thresholding of the vector v with parameter t* > 0.

Example 4.16 Let $\mathbb{R}^{m\times n}$ denote the space of all $m \times n$ real matrices. Recall the nuclear norm $\|\cdot\|_*$ of matrices defined in Example 3.51. We now determine the proximal mapping of $\|\cdot\|_*$. We have

$$\operatorname{prox}_{t\|\cdot\|_*}(Z) = \arg\min_{X\in\mathbb{R}^{m\times n}} \left\{ \|X\|_* + \frac{1}{2t}\|X - Z\|_F^2 \right\}, \quad Z \in \mathbb{R}^{m\times n}.$$

Let $Z = U\Sigma V^T$ be a singular value decomposition of Z. Then, by using the argument in Example 3.51, we can see that

$$\operatorname{prox}_{t\|\cdot\|_*}(Z) = U(\Sigma - tI)_+ V^T$$

which is the *singular value thresholding* of Z with parameter $t > 0$.

By using proximal mapping, we can give a quantitative form of the Fenchel-Young inequality due to [47].

Proposition 4.17 *Let* $f : \mathbb{R}^n \to (-\infty, \infty]$ *be a proper, lower semi-continuous, convex function. Then, for any* $(x, \xi) \in \mathbb{R}^n \times \mathbb{R}^n$ *and* $t > 0$, *there holds*

$$f(x) + f^*(\xi) - \langle \xi, x\rangle \geq \frac{1}{t}\|x - \operatorname{prox}_{tf}(x + t\xi)\|^2.$$

Proof Let $(y, \eta) \in \mathbb{R}^n \times \mathbb{R}^n$ be any pair. By the Fenchel-Young inequality, we have

$$f(x) + f^*(\eta) \geq \langle \eta, x\rangle, \quad f(y) + f^*(\xi) \geq \langle \xi, y\rangle.$$

Adding these two inequalities gives

$$\big(f(x) + f^*(\xi) - \langle \xi, x\rangle\big) + \big(f(y) + f^*(\eta) - \langle \eta, y\rangle\big) \geq \langle \eta - \xi, x - y\rangle.$$

Now, we choose (y, η) such that $f(y) + f^*(\eta) - \langle \eta, y\rangle = 0$. To this end, let $z = \operatorname{prox}_{tf}(x + t\xi)$. By the definition of the proximal mapping, we have $\xi + (x - z)/t \in \partial f(z)$. Thus, by taking $y = z$ and $\eta = \xi + (x - z)/t$, we have $\eta \in \partial f(y)$ and hence $f(y) + f^*(\eta) = \langle \eta, y\rangle$ by Proposition 4.6. Therefore,

$$f(x) + f^*(\xi) - \langle \xi, x\rangle \geq \langle \eta - \xi, x - y\rangle = \frac{1}{t}\|x - z\|^2$$

and the proof is complete. □

We next provide some important properties of the proximal mapping, which will be frequently used.

Lemma 4.18 *The proximal mapping* prox_f *is firmly nonexpansive, i.e.,*

$$\|\mathrm{prox}_f(x) - \mathrm{prox}_f(y)\|^2 \leq \langle x - y, \mathrm{prox}_f(x) - \mathrm{prox}_f(y)\rangle$$

for all $x, y \in \mathbb{R}^n$. *Equivalently, there holds*

$$\|\mathrm{prox}_f(x) - \mathrm{prox}_f(y)\|^2 + \|(I - \mathrm{prox}_f)(x) - (I - \mathrm{prox}_f)(y)\|^2 \leq \|x - y\|^2.$$

Proof Let

$$u = \mathrm{prox}_f(x) = (I + \partial f)^{-1}(x), \qquad v = \mathrm{prox}_f(y) = (I + \partial f)^{-1}(y).$$

Then,

$$x - u \in \partial f(u) \quad \text{and} \quad y - v \in \partial f(v).$$

Since ∂f is monotone, we have

$$0 \leq \langle (x - u) - (y - v), u - v\rangle = \langle x - y, u - v\rangle - \|u - v\|^2,$$

i.e., $\|u - v\|^2 \leq \langle x - y, u - v\rangle$, which is the desired result. □

Lemma 4.19 *Let* $f : \mathbb{R}^n \to (-\infty, \infty]$ *be a proper, lower semi-continuous, convex function, and let* f^* *denote its Legendre-Fenchel conjugate. Then,*

$$x = \mathrm{prox}_f(x) + \mathrm{prox}_{f^*}(x), \qquad \forall x \in \mathbb{R}^n.$$

Proof Let $u = \mathrm{prox}_f(x)$, and then $x - u \in \partial f(u)$. By Theorem 4.9, $u \in \partial f^*(x - u)$, i.e.,

$$x \in (x - u) + \partial f^*(x - u) = (I + \partial f^*)(x - u).$$

This implies that $x - u = (I + \partial f^*)^{-1}(x) = \mathrm{prox}_{f^*}(x)$. Therefore, $x = u + \mathrm{prox}_{f^*}(x)$, which shows the decomposition. □

Lemma 4.19 can be viewed as a generalization of orthogonal decomposition induced by a subspace: Let L be a subspace of $\mathbb{R}^n$ and let $L^\perp$ denote its orthogonal complement. Then, for the indicator function δ_L of L, we have

$$\begin{aligned}(\delta_L)^*(y) &= \sup_{x \in \mathbb{R}^n} \{\langle y, x\rangle - \delta_L(x)\} = \sup_{x \in L} \langle y, x\rangle \\ &= \begin{cases} 0, & \text{if } y \in L^\perp, \\ +\infty, & \text{if } y \notin L^\perp \end{cases} \\ &= \delta_{L^\perp}(y).\end{aligned}$$

By Lemma 4.19 and Example 4.13, we have

$$v = \text{prox}_{\delta_L}(v) + \text{prox}_{\delta_{L^\perp}}(v) = P_L(v) + P_{L^\perp}(v)$$

which is the orthogonal decomposition.

Lemma 4.19 is a simple version of the Moreau decomposition to be considered later. It gives a simple way to obtain the proximal mapping of f in terms of the proximal mapping of f^* and vice versa.

Example 4.20 We calculate prox_f for $f(x) = \|x\|$. By Lemma 4.19, we have $\text{prox}_f(v) = v - \text{prox}_{f^*}(v)$. Note that

$$f^*(y) = \sup_{x \in \mathbb{R}^n} \{\langle y, x\rangle - \|x\|\} = \begin{cases} 0, & \|y\| \le 1, \\ +\infty, & \|y\| > 1, \end{cases}$$

i.e., $f^*(y) = \delta_B(y)$, where $B = \{y : \|y\| \le 1\}$. Consequently,

$$\text{prox}_f(v) = v - \text{prox}_{\delta_B}(v) = v - P_B(v) = v - \min\{1, \|v\|\}\frac{v}{\|v\|},$$

where we used the simple fact $P_B(v) = \min\{1, \|v\|\}\frac{v}{\|v\|}$. □

Definition 4.21 Let $f : \mathbb{R}^n \to (-\infty, \infty]$ be a proper, lower semi-continuous, convex function. Given $t > 0$, we define

$$M_{tf}(v) := \inf_{x \in \mathbb{R}^n} \left\{ f(x) + \frac{1}{2t}\|x - v\|^2 \right\}, \quad v \in \mathbb{R}^n$$

which is called the *Moreau envelope* of f with parameter t.

Example 4.22 Consider $f(x) = |x|$ on $\mathbb{R}$. Then,

$$M_{tf}(v) = \inf_{x \in \mathbb{R}} \left\{ |x| + \frac{1}{2t}(x - v)^2 \right\}.$$

It is known from Example 4.14 that the infimum is achieved at the point $\hat{x} = s_t(v)$. Thus,

$$M_{tf}(v) = |s_t(v)| + \frac{1}{2t}(s_t(v) - v)^2 = \begin{cases} |v| - \frac{t}{2}, & |v| > t, \\ \frac{1}{2t}|v|^2, & |v| \le t. \end{cases}$$

This function is called the *Huber function with parameter t*.

Theorem 4.23 *Let $f : \mathbb{R}^n \to (-\infty, \infty]$ be a proper, lower semi-continuous, convex function and let $t > 0$ be any number. Then:*

(i) $\text{dom}(M_{tf}) = \mathbb{R}^n$.
(ii) *M_{tf} is a convex function and $M_{tf}(v) \le f(v)$ for all $v \in \mathbb{R}^n$.*
(iii) *v^* is a minimizer of f over $\mathbb{R}^n \iff v^*$ is a minimizer of M_{tf} over $\mathbb{R}^n$.*
(iv) *M_{tf} is continuous differentiable and*

$$\nabla M_{tf}(v) = \frac{1}{t}\left(v - \text{prox}_{tf}(v)\right), \quad v \in \mathbb{R}^n.$$

(v) *For any $x, y \in \mathbb{R}^n$, there holds*

$$\|\nabla M_{tf}(x) - \nabla M_{tf}(y)\| \le \frac{1}{t}\|x - y\|,$$

i.e., ∇M_{tf} is Lipschitz continuous with Lipschitz constant $\le 1/t$.

Proof (i) and (ii) are obvious. We now show (iii). Let v^* be a minimizer of M_{tf} over $\mathbb{R}^n$. From the definition of M_{tf}, it follows that

$$M_{tf}(v^*) = f(\text{prox}_{tf}(v^*)) + \frac{1}{2t}\|\text{prox}_{tf}(v^*) - v^*\|^2. \tag{4.3}$$

By (ii) and the minimality, we have $f(\text{prox}_{tf}(v^*)) \ge M_{tf}(\text{prox}_{tf}(v^*)) \ge M_{tf}(v^*)$. It then follows from (4.3) that

$$M_{tf}(v^*) \ge M_{tf}(v^*) + \frac{1}{2t}\|\text{prox}_{tf}(v^*) - v^*\|^2$$

which implies that $\text{prox}_{tf}(v^*) = v^*$. Thus, we can use Lemma 4.12 to conclude that v^* is a minimizer of f over $\mathbb{R}^n$.

Conversely, let v^* be a minimizer of f over $\mathbb{R}^n$. Then,

$$f(x) + \frac{1}{2t}\|x - v\|^2 \ge f(x) \ge f(v^*), \quad \forall x, v \in \mathbb{R}^n.$$

This together with (ii) then implies that

$$M_{tf}(v) = \inf_{x \in \mathbb{R}^n}\left\{f(x) + \frac{1}{2t}\|x - v\|^2\right\} \ge f(v^*) \ge M_{tf}(v^*), \quad \forall v \in \mathbb{R}^n$$

which shows that v^* is a minimizer of M_{tf} over $\mathbb{R}^n$.

(iv) Given $v \in \mathbb{R}^n$, we show that M_{tf} is differentiable at v. To this end, we consider the function

$$g(h) := M_{tf}(v + h) - M_{tf}(v) - \frac{1}{t}\langle v - \text{prox}_{tf}(v), h\rangle, \quad h \in \mathbb{R}^n.$$

It suffices to show that $g(h) = o(\|h\|)$ as $\|h\| \to 0$. Note that

$$M_{tf}(v) = f(\text{prox}_{tf}(v)) + \frac{1}{2t}\|v - \text{prox}_{tf}(v)\|^2,$$

$$M_{tf}(v+h) \le f(\text{prox}_{tf}(v)) + \frac{1}{2t}\|v + h - \text{prox}_{tf}(v)\|^2.$$

Therefore,

$$\begin{aligned} g(h) &\le \frac{1}{2t}\|v+h-\text{prox}_{tf}(v)\|^2 - \frac{1}{2t}\|v-\text{prox}_{tf}(v)\|^2 - \frac{1}{t}\langle v - \text{prox}_{tf}(v), h\rangle \\ &= \frac{1}{2t}\|h\|^2 \end{aligned}$$

for all $h \in \mathbb{R}^n$. Note that g is convex with $g(0) = 0$. We have

$$0 = g(0) = g\left(\frac{1}{2}h + \frac{1}{2}(-h)\right) \le \frac{1}{2}g(h) + \frac{1}{2}g(-h).$$

Consequently, $g(h) \ge -g(-h) \ge -\frac{1}{2t}\|h\|^2$. Thus, $|g(h)| \le \|h\|^2/(2t)$ for all $h \in \mathbb{R}^n$ which shows the differentiability of M_{tf} at v. From the formula for $\nabla M_{tf}(v)$ and the continuity of the proximal mapping, it follows that M_{tf} is continuous differentiable.

(v) For any $x, y \in \mathbb{R}^n$, we have from (iv) that

$$\nabla M_{tf}(x) - \nabla M_{tf}(y) = \frac{1}{t}\left((I - \text{prox}_{tf})(x) - (I - \text{prox}_{tf}((y))\right).$$

By the firm nonexpansiveness of prox_{tf}, see Lemma 4.18, we have

$$\|(I - \text{prox}_{tf})(x) - (I - \text{prox}_{tf})(y)\| \le \|x - y\|.$$

Therefore,

$$\|\nabla M_{tf}(x) - \nabla M_{tf}(y)\| \le \frac{1}{t}\|x - y\|,$$

i.e., ∇M_{tf} is Lipschitz continuous with Lipschitz constant $\le 1/t$. □

Example 4.24 Let C be a nonempty closed convex subset of $\mathbb{R}^n$. Then, the Moreau envelope M_{δ_C} of its indicator function δ_C is

$$M_{\delta_C}(v) = \inf_{x \in C}\left\{\frac{1}{2}\|x - v\|^2\right\} = \frac{1}{2}d_C^2(v).$$

Therefore, according to Theorem 4.23 (iv), d_C^2 is a continuous differentiable function with $\nabla\left(\frac{1}{2}d_C^2\right)(v) = v - \text{prox}_{\delta_C}(v) = v - P_C(v)$. □

We now give the general version of the Moreau decomposition.

Theorem 4.25 (Moreau Decomposition) *Let $f : \mathbb{R}^n \to (-\infty, \infty]$ be a proper, lower semi-continuous, convex function. Then, for any $t > 0$ and $x \in \mathbb{R}^n$, there hold the following:*

(i) $\|x\|^2 = 2t \left\{M_{tf}(x) + M_{f^*/t}(x/t)\right\}$.
(ii) $x = \text{prox}_{tf}(x) + t\text{prox}_{f^*/t}(x/t)$.
(iii) $f(\text{prox}_{tf}(x)) + f^*(\text{prox}_{f^*/t}(x/t)) = \langle \text{prox}_{tf}(x), \text{prox}_{f^*/t}(x/t)\rangle$.

Proof

(i) Since f is proper, lower semi-continuous, and convex, we have $f = f^{**}$ by Theorem 4.8 and thus

$$\begin{aligned} M_{tf}(x) &= \inf_{y\in\mathbb{R}^n} \left\{ f(y) + \frac{1}{2t}\|y - x\|^2 \right\} \\ &= \inf_{y\in\mathbb{R}^n} \sup_{v\in\mathbb{R}^n} \left\{ \langle v, y\rangle - f^*(v) + \frac{1}{2t}\|y - x\|^2 \right\} \\ &\geq \sup_{v\in\mathbb{R}^n} \inf_{y\in\mathbb{R}^n} \left\{ \langle v, y\rangle - f^*(v) + \frac{1}{2t}\|y - x\|^2 \right\} \\ &= \sup_{v\in\mathbb{R}^n} \left\{ -f^*(v) - \frac{t}{2}\|v\|^2 + \langle x, v\rangle \right\}. \end{aligned} \tag{4.4}$$

On the other hand, we know from Theorem 4.23 that M_{tf} is convex and continuous differentiable. By setting $u := \nabla M_{tf}(x)$, we may use the convexity and definition of M_{tf} to obtain

$$M_{tf}(x) \leq M_{tf}(z) - \langle u, z - x\rangle \leq f(y) + \frac{1}{2t}\|y - z\|^2 - \langle u, z - x\rangle$$

for all $y, z \in \mathbb{R}^n$. By taking infimum with respect to z, it follows that

$$\begin{aligned} M_{tf}(x) &\leq f(y) + \inf_{z\in\mathbb{R}^n} \left\{ \frac{1}{2t}\|y - z\|^2 - \langle u, z - x\rangle \right\} \\ &= f(y) - \langle u, y\rangle - \frac{t}{2}\|u\|^2 + \langle x, u\rangle \end{aligned}$$

for all $y \in \mathbb{R}^n$. Now taking infimum with respect to y gives

$$\begin{aligned} M_{tf}(x) &\leq \inf_{y\in\mathbb{R}^n} \{f(y) - \langle u, y\rangle\} - \frac{t}{2}\|u\|^2 + \langle x, u\rangle \\ &= -f^*(u) - \frac{t}{2}\|u\|^2 + \langle x, u\rangle. \end{aligned} \tag{4.5}$$

Combining (4.4) and (4.5), we obtain

$$M_{tf}(x) \le \sup_{v\in\mathbb{R}^n}\left\{-f^*(v) - \frac{t}{2}\|v\|^2 + \langle x, v\rangle\right\} \le M_{tf}(x)$$

which forces

$$\begin{aligned} M_{tf}(x) &= \sup_{v\in\mathbb{R}^n}\left\{-f^*(v) - \frac{t}{2}\|v\|^2 + \langle x, v\rangle\right\} \\ &= -\inf_{v\in\mathbb{R}^n}\left\{f^*(v) + \frac{t}{2}\|v - x/t\|^2\right\} + \frac{1}{2t}\|x\|^2 \\ &= -M_{f^*/t}(x/t) + \frac{1}{2t}\|x\|^2. \end{aligned}$$

(ii) Let $\gamma > 0$ be any number. By using Theorem 4.9, we have

$$\begin{aligned} &y = \mathrm{prox}_{tf}(x) = (I + t\partial f)^{-1}(x) \Longrightarrow x \in y + t\partial f(y) \Longrightarrow \frac{x-y}{t} \in \partial f(y) \\ &\Longrightarrow y \in \partial f^*\left(\frac{x-y}{t}\right) \Longrightarrow \gamma y + \frac{x-y}{t} \in (I + \gamma\partial f^*)\left(\frac{x-y}{t}\right) \\ &\Longrightarrow \frac{x-y}{t} = (I + \gamma\partial f^*)^{-1}\left(\gamma y + \frac{x-y}{t}\right) \\ &\Longrightarrow x = y + t\mathrm{prox}_{\gamma f^*}\left(\gamma y + \frac{x-y}{t}\right). \end{aligned}$$

By taking $\gamma = 1/t$, we obtain the desired decomposition formula.

(iii) Let $y = \mathrm{prox}_{tf}(x) = (I + t\partial f)^{-1}(x)$. Then, $x - y \in t\partial f(y)$. By (ii) we have $x - y = t\mathrm{prox}_{f^*/t}(x/t)$. Thus,

$$\mathrm{prox}_{f^*/t}(x/t) \in \partial f(y).$$

In view of Proposition 4.6, we obtain

$$f(y) + f^*(\mathrm{prox}_{f^*/t}(x/t)) = \langle y, \mathrm{prox}_{f^*/t}(x/t)\rangle$$

which gives the result.

□

The result (i) in Proposition 4.25 actually can be obtained directly from the Fenchel-Rockafellar duality formula stated below, which will be used in the analysis of the augmented Lagrangian method in Chap. 5.

Theorem 4.26 (Fenchel-Rockafellar Duality Formula) *Let $f : \mathbb{R}^n \to (-\infty, \infty]$ and $g : \mathbb{R}^m \to (-\infty, \infty]$ be proper convex functions, and let $A : \mathbb{R}^n \to \mathbb{R}^m$*

be a linear operator. If there is $x_0 \in \mathrm{dom}(f)$ *such that* $Ax_0 \in \mathrm{dom}(g)$ *and* g *is continuous at* Ax_0*, then*

$$\inf_{x\in\mathbb{R}^n} \{f(x) + g(Ax)\} = \sup_{\eta\in\mathbb{R}^m} \left\{-f^*(-A^T\eta) - g^*(\eta)\right\}.$$

Proof We introduce

$$\Phi(x, y) := f(x) + g(Ax + y), \quad (x, y) \in \mathbb{R}^n \times \mathbb{R}^m$$

and

$$h(y) := \inf_{x\in\mathbb{R}^n} \Phi(x, y), \quad y \in \mathbb{R}^m.$$

Note that Φ is a convex function on $\mathbb{R}^n \times \mathbb{R}^m$. Thus, by Example 2.21, h is a convex function on $\mathbb{R}^m$ with

$$h(0) = \inf_{x\in\mathbb{R}^n} \{f(x) + g(Ax)\}$$

and

$$\begin{aligned} h^*(\eta) &= \sup_{y\in\mathbb{R}^m} \left\{\langle\eta, y\rangle - \inf_{x\in\mathbb{R}^n} \Phi(x, y)\right\} \\ &= \sup_{(x,y)\in\mathbb{R}^n\times\mathbb{R}^m} \{\langle\eta, y\rangle - f(x) - g(Ax + y)\} \\ &= \sup_{(x,z)\in\mathbb{R}^n\times\mathbb{R}^m} \{\langle\eta, z - Ax\rangle - f(x) - g(z)\} \\ &= \sup_{x\in\mathbb{R}^n} \left\{\langle -A^*\eta, x\rangle - f(x)\right\} + \sup_{z\in\mathbb{R}^m} \{\langle\eta, z\rangle - g(z)\} \\ &= f^*(-A^*\eta) + g^*(\eta) \end{aligned}$$

for any $\eta \in \mathbb{R}^m$. Therefore, it suffices to show that

$$h(0) = h^{**}(0) := \sup_{\eta\in\mathbb{R}^m} \left\{-h^*(\eta)\right\}. \tag{4.6}$$

By the given condition, we have $h(0) \leq \Phi(x_0, 0) < \infty$. Thus, either $h(0) = -\infty$ or $h(0) \in \mathbb{R}$.

If $h(0) = -\infty$, then by the definition of h^*, we have

$$h^*(\eta) \geq -h(0) = +\infty, \quad \forall \eta \in \mathbb{R}^m$$

which shows (4.6). Therefore, we may assume $h(0) \in \mathbb{R}$. Since $\Phi(x_0, \cdot)$ is continuous at 0, we can find a ball $B_r(0) \subset \mathbb{R}^m$ and $\lambda_0 \in \mathbb{R}$ such that $\Phi(x_0, y) \leq \lambda_0$ for all $y \in B_r(0)$. This implies that

$$h(y) \leq \Phi(x_0, y) \leq \lambda_0, \quad \forall y \in B_r(0).$$

By the convexity of h, we have $h(0) \leq \frac{1}{2}(h(y)+h(-y))$ and hence $h(y) \geq 2h(0) - h(-y) \geq 2h(0) - \lambda_0$ for all $y \in B_r(0)$. This shows that h is bounded on $B_r(0)$, and from Proposition 2.29, it follows that h is continuous at 0. By virtue of Theorem 4.8, we have $h(0) = h^{**}(0)$, which shows (4.6) again. □

4.3 Proximal Point Algorithm

Let $f : \mathbb{R}^n \to (-\infty, \infty]$ be a proper, lower semi-continuous, convex function. Consider the minimization problem:

$$\min_{x \in \mathbb{R}^n} f(x). \tag{4.7}$$

According to Lemma 4.12, x^* is a minimizer of (4.7) if and only if

$$x^* = \operatorname{prox}_{tf}(x^*)$$

for any $t > 0$. This suggests to solve (4.7) by the fixed point iteration

$$x^{k+1} = \operatorname{prox}_{tf}(x^k), \tag{4.8}$$

which is called the *proximal minimization* or the *proximal point algorithm*. The proximal point algorithm was first introduced in [152, 153] by Martinet in 1970 and was then further developed by Rockafellar and many other researchers [43, 94, 192, 193].

The proximal point algorithm can be derived by applying the gradient method to the Moreau envelope. From Theorem 4.23, it follows that M_{tf} is a continuous differentiable convex function whose set of minimizers is identically the same as that of f. Thus, it is natural to use the gradient method

$$x^{k+1} = x^k - t\nabla M_{tf}(x^k) \tag{4.9}$$

to find a minimizer of M_{tf}. By plugging the formula of $\nabla M_{tf}(x^k)$ given in Theorem 4.23 (iv) into (4.9), we then obtain (4.8).

We give one more motivation on the proximal point algorithm by suitable discretization of a gradient flow. Assume that f is a smooth convex function. Then, the minimizer of f can be found as the equilibrium point of the gradient flow:

$$\frac{d}{dt}x(t) = -\nabla f(x(t)).$$

If we use the forward Euler discrete scheme, it results in

$$\frac{x^{k+1} - x^k}{h} = -\nabla f(x^k)$$

with step size $h > 0$. Thus,

$$x^{k+1} = x^k - h\nabla f(x^k)$$

which is the gradient descent method. If we use the backward Euler discrete scheme

$$\frac{x^{k+1} - x^k}{h} = -\nabla f(x^{k+1}),$$

it yields

$$x^{k+1} + h\nabla f(x^{k+1}) = x^k.$$

Consequently,

$$x^{k+1} = (I + h\nabla f)^{-1}(x^k) = \text{prox}_{hf}(x^k)$$

which gives the proximal point algorithm (4.8) in the case of smooth objective functions. This motivation shows gradient steps correspond to forward Euler steps and proximal steps correspond to backward Euler steps. Therefore, we often refer to gradient steps as *forward steps* and proximal steps as *backward steps*.

In the method (4.8), we may let t vary in each step. Thus, we will consider the proximal point algorithm of the form

$$x^{k+1} = \text{prox}_{t_k f}(x^k), \tag{4.10}$$

where t_k are positive numbers.

Example 4.27 Consider the problem of minimizing the quadratic function

$$f(x) = \frac{1}{2}x^T Ax - b^T x,$$

where A is an $n \times n$ positive semi-definite matrix. This problem is equivalent to solving $Ax = b$. The proximal mapping of f is given by

$$\begin{aligned}\operatorname{prox}_{tf}(v) &= \arg\min_{x\in\mathbb{R}^n}\left\{\frac{1}{2}x^TAx - b^Tx + \frac{1}{2t}\|x-v\|^2\right\}\\ &= \left\{x : Ax - b + \frac{1}{t}(x-v) = 0\right\}.\end{aligned}$$

That is,

$$\operatorname{prox}_{tf}(v) = \left(A + \frac{1}{t}I\right)^{-1}\left(b + \frac{1}{t}v\right) = v - \left(A + \frac{1}{t}I\right)^{-1}(Av - b).$$

Therefore, (4.8) becomes the *iterative refinement method* of Wilkinson [224]:

$$x^{k+1} = x^k - \left(A + \frac{1}{t}I\right)^{-1}(Ax^k - b)$$

which is a standard algorithm for solving $Ax = b$. It is useful in case A is singular or has very large condition number. □

Example 4.28 Let A be a matrix of size $m \times n$ and let $b \in \mathbb{R}^m$. Consider the least square problem

$$\min_{x\in\mathbb{R}^n}\left\{f(x) := \frac{1}{2}\|Ax - b\|^2\right\}.$$

By direct calculation, it is easy to see that

$$\operatorname{prox}_{tf}(v) = v - \left(\frac{1}{t}I + A^TA\right)^{-1}A^T(Av - b), \quad \forall v \in \mathbb{R}^n.$$

Therefore, applying the proximal point algorithm (4.10) to solve the least square problem gives the iteration scheme:

$$x^{k+1} = x^k - \left(\frac{1}{t_k}I + A^TA\right)^{-1}A^T(Ax^k - b)$$

which is the nonstationary iterated Tikhonov regularization [98] for solving linear ill-posed inverse problems; when $t_k \equiv t > 0$ for all k, the corresponding method is called Lardy's method [77, 131]. □

For the proximal point algorithm (4.10), we have the following convergence result.

Theorem 4.29 *Let $f : \mathbb{R}^n \to (-\infty, \infty]$ be a proper, lower semi-continuous, convex function. Assume that f has a minimizer. If $t_k > 0$ is such that $\sum_{k=0}^{\infty} t_k = \infty$, then for the method (4.10), there hold the following:*

(i) *$\{f(x^k)\}$ is monotonically decreasing and $f(x^k) \to f_* := \min_{x\in\mathbb{R}^n} f(x)$ as $k \to \infty$.*
(ii) *$\{x^k\}$ converges to a minimizer of f.*

Proof

(i) We first show that $f(x^{k+1}) \leq f(x^k)$ for $k = 0, 1, \cdots$. In fact, by definition, we have

$$x^{k+1} = \operatorname{prox}_{t_k f}(x^k) = (I + t_k \partial f)^{-1}(x^k).$$

Thus, $(x^k - x^{k+1})/t_k \in \partial f(x^{k+1})$. This implies that

$$f(z) - f(x^{k+1}) \geq \left\langle \frac{x^k - x^{k+1}}{t_k}, z - x^{k+1} \right\rangle, \quad \forall z \in \mathbb{R}^n. \tag{4.11}$$

By taking $z = x^k$, it follows that

$$f(x^k) - f(x^{k+1}) \geq \frac{1}{t_k}\|x^k - x^{k+1}\|^2 \geq 0.$$

Now, let z be any minimizer of f. By using (4.11), we have

$$\begin{aligned}\|x^{k+1} - z\|^2 - \|x^k - z\|^2 &= -\|x^k - x^{k+1}\|^2 + 2\langle x^{k+1} - z, x^{k+1} - x^k\rangle \\ &\leq -\|x^k - x^{k+1}\|^2 + 2t_k(f(z) - f(x^{k+1})) \\ &\leq 2t_k(f_* - f(x^{k+1})) \leq 0. \end{aligned} \tag{4.12}$$

This implies that

$$2\sum_{i=0}^{k-1} t_i(f(x^{i+1}) - f_*) \leq \|x_0 - z\|^2 < \infty$$

for all $k \geq 1$. By the monotonicity of $\{f(x^i)\}$, we obtain

$$f(x^k) - f_* \leq \frac{\|x_0 - z\|^2}{2\sum_{i=0}^{k-1} t_i}, \quad k = 1, 2, \cdots$$

which together with $\sum_{i=0}^{\infty} t_i = \infty$ implies that $f(x^k) \to f_*$ as $k \to \infty$.

(ii) From (4.12), it follows that $\{\|x^k - z\|\}$ is decreasing for any minimizer z of f. In particular, $\{x^k\}$ is bounded in $\mathbb{R}^n$ and thus has a subsequence $\{x^{k_l}\}$ that is convergent to some element x^* in $\mathbb{R}^n$. By the lower semi-continuity of f and (i), we have

$$f(x^*) \leq \liminf_{l \to \infty} f(x^{k_l}) = f_*.$$

Therefore, x^* is a minimizer of f and hence $\{\|x^k - x^*\|\}$ is decreasing. Since $\|x^{k_l} - x^*\| \to 0$, we have $\|x^k - x^*\| \to 0$ as $k \to \infty$.

□

We conclude this section by the following useful result from [94], which will be used in the analysis of the augmented Lagrangian method in Chap. 5.

Proposition 4.30 *Assume that $f : \mathbb{R}^n \to (-\infty, \infty]$ is a proper, lower semi-continuous, convex function. Let $\{t_k\}$ be a sequence of positive numbers and let $\{x^k\}$ be defined by the proximal point algorithm (4.10). Then, for any $x \in \mathbb{R}^n$, there holds*

$$f(x^k) - f(x) \leq \frac{\|x - x^0\|^2}{2s_k} - \frac{\|x - x^k\|^2}{2s_k} - \frac{s_k}{2t_{k-1}^2}\|x^k - x^{k-1}\|^2$$

for $k = 1, 2, \cdots$, where $s_k := \sum_{j=0}^{k-1} t_j$.

Proof For any $k \geq 1$, we have by the definition of x^k that $0 \in \partial f(x^k) + (x^k - x^{k-1})/t_{k-1}$. Let $v^k := (x^{k-1} - x^k)/t_{k-1}$. Then, $v^k \in \partial f(x^k)$. Therefore, we may use the convexity of f to obtain

$$\begin{aligned} f(x) - f(x^k) \geq \langle v^k, x - x^k\rangle &= \frac{1}{t_{k-1}}\langle x^{k-1} - x^k, x - x^k\rangle \\ &= \frac{1}{2t_{k-1}}\left(\|x^{k-1} - x^k\|^2 + \|x - x^k\|^2 - \|x - x^{k-1}\|^2\right). \end{aligned} \tag{4.13}$$

Consequently,

$$2t_{k-1}(f(x) - f(x^k)) \geq t_{k-1}^2\|v^k\|^2 + \|x - x^k\|^2 - \|x - x^{k-1}\|^2$$

for all $k \geq 1$. Taking the sum gives

$$2\sum_{j=1}^{k} t_{j-1}(f(x) - f(x^j)) \geq \sum_{j=1}^{k} t_{j-1}^2\|v^j\|^2 + \|x - x^k\|^2 - \|x - x^0\|^2. \tag{4.14}$$

By setting $x = x^{k-1}$ in (4.13), we can obtain

$$f(x^{k-1}) - f(x^k) \geq \frac{1}{t_{k-1}}\|x^{k-1} - x^k\|^2 = t_{k-1}\|v^k\|^2.$$

Multiplying this equation by s_{k-1} and using $s_k - s_{k-1} = t_{k-1}$ with $s_0 = 0$, it gives

$$s_{k-1}f(x^{k-1}) - s_k f(x^k) + t_{k-1}f(x^k) \geq s_{k-1}t_{k-1}\|v^k\|^2.$$

Summing over this inequality and using $s_0 = 0$, we obtain

$$-s_k f(x^k) + \sum_{j=1}^{k} t_{j-1}f(x^j) \geq \sum_{j=1}^{k} s_{j-1}t_{j-1}\|v^j\|^2.$$

Therefore, by combining this with (4.14), it follows that

$$\begin{aligned}2s_k f(x) - 2s_k f(x^k) \geq 2\sum_{j=1}^{k} s_{j-1}t_{j-1}\|v^j\|^2 + \sum_{j=1}^{k} t_{j-1}^2\|v^j\|^2 \\ + \|x - x^k\|^2 - \|x - x^0\|^2. \qquad (4.15)\end{aligned}$$

Recall that $v^k \in \partial f(x^k)$ for $k \geq 1$. By the monotonicity of ∂f, we have

$$0 \leq \langle v^{k+1} - v^k, x^{k+1} - x^k\rangle = -t_k\langle v^{k+1} - v^k, v^{k+1}\rangle.$$

Thus, $\langle v^{k+1} - v^k, v^{k+1}\rangle \leq 0$ which implies that

$$\|v^{k+1}\|^2 \leq \langle v^k, v^{k+1}\rangle \leq \|v^k\|\|v^{k+1}\|$$

and thus $\|v^{k+1}\| \leq \|v^k\|$ for all $k \geq 1$. This shows that $\{\|v^k\|\}$ is monotonically decreasing. Combining this fact with (4.15), we obtain

$$\begin{aligned}2s_k(f(x) - f(x^k)) &\geq \left(\sum_{j=1}^{k} t_{j-1}^2 + 2\sum_{j=1}^{k} s_{j-1}t_{j-1}\right)\|v^k\|^2 + \|x - x^k\|^2 - \|x - x^0\|^2 \\ &= \sum_{j=1}^{k}(s_j^2 - s_{j-1}^2)\|v^k\|^2 + \|x - x^k\|^2 - \|x - x^0\|^2 \\ &= s_k^2\|v^k\|^2 + \|x - x^k\|^2 - \|x - x^0\|^2.\end{aligned}$$

The proof is thus complete. □

4.4 Proximal Gradient Method

In this section, we consider the convex minimization problem:

$$\min_{x\in\mathbb{R}^n} \{F(x) := f(x) + g(x)\}, \tag{4.16}$$

where f, g, and F are assumed to satisfy the following conditions.

Assumption 4.31

(i) $f : \mathbb{R}^n \to \mathbb{R}$ *is convex and differentiable, and* ∇f *is L-Lipschitz continuous, i.e.,*

$$\|\nabla f(x) - \nabla f(y)\| \leq L\|x - y\|, \quad \forall x, y \in \mathbb{R}^n.$$

(ii) $g : \mathbb{R}^n \to (-\infty, \infty]$ *is proper, lower semi-continuous, and convex.*
(iii) $F(x) := f(x) + g(x)$ *has a minimizer over* $\mathbb{R}^n$ *with the finite minimum* F_*.

If the proximal point algorithm (4.10) is used to solve (4.16), at each iteration step, one has to solve the convex minimization problem:

$$x^{k+1} = \arg\min_{x\in\mathbb{R}^n} \left\{ f(x) + g(x) + \frac{1}{2t_k}\|x - x^k\|^2 \right\}. \tag{4.17}$$

Even if both f and g have simple structures, the resolution of (4.17) can be difficult. In order to overcome this inherent difficulty of the minimization problem (4.17), in this section, we will consider the proximal gradient method proposed in [144, 176], which exploits the additively separable structure of (4.16) to decouple f and g in the algorithm formulation.

In order to derive the formulation of the proximal gradient method, we first show that solutions of (4.16) can be interpreted as fixed points of a mapping on $\mathbb{R}^n$. According to Theorem 3.9 and Theorem 3.14, we have

$$\begin{aligned} x^* \text{ is a solution of (4.16)} &\iff 0 \in t\nabla f(x^*) + t\partial g(x^*), \quad \forall t > 0 \\ &\iff x^* - t\nabla f(x^*) \in (I + t\partial g)(x^*) \\ &\iff x^* = (I + t\partial g)^{-1}(x^* - t\nabla f(x^*)). \end{aligned}$$

Therefore, we obtain the following theorem.

Theorem 4.32 x^* *is a minimizer of (4.16) if and only if*

$$x^* = \operatorname{prox}_{tg}(x^* - t\nabla f(x^*)), \quad \forall t > 0.$$

Theorem 4.32 suggests to solve (4.16) by the following fixed point iteration:

$$x^{k+1} = \text{prox}_{t_k g}(x^k - t_k \nabla f(x^k)) \tag{4.18}$$

with variable step size $t_k > 0$. Since each iteration is a combination of a gradient step and a proximal step, this method is called the *proximal gradient method*. Recall that gradient step is a forward step and proximal step is a backward step, the proximal gradient method is also called the *forward-backward splitting algorithm*. The proximal gradient method can be restated as the following algorithm that clearly demonstrates the decoupling structure of the method.

Algorithm 4.33 (Proximal Gradient Algorithm) *Take an initial guess x^0. For $k = 0, 1, \cdots$, repeat*

$$\begin{aligned} y^k &= x^k - t_k \nabla f(x^k); \\ x^{k+1} &= \arg\min_{x \in \mathbb{R}^n} \left\{ g(x) + \frac{1}{2t_k} \|x - y^k\|^2 \right\} \end{aligned}$$

with suitably chosen step size $t_k > 0$.

When $g = \delta_C$, where C is a nonempty closed convex subset of $\mathbb{R}^n$, (4.18) becomes the *projected gradient method*:

$$x^{k+1} = P_C(x^k - t_k \nabla f(x^k))$$

for solving the constrained minimization problem

$$\min_{x \in C} f(x).$$

Therefore, the proximal gradient method generalizes the projected gradient method to solve convex minimization problems of the form (4.16).

There are other motivations for deriving (4.18). We present a derivation using the *majorization-minimization (MM) strategy*. MM algorithms for minimizing a function $\varphi : \mathbb{R}^n \to (-\infty, \infty]$ consist of the following two steps:

(i) *Majorization step*: find $\Phi : \mathbb{R}^n \times \mathbb{R}^n \to (-\infty, \infty]$ that majorizes φ, i.e., for each $z \in \mathbb{R}^n$, there hold

$$\Phi(z, z) = \varphi(z) \quad \text{and} \quad \Phi(x, z) \geq \varphi(x) \quad \forall x \in \mathbb{R}^n.$$

(ii) *Minimization step*: given x^k, define

$$x^{k+1} := \arg\min_{x \in \mathbb{R}^n} \Phi(x, x^k).$$

To apply the MM algorithm to (4.16), let Assumption 4.31 hold, and for $t > 0$, let

$$\hat{f}_t(x, z) := f(z) + \langle \nabla f(z), x - z \rangle + \frac{1}{2t}\|x - z\|^2.$$

If $0 < t \leq 1/L$, one can obtain from Lemma 2.30 that $\hat{f}_t$ is a majorant of f, i.e.,

$$\hat{f}_t(z, z) = f(z) \quad \text{and} \quad \hat{f}_t(x, z) \geq f(x), \quad \forall x, z \in \mathbb{R}^n.$$

Thus, $q_t(x, z) := \hat{f}_t(x, z) + g(x)$ is a majorant of $F := f + g$. Given $t_k > 0$, define

$$x^{k+1} = \arg\min_{x \in \mathbb{R}^n} \left\{ \hat{f}_{t_k}(x, x^k) + g(x) \right\}.$$

Direct calculation shows that

$$\begin{aligned}
x^{k+1} &= \arg\min_{x \in \mathbb{R}^n} \left\{ g(x) + f(x^k) + \langle \nabla f(x^k), x - x^k \rangle + \frac{1}{2t_k}\|x - x^k\|^2 \right\} \\
&= \arg\min_{x \in \mathbb{R}^n} \left\{ g(x) + \frac{1}{2t_k}\left(\|x\|^2 - 2\langle x^k - t_k \nabla f(x^k), x \rangle \right) \right\} \\
&= \arg\min_{x \in \mathbb{R}^n} \left\{ g(x) + \frac{1}{2t_k}\|x - (x^k - t_k \nabla f(x^k))\|^2 \right\} \\
&= \operatorname{prox}_{t_k g}(x^k - t_k \nabla f(x^k))
\end{aligned}$$

which leads to the proximal gradient method (4.18).

For the convergence analysis of the proximal gradient method, we need the following result.

Lemma 4.34 *Let f and g satisfy Assumption 4.31. For $t > 0$ and $u \in \mathbb{R}^n$, define*

$$u^+ := \operatorname{prox}_{tg}(u - t\nabla f(u)).$$

Then, for any $x \in \mathbb{R}^n$, there holds

$$\begin{aligned}
&f(u^+) + g(u^+) \\
&\leq f(x) + g(x) - \frac{1}{t}\langle u - u^+, x - u^+ \rangle + \frac{L}{2}\|u - u^+\|^2 - D_f(x, u)
\end{aligned} \tag{4.19}$$

and, if $0 < t \leq 1/L$, then

$$\begin{aligned}
&f(u^+) + g(u^+) \\
&\leq f(x) + g(x) + \frac{1}{2t}\left(\|u - x\|^2 - \|u^+ - x\|^2 \right) - D_f(x, u),
\end{aligned} \tag{4.20}$$

where $D_f(x, u) := f(x) - f(u) - \langle \nabla f(u), x - u\rangle \geq 0$ is the Bregman distance induced by f at u.

Proof By the definition of u^+, we have

$$0 \in \partial g(u^+) + \frac{1}{t}(u^+ - u + t\nabla f(u))$$

which implies that

$$\frac{u - u^+}{t} - \nabla f(u) \in \partial g(u^+).$$

Therefore, by the convexity of g, we have

$$g(x) \geq g(u^+) + \left\langle \frac{u - u^+}{t} - \nabla f(u), x - u^+ \right\rangle. \tag{4.21}$$

On the other hand, since ∇f is L-Lipschitz continuous, we have from Lemma 2.30 that

$$\begin{aligned} f(u^+) &\leq f(u) + \langle \nabla f(u), u^+ - u\rangle + \frac{L}{2}\|u - u^+\|^2 \\ &= f(x) + \langle \nabla f(u), u^+ - x\rangle - D_f(x, u) + \frac{L}{2}\|u - u^+\|^2 \end{aligned}$$

which implies that

$$f(x) \geq f(u^+) + \langle \nabla f(u), x - u^+\rangle - \frac{L}{2}\|u - u^+\|^2 + D_f(x, u).$$

Adding this inequality with (4.21) gives (4.19). If $0 < t \leq 1/L$, we have from (4.19) that

$$f(u^+)+g(u^+) \leq f(x)+g(x) - \frac{1}{t}\langle u - u^+, x - u^+\rangle + \frac{1}{2t}\|u - u^+\|^2 - D_f(x, u)$$

which is equivalent to (4.20). The nonnegativity of $D_f(x, u)$ follows from the convexity of f and Proposition 2.31. □

Theorem 4.35 *Consider the convex minimization problem (4.16) for which Assumption 4.31 holds. Let $\{x^k\}$ be the sequence generated by Algorithm 4.33. Let $\bar{x}$ denote any solution of (4.16).*

(i) *If the step size $\{t_k\}$ satisfies*

$$0 < \gamma_0 = \inf_{k\geq 0} t_k \leq \sup_{k\geq 0} t_k \leq \gamma_1 < 2/L, \tag{4.22}$$

then $\{x^k\}$ is a bounded sequence whose cluster points are optimal solutions of (4.16), $\{F(x^k)\}$ is monotonically decreasing,

$$0 \le F(x^k) - F_* \le \frac{1}{2\gamma_0 k}\left(\|x^0 - \bar{x}\|^2 + \frac{2L\gamma_1^2}{2 - L\gamma_1}(F(x^0) - F_*)\right) \tag{4.23}$$

for all $k \ge 1$, and $F(x^k) - F_* = o(1/k)$ as $k \to \infty$.

(ii) *If the step size $\{t_k\}$ satisfies*

$$0 < \gamma_0 = \inf_{k\ge 0} t_k \le \sup_{k\ge 0} t_k \le 1/L, \tag{4.24}$$

then $\{\|x^{k+1} - \bar{x}\|\}$ is monotonically decreasing,

$$0 \le F(x^k) - F_* \le \frac{\|x^0 - x^*\|^2}{2\gamma_0 k} \tag{4.25}$$

for all $k \ge 1$ and $\{x^k\}$ converges to a minimizer of F over $\mathbb{R}^n$.

Proof

(i) As $x^{k+1} = \operatorname{prox}_{t_k g}(x^k - t_k \nabla f(x^k))$, we may use (4.19) in Lemma 4.34 with $x = u = x^k$, $u^+ = x^{k+1}$, and $t = t_k$ to obtain

$$F(x^{k+1}) - F(x^k) \le -\left(\frac{1}{t_k} - \frac{L}{2}\right)\|x^{k+1} - x^k\|^2. \tag{4.26}$$

According to the condition (4.22), we have $1/t_k - L/2 > 0$. Thus, $\{F(x^k)\}$ is monotonically decreasing.

Next, by using again (4.19) in Lemma 4.34 with $x = \bar{x}$, $u = x^k$, and $u^+ = x^{k+1}$, we obtain

$$\begin{aligned} F(x^{k+1}) &\le F_* - \frac{1}{t_k}\langle x^k - x^{k+1}, \bar{x} - x^{k+1}\rangle + \frac{L}{2}\|x^{k+1} - x^k\|^2 \\ &= F_* + \frac{1}{2t_k}\left(\|x^k - \bar{x}\|^2 - \|x^{k+1} - \bar{x}\|^2\right) + \left(\frac{L}{2} - \frac{1}{2t_k}\right)\|x^{k+1} - x^k\|^2. \end{aligned}$$

Consequently,

$$\|x^{k+1} - \bar{x}\|^2 - \|x^k - \bar{x}\|^2 \le 2t_k(F_* - F(x^{k+1})) + (Lt_k - 1)\|x^{k+1} - x^k\|^2. \tag{4.27}$$

With the help of (4.26), we then obtain

$$\|x^{k+1}-\bar{x}\|^2-\|x^k-\bar{x}\|^2 \le 2t_k\left(F_*-F(x^{k+1})\right)+\frac{2Lt_k^2}{2-Lt_k}(F(x^k)-F(x^{k+1})).$$

Note that $F_* - F(x^{k+1}) \le 0$ and $F(x^k) - F(x^{k+1}) \ge 0$, and we may use the condition (4.22) to derive

$$\|x^{k+1}-\bar{x}\|^2-\|x^k-\bar{x}\|^2 \le 2\gamma_0\left(F_*-F(x^{k+1})\right)+\frac{2L\gamma_1^2}{2-L\gamma_1}(F(x^k)-F(x^{k+1})).$$

By summing both sides, we obtain

$$\begin{aligned}
&\|x^{k+1}-\bar{x}\|^2+2\gamma_0\sum_{j=0}^{k}(F(x^{j+1})-F_*)\\
&\le \|x^0-\bar{x}\|^2+\frac{2L\gamma_1^2}{2-L\gamma_1}\sum_{j=0}^{k}(F(x^j)-F(x^{j+1}))\\
&= \|x^0-\bar{x}\|^2+\frac{2L\gamma_1^2}{2-L\gamma_1}(F(x^0)-F(x^{k+1}))\\
&\le \|x^0-\bar{x}\|^2+\frac{2L\gamma_1^2}{2-L\gamma_1}(F(x^0)-F_*).
\end{aligned}$$

This shows that $\{x^k\}$ is a bounded sequence and

$$2\gamma_0\sum_{j=0}^{k}(F(x^{j+1})-F_*) \le \|x^0-\bar{x}\|^2+\frac{2L\gamma_1^2}{2-L\gamma_1}(F(x^0)-F_*).$$

By the monotonicity of $F(x^k) - F_*$, we thus obtain (4.23) and $F(x^k) - F_* = o(1/k)$ as $k \to \infty$.

Let x^* be a cluster point of $\{x^k\}$. Then, there is a subsequence $\{k_j\}$ of integers such that $x^{k_j} \to x^*$ as $j \to \infty$. By virtue of the lower semi-continuity of F, we then obtain

$$F(x^*) \le \liminf_{j\to\infty} F(x^{k_j}) = F_* \le F(x^*).$$

Thus, $F(x^*) = F_*$, i.e., x^* is an optimal solution of (4.16).

(ii) When the step size $\{t_k\}$ satisfies (4.24), it follows from (4.27) that $\{\|x^k - \bar{x}\|\}$ is monotonically decreasing and

$$2\gamma_0(F(x^{k+1})-F_*) \le \|x^k-\bar{x}\|^2-\|x^{k+1}-\bar{x}\|^2.$$

Summing over this inequality gives

$$2\gamma_0 \sum_{i=0}^{k-1} \left(F(x^{i+1}) - F_* \right) \le \|x^0 - \bar{x}\|^2 - \|x^k - \bar{x}\|^2 \le \|x^0 - \bar{x}\|^2.$$

Since $\{F(x^i)\}$ is monotonically decreasing, we obtain

$$2\gamma_0 k(F(x^k) - F_*) \le 2\gamma_0 \sum_{i=0}^{k-1} (F(x^{i+1}) - F_*) \le \|x^0 - \bar{x}\|^2$$

for $k \ge 1$, which implies the estimate (4.25).

Finally, we show that $\{x^k\}$ converges to a minimizer of F. In (i) we have shown that $\{x^k\}$ has a subsequence $\{x^{k_j}\}$ converging to a minimizer x^* of F. By the monotonicity of $\{\|x_k - x^*\|\}$, we must have $\lim_{k\to\infty} \|x^k - x^*\| = \lim_{j\to\infty} \|x^{k_j} - x^*\| = 0$.

□

When f is strongly convex, the convergence result in Theorem 4.35 can be improved. Actually, we have the following linear convergence result.

Theorem 4.36 *Consider the convex minimization problem (4.16) for which Assumption 4.31 holds. Assume further that f is μ-strongly convex for some $\mu > 0$. Then, for the sequence $\{x^k\}$ generated by Algorithm 4.33 with $0 < t_k = t \le 1/L$, there hold*

$$\|x^{k+1} - x^*\|^2 \le (1 - 2\mu t)\|x^k - x^*\|^2 \tag{4.28}$$

and

$$F(x^{k+1}) - F(x^*) \le (1 - 2\mu t) \left(F(x^k) - F(x^*) \right) \tag{4.29}$$

for all integers $k \ge 0$, where x^* denotes the unique solution of (4.16).

Proof Since $0 < t \le 1/L$, by using (4.20) in Lemma 4.34 with $x = x^*$, $u = x^k$ and $u^+ = x^{k+1}$, we can obtain

$$F(x^{k+1}) \le F(x^*) + \frac{1}{2t} \left(\|x^k - x^*\|^2 - \|x^{k+1} - x^*\|^2 \right) - D_f(x^*, x^k).$$

Since f is μ-strongly convex, we have $D_f(x^*, x^k) \ge \mu \|x^k - x^*\|^2$. Therefore,

$$F(x^{k+1}) - F(x^*) \le \left(\frac{1}{2t} - \mu \right) \|x^k - x^*\|^2 - \frac{1}{2t}\|x^{k+1} - x^*\|^2$$

which together with $F(x^{k+1}) - F(x^*) \geq 0$ implies that

$$\|x^{k+1} - x^*\|^2 \leq (1 - 2\mu t)\|x^k - x^*\|^2$$

and thus (4.28) follows.

Next, we show (4.29). By using again (4.20) in Lemma 4.34 with $u = x^k$ and $u^+ = x^{k+1}$, we have for any x that

$$F(x^{k+1}) \leq F(x) + \frac{1}{2t}\left(\|x^k - x\|^2 - \|x^{k+1} - x\|^2\right) - D_f(x, x^k).$$

Using the μ-strong convexity of f, we further have

$$F(x^{k+1}) \leq F(x) + \left(\frac{1}{2t} - \mu\right)\|x^k - x\|^2.$$

Now, we take $x = x^k + 2\mu t(x^* - x^k)$. Noting that $0 < \mu \leq L/2$, we have $0 < 2\mu t \leq 1$. Thus, by the μ-strongly convexity of F, we can obtain

$$\begin{aligned} F(x^{k+1}) &\leq F\left(x^k + 2\mu t(x^* - x^k)\right) + \mu(2\mu t)(1 - 2\mu t)\|x^* - x^k\|^2 \\ &\leq 2\mu t F(x^*) + (1 - 2\mu t)F(x^k) \end{aligned}$$

which implies (4.29). The proof is thus complete. □

Remark 4.37 For further linear convergence results on the proximal gradient method, one may refer to Theorem 8.51.

Example 4.38 Let us consider applying the proximal gradient method to solve the lasso problem:

$$\min_{x \in \mathbb{R}^n} \left\{\frac{1}{2}\|Ax - b\|^2 + \lambda\|x\|_1\right\},$$

where $b \in \mathbb{R}^m$, A is an $m \times n$ matrix, and $\lambda > 0$ is a parameter. This problem has the form (4.16) with

$$f(x) = \frac{1}{2}\|Ax - b\|^2 \quad \text{and} \quad g(x) = \lambda\|x\|_1.$$

Note that $\nabla f(x) = A^T(Ax - b)$, which means that ∇f is Lipschitz continuous with Lipschitz constant $\lambda_{\max}(A^T A)$. Note also that

$$\text{prox}_{tg}(x) = \arg\min_{z \in \mathbb{R}^n}\left\{\|z\|_1 + \frac{1}{2\lambda t}\|z - x\|^2\right\} = S_{\lambda t}(x),$$

where S_λ denotes the soft thresholding; see Example 4.15. Thus, the proximal gradient method (4.18) becomes

$$x^{k+1} = S_{\lambda t_k}\left(x^k - t_k A^T(Ax^k - b)\right)$$

which is called the iterative soft thresholding algorithm (ISTA) [59, 65]. By Theorem 4.35, ISTA is convergent with constant step size $0 < t_k = t \leq 1/\lambda_{\max}(A^T A)$.

Example 4.39 Given two closed convex sets C and D in $\mathbb{R}^n$, assume that the Euclidean distance

$$d(C, D) := \inf_{x\in D, y\in C} d(x, y) = \inf_{x\in D} d_C(x)$$

between C and D is realized, and we want to find a point $y^* \in C$ and $x^* \in D$ such that $d(y^*, x^*) = d(C, D)$. This amounts to solving the constrained convex minimization problem:

$$\min_{x\in D} \left\{\frac{1}{2} d_C^2(x)\right\}, \tag{4.30}$$

where $d_C(x)$ denotes the Euclidean distance of x to C, i.e.,

$$d_C(x) = \min_{z\in C} \|z - x\| = \|P_C(x) - x\|.$$

Here, P_C denotes the Euclidean projection onto C. Let δ_D denote the indicator function of D, and then (4.30) takes the form (4.16) with $f(x) = \frac{1}{2} d_C^2(x)$ and $g(x) = \delta_D(x)$. According to Example 4.24, f is a continuous differentiable convex function with gradient

$$\nabla f(x) = x - P_C(x)$$

which is Lipschitz continuous with Lipschitz constant $L = 1$. Therefore, applying the proximal gradient method to solve (4.30) yields the iteration scheme:

$$x^{k+1} = P_D\left(x^k - t_k(x^k - P_C(x^k))\right).$$

Taking the step size $t_k \equiv 1$ for all k gives $x^{k+1} = P_D P_C(x^k)$, which can be equivalently stated as

$$y^k = P_C(x^k), \quad x^{k+1} = P_D(y^k), \quad k = 0, 1, \cdots. \tag{4.31}$$

This is the famous *alternating projection method*. According to Theorem 4.35, for the sequence $\{x^k\}$ and $\{y^k\}$ defined by (4.31) from any initial guess x^0, we have

$x^k \to x^*$ and $y^k \to y^*$ for some $x^* \in D$ and $y^* \in C$ and $d_C^2(x^k) \to d^2(C, D)$ as $k \to \infty$. Note that

$$d_C(x^k) = \|P_C(x^k) - x^k\| = \|y^k - x^k\|,$$

and we therefore have $\|y^* - x^*\| = \lim_{k\to\infty} d_C(x^k) = d(C, D)$. This is the convergence result on alternating projection method proved in [56] by a different approach. □

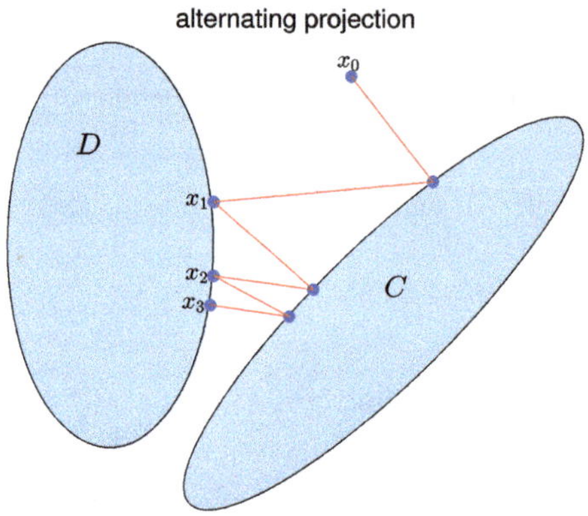

Example 4.40 Let A be a matrix of size $m \times n$. Let C and Q be closed convex subsets in $\mathbb{R}^n$ and $\mathbb{R}^m$, respectively. Consider the problem of finding $x \in C$ such that $Ax \in Q$, which is called the split feasibility problem. This problem was proposed in [48] as a model for solving certain inverse problems. For any $y \in \mathbb{R}^m$, let $d_Q(y)$ denote the Euclidean distance of y to Q. Then, the split feasibility problem is equivalent to the minimization problem:

$$\min_{x\in C} \left\{ \frac{1}{2} d_Q^2(Ax) := \frac{1}{2}\|Ax - P_Q(Ax)\|^2 \right\}$$

which has the form (4.16) with $f(x) = \frac{1}{2} d_Q^2(Ax)$ and $g(x) = \delta_C(x)$. According to Example 4.24, f is a continuous differentiable convex function and its gradient is given by $\nabla f(x) = A^T(I - P_Q)Ax$, which is Lipschitz continuous with Lipschitz constant $\|A\|^2$. Therefore, we may solve it by the proximal gradient method, leading to the iterative scheme:

$$x^{k+1} = P_C\left(x^k - tA^T(I - P_Q)Ax^k\right)$$

which is the CQ algorithm proposed in [45]. According to Theorem 4.35, the CQ algorithm is convergent as long as $0 < t \le 1/\|A\|^2$.

4.5 Accelerated Proximal Gradient Methods

In the previous section, we considered the proximal gradient method, i.e., Algorithm 4.33, for solving the minimization problem (4.16). It has been shown in Theorem 4.35 that this method has the convergence rate $o(1/k)$ in the sense that

$$F(x^k) - F_* = o(1/k),$$

where $F_* := \min_{x\in\mathbb{R}^n} F(x)$. The proximal gradient method can be modified to achieve faster convergence rate using Nesterov's acceleration strategy, which was first introduced in [164] to accelerate the gradient method for solving smooth convex minimization problems. This strategy was later incorporated into the proximal gradient method in [18] that leads to a significant improvement on convergence rates.

In this section, we will consider the following over-relaxation version of the proximal gradient method:

$$\begin{aligned} y^k &= x^k + \alpha_k(x^k - x^{k-1}), \\ x^{k+1} &= \text{prox}_{sg}(y^k - s\nabla f(y^k)) \end{aligned} \tag{4.32}$$

with given $x^{-1} = x^0 \in \mathbb{R}^n$, where $0 < s \le 1/L$ and $\{\alpha_k\}$ is a suitably chosen sequence of numbers. We will consider the choice

$$\alpha_k = \frac{t_k - 1}{t_{k+1}}, \tag{4.33}$$

where $\{t_k\}$ is a sequence of numbers satisfying

$$t_k \ge 1 \quad \text{and} \quad t_k^2 - t_k \le t_{k-1}^2 \quad \text{for all } k \ge 1. \tag{4.34}$$

For the analysis of the method (4.32), we will use Lemma 4.34. Let x^* be any solution of (4.16). By using (4.20) in Lemma 4.34 with

$$u^+ = x^{k+1}, \quad u = y^k \quad \text{and} \quad x = \left(1 - \frac{1}{t_{k+1}}\right)x^k + \frac{1}{t_{k+1}}x^*$$

and the convexity of F, we can derive that

$$\begin{aligned} F(x^{k+1}) \le F\left(\left(1-\frac{1}{t_{k+1}}\right)x^k+\frac{1}{t_{k+1}}x^*\right) + \frac{1}{2s}\left\|\left(1-\frac{1}{t_{k+1}}\right)x^k+\frac{1}{t_{k+1}}x^* - y^k\right\|^2 \\ - \frac{1}{2s}\left\|\left(1 - \frac{1}{t_{k+1}}\right)x^k + \frac{1}{t_{k+1}}x^* - x^{k+1}\right\|^2 \end{aligned}$$

$$\leq \left(1-\frac{1}{t_{k+1}}\right) F(x^k)+\frac{1}{t_{k+1}} F(x^*)+\frac{1}{2st_{k+1}^2} \left\|x^*-x^k - t_{k+1}(y^k - x^k)\right\|^2$$
$$- \frac{1}{2st_{k+1}^2} \left\|x^* - x^k - t_{k+1}(x^{k+1} - x^k)\right\|^2.$$

Let

$$u^k := x^{k-1} + t_k(x^k - x^{k-1}). \tag{4.35}$$

Then, by using $y^k = x^k + \frac{t_k-1}{t_{k+1}}(x^k - x^{k-1})$, we can see that

$$x^k + t_{k+1}(y^k - x^k) = x^{k-1} + t_k(x^k - x^{k-1}) = u^k.$$

Therefore,

$$F(x^{k+1}) \leq \left(1 - \frac{1}{t_{k+1}}\right) F(x^k) + \frac{1}{t_{k+1}} F(x^*) + \frac{1}{2st_{k+1}^2} \left\|x^* - u^k\right\|^2$$
$$- \frac{1}{2st_{k+1}^2} \left\|x^* - u^{k+1}\right\|^2.$$

This implies that

$$F(x^{k+1}) - F_* - \left(1 - \frac{1}{t_{k+1}}\right) (F(x^k) - F_*)$$
$$\leq \frac{1}{2st_{k+1}^2} \left\|x^* - u^k\right\|^2 - \frac{1}{2st_{k+1}^2} \left\|x^* - u^{k+1}\right\|^2.$$

We therefore obtain the following result.

Proposition 4.41 *Consider the convex minimization problem (4.16) for which Assumption 4.31 holds. Let x^* be any solution of (4.16). Consider the method (4.32) with $\{\alpha_k\}$ given by (4.33), where $\{t_k\}$ is chosen to satisfy (4.34). If $0 < s \leq 1/L$, then for any $k \geq 0$, there holds*

$$2st_{k+1}^2(F(x^{k+1}) - F_*) + \|x^* - u^{k+1}\|^2$$
$$\leq 2s(t_k^2 + \rho_k)(F(x^k) - F_*) + \|x^* - u^k\|^2, \tag{4.36}$$

where $\rho_k := t_{k+1}^2 - t_{k+1} - t_k^2$.

We now give some special choices of $\{t_k\}$ such that the corresponding method (4.32) yields faster convergence rate than the proximal gradient method in terms of the objective error.

Example 4.42 We first consider the choice

$$t_k = \frac{k+\alpha-1}{\alpha} \qquad \text{for all } k \geq 0$$

with $\alpha \geq 2$. Then,

$$\alpha_k = \frac{t_k - 1}{t_{k+1}} = \frac{k-1}{k+\alpha}.$$

Therefore, the method (4.32) becomes the form

$$\begin{aligned} y^k &= x^k + \frac{k-1}{k+\alpha}(x^k - x^{k-1}), \\ x^{k+1} &= \text{prox}_{sg}(y^k - s\nabla f(y^k)), \end{aligned} \tag{4.37}$$

where $x^{-1} = x^0$. Note that for $k \geq 0$, we have

$$\rho_k = t_{k+1}^2 - t_{k+1} - t_k^2 = -\frac{1}{\alpha^2}[(\alpha-2)(k+\alpha)+1] < 0 \tag{4.38}$$

which shows that $\{t_k\}$ satisfies (4.34). Thus, we may use Proposition 4.41 to conclude the following result.

Theorem 4.43 *Consider the convex minimization problem (4.16) for which Assumption 4.31 holds. Let x^* be any solution of (4.16). Let $\{x^k\}$ be defined by (4.37) with $0 < s \leq 1/L$ and $\alpha \geq 2$. Then,*

$$0 \leq F(x^k) - F_* \leq \frac{\alpha^2\|x^0 - x^*\|^2}{2s(k+\alpha-1)^2}, \qquad \forall k \geq 1.$$

Proof Since $\rho_k \leq 0$, we may use Proposition 4.41 to conclude for $k \geq 1$ that

$$\begin{aligned} 2st_k^2(F(x^k) - F_*) + \|u^k - x^*\|^2 &\leq 2st_1^2(F(x^1) - F_*) + \|u^1 - x^*\|^2 \\ &\leq 2s(t_0^2 + \rho_0)(F(x^0) - F_*) + \|u^0 - x^*\|^2. \end{aligned}$$

Noting that $t_0^2 + \rho_0 = 0$ and $u^0 = x^0$, we thus have

$$2st_k^2(F(x^k) - F_*) + \|u^k - x^*\|^2 \leq \|x^0 - x^*\|^2.$$

Consequently,

$$0 \leq F(x^k) - F_* \leq \frac{\|x^0 - x^*\|^2}{2st_k^2} = \frac{\alpha^2\|x^0 - x^*\|^2}{2s(k+\alpha-1)^2}.$$

The proof is complete. □

Example 4.44 Next, we consider the choice of $\{t_k\}$ defined by

$$t_0 = 0 \quad \text{and} \quad t_{k+1} = \frac{1 + \sqrt{1 + 4t_k^2}}{2} \quad \text{for } k \geq 0. \tag{4.39}$$

One can show by induction that

$$t_k \geq \frac{k+1}{2} \quad \text{for } k \geq 1. \tag{4.40}$$

Indeed, this is true for $k = 1$ because $t_1 = 1$. If it is true for some $k \geq 1$, then

$$t_{k+1} \geq \frac{1 + \sqrt{1 + (k+1)^2}}{2} \geq \frac{1 + (k+1)}{2} = \frac{k+2}{2}.$$

Consequently, we may use Proposition 4.41 to conclude the following result.

Theorem 4.45 *Consider the convex minimization problem (4.16) for which Assumption 4.31 holds. Let x^* be any solution of (4.16). Let $\{x^k\}$ be defined by (4.32) with $\{\alpha_k\}$ given by (4.33), where $\{t_k\}$ is defined by (4.39). If $0 < s \leq 1/L$, then there holds*

$$0 \leq F(x^k) - F_* \leq \frac{2\|x^0 - x^*\|^2}{s(k+1)^2}, \qquad \forall k \geq 1.$$

Proof By the choice of $\{t_k\}$, we have $\rho_k = 0$ for all $k \geq 0$. Thus, it follows from Proposition 4.41 that

$$\begin{aligned} 2st_k^2(F(x^k) - F_*) + \|u^k - x^*\|^2 &\leq 2st_0^2(F(x^0) - F_*) + \|u^0 - x^*\|^2 \\ &= \|x^0 - x^*\|^2. \end{aligned}$$

In view of (4.40), we can obtain the desired estimate. □

Theorems 4.43 and 4.45 establish convergence rate $O(k^{-2})$ in terms of objective error for the accelerated proximal gradient method (4.32) with particular choices of $\{t_k\}$. These results however do not imply the convergence of the iterative sequence $\{x^k\}$ automatically in general unless F admits stronger convexity. Progress toward the convergence of iterates has been made in [9, 49], which shows that for the method (4.37) with $\alpha > 2$, the iterative sequence $\{x^k\}$ is indeed convergent under general conditions.

In the following, we will focus on the method (4.37) with $\alpha > 2$ and show that the rate in Theorem 4.43 can be improved to

$$F(x^k) - F_* = o(k^{-2})$$

and the sequence $\{x^k\}$ converges to a solution of (4.16). We will follow the arguments in [9]. We first introduce

$$\begin{aligned}\mathcal{E}_k &:= 2st_k^2(F(x^k) - F_*) + \|u^k - x^*\|^2 \\ &= \frac{2s(k+\alpha-1)^2}{\alpha^2}(F(x^k) - F_*) + \|u^k - x^*\|^2.\end{aligned}$$

By virtue of Proposition 4.41, we can obtain the following result.

Lemma 4.46 *Let Assumption 4.31 hold. Consider the method (4.37) with $\alpha \geq 2$ and $0 < s \leq 1/L$. The following properties hold.*

(i) *$\{\mathcal{E}_k\}$ is monotonically decreasing and $\lim_{k\to\infty} \mathcal{E}_k$ exists.*
(ii) *$\mathcal{E}_k \leq \mathcal{E}_0$ and thus*

$$F(x^k) - F_* \leq \frac{\alpha^2}{2s(k+\alpha-1)^2}\mathcal{E}_0, \quad \|u^k - x^*\| \leq \mathcal{E}_0.$$

(iii) *If $\alpha > 2$, then*

$$\sum_{k=0}^{\infty} k(F(x^k) - F_*) \leq \frac{\alpha^2}{2s(\alpha-2)}\mathcal{E}_0.$$

Proof From (4.38), we can see that $\rho_k \leq -(\alpha-2)k/\alpha^2$. Since $\alpha \geq 2$ and $F(x^k) \geq F_*$, it follows from (4.36) that

$$\mathcal{E}_{k+1} + \frac{2s(\alpha-2)k}{\alpha^2}(F(x^k) - F_*) \leq \mathcal{E}_k. \tag{4.41}$$

This implies (i) and (ii). By summing (4.41) over k, we can obtain (iii). □

In order to proceed further, we use (4.20) in Lemma 4.34 with $u^+ = x^{k+1}$, $u = y^k$, and $x = x^k$ to obtain

$$F(x^{k+1}) \leq F(x^k) - \frac{1}{2s}\|x^{k+1} - x^k\|^2 + \frac{1}{2s}\|y^k - x^k\|^2.$$

Thus, using the definition of y^k, we have

$$F(x^{k+1}) + \frac{1}{2s}\|x^{k+1} - x^k\|^2 \leq F(x^k) + \frac{(k-1)^2}{2s(k+\alpha)^2}\|x^k - x^{k-1}\|^2. \tag{4.42}$$

Let

$$w_k := F(x^k) - F_*, \quad d_k := \frac{1}{2s}\|x^k - x^{k-1}\|^2.$$

It then follows from (4.42) that

$$w_{k+1} + d_{k+1} \le w_k + \frac{(k-1)^2}{(k+\alpha)^2} d_k \le w_k + \frac{(k-1)^2}{(k+2)^2} d_k.$$

Consequently,

$$(k+2)^2 d_{k+1} - (k-1)^2 d_k \le (k+2)^2 (w_k - w_{k+1})$$

which can be equivalently written as

$$\begin{aligned} k^2 d_{k+1} - (k-1)^2 d_k + 4(k+1) d_{k+1} \le (k+1)^2 w_k - (k+2)^2 w_{k+1} \\ + (2k+3) w_k. \end{aligned} \tag{4.43}$$

Summing this inequality over k gives

$$4 \sum_{k=2}^{\infty} k d_k \le 4 w_1 + \sum_{k=1}^{\infty} (2k+3) w_k \le 4 w_1 + 5 \sum_{k=1}^{\infty} k w_k.$$

When $\alpha > 2$, we may use Lemma 4.46 to obtain

$$4 \sum_{k=2}^{\infty} k d_k \le \frac{2}{s} \mathcal{E}_1 + \frac{5\alpha^2}{2s(\alpha-2)} \mathcal{E}_0 \le \frac{2}{s} \mathcal{E}_0 + \frac{5\alpha^2}{2s(\alpha-2)} \mathcal{E}_0.$$

We therefore obtain the following lemma.

Lemma 4.47 *Let Assumption 4.31 hold. Consider the method (4.37) with* $0 < s \le 1/L$. *If* $\alpha > 2$, *then*

$$\sum_{k=2}^{\infty} k \|x^k - x^{k-1}\|^2 \le C_\alpha \mathcal{E}_0 < \infty,$$

where C_α *is a constant depending only on* α *and* s.

Lemma 4.48 *Let Assumption 4.31 hold. Consider the method (4.37) with* $0 < s \le 1/L$. *If* $\alpha > 2$, *then*

$$\lim_{k \to \infty} \left[k^2 \|x^k - x^{k-1}\|^2 + k^2 (F(x^k) - F_*) \right]$$

exists.

Proof From (4.43), it follows that

$$\left[k^2 d_{k+1} + (k+2)^2 w_{k+1} \right] - \left[(k-1)^2 d_k + (k+1)^2 w_k \right] \le 5 k w_k.$$

Note that $(k-1)^2 d_k + (k+1)^2 w_k \geq 0$ and Lemma 4.46 shows that $\sum_k k w_k < \infty$. We may use Lemma 3.20 to conclude that

$$\lim_{k\to\infty} [k^2 d_{k+1} + (k+2)^2 w_{k+1}]$$

exists. Since Lemmas 4.46 and 4.47 imply that

$$k w_k \to 0 \quad \text{and} \quad k d_k \to 0 \quad \text{as } k \to \infty,$$

we can conclude that $\lim_{k\to\infty} \left[k^2 d_k + k^2 w_k\right]$ must exist. □

Theorem 4.49 *Let Assumption 4.31 hold. Consider the method (4.37) with* $0 < s \leq 1/L$. *If* $\alpha > 2$, *then*

$$F(x^k) - F_* = o(k^{-2}) \quad \textit{and} \quad \|x^{k+1} - x^k\| = o(k^{-1})$$

as $k \to \infty$.

Proof In view of Lemmas 4.46 and 4.47, we have

$$\sum_{k=1}^{\infty} \frac{1}{k} \left[k^2 \|x^k - x^{k-1}\|^2 + k^2 (F(x^k) - F_*)\right] < \infty.$$

This together with Lemma 4.48 implies that

$$\lim_{k\to\infty} \left[k^2 \|x^k - x^{k-1}\|^2 + k^2 (F(x^k) - F_*)\right] = 0$$

which completes the proof. □

Theorem 4.50 *Let Assumption 4.31 hold. Let* $\{x^k\}$ *be the sequence defined by the method (4.37) with* $\alpha > 2$ *and* $0 < s \leq 1/L$. *Then,* $\{x^k\}$ *converges to a minimizer of* F *over* $\mathbb{R}^n$.

Proof Let x^* be any minimizer of F. Recalling the definition (4.35) of u^k, we have

$$u^k = x^k + (t_k - 1)(x^k - x^{k-1}) = x^k + \frac{k-1}{\alpha}(x^k - x^{k-1}). \tag{4.44}$$

Consequently,

$$\begin{aligned} \|u^k - x^*\|^2 &= \frac{(k-1)^2}{\alpha^2} \|x^k - x^{k-1}\|^2 + \|x^k - x^*\|^2 \\ &\quad + \frac{2(k-1)}{\alpha} \langle x^k - x^*, x^k - x^{k-1} \rangle \end{aligned}$$

$$
\begin{aligned}
&= \frac{(k-1)^2}{\alpha^2}\|x^k - x^{k-1}\|^2 + \|x^k - x^*\|^2 \\
&\quad + \frac{k-1}{\alpha}\left(\|x^k - x^*\|^2 - \|x^{k-1} - x^*\|^2 + \|x^k - x^{k-1}\|^2\right) \\
&= \left(\frac{(k-1)^2}{\alpha^2} + \frac{k-1}{\alpha}\right)\|x^k - x^{k-1}\|^2 + \|x^k - x^*\|^2 \\
&\quad + \frac{k-1}{\alpha}\left(\|x^k - x^*\|^2 - \|x^{k-1} - x^*\|^2\right).
\end{aligned}
$$

Let

$$
\delta_k := \frac{k-1}{\alpha}\left(\|x^k - x^*\|^2 - \|x^{k-1} - x^*\|^2\right) + \|x^k - x^*\|^2.
$$

Then,

$$
\|u^k - x^*\|^2 = \left(\frac{(k-1)^2}{\alpha^2} + \frac{k-1}{\alpha}\right)\|x^k - x^{k-1}\|^2 + \delta_k. \tag{4.45}
$$

By the boundedness of $\{u^k\}$ given in Lemma 4.46 and the boundedness of $\{k^2\|x^k - x^{k-1}\|^2\}$ given in Theorem 4.49, we can see from (4.44) that $\{x^k\}$ is bounded. Observing that

$$
\delta_k \leq \frac{k-1}{\alpha}\|x^k - x^{k-1}\|\left(\|x^k - x^*\| + \|x^{k-1} - x^*\|\right) + \|x^k - x^*\|^2,
$$

we can further conclude that $\{\delta_k\}$ is a bounded sequence. Let $h_k := \|x^k - x^*\|^2$. It then follows that

$$
\begin{aligned}
\delta_{k+1} - \delta_k &= h_{k+1} - h_k + \frac{k}{\alpha}(h_{k+1} - h_k) - \frac{k-1}{\alpha}(h_k - h_{k-1}) \\
&= \frac{k+\alpha}{\alpha}(h_{k+1} - h_k) - \frac{k-1}{\alpha}(h_k - h_{k-1}).
\end{aligned} \tag{4.46}
$$

On the other hand, by using (4.20) in Lemma 4.34 with $u^+ = x^{k+1}$, $u = y^k$, and $x = x^*$, we have

$$
0 \leq F(x^{k+1}) - F_* \leq \frac{1}{2s}\left(\|y^k - x^*\|^2 - \|x^{k+1} - x^*\|^2\right).
$$

Thus,

$$
\begin{aligned}
\|x^{k+1} - x^*\|^2 \le \|y^k - x^*\|^2 &= \left\| x^k - x^* + \frac{k-1}{k+\alpha}(x^k - x^{k-1}) \right\|^2 \\
&= \|x^k - x^*\|^2 + \frac{(k-1)^2}{(k+\alpha)^2}\|x^k - x^{k-1}\|^2 \\
&\quad + \frac{2(k-1)}{k+\alpha}\langle x^k - x^*, x^k - x^{k-1}\rangle \\
&= \|x^k - x^*\|^2 + \left(\frac{(k-1)^2}{(k+\alpha)^2} + \frac{k-1}{k+\alpha}\right)\|x^k - x^{k-1}\|^2 \\
&\quad + \frac{k-1}{k+\alpha}\left(\|x^k - x^*\|^2 - \|x^{k-1} - x^*\|^2\right) \\
&\le \|x^k - x^*\|^2 + 2\|x^k - x^{k-1}\|^2 \\
&\quad + \frac{k-1}{k+\alpha}\left(\|x^k - x^*\|^2 - \|x^{k-1} - x^*\|^2\right).
\end{aligned}
$$

This shows that

$$
(k+\alpha)(h_{k+1} - h_k) - (k-1)(h_k - h_{k-1}) \le 2(k+\alpha)\|x^k - x^{k-1}\|^2.
$$

Combining this with (4.46), we therefore obtain

$$
\delta_{k+1} - \delta_k \le \frac{2(k+\alpha)}{\alpha}\|x^k - x^{k-1}\|^2.
$$

Since Lemma 4.47 implies that $\sum_k (k+\alpha)\|x^k - x^{k-1}\|^2 < \infty$ and since $\{\delta_k\}$ is bounded, we may use Lemma 3.20 to conclude that $\{\delta_k\}$ is convergent. It then follows from (4.45) and Theorem 4.49 that $\lim_{k\to\infty}\|u^k - x^*\|$ exists. Recall that

$$
u^k - x^* = (x^k - x^*) + \frac{k-1}{\alpha}(x^k - x^{k-1}),
$$

and by using again Theorem 4.49, we can see that $\lim_{k\to\infty}\|x^k - x^*\|$ exists for any minimizer x^* of F. This in particular implies that $\{x^k\}$ has a convergent subsequence $\{x^{k_j}\}$ with $x^{k_j} \to \bar{x}$ as $j \to \infty$. By the lower semi-continuity of F and Theorem 4.49, we have

$$
F(\bar{x}) \le \liminf_{j\to\infty} F(x^{k_j}) = F_* \le F(\bar{x}).
$$

Thus, $F(\bar{x}) = F_*$ and $\bar{x}$ is a minimizer of F. Consequently, $\lim_{k\to\infty} \|x^k - \bar{x}\|$ exists with

$$\lim_{k\to\infty} \|x^k - \bar{x}\| = \lim_{j\to\infty} \|x^{k_j} - \bar{x}\| = 0.$$

The proof is thus complete. □

Next, we will show that, if f is strongly convex, the method (4.32) with suitable choice of s and α_k can converge linearly. We will consider the constant choice of α_k with the form $\alpha_k = (1-t)/(1+t)$; the corresponding method takes the form

$$\begin{aligned} y^k &= x^k + \frac{1-t}{1+t}(x^k - x^{k-1}), \\ x^{k+1} &= \mathrm{prox}_{sg}(y^k - s\nabla f(y^k)) \end{aligned} \tag{4.47}$$

with the given initial guess $x^0 = x^{-1}$, where $s > 0$ and $t \in (0, 1)$ need to be chosen carefully. We have the following result ([16, 51, 165]).

Theorem 4.51 *Consider the convex minimization problem (4.16) for which Assumption 4.31 holds. Assume further that f is μ-strongly convex for some $\mu > 0$. Consider the method (4.47) with $0 < s \le 1/L$ and $t = \sqrt{2s\mu}$. There holds*

$$F(x^k) - F(x^*) \le (1-t)^k \left(F(x^0) - F(x^*) + \mu\|x^0 - x^*\|^2\right)$$

for all integers $k \ge 0$, where x^* is the unique solution of (4.16).

Proof Since $0 < s \le 1/L$, by using (4.20) in Lemma 4.34 with $u = y^k$ and $u^+ = x^{k+1}$, we have for any x that

$$F(x^{k+1}) \le F(x) + \frac{1}{2s}\left(\|x - y^k\|^2 - \|x - x^{k+1}\|^2\right) - D_f(x, y^k)$$

which together with the μ-strong convexity of f gives

$$F(x^{k+1}) \le F(x) + \left(\frac{1}{2s} - \mu\right)\|x - y^k\|^2 - \frac{1}{2s}\|x - x^{k+1}\|^2.$$

Now, we may take $x = tx^* + (1-t)x^k$. Noting that $\mu \le L/2$ and $0 < s \le 1/L$, we have $0 < t \le 1$. Thus, we may use the μ-strong convexity of F to obtain

$$\begin{aligned} F(x^{k+1}) \le\; & F\left(tx^* + (1-t)x^k\right) - \frac{1}{2s}\left\|tx^* + (1-t)x^k - x^{k+1}\right\|^2 \\ & + \frac{1-2s\mu}{2s}\left\|tx^* + (1-t)x^k - y^k\right\|^2 \\ \le\; & tF(x^*) + (1-t)F(x^k) - \mu t(1-t)\|x^* - x^k\|^2 \end{aligned}$$

$$
\begin{aligned}
&- \frac{1}{2s} \left\| tx^* + (1-t)x^k - x^{k+1} \right\|^2 \\
&+ \frac{1-2s\mu}{2s} \left\| tx^* + (1-t)x^k - y^k \right\|^2 .
\end{aligned}
$$

Therefore,

$$
\begin{aligned}
&F(x^{k+1}) - F(x^*) + \frac{1}{2s} \left\| tx^* + (1-t)x^k - x^{k+1} \right\|^2 \\
&\le (1-t)\left(F(x^k) - F(x^*)\right) + \frac{1-2s\mu}{2s} \left\| t(x^* - x^k) + (x^k - y^k) \right\|^2 \\
&- \mu t(1-t)\|x^* - x^k\|^2 .
\end{aligned}
$$

By using the polarization identity and $2s\mu = t^2$, we have

$$
\begin{aligned}
&\frac{1-2s\mu}{2s} \left\| t(x^* - x^k) + (x^k - y^k) \right\|^2 - \mu t(1-t)\|x^* - x^k\|^2 \\
&= \frac{1-t^2}{2s} \left(\|t(x^* - x^k) + (x^k - y^k)\|^2 - \frac{t^3}{1+t}\|x^* - x^k\|^2 \right) \\
&= \frac{1-t^2}{2s} \left(\|x^k - y^k\|^2 + 2t\langle x^* - x^k, x^k - y^k\rangle + \frac{t^2}{1+t}\|x^* - x^k\|^2 \right) \\
&= \frac{1-t^2}{2s} \left(\frac{t^2}{1+t} \left\| x^* - x^k + \frac{1+t}{t}(x^k - y^k) \right\|^2 - t\|x^k - y^k\|^2 \right) \\
&\le \frac{t^2(1-t)}{2s} \left\| x^* - x^k + \frac{1+t}{t}(x^k - y^k) \right\|^2 .
\end{aligned}
$$

Consequently,

$$
\begin{aligned}
&F(x^{k+1}) - F(x^*) + \frac{1}{2s} \left\| tx^* + (1-t)x^k - x^{k+1} \right\|^2 \\
&\le (1-t)\left(F(x^k) - F(x^*)\right) + \frac{t^2(1-t)}{2s} \left\| x^* - x^k + \frac{1+t}{t}(x^k - y^k) \right\|^2 .
\end{aligned}
$$

By the definition of y^k, we then obtain

$$
\begin{aligned}
&F(x^{k+1}) - F(x^*) + \frac{1}{2s} \left\| tx^* + (1-t)x^k - x^{k+1} \right\|^2 \\
&\le (1-t)\left(F(x^k) - F(x^*)\right) + \frac{1-t}{2s} \left\| t(x^* - x^k) - (1-t)(x^k - x^{k-1}) \right\|^2 \\
&= (1-t)\left(F(x^k) - F(x^*) + \frac{1}{2s} \left\| tx^* + (1-t)x^{k-1} - x^k \right\|^2 \right).
\end{aligned}
$$

Recursively using this inequality and noting $x^{-1} = x^0$, it follows that

$$F(x^k) - F(x^*) \le (1-t)^k \left(F(x^0) - F(x^*) + \frac{1}{2s} \left\| t(x^* - x^0) \right\|^2 \right)$$

which shows the desired estimate. □

Theorem 4.51 provides a linear convergence result for the method (4.32) with $\alpha_k = (1-t)/(1+t)$ and $t = \sqrt{2s\mu}$ when f is μ-strongly convex. This method requires the information on μ, which is generally hard to acquire. For the method (4.32) with $\alpha_k = (k-1)/(k+\alpha)$, i.e., the method (4.37), which is independent of μ, we have shown its acceleration effect in general situation. A natural question arises: does this method automatically achieve linear convergence when f is strongly convex? This question was affirmatively answered in [142] based on a delicate Lyapunov-type argument, which we will elaborate on below. The key insight lies in the construction of an appropriate Lyapunov function.

In the following, in addition to Assumption 4.31, we assume f is μ-strongly convex with $\mu > 0$ and $0 < s < 1/L$, and let x^* be the unique minimizer of (4.16). We take

$$\mathcal{E}(k) := \mathcal{E}_1(k) + \mathcal{E}_2(k) + \mathcal{E}_3(k),$$

where

$$\begin{aligned}
\mathcal{E}_1(k) &:= \frac{1}{2} \left\| (k-1)(x^k - x^{k-1}) + \alpha(x^k - x^*) \right\|^2, \\
\mathcal{E}_2(k) &:= \frac{\tau(k)}{2} \|x^k - x^{k-1}\|^2, \\
\mathcal{E}_3(k) &:= s\gamma(k) \left(F(x^k) - F(x^*) \right)
\end{aligned}$$

with the nonnegative coefficients $\tau(k)$ and $\gamma(k)$ to be chosen shortly. We need to estimate $\mathcal{E}(k+1) - \mathcal{E}(k)$. To this end, let us consider $\mathcal{E}_i(k+1) - \mathcal{E}_i(k)$ for $i = 1, 2, 3$. By noting that

$$(k-1)(x^k - x^{k-1}) = (k+\alpha)(y^k - x^k),$$

we have

$$\begin{aligned}
&\mathcal{E}_1(k+1) - \mathcal{E}_1(k) \\
&= \frac{1}{2} \left\| k(x^{k+1} - x^k) + \alpha(x^{k+1} - x^*) \right\|^2 - \frac{1}{2} \left\| (k+\alpha)(y^k - x^k) + \alpha(x^k - x^*) \right\|^2 \\
&= \frac{1}{2} \left\| k(x^{k+1} - x^k) + \alpha(x^{k+1} - x^*) \right\|^2
\end{aligned}$$

$$
\begin{aligned}
&- \frac{1}{2} \left\| k(x^{k+1} - x^k) + \alpha(x^{k+1} - x^*) + (k+\alpha)(y^k - x^{k+1}) \right\|^2 \\
&= - \left\langle k(x^{k+1} - x^k) + \alpha(x^{k+1} - x^*), \, (k+\alpha)(y^k - x^{k+1}) \right\rangle - \frac{(k+\alpha)^2}{2} \|y^k - x^{k+1}\|^2 \\
&= -k(k+\alpha)\langle y^k - x^{k+1}, x^{k+1} - x^k \rangle - \alpha(k+\alpha)\langle y^k - x^{k+1}, x^{k+1} - x^* \rangle \\
&\quad - \frac{(k+\alpha)^2}{2} \|y^k - x^{k+1}\|^2 \\
&= -k(k+\alpha)\langle y^k - x^{k+1}, x^{k+1} - x^k \rangle - \alpha(k+\alpha)\langle y^k - x^{k+1}, y^k - x^* \rangle \\
&\quad - \frac{1}{2}(k-\alpha)(k+\alpha)\|y^k - x^{k+1}\|^2. \qquad (4.48)
\end{aligned}
$$

For $\mathcal{E}_2(k+1) - \mathcal{E}_2(k)$, we have

$$
\begin{aligned}
\mathcal{E}_2(k+1) - \mathcal{E}_2(k) &= \frac{\tau(k+1)}{2} \|x^{k+1} - x^k\|^2 - \frac{\tau(k)}{2} \left(\frac{k+\alpha}{k-1} \right)^2 \|y^k - x^k\|^2 \\
&= \frac{1}{2} \left(\tau(k+1) - \tau(k) \left(\frac{k+\alpha}{k-1} \right)^2 \right) \|x^{k+1} - x^k\|^2 \\
&\quad - \tau(k) \left(\frac{k+\alpha}{k-1} \right)^2 \langle y^k - x^{k+1}, x^{k+1} - x^k \rangle \\
&\quad - \frac{1}{2} \tau(k) \left(\frac{k+\alpha}{k-1} \right)^2 \|y^k - x^{k+1}\|^2.
\end{aligned}
$$

By choosing $\tau(k) = (k-1)^2$, we then obtain

$$
\begin{aligned}
\mathcal{E}_2(k+1) - \mathcal{E}_2(k) &= \frac{1}{2} \left(k^2 - (k+\alpha)^2 \right) \|x^{k+1} - x^k\|^2 \\
&\quad - (k+\alpha)^2 \langle y^k - x^{k+1}, x^{k+1} - x^k \rangle \\
&\quad - \frac{1}{2}(k+\alpha)^2 \|y^k - x^{k+1}\|^2. \qquad (4.49)
\end{aligned}
$$

For $\mathcal{E}_3(k+1) - \mathcal{E}_3(k)$, we write

$$
\begin{aligned}
&\mathcal{E}_3(k+1) - \mathcal{E}_3(k) \\
&= s\gamma(k) \left(F(x^{k+1}) - F(x^k) \right) + s\left(\gamma(k+1) - \gamma(k) \right) \left(F(x^{k+1}) - F(x^*) \right).
\end{aligned}
$$

We need to estimate $F(x^{k+1}) - F(x^k)$. According to (4.19) in Lemma 4.34 and the μ-convexity of f, we have

$$\begin{aligned} F(x^{k+1}) - F(x^k) &\le \frac{1}{s}\langle y^k - x^{k+1}, x^{k+1} - x^k\rangle + \frac{L}{2}\|y^k - x^{k+1}\|^2 - \mu\|y^k - x^k\|^2 \\ &= \left(\frac{1}{s} - 2\mu\right)\langle y^k - x^{k+1}, x^{k+1} - x^k\rangle \\ &\quad + \frac{L-2\mu}{2}\|y^k - x^{k+1}\|^2 - \mu\|x^{k+1} - x^k\|^2. \end{aligned}$$

Therefore,

$$\begin{aligned} \mathcal{E}_3(k+1) - \mathcal{E}_3(k) &\le s\gamma(k)\left(\frac{1}{s} - 2\mu\right)\langle y^k - x^{k+1}, x^{k+1} - x^k\rangle \\ &\quad + \frac{(L-2\mu)s}{2}\gamma(k)\|y^k - x^{k+1}\|^2 - \mu s\gamma(k)\|x^{k+1} - x^k\|^2 \\ &\quad + s\left(\gamma(k+1) - \gamma(k)\right)\left(F(x^{k+1}) - F(x^*)\right). \end{aligned} \tag{4.50}$$

Note that all the right-hand sides of (4.48), (4.49), and (4.50) contain $\langle y^k - x^{k+1}, x^{k+1} - x^k\rangle$. We hope this term can be cancelled out after adding these three equations. This leads us to choose $\gamma(k)$ such that

$$-k(k+\alpha) - (k+\alpha)^2 + \gamma(k)(1 - 2s\mu) = 0,$$

i.e.,

$$\gamma(k) = \frac{(k+\alpha)(2k+\alpha)}{1 - 2\mu s}.$$

Note that $2\mu \le L$. Thus, by using $0 < s < 1/L$, we can see that $\gamma(k) > 0$ for large k. Consequently, the Lyapunov function $\mathcal{E}(k)$ takes the form

$$\begin{aligned} \mathcal{E}(k) &= \frac{1}{2}\|(k-1)(x^k - x^{k-1}) + \alpha(x^k - x^*)\|^2 + \frac{(k-1)^2}{2}\|x^k - x^{k-1}\|^2 \\ &\quad + \frac{s(k+\alpha)(2k+\alpha)}{1 - 2\mu s}\left(F(x^k) - F(x^*)\right). \end{aligned}$$

and

$$\begin{aligned} &\mathcal{E}(k+1) - \mathcal{E}(k) \\ &\le -\alpha(k+\alpha)\langle y^k - x^{k+1}, y^k - x^*\rangle - \frac{1}{2}(k-\alpha)(k+\alpha)\|y^k - x^{k+1}\|^2 \end{aligned}$$

$$
\begin{aligned}
&+\frac{1}{2}(k^2-(k+\alpha)^2)\|x^{k+1}-x^k\|^2-\frac{1}{2}(k+\alpha)^2\|y^k-x^{k+1}\|^2\\
&+\frac{s(k+\alpha)(2k+\alpha)}{1-2\mu s}\left(\frac{L-2\mu}{2}\|y^k-x^{k+1}\|^2-\mu\|x^{k+1}-x^k\|^2\right)\\
&+\frac{s}{1-2\mu s}\left((k+\alpha+1)(2k+\alpha+2)-(k+\alpha)(2k+\alpha)\right)\left(F(x^{k+1})-F(x^*)\right)\\
=&-\alpha(k+\alpha)\langle y^k-x^{k+1},y^k-x^*\rangle-\frac{\alpha(2k+\alpha)}{2}\|x^{k+1}-x^k\|^2\\
&-k(k+\alpha)\|y^k-x^{k+1}\|^2+\frac{s(4k+3\alpha+2)}{1-2\mu s}\left(F(x^{k+1})-F(x^*)\right)\\
&+\frac{s(k+\alpha)(2k+\alpha)}{1-2\mu s}\left(\frac{L-2\mu}{2}\|y^k-x^{k+1}\|^2-\mu\|x^{k+1}-x^k\|^2\right).
\end{aligned}
\tag{4.51}
$$

In order to deal with the term $\langle y^k-x^{k+1},y^k-x^*\rangle$, we use again (4.19) in Lemma 4.34 and the μ-strong convexity of f to obtain

$$
\begin{aligned}
F(x^{k+1})-F(x^*)&\le-\frac{1}{s}\langle y^k-x^{k+1},x^*-x^{k+1}\rangle+\frac{L}{2}\|y^k-x^{k+1}\|^2-\mu\|y^k-x^*\|^2\\
&=\frac{1}{s}\langle y^k-x^{k+1},y^k-x^*\rangle-\left(\frac{1}{s}-\frac{L}{2}\right)\|y^k-x^{k+1}\|^2\\
&\quad-\mu\|y^k-x^*\|^2.
\end{aligned}
$$

Since $0<s<1/L$, we have $1/s-L/2\ge 1/(2s)$, and by (4.20) in Lemma 4.34, we have

$$
\|y^k-x^*\|^2-\|x^{k+1}-x^*\|^2\ge 2s\left(F(x^{k+1})-F(x^*)\right)\ge 0.
$$

Thus,

$$
\langle y^k-x^{k+1},y^k-x^*\rangle\ge s\left(F(x^{k+1})-F(x^*)\right)+\frac{1}{2}\|y^k-x^{k+1}\|^2+\mu s\|x^{k+1}-x^*\|^2.
$$

Combining this with (4.51), we therefore obtain

$$
\begin{aligned}
\mathcal{E}(k+1)-\mathcal{E}(k)\le{}& s\left(\frac{4k+3\alpha+2}{1-2\mu s}-\alpha(k+\alpha)\right)\left(F(x^{k+1})-F(x^*)\right)\\
&-\frac{1}{2}\left(\frac{2\mu s}{1-2\mu s}(k+\alpha)(2k+\alpha)+\alpha(2k+\alpha)\right)\|x^{k+1}-x^k\|^2\\
&-\frac{1-Ls}{2(1-2\mu s)}(k+\alpha)(2k+\alpha)\|y^k-x^{k+1}\|^2\\
&-\mu s\alpha(k+\alpha)\|x^{k+1}-x^*\|^2.
\end{aligned}
\tag{4.52}
$$

Lemma 4.52 *Let f and g satisfy Assumption 4.31 with $F := f + g$. For $0 < s \leq 2/L$ and $u \in \mathbb{R}^n$, define*

$$u^+ := \text{prox}_{sg}(u - s\nabla f(u)).$$

If f is μ-strongly convex, then

$$\|u^+ - u\|^2 \geq 4\mu s^2 \left(F(u^+) - \inf_{x\in\mathbb{R}^n} F(x) \right).$$

Proof By using (4.19) in Lemma 4.34 and the μ-strong convexity of f, we have for any $x \in \mathbb{R}^n$ that

$$F(x) \geq F(u^+) + \frac{1}{s}\langle u - u^+, x - u^+\rangle - \frac{L}{2}\|u - u^+\|^2 + \mu\|x - u\|^2.$$

Note that the right-hand side is quadratic in x and achieves its minimum at $x = u - \frac{1}{2\mu s}(u - u^+)$. Plugging this x into the right-hand side, we can obtain

$$\begin{aligned} F(x) &\geq F(u^+) - \frac{1}{4\mu s^2}\|u - u^+\|^2 + \left(\frac{1}{s} - \frac{L}{2}\right)\|u - u^+\|^2 \\ &\geq F(u^+) - \frac{1}{4\mu s^2}\|u - u^+\|^2 \end{aligned}$$

because $0 < s \leq 2/L$. Thus, $\|u - u^+\|^2 \geq 4\mu s^2(F(u^+) - F(x))$ for all $x \in \mathbb{R}^n$. The proof is therefore complete. □

Based on Lemma 4.52 and (4.52), we are now ready to prove the following linear convergence result [142].

Theorem 4.53 *For the convex minimization problem (4.16) let Assumption 4.31 hold. Consider the method (4.37) with $0 < s < 1/L$ and $\alpha \geq 0$. If f is μ-strongly convex for some $\mu > 0$, then there exists an integer k_0 such that*

$$F(x^k) - F(x^*) \leq \frac{(1 - 2\mu s)\mathcal{E}(k_0)}{s(k + \alpha)(2k + \alpha)[1 + \mu s(1 - Ls)/2]^{k-k_0}}$$

for all $k \geq k_0$, where x^* is unique solution of (4.16).

Proof According to Lemma 4.52, we have

$$\|y^k - x^{k+1}\|^2 \geq 4\mu s^2 \left(F(x^{k+1}) - F(x^*)\right).$$

Thus, it follows from (4.52) that

$$
\begin{aligned}
&\mathcal{E}(k+1)-\mathcal{E}(k) \\
&\leq -s\left(\frac{\mu s(1-Ls)}{1-2\mu s}(k+\alpha)(2k+\alpha)-\frac{4k+3\alpha+2}{1-2\mu s}+\alpha(k+\alpha)\right) \\
&\times\left(F(x^{k+1})-F(x^*)\right) \\
&-\frac{1}{2}\left(\frac{2\mu s}{1-2\mu s}(k+\alpha)(2k+\alpha)+\alpha(2k+\alpha)\right)\|x^{k+1}-x^k\|^2 \\
&-\mu s\alpha(k+\alpha)\|x^{k+1}-x^*\|^2 \\
&-\frac{1-Ls}{4(1-2\mu s)}(k+\alpha)(2k+\alpha)\|y^k-x^{k+1}\|^2.
\end{aligned}
\tag{4.53}
$$

Next, we will demonstrate that the first three terms on the right-hand side of (4.53) can be replaced by a suitable negative multiple of $\mathcal{E}(k+1)$. To see this, we note that

$$
\begin{aligned}
\mathcal{E}(k+1) &= \frac{s(k+1+\alpha)(2k+\alpha+2)}{1-2\mu s}\left(F(x^{k+1})-F(x^*)\right) \\
&\quad+\frac{1}{2}\|k(x^{k+1}-x^k)+\alpha(x^{k+1}-x^*)\|^2+\frac{k^2}{2}\|x^{k+1}-x^k\|^2 \\
&= \frac{s(k+1+\alpha)(2k+\alpha+2)}{1-2\mu s}\left(F(x^{k+1})-F(x^*)\right)+k^2\|x^{k+1}-x^k\|^2 \\
&\quad+\frac{\alpha^2}{2}\|x^{k+1}-x^*\|^2+\langle k(x^{k+1}-x^k),\alpha(x^{k+1}-x^*)\rangle \\
&\leq \frac{s(k+1+\alpha)(2k+\alpha+2)}{1-2\mu s}\left(F(x^{k+1})-F(x^*)\right) \\
&\quad+\frac{3k^2}{2}\|x^{k+1}-x^k\|^2+\alpha^2\|x^{k+1}-x^*\|^2.
\end{aligned}
$$

Comparing this with the right-hand side of (4.53) and noting that there exists an integer k_0 such that

$$
\begin{aligned}
&\frac{\mu s(1-Ls)}{1-2\mu s}(k+\alpha)(2k+\alpha)-\frac{4k+3\alpha+2}{1-2\mu s}+\alpha(k+\alpha) \\
&\geq \frac{\mu s(1-Ls)}{2(1-2\mu s)}(k+1+\alpha)(2k+\alpha+2)
\end{aligned}
$$

for $k \geq k_0$, we thus obtain for $k \geq k_0$ that

$$\begin{aligned}\mathcal{E}(k+1) - \mathcal{E}(k) \leq & -\frac{\mu s^2(1-Ls)}{2(1-2\mu s)}(k+1+\alpha)(2k+\alpha+2)\left(F(x^{k+1}) - F(x^*)\right) \\ & - 2\mu s k^2\|x^{k+1} - x^k\|^2 - \mu s \alpha^2\|x^{k+1} - x^*\|^2 \\ & - \frac{1-Ls}{4(1-2\mu s)}(k+\alpha)(2k+\alpha)\|y^k - x^{k+1}\|^2 \\ \leq & -\frac{(1-Ls)\mu s}{2}\mathcal{E}(k+1) \\ & - \frac{1-Ls}{4(1-2\mu s)}(k+\alpha)(2k+\alpha)\|y^k - x^{k+1}\|^2.\end{aligned}$$

This in particular implies

$$\mathcal{E}(k+1) \leq \frac{1}{1+\mu s(1-Ls)/2}\mathcal{E}(k), \quad \forall k \geq k_0$$

and hence

$$\mathcal{E}(k) \leq \frac{\mathcal{E}(k_0)}{[1+\mu s(1-Ls)/2]^{k-k_0}}, \quad \forall k \geq k_0.$$

By invoking the definition of $\mathcal{E}(k)$, we finally obtain

$$\frac{s(k+\alpha)(2k+\alpha)}{1-2\mu s}\left(F(x^k) - F(x^*)\right) \leq \mathcal{E}(k) \leq \frac{\mathcal{E}(k_0)}{[1+\mu s(1-Ls)/2]^{k-k_0}}$$

for $k \geq k_0$, which completes the proof. □

4.6 Douglas-Rachford Splitting Method

In this section, we continue to consider the minimization problem (4.16), i.e.,

$$\min_{x \in \mathbb{R}^n} \{f(x) + g(x)\}, \tag{4.54}$$

where $f : \mathbb{R}^n \to (-\infty, \infty]$ and $g : \mathbb{R}^n \to (-\infty, \infty]$ are convex functions. However, different from the previous two sections, neither f nor g is assumed to be differentiable. Instead, we will use the following condition.

Assumption 4.54 *Both $f : \mathbb{R}^n \to (-\infty, \infty]$ and $g : \mathbb{R}^n \to (-\infty, \infty]$ are proper, lower semi-continuous, and convex, and the problem (4.54) has an optimal solution.*

Moreover, there is an $x_0 \in dom(f) \cap dom(g)$ such that either f or g is continuous at x_0.

We will derive a splitting method, the Douglas-Rachford splitting method, by interpreting solutions of (4.16) as fixed points of a mapping on $\mathbb{R}^n$. Under Assumption 4.54, we have from Theorem 3.14 that

$$\partial(f+g)(x) = \partial f(x) + \partial g(x), \quad \forall x \in \mathbb{R}^n.$$

Therefore, according to Theorem 3.14, we have for any $t > 0$ that

$$\begin{aligned} x^* \in \arg\min_{x \in \mathbb{R}^n}\{f(x)+g(x)\} &\Longleftrightarrow 0 \in \partial(f+g)(x^*) \Longleftrightarrow 0 \in t\partial f(x^*) + t\partial g(x^*) \\ &\Longleftrightarrow \text{there is } u^* \in t\partial g(x^*) \text{ such that } -u^* \in t\partial f(x^*). \end{aligned}$$

Thus, if x^* is an optimal solution of (4.54), then there is $u^* \in \mathbb{R}^n$ such that

$$x^* - u^* \in (I + t\partial f)(x^*) \quad \text{and} \quad x^* + u^* \in (I + t\partial g)(x^*).$$

Consequently,

$$\begin{aligned} x^* &= (I + t\partial f)^{-1}(x^* - u^*) = \text{prox}_{tf}(x^* - u^*), \\ x^* &= (I + t\partial g)^{-1}(x^* + u^*) = \text{prox}_{tg}(x^* + u^*). \end{aligned}$$

Let $z^* = x^* - u^*$. Then, $x^* = \text{prox}_{tf}(z^*)$ and $u^* = x^* - z^*$. So

$$\text{prox}_{tf}(z^*) = x^* = \text{prox}_{tg}(2x^* - z^*) = \text{prox}_{tg}(2\text{prox}_{tf}(z^*) - z^*).$$

By introducing the function

$$F(z) := z + \text{prox}_{tg}(2\text{prox}_{tf}(z) - z) - \text{prox}_{tf}(z) \tag{4.55}$$

on $\mathbb{R}^n$, we can state the following conclusion.

Lemma 4.55 *Let Assumption 4.54 hold and let $t > 0$ be any number:*

(i) *If x^* is an optimal solution of (4.54), then there is u^* such that $z^* := x^* - u^*$ is a fixed point of F.*
(ii) *If z^* is a fixed point of F, then $x^* := \text{prox}_{tf}(z^*)$ is an optimal solution of (4.54)*

Proof (i) has been proved above. We only need to show (ii). Let z^* be a fixed point of F, i.e. $z^* = F(z^*)$. Then,

$$\text{prox}_{tf}(z^*) = \text{prox}_{tg}(2\text{prox}_{tf}(z^*) - z^*).$$

Let $x^* := \text{prox}_{tf}(z^*)$ and $u^* := x^* - z^*$. Then, $z^* = x^* - u^*$ and

$$x^* = \text{prox}_{tf}(x^* - u^*) = (I + t\partial f)^{-1}(x^* - u^*),$$
$$x^* = \text{prox}_{tg}(x^* + u^*) = (I + t\partial g)^{-1}(x^* + u^*).$$

Consequently, $-u^* \in t\partial f(x^*)$ and $u^* \in t\partial g(x^*)$. This implies that

$$0 \in \partial f(x^*) + \partial g(x^*) \subset \partial(f + g)(x^*).$$

Therefore, x^* is an optimal solution of (4.54). □

To find a fixed point of F, we may use the fixed point iteration

$$z^{k+1} = F(z^k)$$

to define a sequence $\{z^k\}$ with the hope that $\{z^k\}$ converges to a fixed point of F. By the definition of z^{k+1}, we have

$$z^{k+1} = z^k + \text{prox}_{tg}(2\text{prox}_{tf}(z^k) - z^k) - \text{prox}_{tf}(z^k).$$

This can be written into the following equivalent form, which is called the *Douglas-Rachford splitting method* [72, 75, 144]:

Algorithm 4.56 *Let $t > 0$ be a fixed number. Take an initial guess z^0 and define*

$$\begin{aligned} x^{k+1} &= \text{prox}_{tf}(z^k), \\ y^{k+1} &= \text{prox}_{tg}(2x^{k+1} - z^k), \\ z^{k+1} &= z^k + (y^{k+1} - x^{k+1}) \end{aligned}$$

for $k = 0, 1, \cdots$.

Example 4.57 As an application of the Douglas-Rachford splitting method, let us consider the constrained convex optimization problem:

$$\min_{x \in \mathbb{R}^n} \{\|x\|_1 : Ax = b\}, \tag{4.56}$$

where A is an $m \times n$ matrix with $m \ll n$ and $b \in \mathbb{R}^m$. We assume AA^T is invertible. Let $L = \{x \in \mathbb{R}^n : Ax = b\}$ and let δ_L be the indicator function of L. Then, (4.56) is equivalent to the unconstrained minimization problem:

$$\min_{x \in \mathbb{R}^n} \{\|x\|_1 + \delta_L(x)\}$$

which has the form (4.54) with $f(x) = \delta_L(x)$ and $g(x) = \|x\|_1$. Let P_L denote the projection of $\mathbb{R}^n$ onto L. Then, for any $v \in \mathbb{R}^n$, we have

$$P_L(v) = v + A^T(AA^T)^{-1}(b - Av).$$

Applying the Douglas-Rachford splitting method, i.e., Algorithm 4.56, to the unconstrained problem, we obtain the iteration formula:

$$\begin{aligned} x^{k+1} &= z^k + A^T(AA^T)^{-1}(b - Az^k), \\ y^{k+1} &= S_t(2x^{k+1} - z^k), \\ z^{k+1} &= z^k + y^{k+1} - x^{k+1}, \end{aligned}$$

where $S_t(\cdot)$ denotes the soft thresholding operator. □

The Douglas-Rachford splitting method can be formulated in other equivalent forms. In Algorithm 4.56, if we start from y, then z, and then x and repeat the loop, we can obtain

$$\begin{aligned} y^{k+1} &= \operatorname{prox}_{tg}(2x^k - z^k), \\ z^{k+1} &= z^k + (y^{k+1} - x^k), \\ x^{k+1} &= \operatorname{prox}_{tf}(z^{k+1}). \end{aligned}$$

Equivalently, we have

$$\begin{aligned} y^{k+1} &= \operatorname{prox}_{tg}(2x^k - z^k), \\ x^{k+1} &= \operatorname{prox}_{tf}(z^k + y^{k+1} - x^k), \\ z^{k+1} &= z^k + y^{k+1} - x^k. \end{aligned}$$

Let $w^k = z^k - x^k$. Then,

$$w^{k+1} = z^{k+1} - x^{k+1} = z^k + y^{k+1} - x^k - x^{k+1} = w^k + y^{k+1} - x^{k+1}.$$

We thus obtain the following alternative form of the Douglas-Rachford splitting method.

Algorithm 4.58 *Let $t > 0$ be a given number. Take any initial guess x^0 and w^0, and for $k \geq 0$, define*

$$\begin{aligned} y^{k+1} &= \operatorname{prox}_{tg}(x^k - w^k), \\ x^{k+1} &= \operatorname{prox}_{tf}(y^{k+1} + w^k), \\ w^{k+1} &= w^k + y^{k+1} - x^{k+1}. \end{aligned}$$

In Chap. 6, we will use Algorithm 4.58 to provide an alternative derivation of the alternating direction method of multipliers.

Now, we turn to the convergence analysis of the Douglas-Rachford splitting method formulated in Algorithm 4.56. Consider the function F defined in (4.55). Recall that for the sequence $\{x^k, y^k, z^k\}$ defined by Algorithm 4.56, we have $z^{k+1} = F(z^k)$. Thus, it is necessary to investigate the properties of F.

Lemma 4.59

(i) *The function F is firmly nonexpansive, i.e.,*

$$\langle F(z_2) - F(z_1), z_2 - z_1 \rangle \geq \|F(z_2) - F(z_1)\|^2, \quad \forall z_1, z_2 \in \mathbb{R}^n.$$

(ii) *The function $G(z) := z - F(z)$ is also firmly nonexpansive.*

Proof

(i) For any $z_1, z_2 \in \mathbb{R}^n$, let

$$x_1 = \text{prox}_{tf}(z_1), \qquad y_1 = \text{prox}_{tg}(2x_1 - z_1),$$
$$x_2 = \text{prox}_{tf}(z_2), \qquad y_2 = \text{prox}_{tg}(2x_2 - z_2).$$

We have

$$F(z_1) = z_1 + y_1 - x_1, \qquad F(z_2) = z_2 + y_2 - x_2.$$

In view of Lemma 4.18, we have

$$\langle x_2 - x_1, z_2 - z_1 \rangle \geq \|x_2 - x_1\|^2,$$
$$\langle y_2 - y_1, (2x_2 - z_2) - (2x_1 - z_1) \rangle \geq \|y_2 - y_1\|^2.$$

Therefore,

$$\begin{aligned}
&\langle F(z_2) - F(z_1), z_2 - z_1 \rangle - \|F(z_2) - F(z_1)\|^2 \\
&= \langle F(z_2) - F(z_1), (z_2 - F(z_2)) - (z_1 - F(z_1)) \rangle \\
&= \langle (z_2 + y_2 - x_2) - (z_1 + y_1 - x_1), x_2 - y_2 - x_1 + y_1 \rangle \\
&= - \|y_2 - x_2 - y_1 + x_1\|^2 + \langle z_2 - z_1, x_2 - x_1 - y_2 + y_1 \rangle \\
&= - \|y_2 - x_2 - y_1 + x_1\|^2 + \langle x_2 - x_1, z_2 - z_1 \rangle \\
&\quad + \langle y_2 - y_1, (2x_2 - z_2) - (2x_1 - z_1) \rangle - 2\langle x_2 - x_1, y_2 - y_1 \rangle \\
&\geq - \|y_2 - x_2 - y_1 + x_1\|^2 + \|x_2 - x_1\|^2 + \|y_2 - y_1\|^2 - 2\langle x_2 - x_1, y_2 - y_1 \rangle \\
&= - \|y_2 - x_2 - y_1 + x_1\|^2 + \|y_2 - y_1 - x_2 + x_1\|^2 \\
&= 0.
\end{aligned}$$

(ii) Next, we show the firmly nonexpansiveness of $G(z) = z - F(z)$. For any $z_1, z_2 \in \mathbb{R}^n$, we have from (i) that

$$\begin{aligned}
&\langle G(z_2) - G(z_1), z_2 - z_1\rangle \\
&= \langle G(z_2) - G(z_1), F(z_2) + G(z_2) - F(z_1) - G(z_1)\rangle \\
&= \|G(z_2) - G(z_1)\|^2 + \langle F(z_2) - F(z_1), G(z_2) - G(z_1)\rangle \\
&= \|G(z_2) - G(z_1)\|^2 + \langle F(z_2) - F(z_1), z_2 - F(z_2) - z_1 + F(z_1)\rangle \\
&= \|G(z_2) - G(z_1)\|^2 + \langle F(z_2) - F(z_1), z_2 - z_1\rangle - \|F(z_2) - F(z_1)\|^2 \\
&\geq \|G(z_2) - G(z_1)\|^2.
\end{aligned}$$

The proof is therefore complete. □

We are now ready to prove the convergence of the Douglas-Rachford splitting method.

Theorem 4.60 *Let Assumption 4.54 hold and let $t > 0$ be any number. Let $\{x^k, y^k, z^k\}$ be defined by Algorithm 4.56. Then,*

(i) *For any fixed point $\hat{z}$ of F, $\{\|z^k - \hat{z}\|\}$ is monotonically decreasing.*
(ii) *$\{\|z^{k+1} - z^k\|\}$ is monotonically decreasing and*

$$\|z^{k+1} - z^k\| \leq \frac{\|z^0 - \hat{z}\|}{\sqrt{k+1}}$$

for all $k \geq 0$ and $\|z^{k+1} - z^k\| = o(1/\sqrt{k+1})$ as $k \to \infty$, where $\hat{z}$ is any fixed point of F.
(iii) *$\{z^k\}$ converges to a fixed point z^* of F.*
(iv) *$\{x^k\}$ converges to the optimal solution $x^* := \mathrm{prox}_{tf}(z^*)$ of (4.54).*

Proof In view of Lemma 4.55, F has a fixed point. Let $\hat{z}$ be any fixed point of F. Noting that $z^{k+1} - z^k = -G(z^k)$, we have

$$\begin{aligned}
\|z^{k+1} - \hat{z}\|^2 - \|z^k - \hat{z}\|^2 &= 2\langle z^{k+1} - z^k, z^k - \hat{z}\rangle + \|z^{k+1} - z^k\|^2 \\
&= -2\langle G(z^k), z^k - \hat{z}\rangle + \|G(z^k)\|^2.
\end{aligned}$$

By using $G(\hat{z}) = 0$ and Lemma 4.59, we have

$$\begin{aligned}
\|z^{k+1} - \hat{z}\|^2 - \|z^k - \hat{z}\|^2 &= -2\langle G(z^k) - G(\hat{z}), z^k - \hat{z}\rangle + \|G(z^k) - G(\hat{z})\|^2 \\
&\leq -2\|G(z^k) - G(\hat{z})\|^2 + \|G(z^k) - G(\hat{z})\|^2 \\
&= -\|G(z^k) - G(\hat{z})\|^2 = -\|G(z^k)\|^2. \qquad (4.57)
\end{aligned}$$

This implies that $\{\|z^k - \hat{z}\|\}$ is monotonically decreasing and hence (i) follows. Moreover, from (4.57), it follows that

$$\sum_{k=0}^{\infty} \|z^{k+1} - z^k\|^2 = \sum_{k=0}^{\infty} \|G(z^k)\|^2 \leq \|z^0 - \hat{z}\|^2 < \infty. \tag{4.58}$$

By using the nonexpansiveness of F given in Lemma 4.59, we have

$$\|z^{k+1} - z^k\| = \|F(z^k) - F(z^{k-1})\| \leq \|z^k - z^{k-1}\|$$

which shows that $\{\|z^{k+1} - z^k\|\}$ is monotonically decreasing. Consequently, it follows from (4.58) that

$$(k+1)\|z^{k+1} - z^k\|^2 \leq \sum_{j=0}^{k} \|z^{j+1} - z^j\|^2 \leq \|z^0 - \hat{z}\|$$

and

$$\frac{k+2}{2}\|z^{k+1} - z^k\|^2 \leq \sum_{j=[k/2]}^{k} \|z^{j+1} - z^j\|^2 \to 0 \quad \text{as } k \to \infty.$$

We therefore obtain (ii). From (4.58), we can see that $G(z^k) \to 0$ as $k \to \infty$. Since (i) implies that $\{z^k\}$ is bounded, we can pick a convergent subsequence $\{z^{k_j}\}$ whose limit is denoted by z^*. Then,

$$G(z^*) = \lim_{j\to\infty} G(z^{k_j}) = 0$$

which shows that z^* is a fixed point of F. Thus, by (i), $\{\|z^k - z^*\|\}$ is monotonically decreasing, and we can conclude that

$$\lim_{k\to\infty} \|z^k - z^*\| = \lim_{j\to\infty} \|z^{k_j} - z^*\| = 0.$$

That is, $\{z^k\}$ converges to a fixed point z^* of F, which shows (iii). In view of Lemma 4.55, $x^* = \text{prox}_{tf}(z^*)$ is an optimal solution of (4.54). Recall that $x^k = \text{prox}_{tf}(z^k)$, and by the continuity of prox_{tf}, we have $x^k \to x^*$ as $k \to \infty$, which shows (iv). □

Next, we will present convergence rate results for the Douglas-Rachford splitting method in terms of the objective error. We will follow [66] to provide both ergodic and non-ergodic convergence rate results. The following lemma is crucial for the analysis.

Lemma 4.61 *Let Assumption 4.54 hold and let $t > 0$ be any number. Let z^* be a fixed point of F and let $x^* = \operatorname{prox}_{tf}(z^*)$. Then,*

$$f(x) + g(y) - f(x^*) - g(x^*) \geq \frac{1}{t}\langle z^* - x^*, x - y\rangle \tag{4.59}$$

for all x and y. Furthermore, for the sequence $\{x^k, y^k, z^k\}$ defined by Algorithm 4.56, there holds

$$f(x^k) + g(y^k) - f(x^*) - g(x^*) \leq \frac{1}{t}\langle z^k - z^{k-1}, x^* - z^k\rangle. \tag{4.60}$$

Proof Since z^* is a fixed point of F and $x^* = \operatorname{prox}_{tf}(z^*)$, we have $x^* = \operatorname{prox}_{tg}(2x^* - z^*)$. Thus,

$$\frac{z^* - x^*}{t} \in \partial f(x^*) \quad \text{and} \quad \frac{x^* - z^*}{t} \in \partial g(x^*).$$

By the convexity of f and g, we then obtain for any x and y that

$$f(x) - f(x^*) \geq \frac{1}{t}\langle z^* - x^*, x - x^*\rangle \quad \text{and} \quad g(y) - g(x^*) \geq \frac{1}{t}\langle x^* - z^*, y - x^*\rangle.$$

Adding these two inequalities gives (4.59). Next, according to the definition of x^k, y^k, and z^k, we have

$$\frac{z^{k-1} - x^k}{t} \in \partial f(x^k) \quad \text{and} \quad \frac{x^k - z^k}{t} = \frac{2x^k - z^{k-1} - y^k}{t} \in \partial g(y^k).$$

Thus, by using again the convexity of f and g, we have

$$\begin{aligned} f(x^*) - f(x^k) &\geq \frac{1}{t}\langle z^{k-1} - x^k, x^* - x^k\rangle, \\ g(x^*) - g(y^k) &\geq \frac{1}{t}\langle x^k - z^k, x^* - y^k\rangle. \end{aligned}$$

Adding these two inequalities and using $y^k = z^k - z^{k-1} + x^k$, it gives

$$\begin{aligned} &t[f(x^k) + g(y^k) - f(x^*) - g(x^*)] \\ &\quad \leq \langle x^k - z^{k-1}, x^* - x^k\rangle + \langle z^k - x^k, x^* - y^k\rangle \\ &\quad = \langle x^k - z^{k-1}, x^* - x^k\rangle + \langle z^k - x^k, x^* - x^k + z^{k-1} - z^k\rangle \\ &\quad = \langle z^k - z^{k-1}, x^* - x^k\rangle + \langle z^k - x^k, z^{k-1} - z^k\rangle \\ &\quad = \langle z^k - z^{k-1}, x^* - z^k\rangle \end{aligned}$$

which shows (4.60). □

Theorem 4.62 (Ergodic Convergence Rate) *Let Assumption 4.54 hold and let $t > 0$ be any number. Let z^* be a fixed point of F and set $x^* := \operatorname{prox}_{tf}(z^*)$. Let $\{x^k, y^k, z^k\}$ be defined by Algorithm 4.56. For $k \geq 1$ let*

$$\bar{x}^k := \frac{1}{k}\sum_{i=1}^{k} x^i \quad \text{and} \quad \bar{y}^k := \frac{1}{k}\sum_{i=1}^{k} y^i.$$

Then,

$$-\frac{2\|z^0 - z^*\|\|z^* - x^*\|}{tk} \leq f(\bar{x}^k) + g(\bar{y}^k) - f(x^*) - g(x^*) \leq \frac{\|z^0 - x^*\|^2}{2tk} \tag{4.61}$$

for all $k \geq 1$. If in addition g is L-Lipschitz continuous on the closed ball $\overline{B_r(x^*)}$ with $r := \|z^0 - z^*\|$, then

$$0 \leq f(\bar{x}^k) + g(\bar{x}^k) - f(x^*) - g(x^*) \leq \frac{\|z^0 - x^*\|^2}{2tk} + \frac{2L\|z^0 - z^*\|}{k} \tag{4.62}$$

for all $k \geq 1$.

Proof Since $z^i - z^{i-1} = y^i - x^i$ for $i \geq 1$, we have

$$\bar{x}^k - \bar{y}^k = \frac{1}{k}\sum_{i=1}^{k}(x^i - y^i) = \frac{1}{k}\sum_{i=1}^{k}(z^{i-1} - z^i) = \frac{1}{k}(z^0 - z^k).$$

According to Theorem 4.60 (i), we have

$$\|\bar{x}^k - \bar{y}^k\| \leq \frac{\|z^k - z^0\|}{k} \leq \frac{\|z^k - z^*\| + \|z^0 - z^*\|}{k} \leq \frac{2\|z^0 - z^*\|}{k}. \tag{4.63}$$

By taking $x = \bar{x}^k$ and $y = \bar{y}^k$ in (4.59) and using (4.63), we thus obtain the first inequality in (4.61). On the other hand, by using (4.60), we have

$$\begin{aligned}
&t[f(x^k) + g(y^k) - f(x^*) - g(x^*)] \\
&\leq \langle z^k - z^{k-1}, x^* - z^k \rangle = \frac{1}{2}\left(\|z^{k-1} - x^*\|^2 - \|z^k - x^*\|^2 - \|z^k - z^{k-1}\|^2\right) \\
&\leq \frac{1}{2}\left(\|z^{k-1} - x^*\|^2 - \|z^k - x^*\|^2\right).
\end{aligned}$$

Therefore, it follows from the convexity of f and g that

$$
\begin{aligned}
t[f(\bar{x}^k) + g(\bar{y}^k) - f(x^*) - g(x^*)] &\le \frac{t}{k}\sum_{i=1}^{k}[f(x^i) + g(y^i) - f(x^*) - g(x^*)] \\
&\le \frac{1}{2k}\sum_{i=1}^{k}(\|z^{i-1} - x^*\|^2 - \|z^i - x^*\|^2) \\
&\le \frac{\|z^0 - x^*\|^2}{2k}
\end{aligned}
$$

which shows the second inequality in (4.61).

Next, we show (4.62). We first show that $x^k, y^k \in \overline{B_r(x^*)}$ for all $k \ge 1$. By the firmly nonexpansiveness of proximal mappings, see Lemma 4.18, and Theorem 4.60, then we have

$$
\|x^k - x^*\| = \|\text{prox}_{tf}(z^{k-1}) - \text{prox}_{tf}(z^*)\| \le \|z^{k-1} - z^*\| \le \|z^0 - z^*\|
$$

and

$$
\begin{aligned}
\|y^k - x^*\|^2 &= \|\text{prox}_{tg}(2\text{prox}_{tf}(z^{k-1}) - z^{k-1}) - \text{prox}_{tg}(2\text{prox}_{tf}(z^*) - z^*)\|^2 \\
&\le \|2(\text{prox}_{tf}(z^{k-1}) - \text{prox}_{tf}(z^*)) - (z^{k-1} - z^*)\|^2 \\
&= 4\|\text{prox}_{tf}(z^{k-1}) - \text{prox}_{tf}(z^*)\|^2 + \|z^{k-1} - z^*\|^2 \\
&\quad - 4\langle \text{prox}_{tf}(z^{k-1}) - \text{prox}_{tf}(z^*), z^{k-1} - z^*\rangle \\
&\le \|z^{k-1} - z^*\|^2 \le \|z^0 - z^*\|^2.
\end{aligned}
$$

Thus, $x^k, y^k \in \overline{B_r(x^*)}$ and hence $\bar{x}_k, \bar{y}_k \in \overline{B_r(x^*)}$ for all $k \ge 1$. Therefore, it follows from (4.61), the L-Lipschitz continuity of g over $\overline{B_r(x^*)}$ and (4.63) that

$$
\begin{aligned}
0 &\le f(\bar{x}^k) + g(\bar{x}^k) - f(x^*) - g(x^*) \\
&\le \left[f(\bar{x}^k) + g(\bar{y}^k) - f(x^*) - g(x^*)\right] + \left[g(\bar{x}^k) - g(\bar{y}^k)\right] \\
&\le \frac{\|z^0 - x^*\|^2}{2tk} + L\|\bar{x}^k - \bar{y}^k\| \\
&\le \frac{\|z^0 - x^*\|^2}{2tk} + \frac{2L\|z^0 - z^*\|}{k}
\end{aligned}
$$

which completes the proof of (4.62). □

Theorem 4.63 (Non-ergodic Convergence Rate) *Let Assumption 4.54 be fulfilled and let $t > 0$ be any number. Let z^* be a fixed point of F and set $x^* := \mathrm{prox}_{tf}(z^*)$. Let $\{x^k, y^k, z^k\}$ be defined by Algorithm 4.56. Then,*

$$\left| f(x^k) + g(y^k) - f(x^*) - g(x^*) \right| \le \frac{(\|z^0 - x^*\| + \|z^* - x^*\|)\|z^0 - z^*\|}{t\sqrt{k}} \tag{4.64}$$

for all $k \ge 1$ and

$$f(x^k) + g(y^k) - f(x^*) - g(x^*) = o\left(1/\sqrt{k}\right) \quad \text{as } k \to \infty.$$

If in addition g is L-Lipschitz continuous on the closed ball $\overline{B_r(x^*)}$ with $r := \|z^0 - z^*\|$, then

$$0 \le f(x^k) + g(x^k) - f(x^*) - g(x^*) \le \frac{(\|z^0 - x^*\| + \|z^* - x^*\| + tL)\|z^0 - z^*\|}{t\sqrt{k}} \tag{4.65}$$

for all $k \ge 1$ and $f(x^k) + g(x^k) - f(x^*) - g(x^*) = o\left(1/\sqrt{k}\right)$ as $k \to \infty$.

Proof By using (4.60) and Theorem 4.60 (i), we have

$$\begin{aligned} & t\left[f(x^k) + g(y^k) - f(x^*) - g(x^*)\right] \\ & \le \langle z^{k-1} - z^k, z^k - x^* \rangle \le \|z^k - x^*\|\|z^k - z^{k-1}\| \\ & \le (\|z^k - z^*\| + \|z^* - x^*\|)\|z^k - z^{k-1}\| \\ & \le (\|z^0 - z^*\| + \|z^* - x^*\|)\|z^k - z^{k-1}\|. \end{aligned}$$

By using (4.59) with $x = x^k$ and $y = y^k$ and the relation $z^k - z^{k-1} = y^k - x^k$, we also have

$$t\left[f(x^k) + g(y^k) - f(x^*) - g(x^*)\right] \ge -\|z^* - x^*\|\|z^k - z^{k-1}\|.$$

Therefore,

$$t\left|f(x^k) + g(y^k) - f(x^*) - g(x^*)\right| \le (\|z^0 - z^*\| + \|z^* - x^*\|)\|z^k - z^{k-1}\|.$$

According to Theorem 4.60 (ii), we thus obtain the first part. For the second part, recall that $x^k, y^k \in \overline{B_r(x^*)}$, which has been shown in the proof of Theorem 4.62.

Consequently, by using the assumed condition on g, we can obtain

$$
\begin{aligned}
0 &\le f(x^k) + g(x^k) - f(x^*) - g(x^*) \\
&\le [f(x^k) + g(y^k) - f(x^*) - g(x^*)] + [g(x^k) - g(y^k)] \\
&\le \frac{1}{t}(\|z^0 - z^*\| + \|z^* - x^*\|)\|z^k - z^{k-1}\| + L\|x^k - y^k\| \\
&= \frac{1}{t}(\|z^0 - z^*\| + \|z^* - x^*\| + tL)\|z^k - z^{k-1}\|
\end{aligned}
$$

which together with Theorem 4.60 (ii) completes the proof. □

Problems

4.1 A function $p : \mathbb{R}^n \to [-\infty, \infty]$ is said to be sublinear if it is subadditive and positively homogeneous, i.e.,

$$p(\tilde{x} + x) \le p(\tilde{x}) + p(x) \quad \text{and} \quad p(\alpha x) = \alpha p(x)$$

for all $\tilde{x}, x \in \mathbb{R}^n$ and $\alpha \ge 0$. For a proper, lower semi-continuous, sublinear function $p : \mathbb{R}^n \to [-\infty, \infty]$, show that the set

$$A := \{\xi \in \mathbb{R}^n : \langle \xi, x \rangle \le p(x) \text{ for all } x \in \mathbb{R}^n\}$$

is nonempty and

$$p(x) = \sup\{\langle \xi, x \rangle : \xi \in A\}. \quad \forall x \in \mathbb{R}^n.$$

(Hint: use Proposition 4.2 and Theorem 4.8.)

4.2 Consider the convex function

$$f(x) = \begin{cases} -\log x, & \text{if } x > 0, \\ +\infty, & \text{if } x \le 0. \end{cases}$$

For every $v \in \mathbb{R}$ and $\lambda > 0$, determine $\text{prox}_{\lambda f}(v)$.

4.3 Let C be a nonempty closed convex set in $\mathbb{R}^n$, and consider the function $f(x) = \frac{1}{2}\|x\|_2^2 + \delta_C(x)$, where δ_C denotes the indicator function of C. Show that for any $v \in \mathbb{R}^n$ and $\lambda > 0$, there holds

$$\text{prox}_{\lambda f}(v) = P_C\left(\frac{v}{\lambda + 1}\right),$$

where P_C denotes the projection onto C.

4.4 Consider the function $f(x) = \|x\|_\infty$ on $\mathbb{R}^n$. Show that

$$\text{prox}_f(v) = v - P_C(v) \quad \text{for all } v \in \mathbb{R}^n,$$

where $C := \{x \in \mathbb{R}^n : \|x\|_1 \leq 1\}$.

4.5 Let $f : \mathbb{R}^n \to (-\infty, \infty]$ be a proper, lower semi-continuous, convex function with the modulus of convexity $\mu \geq 0$. Show that for any $t > 0$, there holds

$$\langle x - y, \text{prox}_{tf}(x) - \text{prox}_{tf}(y)\rangle \geq (1 + 2\mu t)\|\text{prox}_{tf}(x) - \text{prox}_{tf}(y)\|^2$$

for all $x, y \in \mathbb{R}^n$.

4.6

(i) Given two convex functions $f, g : \mathbb{R}^n \to (-\infty, \infty]$, we define

$$(f\Box g)(x) := \inf_{y\in\mathbb{R}^n} \{f(y) + g(x - y)\}, \quad x \in \mathbb{R}^n$$

which is called the infimal convolution of f and g. Show that for the Legendre-Fenchel conjugate, there holds $(f\Box g)^* = f^* + g^*$.

(ii) Given a parameter $\mu > 0$, we define $H_\mu(x) := \left(\|\cdot\|_1\Box\frac{1}{2\mu}\|\cdot\|_2^2\right)(x)$, i.e.,

$$H_\mu(x) = \inf_{y\in\mathbb{R}^n} \left\{\|y\|_1 + \frac{1}{2\mu}\|x - y\|_2^2\right\}, \quad x \in \mathbb{R}^n$$

which is called the Huber function with parameter $\mu > 0$. Show that

$$H_\mu(x) = \sum_{i=1}^n h_\mu(x_i), \quad \text{where } h_\mu(t) = \begin{cases} |t| - \frac{\mu}{2} & \text{if } |t| > \mu, \\ \frac{1}{2\mu}t^2 & \text{if } |t| \leq \mu. \end{cases}$$

(iii) Use (i) to show that the Legendre-Fenchel conjugate of H_μ is given by

$$H_\mu^*(\xi) = \delta_{B_\infty}(\xi) + \frac{\mu}{2}\|\xi\|_2^2, \quad \forall \xi \in \mathbb{R}^n,$$

where $B_\infty := \{\xi \in \mathbb{R}^n : \|\xi\|_\infty \leq 1\}$ and δ_{B_∞} is the indicator function of B_∞.

(iv) For any parameter $\lambda > 0$, show that

$$\text{prox}_{\lambda H_\mu}(v) = v - \lambda P_{B_\infty}\left(\frac{v}{\lambda + \mu}\right) = v - \frac{\lambda v}{\max\{|v|, \lambda + \mu\}},$$

where $|\cdot|$ and max are operations on components.

4.7 Let $f : \mathbb{R}^n \to (-\infty, \infty]$ and $g : \mathbb{R}^n \to (-\infty, \infty]$ be two proper, lower semi-continuous, convex functions. If $M_{tf} = M_{tg}$ for some $t > 0$, show that $f = g$.

4.8 Consider the total variation denoising problem given in Problem 3.8. Following [202], we will use the majorization-minimization (MM) strategy to develop an iterative algorithm for solving this denoising problem. Assuming x^k is the current iterate, the MM strategy is to find a simpler function $F_k(x)$ such that $F_k(x^k) = F(x^k)$ and $F_k(x) \geq F(x)$ for all x and then use the minimizer of $F_k(x)$ over $\mathbb{R}^n$ as the next iterate x^{k+1}.

(i) Assume that every component of Dx^k is nonzero and define the diagonal matrix:

$$\Lambda_k = \begin{pmatrix} |(Dx^k)_1| & & \\ & \ddots & \\ & & |(Dx^k)_{n-1}| \end{pmatrix},$$

where $(Dx^k)_i$ denotes the i-th component of Dx^k. Show that the function

$$F_k(x) = \frac{1}{2}\|x - b\|_2^2 + \frac{1}{2}\lambda \left(\|Dx^k\|_1 + x^T D^T \Lambda_k^{-1} D x \right)$$

is a majorant of $F(x)$, i.e., $F_k(x^k) = F(x^k)$ and $F_k(x) \geq F(x)$ for all x.

(ii) Let x^{k+1} denote the minimizer of $F_k(x)$ over $\mathbb{R}^n$. Show that

$$x^{k+1} = \left(I + \lambda D^T \Lambda_k^{-1} D \right)^{-1} b.$$

(iii) Use the Sherman-Morrison-Woodbury formula:

$$(A + UCV)^{-1} = A^{-1} - A^{-1}U \left(C^{-1} + VA^{-1}U \right)^{-1} VA$$

to show that

$$x^{k+1} = b - D^T \left(\frac{1}{\lambda}\Lambda_k + DD^T \right)^{-1} Db.$$

This is our iterative algorithm to solve the total variation denoising problem.

(iv) Write a code to implement the algorithm obtained in (iii) and test your code in the following way: create a piecewise constant signal x^* of length $n = 4000$ with at least 10 discontinuous points, and obtain noisy data $b = x^* + \texttt{randn}(n, 1)$ by adding to x^* a Gaussian white noise with standard deviation 1. Use total variation denoising with $\lambda = 10, 20, 40$ to reconstruct x^*, perform 10, 20, and 100 iterations, plot your results, and make your comments.

4.9 By modifying the lasso problem, we consider the minimization problem:

$$\min_{x \in \mathbb{R}^n} \left\{ \frac{1}{2}\|Ax - b\|_2^2 + \lambda \left(\|x\|_1 + \frac{\alpha}{2}\|x\|_2^2 \right) \right\},$$

where $\lambda > 0$, $\alpha > 0$, $b \in \mathbb{R}^m$, and A is an $m \times n$ matrix with $m \ll n$. This is the elastic-net regularization for solving $Ax = b$. When applying the proximal gradient method to solve this minimization problem, there are two different splitting strategies:

(i) If we take $f(x) = \frac{1}{2}\|Ax - b\|_2^2$ and $g(x) = \lambda \left(\|x\|_1 + \frac{\alpha}{2}\|x\|_2^2 \right)$, show that the proximal gradient method results in the iterative algorithm:

$$x^{k+1} = S_{\frac{\lambda t_k}{\lambda \alpha t_k + 1}} \left(\frac{1}{\lambda \alpha t_k + 1} \left(x^k - t_k A^T (Ax^k - b) \right) \right),$$

where $S_\lambda(\cdot)$ denotes the soft thresholding operator.

(ii) If we take $f(x) = \frac{1}{2}\|Ax - b\|_2^2 + \frac{\lambda\alpha}{2}\|x\|_2^2$ and $g(x) = \lambda\|x\|_1$, show that the proximal gradient method results in the iterative algorithm:

$$x^{k+1} = S_{\lambda t_k} \left(x^k - t_k \left(A^T (Ax^k - b) + \lambda \alpha x^k \right) \right).$$

(iii) Write codes to implement these two algorithms and test your codes in the following way: use A = `randn(m, n)` to generate a matrix of size $m = 400$ and $n = 600$, and pick $x^* \in \mathbb{R}^n$ with $\|x\|_0 = 100$; you may produce such x^* by performing the following commands:

```
x_exact = zeros(n,1);
ind = randperm(n,100);
x0 = sign(randn(n,1));
x_exact(ind) = x0(ind);
```

Next, compute Ax^* and add noise $\varepsilon = 0.1 *$ `randn(m, 1)` to obtain b. When running the two algorithms, take $\lambda = 5$ and $t_k = 1/\lambda_{\max}(A^T A)$; the value $\lambda_{\max}(A^T A)$ can be obtained using `eigs(A'*A,1)` in `Matlab`. Run your codes 2000 iterations for the values $\alpha = 0, 0.01, 0.1, 1, 10$. Plot your results and make comments.

4.10 In Sect. 4.5, we have considered the accelerated proximal gradient method for solving the convex minimization problem (4.16) for which Assumption 4.31 holds. For this method, the objective function value is not necessarily decreasing along the iteration. A monotone version of the accelerated proximal gradient method—MFISTA—was proposed in [18], which can be stated as follows: pick $x^{-1} = x^0$, and set $z^0 = x^0$. For $k = 0, 1, \cdots$, do

$$
\begin{aligned}
y^k &= x^k + \frac{k+\alpha-1}{k+\alpha}(z^k - x^k) + \frac{k-1}{k+\alpha}(x^k - x^{k-1}),\\
z^{k+1} &= \text{prox}_{sg}(y^k - s\nabla f(y^k)),\\
x^{k+1} &= \begin{cases} z^{k+1} & \text{if } F(z^{k+1}) \le F(x^k),\\ x^k & \text{if } F(z^{k+1}) > F(x^k). \end{cases}
\end{aligned} \tag{4.66}
$$

where $\alpha \ge 2$ and $s > 0$. Under Assumption 4.31, show that if $0 < s \le 1/L$, then

$$F(x^k) - F(x^*) \le \frac{\alpha^2\|x^0 - x^*\|^2}{2s(k+\alpha-1)^2}, \quad \forall k \ge 0$$

by adapting the proof of Theorem 4.43, where x^* denotes any minimizer of F.

4.11 Consider an operator $T : \mathbb{R}^n \to \mathbb{R}^n$. It is called nonexpansive if

$$\|T(x) - T(y)\| \le \|x - y\|, \quad \forall x, y \in \mathbb{R}^n,$$

and it is called firmly nonexpansive if

$$\langle x - y, T(x) - T(y)\rangle \ge \|T(x) - T(y)\|^2, \quad \forall x, y \in \mathbb{R}^n.$$

Show that an operator $T : \mathbb{R}^n \to \mathbb{R}^n$ is firmly nonexpansive if and only if $2T - I$ is nonexpansive.

4.12 For the convex minimization problem (4.54) satisfying Assumption 4.54, we have considered the Douglas-Rachford splitting method, i.e., Algorithm 4.56, which is deduced from the fixed point iteration $z^{k+1} = F(z^k)$ with F defined by (4.55). Alternatively, if we introduce the mapping $\tilde{F} : \mathbb{R}^n \to \mathbb{R}^n$ by

$$\tilde{F}(z) := z + 2\left(\text{prox}_{tg}(2\text{prox}_{tf}(z) - z) - \text{prox}_{tf}(z)\right)$$

and consider the fixed point iteration $z^{k+1} = \tilde{F}(z^k)$, it leads to the Peaceman-Rachford splitting method [144, 179]:

$$
\begin{aligned}
x^{k+1} &= \text{prox}_{tf}(z^k),\\
y^{k+1} &= \text{prox}_{tg}(2x^{k+1} - z^k),\\
z^{k+1} &= z^k + 2(y^{k+1} - x^{k+1}),
\end{aligned}
$$

where $t > 0$ is a given number. Show that

$$z^{k+1} = (2\text{prox}_{tg} - I) \circ (2\text{prox}_{tf} - I)(z^k).$$

If f is strongly convex with modulus of convexity $\mu > 0$, based on this equation, Lemma 4.18, Problem 4.11, and Problem 4.5 to show that

$$\|z^{k+1} - z^*\|^2 \leq \|z^k - z^*\|^2 - 8\mu t\|x^k - x^*\|^2$$

and to conclude that $\|x^k - x^*\| \to 0$ as $k \to \infty$, where z^* is a fixed point of $\tilde{F}$ and $x^* = \text{prox}_{tf}(z^*)$ is the unique solution of (4.54).

Chapter 5
Karush-Kuhn-Tucker Theory and Lagrangian Duality

In this chapter, we first develop the Karush-Kuhn-Tucker (KKT) theory for nonsmooth convex minimization problems. This is achieved by leveraging the characterization of the subdifferential of convex functions through their directional derivatives, as well as the formula for the subdifferential of maximum functions. We then present several applications of the KKT theory, including the derivation of the well-known Hoffman error bound. Next, we explore Lagrangian duality theory and its various applications. In particular, we investigate the dual proximal gradient method and the augmented Lagrangian method.

5.1 Directional Derivative and Subdifferential

In this section, we will give a characterization of subdifferentials of convex functions in terms of directional derivatives. Let us start with the definition of directional derivatives for general proper functions.

Definition 5.1 Let $f : \mathbb{R}^n \to (-\infty, \infty]$ be a proper function. Given $x \in \text{dom}(f)$ and a vector $v \in \mathbb{R}^n$, we define

$$f'(x; v) := \lim_{t \searrow 0} \frac{f(x + tv) - f(x)}{t}$$

if the limit exists and call it the *directional derivative of f at x in the direction v.*

From the definition of directional derivative, it follows immediately that $f'(x; v)$ is *positively homogeneous* in v, i.e., $f'(x; \lambda v) = \lambda f'(x; v)$ for any $\lambda \geq 0$. When f is a convex function, one can show that, for any $x \in \text{dom}(f)$, the function

$$t \to \frac{f(x + tv) - f(x)}{t}, \qquad t > 0$$

Q. Jin, *Lectures on Nonsmooth Optimization*, Texts in Applied Mathematics 82,
https://doi.org/10.1007/978-3-031-91417-1_5

is increasing. Indeed, by convexity, for any $0 < t_1 \leq t_2$, we have

$$f(x + t_1 v) = f\left(\left(1 - \frac{t_1}{t_2}\right)x + \frac{t_1}{t_2}(x + t_2 v)\right) \leq \left(1 - \frac{t_1}{t_2}\right)f(x) + \frac{t_1}{t_2}f(x + t_2 v)$$

which implies that

$$\frac{f(x + t_1 v) - f(x)}{t_1} \leq \frac{f(x + t_2 v) - f(x)}{t_2}.$$

Thus, for a proper convex function $f : \mathbb{R}^n \to (-\infty, \infty]$, there holds

$$f'(x; v) = \inf_{t>0} \frac{f(x + tv) - f(x)}{t}, \qquad x \in \mathrm{dom}(f). \tag{5.1}$$

This in particular implies

$$f(x + v) - f(x) \geq f'(x; v), \qquad \forall x \in \mathrm{dom}(f),\ v \in \mathbb{R}^n. \tag{5.2}$$

Lemma 5.2 *Let $f : \mathbb{R}^n \to (-\infty, \infty]$ be a proper convex function. For each $x \in \mathrm{dom}(f)$, the function $v \to f'(x; v)$ is convex on $\mathbb{R}^n$.*

Proof Let $u, v \in \mathbb{R}^n$ and $0 \leq \lambda \leq 1$. According to the definition of directional derivative and the convexity of f, we have

$$\begin{aligned}
f'(x; \lambda u + (1 - \lambda)v) &= \lim_{t \searrow 0} \frac{f(x + t(\lambda u + (1 - \lambda)v)) - f(x)}{t} \\
&\leq \lim_{t \searrow 0} \left(\lambda \frac{f(x + tu) - f(x)}{t} + (1 - \lambda)\frac{f(x + tv) - f(x)}{t}\right) \\
&= \lambda \lim_{t \searrow 0} \frac{f(x + tu) - f(x)}{t} + (1 - \lambda) \lim_{t \searrow 0} \frac{f(x + tv) - f(x)}{t} \\
&= \lambda f'(x, u) + (1 - \lambda) f'(x, v)
\end{aligned}$$

which shows the convexity of the function $v \to f'(x; v)$ on $\mathbb{R}^n$. □

Proposition 5.3 *Let $f : \mathbb{R}^n \to (-\infty, \infty]$ be proper and convex. If f is continuous at $x \in \mathrm{dom}(f)$, then*

$$\partial f(x) = \left\{\xi \in \mathbb{R}^n : \langle \xi, v \rangle \leq f'(x; v) \text{ for all } v \in \mathbb{R}^n\right\} \tag{5.3}$$

and

$$f'(x; v) = \sup\left\{\langle \xi, v \rangle : \xi \in \partial f(x)\right\}, \quad \forall v \in \mathbb{R}^n. \tag{5.4}$$

Proof Since f is continuous at x, $\partial f(x) \neq \emptyset$ by Theorem 3.6. For any $\xi \in \partial f(x)$, we have

$$\frac{f(x+tv)-f(x)}{t} \geq \frac{1}{t}\langle \xi, tv\rangle = \langle \xi, v\rangle, \quad \forall v \in \mathbb{R}^n$$

for any $t > 0$. Letting $t \to 0$ gives $f'(x; v) \geq \langle \xi, v\rangle$ for all $v \in \mathbb{R}^n$, which shows that $\partial f(x)$ is contained in the set on the right-hand side of (5.3). Conversely, let ξ be an element from the set on the right-hand side of (5.3). Then, by using (5.2), we have

$$\langle \xi, y-x\rangle \leq f'(x; y-x) \leq f(y) - f(x), \quad \forall y \in \mathbb{R}^n$$

and consequently $\xi \in \partial f(x)$. We thus obtain (5.3).

Next, we prove (5.4). It suffices to show that for any $v \in \mathbb{R}^n$, there is $\bar{\xi} \in \partial f(x)$ with $f'(x; v) = \langle \bar{\xi}, v\rangle$. Consider the function $u \to f'(x; u)$. From (5.3), it follows that $f'(x; u) > -\infty$ for all u. By the continuity of f at x and

$$f'(x; u) \leq \frac{f(x+tu)-f(x)}{t},$$

we have $f'(x; u) < \infty$ for all u. By Lemma 5.2, $u \to f'(x; u)$ is convex. Therefore, it follows from Theorem 3.6 that $u \to f'(x; u)$ is subdifferentiable at all u. Let $\bar{\xi}$ be a subgradient of $u \to f'(x; u)$ at v. Then,

$$\lambda f'(x; u) = f'(x; \lambda u) \geq f'(x; v) + \langle \bar{\xi}, \lambda u - v\rangle \tag{5.5}$$

and thus

$$f'(x; u) \geq \langle \bar{\xi}, u\rangle + \frac{f'(x; v) - \langle \bar{\xi}, v\rangle}{\lambda}$$

for all $\lambda > 0$ and $u \in \mathbb{R}^n$. Letting $\lambda \nearrow \infty$ gives $f'(x; u) \geq \langle \bar{\xi}, u\rangle$ for all u. Therefore, it follows from (5.2) that

$$f(x+u) - f(x) \geq f'(x; u) \geq \langle \bar{\xi}, u\rangle, \quad \forall u$$

which implies $\bar{\xi} \in \partial f(x)$. Next, taking $\lambda = 0$ in (5.5) gives $0 \geq f'(x; v) - \langle \bar{\xi}, v\rangle$. Since $\bar{\xi} \in \partial f(x)$, we always have $\langle \bar{\xi}, v\rangle \leq f'(x; v)$. Hence, $f'(x; v) = \langle \bar{\xi}, v\rangle$. □

Corollary 5.4 *Let $f : \mathbb{R}^n \to (-\infty, \infty]$ be convex. If f is continuous at x and $\partial f(x) = \{\xi\}$ is a singleton, then f is differentiable at x with $\nabla f(x) = \xi$.*

Proof Since $\partial f(x) = \{\xi\}$ is a singleton, according to Proposition 5.3, we have $f'(x; v) = \langle \xi, v\rangle$ for all $v \in \mathbb{R}^n$. Assume that f is not differentiable at x_0. Then, there exist $\varepsilon_0 > 0$ and a sequence $\{h_k\} \subset \mathbb{R}^n$ with $h_k \to 0$ such that

$$\frac{|f(x+h_k) - f(x) - \langle \xi, h_k \rangle|}{\|h_k\|} \geq \varepsilon_0$$

for all k. Let $t_k = \|h_k\|$ and $v_k = h_k/\|h_k\|$. Then, $\|v_k\| = 1$ and hence there is $v \in \mathbb{R}^n$ such that $v_k \to v$ as $k \to \infty$ by passing to a subsequence if necessary. Thus,

$$\begin{aligned}\varepsilon_0 &\leq \frac{|f(x+t_k v_k) - f(x) - t_k\langle \xi, v_k \rangle|}{t_k} \\ &\leq \frac{|f(x+t_k v_k) - f(x+t_k v)|}{t_k} + |\langle \xi, v_k - v \rangle| \\ &\quad + \frac{|f(x+t_k v) - f(x) - t_k\langle \xi, v \rangle|}{t_k}.\end{aligned}$$

Since f is convex and is continuous at x, f must be locally Lipschitz continuous around x, i.e., there is a neighborhood U of x and a constant L such that $|f(y) - f(z)| \leq L\|y - z\|$ for all $y, z \in U$. Consequently,

$$\varepsilon_0 \leq (L + \|\xi\|)\|v_k - v\| + \left| \frac{f(x+t_k v) - f(x)}{t_k} - \langle \xi, v \rangle \right| \to 0$$

as $k \to \infty$ which is a contradiction. □

5.2 Strong Convexity and Smoothness

In this section, we will use Corollary 5.4 to show that, for any strongly convex function, its Legendre-Fenchel conjugate is differentiable everywhere with Lipschitz continuous gradient.

Lemma 5.5 *Let $f : \mathbb{R}^n \to (-\infty, \infty]$ be proper, lower semi-continuous, and strongly convex with modulus of convexity $\mu > 0$. Then,*

$$\liminf_{\|x\| \to \infty} \frac{f(x)}{\|x\|^2} \geq \frac{\mu}{2}.$$

Proof Under the given conditions on f, it follows from Lemma 2.23 that there exists $(\xi, \gamma) \in \mathbb{R}^n \times \mathbb{R}$ such that $f(x) \geq \langle \xi, x \rangle + \gamma$ for all $x \in \mathbb{R}^n$. Since f is proper, there is $\bar{x}$ such that $f(\bar{x})$ is finite. Therefore, for any $x \in \mathbb{R}^n$, by the strong convexity of f, we have

$$\gamma + \left\langle \xi, \frac{1}{2}(\bar{x} + x) \right\rangle \leq f\left(\frac{1}{2}\bar{x} + \frac{1}{2}x\right) \leq \frac{1}{2}f(\bar{x}) + \frac{1}{2}f(x) - \frac{\mu}{4}\|x - \bar{x}\|^2$$

which implies that

$$f(x) \geq \frac{\mu}{2}\|x - \bar{x}\|^2 + \langle \xi, x \rangle - f(\bar{x}) + \langle \xi, \bar{x} \rangle + 2\gamma.$$

The proof is complete. □

Lemma 5.6 *Let* $f : \mathbb{R}^n \to (-\infty, \infty]$ *be proper, lower semi-continuous, and strongly convex. Then,* f^* *is finite everywhere.*

Proof From Lemma 5.5, it follows that f coercive. Thus, we may use Corollary 1.15 and Proposition 2.28 to conclude that f possesses a unique minimizer x^* over $\mathbb{R}^n$. Since $0 \in \partial f(x^*)$, it follows from Proposition 3.12 that

$$f(x) \geq f(x^*) + \mu\|x - x^*\|^2, \quad \forall x \in \mathbb{R}^n,$$

where $\mu > 0$ denotes the modulus of convexity of f. Therefore, for any $\xi \in \mathbb{R}^n$, we have

$$\begin{aligned}
f^*(\xi) &= \sup_{x\in\mathbb{R}^n} \{\langle \xi, x \rangle - f(x)\} \leq \sup_{x\in\mathbb{R}^n} \left\{\langle \xi, x \rangle - f(x^*) - \mu\|x - x^*\|^2\right\} \\
&= \sup_{x\in\mathbb{R}^n} \left\{\langle \xi, x - x^* \rangle - \mu\|x - x^*\|^2\right\} - f(x^*) + \langle \xi, x^* \rangle \\
&\leq \sup_{x\in\mathbb{R}^n} \left\{\|\xi\|\|x - x^*\| - \mu\|x - x^*\|^2\right\} - f(x^*) + \langle \xi, x^* \rangle \\
&= \sup_{s\geq 0} \left\{\|\xi\|s - \mu s^2\right\} - f(x^*) + \langle \xi, x^* \rangle \\
&= \frac{\|\xi\|^2}{4\mu} - f(x^*) + \langle \xi, x^* \rangle < \infty
\end{aligned}$$

which completes the proof. □

Lemma 5.7 *Let* $f : \mathbb{R}^n \to (-\infty, \infty]$ *be proper, lower semi-continuous, and strongly convex. Then,* $\partial f^*(\xi)$ *is a singleton for each* $\xi \in \mathbb{R}^n$.

Proof According to Lemma 5.6 and the convexity of f^*, it follows from Proposition 2.29 that f^* is continuous everywhere. Thus, by Theorem 3.6, $\partial f^*(\xi) \neq \emptyset$ for each ξ. Assume now that there exist $x_1, x_2 \in \partial f^*(\xi)$. Then, $\xi \in \partial f(x_1) \cap \partial f(x_2)$ by Theorem 4.9. Thus, by Proposition 4.6, we have

$$f(x_i) + f^*(\xi) = \langle \xi, x_i \rangle, \qquad i = 1, 2.$$

Therefore, for $0 < t < 1$, we have by the Fenchel-Young inequality and the strong convexity of f that

$$\begin{aligned}&tf(x_1) + (1-t)f(x_2) + f^*(\xi)\\&= \langle \xi, tx_1 + (1-t)x_2\rangle \le f^*(\xi) + f(tx_1 + (1-t)x_2)\\&\le f^*(\xi) + tf(x_1) + (1-t)f(x_2) - \mu t(1-t)\|x_1 - x_2\|^2,\end{aligned}$$

where $\mu > 0$ is the modulus of convexity of f. This implies $\|x_1 - x_2\|^2 \le 0$ and hence $x_1 = x_2$. □

Now, we are ready to show the following important result on f^* for a strongly convex function f.

Theorem 5.8 *Let $f : \mathbb{R}^n \to (-\infty, \infty]$ be proper, lower semi-continuous, and strongly convex with modulus of convexity $\mu > 0$. Then, f^* is differentiable everywhere and*

$$\|\nabla f^*(\xi) - \nabla f^*(\eta)\| \le \frac{\|\xi - \eta\|}{2\mu}$$

for all $\xi, \eta \in \mathbb{R}^n$.

Proof The differentiability of f^* follows from Lemma 5.7 and Corollary 5.4. Let $x = \nabla f^*(\xi)$ and $y = \nabla f^*(\eta)$. Then, $\xi \in \partial f(x)$ and $\eta \in \partial f(y)$. By Proposition 3.12, we have

$$2\mu\|x - y\|^2 \le \langle \xi - \eta, x - y\rangle \le \|\xi - \eta\|\|x - y\|$$

which implies $2\mu\|x - y\| \le \|\xi - \eta\|$. □

The next result provides a converse result to Theorem 5.8.

Theorem 5.9 *If $f : \mathbb{R}^n \to \mathbb{R}$ is convex differentiable with an L-Lipschitz continuous gradient ∇f for some $L > 0$, then $f^* : \mathbb{R}^n \to (-\infty, \infty]$ is strongly convex with modulus of convexity $\ge 1/(2L)$.*

Proof For any $x, y \in \mathbb{R}^n$ and $0 < t < 1$, let $x_t := (1-t)x + ty$. By the conditions on f, we have

$$f(x) \le f(x_t) + \langle \nabla f(x_t), x - x_t\rangle + \frac{L}{2}\|x - x_t\|^2,$$

$$f(y) \le f(x_t) + \langle \nabla f(x_t), y - x_t\rangle + \frac{L}{2}\|y - x_t\|^2.$$

Note that $x - x_t = t(x - y)$ and $y - x_t = (1 - t)(y - x)$. We have

$$f(x) \leq f(x_t) + t\langle \nabla f(x_t), x - y\rangle + \frac{L}{2}t^2\|x - y\|^2,$$

$$f(y) \leq f(x_t) + (1 - t)\langle \nabla f(x_t), y - x\rangle + \frac{L}{2}(1 - t)^2\|x - y\|^2.$$

Multiplying the first equation by $(1 - t)$ and the second equation by t and then adding them, we obtain

$$(1 - t)f(x) + tf(y) \leq f(x_t) + \frac{L}{2}t(1 - t)\|x - y\|^2.$$

Consequently,

$$(1 - t)f(x - ty) + tf(x + (1 - t)y) \leq f(x) + \frac{L}{2}t(1 - t)\|y\|^2 \tag{5.6}$$

for all $x, y \in \mathbb{R}^n$ and $0 < t < 1$.

To show the strong convexity of f^*, let $\xi, \eta \in \mathbb{R}^n$ and $0 < t < 1$. By the Fenchel-Young inequality and (5.6), we have for any $x, y \in \mathbb{R}^n$ that

$$\begin{aligned}
&(1 - t)f^*(\xi) + tf^*(\eta) \\
&\geq (1 - t)\left(\langle \xi, x - ty\rangle - f(x - ty)\right) + t\left(\langle \eta, x + (1 - t)y\rangle - f(x + (1 - t)y)\right) \\
&= \langle (1 - t)\xi + t\eta, x\rangle + t(1 - t)\langle \eta - \xi, y\rangle \\
&\quad - \left((1 - t)f(x - ty) + tf(x + (1 - t)y)\right) \\
&\geq \langle (1 - t)\xi + t\eta, x\rangle + t(1 - t)\langle \eta - \xi, y\rangle - f(x) - \frac{L}{2}t(1 - t)\|y\|^2.
\end{aligned}$$

Therefore,

$$\begin{aligned}
&(1 - t)f^*(\xi) + tf^*(\eta) \\
&\geq \sup_{x,y\in\mathbb{R}^n}\left\{\langle (1 - t)\xi + t\eta, x\rangle - f(x) + t(1 - t)\langle \eta - \xi, y\rangle - \frac{L}{2}t(1 - t)\|y\|^2\right\} \\
&= f^*((1 - t)\xi + t\eta) + t(1 - t)\sup_{y\in\mathbb{R}^n}\left\{\langle \eta - \xi, y\rangle - \frac{L}{2}\|y\|^2\right\} \\
&= f^*((1 - t)\xi + t\eta) + \frac{1}{2L}t(1 - t)\|\xi - \eta\|^2.
\end{aligned}$$

The proof is complete. □

5.3 Subdifferential of Maximum Functions

Let $f_1, \cdots, f_m : \mathbb{R}^n \to (-\infty, \infty]$ be convex functions, and we define

$$f(x) := \max\{f_1(x), \cdots, f_m(x)\}, \quad x \in \mathbb{R}^n \tag{5.7}$$

which is also a convex function from $\mathbb{R}^n$ to $(-\infty, \infty]$. Since the max is taken for finite many values, for each $x \in \mathbb{R}^n$, there must exists an $i \in \{1, \cdots, m\}$ such that $f_i(x) = f(x)$. Let

$$I(x) = \{i \in \{1, \cdots, m\} : f_i(x) = f(x)\}$$

which is called the active set of x. For any $\xi \in \partial f_i(x)$ with $i \in I(x)$, we have

$$f(y) \geq f_i(y) \geq f_i(x) + \langle \xi, y - x\rangle = f(x) + \langle \xi, y - x\rangle$$

for all $y \in \mathbb{R}^n$. This shows that $\partial f_i(x) \subset \partial f(x)$ for all $i \in I(x)$. Since $\partial f(x)$ is a convex set, we therefore have

$$\operatorname{conv}\left(\bigcup_{i \in I(x)} \partial f_i(x)\right) \subset \partial f(x). \tag{5.8}$$

The following result shows that the inclusion in (5.8) can be turned into an equality for those x from the intersection of the interior of all $\operatorname{dom}(f_i)$.

Theorem 5.10 (Dubovitskii-Milyutin) *Let $f_1, \cdots, f_m : \mathbb{R}^n \to (-\infty, \infty]$ be convex functions and define f by (5.7). For $x_0 \in \bigcap_{i=1}^m \operatorname{int}(\operatorname{dom}(f_i))$, there holds*

$$\partial f(x_0) = \operatorname{conv}\left(\bigcup_{i \in I(x_0)} \partial f_i(x_0)\right). \tag{5.9}$$

Proof Let C denote the right-hand side of (5.9). We prove that $\partial f(x_0) \subset C$. We will use a contradiction argument and assume there is $\xi_0 \in \partial f(x_0)$ with $\xi_0 \notin C$. Since $x_0 \in \bigcap_{i=1}^m \operatorname{int}(\operatorname{dom}(f_i))$, it follows from Proposition 2.29 that each f_i is locally Lipschitz continuous around x_0 and thus, by Lemma 3.19, each $\partial f_i(x_0)$ is bounded and hence is compact. By virtue of Corollary 2.6, we can conclude that C is compact and convex. Therefore, by Theorem 2.18, there is $v \in \mathbb{R}^n$ such that

$$\langle \xi_0, v\rangle > \sup_{\xi \in C}\langle \xi, v\rangle = \max_{i \in I(x_0)} \sup_{\xi \in \partial f_i(x_0)} \langle \xi, v\rangle.$$

In view of Proposition 5.3, we have

$$\langle \xi_0, v\rangle > \max_{i\in I(x_0)} f_i'(x_0; v). \tag{5.10}$$

Because $x_0 \in \cap_{i=1}^m \text{int}(\text{dom}(f_i))$, we have $x_0 \in \text{int}(\text{dom}(f))$ and there is $\bar{t} > 0$ such that $x + tv \in \text{dom}(f)$ for $0 \le t < \bar{t}$. For each $0 < t < \bar{t}$, we can find $i(t) \in \{1, \cdots, m\}$ such that

$$f_{i(t)}(x_0 + tv) = f(x_0 + tv).$$

Because m is finite, there is an index $j \in \{1, \cdots, m\}$ that appears infinitely many times among $\{i(t) : 0 < t < \bar{t}\}$. Thus, there is a sequence $\{t_k\}$ with $0 < t_k < \bar{t}$ and $t_k \to 0$ such that $i(t_k) = j$ for all k. Consequently,

$$f_j(x_0 + t_k v) = f(x_0 + t_k v), \quad \forall k$$

Since f and f_j are continuous at x_0, taking $k \to \infty$ gives $f_j(x_0) = f(x_0)$, which shows $j \in I(x_0)$. Now, using $\xi_0 \in \partial f(x_0)$, we have

$$\frac{f_j(x_0 + t_k v) - f_j(x_0)}{t_k} = \frac{f(x_0 + t_k v) - f(x_0)}{t_k} \ge \frac{\langle \xi_0, t_k v\rangle}{t_k} = \langle \xi_0, v\rangle.$$

This shows that

$$f_j'(x_0; v) = \lim_{k\to\infty} \frac{f_j(x_0 + t_k v) - f_j(x_0)}{t_k} \ge \langle \xi_0, v\rangle.$$

Therefore, by using (5.10), we obtain

$$\langle \xi_0, v\rangle > \max_{i\in I(x_0)} f_i'(x_0; v) \ge f_j'(x_0; v) \ge \langle \xi_0, v\rangle$$

which is a contradiction. □

Example 5.11 Let $a_1, \cdots, a_m \in \mathbb{R}^n$ and $b_1, \cdots, b_m \in \mathbb{R}$. Consider the function

$$f(x) = \max\{\langle a_1, x\rangle + b_1, \cdots, \langle a_m, x\rangle + b_m\}.$$

Here, each $f_i(x) = \langle a_i, x\rangle + b_i$ is differentiable with $\partial f_i(x) = \{a_i\}$. Thus, by Theorem 5.10, we have

$$\partial f(x) = \left\{ \sum_{i\in I(x)} t_i a_i : t_i \ge 0, \sum_{i\in I(x)} t_i = 1 \right\}$$

for all $x \in \mathbb{R}^n$, where $I(x) = \{i : 1 \le i \le m,\ f(x) = \langle a_i, x\rangle + b_i\}$.

5.4 Karush-Kuhn-Tucker Theory

Consider the constrained convex minimization problem:

$$\begin{aligned} \text{minimize} \quad & f(x) \\ \text{subject to} \quad & g_1(x) \leq 0, \ \cdots, \ g_p(x) \leq 0, \\ & Ax = b, \end{aligned} \tag{5.11}$$

where $f, g_1, \cdots, g_p : \mathbb{R}^n \to (-\infty, \infty]$ are proper convex functions, A is an $m \times n$ matrix, and $b \in \mathbb{R}^m$. Any point x satisfying $g_1(x) \leq 0, \cdots, g_p(x) \leq 0$ and $Ax = b$ is called a *feasible point* of (5.11).

We will give necessary and sufficient conditions to guarantee a feasible point to be an optimal solution of (5.11). Assuming x^* is an optimal solution of (5.11), consider the convex function:

$$H(x) := \max\{f(x) - f(x^*), g_1(x), \cdots, g_p(x)\} + \delta_L(x),$$

where δ_L is the indicator function of

$$L := \{x : Ax = b\}.$$

Then, $H(x^*) = 0$ and $H(x) \geq 0$ for all $x \in \mathbb{R}^n$. Thus, x^* is a minimizer of H over $\mathbb{R}^n$. If the function

$$\phi(x) = \max\{f(x) - f(x^*), g_1(x), \cdots, g_p(x)\}$$

is continuous at some point in L, then by Theorems 3.9 and 3.14, we have

$$0 \in \partial H(x^*) = \partial\phi(x^*) + \partial\delta_L(x^*).$$

Recalling from Example 3.5 that $\partial\delta_L(x) = \text{Ran}(A^T)$, we have

$$0 \in \partial\phi(x^*) + \text{Ran}(A^T).$$

If, in addition, $x^* \in \text{int}(\text{dom}(f)) \cap \left(\cap_{i=1}^p \text{int}(\text{dom}(g_i))\right)$, we have from Theorem 5.10 that

$$0 \in \text{conv}\left(\partial f(x^*) \bigcup \left(\bigcup_{i \in I(x^*)} \partial g_i(x^*)\right)\right) + \text{Ran}(A^T),$$

where $I(x^*) = \{i : 1 \le i \le p, g_i(x^*) = 0\}$. By the definition of convex hull, there exist $\mu^* \in \mathbb{R}^m$, $\lambda_0^* \ge 0$, and $\lambda_i^* \ge 0$, for $i \in I(x^*)$ with $\lambda_0^* + \sum_{i \in I(x^*)} \lambda_i^* = 1$ such that

$$0 \in \lambda_0^* \partial f(x^*) + \sum_{i \in I(x^*)} \lambda_i^* \partial g_i(x^*) + A^T \mu^*.$$

Clearly, $\lambda_i^* g_i(x^*) = 0$ for $i \in I(x^*)$. If we define $\lambda_i^* = 0$ for $i \in \{1, \cdots, p\} \setminus I(x^*)$, then $\lambda_i^* \ge 0$, $\lambda_i^* g_i(x^*) = 0$ for all $i = 1, \cdots, p$ and

$$0 \in \lambda_0^* \partial f(x^*) + \sum_{i=1}^{p} \lambda_i^* \partial g_i(x^*) + A^T \mu^*.$$

Under certain conditions, we can take $\lambda_0^* = 1$. One of most important conditions is *Slater's qualification condition*: there exists $\bar{x} \in \mathbb{R}^n$ such that

$$A\bar{x} - b = 0, \quad g_1(\bar{x}) < 0, \cdots, g_p(\bar{x}) < 0.$$

This leads to the following theorem:

Theorem 5.12 *Consider the problem (5.11). If x^* is an optimal solution of (5.11) such that $f, g_1, \cdots, g_p$ are continuous at x^*, then there exists nonzero $(\lambda_0^*, \lambda_1^*, \cdots, \lambda_p^*)$ and $\mu^* \in \mathbb{R}^m$ such that*

$$\begin{aligned} & 0 \in \lambda_0^* \partial f(x^*) + \sum_{i=1}^{p} \lambda_i^* \partial g_i(x^*) + A^T \mu^*, \\ & \lambda_i^* g_i(x^*) = 0, \quad i = 1, \cdots, p, \\ & \lambda_i^* \ge 0, \quad i = 0, \cdots, p, \\ & Ax^* = b, \quad g_i(x^*) \le 0, \quad i = 1, \cdots, p. \end{aligned}$$

If, in addition, Slater's qualification condition holds, we can take $\lambda_0^* = 1$.

Proof It remains only to show that, if Slater's condition holds, we can take $\lambda_0^* = 1$. It suffices to show that $\lambda_0^* \ne 0$. If $\lambda_0^* = 0$, then

$$\begin{aligned} 0 &\in \lambda_1^* \partial g_1(x^*) + \cdots + \lambda_p^* \partial g_p(x^*) + \mathrm{Ran}(A^T) \\ &\subset \partial \left(\lambda_1^* g_1 + \cdots \lambda_p^* g_p + \delta_L \right)(x^*). \end{aligned}$$

Thus, x^* is a minimizer of $\lambda_1^* g_1 + \cdots + \lambda_p^* g_p + \delta_L$. So

$$0 = \lambda_1^* g_1(x^*) + \cdots + \lambda_p^* g_p(x^*) + \delta_L(x^*) \le \lambda_1^* g_1(\bar{x}) + \cdots + \lambda_p^* g_p(\bar{x}) < 0$$

by Slater's condition because at least one of $\lambda_1^*, \cdots, \lambda_p^*$ is positive. This is a contradiction. □

The following result gives the converse of Theorem 5.12.

Theorem 5.13 *Consider the problem (5.11). For a point $x^* \in \mathbb{R}^n$, if there exist $\mu^* \in \mathbb{R}^m$ and numbers $\lambda_1^*, \cdots, \lambda_p^*$ such that*

$$\begin{aligned} &0 \in \partial f(x^*) + \sum_{i=1}^{p} \lambda_i^* \partial g_i(x^*) + A^T \mu^*, \\ &\lambda_i^* g_i(x^*) = 0, \quad i = 1, \cdots, p, \\ &\lambda_i^* \geq 0, \quad i = 1, \cdots, p, \\ &Ax^* = b, \quad g_i(x^*) \leq 0, \quad i = 1, \cdots, p, \end{aligned}$$

then x^* is an optimal solution of (5.11).

Proof By the given conditions, we have

$$0 \in \partial f(x^*) + \sum_{i=1}^{p} \lambda_i^* \partial g_i(x^*) + A^T \mu^* \subset \partial \left(f + \sum_{i=1}^{p} \lambda_i^* g_i + \langle \mu^*, A \cdot - b \rangle \right) (x^*)$$

which implies for all $x \in \mathbb{R}^n$ that

$$\begin{aligned} f(x) + \sum_{i=1}^{p} \lambda_i^* g_i(x) + \langle \mu^*, Ax - b \rangle &\geq f(x^*) + \sum_{i=1}^{p} \lambda_i^* g_i(x^*) + \langle \mu^*, Ax^* - b \rangle \\ &= f(x^*). \end{aligned}$$

For any feasible point x, we have $Ax - b = 0$ and $g_i(x) \leq 0$. Thus, $\sum_{i=1}^{p} \lambda_i^* g_i(x) \leq 0$ and so $f(x) \geq f(x^*)$ for all feasible x. Therefore, x^* is an optimal solution of (5.11). □

The equations in Theorem 5.13 are called *Karush-Kuhn-Tucker (KKT) conditions*. Any point $(x^*, \lambda^*, \mu^*) \in \mathbb{R}^n \times \mathbb{R}^p \times \mathbb{R}^m$ satisfying the KKT conditions is called a *KKT point*. The function

$$L(x, \lambda, \mu) := f(x) + \sum_{i=1}^{p} \lambda_i g_i(x) + \langle \mu, Ax - b \rangle$$

appearing in the proof of Theorem 5.13 is called the *Lagrangian function*, where x are called *primal variables* and μ, λ are called *dual variables*.

Example 5.14 Finding the Euclidean projection of a vector $v \in \mathbb{R}^n$ onto the unit simplex can be stated as the convex minimization problem:

$$\min \left\{ \frac{1}{2}\|x - v\|^2 : x \geq 0 \text{ and } \sum_{i=1}^{n} x_i = 1 \right\}. \tag{5.12}$$

The corresponding Lagrangian function is

$$L(x, \lambda, \mu) := \frac{1}{2}\|x - v\|^2 - \langle \lambda, x \rangle - \mu \left(\sum_{i=1}^{n} x_i - 1 \right).$$

Let x^* denote the unique solution of (5.12). It follows from Theorem 5.12 that there exist $\lambda^* \in \mathbb{R}^n$ and $\mu^* \in \mathbb{R}$ such that (x^*, λ^*, μ^*) satisfies the KKT conditions:

$$\begin{aligned} x_i^* - v_i - \lambda_i^* - \mu^* = 0, &\quad i = 1, \cdots, n, \\ \sum_{i=1}^{n} x_i^* = 1, \ x_i^* \geq 0, &\quad i = 1, \cdots, n, \\ \lambda_i^* \geq 0, \ \lambda_i^* x_i^* = 0, &\quad i = 1, \cdots, n. \end{aligned}$$

In order to determine x^* from these equations, we assume the components of v are sorted in a descending order, i.e., $v_1 \geq v_2 \geq \cdots \geq v_n$ and let ρ be the number of positive components in x^*. If $x_i^* > 0$, then we have $\lambda_i^* = 0$ and hence $x_i^* = v_i + \mu^* > 0$, while if $x_i^* = 0$, then $v_i + \mu^* = -\lambda_i^* \leq 0$. This in particular shows that the components of x^* are in descending order with

$$x_1^* \geq \cdots \geq x_\rho^* > x_{\rho+1}^* = \cdots = x_n^* = 0.$$

By the equation $\sum_{i=1}^{n} x_i^* = 1$, we can obtain

$$1 = \sum_{i=1}^{\rho} x_i^* = \sum_{i=1}^{\rho} (v_i + \mu^*) = \sum_{i=1}^{\rho} v_i + \rho\mu^*$$

which gives

$$\mu^* = \frac{1}{\rho} \left(1 - \sum_{i=1}^{\rho} v_i \right).$$

Therefore, once we know ρ, we can obtain μ^* by the above equation, and consequently, we can determine λ^* and x^* completely; in particular, we have

$$x_i^* = v_i + \mu^* = v_i + \frac{1}{\rho}\left(1 - \sum_{i=1}^{\rho} v_i\right), \quad i = 1, \cdots, \rho.$$

It remains to determine ρ. Recall that $v_i + \mu^* > 0$ for $i \leq \rho$ and $v_i + \mu^* \leq 0$ for $i > \rho$. Therefore, for $1 \leq j \leq \rho$, we have

$$\begin{aligned} v_j + \frac{1}{j}\left(1 - \sum_{i=1}^{j} v_i\right) &= v_j + \frac{1}{j}\left(1 - \sum_{i=1}^{\rho} v_i + \sum_{i=j+1}^{\rho} v_i\right) \\ &= v_j + \frac{1}{j}\left(\rho\mu^* + \sum_{i=j+1}^{\rho} v_i\right) \\ &= (v_j + \mu^*) + \frac{1}{j}\sum_{i=j+1}^{\rho} (v_i + \mu^*) > 0, \end{aligned}$$

and for $j > \rho$, we have

$$\begin{aligned} v_j + \frac{1}{j}\left(1 - \sum_{i=1}^{j} v_i\right) &= v_j + \frac{1}{j}\left(1 - \sum_{i=1}^{\rho} v_i - \sum_{i=\rho+1}^{j} v_i\right) \\ &= v_j + \frac{1}{j}\left(\rho\mu^* - \sum_{i=\rho+1}^{j} v_i\right) \\ &= \frac{1}{j}\left(\rho(v_j + \mu^*) + \sum_{i=\rho+1}^{j} (v_j - v_i)\right) \leq 0. \end{aligned}$$

This shows that ρ must be the largest integer such that $v_\rho + \frac{1}{\rho}\left(1 - \sum_{i=1}^{\rho} v_i\right) > 0$. Putting together the above analysis, we obtain the following algorithm for determining the Euclidean projection onto the unit simplex [44, 203, 221]:

Algorithm 5.15 (Euclidean Projection onto Unit Simplex) *Given $v \in \mathbb{R}^n$, do the following:*

(i) *Sort the components of v in a descending order to form the new vector u.*
(ii) *Find the largest integer ρ such that $u_\rho + \frac{1}{\rho}\left(1 - \sum_{i=1}^{\rho} u_i\right) > 0$.*
(iii) *Define $\mu^* = \frac{1}{\rho}\left(1 - \sum_{i=1}^{\rho} u_i\right)$ and set $x^* = \max\{v + \mu^*, 0\}$.*

5.5 Hoffman's Error Bound and Linear Regularity

In this section, we will use the KKT theory to prove the famous error bound result of Hoffman [106], which demonstrates that the distance from any point to the solution set of a linear system can be controlled by the residual. More precisely, we have the following result.

Theorem 5.16 (Hoffman's Error Bound) *Let A be a nonzero $m \times n$ matrix. For each $b \in \mathbb{R}^m$, let*

$$\Phi(b) := \left\{x \in \mathbb{R}^n : Ax \leq b\right\}.$$

Then, there exists a constant κ depending only on A such that

$$d(x, \Phi(b)) \leq \kappa \|(Ax - b)_+\|$$

for all $x \in \mathbb{R}^n$ and $b \in \mathbb{R}^m$ with $\Phi(b) \neq \emptyset$.

Proof Let $b \in \mathbb{R}^m$ be such that $\Phi(b) \neq \emptyset$. Note that $\Phi(b)$ is a closed convex set in $\mathbb{R}^n$. Let $x \in \mathbb{R}^n$ be any point. If $x \in \Phi(b)$, then the result holds trivially. So we may assume $x \notin \Phi(b)$. Let $\hat{x}$ be the unique projection of x onto $\Phi(b)$, i.e.,

$$\hat{x} := \arg\min_{z \in \mathbb{R}^n} \left\{\frac{1}{2}\|z - x\|^2 : Az \leq b\right\}.$$

By the KKT theory, there exists $\lambda \in \mathbb{R}^m$ such that

$$\begin{aligned} &\hat{x} - x + A^T\lambda = 0, \\ &\lambda \geq 0, \quad \langle \lambda, A\hat{x} - b\rangle = 0, \\ &A\hat{x} - b \leq 0. \end{aligned} \tag{5.13}$$

Since $\hat{x} \neq x$, we have $\lambda \neq 0$, i.e., λ has positive components. Since there may have many λ satisfying (5.13), we use λ^* to denote a one such that $\text{supp}(\lambda^*)$ has the smallest cardinality, where, for each $\lambda \in \mathbb{R}^m$, $\text{supp}(\lambda)$ denotes its support, i.e.,

$$\text{supp}(\lambda) := \{i : i \in \{1, \cdots, m\} \text{ and } \lambda_i \neq 0\}.$$

Let $I := \text{supp}(\lambda^*)$ and let a_i^T be the i-th row of A. Then, from (5.13), it follows that $\langle a_i, \hat{x}\rangle - b_i = 0$ for $i \in I$ and $\lambda_i^* = 0$ for $i \notin I$.

Now, we claim that $a_i, i \in I$ must be linearly independent. If it is not true, then there exist $\mu_i, i \in I$, not all zeros, such that $\sum_{i \in I} \mu_i a_i = 0$. By changing μ_i to $-\mu_i$ for all $i \in I$ if necessary, we may assume $\min_{i \in I} \mu_i < 0$. Thus, there exists $t > 0$ such that

$$\lambda_i^* + t\mu_i \geq 0 \text{ for all } i \in I \quad \text{and} \quad \min_{i \in I}(\lambda_i^* + t\mu_i) = 0.$$

Define $\hat{\lambda}$ by setting $\hat{\lambda}_i = \lambda_i^* + t\mu_i$ if $i \in I$ and $\hat{\lambda}_i = 0$ if $i \notin I$. Then, $\hat{\lambda}$ still satisfies (5.13). However, $\text{supp}(\hat{\lambda})$ is smaller than $\text{supp}(\lambda^*)$, contradicting the choice of λ^*.

Let A_I, λ_I^*, and b_I denote respectively the matrix and vectors formed from A, λ^*, and b by deleting the rows with indices not in I. Then, A_I has full row rank and

$$x - \hat{x} = A^T \lambda^* = A_I^T \lambda_I^*. \tag{5.14}$$

Thus, by (5.13), we have

$$\begin{aligned}
\|x - \hat{x}\|^2 &= \langle A_I^T \lambda_I^*, x - \hat{x} \rangle = \langle \lambda_I^*, A_I(x - \hat{x}) \rangle \\
&= \langle \lambda_I^*, A_I x - b_I \rangle - \langle \lambda_I^*, A_I \hat{x} - b_I \rangle \\
&= \langle \lambda_I^*, A_I x - b_I \rangle \le \langle \lambda_I^*, (A_I x - b_I)_+ \rangle \\
&\le \|\lambda_I^*\| \|(A_I x - b_I)_+\| \\
&\le \|\lambda_I^*\| \|(Ax - b)_+\|.
\end{aligned}$$

Since A_I has full row rank, $A_I A_I^T$ is invertible. It then follows from (5.14) that $\lambda_I^* = (A_I A_I^T)^{-1} A_I (x - \hat{x})$, which gives

$$\|\lambda_I^*\| \le \|(A_I A_I^T)^{-1} A_I\| \|x - \hat{x}\|.$$

Therefore, by defining

$$\kappa := \max \left\{ \|(A_I A_I^T)^{-1} A_I\| : I \subset \{1, \cdots, m\} \text{ and } A_I \text{ has full row rank} \right\}.$$

Then, κ is a finite number depending only on A and $\|\lambda_I^*\| \le \kappa \|x - \hat{x}\|$. Consequently,

$$\|x - \hat{x}\|^2 \le \kappa \|x - \hat{x}\| \|(Ax - b)_+\|$$

and hence

$$d(x, \Phi(b)) = \|x - \hat{x}\| \le \kappa \|(Ax - b)_+\|.$$

The result is thus proved. □

Hoffman's error bound result has far-reaching impact on the development of sensitivity analysis in optimization; see [37]. In the following, we will use Theorem 5.16 to show the linear regularity of polyhedral sets, which enables us to establish linear convergence of some cyclic projection algorithms.

Definition 5.17 Let $C_1, \cdots, C_N$ be closed convex subsets in $\mathbb{R}^n$ with nonempty intersection.

(i) They are called *linearly regular* if there is a constant $\kappa > 0$ such that

$$d\left(x, \bigcap_{i=1}^{N} C_i\right) \leq \kappa \max_{1\leq i\leq N} d(x, C_i), \quad \forall x \in \mathbb{R}^n. \tag{5.15}$$

(ii) They are called *bounded linearly regular* if for any bounded set $D \subset \mathbb{R}^n$, there is a constant κ such that

$$d\left(x, \bigcap_{i=1}^{N} C_i\right) \leq \kappa \max_{1\leq i\leq N} d(x, C_i), \quad \forall x \in D. \tag{5.16}$$

Example 5.18 Let $C_1, \cdots, C_N$ be closed convex subsets in $\mathbb{R}^n$ such that their intersection $C := \bigcap_{i=1}^{N} C_i$ has nonempty interior, i.e., there exist $\bar{x} \in \mathbb{R}^n$ and $r > 0$ such that $B_r(\bar{x}) \subset C$. Then, $C_1, \cdots, C_N$ are bounded linearly regular, more precisely,

$$d(x, C) \leq \left(1 + \frac{2\|x - \bar{x}\|}{r}\right)^{N-1} \max_{1\leq i\leq N} d(x, C_i), \quad \forall x \in \mathbb{R}^n.$$

To see this, we first consider the case $N = 2$. Let $C := C_1 \cap C_2$, and for any fixed $x \in \mathbb{R}^n$, let $d := 2\max\{d(x, C_1), d(x, C_2)\}$. By the 1-Lipschitz continuity of the function $d(\cdot, C_2)$, we have

$$\left|d(P_{C_1}(x), C_2) - d(x, C_2)\right| \leq \|P_{C_1}(x) - x\| = d(x, C_1).$$

Thus,

$$d(P_{C_1}(x), C_2) \leq d(x, C_2) + d(x, C_1) \leq d.$$

Now, we take $y = \bar{x} + \frac{r}{d}(P_{C_1}(x) - P_{C_2}P_{C_1}(x))$. We have

$$\|y - \bar{x}\| = \frac{r}{d}\|P_{C_1}(x) - P_{C_2}P_{C_1}(x)\| = \frac{r}{d}d(P_{C_1}(x), C_2) \leq \frac{r}{d}d = r$$

which implies $y \in B_r(\bar{x}) \subset C$. Next, we choose $z \in \mathbb{R}^n$ to be the intersection point of the two line segments $[\bar{x}, P_{C_1}(x)]$ and $[y, P_{C_2}P_{C_1}(x)]$. By direct calculation, it is easy to see that

$$z = \frac{d}{d+r}y + \frac{r}{d+r}P_{C_2}P_{C_1}(x) = \frac{d}{d+r}\bar{x} + \frac{r}{d+r}P_{C_1}(x).$$

Consequently, $z \in C_1 \cap C_2$. Therefore,

$$\begin{aligned} d(x, C) &\le \|x - z\| \le \|x - P_{C_1}(x)\| + \|z - P_{C_1}(x)\| \\ &= d(x, C_1) + \frac{d}{d+r}\|\bar{x} - P_{C_1}(x)\| \le \frac{d}{2} + \frac{d}{d+r}\|P_{C_1}(\bar{x}) - P_{C_1}(x)\| \\ &\le \frac{d}{2} + \frac{d}{d+r}\|\bar{x} - x\| \le \left(\frac{1}{2} + \frac{\|x - \bar{x}\|}{r}\right) d \\ &= \left(1 + \frac{2\|x - \bar{x}\|}{r}\right) \max\{d(x, C_1), d(x, C_2)\}. \end{aligned}$$

For the general case $N \ge 2$, we can recursively use the above result to obtain

$$\begin{aligned} d\left(x, \bigcap_{i=1}^{N} C_i\right) &\le \left(1 + \frac{2\|x - \bar{x}\|}{r}\right) \max\left\{d\left(x, \bigcap_{i=1}^{N-1} C_i\right), d(x, C_N)\right\} \le \cdots \\ &\le \left(1 + \frac{2\|x - \bar{x}\|}{r}\right)^{N-1} \max_{1 \le i \le N} d(x, C_i) \end{aligned}$$

for any $x \in \mathbb{R}^n$, which shows the desired estimate.

Proposition 5.19 *Let $C_1, \cdots, C_N$ be (bounded) linearly regular closed convex subsets in $\mathbb{R}^n$. Let $I_1, \cdots, I_M$ be a disjoint decomposition of $\{1, \cdots, N\}$, i.e., $I_i \neq \emptyset$ for each i, $I_i \cap I_j = \emptyset$ when $i \neq j$, and $\bigcup_{i=1}^{M} I_i = \{1, \cdots, N\}$. Define*

$$D_i := \bigcap_{j \in I_i} C_j, \quad i = 1, \cdots, M.$$

Then, $D_1, \cdots, D_M$ are (bounded) linearly regular.

Proof We only show the result for the linearly regular case; the bounded linearly regular case can be proved by the same argument. Since $C_1, \cdots, C_N$ are linearly regular, there is a constant $\kappa > 0$ such that (5.15) holds for all $x \in \mathbb{R}^n$. Noting that $\bigcap_{i=1}^{M} D_i = \bigcap_{j=1}^{N} C_j$ and $I_1, \cdots, I_M$ form a disjoint decomposition of $\{1, \cdots, N\}$, we have for any $x \in \mathbb{R}^n$ that

$$\begin{aligned} d\left(x, \bigcap_{i=1}^{M} D_i\right) = d\left(x, \bigcap_{j=1}^{N} C_j\right) &\le \kappa \max_{1 \le j \le N} d(x, C_j) \\ &= \kappa \max_{1 \le i \le M} \max_{j \in I_i} d(x, C_j) \\ &\le \kappa \max_{1 \le i \le M} d(x, D_i) \end{aligned}$$

which shows that $D_1, \cdots, D_M$ are linearly regular. □

Based on Proposition 5.19 and Theorem 5.16, we now provide examples of linear regular sets by showing that any finite collection of polyhedral convex sets are linearly regular. A set D in $\mathbb{R}^n$ is called *polyhedral convex* if it is the intersection of a finite number of closed half spaces, i.e.,

$$D = \left\{x \in \mathbb{R}^n : Ax \leq b\right\},$$

where A is an $m \times n$ matrix and $b \in \mathbb{R}^m$.

Corollary 5.20 *Any finite many polyhedral convex sets $C_1, \cdots, C_N$ in $\mathbb{R}^n$ with nonempty intersection are linearly regular.*

Proof Note that for any closed half-space $H := \{x \in \mathbb{R}^n : \langle a, x\rangle \leq \alpha\}$ and any $x \in \mathbb{R}^n$, there holds

$$d(x, H) = \frac{1}{\|a\|}(\langle a, x\rangle - \alpha)_+.$$

This fact and Hoffman's error bound show that any finite family of half-spaces are linearly regular. Since any polyhedral convex set is an intersection of finitely many half-spaces, it follows from Proposition 5.19 that any finite family of polyhedral convex sets are linearly regular. □

Given finite many closed convex sets with nonempty intersection, we may use a cyclic projection algorithm to find a point in their intersection; see Example 3.34. The following result shows that the algorithm enjoys a linear convergence property if they are bounded linearly regular ([13]).

Theorem 5.21 *Let $C_1, \cdots, C_N$ be bounded linearly regular, closed, convex subsets in $\mathbb{R}^n$ with nonempty intersection $C := \bigcap_{i=1}^N C_i$, where $N \geq 2$. Consider the cyclic projection algorithm:*

$$x^{k+1} = P_{C_{i_k}}(x^k), \quad k = 0, 1 \cdots,$$

where $i_k = k \pmod N) + 1$. Then, for any initial guess x^0, the sequence $\{x^k\}$ linearly converges to an element $x^* \in C$, i.e., there exist a number $\theta \in (0, 1)$ and a constant M such that

$$\|x^k - x^*\| \leq M\theta^k d(x_0, C)$$

for all $k \geq 0$

Proof For any $c \in C$, we have by the polarization identity that

$$\|x^k - c\|^2 - \|x^{k+1} - c\|^2 = \|x^k - x^{k+1}\|^2 - 2\langle x^k - x^{k+1}, c - x^{k+1}\rangle.$$

Since $x^{k+1} = P_{C_{i_k}}(x^k)$ and $c \in C \subset C_{i_k}$, it follows from (2.2) that $\langle x^k - x^{k+1}, c - x^{k+1}\rangle \le 0$. Therefore,

$$\|x^k - c\|^2 - \|x^{k+1} - c\|^2 \ge \|x^k - x^{k+1}\|^2 \tag{5.17}$$

which in particular implies that

$$\|x^{k+1} - c\|^2 \le \|x^k - c\|^2. \tag{5.18}$$

Now, let $i \in \{1, \cdots, N\}$ be any fixed index. By the triangle inequality, we have for any integer $p \ge 0$ that

$$\begin{aligned} d(x^{pN}, C_i) &\le \|x^{pN} - x^{pN+i-1}\| + d(x^{pN+i-1}, C_i) \\ &\le \sum_{j=0}^{i-2} \|x^{pN+j} - x^{pN+j+1}\| + d(x^{pN+i-1}, C_i). \end{aligned}$$

Note that

$$d(x^{pN+i-1}, C_i) = \|x^{pN+i-1} - P_{C_i}(x^{pN+i-1})\| = \|x^{pN+i-1} - x^{pN+i}\|.$$

Therefore, by using the Cauchy-Schwarz inequality, (5.17) and (5.18), we can obtain for any $c \in C$ that

$$\begin{aligned} d^2(x^{pN}, C_i) &\le \left(\sum_{j=0}^{i-1} \|x^{pN+j} - x^{pN+j+1}\|\right)^2 \le i \sum_{j=0}^{i-1} \|x^{pN+j} - x^{pN+j+1}\|^2 \\ &\le i\left(\|x^{pN} - c\|^2 - \|x^{pN+i} - c\|^2\right) \\ &\le N\left(\|x^{pN} - c\|^2 - \|x^{(p+1)N} - c\|^2\right). \end{aligned}$$

Taking $c = P_C(x^{pN})$ gives

$$d^2(x^{pN}, C_i) \le N\left(d^2(x^{pN}, C) - d^2(x^{(p+1)N}, C)\right).$$

According to (5.18), $\{x^k\}$ is bounded. Since $C_1, \cdots, C_N$ are bounded linearly regular, there is a constant $\kappa \ge 1$ such that

$$d(x^k, C) \le \kappa \max_{1\le i\le N} d(x^k, C_i), \quad \forall k.$$

Thus,

$$d^2(x^{pN}, C) \leq N\kappa^2 \left(d^2(x^{pN}, C) - d^2(x^{(p+1)N}, C)\right)$$

which implies

$$d(x^{(p+1)N}, C) \leq \tau d(x^{pN}, C), \tag{5.19}$$

where $\tau := \sqrt{1 - 1/(N\kappa^2)} \in (0, 1)$. Consequently,

$$d(x^{pN}, C) \to 0 \quad \text{as } p \to \infty. \tag{5.20}$$

Note that for any integers $0 \leq p < q$, we have from (5.18) that

$$\begin{aligned} \|x^{pN} - x^{qN}\| &\leq \|x^{pN} - P_C(x^{pN})\| + \|x^{qN} - P_C(x^{pN})\| \\ &\leq 2\|x^{pN} - P_C(x^{pN})\| = 2d(x^{pN}, C). \end{aligned} \tag{5.21}$$

This together with (5.20) shows that $\{x^{pN}\}$ is a Cauchy sequence and hence $x^{pN} \to x^*$ for some $x^* \in \mathbb{R}^n$. According to (5.20), we have $d(x^*, C) = 0$ and thus $x^* \in C$ by the closedness of C. Furthermore, by taking $q \to \infty$ in (5.21) and using (5.19), we can obtain

$$\|x^{pN} - x^*\| \leq 2d(x^{pN}, C) \leq 2\tau^p d(x^0, C) \tag{5.22}$$

which shows the linear convergence of $\{x^{pN}\}$ to $x^* \in C$.

Finally, we show that $\{x^k\}$ converges to x^* linearly. For any integer $k \geq 0$, we write $k = pN+i$ for some integers $p \geq 0$ and $0 \leq i \leq N-1$. Then, by using (5.18) and (5.22), we have

$$\begin{aligned} \|x^k - x^*\| \leq \|x^{pN} - x^*\| &\leq 2\tau^p d(x^0, C) \\ &= 2\tau^{(k-i)/N} d(x^0, C) \\ &\leq M\tau^{k/N} d(x^0, C), \end{aligned}$$

where $M := 2\tau^{-(N-1)/N}$. The proof is complete. □

As an application, we consider solving the linear consistent system $Ax = b$, where A is an $m \times n$ matrix and $b \in \mathbb{R}^m$. Let a_i^T denote the i-th row of A. In Chap. 3, we have considered the greedy Kaczmarz algorithm; see Algorithm 3.28, in which at each iteration step, one has to construct x^{k+1} from x^k by determining a hyperplane

$$H_i := \{x \in \mathbb{R}^n : \langle a_i, x\rangle = b_i\}$$

that is farthest to the current iterate x^k. This makes the implementation of Algorithm 3.28 very expensive. The workload can be relieved by considering the cyclic Kaczmarz algorithm of the following form.

Algorithm 5.22 *Take an initial guess $x^0 \in \mathbb{R}^n$. For $k = 0, 1, \cdots$, define*

$$x^{k+1} = x^k + \frac{b_{i_k} - \langle a_{i_k}, x^k \rangle}{\|a_{i_k}\|^2} a_{i_k},$$

where $i_k \in \{1, \cdots, m\}$ is given by $i_k = k \pmod m + 1$.

According to Corollary 5.20, the hyperplanes $H_1, \cdots, H_m$ are linearly regular. Therefore, by virtue of Theorem 5.21, we can obtain the following result on Algorithm 5.22.

Corollary 5.23 *Consider Algorithm 5.22 for solving the linear consistent system $Ax = b$. Then, for any initial guess x^0, the sequence $\{x^k\}$ converges linearly to a solution of $Ax = b$.*

Although Algorithm 5.22 exhibits linear convergence, its speed depends critically on the order in which the equations are processed. To enhance performance, one may consider reshuffling the equations before applying Algorithm 5.22 to minimize the likelihood of successive hyperplanes forming small angles.

5.6 Lagrangian Duality Theory

Consider again the constrained convex optimization problem (5.11). We use Ω to denote the feasible set of (5.11), i.e.,

$$\Omega = \{x \in \mathbb{R}^n : g_1(x) \le 0, \ \cdots, \ g_p(x) \le 0, \ Ax = b\}.$$

The associated Lagrangian function is

$$L(x; \lambda) = f(x) + \sum_{i=1}^{p} \lambda_i g_i(x) + \langle \mu, Ax - b \rangle,$$

where $x \in \mathbb{R}^n$, $\lambda \in \mathbb{R}^p_+$, and $\mu \in \mathbb{R}^m$. We have the following observation.

Lemma 5.24 *There holds*

$$\sup_{(\lambda,\mu) \in \mathbb{R}^p_+ \times \mathbb{R}^m} L(x; \lambda, \mu) = \begin{cases} f(x), & x \in \Omega \\ +\infty, & \text{otherwise.} \end{cases}$$

Proof We first have

$$\sup_{(\lambda,\mu)\in\mathbb{R}_+^p\times\mathbb{R}^m} L(x;\lambda,\mu) = f(x) + \sup_{\lambda\in\mathbb{R}_+^p}\sum_{i=1}^p \lambda_i g_i(x) + \sup_{\mu\in\mathbb{R}^m}\langle\mu, Ax-b\rangle.$$

It is easy to show that

$$\sup_{\mu\in\mathbb{R}^m}\langle\mu, Ax-b\rangle = \begin{cases} 0, & \text{if } Ax=b,\\ +\infty, & \text{otherwise}\end{cases}$$

and

$$\sup_{\lambda\in\mathbb{R}_+^p}\sum_{i=1}^p \lambda_i g_i(x) = \begin{cases} 0, & \text{if } g_1(x)\le 0,\cdots,g_p(x)\le 0,\\ +\infty, & \text{otherwise.}\end{cases}$$

Combining the above three equations, we complete the proof. □

By Lemma 5.24, the problem (5.11) can be equivalently stated as

$$\inf_{x\in\mathbb{R}^n}\sup_{(\lambda,\mu)\in\mathbb{R}_+^p\times\mathbb{R}^m} L(x;\lambda,\mu).$$

Let p^* denote the optimal value of the problem (5.11). Then,

$$p^* \ge \inf_{x\in\mathbb{R}^n} L(x;\lambda,\mu), \quad \forall(\lambda,\mu)\in\mathbb{R}_+^p\times\mathbb{R}^m.$$

Definition 5.25 (Lagrangian Dual Function) The function

$$h(\lambda,\mu) := \inf_{x\in\mathbb{R}^n} L(x;\lambda,\mu), \quad (\lambda,\mu)\in\mathbb{R}_+^p\times\mathbb{R}^m$$

is called the *Lagrangian dual function* associated with (5.11).

Therefore, for every $(\lambda,\mu)\in\mathbb{R}_+^p\times\mathbb{R}^m$, $h(\lambda,\mu)$ gives a lower bound on the optimal value p^* of (5.11). So, what is the greatest lower bound on p^* we can get from the Lagrangian dual function $h(\lambda,\mu)$? This naturally leads to consider the optimization problem:

$$\begin{aligned} &\text{maximize} && h(\lambda,\mu)\\ &\text{subject to} && \lambda\in\mathbb{R}_+^p,\ \mu\in\mathbb{R}^m. \end{aligned} \tag{5.23}$$

We call this maximization problem the *dual problem* of (5.11). Correspondingly, the original problem (5.11) is called the *primal problem*. We also call x the *primal variables* and the Lagrangian multipliers λ, μ the *dual variables*.

Example 5.26 Given $b \in \mathbb{R}^m$, $c \in \mathbb{R}^n$, and an $m \times n$ matrix A. Consider the linear programming problem:

$$\min\{\langle c, x\rangle : Ax = b \text{ and } x \geq 0\}. \tag{5.24}$$

The associated Lagrangian function is

$$L(x; \lambda, \mu) = \langle c, x\rangle - \langle \lambda, Ax - b\rangle - \langle \mu, x\rangle,$$

where $\lambda \in \mathbb{R}^m$ and $\mu \geq 0$. The Lagrangian dual function is given by

$$\begin{aligned} h(\lambda, \mu) &= \inf_{x\in\mathbb{R}^n} L(x; \lambda, \mu) = \inf_{x\in\mathbb{R}^n} \{\langle c, x\rangle - \langle \lambda, Ax - b\rangle - \langle \mu, x\rangle\} \\ &= \inf_{x\in\mathbb{R}^n} \left\{\langle c - A^T\lambda - \mu, x\rangle + \langle b, \lambda\rangle\right\} \\ &= \begin{cases} \langle b, \lambda\rangle, \text{ if } c - A^T\lambda - \mu = 0, \\ -\infty, \quad \text{otherwise.} \end{cases} \end{aligned}$$

Thus, the dual problem is

$$\max\{\langle b, \lambda\rangle : \mu + A^T\lambda = c \text{ and } \mu \geq 0\}$$

which is equivalent to

$$\max\{\langle b, \lambda\rangle : A^T\lambda \leq c\}. \tag{5.25}$$

Example 5.27 Given an $m \times n$ matrix A and $b \in \mathbb{R}^m$, consider the problem:

$$\min\{\|x\|_1 : Ax = b\}.$$

The associated Lagrangian function is

$$L(x; \lambda) = \|x\|_1 + \langle \lambda, Ax - b\rangle, \quad x \in \mathbb{R}^n, \lambda \in \mathbb{R}^m.$$

Let $f(x) = \|x\|_1$. The Lagrangian dual function is given by

$$\begin{aligned} h(\lambda) &= \inf_{x\in\mathbb{R}^n} \{\|x\|_1 + \langle \lambda, Ax - b\rangle\} = -\sup_{x\in\mathbb{R}^n} \left\{\langle -A^T\lambda, x\rangle - f(x)\right\} - \langle b, \lambda\rangle \\ &= -f^*(-A^T\lambda) - \langle b, \lambda\rangle, \end{aligned}$$

where f^* denotes the Legendre-Fenchel conjugate of f. In view of Example 4.5, we have

$$h(\lambda) = \begin{cases} -\langle b, \lambda \rangle, & \text{if } \|A^T\lambda\|_\infty \le 1, \\ -\infty, & \text{if } \|A^T\lambda\|_\infty > 1. \end{cases}$$

Therefore, the dual problem is

$$\max\{-\langle b, \lambda \rangle : \|A^T\lambda\|_\infty \le 1\}$$

which can be written equivalently as

$$\min\{\langle b, \lambda \rangle : \|A^T\lambda\|_\infty \le 1\}.$$

Return to the dual problem (5.23). Recall that $p^* \ge h(\lambda, \mu)$ for all $(\lambda, \mu) \in \mathbb{R}^p_+ \times \mathbb{R}^m$. Let d^* be the optimal value of the dual problem (5.23). Then, $d^* \le p^*$, i.e.,

$$\sup_{(\lambda,\mu)\in\mathbb{R}^p_+\times\mathbb{R}^m} \inf_{x\in\mathbb{R}^n} L(x; \lambda, \mu) \le \inf_{x\in\mathbb{R}^n} \sup_{(\lambda,\mu)\in\mathbb{R}^p_+\times\mathbb{R}^m} L(x; \lambda, \mu). \tag{5.26}$$

This is called the *weak duality*. The value $p^* - d^*$ is called the *duality gap*.

A natural question is under what conditions the inequality in (5.26) becomes equality, i.e., there is no duality gap. Any such a result is called a *strong duality* result. We will introduce the concept of saddle point and use the Karush-Kuhn-Tucker theory to answer this question.

Definition 5.28 (Saddle Point) A point $(x^*; \lambda^*, \mu^*) \in \mathbb{R}^n \times \mathbb{R}^p_+ \times \mathbb{R}^m$ is called a *saddle point* of the Lagrangian function L if

$$L(x^*; \lambda, \mu) \le L(x^*, \lambda^*, \mu^*) \le L(x, \lambda^*, \mu^*)$$

for all $x \in \mathbb{R}^n$ and $(\lambda, \mu) \in \mathbb{R}^p_+ \times \mathbb{R}^m$.

There holds the following result.

Theorem 5.29 *Let $L(x; \lambda, \mu)$ be the Lagrangian function associated with (5.11). For $(x^*; \lambda^*, \mu^*) \in \mathbb{R}^n \times \mathbb{R}^p_+ \times \mathbb{R}^m$, the following two conditions are equivalent:*

(i) *$(x^*; \lambda^*, \mu^*)$ is a saddle point of L.*
(ii) *x^* is an optimal solution of the primal problem (5.11), (λ^*, μ^*) is an optimal solution of the dual problem (5.23), and $p^* = d^*$, where p^* and d^* denote the optimal values of the primal and dual problems, respectively.*

In particular, if one of the conditions holds, then

$$\sup_{(\lambda,\mu)\in\mathbb{R}^p_+\times\mathbb{R}^m} \inf_{x\in\mathbb{R}^n} L(x; \lambda, \mu) = L(x^*; \lambda^*, \mu^*) = \inf_{x\in\mathbb{R}^n} \sup_{(\lambda,\mu)\in\mathbb{R}^p_+\times\mathbb{R}^m} L(x; \lambda, \mu).$$

Proof (i) $\Longrightarrow$ (ii). Since $(x^*; \lambda^*, \mu^*)$ is a saddle point of L, by using Lemma 5.24, we have

$$L(x^*; \lambda^*, \mu^*) = \sup_{(\lambda,\mu)\in\mathbb{R}_+^p\times\mathbb{R}^m} L(x^*; \lambda, \mu) = f(x^*) + \delta_\Omega(x^*) \geq \inf_{x\in\Omega} f(x),$$

$$L(x^*; \lambda^*, \mu^*) = \inf_{x\in\mathbb{R}^n} L(x; \lambda^*, \mu^*) = h(\lambda^*, \mu^*) \leq \sup_{(\lambda,\mu)\in\mathbb{R}_+^p\times\mathbb{R}^m} h(\lambda, \mu).$$

By virtue of the weak duality (5.26), it follows that

$$\begin{aligned} L(x^*; \lambda^*, \mu^*) &= f(x^*) + \delta_\Omega(x^*) \geq \inf_{x\in\Omega} f(x) \geq \sup_{(\lambda,\mu)\in\mathbb{R}_+^p\times\mathbb{R}^m} h(\lambda, \mu) \\ &\geq h(\lambda^*, \mu^*) = L(x^*; \lambda^*, \mu^*). \end{aligned}$$

Thus, all the inequalities in the above equation must be equalities. The desired conclusions thus follow.

(ii) $\Longrightarrow$ (i). By the given conditions and Lemma 5.24, we have

$$\begin{aligned} h(\lambda^*, \mu^*) &= \inf_{x\in\mathbb{R}^n} L(x; \lambda^*, \mu^*) \leq L(x^*; \lambda^*, \mu^*) \leq \sup_{(\lambda,\mu)\in\mathbb{R}_+^p\times\mathbb{R}^m} L(x^*; \lambda, \mu) \\ &= f(x^*) + \delta_\Omega(x^*) = f(x^*) = h(\lambda^*, \mu^*). \end{aligned}$$

Therefore,

$$\begin{aligned} h(\lambda^*, \mu^*) &= \inf_{x\in\mathbb{R}^n} L(x; \lambda^*, \mu^*) = L(x^*; \lambda^*, \mu^*) \\ &= \sup_{(\lambda,\mu)\in\mathbb{R}_+^p\times\mathbb{R}^m} L(x^*; \lambda, \mu) = f(x^*) \end{aligned}$$

which shows that $(x^*; \lambda^*, \mu^*)$ is a saddle point of L. □

Now, we discuss the existence of saddle points for the Lagrangian function L in terms of conditions on the problem (5.11) itself. We use the KKT theory.

Assume Slater's condition is satisfied. If x^* is an optimal solution of (5.11) such that $f_1, g_1, \cdots, g_p$ are continuous at x^*, then Theorem 5.12 implies the existence of $\lambda^* = (\lambda_1^*, \cdots, \lambda_p^*) \in \mathbb{R}^p$ and $\mu^* \in \mathbb{R}^m$ such that

$$0 \in \partial f(x^*) + \sum_{i=1}^{p} \lambda_i^* \partial g_i(x^*) + A^T \mu^*, \tag{5.27}$$

$$\lambda_i^* g_i(x^*) = 0, \quad i = 1, \cdots, p, \tag{5.28}$$

$$\lambda_i^* \geq 0, \quad i = 1, \cdots, p, \tag{5.29}$$

$$Ax^* = b, \quad g_i(x^*) \leq 0, \quad i = 1, \cdots, p. \tag{5.30}$$

In view of (5.27), we have

$$0 \in \partial f(x^*) + \sum_{i=1}^{p} \lambda_i^* \partial g_i(x^*) + A^T \mu^* \subset \partial\Big(f + \sum_{i=1}^{p} \lambda_i^* g_i + \langle \mu^*, A \cdot -b\rangle\Big)(x^*).$$

This implies that

$$\begin{aligned} L(x^*; \lambda^*, \mu^*) &= f(x^*) + \sum_{i=1}^{p} \lambda_i^* g_i(x^*) + \langle \mu^*, Ax^* - b\rangle \\ &\le f(x) + \sum_{i=1}^{p} \lambda_i^* g_i(x) + \langle \mu^*, Ax - b\rangle \\ &= L(x; \lambda^*, \mu^*) \end{aligned}$$

for all $x \in \mathbb{R}^n$. On the other hand, by using (5.28), (5.29), and (5.30), we have

$$L(x^*; \lambda^*, \mu^*) = f(x^*) + \sum_{i=1}^{p} \lambda_i^* g_i(x^*) + \langle \mu^*, Ax^* - b\rangle = f(x^*)$$

and

$$L(x^*; \lambda, \mu) = f(x^*) + \sum_{i=1}^{p} \lambda_i g_i(x^*) + \langle \mu, Ax^* - b\rangle \le f(x^*)$$

for all $(\lambda, \mu) \in \mathbb{R}^p_+ \times \mathbb{R}^m$. Therefore,

$$L(x^*; \lambda, \mu) \le L(x^*; \lambda^*, \mu^*), \quad \forall (\lambda, \mu) \in \mathbb{R}^p_+ \times \mathbb{R}^m.$$

In summary, we obtain the following result.

Theorem 5.30 *Consider the constrained convex minimization problem (5.11) and assume that Slater's condition is satisfied. If x^* is an optimal solution of (5.11) such that $f_1, g_1, \cdots, g_p$ are continuous at x^*, then there exist $\lambda^* \in \mathbb{R}^p_+$ and $\mu^* \in \mathbb{R}^m$ such that $(x^*; \lambda^*, \mu^*)$ is a saddle point of the Lagrangian function L, i.e.,*

$$L(x^*; \lambda, \mu) \le L(x^*, \lambda^*, \mu^*) \le L(x, \lambda^*, \mu^*)$$

for all $x \in \mathbb{R}^n$ and $(\lambda, \mu) \in \mathbb{R}^p_+ \times \mathbb{R}^m$.

We conclude this section by proving a duality result for linear programming in which Slater's condition can be dropped. We need the following lemma.

Lemma 5.31 *Let B be an $m \times n$ matrix. Then, $K := \{B\lambda : \lambda \geq 0\}$ is a closed convex cone.*

Proof We will only prove the closedness since other assertions are trivial. Given an index set $I \subset \{1, \cdots, m\}$, we will use B_I to denote the submatrix of B formed by those columns with indices in I. Given a vector λ, we will also use λ_I to denote the subvector obtained from λ by retaining those components with indices in I.

We first show that for any nonzero $x \in K$, there is an index set $I \subset \{1, \cdots, m\}$ such that $x = B_I\mu$ for some $\mu > 0$ and B_I has full column rank. To see this, we first write $x = B\lambda$ for some $\lambda \geq 0$. Let $J := \{i : \lambda_i > 0\}$. Then, $x = B_J\lambda_J$. If B_J is of full column rank, we are done. Otherwise, there is a nonzero λ_0 such that $B_J\lambda_0 = 0$; by replacing λ_0 by $-\lambda_0$ if necessary, we may assume λ_0 has at least one positive component. Then, $x = B_J(\lambda_J - s\lambda_0)$ for any $s \geq 0$. Take $s > 0$ to be the smallest number such that $\lambda' := \lambda_J - s\lambda_0 \geq 0$ and at least one component of λ' is zero. Define $J' := \{i \in J : \lambda'_i > 0\}$. Then, J' is a proper subset of J and $x = B_{J'}\lambda'_{J'}$. Continuing this procedure with finite many steps, we can obtain the desired index set I and the vector μ.

Now, let $\{x^k\} \subset K$ be a sequence such that $x^k \to \bar{x}$ as $k \to \infty$. We may assume $\bar{x} \neq 0$. Then, $x^k \neq 0$ for large k. By the above argument, each x^k can be written as $x^k = B_{I_k}\mu^k$, where $I_k \subset \{1, \cdots, m\}$, B_{I_k} is of full column rank, and $\mu^k \geq 0$. Since $\{1, \cdots, m\}$ has only finitely many subsets, there exist an index set I and a subsequence $\{k_l\}$ of integers such that $I_{k_l} = I$ for all l. Consequently,

$$x^{k_l} = B_{I_{k_l}}\mu^{k_l} = B_I\mu^{k_l}.$$

Because B_I has full column rank, $B_I^T B_I$ is invertible and thus

$$\mu^{k_l} = (B_I^T B_I)^{-1} B_I^T x^{k_l}.$$

This together with $x^{k_l} \to \bar{x}$ shows that $\mu^{k_l} \to \bar{\mu} := (B_I^T B_I)^{-1} B_I^T \bar{x}$ as $l \to \infty$. Furthermore, $\mu^{k_l} \geq 0$ implies $\bar{\mu} \geq 0$. Finally, by taking $l \to \infty$ in the equation $x^{k_l} = B_I\mu^{k_l}$, we obtain $\bar{x} = B_I\bar{\mu}$ with $\bar{\mu} \geq 0$. Therefore, $\bar{x} \in K$. □

Now, we consider the linear programming problem (5.24) whose dual problem is given by (5.25); see Example 5.26. We have the following result.

Theorem 5.32 *Consider the linear programming (5.24) and its dual problem (5.25). If either of them has a finite optimal solution, so does the other and the corresponding values of the objective functions are equal. Consequently, if x^* is an optimal solution of (5.24), then (5.25) has an optimal solution λ^* such that*

$$Ax^* = b, \quad x^* \geq 0, \quad A^T\lambda^* \leq c \quad \textit{and} \quad \langle c, x^*\rangle = \langle b, \lambda^*\rangle.$$

Proof Assume the primal problem (5.24) has a finite optimal solution with optimum γ. Consider the set

$$K := \{(u, r) \in \mathbb{R}^m \times \mathbb{R} : u = tb - Ax,\ r = t\gamma - \langle c, x\rangle,\ x \geq 0,\ t \geq 0\}$$

which is a closed convex cone; see Lemma 5.31. We claim that $(0, 1) \notin K$. If not, then there exist $x_0 \geq 0$ and $t_0 \geq 0$ such that

$$0 = t_0 b - Ax_0 \quad \text{and} \quad 1 = t_0\gamma - \langle c, x_0\rangle.$$

If $t_0 > 0$, then x_0/t_0 is a feasible point of (5.24) and hence $1/t_0 = \gamma - \langle c, x_0/t_0\rangle \leq 0$, which is absurd. If $t_0 = 0$, then $Ax_0 = 0$ and $\langle c, x_0\rangle = -1$. Let x be a feasible point of (5.24). Then, $x + sx_0$ is also a feasible point of (5.24) for any $s \geq 0$. Since $\langle c, x + sx_0\rangle = \langle c, x\rangle - s \to -\infty$ as $s \to \infty$, we obtain a contradiction to the assumption that (5.24) has a finite optimal solution.

Since $(0, 1) \notin K$, we may use Corollary 2.18 to strictly separate $(0, 1)$ and K by a hyperplane, i.e., there exists nonzero $(\xi, \alpha) \in \mathbb{R}^m \times \mathbb{R}$ such that

$$\alpha < c := \inf\{\langle \xi, u\rangle + \alpha r : (u, r) \in K\}. \tag{5.31}$$

Since $(0, 0) \in K$, we have $c \leq 0$. Since K is a cone, for any $(u, r) \in K$, we have $(su, sr) \in K$ for any $s \geq 0$. Thus,

$$s(\langle \xi, u\rangle + \alpha r) = \langle \xi, su\rangle + \alpha(sr) > \alpha$$

which implies $\langle \xi, u\rangle + \alpha r \geq \alpha/s$ for any $(u, r) \in K$ and $s > 0$. Thus, $c \geq 0$. Consequently, $c = 0$ and $\alpha < 0$. By rescaling, we may assume $\alpha = -1$. It then follows from (5.31) that $\langle \xi, u\rangle - r \geq 0$ for all $(u, r) \in K$, i.e.,

$$t(\langle \xi, b\rangle - \gamma) - (\langle \xi, Ax\rangle - \langle c, x\rangle) \geq 0, \quad \forall x \geq 0,\ t \geq 0. \tag{5.32}$$

Taking $t = 0$ in (5.32) gives $\langle A^T\xi - c, x\rangle \leq 0$ for all $x \geq 0$, which implies $A^T\xi \leq c$, i.e., ξ is a feasible point of the dual problem (5.25). By the weak duality, we have $\langle b, \xi\rangle \leq \gamma$. On the other hand, taking $x = 0$ and $t = 1$ in (5.32) gives $\langle b, \xi\rangle \geq \gamma$. Therefore, $\langle b, \xi\rangle = \gamma$ and ξ is an optimal solution of the dual problem (5.25). □

As an application, we now prove that the distance between two disjoint polyhedral convex sets must be positive.

Proposition 5.33 *Let $P = \{x \in \mathbb{R}^n : Ax \leq b\}$ and $Q = \{x \in \mathbb{R}^n : Cx \leq d\}$ be two polyhedral convex sets. If $P \cap Q = \emptyset$, then P and Q can be strongly separated by hyperplanes and hence $d(P, Q) > 0$.*

Proof Consider the following two linear programming problems:

$$\max\{\langle 0, x\rangle : Ax \leq b,\ Cx \leq d\}$$

and

$$\min\{\langle b, \lambda\rangle + \langle d, \mu\rangle : A^T\lambda + C^T\mu = 0,\ \lambda \geq 0,\ \mu \geq 0\}. \tag{5.33}$$

It is easy to see that they are dual to each other. We claim that there must exist (λ^*, μ^*) such that

$$\alpha := \langle b, \lambda^*\rangle + \langle d, \mu^*\rangle < 0, \quad A^T\lambda^* + C^T\mu^* = 0, \quad \lambda^* \geq 0, \quad \mu^* \geq 0. \tag{5.34}$$

If not, then (5.33) attains its minimum 0 at the feasible point (0, 0). According to Theorem 5.32, there must exist x^* such that $Ax^* \leq b$ and $Cx^* \leq d$, contradicting the assumption $P \cap Q = \emptyset$. Based on (5.34), we have for any $x \in P$ and $y \in Q$ that

$$\begin{aligned} \alpha &= \langle b, \lambda^*\rangle + \langle d, \mu^*\rangle \geq \langle Ax, \lambda^*\rangle + \langle Cy, \mu^*\rangle \\ &= \langle A^T\lambda^*, x\rangle + \langle C^T\mu^*, y\rangle = \langle \xi, x - y\rangle, \end{aligned}$$

where $\xi := A^T\lambda^*$. This shows that

$$\sup_{x\in P}\langle \xi, x\rangle \leq \alpha + \inf_{y\in Q}\langle \xi, y\rangle < \inf_{y\in Q}\langle \xi, y\rangle$$

which means P and Q can be strongly separated by hyperplanes and hence $d(P, Q) > 0$. □

5.7 Minimax Theorem

In this section, we will generalize the discussion in the previous section and develop a duality theory in a more general setting. Let $\Phi : X \times \Lambda \to \mathbb{R}$ be a function with $X \subset \mathbb{R}^n$ and $\Lambda \subset \mathbb{R}^m$. We introduce the functions

$$f(x) := \sup_{\lambda\in\Lambda} \Phi(x, \lambda), \qquad h(\lambda) := \inf_{x\in X} \Phi(x, \lambda) \tag{5.35}$$

which are called *primal function* and *dual function*, respectively. Correspondingly, we introduce the optimization problems:

$$\min_{x\in X} f(x) \qquad \text{and} \qquad \max_{\lambda\in\Lambda} h(\lambda)$$

which are called *primal problem* and *dual problem*, respectively. By the definitions of f and h, we have

$$f(x) \geq h(\lambda) \quad \text{for all } x \in X \text{ and } \lambda \in \Lambda.$$

Thus, by letting p^* and d^* be the optimal values of the primal problem and dual problem, respectively, we have $p^* \geq d^*$, i.e.,

$$\sup_{\lambda \in \Lambda} \inf_{x \in X} \Phi(x, \lambda) \leq \inf_{x \in X} \sup_{\lambda \in \Lambda} \Phi(x, \lambda)$$

which is called the *weak duality* result. The value $p^* - d^*$ is called the *duality gap*.

A natural question is under what conditions we have $p^* = d^*$.

Definition 5.34 (Saddle Points) A point $(x^*, \lambda^*) \in X \times \Lambda$ is called a saddle point of Φ if

$$\Phi(x^*, \lambda) \leq \Phi(x^*, \lambda^*) \leq \Phi(x, \lambda^*), \quad \forall (x, \lambda) \in X \times \Lambda.$$

Theorem 5.35 *For $(x^*, \lambda^*) \in X \times \Lambda$, the following two conditions are equivalent:*

(i) *(x^*, λ^*) is a saddle point of Φ.*
(ii) *f attains its minimum at x^*, h attains its maximum at λ^*, and $\inf_{x \in X} f(x) = \sup_{\lambda \in \Lambda} h(\lambda)$.*

In particular, if one of the conditions holds, then

$$\sup_{\lambda \in \Lambda} \inf_{x \in X} \Phi(x, \lambda) = \Phi(x^*, \lambda^*) = \inf_{x \in X} \sup_{\lambda \in \Lambda} \Phi(x, \lambda).$$

Proof The proof follows the same lines as in the proof of Theorem 5.29. □

Corollary 5.36 *For a function $\Phi : X \times \Lambda \to \mathbb{R}$, define f and h as in (5.35). Assume that f attains its minimum at $x^* \in X$, h attains its maximum at $\lambda^* \in \Lambda$, and*

$$\min_{x \in X} f(x) = \max_{\lambda \in \Lambda} h(\lambda).$$

Assume further that for each $\lambda \in \Lambda$, there is a unique $x_\lambda \in X$ with $h(\lambda) = \Phi(x_\lambda, \lambda)$. Then, $x_{\lambda^*} = x^*$, i.e., x_{λ^*} is the optimal solution of $\min_{x \in X} f(x)$.

Proof By the given conditions and Theorem 5.35, (x^*, λ^*) is a saddle point of Φ, and in particular

$$\Phi(x^*, \lambda^*) = \max_{\lambda \in \Lambda} h(\lambda) = h(\lambda^*).$$

Since $x_{\lambda^*} \in X$ is the unique element with $h(\lambda^*) = \Phi(x_{\lambda^*}, \lambda^*)$, we must have $x^* = x_{\lambda^*}$. □

Corollary 5.36 shows how to determine the optimal solution of the primal problem by finding the solution of the dual problem.

Next, we will provide some results to guarantee the strong duality, i.e., $p^* = d^*$, in terms of conditions on Φ.

Theorem 5.37 *Let $X \subset \mathbb{R}^n$ be a compact convex set and let $\Lambda \subset \mathbb{R}^m$ be a convex set. Let $\Phi : X \times \Lambda \to (-\infty, \infty)$ be such that:*

(i) *$x \to \Phi(x, \lambda)$ is convex and lower semi-continuous for each $\lambda \in \Lambda$,*
(ii) *$\lambda \to \Phi(x, \lambda)$ is concave for each $x \in X$.*

Then, there holds

$$\inf_{x \in X} \sup_{\lambda \in \Lambda} \Phi(x, \lambda) = \sup_{\lambda \in \Lambda} \inf_{x \in X} \Phi(x, \lambda).$$

Replacing Φ by $-\Phi$, we can obtain the following equivalent result.

Theorem 5.38 *Let $X \subset \mathbb{R}^n$ be a convex set and let $\Lambda \subset \mathbb{R}^m$ be a compact convex set. Let $\Phi : X \times \Lambda \to (-\infty, \infty)$ be such that:*

(i) *$x \to \Phi(x, \lambda)$ is convex for each $\lambda \in \Lambda$,*
(ii) *$\lambda \to \Phi(x, \lambda)$ is concave and upper semi-continuous for each $x \in X$.*

Then, there holds

$$\inf_{x \in X} \sup_{\lambda \in \Lambda} \Phi(x, \lambda) = \sup_{\lambda \in \Lambda} \inf_{x \in X} \Phi(x, \lambda).$$

We will follow [209] to provide a proof of Theorem 5.37. Let us start with the following lemma.

Lemma 5.39 *Let $X \subset \mathbb{R}^n$ be a compact set and let $\mathcal{F}$ be a set consisting of lower semi-continuous functions defined on X. Then, the following are equivalent:*

(i) *For any finite subset $\mathcal{G} \subset \mathcal{F}$ and any $\alpha \in \mathbb{R}$ with*

$$\alpha < \min_{x \in X} \max_{g \in \mathcal{G}} g(x),$$

there exists $h \in \mathcal{F}$ with $\alpha \leq \min_{x \in X} h(x)$.

(ii) $\sup_{f \in \mathcal{F}} \min_{x \in X} f(x) = \min_{x \in X} \sup_{f \in \mathcal{F}} f(x).$

Proof (i) $\Longrightarrow$ (ii). For any $\alpha < \min_{x \in X} \sup_{f \in \mathcal{F}} f(x)$, define

$$A(f) = \{x \in X : f(x) \leq \alpha\}, \quad \forall f \in \mathcal{F}.$$

Then, $\bigcap_{f \in \mathcal{F}} A(f) = \emptyset$ and thus

$$X \subset \bigcup_{f \in \mathcal{F}} A(f)^c.$$

Since each $f \in \mathcal{F}$ is lower semi-continuous, $A(f)$ is closed and hence $A(f)^c$ is open. Because X is compact, we can find a finite subset $\mathcal{G} \subset \mathcal{F}$ such that $X \subset \bigcup_{f \in \mathcal{G}} A(f)^c$ and hence

$$\bigcap_{f \in \mathcal{G}} A(f) = \emptyset.$$

Consequently,

$$\max_{f \in \mathcal{G}} f(x) > \alpha, \quad \forall x \in X.$$

Note that $x \to \max_{f \in \mathcal{G}} f(x)$ is lower semi-continuous and X is compact; thus, there is $\bar{x} \in X$ such that

$$\min_{x \in X} \max_{f \in \mathcal{G}} f(x) = \max_{f \in \mathcal{G}} f(\bar{x}) > \alpha.$$

By (i), there is $h \in \mathcal{F}$ such that

$$\alpha \leq \min_{x \in X} h(x) \leq \sup_{f \in \mathcal{F}} \min_{x \in X} f(x).$$

By the arbitrariness of α, we obtain

$$\min_{x \in X} \sup_{f \in \mathcal{F}} f(x) \leq \sup_{f \in \mathcal{F}} \min_{x \in X} f(x)$$

and hence (ii) since the opposite inequality holds trivially.

(ii) $\Longrightarrow$ (i). By (ii), for any $\alpha \in \mathbb{R}$ and any finite subset $\mathcal{G} \subset \mathcal{F}$ satisfying $\alpha < \min_{x \in X} \max_{g \in \mathcal{G}} g(x)$, we have

$$\alpha < \min_{x \in X} \sup_{f \in \mathcal{F}} f(x) = \sup_{f \in \mathcal{F}} \min_{x \in X} f(x).$$

Thus, there exists $h \in \mathcal{F}$ such that $\alpha \leq \min_{x \in X} h(x)$. □

Proof of Theorem 5.37 Let $\mathcal{F} = \{\Phi(\cdot, \lambda) : \lambda \in \Lambda\}$. It suffices to show that (i) of Lemma 5.39 holds. Take any finite subset $\{\lambda_1, \cdots, \lambda_n\} \subset \Lambda$ and any $\alpha \in \mathbb{R}$ with

$$\alpha < \min_{x \in X} \max_{1 \leq i \leq n} \Phi(x, \lambda_i). \tag{5.36}$$

Let $P \subset \mathbb{R}^n$ be the convex hull of the set:

$$\{(\Phi(x, \lambda_1), \cdots, \Phi(x, \lambda_n)) : x \in X\}.$$

Let $Q \subset \mathbb{R}^n$ be the cone:

$$Q = \{(z_1, \cdots, z_n) : z_i \leq \alpha,\ 1 \leq i \leq n\}.$$

We claim that $P \cap Q = \emptyset$. Indeed, any $z \in P$ is of the form

$$z = \sum_{j=1}^{m} \beta_j (\Phi(x_j, \lambda_1), \cdots, \Phi(x_j, \lambda_n))$$

for some $m \geq 1$, where $x_j \in X$, $\beta_j \geq 0$ for $j = 1, \cdots, m$ and $\sum_{j=1}^{m} \beta_j = 1$. Set $x_0 = \sum_{j=1}^{m} \beta_j x_j$. Then, $x_0 \in X$. According to (5.36), there exists $i \in \{1, \cdots, n\}$ such that $\Phi(x_0, \lambda_i) > \alpha$. Since $x \to \Phi(x, \lambda_i)$ is convex, the i-th component z_i of z satisfies

$$z_i = \sum_{j=1}^{m} \beta_j \Phi(x_j, \lambda_i) \geq \Phi(x_0, \lambda_i) > \alpha.$$

Thus, $z \notin Q$.

The disjoint convex sets P and Q can be separated by a hyperplane, i.e., there is a nonzero vector $\gamma = (\gamma_1, \cdots, \gamma_n)$ with

$$\sup_{z \in Q} \langle \gamma, z \rangle \leq \inf_{w \in P} \langle \gamma, w \rangle.$$

We must have $\gamma_i \geq 0$ for $i = 1, \cdots, n$, for otherwise we have $\gamma_i < 0$ for some i, and we take $z \in Q$ with $z_i \to -\infty$ to get a contradiction. We may assume $\sum_{i=1}^{n} \gamma_i = 1$. Define $\lambda_0 = \sum_{i=1}^{n} \gamma_i \lambda_i$. By the concavity of $\lambda \to \Phi(x, \lambda)$, we have

$$\Phi(x, \lambda_0) \geq \sum_{i=1}^{n} \gamma_i \Phi(x, \lambda_i) \geq \inf_{w \in P} \langle \gamma, w \rangle \geq \sup_{z \in Q} \langle \gamma, z \rangle = \sup_{z \in Q} \sum_{i=1}^{n} \gamma_i z_i = \alpha, \quad \forall x \in X$$

showing that (i) of Lemma 5.39 holds. □

Example 5.40 (Projection Onto ℓ^1-Balls) Finding the Euclidean projection of a vector $v \in \mathbb{R}^n$ onto the ℓ^1-ball $\{x \in \mathbb{R}^n : \|x\|_1 \leq r\}$ of radius $r > 0$ is equivalent to solving the optimization problem:

$$\begin{cases} \text{minimize} & \frac{1}{2}\|x - v\|_2^2 \\ \text{subject to} & \|x\|_1 \leq r. \end{cases} \tag{5.37}$$

When $\|v\|_1 \leq r$, $x = v$ is the obvious solution. In the following, we assume that $\|v\|_1 > r$. Note that the associate Lagrangian function is

$$L(x, \lambda) = \frac{1}{2}\|x - v\|_2^2 + \lambda(\|x\|_1 - r).$$

Thus, the Lagrangian dual function is

$$h(\lambda) = \inf_{x \in \mathbb{R}^n} \left\{ \frac{1}{2}\|x - v\|_2^2 + \lambda(\|x\|_1 - r) \right\} = \sum_{i=1}^{n} h_i(\lambda) - \lambda r,$$

where

$$h_i(\lambda) = \min_{x_i \in \mathbb{R}} \left\{ \frac{1}{2}(x_i - v_i)^2 + \lambda|x_i| \right\}.$$

According to Example 4.14, with $s_\lambda(v_i)$ denoting the soft-thresholding of v_i with parameter λ, we can derive by a straightforward calculation that

$$h_i(\lambda) = \frac{1}{2}(s_\lambda(v_i) - v_i)^2 + \lambda|s_\lambda(v_i)| = \begin{cases} -\frac{1}{2}\lambda^2 + \lambda|v_i| & \text{if } \lambda < |v_i| \\ \frac{1}{2}v_i^2 & \text{if } \lambda \geq |v_i|. \end{cases}$$

Therefore, the dual problem of (5.37) is given by

$$\begin{cases} \text{maximize} & h(\lambda) \\ \text{subject to} & \lambda \geq 0 \end{cases}$$

which is a one-dimensional problem. It is clear that Slater's condition holds for (5.37). Thus, it follows from Theorems 5.30 and 5.29 that the strong duality holds for this minimization problem. Furthermore, for each $\lambda \geq 0$, there is a unique $x_\lambda \in \mathbb{R}^n$ such that $h(\lambda) = L(x_\lambda, \lambda)$ which is given by $x_\lambda = S_\lambda(v)$, where $S_\lambda(v)$ denotes the soft thresholding of v. According to Corollary 5.36, if λ^* is the solution of the dual problem, then $x^* := S_{\lambda^*}(v)$ is the projection of v onto $\{x : \|x\|_1 \leq r\}$.

We need to determine the optimal solution λ^* of the dual problem. Note that $h(0) = 0$. By using $\|v\|_1 > r$, we also have

$$h(\lambda) = \sum_{i:v_i \neq 0} h_i(\lambda) - \lambda r = -\frac{1}{2}\lambda^2|\{i : v_i \neq 0\}| + \lambda(\|v\|_1 - r) > 0$$

for sufficiently small $\lambda > 0$. Therefore, we must have $\lambda^* > 0$. Noting that h is a differentiable function on $(0, \infty)$, we thus have $h'(\lambda^*) = 0$. Direct calculation shows that

$$h'(\lambda) = \sum_{i=1}^{n} \max\{|v_i| - \lambda, 0\} - r.$$

Therefore, λ^* must be the root of the equation:

$$\sum_{i=1}^{n} \max\{|v_i| - \lambda, 0\} - r = 0.$$

Sort the components of v in magnitude in descending order to form a vector u such that

$$|u_1| \geq |u_2| \geq \cdots \geq |u_n|.$$

Note that $h'(|u_1|) = -r < 0$. Let $j \in \{1, \cdots, n\}$ be the largest integer such that $h'(|u_j|) < 0$. Then, $0 < \lambda^* < |u_j|$ and thus

$$0 = \sum_{i=1}^{j} \max\{|u_i| - \lambda^*, 0\} - r = \sum_{i=1}^{j} |u_i| - r - j\lambda^*.$$

Therefore,

$$\lambda^* = \frac{1}{j}\left(\sum_{i=1}^{j} |u_i| - r\right) \tag{5.38}$$

and consequently $x^* = S_{\lambda^*}(v)$ is the projection of v onto $\{x : \|x\|_1 \leq r\}$. This leads to the following algorithm.

Algorithm 5.41 (Projection Onto ℓ^1-Balls) *Given $v \in \mathbb{R}^n$ and $r > 0$, if $\|v\|_1 \leq r$, set $x^* = v$. If $\|v\|_1 > r$, do the following:*

(i) *Sort the components of v in magnitude in descending order to form a new vector u with $|u_1| \geq |u_2| \geq \cdots \geq |u_n|$.*
(ii) *Find the largest integer $j \in \{1, \cdots, n\}$ such that $h'(|u_j|) < 0$.*
(iii) *Define λ^* by (5.38) and set $x^* := S_{\lambda^*}(v)$.*

5.8 Dual Proximal Gradient Method

We consider the minimization problem of the form

$$\min_{x \in \mathbb{R}^n} \{f(x) + g(Ax)\}, \tag{5.39}$$

where f, g, and A satisfy the following conditions.

Assumption 5.42

(i) $f : \mathbb{R}^n \to (-\infty, \infty]$ *is proper, lower semi-continuous, and strongly convex with modulus of convexity* $\mu > 0$.
(ii) $g : \mathbb{R}^m \to (-\infty, \infty]$ *is proper, lower semi-continuous, and convex.*
(iii) $A : \mathbb{R}^n \to \mathbb{R}^m$ *is a linear operator, and there is* $x_0 \in \mathrm{dom}(f)$ *such that* g *is continuous at* Ax_0.

Under Assumption 5.42, we may use Lemma 5.5 and Corollary 1.15 to conclude that (5.39) has a unique solution x^*. Since neither f nor g is assumed to be differentiable, we cannot use the proximal gradient method to solve it. Nevertheless, it turns out that its dual problem can be solved by a proximal gradient method as will be shown below.

To determine its dual problem, we reformulate (5.39) equivalently as

$$\min_{x\in\mathbb{R}^n, y\in\mathbb{R}^m} \{f(x) + g(y) : Ax - y = 0\}.$$

The Lagrangian function is

$$L(x, y, \lambda) = f(x) + g(y) + \langle \lambda, Ax - y\rangle, \quad x \in \mathbb{R}^n, y \in \mathbb{R}^m, \lambda \in \mathbb{R}^m$$

and hence the dual function is

$$\begin{aligned}\inf_{x\in\mathbb{R}^n, y\in\mathbb{R}^m} L(x, y, \lambda) &= \inf_{x\in\mathbb{R}^n} \left\{f(x) + \langle A^T\lambda, x\rangle\right\} + \inf_{y\in\mathbb{R}^m} \{g(y) - \langle \lambda, y\rangle\} \\ &= -\sup_{x\in\mathbb{R}^n} \left\{\langle -A^T\lambda, x\rangle - f(x)\right\} - \sup_{y\in\mathbb{R}^m} \{\langle \lambda, y\rangle - g(y)\} \\ &= -f^*(-A^T\lambda) - g^*(\lambda),\end{aligned}$$

where f^* and g^* denote the Legendre-Fenchel conjugates of f and g, respectively. Thus, the dual problem is

$$\min_{\lambda\in\mathbb{R}^m} \left\{d(\lambda) := f^*(-A^T\lambda) + g^*(\lambda)\right\}. \tag{5.40}$$

Since f is strongly convex, it follows from Theorem 5.8 that f^* is continuous differentiable and so is $f^*(-A^T\lambda)$ with gradient $-A\nabla f^*(-A^T\lambda)$; moreover,

$$\begin{aligned}\|(-A\nabla f^*(-A^T\hat{\lambda}))-(-A\nabla f^*(-A^T\lambda))\| &\le \|A\|\|\nabla f^*(-A^T\hat{\lambda})-\nabla f^*(-A^T\lambda)\| \\ &\le \frac{\|A\|}{2\mu}\|A^T\hat{\lambda} - A^T\lambda\| \\ &\le \frac{\|A\|^2}{2\mu}\|\hat{\lambda} - \lambda\|\end{aligned}$$

for all $\hat{\lambda}, \lambda \in \mathbb{R}^m$. Thus, we may use the proximal gradient method to solve (5.40), leading to the algorithm:

$$\lambda^{k+1} = \text{prox}_{tg^*}\left(\lambda^k + tA\nabla f^*(-A^T\lambda^k)\right).$$

Let $x^k = \nabla f^*(-A^T\lambda^k)$. Then, $-A^T\lambda^k \in \partial f(x^k)$, which implies that

$$x^k = \arg\min_{x\in\mathbb{R}^n}\left\{f(x) + \langle\lambda^k, Ax\rangle\right\}.$$

Therefore, we obtain the following dual proximal gradient method for solving (5.39) in case prox_{tg^*} can be easily calculated.

Algorithm 5.43 *Take an initial guess $\lambda^0 \in \mathbb{R}^m$. For $k \geq 0$, do*

$$\begin{aligned} x^k &= \arg\min_{x\in\mathbb{R}^n}\left\{f(x) + \langle\lambda^k, Ax\rangle\right\}, \\ \lambda^{k+1} &= \text{prox}_{tg^*}\left(\lambda^k + tAx^k\right), \end{aligned}$$

where $t > 0$ is a step size.

If the proximal mapping prox_{tg^*} is difficult to calculate, we may use the Moreau decomposition formula, i.e., Theorem 4.25, to obtain

$$\lambda^{k+1} = \text{prox}_{tg^*}\left(\lambda^k + tAx^k\right) = \lambda^k + tAx^k - t\text{prox}_{g/t}(\lambda^k/t + Ax^k).$$

Let $y^k = \text{prox}_{g/t}(\lambda^k/t + Ax^k)$. Then,

$$y^k = \arg\min_{y\in\mathbb{R}^m}\left\{g(y) + \langle\lambda^k, Ax^k - y\rangle + \frac{t}{2}\|Ax^k - y\|^2\right\}.$$

Consequently, we obtain the following alternative version of the dual proximal gradient method.

Algorithm 5.44 *Take an initial guess $\lambda^0 \in \mathbb{R}^m$. For $k \geq 0$, do*

$$\begin{aligned} x^k &= \arg\min_{x\in\mathbb{R}^n}\left\{f(x) + \langle\lambda^k, Ax\rangle\right\}, \\ y^k &= \arg\min_{y\in\mathbb{R}^m}\left\{g(y) + \langle\lambda^k, Ax^k - y\rangle + \frac{t}{2}\|Ax^k - y\|^2\right\}, \\ \lambda^{k+1} &= \lambda^k + t(Ax^k - y^k), \end{aligned}$$

where $t > 0$ is a step size.

We have the following convergence result concerning the dual proximal gradient method.

Theorem 5.45 *Let Assumption 5.42 hold and let x^* be the unique minimizer of (5.39). Then, for Algorithm 5.44 with $0 < t \leq 2\mu/\|A\|^2$, there holds*

$$\|x^k - x^*\| = O(1/\sqrt{k}).$$

Proof Under Assumption 5.42, we have from Theorem 3.9, Theorem 3.13, and Theorem 3.14 that

$$0 \in \partial(f + g(A\cdot))(x^*) = \partial f(x^*) + A^T \partial g(Ax^*).$$

This implies that there is $\lambda^* \in \partial g(Ax^*)$ such that $-A^T\lambda^* \in \partial f(x^*)$. By the definition of x^k, we have $-A^T\lambda^k \in \partial f(x^k)$. Let

$$\Delta_k := f(x^*) - f(x^k) - \langle -A^T\lambda^k, x^* - x^k\rangle.$$

It follows from Proposition 3.12 that

$$\Delta_k \geq \mu\|x^k - x^*\|^2.$$

On the other hand, by using $-A^T\lambda^k \in \partial f(x^k)$ and $-A^T\lambda^* \in \partial f(x^*)$, we have from Proposition 4.6 that

$$f(x^k) + f^*(-A^T\lambda^k) = \langle -A^T\lambda^k, x^k\rangle,$$
$$f(x^*) + f^*(-A^T\lambda^*) = \langle -A^T\lambda^*, x^*\rangle.$$

Therefore,

$$\begin{aligned}\Delta_k &= \left[\langle -A^T\lambda^*, x^*\rangle - f^*(-A^T\lambda^*)\right] - \left[\langle -A^T\lambda^k, x^k\rangle - f^*(-A^T\lambda^k)\right] \\ &\quad - \langle -A^T\lambda^k, x^* - x^k\rangle \\ &= f^*(-A^T\lambda^k) - f^*(-A^T\lambda^*) + \langle A^T(\lambda^k - \lambda^*), x^*\rangle \\ &= d(\lambda^k) - d(\lambda^*) - \left[g^*(\lambda^k) - g^*(\lambda^*) - \langle Ax^*, \lambda^k - \lambda^*\rangle\right].\end{aligned}$$

Since $\lambda^* \in \partial g(Ax^*)$ implies $Ax^* \in \partial g^*(\lambda^*)$, we have

$$g^*(\lambda^k) - g^*(\lambda^*) - \langle Ax^*, \lambda^k - \lambda^*\rangle \geq 0.$$

Thus,

$$\mu\|x^k - x^*\|^2 \leq \Delta_k \leq d(\lambda^k) - d(\lambda^*).$$

Since $\lambda^* \in \partial g(Ax^*)$ and $-A^T\lambda^* \in \partial f(x^*)$, we have $x^* = \nabla f^*(-A^T\lambda^*)$ and $Ax^* \in \partial g^*(\lambda^*)$. Hence,

$$0 \in -Ax^* + \partial g(\lambda^*) = -A\nabla f^*(-A^T\lambda^*) + \partial g^*(\lambda^*) = \partial d(\lambda^*)$$

which shows that λ^* is a minimizer of d over $\mathbb{R}^m$. Since $\{\lambda^k\}$ is obtained by applying the proximal gradient method to the dual problem, from Theorem 4.35, it follows that

$$d(\lambda^k) - d(\lambda^*) = O(1/k).$$

Consequently, $\|x^k - x^*\|^2 = O(1/k)$. □

In order to produce a method with faster convergence speed, we may apply the accelerated proximal gradient method to the dual problem and obtain

$$\begin{aligned}
\hat{\lambda}^k &= \lambda^k + \frac{k-1}{k+2}(\lambda^k - \lambda^{k-1}),\\
\lambda^{k+1} &= \text{prox}_{tg^*}\left(\hat{\lambda}^k + tA\nabla f^*(-A^T\hat{\lambda}^k)\right).
\end{aligned}$$

We then output

$$x^{k+1} := \nabla f^*(-A^T\lambda^{k+1})$$

as an approximate solution of (5.39). By introducing $\hat{x}^k := \nabla f^*(-A^T\hat{\lambda}^k)$, we thus obtain the following accelerated dual proximal gradient method for solving (5.39); see [21].

Algorithm 5.46 *Take an initial guess $\lambda^0 \in \mathbb{R}^m$ and set $\lambda^{-1} = \lambda^0$. For $k \geq 0$, do*

$$\begin{aligned}
\hat{\lambda}^k &= \lambda^k + \frac{k-1}{k+2}(\lambda^k - \lambda^{k-1}),\\
\hat{x}^k &= \arg\min_{x\in\mathbb{R}^n}\left\{f(x) + \langle\hat{\lambda}^k, Ax\rangle\right\},\\
\lambda^{k+1} &= \text{prox}_{tg^*}\left(\hat{\lambda}^k + tA\hat{x}^k\right),\\
x^{k+1} &= \arg\min_{x\in\mathbb{R}^n}\left\{f(x) + \langle\lambda^{k+1}, Ax\rangle\right\},
\end{aligned}$$

where $t > 0$ is a step size.

In case prox_{tg^*} is not easy to calculate, we may use the Moreau decomposition formula to write $\lambda^{k+1} = \hat{\lambda}^k + t(A\hat{x}^k - \hat{y}^k)$, where

$$\hat{y}^k := \text{prox}_{g/t}(\hat{\lambda}^k/t + A\hat{x}^k).$$

This leads to the following alternative version of the accelerated proximal gradient method for solving (5.39).

Algorithm 5.47 *Take an initial guess* $\lambda^0 \in \mathbb{R}^m$ *and set* $\lambda^{-1} = \lambda^0$. *For* $k \geq 0$, *do*

$$
\begin{aligned}
\hat{\lambda}^k &= \lambda^k + \frac{k-1}{k+2}(\lambda^k - \lambda^{k-1}), \\
\hat{x}^k &= \arg\min_{x \in \mathbb{R}^n} \left\{ f(x) + \langle \hat{\lambda}^k, Ax \rangle \right\}, \\
\hat{y}^k &= \arg\min_{y \in \mathbb{R}^m} \left\{ g(y) + \langle \hat{\lambda}^k, A\hat{x}^k - y \rangle + \frac{t}{2}\|A\hat{x}^k - y\|^2 \right\}, \\
\lambda^{k+1} &= \hat{\lambda}^k + t(A\hat{x}^k - \hat{y}^k), \\
x^{k+1} &= \arg\min_{x \in \mathbb{R}^n} \left\{ f(x) + \langle \lambda^{k+1}, Ax \rangle \right\},
\end{aligned}
$$

where $t > 0$ *is a step size.*

There holds the following convergence result concerning the accelerated dual proximal gradient method for solving (5.39).

Theorem 5.48 *Let Assumption 5.42 hold and let* x^* *be the unique minimizer of (5.39). Then, for Algorithm 5.47 with* $0 < t \leq 2\mu/\|A\|^2$, *there holds*

$$\|x^k - x^*\| = O(1/k).$$

Proof As in the proof of Theorem 5.45, let λ^* be such that $\lambda^* \in \partial g(Ax^*)$ and $-A^T\lambda^* \in \partial f(x^*)$. Then, λ^* is a minimizer of the dual function d over $\mathbb{R}^m$, and, by the fact $-A^T\lambda^k \in \partial f(x^k)$ for $k \geq 1$, we have

$$\mu\|x^k - x^*\|^2 \leq d(\lambda^k) - d(\lambda^*)$$

Since $\{\lambda^k\}$ is obtained by applying the accelerated proximal gradient method to the dual problem, we have from Theorem 4.43 that

$$d(\lambda^k) - d(\lambda^*) = O(1/k^2).$$

Consequently, $\|x^k - x^*\| = O(1/k)$. □

Now, we provide various applications of the (accelerated) dual proximal gradient method.

Example 5.49 Consider the convex minimization problem:

$$
\begin{aligned}
&\text{minimize} \quad f(x) \\
&\text{subject to} \quad Ax = b \quad \text{and} \quad Cx \leq d,
\end{aligned}
\tag{5.41}
$$

where A and C are matrices of size $m \times n$ and $p \times n$, respectively, $b \in \mathbb{R}^m$, $d \in \mathbb{R}^p$, and $f : \mathbb{R}^n \to (-\infty, \infty]$ is a proper, lower semi-continuous, strongly convex function. Let $\Omega := \{z \in \mathbb{R}^p : z \leq 0\}$ and define the linear map $L : \mathbb{R}^n \to \mathbb{R}^m \times \mathbb{R}^p$ by

$$Lx := (Ax, Cx), \quad \forall x \in \mathbb{R}^n.$$

Then, (5.41) can be written in the form (5.39) with

$$g(y, z) = \delta_{\{b\}}(y) + \delta_\Omega(z - d).$$

Note that $L^T : \mathbb{R}^m \times \mathbb{R}^p \to \mathbb{R}^n$ is given by

$$L^T(y, z) = A^T y + C^T z, \quad \forall (y, z) \in \mathbb{R}^m \times \mathbb{R}^p.$$

Moreover,

$$\mathrm{prox}_{g/t}(y, z) = (b, d + P_\Omega(z - d)), \quad \forall (y, z) \in \mathbb{R}^m \times \mathbb{R}^p.$$

Therefore, by applying Algorithm 5.44 to solve (5.41), after cancelling y^k, we obtain the following algorithm:

$$\begin{aligned}
x^k &= \arg\min_{x \in \mathbb{R}^n} \left\{ f(x) + \langle \lambda^k, Ax \rangle + \langle \mu^k, Cx \rangle \right\}, \\
\lambda^{k+1} &= \lambda^k + t(Ax^k - b), \\
\mu^{k+1} &= [\mu^k + t(Cx^k - d)]_+,
\end{aligned}$$

where for each $z \in \mathbb{R}^p$, we used $z - P_\Omega(z) = z_+$ with z_+ denoting the projection of z onto the first quadrant. Applying Algorithm 5.47 to solve (5.41) gives the following accelerated algorithm:

$$\begin{aligned}
\hat{\lambda}^k &= \lambda^k + \frac{k-1}{k+2}(\lambda^k - \lambda^{k-1}), \\
\hat{\mu}^k &= \mu^k + \frac{k-1}{k+2}(\mu^k - \mu^{k-1}), \\
\hat{x}^k &= \arg\min_{x \in \mathbb{R}^n} \left\{ f(x) + \langle \hat{\lambda}^k, Ax \rangle + \langle \hat{\mu}^k, Cx \rangle \right\}, \\
\lambda^{k+1} &= \hat{\lambda}^k + t(A\hat{x}^k - b), \\
\mu^{k+1} &= [\hat{\mu}^k + t(C\hat{x}^k - d)]_+, \\
x^{k+1} &= \arg\min_{x \in \mathbb{R}^n} \left\{ f(x) + \langle \lambda^{k+1}, Ax \rangle + \langle \mu^{k+1}, Cx \rangle \right\}.
\end{aligned}$$

Note that

$$\|Lx\|^2 = \|Ax\|^2 + \|Cx\|^2 \le (\|A\|^2 + \|C\|^2)\|x\|^2.$$

We have $\|L\|^2 \le \|A\|^2 + \|C\|^2$. Thus, the above two methods with $0 < t \le 2\mu/(\|A\|^2 + \|C\|^2)$ possess the convergence rates given in Theorem 5.45 and Theorem 5.48, respectively, where $\mu > 0$ is the modulus of convexity of f.

Example 5.50 (One-Dimensional Total Variation Denoising) Given a noisy data $b \in \mathbb{R}^n$, consider the total variation denoising problem:

$$\hat{x} = \arg\min_{x\in\mathbb{R}^n} \left\{ F(x) = \frac{1}{2}\|x - b\|_2^2 + \alpha \sum_{i=1}^{n-1} |x_{i+1} - x_i| \right\}. \tag{5.42}$$

It can be written as

$$\hat{x} = \arg\min_{x\in\mathbb{R}^n} \left\{ F(x) = \frac{1}{2}\|x - b\|_2^2 + \alpha \|Dx\|_1 \right\},$$

where $D : \mathbb{R}^n \to \mathbb{R}^{n-1}$ is the linear operator defined by

$$Dx := (x_2 - x_1, \cdots, x_n - x_{n-1})^T, \quad \forall x \in \mathbb{R}^n.$$

It is easy to see that $D^T : \mathbb{R}^{n-1} \to \mathbb{R}^n$, the transpose of D, is given by

$$D^T\lambda = (\lambda_0 - \lambda_1, \lambda_1 - \lambda_2, \cdots, \lambda_{n-1} - \lambda_n)^T, \quad \forall \lambda \in \mathbb{R}^{n-1},$$

where we set $\lambda_0 = \lambda_n = 0$.

The problem (5.42) takes the form (5.39) with

$$f(x) = \frac{1}{2}\|x - b\|^2 \quad \text{and} \quad g(y) = \alpha\|y\|_1.$$

Applying the dual proximal gradient method given in Algorithm 5.43 to solve (5.42) gives

$$x^k = b - D^T\lambda^k, \quad \lambda^{k+1} = \text{prox}_{tg^*}(\lambda^k + tDx^k).$$

Since $g(y) = \alpha\|y\|_1$, g^* is the indicator function of the convex set,

$$\{v \in \mathbb{R}^m : \|v\|_\infty \le \alpha\}$$

and hence prox_{tg^*} is the Euclidean projection onto this set. Therefore,

$$\lambda^{k+1} = \max\left\{\min\left\{\lambda^k + tDx^k, \alpha\right\}, -\alpha\right\}.$$

Thus, we obtain the algorithm:

$$\begin{aligned} x^k &= b - D^T\lambda^k, \\ \lambda^{k+1} &= \max\left\{\min\left\{\lambda^k + tDx^k, \alpha\right\}, -\alpha\right\} \end{aligned}$$

with $0 < t \le 1/\|D\|^2$. Since

$$\|Dx\|^2 = \sum_{i=1}^{n-1}(x_{i+1} - x_i)^2 \le 2\sum_{i=1}^{n-1}(x_{i+1}^2 + x_i^2) \le 4\|x\|^2$$

for any $x \in \mathbb{R}^n$, we have $\|D\|^2 \le 4$. Therefore, we may take $t = 1/4$ and obtain the following algorithm:

Algorithm 5.51 *Let $\lambda^0 = 0$ and for $k \ge 0$, do*

$$\begin{aligned} x^k &= b - D^T\lambda^k, \\ \lambda^{k+1} &= \max\left(\min\left(\lambda^k + \frac{1}{4}Dx^k, \alpha\right), -\alpha\right) \end{aligned}$$

until a stopping rule is met. The last x^k is used as an approximation.

By using the accelerated dual proximal gradient method given in Algorithm 5.46, we can obtain the following accelerated algorithm:

Algorithm 5.52 *Let $\lambda^0 = \lambda^{-1} = 0$. For $k \ge 0$, do*

$$\begin{aligned} \hat{\lambda}^k &= \lambda^k + \frac{k-1}{k+2}(\lambda^k - \lambda^{k-1}), \\ \hat{x}^k &= b - D^T\hat{\lambda}^k, \\ \lambda^{k+1} &= \max\left(\min\left(\hat{\lambda}^k + \frac{1}{4}D\hat{x}^k, \alpha\right), -\alpha\right), \\ x^{k+1} &= b - D^T\lambda^{k+1} \end{aligned}$$

until a stopping criterion is satisfied. We then use the last x^k as an approximate solution.

Example 5.53 (Two-Dimensional Total Variation Denoising) Consider an image of size $m \times n$ corrupted by Gaussian noise. Assume, instead of the original image, only a noisy image b is available. Can we reconstruct the original image from b as accurate as possible? Many models have been proposed to answer this question. The total variation model answers this question by solving the minimization problem:

$$\min_{x\in\mathbb{R}^{m\times n}}\left\{F(x) := \frac{1}{2}\|x-b\|_F^2 + \alpha|x|_{TV}\right\}, \tag{5.43}$$

where $\|x\|_F$ and $|x|_{TV}$ denote the Frobenius norm and the total variation of x, respectively, i.e.,

$$\|x\|_F^2 = \sum_{i=1}^{m}\sum_{j=1}^{n}|x_{i,j}|^2$$

and

$$\begin{aligned}|x|_{TV} = &\sum_{i=1}^{m-1}\sum_{j=1}^{n-1}\sqrt{(x_{i,j}-x_{i+1,j})^2+(x_{i,j}-x_{i,j+1})^2}\\ &+\sum_{i=1}^{m-1}|x_{i,n}-x_{i+1,n}|+\sum_{j=1}^{n-1}|x_{m,j}-x_{m,j+1}|.\end{aligned}$$

We introduce the operator $D:\mathbb{R}^{m\times n}\to\mathbb{R}^{(m-1)\times n}\times\mathbb{R}^{m\times(n-1)}$ by

$$Dx = (p,q) \quad \text{for each } x\in\mathbb{R}^{m\times n},$$

where $p\in\mathbb{R}^{(m-1)\times n}$ and $q\in\mathbb{R}^{m\times(n-1)}$ are matrices defined by

$$\begin{aligned}p_{i,j} &= x_{i,j}-x_{i+1,j}, \quad i=1,\cdots,m-1;\ j=1,\cdots,n,\\ q_{i,j} &= x_{i,j}-x_{i,j+1}, \quad i=1,\cdots,m;\ j=1,\cdots,n-1,\end{aligned}$$

We also introduce the function $g:\mathbb{R}^{(m-1)\times n}\times\mathbb{R}^{m\times(n-1)}\to\mathbb{R}$ by

$$g(p,q) = \alpha\sum_{i=1}^{m-1}\sum_{j=1}^{n-1}\sqrt{p_{i,j}^2+q_{i,j}^2}+\alpha\sum_{i=1}^{m-1}|p_{i,n}|+\alpha\sum_{j=1}^{n-1}|q_{m,j}|$$

for any $p\in\mathbb{R}^{(m-1)\times n}$ and $q\in\mathbb{R}^{m\times(n-1)}$. Then, the minimization problem (5.43) can be written as

$$\min_{x\in\mathbb{R}^{m\times n}}\left\{\frac{1}{2}\|x-b\|_F^2+g(Dx)\right\}$$

which is of the form (5.39) with $f(x) = \|x - b\|_F^2/2$. Thus, the dual proximal gradient method becomes

$$\begin{aligned} x^k &= b - D^T(p^k, q^k), \\ (p^{k+1}, q^{k+1}) &= \operatorname{prox}_{tg^*}\left((p^k, q^k) + tDx^k\right). \end{aligned} \tag{5.44}$$

According to the definition of g, it is easy to see that the Legendre-Fenchel conjugate of g is the indicator function of the convex set C_α, where C_α is the set consisting of matrix pairs $(p, q) \in \mathbb{R}^{(m-1)\times n} \times \mathbb{R}^{m\times(n-1)}$ satisfying

$$\begin{aligned} &p_{i,j}^2 + q_{i,j}^2 \le \alpha^2, \quad i = 1, \cdots, m-1,\ j = 1, \cdots, n-1, \\ &|p_{i,n}| \le \alpha, \quad i = 1, \cdots, m-1, \\ &|q_{m,j}| \le \alpha, \quad j = 1, \cdots, n-1. \end{aligned}$$

Consequently, $\operatorname{prox}_{tg^*}((r, s)) = P_{C_\alpha}(r, s)$, where P_{C_α} denotes the Euclidean projection onto C_α. Thus, the method (5.44) becomes

$$\begin{aligned} x^k &= b - D^T(p^k, q^k), \\ (p^{k+1}, q^{k+1}) &= P_{C_\alpha}\left((p^k, q^k) + tDx^k\right). \end{aligned} \tag{5.45}$$

In order to carry out the numerical implementation, we need to determine D^T and P_{C_α}. By definition, we have for $(p, q) \in \mathbb{R}^{(m-1)\times n} \times \mathbb{R}^{m\times(n-1)}$ and $x \in \mathbb{R}^{m\times n}$ that

$$\begin{aligned} \langle D^T(p, q), x\rangle &= \langle (p, q), Dx\rangle \\ &= \sum_{i=1}^{m-1}\sum_{j=1}^{n} p_{i,j}(x_{i,j} - x_{i+1,j}) + \sum_{i=1}^{m}\sum_{j=1}^{n-1} q_{i,j}(x_{i,j} - x_{i,j+1}) \\ &= \sum_{i=1}^{m}\sum_{j=1}^{n}(p_{i,j} + q_{i,j} - p_{i-1,j} - q_{i,j-1})x_{i,j}, \end{aligned}$$

and therefore $D^T : \mathbb{R}^{(m-1)\times n} \times \mathbb{R}^{m\times(n-1)} \to \mathbb{R}^{m\times n}$ is given by

$$D^T(p, q)_{i,j} = p_{i,j} + q_{i,j} - p_{i-1,j} - q_{i,j-1},$$

where we set $p_{0,j} = q_{i,0} = p_{m,j} = q_{i,n} = 0$ for $i = 1, \cdots, m$ and $j = 1, \cdots, n$. Using the definition of C_α, it is easy to see that $P_{C_\alpha}(r, s) = (p, q)$, where

$$p_{i,j} = \begin{cases} \min\left\{1, \frac{\alpha}{\sqrt{r_{i,j}^2 + s_{i,j}^2}}\right\} r_{i,j}, & i = 1, \cdots, m-1,\ j = 1, \cdots, n-1, \\ \min\left\{1, \frac{\alpha}{|r_{i,n}|}\right\} r_{i,n}, & i = 1, \cdots, m-1 \end{cases}$$

and

$$q_{i,j} = \begin{cases} \min\left\{1, \frac{\alpha}{\sqrt{r_{i,j}^2+s_{i,j}^2}}\right\} s_{i,j}, \; i = 1, \cdots, m-1, \; j = 1, \cdots, n-1, \\ \min\left\{1, \frac{\alpha}{|s_{m,j}|}\right\} s_{m,j}, \qquad j = 1, \cdots, n-1. \end{cases}$$

In order to determine the step size, we need to estimate $\|D\|$. Note that

$$\begin{aligned} \|Dx\|_F^2 &= \sum_{i=1}^{m-1}\sum_{j=1}^{n}(x_{i,j} - x_{i+1,j})^2 + \sum_{i=1}^{m}\sum_{j=1}^{n-1}(x_{i,j} - x_{i,j+1})^2 \\ &\le 2\sum_{i=1}^{m-1}\sum_{j=1}^{n}(|x_{i,j}|^2 + |x_{i+1,j}|^2) + 2\sum_{i=1}^{m}\sum_{j=1}^{n-1}(|x_{i,j}|^2 + |x_{i,j+1}|^2) \\ &\le 8\sum_{i=1}^{m}\sum_{j=1}^{n} |x_{i,j}|^2 = 8\|x\|_F^2. \end{aligned}$$

Thus, $\|D\|^2 \le 8$. By taking $t = 1/8$, we obtain from Algorithm 5.43 the following dual proximal gradient method for solving (5.43).

Algorithm 5.54 *Let $(p^0, q^0) = (0_{(m-1)\times n}, 0_{m\times(n-1)})$. For $k \ge 0$, calculate*

$$\begin{aligned} x^k &= b - D^T(p^k, q^k), \\ (p^{k+1}, q^{k+1}) &= P_{C_\alpha}\left((p^k, q^k) + \frac{1}{8}Dx^k\right) \end{aligned}$$

until a stopping rule is satisfied. Take the last x^k as an approximation.

By using the same reasoning, we obtain from Algorithm 5.46 the following accelerated algorithm ([19]).

Algorithm 5.55 *Let $(p^0, q^0) = (p^{-1}, q^{-1}) = (0_{(m-1)\times n}, 0_{m\times(n-1)})$. For $k \ge 0$, calculate*

$$\begin{aligned} (\hat{p}^k, \hat{q}^k) &= (p^k, q^k) + \frac{k-1}{k+2}(p^k - p^{k-1}, q^k - q^{k-1}), \\ \hat{x}^k &= b - D^T(\hat{p}^k, \hat{q}^k), \\ (p^{k+1}, q^{k+1}) &= P_{C_\alpha}\left((\hat{p}^k, \hat{q}^k) + \frac{1}{8}D\hat{x}^k\right), \\ x^{k+1} &= b - D^T(p^{k+1}, q^{k+1}) \end{aligned}$$

until a stopping rule is satisfied. Take the last x^k as an approximation.

5.9 Augmented Lagrangian Method

We consider the minimization problem of the form

$$\min\{f(x) : Ax = b\}, \tag{5.46}$$

where A is an $m \times n$ matrix, $b \in \mathbb{R}^m$, and $f : \mathbb{R}^n \to (-\infty, \infty]$ is a proper, lower semi-continuous, convex function. Applying a dual (sub)gradient method to solve (5.46) leads to an algorithm of the form

$$\begin{aligned} x^k &\in \arg\min_{x\in\mathbb{R}^n} \left\{ f(x) + \langle \lambda^k, Ax - b\rangle \right\}, \\ \lambda^{k+1} &= \lambda^k + t(Ax^k - b); \end{aligned} \tag{5.47}$$

see Example 5.49. Although this method preserves the decomposability of the function f, its convergence analysis requires f to be strongly convex. When f does not have enough convexity, even the definition of x^k can be problematic. To solve this issue, in the following, we introduce the *augmented Lagrangian method.*

To derive the augmented Lagrangian method, we apply the proximal point algorithm to the dual problem:

$$\min_{\lambda\in\mathbb{R}^m} \left\{ d(\lambda) := f^*(-A^T\lambda) + \langle \lambda, b\rangle \right\} \tag{5.48}$$

of (5.46). The updating formula is

$$\lambda^k = \arg\min_{\lambda\in\mathbb{R}^m} \left\{ f^*(-A^T\lambda) + \langle b, \lambda\rangle + \frac{1}{2t_k}\|\lambda - \lambda^{k-1}\|^2 \right\},$$

where $t_k > 0$ is a step size. Thus, by the first-order optimality condition on λ^k, we have

$$0 \in -A\partial f^*(-A^T\lambda^k) + b + \frac{1}{t_k}(\lambda^k - \lambda^{k-1}),$$

where we assumed $\partial(f^* \circ (-A^T))(\lambda) = -A\partial f^*(-A^T\lambda)$, which may not be true in general; this condition is only used in motivating the method and will not be needed in the forthcoming convergence analysis. The above equation means that there exists $x^k \in \partial f^*(-A^T\lambda^k)$ such that

$$\lambda^k = \lambda^{k-1} + t_k(Ax^k - b). \tag{5.49}$$

By using Theorem 4.9, we have

$$x^k \in \partial f^*(-A^T\lambda^k) \Longleftrightarrow -A^T\lambda^k \in \partial f(x^k) \Longleftrightarrow 0 \in \partial f(x^k) + A^T\lambda^k.$$

From (5.49), it then follows that

$$\begin{aligned} 0 &\in \partial f(x^k) + A^T\lambda^{k-1} + t_k A^T(Ax^k - b) \\ &= \partial\left(f + \langle \lambda^{k-1}, A\cdot - b\rangle + \frac{t_k}{2}\|A\cdot - b\|^2\right)(x^k) \end{aligned}$$

which implies that

$$x^k \in \arg\min_{x\in\mathbb{R}^n}\left\{f(x) + \langle \lambda^{k-1}, Ax - b\rangle + \frac{t_k}{2}\|Ax - b\|^2\right\}.$$

This together with (5.49) motivates the following algorithm.

Algorithm 5.56 (The Augmented Lagrangian Method) *Take an initial guess $\lambda^0 \in \mathbb{R}^m$ and define*

$$\begin{aligned} x^k &\in \arg\min_{x\in\mathbb{R}^n}\left\{f(x) + \langle \lambda^{k-1}, Ax - b\rangle + \frac{t_k}{2}\|Ax - b\|^2\right\}, \\ \lambda^k &= \lambda^{k-1} + t_k(Ax^k - b) \end{aligned}$$

for $k = 1, 2, \cdots$, where $t_k > 0$ are step sizes.

This algorithm can be viewed as a modification of the method (5.47) by adding the term $\frac{t_k}{2}\|Ax - b\|^2$ in the definition of x^k to augment the Lagrangian function $L(x, \lambda^{k-1}) = f(x) + \langle \lambda^{k-1}, Ax - b\rangle$. It is this reason that the algorithm is called the *augmented Lagrangian method*. This method is also called *the method of multipliers*; it was proposed by Hestenes [104] and Powell [183] independently in 1969.

For the convergence analysis of the augmented Lagrangian method, we make the following assumption.

Assumption 5.57

(i) *$f : \mathbb{R}^n \to (-\infty, \infty]$ is proper, lower semi-continuous, and convex.*
(ii) *The equation $Ax = b$ has a solution in* dom(f).
(iii) *For each $\rho > 0$, the set $\{x \in \mathbb{R}^n : \|Ax\| + f(x) \le \rho\}$ is bounded.*

Under Assumption 5.57, it follows from Theorem 1.12 that (5.46) has a global minimizer and Algorithm 5.56 is well-defined, i.e., for each $k \ge 1$, the minimization problem defining x^k always has a minimizer; we take x^k to be any one of the minimizers.

Lemma 5.58 *Let Assumption 5.57 hold and let* $\{t_k\}$ *be a sequence of positive numbers. Let* $\{x^k\}$ *and* $\{\lambda^k\}$ *be defined by Algorithm 5.56. Then,* $\xi^k := -A^T\lambda^k \in \partial f(x^k)$ *and*

$$\|Ax^{k+1} - b\| \le \|Ax^k - b\|$$

for all $k \ge 1$*. Let* $x^* \in \mathrm{dom}(f)$ *be any element satisfying* $Ax^* = b$*. Then,*

$$D_f^{\xi^{k+1}}(x^*, x^{k+1}) \le D_f^{\xi^k}(x^*, x^k)$$

for all $k \ge 1$*, where* $D_f^{\xi^k}(x, x^k)$ *denotes the Bregman distance induced by* f *at* x^k *in the direction* ξ^k*.*

Proof According to the definition of x^k and Theorem 3.14, we have

$$\begin{aligned}0 &\in \partial f(x^k) + A^T\lambda^{k-1} + t_k A^T(Ax^k - b)\\ &= \partial f(x^k) + A^T(\lambda^{k-1} + t_k(Ax^k - b)).\end{aligned}$$

By using the equation $\lambda^k = \lambda^{k-1} + t_k(Ax^k - b)$, we then have $0 \in \partial f(x^k) + A^T\lambda^k$, which shows $\xi^k := -A^T\lambda^k \in \partial f(x^k)$. From the monotonicity of ∂f, it follows that

$$\begin{aligned}0 &\le \langle \xi^{k+1} - \xi^k, x^{k+1} - x^k\rangle = -\langle \lambda^{k+1} - \lambda^k, Ax^{k+1} - Ax^k\rangle\\ &= -t_{k+1}\langle Ax^{k+1} - b, Ax^{k+1} - Ax^k\rangle\end{aligned}$$

By virtue of the polarization identity and $t_{k+1} > 0$, we can further obtain

$$0 \le -\frac{1}{2}\left(\|Ax^{k+1} - b\|^2 - \|Ax^k - b\|^2 + \|A(x^{k+1} - x^k)\|^2\right).$$

Therefore,

$$\|Ax^{k+1} - b\|^2 + \|A(x^{k+1} - x^k)\|^2 \le \|Ax^k - b\|^2$$

which implies that $\|Ax^{k+1} - b\| \le \|Ax^k - b\|$.

Finally, by using Lemma 3.11 and nonnegativity of the Bregman distance, we have

$$\begin{aligned}D_f^{\xi^{k+1}}(x^*, x^{k+1}) - D_f^{\xi^k}(x^*, x^k) &= -D_f^{\xi^k}(x^{k+1}, x^k) + \langle \xi^{k+1} - \xi^k, x^{k+1} - x^*\rangle\\ &\le \langle \xi^{k+1} - \xi^k, x^{k+1} - x^*\rangle\\ &= -\langle \lambda^{k+1} - \lambda^k, Ax^{k+1} - b\rangle\\ &= -t_{k+1}\|Ax^{k+1} - b\|^2 \le 0.\end{aligned}$$

The proof is complete. □

Recall that we have derived the augmented Lagrangian method from an application of the proximal point algorithm to the dual problem (5.48); the derivation requires $\partial(f^* \circ (-A^T))(\lambda) = -A\partial f^*(-A^T\lambda)$. The following result, due to [192], shows that the sequence $\{\lambda^k\}$ defined by Algorithm 5.56 is exactly the sequence obtained by applying the proximal point algorithm to the dual problem without relying on this extra requirement on f^*.

Lemma 5.59 *Let Assumption 5.57 hold. For Algorithm 5.56, there holds*

$$\lambda^k = \arg\min_{\lambda \in \mathbb{R}^n} \left\{ f^*(-A^T\lambda) + \langle b, \lambda \rangle + \frac{1}{2t_k}\|\lambda - \lambda^{k-1}\|^2 \right\}$$

for all $k = 1, 2, \cdots$.

Proof According to Lemma 5.58, we have $-A^T\lambda^k \in \partial f(x^k)$. Thus, it follows from Theorem 4.9 that $x^k \in \partial f^*(-A^T\lambda^k)$. Consequently,

$$-Ax^k \in -A\partial f^*(-A^T\lambda^k) \subset \partial\left(f^* \circ (-A^T)\right)(\lambda^k).$$

Since $-Ax^k = -b - (\lambda^k - \lambda^{k-1})/t_k$, we have

$$-b - \frac{\lambda^k - \lambda^{k-1}}{t_k} \in \partial\left(f^* \circ (-A^T)\right)(\lambda^k).$$

This implies that

$$0 \in \partial\left(f^* \circ (-A^T) + \langle b, \cdot\rangle + \frac{1}{2t_k}\|\cdot - \lambda^{k-1}\|^2\right)(\lambda^k)$$

which shows the desired conclusion. □

Based on Lemma 5.59, we may use Proposition 4.30 to obtain immediately the following result.

Lemma 5.60 *Let Assumption 5.57 hold and let $d(\lambda) := f^*(-A^T\lambda) + \langle b, \lambda\rangle$. Then, for any $\lambda \in \mathbb{R}^m$, there holds*

$$d(\lambda^k) - d(\lambda) \le \frac{\|\lambda\|^2}{2s_k} - \frac{\|\lambda - \lambda^k\|^2}{2s_k} - \frac{s_k}{2t_k^2}\|\lambda^k - \lambda^{k-1}\|^2$$

for $k = 1, 2, \cdots$, where $s_k := \sum_{j=1}^k t_j$.

Using Lemma 5.60, we can obtain the following estimate concerning the augmented Lagrangian method.

Lemma 5.61 *Let Assumption 5.57 hold and consider Algorithm 5.56. Then, for any* $\lambda \in \mathbb{R}^m$*, there holds*

$$s_k\|Ax^k - b\|^2 + \frac{\|\lambda^k\|^2}{4s_k} \le d(\lambda) - d(\lambda^k) + \frac{\|\lambda\|^2}{s_k}$$

for all $k \ge 1$.

Proof By using Lemma 5.60 and the fact $\lambda^k - \lambda^{k-1} = t_k(Ax^k - b)$, we can obtain

$$s_k\|Ax^k - b\|^2 + \frac{\|\lambda - \lambda^k\|^2}{2s_k} \le d(\lambda) - d(\lambda^k) + \frac{\|\lambda\|^2}{2s_k}.$$

This together with the inequality $\|\lambda^k\|^2 \le 2(\|\lambda - \lambda^k\|^2 + \|\lambda\|^2)$ then shows the conclusion. □

Now, we are ready to prove the following convergence result on Algorithm 5.56.

Theorem 5.62 *Let Assumption 5.57 hold and let* x^* *be a solution of (5.46). Let* $\{t_k\}$ *be a sequence of positive numbers such that* $s_k := \sum_{j=1}^k t_j \to \infty$ *as* $k \to \infty$*. Let* $\{x^k\}$ *and* $\{\lambda^k\}$ *be defined by Algorithm 5.56 and let* $\xi^k := -A^*\lambda^k$*. Then,*

$$s_k\|Ax^k - b\|^2 \to 0, \qquad f(x^k) \to f(x^*) \quad \text{and} \quad D_f^{\xi^k}(x^*, x^k) \to 0$$

as $k \to \infty$*. Moreover,* $\{x^k\}$ *is a bounded sequence, and any cluster point of* $\{x^k\}$ *is a solution of (5.46).*

Proof By the Fenchel-Young inequality and $Ax^* = b$, we have for any $\lambda \in \mathbb{R}^m$ that

$$d(\lambda) = f^*(-A^T\lambda) - \langle -A^T\lambda, x^*\rangle \ge -f(x^*)$$

which implies that $\inf_{\lambda\in\mathbb{R}^m} d(\lambda) \ge -f(x^*) > -\infty$. Thus, for any $\varepsilon > 0$, we can find $\lambda_\varepsilon \in \mathbb{R}^m$ such that

$$d(\lambda_\varepsilon) \le \inf_{\lambda\in\mathbb{R}^m} d(\lambda) + \varepsilon.$$

Consequently, $d(\lambda_\varepsilon) \le d(\lambda^k) + \varepsilon$. Therefore, by taking $\lambda = \lambda_\varepsilon$ in the estimate given in Lemma 5.61, we have

$$s_k\|Ax^k - b\|^2 + \frac{\|\lambda^k\|^2}{4s_k} \le \varepsilon + \frac{\|\lambda_\varepsilon\|^2}{s_k}.$$

Since $s_k \to \infty$ as $k \to \infty$, we can obtain $\limsup_{k\to\infty} s_k\|Ax^k - b\|^2 \leq \varepsilon$. Because $\varepsilon > 0$ can be arbitrarily small, we thus have $s_k\|Ax^k - b\|^2 \to 0$ as $k \to \infty$. From the above inequality, we also have

$$\|\lambda^k\|\|Ax^k - b\| \leq s_k\|Ax^k - b\|^2 + \frac{\|\lambda^k\|^2}{4s_k} \leq \varepsilon + \frac{\|\lambda_\varepsilon\|^2}{s_k}.$$

Therefore, $\limsup_{k\to\infty} \|\lambda^k\|\|Ax^k - b\| \leq \varepsilon$. Since $\varepsilon > 0$ is arbitrary, this implies that

$$\lim_{k\to\infty} \|\lambda^k\|\|Ax^k - b\| = 0. \tag{5.50}$$

Recall that $-A^T\lambda^k \in \partial f(x^k)$. We have

$$\begin{aligned} f(x^k) &\leq f(x^*) + \langle A^T\lambda^k, x^* - x^k\rangle \\ &= f(x^*) + \langle \lambda^k, b - Ax^k\rangle \\ &\leq f(x^*) + \|\lambda^k\|\|b - Ax^k\|. \end{aligned}$$

By virtue of (5.50), we thus obtain

$$\limsup_{k\to\infty} f(x^k) \leq f(x^*).$$

Therefore, we can find a constant ρ such that $\|Ax^k\| + f(x^k) \leq \rho$ for all k. Consequently, we may use Assumption 5.57 (iii) to conclude that $\{x^k\}$ is a bounded sequence.

Let $\bar{x}$ be any cluster point of $\{x^k\}$, i.e., there is a subsequence $\{x^{k_l}\}$ of $\{x^k\}$ such that $x^{k_l} \to \bar{x}$ as $l \to \infty$. Then,

$$\|A\bar{x} - b\| = \lim_{l\to\infty} \|Ax^{k_l} - b\| = 0,$$

and by the lower semi-continuity of f, we have

$$f(\bar{x}) \leq \liminf_{l\to\infty} f(x^{k_l}) \leq \limsup_{l\to\infty} f(x^{k_l}) \leq f(x^*).$$

This shows that $A\bar{x} = b$ and thus $\bar{x}$ is a solution of (5.46) with

$$\lim_{l\to\infty} f(x^{k_l}) = f(x^*) = f(\bar{x}). \tag{5.51}$$

The above argument actually shows that for any subsequence $\{x^{k_l}\}$ of $\{x^k\}$, there is a subsequence, still denoted as $\{x^{k_l}\}$, such that (5.51) holds. By a subsequence-subsequence argument, we therefore obtain $\lim_{k\to\infty} f(x^k) = f(x^*)$. Finally, by using (5.50), we have

$$D_f^{\xi^k}(x^*, x^k) = f(x^*) - f(x^k) - \langle \lambda^k, Ax^k - b\rangle \to 0$$

as $k \to \infty$. The proof is complete. □

In the next result, we will improve the convergence result in Theorem 5.62 by deriving convergence rates when the problem (5.46) has a KKT point.

Theorem 5.63 *Let Assumption 5.57 hold and assume that the problem (5.46) has a KKT point (x^*, λ^*), i.e., $Ax^* = b$ and $-A^T\lambda^* \in \partial f(x^*)$. Consider Algorithm 5.56 with $\{t_k\}$ being a sequence of positive numbers. Then,*

$$\begin{aligned} \|Ax^k - b\| &\le \|\lambda^*\| s_k^{-1}, \\ |f(x^k) - f(x^*)| &\le 2\|\lambda^*\|^2 s_k^{-1}, \\ D_f^{-A^T\lambda^k}(x^*, x^k) &\le 3\|\lambda^*\|^2 s_k^{-1} \end{aligned}$$

for all $k \ge 1$, where $s_k := \sum_{j=1}^k t_j$.

Proof Recall that $-A^T\lambda^k \in \partial f(x^k)$. We may use $Ax^* = b$ and the Fenchel-Young inequality to obtain

$$d(\lambda^k) = f^*(-A^T\lambda^k) - \langle -A^T\lambda^k, x^*\rangle \ge -f(x^*).$$

Thus, it follows from Lemma 5.61 and the definition of $d(\lambda)$ that

$$s_k\|Ax^k - b\|^2 + \frac{\|\lambda^k\|^2}{4s_k} \le f^*(-A^T\lambda) + \langle \lambda, b\rangle + f(x^*) + \frac{\|\lambda\|^2}{s_k}$$

for all $\lambda \in \mathbb{R}^m$. Consequently,

$$\begin{aligned} s_k\|Ax^k - b\|^2 + \frac{\|\lambda^k\|^2}{4s_k} &\le f(x^*) + \inf_{\lambda\in\mathbb{R}^m}\left\{f^*(-A^T\lambda) + \langle \lambda, b\rangle + \frac{\|\lambda\|^2}{s_k}\right\} \\ &= f(x^*) - \sup_{\lambda\in\mathbb{R}^m}\left\{-f^*(-A^T\lambda) - \langle \lambda, b\rangle - \frac{\|\lambda\|^2}{s_k}\right\}. \end{aligned}$$

By virtue of the Fenchel-Rockafellar duality formula given in Theorem 4.26, we have

$$\sup_{\lambda\in\mathbb{R}^m}\left\{-f^*(-A^T\lambda) - \langle \lambda, b\rangle - \frac{\|\lambda\|^2}{s_k}\right\} = \inf_{x\in\mathbb{R}^n}\left\{f(x) + \frac{1}{4}s_k\|Ax - b\|^2\right\}.$$

Therefore,

$$s_k\|Ax^k - b\|^2 + \frac{\|\lambda^k\|^2}{4s_k} \le \sup_{x\in\mathbb{R}^n}\left\{f(x^*) - f(x) - \frac{1}{4}s_k\|Ax - b\|^2\right\}.$$

Since $-A^T\lambda^* \in \partial f(x^*)$, we have

$$f(x^*) - f(x) \le \langle \lambda^*, Ax - b\rangle, \quad \forall x \in \mathbb{R}^n \tag{5.52}$$

and hence

$$\begin{aligned} s_k\|Ax^k - b\|^2 + \frac{\|\lambda^k\|^2}{4s_k} &\le \sup_{x\in\mathbb{R}^n}\left\{\langle \lambda^*, Ax - b\rangle - \frac{1}{4}s_k\|Ax - b\|^2\right\} \\ &\le \sup_{x\in\mathbb{R}^n}\left\{\|\lambda^*\|\|Ax - b\| - \frac{1}{4}s_k\|Ax - b\|^2\right\} \\ &\le \sup_{\tau\ge 0}\left\{\|\lambda^*\|\tau - \frac{1}{4}s_k\tau^2\right\} \\ &= \frac{\|\lambda^*\|^2}{s_k} \end{aligned}$$

This in particular implies that

$$\|Ax^k - b\| \le \frac{\|\lambda^*\|}{s_k} \qquad \text{and} \qquad \|\lambda^k\| \le 2\|\lambda^*\|.$$

Using again $-A^T\lambda^k \in \partial f(x^k)$, we have $f(x^k) - f(x^*) \le \langle \lambda^k, b - Ax^k\rangle$. This together with (5.52) with $x = x^k$ shows that

$$|f(x^k) - f(x^*)| \le \max\{\|\lambda^*\|, \|\lambda^k\|\}\|Ax^k - b\| \le \frac{2\|\lambda^*\|^2}{s_k}.$$

Finally, using again (5.52) with $x = x^k$, we have

$$\begin{aligned} D_f^{-A^T\lambda^k}(x^*, x^k) &= f(x^*) - f(x^k) + \langle \lambda^k, b - Ax^k\rangle \\ &\le \langle \lambda^* - \lambda^k, Ax^k - b\rangle \\ &\le (\|\lambda^*\| + \|\lambda^k\|)\|Ax^k - b\| \\ &\le \frac{3\|\lambda^*\|^2}{s_k}. \end{aligned}$$

The proof is therefore complete. □

Finally, we conclude this section by mentioning that the augmented Lagrangian method has been extended for solving ill-posed inverse problems in various contexts; see [84–86, 116, 119, 172], for instance.

Problems

5.1 Let $f : \mathbb{R}^n \to \mathbb{R}$ be convex and let $\varphi : \mathbb{R} \to \mathbb{R}$ be differentiable, increasing, and convex. Show that $\varphi \circ f : \mathbb{R}^n \to \mathbb{R}$ is convex and its directional derivative and subdifferential at $x \in \mathbb{R}^n$ are given, respectively, by

$$(\varphi \circ f)'(x; v) = \varphi'(f(x)) f'(x; v), \quad \forall v \in \mathbb{R}^n$$

and

$$\partial(\varphi \circ f)(x) = \varphi'(f(x)) \partial f(x), \quad \forall x \in \mathbb{R}^n.$$

5.2 Let $f : \mathbb{R}^n \to (-\infty, \infty]$ be a strong convex function with modulus of convexity $\mu > 0$. Show that for any $x, y \in \text{dom}(\partial f)$ and $\xi \in \partial f(x)$, $\eta \in \partial f(y)$ there holds

$$D_f^{\xi}(y, x) \le \frac{1}{4\mu} \|\xi - \eta\|^2.$$

5.3 Let $A \in \mathbb{R}^{m \times n}$ have full row rank and $b \in \text{Ran}(A)$. Use the KKT theory to show that, for any $v \in \mathbb{R}^n$, its Euclidean projection $\hat{v}$ onto the set $\{x \in \mathbb{R}^n : Ax = b\}$ is given by

$$\hat{v} = v - A^T (AA^T)^{-1} (Av - b).$$

5.4 Let $\gamma_1, \cdots, \gamma_n$ be given positive numbers. Consider the constrained minimization problem:

$$\min \left\{ -\sum_{i=1}^n \log(\gamma_i + x_i) : \sum_{i=1}^n x_i = 1 \text{ and } x_i \ge 0 \text{ for } i = 1, \cdots, n \right\}.$$

Use the KKT theory to show that its optimal solution $\hat{x} = (\hat{x}_1, \cdots, \hat{x}_n)$ is given by

$$\hat{x}_i = \max\left\{ \frac{1}{\mu} - \gamma_i, 0 \right\}, \quad i = 1, \cdots, n,$$

where μ is a positive number satisfying

$$\sum_{i=1}^{n} \max\left\{\frac{1}{\mu} - \gamma_i, 0\right\} = 1.$$

5.5 Let $A \in \mathbb{R}^{m\times n}$ and $B \in \mathbb{R}^{p\times n}$ be nonzero matrices. For any $\alpha \in \mathbb{R}^m$ and $\beta \in \mathbb{R}^p$, let

$$\Psi(\alpha, \beta) := \left\{x \in \mathbb{R}^n : Ax \leq \alpha \text{ and } Bx = \beta\right\}.$$

Use Theorem 5.16 to show that there exists a constant κ depending only on A and B such that

$$d(x, \Psi(\alpha, \beta)) \leq \kappa\left(\|(Ax - \alpha)_+\| + \|Bx - \beta\|\right)$$

for all $x \in \mathbb{R}^n$ and $(\alpha, \beta) \in \mathbb{R}^m \times \mathbb{R}^p$ with $\Psi(\alpha, \beta) \neq \emptyset$.

5.6 Let $C_1, \cdots, C_N$ be closed convex subsets in $\mathbb{R}^n$ with nonempty intersection. If for each $2 \leq k \leq N$ the pair of the sets $(\cap_{i=1}^{k-1} C_i, C_k)$ is (bounded) linear regular, show that $C_1, \cdots, C_N$ are (bounded) linear regular.

5.7 Let $C_1, \cdots, C_p$ be closed convex sets in $\mathbb{R}^n$ with

$$C := \bigcap_{i=1}^{p} C_i \neq \emptyset.$$

Given a point $v \in \mathbb{R}^n$, we consider the problem of finding its Euclidean projection onto C. This problem can be formulated as

$$\min_{x\in\mathbb{R}^n} \left\{\frac{1}{2}\|x - v\|^2 + \delta_{C_1}(x) + \cdots + \delta_{C_p}(x)\right\}, \tag{5.53}$$

where δ_{C_i} denotes the indicator function of C_i for each $i = 1, \cdots, p$. Use the dual gradient method to derive the following algorithm for solving (5.53):

$$\begin{cases} x^k = v - \dfrac{1}{p}\displaystyle\sum_{i=1}^{p} \xi_i^k, \\ \xi_i^{k+1} = \xi_i^k + x^k - P_{C_i}(\xi_i^k + x^k), \quad i = 1, \cdots, p, \end{cases}$$

where P_{C_i} denotes the Euclidean projection onto C_i. (Hint: Introducing

$$g(z_1, \cdots, z_p) := \delta_{C_1}(z_1) + \cdots + \delta_{C_p}(z_p)$$

on $\mathbb{R}^n \times \cdots \times \mathbb{R}^n$ and the linear operator $A : \mathbb{R}^n \to \mathbb{R}^n \times \cdots \times \mathbb{R}^n$ by $Ax = (x, \cdots, x)$ to reformulate (5.53) as

$$\min_{x \in \mathbb{R}^n} \left\{ \frac{1}{2}\|x - v\|^2 + g(Ax) \right\}$$

and then applying the dual gradient method. You may need to rescale the dual variables and justify the use of the step size $t = 1/p$.)

5.8 Let A be an $m \times n$ matrix with $m \ll n$ and $b \in \mathrm{Ran}(A) \subset \mathbb{R}^m$. A sparse solution of $Ax = b$ can be determined by finding the unique solution $x^\dagger$ of the minimization problem:

$$\min \left\{ \alpha\|x\|_1 + \frac{1}{2}\|x\|^2 : Ax = b \right\} \tag{5.54}$$

with a large parameter $\alpha > 0$.

(i) Use the dual gradient method to derive the following iteration scheme for solving (5.54):

$$x^k = \mathrm{sign}(\xi^k) \cdot \max\left\{|\xi^k| - \alpha, 0\right\}, \quad \xi^{k+1} = \xi^k - tA^T(Ax^k - b)$$

with $\xi^0 = 0$, and show that, when $0 < t \le 1/\|A\|^2$, there holds $\|x^k - x^\dagger\| = O(1/\sqrt{k})$.

(ii) Use the accelerated dual gradient method to derive the following iteration scheme for solving (5.54):

$$\hat{\xi}^k = \xi^k + \frac{k-1}{k+2}(\xi^k - \xi^{k-1}), \qquad \hat{x}^k = \mathrm{sign}(\hat{\xi}^k) \cdot \max\left\{|\hat{\xi}^k| - \alpha, 0\right\},$$

$$\xi^{k+1} = \hat{\xi}^k - tA^T(A\hat{x}^k - b), \qquad x^{k+1} = \mathrm{sign}(\xi^{k+1}) \cdot \max\left\{|\xi^{k+1}| - \alpha, 0\right\}$$

with $\xi^{-1} = \xi^0 = 0$, and show that, when $0 < t \le 1/\|A\|^2$, there holds $\|x^k - x^\dagger\| = O(1/k)$.

Chapter 6
ADMM: Alternating Direction Method of Multipliers

The alternating direction method of multipliers (ADMM) is a versatile approach for solving multi-block separable optimization problems. This chapter focuses on ADMM for two-block convex separable optimization problems with equality constraints. We establish the convergence of ADMM and its proximal variant, deriving both ergodic and non-ergodic convergence rates. Additionally, we analyze the linear convergence of proximal ADMM and discuss an accelerated ADMM. For a broader discussion on the convergence of ADMM for multi-block separable optimization problems, we refer the reader to the survey paper [97].

6.1 The Method

We consider the convex minimization problem:

$$\begin{aligned} &\text{minimize} \quad && H(x, y) := f(x) + g(y) \\ &\text{subject to} \quad && Ax + By = c, \end{aligned} \tag{6.1}$$

where $f : \mathbb{R}^n \to (-\infty, \infty]$ and $g : \mathbb{R}^m \to (-\infty, \infty]$ are proper, lower semicontinuous, convex functions, A and B are matrices of size $p \times n$ and $p \times m$, respectively, and $c \in \mathbb{R}^p$. The associated Lagrangian function for (6.1) is

$$L(x, y, \lambda) = f(x) + g(y) + \langle \lambda, Ax + By - c \rangle.$$

If we use the augmented Lagrangian method to solve (6.1), the algorithm takes the form

$$\begin{aligned} (x^{k+1}, y^{k+1}) &= \arg \min_{(x,y) \in \mathbb{R}^n \times \mathbb{R}^m} L_\rho(x, y, \lambda^k), \\ \lambda^{k+1} &= \lambda^k + \rho(Ax^{k+1} + By^{k+1} - c), \end{aligned}$$

Q. Jin, *Lectures on Nonsmooth Optimization*, Texts in Applied Mathematics 82,
https://doi.org/10.1007/978-3-031-91417-1_6

where

$$L_\rho(x, y, \lambda) := f(x) + g(y) + \langle \lambda, Ax + By - c \rangle + \frac{\rho}{2}\|Ax + By - c\|^2$$

is the augmented Lagrangian function with parameter $\rho > 0$. Therefore, finding (x^{k+1}, y^{k+1}) requires to solving a minimization problem which couples f and g. Even if f and g have simple structure, solving this minimization problem could be very difficult and thus makes the computation expensive.

One may consider defining x^{k+1} and y^{k+1} alternatively by decoupling f and g into separate minimization problems to make the computation easier. This leads to the alternating direction method of multipliers (ADMM).

Algorithm 6.1 (ADMM) *Take initial y^0 and λ^0. For $k \geq 0$, define*

$$\begin{aligned} x^{k+1} &= \arg\min_{x \in \mathbb{R}^n} L_\rho(x, y^k, \lambda^k), \\ y^{k+1} &= \arg\min_{y \in \mathbb{R}^m} L_\rho(x^{k+1}, y, \lambda^k), \\ \lambda^{k+1} &= \lambda^k + \rho(Ax^{k+1} + By^{k+1} - c). \end{aligned}$$

According to the expression of the augmented Lagrangian function, by dropping the constant terms in the minimization problems, the ADMM can be restated as

$$\begin{aligned} x^{k+1} &= \arg\min_{x \in \mathbb{R}^n} \left\{ f(x) + \langle \lambda^k, Ax \rangle + \frac{\rho}{2}\|Ax + By^k - c\|^2 \right\}, \\ y^{k+1} &= \arg\min_{y \in \mathbb{R}^m} \left\{ g(y) + \langle \lambda^k, By \rangle + \frac{\rho}{2}\|Ax^{k+1} + By - c\|^2 \right\}, \\ \lambda^{k+1} &= \lambda^k + \rho(Ax^{k+1} + By^{k+1} - c). \end{aligned} \tag{6.2}$$

This ADMM algorithm was proposed in [88, 90] around the mid-1970 and was analyzed in [75, 87, 144]. It has been widely used in solving structured optimization problems due to its decomposability and superior flexibility. Recently, it has been revisited and popularized in modern signal/image processing, statistics, machine learning, and so on; see [41, 97, 113, 227] and the references therein.

The ADMM algorithm can be derived by applying the Douglas-Rachford splitting method to the dual problem of (6.1). We will provide this derivation under the additional condition:

$$\partial(f^* \circ A^T)(\lambda) = A\partial f^*(A^T\lambda) \quad \text{and} \quad \partial(g^* \circ B^T)(\lambda) = B\partial g^*(B^T\lambda)$$

for all $\lambda \in \mathbb{R}^p$ which is not required in the convergence analysis of Algorithm 6.1 in the forthcoming sections. To see the derivation, we note that, for the minimization problem (6.1), the dual function is

$$\begin{aligned} h(\lambda) &= \inf_{x\in\mathbb{R}^n, y\in\mathbb{R}^m} \{f(x) + g(y) + \langle \lambda, Ax + By - c\rangle\} \\ &= \inf_{x\in\mathbb{R}^n} \left\{ f(x) + \langle A^T\lambda, x\rangle \right\} + \inf_{y\in\mathbb{R}^m} \left\{ g(y) + \langle B^T\lambda, y\rangle \right\} - \langle \lambda, c\rangle \\ &= -\sup_{x\in\mathbb{R}^n} \left\{ \langle -A^T\lambda, x\rangle - f(x) \right\} - \sup_{y\in\mathbb{R}^m} \left\{ \langle -B^T\lambda, y\rangle - g(y) \right\} - \langle \lambda, c\rangle \\ &= -f^*(-A^T\lambda) - g^*(-B^T\lambda) - \langle c, \lambda\rangle. \end{aligned}$$

Thus, the dual problem is

$$\max_{\lambda\in\mathbb{R}^p} \left\{ -f^*(-A^T\lambda) - g^*(-B^T\lambda) - \langle c, \lambda\rangle \right\}$$

or equivalently

$$\min_{\lambda\in\mathbb{R}^p} \left\{ f^*(-A^T\lambda) + g^*(-B^T\lambda) + \langle c, \lambda\rangle \right\}.$$

Let

$$h_1(\lambda) = f^*(-A^T\lambda) + \langle c, \lambda\rangle \quad \text{and} \quad h_2(\lambda) = g^*(-B^T\lambda).$$

By applying the Douglas-Rachford splitting method described in Algorithm 4.58 to this dual problem, we obtain

$$\begin{aligned} \mu^{k+1} &= \mathrm{prox}_{\rho h_1}(\lambda^k - w^k), \\ \lambda^{k+1} &= \mathrm{prox}_{\rho h_2}(\mu^{k+1} + w^k), \\ w^{k+1} &= w^k + \mu^{k+1} - \lambda^{k+1}. \end{aligned}$$

From the definition of μ^{k+1}, we have

$$\lambda^k - w^k \in \mu^{k+1} + \rho\partial h_1(\mu^{k+1}) = \mu^{k+1} + \rho\left(c - A\partial f^*(-A^T\mu^{k+1})\right).$$

This implies that

$$\mu^{k+1} = \lambda^k - w^k + \rho(Ax^{k+1} - c)$$

for some $x^{k+1} \in \partial f^*(-A^T\mu^{k+1})$. In view of Theorems 4.9 and 3.14, we have

$$\begin{aligned} x^{k+1} \in \partial f^*(-A^T\mu^{k+1}) &\Longleftrightarrow -A^T\mu^{k+1} \in \partial f(x^{k+1}) \Longleftrightarrow 0 \in \partial f(x^{k+1}) + A^T\mu^{k+1} \\ &\Longleftrightarrow 0 \in \partial f(x^{k+1}) + A^T\left(\lambda^k - w^k + \rho(Ax^{k+1} - c)\right) \\ &\Longleftrightarrow 0 \in \partial\left(f + \langle \lambda^k, A\,\cdot\rangle + \frac{\rho}{2}\|A\cdot - c - w^k/\rho\|^2\right)(x^{k+1}). \end{aligned}$$

Therefore, μ^{k+1} can be determined by the following two steps:

$$\begin{aligned} x^{k+1} &= \arg\min_{x\in\mathbb{R}^n}\left\{f(x) + \langle \lambda^k, Ax\rangle + \frac{\rho}{2}\|Ax - c - w^k/\rho\|^2\right\}, \\ \mu^{k+1} &= \lambda^k - w^k + \rho(Ax^{k+1} - c). \end{aligned} \tag{6.3}$$

Next, from the definition of λ^{k+1}, we have

$$\mu^{k+1} + w^k \in \lambda^{k+1} + \rho\partial h_2(\lambda^{k+1}) = \lambda^{k+1} - \rho B\partial g^*(-B^T\lambda^{k+1}).$$

This implies that

$$\lambda^{k+1} = \mu^{k+1} + w^k + \rho By^{k+1}$$

for some $y^{k+1} \in \partial g^*(-B^T\lambda^{k+1})$. Using again Theorem 4.9 and (6.3), it yields

$$\begin{aligned} y^{k+1} \in \partial g^*(-B^T\lambda^{k+1}) &\Longleftrightarrow -B^T\lambda^{k+1} \in \partial g(y^{k+1}) \Longleftrightarrow 0 \in \partial g(y^{k+1}) + B^T\lambda^{k+1} \\ &\Longleftrightarrow 0 \in \partial g(y^{k+1}) + B^T(\mu^{k+1} + w^k + \rho By^{k+1}) \\ &\Longleftrightarrow 0 \in \partial g(y^{k+1}) + B^T\left(\lambda^k + \rho(Ax^{k+1} + By^{k+1} - c)\right). \end{aligned}$$

Therefore, λ^{k+1} can be determined by the following two steps:

$$\begin{aligned} y^{k+1} &= \arg\min_{y\in\mathbb{R}^m}\left\{g(y) + \langle \lambda^k, By\rangle + \frac{\rho}{2}\|Ax^{k+1} + By - c\|^2\right\}, \\ \lambda^{k+1} &= \mu^{k+1} + w^k + \rho By^{k+1}. \end{aligned} \tag{6.4}$$

Finally, for w^{k+1}, we have from (6.4) that

$$w^{k+1} = w^k + \mu^{k+1} - \lambda^{k+1} = -\rho By^{k+1}. \tag{6.5}$$

Combining (6.3), (6.4), and (6.5), we obtain the formulation (6.2) of the alternating direction method of multipliers.

The implementation of Algorithm 6.1 requires to determine x^{k+1} and y^{k+1} by solving two minimization problems during each iteration. According to the formulation (6.2), by completing squares, we may restate the algorithm in the following form:

$$\begin{aligned} x^{k+1} &= \arg\min_{x\in\mathbb{R}^n}\left\{f(x)+\frac{\rho}{2}\left\|Ax+By^k-c+\frac{1}{\rho}\lambda^k\right\|^2\right\},\\ y^{k+1} &= \arg\min_{y\in\mathbb{R}^m}\left\{g(y)+\frac{\rho}{2}\left\|Ax^{k+1}+By-c+\frac{1}{\rho}\lambda^k\right\|^2\right\},\\ \lambda^{k+1} &= \lambda^k+\rho(Ax^{k+1}+By^{k+1}-c). \end{aligned} \tag{6.6}$$

Although f and g may have the special structure so that their proximal mappings are easily calculated, solving the minimization problem in (6.6) is highly nontrivial expect that A and B are diagonal matrices or f and g are quadratic functions. In order to avoid this implementation issue, one may consider to add certain terms to the x-subproblem and y-subproblem in (6.6) to remove the terms $\|Ax\|^2$ and $\|By\|^2$. For any positive semi-definite matrix M, we will use the notation

$$\|z\|_M^2 := z^T Mz.$$

By taking two positive semi-definite matrices P of size $n\times n$ and Q of size $m\times m$, we may add the terms $\frac{1}{2}\|x-x^k\|_P^2$ and $\frac{1}{2}\|y-y^k\|_Q^2$ to the x- and y-subproblems in (6.6), respectively, to obtain the following proximal alternating direction method of multipliers [16, 69, 99, 100, 114, 233].

Algorithm 6.2 (Proximal ADMM) *Take initial guess y^0 and λ^0. For $k=0,1,\cdots$ do*

$$\begin{aligned} x^{k+1} &\in \arg\min_{x\in\mathbb{R}^n}\left\{f(x)+\frac{\rho}{2}\left\|Ax+By^k-c+\frac{1}{\rho}\lambda^k\right\|^2+\frac{1}{2}\|x-x^k\|_P^2\right\},\\ y^{k+1} &\in \arg\min_{y\in\mathbb{R}^m}\left\{g(y)+\frac{\rho}{2}\left\|Ax^{k+1}+By-c+\frac{1}{\rho}\lambda^k\right\|^2+\frac{1}{2}\|y-y^k\|_Q^2\right\},\\ \lambda^{k+1} &= \lambda^k+\rho(Ax^{k+1}+By^{k+1}-c). \end{aligned}$$

Let us illustrate how to choose P and Q to make the x- and y-subproblems easier to solve. For the x-subproblem, we have

$$\begin{aligned} x^{k+1} &\in \arg\min_{x\in\mathbb{R}^n}\left\{f(x)+\frac{\rho}{2}\|Ax\|^2+\langle Ax,\rho(By^k-c)+\lambda^k\rangle+\frac{1}{2}\|x\|_P^2-x^TPx^k\right\}\\ &= \arg\min_{x\in\mathbb{R}^n}\left\{f(x)+\frac{1}{2}x^T(\rho A^TA+P)x+\langle x, A^T[\lambda^k+\rho(By^k-c)]-Px^k\rangle\right\}. \end{aligned}$$

If we choose

$$P = \frac{1}{\tau} I - \rho A^T A,$$

where the constant $\tau > 0$ is chosen so that $P \succeq 0$, then

$$\begin{aligned} x^{k+1} &\in \arg\min_{x \in \mathbb{R}^n} \left\{ f(x) + \frac{1}{2\tau} \|x\|^2 + \langle x, A^T [\lambda^k + \rho(By^k - c)] - Px^k \rangle \right\} \\ &= \arg\min_{x \in \mathbb{R}^n} \left\{ f(x) + \frac{1}{2\tau} \|x + \tau A^T [\lambda^k + \rho(By^k - c)] - \tau Px^k\|^2 \right\} \\ &= \arg\min_{x \in \mathbb{R}^n} \left\{ f(x) + \frac{1}{2\tau} \|x - (x^k - \tau A^T (\lambda^k + \rho(Ax^k + By^k - c)))\|^2 \right\} \\ &= \text{prox}_{\tau f}(x^k - \tau A^T (\lambda^k + \rho(Ax^k + By^k - c))). \end{aligned}$$

Therefore, the resolution of x-subproblem reduces to finding the proximal mapping of f. By a similar argument, one can see that the y-subproblems can be reduced to finding the proximal mapping of g under a suitable choice of Q.

We conclude this section by providing various examples to illustrate the versatility of ADMM and proximal ADMM.

Example 6.3 Consider the ℓ^1-robust estimation problem:

$$\min_{x \in \mathbb{R}^n} \|Ax - b\|_1,$$

where A is an $m \times n$ matrix and $b \in \mathbb{R}^m$. It can be written as the equivalent form

$$\min_{x \in \mathbb{R}^n, y \in \mathbb{R}^m} \{\|y\|_1 : Ax - y - b = 0\}.$$

The corresponding augmented Lagrangian function is

$$L_\rho(x, y; \lambda) = \|y\|_1 + \langle \lambda, Ax - y - b \rangle + \frac{\rho}{2} \|Ax - y - b\|^2.$$

By applying Algorithm 6.2 with $Q = 0$ and a positive semi-definite matrix P, we obtain

$$\begin{aligned} x^{k+1} &= \arg\min_{x \in \mathbb{R}^n} \left\{ \langle \lambda^k, Ax \rangle + \frac{\rho}{2} \|Ax - y^k - b\|^2 + \frac{1}{2} \|x - x^k\|_P^2 \right\} \\ y^{k+1} &= \arg\min_{y \in \mathbb{R}^m} \left\{ \|y\|_1 - \langle \lambda^k, y \rangle + \frac{\rho}{2} \|Ax^{k+1} - y - b\|^2 \right\} \\ \lambda^{k+1} &= \lambda^k + \rho(Ax^{k+1} - y^{k+1} - b). \end{aligned}$$

After simplification, it becomes the algorithm:

$$\begin{aligned} x^{k+1} &= (P+\rho A^T A)^{-1}\left(\rho A^T(y^k+b-\lambda^k/\rho)+Px^k\right), \\ y^{k+1} &= S_{1/\rho}(Ax^{k+1}-b+\lambda^k/\rho), \\ \lambda^{k+1} &= \lambda^k+\rho(Ax^{k+1}-y^{k+1}-b), \end{aligned} \tag{6.7}$$

where $S_t(\cdot)$ denotes the soft thresholding function. If $A^T A$ is invertible and linear systems related to $A^T A$ are easy to solve, we may take $P=0$ and the x-subproblem in (6.7) becomes

$$x^{k+1}=(A^T A)^{-1}A^T(y^k+b-\lambda^k/\rho).$$

Otherwise, we may take $P=\frac{1}{\tau}I-\rho A^T A$ with $0<\tau\le 1/(\rho\|A\|^2)$, then the x-subproblem in (6.7) becomes

$$x^{k+1}=x^k-\tau\rho A^T\left(Ax^k-y^k-b+\lambda^k/\rho\right).$$

Example 6.4 Consider the convex constrained minimization problem:

$$\begin{cases} \text{minimize} & f(x) \\ \text{subject to} & x\in C, \end{cases} \tag{6.8}$$

where $f:\mathbb{R}^n\to(-\infty,\infty]$ is a proper, lower semicontinuous, convex function and $C\subset\mathbb{R}^n$ is a closed convex set. Let δ_C be the indicator function of C. The above problem can be reformulated as

$$\min_{x\in\mathbb{R}^n}\{f(x)+\delta_C(x)\}.$$

By introducing an additional variable, it can be transformed into the following equivalent form:

$$\min\{f(x)+\delta_C(y):x-y=0\}.$$

Applying Algorithm 6.1 gives

$$\begin{aligned} x^{k+1} &= \arg\min_{x\in\mathbb{R}^n}\left\{f(x)+\langle\lambda^k,x\rangle+\frac{\rho}{2}\|x-y^k\|^2\right\} \\ y^{k+1} &= \arg\min_{y\in\mathbb{R}^n}\left\{\delta_C(y)-\langle\lambda^k,y\rangle+\frac{\rho}{2}\|x^{k+1}-y\|^2\right\} \\ \lambda^{k+1} &= \lambda^k+\rho(x^{k+1}-y^{k+1}). \end{aligned}$$

After simplification, it becomes the method

$$\begin{aligned} x^{k+1} &= \arg\min_{x\in\mathbb{R}^n} \left\{ f(x) + \frac{\rho}{2}\|x-(y^k-\lambda^k/\rho)\|^2 \right\}, \\ y^{k+1} &= P_C(x^{k+1}+\lambda^k/\rho), \\ \lambda^{k+1} &= \lambda^k + \rho(x^{k+1}-y^{k+1}). \end{aligned} \tag{6.9}$$

Now, we apply this method to solve the minimization problem:

$$\min\{\|x\|_1 : Ax = b\}, \tag{6.10}$$

where A is an $m \times n$ matrix and $b \in \mathbb{R}^m$. Problem (6.10) takes the form (6.8) with $f(x) = \|x\|_1$ and $C = \{x \in \mathbb{R}^n : Ax = b\}$. Assuming AA^T is invertible, then

$$P_C(v) = v + A^T(AA^T)^{-1}(b - Av), \quad \forall v \in \mathbb{R}^n.$$

An application of (6.9) leads to the following algorithm:

$$\begin{aligned} x^{k+1} &= S_{1/\rho}(y^k - \lambda^k/\rho), \\ y^{k+1} &= (x^{k+1}+\lambda^k/\rho) + A^T(AA^T)^{-1}(b - A(x^{k+1}+\lambda^k/\rho)), \\ \lambda^{k+1} &= \lambda^k + \rho(x^{k+1}-y^{k+1}) \end{aligned}$$

for solving (6.10) whenever AA^T is invertible.

Example 6.5 We next use Algorithm 6.2 to formulate an alternative algorithm for solving (6.10). By introducing an additional variable, we may reformulate (6.10) into the equivalent form:

$$\min\{\|y\|_1 : Ax = b \text{ and } x - y = 0\}. \tag{6.11}$$

The corresponding augmented Lagrangian function is

$$L_\rho(x, y; \lambda, \mu) = \|y\|_1 + \langle \lambda, Ax - b\rangle + \langle \mu, x - y\rangle + \frac{\rho}{2}(\|Ax - b\|^2 + \|x - y\|^2).$$

Thus, applying Algorithm 6.2 with $P \succeq 0$ and $Q = 0$ to (6.11) gives

$$\begin{aligned} x^{k+1} &= \arg\min_{x\in\mathbb{R}^n} \left\{ \langle\lambda^k, Ax\rangle + \langle\mu^k, x\rangle + \frac{\rho}{2}(\|Ax-b\|^2 + \|x-y^k\|^2) + \frac{1}{2}\|x-x^k\|_P^2 \right\}, \\ y^{k+1} &= \arg\min_{y\in\mathbb{R}^n} \left\{ \|y\|_1 - \langle\mu^k, y\rangle + \frac{\rho}{2}\|x^{k+1}-y\|^2 \right\}, \\ \lambda^{k+1} &= \lambda^k + \rho(Ax^{k+1} - b), \\ \mu^{k+1} &= \mu^k + \rho(x^{k+1} - y^{k+1}) \end{aligned}$$

which after simplification becomes the algorithm

$$\begin{aligned}
x^{k+1} &= (I + A^T A + P/\rho)^{-1}(A^T b + y^k - (A^T \lambda^k + \mu^k - P x^k)/\rho), \\
y^{k+1} &= S_{1/\rho}(x^{k+1} + \mu^k/\rho), \\
\lambda^{k+1} &= \lambda^k + \rho(Ax^{k+1} - b), \\
\mu^{k+1} &= \mu^k + \rho(x^{k+1} - y^{k+1}).
\end{aligned}$$

In case the matrix $(I + A^T A)^{-1}$ is easy to handle, we may take $P = 0$ and the x-subproblem becomes

$$x^{k+1} = (I + A^T A)^{-1}(A^T b + y^k - (A^T \lambda^k + \mu^k)/\rho).$$

Otherwise, we may take $P = \rho(\tau I - A^T A)$ with $\tau \geq \|A\|^2$ and the x-subproblem then becomes

$$x^{k+1} = x^k - \frac{1}{1+\tau}\left(A^T(Ax^k - b) + (x^k - y^k) + (A^T \lambda^k + \mu^k)/\rho\right).$$

Example 6.6 Consider the unconstrained convex minimization problem:

$$\min_{x\in\mathbb{R}^n} \{f(Ax) + g(Dx)\},$$

where A and D are matrices of size $m \times n$ and $p \times n$, respectively, and $f : \mathbb{R}^m \to (-\infty, \infty]$ and $g : \mathbb{R}^p \to (-\infty, \infty]$ are proper convex functions. This problem appears in many application areas such as imaging, compressive sensing, inverse problems, and so on. We assume that the proximal mappings of f and g are easy to be computed. This problem can be written equivalently as

$$\min \{f(y) + g(z) : Ax - y = 0 \text{ and } Dx - z = 0\}.$$

The corresponding augmented Lagrangian function is

$$\begin{aligned}
L_\rho(x, y, z; \lambda, \mu) = {} & f(y) + g(z) + \langle \lambda, Ax - y\rangle + \langle \mu, Dx - z\rangle \\
& + \frac{\rho}{2}\left(\|Ax - y\|^2 + \|Dx - z\|^2\right).
\end{aligned}$$

Applying Algorithm 6.2 with $P \succeq 0$ gives

$$\begin{aligned}
x^{k+1} = \arg\min_{x\in\mathbb{R}^n} \Bigg\{ & \langle A^T \lambda^k + D^T \mu^k, x\rangle + \frac{\rho}{2}(\|Ax - y^k\|^2 + \|Dx - z^k\|^2) \\
& + \frac{1}{2}\|x - x^k\|_P^2 \Bigg\},
\end{aligned}$$

$$
\begin{aligned}
y^{k+1} &= \arg\min_{y\in\mathbb{R}^m} \left\{ f(y) - \langle \lambda^k, y\rangle + \frac{\rho}{2}\|Ax^{k+1} - y\|^2 \right\}, \\
z^{k+1} &= \arg\min_{z\in\mathbb{R}^p} \left\{ g(z) - \langle \mu^k, z\rangle + \frac{\rho}{2}\|Dx^{k+1} - z\|^2 \right\}, \\
\lambda^{k+1} &= \lambda^k + \rho(Ax^{k+1} - y^{k+1}), \\
\mu^{k+1} &= \mu^k + \rho(Dx^{k+1} - z^{k+1}).
\end{aligned}
$$

After simplification, this leads to the algorithm:

$$
\begin{aligned}
x^{k+1} &= (A^TA + D^TD + P/\rho)^{-1}\left(A^Ty^k + D^Tz^k - (A^T\lambda^k + D^T\mu^k - Px^k)/\rho\right), \\
y^{k+1} &= \operatorname{prox}_{f/\rho}(Ax^{k+1} + \lambda^k/\rho), \\
z^{k+1} &= \operatorname{prox}_{g/\rho}(Dx^{k+1} + \mu^k/\rho), \\
\lambda^{k+1} &= \lambda^k + \rho(Ax^{k+1} - y^{k+1}), \\
\mu^{k+1} &= \mu^k + \rho(Dx^{k+1} - z^{k+1}).
\end{aligned}
$$

Note that the x-subproblem in the above algorithm involves the inversion of the matrix $A^TA + D^TD + P/\rho$. According to the properties of A and D, a proper choice of P can make the computation of the update x^{k+1} easy. If the inversion of the matrix $A^TA + D^TD$ can be done by a fast computational method, we may simply take $P = 0$. Otherwise, if one of them, say D, has good properties so that the inversion of $tI + D^TD$ can be treated by a fast method for any $t > 0$, we may take $P = \rho(\tau I - A^TA)$ with $\tau \geq \|A\|^2$ which leads to an update formula of x^{k+1} involving only the inversion of $\tau I + D^TD$. If no fast computation method is expected to either A or D, we may take $P = \rho(\tau I - A^TA - D^TD)$ with a proper choice of $\tau > 0$ so that $P \succeq 0$, then the update formula of x^{k+1} can avoid inversion of any matrices.

6.2 Convergence

The analysis of Algorithm 6.2 is based on the following two assumptions:

Assumption 6.7 *$f : \mathbb{R}^n \to (-\infty, \infty]$ and $g : \mathbb{R}^m \to (-\infty, \infty]$ are proper, lower semicontinuous, convex functions.*

Assumption 6.8 *The problem (6.1) has a Karush-Kuhn-Tucker (KKT) point, i.e., there exists $(x^*, y^*, \lambda^*) \in \mathbb{R}^n \times \mathbb{R}^m \times \mathbb{R}^p$ such that*

$$
-A^T\lambda^* \in \partial f(x^*), \quad -B^T\lambda^* \in \partial g(y^*), \quad Ax^* + By^* = c.
$$

It should be mentioned that, to guarantee the alternating direction method of multipliers to be well-defined, certain additional conditions need to be imposed to ensure that the x- and y-subproblems do have minimizers. Since the well-definedness of ADMM can be easily seen in concrete applications, we will not state these conditions explicitly to make the presentation more succinct.

In the following, we will use

$$H_* := \min\{H(x, y) := f(x) + g(y) : Ax + By = c\}$$

to denote the optimal value of the problem (6.1). Clearly, $H_* = H(x^*, y^*)$ for any KKT point (x^*, y^*, λ^*) of (6.1).

We will perform the convergence analysis of Algorithm 6.2 by using the optimality conditions on x^{k+1} and y^{k+1} which read as

$$\begin{aligned} -A^T(\lambda^k + \rho(Ax^{k+1} + By^k - c)) - P(x^{k+1} - x^k) &\in \partial f(x^{k+1}), \\ -B^T(\lambda^k + \rho(Ax^{k+1} + By^{k+1} - c)) - Q(y^{k+1} - y^k) &\in \partial g(y^{k+1}). \end{aligned} \tag{6.12}$$

For simplicity of exposition, we set

$$\begin{aligned} &\tilde{\lambda}^{k+1} := \lambda^k + \rho(Ax^{k+1} + By^k - c), \quad \widehat{Q} := Q + \rho B^T B, \\ &\Delta x^k = x^k - x^{k-1}, \quad \Delta y^k = y^k - y^{k-1}, \quad \Delta\lambda^k = \lambda^k - \lambda^{k-1}. \end{aligned}$$

Then, it follows from (6.12) that

$$-A^T\tilde{\lambda}^{k+1} - P\Delta x^{k+1} \in \partial f(x^{k+1}), \quad -B^T\tilde{\lambda}^{k+1} - \widehat{Q}\Delta y^{k+1} \in \partial g(y^{k+1}). \tag{6.13}$$

Let μ_f and μ_g denote the modulus of convexity of f and g, respectively. By using the convexity of f and g and (6.13), we can obtain for any $x \in \mathbb{R}^n$ and $y \in \mathbb{R}^m$ that

$$\begin{aligned} \mu_f\|x^{k+1} - x\|^2 &\le f(x) - f(x^{k+1}) + \langle A^T\tilde{\lambda}^{k+1} + P\Delta x^{k+1}, x - x^{k+1}\rangle, \\ \mu_g\|y^{k+1} - y\|^2 &\le g(y) - g(y^{k+1}) + \langle B^T\tilde{\lambda}^{k+1} + \widehat{Q}\Delta y^{k+1}, y - y^{k+1}\rangle. \end{aligned}$$

Adding these two inequalities gives

$$\begin{aligned} &\mu_f\|x^{k+1} - x\|^2 + \mu_g\|y^{k+1} - y\|^2 \\ &\quad \le H(x, y) - H(x^{k+1}, y^{k+1}) + \langle\tilde{\lambda}^{k+1}, A(x - x^{k+1}) + B(y - y^{k+1})\rangle \\ &\qquad + \langle P\Delta x^{k+1}, x - x^{k+1}\rangle + \langle\widehat{Q}\Delta y^{k+1}, y - y^{k+1}\rangle. \end{aligned}$$

Note that

$$\langle P\Delta x^{k+1}, x - x^{k+1}\rangle = \frac{1}{2}\left(\|x^k - x\|_P^2 - \|x^{k+1} - x\|_P^2 - \|\Delta x^{k+1}\|_P^2\right),$$

$$\langle \widehat{Q}\Delta y^{k+1}, y - y^{k+1}\rangle = \frac{1}{2}\left(\|y^k - y\|_{\widehat{Q}}^2 - \|y^{k+1} - y\|_{\widehat{Q}}^2 - \|\Delta y^{k+1}\|_{\widehat{Q}}^2\right).$$

We therefore have for any $\lambda \in \mathbb{R}^p$ that

$$\begin{aligned}
&\mu_f\|x^{k+1} - x\|^2 + \mu_g\|y^{k+1} - y\|^2 \\
&\quad \le H(x, y) - H(x^{k+1}, y^{k+1}) + \langle \tilde{\lambda}^{k+1}, Ax + By - c\rangle \\
&\quad\quad - \langle \lambda, Ax^{k+1} + By^{k+1} - c\rangle + \langle \lambda - \tilde{\lambda}^{k+1}, Ax^{k+1} + By^{k+1} - c\rangle \\
&\quad\quad + \frac{1}{2}\left(\|x^k - x\|_P^2 - \|x^{k+1} - x\|_P^2 - \|\Delta x^{k+1}\|_P^2\right) \\
&\quad\quad + \frac{1}{2}\left(\|y^k - y\|_{\widehat{Q}}^2 - \|y^{k+1} - y\|_{\widehat{Q}}^2 - \|\Delta y^{k+1}\|_{\widehat{Q}}^2\right).
\end{aligned} \tag{6.14}$$

Recall that $\lambda^{k+1} - \lambda^k = \rho(Ax^{k+1} + By^{k+1} - c)$, we have

$$\begin{aligned}
&\langle \lambda - \tilde{\lambda}^{k+1}, Ax^{k+1} + By^{k+1} - c\rangle = \frac{1}{\rho}\langle \lambda - \tilde{\lambda}^{k+1}, \lambda^{k+1} - \lambda^k\rangle \\
&\quad = \frac{1}{2\rho}\left(\|\lambda^k - \lambda\|^2 - \|\lambda^{k+1} - \lambda\|^2 - \|\lambda^k - \tilde{\lambda}^{k+1}\|^2 + \|\lambda^{k+1} - \tilde{\lambda}^{k+1}\|^2\right).
\end{aligned}$$

By the definition of $\tilde{\lambda}^{k+1}$, we have

$$\tilde{\lambda}^{k+1} - \lambda^k = \rho(Ax^{k+1} + By^k - c), \quad \lambda^{k+1} - \tilde{\lambda}^{k+1} = \rho B\Delta y^{k+1}.$$

Therefore

$$\begin{aligned}
&\langle \lambda - \tilde{\lambda}^{k+1}, Ax^{k+1} + By^{k+1} - c\rangle \\
&= \frac{1}{2\rho}\left(\|\lambda^k - \lambda\|^2 - \|\lambda^{k+1} - \lambda\|^2\right) - \frac{\rho}{2}\|Ax^{k+1} + By^k - c\|^2 + \frac{\rho}{2}\|B\Delta y^{k+1}\|^2.
\end{aligned}$$

Combining this equation with (6.14) gives

$$\begin{aligned}
&\mu_f\|x^{k+1} - x\|^2 + \mu_g\|y^{k+1} - y\|^2 \\
&\le H(x, y) - H(x^{k+1}, y^{k+1}) + \langle \tilde{\lambda}^{k+1}, Ax + By - c\rangle - \langle \lambda, Ax^{k+1} + By^{k+1} - c\rangle \\
&\quad + \frac{1}{2}\left(\|x^k - x\|_P^2 - \|x^{k+1} - x\|_P^2 - \|\Delta x^{k+1}\|_P^2\right)
\end{aligned}$$

$$+\frac{1}{2}\left(\|y^k-y\|_{\widehat{Q}}^2-\|y^{k+1}-y\|_{\widehat{Q}}^2-\|\Delta y^{k+1}\|_{\widehat{Q}}^2\right)+\frac{\rho}{2}\|B\Delta y^{k+1}\|^2$$
$$+\frac{1}{2\rho}\left(\|\lambda^k-\lambda\|^2-\|\lambda^{k+1}-\lambda\|^2\right)-\frac{\rho}{2}\|Ax^{k+1}+By^k-c\|^2. \tag{6.15}$$

For ease of exposition, we introduce the notation:

$$u=\begin{pmatrix}x\\y\\\lambda\end{pmatrix},\quad u^k=\begin{pmatrix}x^k\\y^k\\\lambda^k\end{pmatrix},\quad u^*=\begin{pmatrix}x^*\\y^*\\\lambda^*\end{pmatrix},\quad G=\begin{pmatrix}P&&\\&\hat{Q}&\\&&\frac{1}{\rho}I\end{pmatrix} \tag{6.16}$$

Then

$$\|u^k-u\|_G^2=\|x^k-x\|_P^2+\|y^k-y\|_{\widehat{Q}}^2+\frac{1}{\rho}\|\lambda^k-\lambda\|^2.$$

Therefore, we may use (6.15) to obtain

$$\mu_f\|x^{k+1}-x\|^2+\mu_g\|y^{k+1}-y\|^2$$
$$\leq H(x,y)-H(x^{k+1},y^{k+1})+\langle\tilde{\lambda}^{k+1},Ax+By-c\rangle-\langle\lambda,Ax^{k+1}+By^{k+1}-c\rangle$$
$$+\frac{1}{2}\left(\|u^k-u\|_G^2-\|u^{k+1}-u\|_G^2\right)-\frac{\rho}{2}\|Ax^{k+1}+By^k-c\|^2$$
$$-\frac{1}{2}\|\Delta x^{k+1}\|_P^2-\frac{1}{2}\|\Delta y^{k+1}\|_{\widehat{Q}}^2+\frac{\rho}{2}\|B\Delta y^{k+1}\|^2. \tag{6.17}$$

Note that

$$\rho\|B\Delta y^{k+1}\|^2-\|\Delta y^{k+1}\|_{\widehat{Q}}^2=-\|\Delta y^{k+1}\|_Q^2.$$

We therefore obtain from (6.17) the following result.

Proposition 6.9 *Let Assumption 6.7 hold. Consider Algorithm 6.2. There holds*

$$\mu_f\|x^{k+1}-x\|^2+\mu_g\|y^{k+1}-y\|^2$$
$$\leq H(x,y)-H(x^{k+1},y^{k+1})+\langle\tilde{\lambda}^{k+1},Ax+By-c\rangle-\langle\lambda,Ax^{k+1}+By^{k+1}-c\rangle$$
$$+\frac{1}{2}\left(\|u^k-u\|_G^2-\|u^{k+1}-u\|_G^2\right)-\frac{\rho}{2}\|Ax^{k+1}+By^k-c\|^2$$
$$-\frac{1}{2}\|\Delta x^{k+1}\|_P^2-\frac{1}{2}\|\Delta y^{k+1}\|_Q^2$$

for all $x\in\mathbb{R}^n$, $y\in\mathbb{R}^m$ *and* $\lambda\in\mathbb{R}^p$.

This result will play the fundamental importance in the following analysis, and it will be used to derive various convergence and convergence rate results.

Let us first use Proposition 6.9 to derive some convergence results concerning the sequence $\{(x^k, y^k, \lambda^k)\}$ produced by Algorithm 6.2. Let (x^*, y^*, λ^*) be a KKT point of (6.1). By taking $(x, y, \lambda) = (x^*, y^*, \lambda^*)$ in Proposition 6.9, we obtain

$$\begin{aligned}
&\mu_f \|x^{k+1} - x^*\|^2 + \mu_g \|y^{k+1} - y^*\|^2 \\
&\leq H_* - H(x^{k+1}, y^{k+1}) - \langle \lambda^*, Ax^{k+1} + By^{k+1} - c\rangle - \frac{\rho}{2}\|Ax^{k+1} + By^k - c\|^2 \\
&+ \frac{1}{2}\left(\|u^k - u^*\|_G^2 - \|u^{k+1} - u^*\|_G^2\right) - \frac{1}{2}\|\Delta x^{k+1}\|_P^2 - \frac{1}{2}\|\Delta y^{k+1}\|_Q^2. \qquad (6.18)
\end{aligned}$$

By using $-A^T\lambda^* \in \partial f(x^*)$, $-B^T\lambda^* \in \partial g(y^*)$, and the convexity of f and g, we have

$$\begin{aligned}
\mu_f \|x^{k+1} - x^*\|^2 &\leq f(x^{k+1}) - f(x^*) + \langle \lambda^*, A(x^{k+1} - x^*)\rangle, \\
\mu_g \|y^{k+1} - y^*\|^2 &\leq g(y^{k+1}) - g(y^*) + \langle \lambda^*, B(y^{k+1} - y^*)\rangle.
\end{aligned}$$

Adding these two equations and using $Ax^* + By^* - c = 0$, it follows that

$$\begin{aligned}
&\mu_f \|x^{k+1} - x^*\|^2 + \mu_g \|y^{k+1} - y^*\|^2 \\
&\leq H(x^{k+1}, y^{k+1}) - H_* + \langle \lambda^*, Ax^{k+1} + By^{k+1} - c\rangle. \qquad (6.19)
\end{aligned}$$

Combining (6.19) with (6.18) gives

$$\begin{aligned}
2\mu_f \|x^{k+1} - x^*\|^2 + 2\mu_g \|y^{k+1} - y^*\|^2 \leq &\frac{1}{2}\left(\|u^k - u^*\|_G^2 - \|u^{k+1} - u^*\|_G^2\right) \\
&- \frac{1}{2}\|\Delta x^{k+1}\|_P^2 - \frac{1}{2}\|\Delta y^{k+1}\|_Q^2 \\
&- \frac{\rho}{2}\|Ax^{k+1} + By^k - c\|^2. \qquad (6.20)
\end{aligned}$$

From (6.20), we can immediately obtain the following result.

Theorem 6.10 *Let Assumptions 6.7 and 6.8 hold. Consider Algorithm 6.2. Let (x^*, y^*, λ^*) be any KKT point of (6.1).*

(i) *The sequence $\{\|u^k - u^*\|_G^2\}$ is monotonically decreasing.*
(ii) *If $\mu_f > 0$ and $\mu_g > 0$, i.e., both f and g are strongly convex, then, for any initial guess y^0 and λ^0 and any $\rho > 0$, there hold $\|x^k - x^*\| \to 0$ and $\|y^k - y^*\| \to 0$ as $k \to \infty$.*

Next, we will prove some convergence results without assuming strong convexity on f and g. In order to use (6.20) further, we need to treat the term $\|Ax^{k+1}+By^k - c\|^2$. Note that

$$Ax^{k+1} + By^k - c = \frac{1}{\rho}\Delta\lambda^{k+1} - B\Delta y^{k+1}.$$

We have

$$\|Ax^{k+1} + By^k - c\| = \frac{1}{\rho^2}\|\Delta\lambda^{k+1}\|^2 + \|B\Delta y^{k+1}\|^2 - \frac{2}{\rho}\langle\Delta\lambda^{k+1}, B\Delta y^{k+1}\rangle.$$

Therefore, it follows from (6.20) that

$$\begin{aligned}&2\mu_f\|x^{k+1} - x^*\|^2 + 2\mu_g\|y^{k+1} - y^*\|^2\\&< \frac{1}{2}\left(\|u^k - u^*\|_G^2 - \|u^{k+1} - u^*\|_G^2 - \|u^{k+1} - u^k\|_G^2\right) + \langle\Delta\lambda^{k+1}, B\Delta y^{k+1}\rangle.\end{aligned} \tag{6.21}$$

Recall from (6.12) that

$$-B^T\lambda^{k+1} - Q\Delta y^{k+1} \in \partial g(y^{k+1}), \quad -B^T\lambda^k - Q\Delta y^k \in \partial g(y^k).$$

Therefore, we may use the monotonicity of subdifferential to conclude that

$$\begin{aligned}2\mu_g\|\Delta y^{k+1}\|^2 &\leq \langle Q\Delta y^k - Q\Delta y^{k+1} - B^T\Delta\lambda^{k+1}, \Delta y^{k+1}\rangle\\&= \frac{1}{2}\left(\|\Delta y^k\|_Q^2 - \|\Delta y^{k+1}\|_Q^2 - \|\Delta y^k - \Delta y^{k+1}\|_Q^2\right)\\&\quad - \langle\Delta\lambda^{k+1}, B\Delta y^{k+1}\rangle\\&\leq \frac{1}{2}\left(\|\Delta y^k\|_Q^2 - \|\Delta y^{k+1}\|_Q^2\right) - \langle\Delta\lambda^{k+1}, B\Delta y^{k+1}\rangle.\end{aligned}$$

Adding this inequality and (6.21), we thus obtain the following result.

Lemma 6.11 *Let Assumptions 6.7 and 6.8 hold. Consider Algorithm 6.2. Let* $u^* = (x^*; y^*; \lambda^*)$ *be any KKT point of (6.1). Then, there holds*

$$\begin{aligned}&2\mu_f\|x^{k+1} - x^*\|^2 + 2\mu_g\|y^{k+1} - y^*\|^2 + 2\mu_g\|\Delta y^{k+1}\|^2 + \frac{1}{2}\|u^{k+1} - u^k\|_G^2\\&\leq \frac{1}{2}\left(\|u^k - u^*\|_G^2 + \|\Delta y^k\|_Q^2\right) - \frac{1}{2}\left(\|u^{k+1} - u^*\|_G^2 + \|\Delta y^{k+1}\|_Q^2\right)\end{aligned} \tag{6.22}$$

for all $k \geq 1$.

Now, we are ready to prove the following convergence result.

Theorem 6.12 *Let Assumptions 6.7 and 6.8 hold. Consider Algorithm 6.2. There hold*

$$\begin{aligned} &H(x^k, y^k) \to H_*,\ Ax^k + By^k - c \to 0, \quad \lambda^{k+1} - \lambda^k \to 0, \\ &P(x^{k+1} - x^k) \to 0,\ B(y^{k+1} - y^k) \to 0, \quad Q(y^{k+1} - y^k) \to 0 \end{aligned} \tag{6.23}$$

as $k \to \infty$. *If* $\{(x^k, y^k, \lambda^k)\}$ *is bounded, then any cluster point is a KKT point of (6.1); moreover, there is a KKT point* $\bar{u} = (\bar{x}, \bar{y}, \bar{\lambda})$ *such that* $\|u^k - \bar{u}\|_G^2 \to 0$ *as* $k \to \infty$.

Proof Let $u^* = (x^*, y^*, \lambda^*)$ be any KKT point of (6.1). According to Lemma 6.11, we have

$$\|u^{k+1} - u^k\|_G^2 \le \left(\|u^k - u^*\|_G^2 + \|\Delta y^k\|_Q^2\right) - \left(\|u^{k+1} - u^*\|_G^2 + \|\Delta y^{k+1}\|_Q^2\right).$$

This shows that $\{\|u^k - u^*\|_G^2 + \|\Delta y^k\|_Q^2\}$ is monotonically decreasing and $\|u^{k+1} - u^k\|_G^2 \to 0$ which means that

$$\Delta\lambda^k \to 0, \quad P\Delta x^k \to 0, \quad B\Delta y^k \to 0, \quad Q\Delta y^k \to 0 \tag{6.24}$$

as $k \to \infty$. Since $\Delta\lambda^k = \rho(Ax^k + By^k - c)$, we therefore have $Ax^k + By^k - c \to 0$ as $k \to \infty$. Next, we show that $H(x^k, y^k) \to H_*$ as $k \to \infty$. According to (6.19), we have

$$H(x^k, y^k) - H_* \ge -\langle \lambda^*, Ax^k + By^k - c\rangle.$$

Therefore

$$\liminf_{k\to\infty} H(x^k, y^k) \ge H_*.$$

On the other hand, by using (6.18), we also have

$$\begin{aligned} H(x^{k+1}, y^{k+1}) - H_* \le\ & \frac{1}{2}\left(\|u^k - u^*\|_G^2 - \|u^{k+1} - u^*\|_G^2\right) \\ & - \langle \lambda^*, Ax^{k+1} + By^{k+1} - c\rangle. \end{aligned}$$

From Theorem 6.10, we know that $\{\|u^k - u^*\|_G^2\}$ is monotonically decreasing. Consequently, we can obtain

$$\limsup_{k\to\infty} H(x^k, y^k) \le H_*.$$

Therefore, $H(x^k, y^k) \to H_*$ as $k \to \infty$.

Now, we assume $\{(x^k, y^k, \lambda^k)\}$ is bounded and let $(\bar{x}, \bar{y}, \bar{z})$ be a cluster point. Then, there is a subsequence $\{(x^{k_j}, y^{k_j}, \lambda^{k_j})\}$ such that

$$(x^{k_j}, y^{k_j}, \lambda^{k_j}) \to (\bar{x}, \bar{y}, \bar{\lambda}) \quad \text{as } j \to \infty.$$

We will show that $(\bar{x}, \bar{y}, \bar{\lambda})$ is a KKT point. By using the second equation in (6.23), we immediately obtain $A\bar{x} + B\bar{y} - c = 0$. By using (6.12), we have

$$\begin{aligned} -A^T(\lambda^{k_j-1} + \rho(Ax^{k_j} + By^{k_j-1} - c)) - P(x^{k_j} - x^{k_j-1}) &\in \partial f(x^{k_j}), \\ -B^T(\lambda^{k_j-1} + \rho(Ax^{k_j} + By^{k_j} - c)) - Q(y^{k_j} - y^{k_j-1}) &\in \partial g(y^{k_j}). \end{aligned} \tag{6.25}$$

By taking $j \to \infty$, using (6.24) and the closedness of subdifferential, we can obtain $-A^T\bar{\lambda} \in \partial f(\bar{x})$ and $-B^T\bar{\lambda} \in \partial g(\bar{y})$. Thus, $\bar{u} := (\bar{x}, \bar{y}, \bar{\lambda})$ is a KKT point. Finally, as $\{\|u^k - \bar{u}\|_G^2\}$ is monotonically decreasing and $\|u^{k_j} - \bar{u}\|_G^2 \to 0$ as $j \to \infty$, we must have $\|u^k - \bar{u}\|_G^2 \to 0$ as $k \to \infty$. □

Remark 6.13 In Theorem 6.12, we obtain the convergence $\|u^k - u^*\|_G^2 \to 0$ as $k \to \infty$ under the condition that the sequence $\{u^k\}$ is bounded. According to Theorem 6.10, $\{\|u^k\|_G^2\}$ is bounded which implies the boundedness of $\{\lambda^k\}$. In the following, we will provide sufficient conditions to guarantee the boundedness of $\{(x^k, y^k)\}$.

(i) From (6.18), it follows that $\{\mu_f\|x^k\|^2 + \mu_g\|y^k\|^2 + \|u^k\|_G^2\}$ is bounded. By the definition of G, this in particular implies the boundedness of $\{\lambda^k\}$ and $\{By^k\}$. Consequently, it follows from $\Delta\lambda^k = \rho(Ax^k + By^k - c)$ that $\{Ax^k\}$ is bounded. Putting the above together, we can conclude that both $\{(\mu_f I + P + A^TA)x^k\}$ and $\{(\mu_g I + Q + B^TB)y^k\}$ are bounded. Therefore, if both the matrices

$$\mu_f I + P + A^TA \quad \text{and} \quad \mu_g I + Q + B^TB$$

are invertible, we can conclude the boundedness of $\{x^k\}$ and $\{y^k\}$.

(ii) If there exist $\beta > H_*$ and $\sigma > 0$ such that the set

$$\{(x, y) \in \mathcal{X} \times \mathcal{Y} : H(x, y) \le \beta \text{ and } \|Ax + By - c\| \le \sigma\}$$

is bounded, then $\{(x^k, y^k)\}$ is bounded. In fact, since $H(x^k, y^k) \to H_*$ and $Ax^k + By^k - c \to 0$ as shown in Theorem 6.12, the sequence $\{(x^k, y^k)\}$ is contained in the above set except for finite many terms. Thus, $\{(x^k, y^k)\}$ is bounded.

6.3 Ergodic Convergence Rate

By virtue of Proposition 6.9, we will obtain a convergence rate result for Algorithm 6.2 in the ergodic sense. That is, by setting

$$x^{(\ell)} = \frac{1}{\ell}\sum_{k=1}^{\ell} x^k, \quad y^{(\ell)} = \frac{1}{\ell}\sum_{k=1}^{\ell} y^k, \quad \tilde{\lambda}^{(\ell)} = \frac{1}{\ell}\sum_{k=1}^{\ell} \tilde{\lambda}^k,$$

we will derive some convergence rate results with respect to $x^{(\ell)}$ and $y^{(\ell)}$ in terms of the objective error and the constraint error. To see this, we sum the inequality in Proposition 6.9 over k from $k = 0$ to $k = \ell - 1$ and use the convexity of H to obtain

$$\begin{aligned}
0 &\le \ell H(x, y) - \sum_{k=0}^{\ell-1} H(x^{k+1}, y^{k+1}) + \left\langle \sum_{k=0}^{\ell-1} \tilde{\lambda}^{k+1}, Ax + By - c \right\rangle \\
&\quad - \left\langle \lambda, A\left(\sum_{k=0}^{\ell-1} x^{k+1}\right) + B\left(\sum_{k=0}^{\ell-1} y^{k+1}\right) - \ell c \right\rangle + \frac{1}{2}\|u^0 - u\|_G^2 \\
&\le \ell H(x, y) - \ell H\left(\frac{1}{\ell}\sum_{k=0}^{\ell-1} x^{k+1}, \frac{1}{\ell}\sum_{k=0}^{\ell-1} y^{k+1}\right) + \ell\langle \tilde{\lambda}^{(\ell)}, Ax + By - c\rangle \\
&\quad - \ell\langle \lambda, Ax^{(\ell)} + By^{(\ell)} - c\rangle + \frac{1}{2}\|u^0 - u\|_G^2 \\
&= \ell H(x, y) - \ell H(x^{(\ell)}, y^{(\ell)}) + \ell\langle \tilde{\lambda}^{(\ell)}, Ax + By - c\rangle \\
&\quad - \ell\langle \lambda, Ax^{(\ell)} + By^{(\ell)} - c\rangle + \frac{1}{2}\|u^0 - u\|_G^2.
\end{aligned}$$

Let (x^*, y^*, λ^*) be any KKT point of (6.1). Dividing both sides of the above equation by ℓ, then taking $x = x^*$ and $y = y^*$, and using $Ax^* + By^* - c = 0$, we can obtain

$$\begin{aligned}
&H(x^{(\ell)}, y^{(\ell)}) - H_* + \langle \lambda, Ax^{(\ell)} + By^{(\ell)} - c\rangle \\
&\quad \le \frac{1}{2\ell}\left(\|x^0 - x^*\|_P^2 + \|y^0 - y^*\|_Q^2 + \frac{1}{\rho}\|\lambda^0 - \lambda\|^2\right)
\end{aligned} \tag{6.26}$$

for any $\lambda \in \mathbb{R}^p$. With the help of this inequality, we can obtain the following ergodic convergence rate [99].

Theorem 6.14 (Ergodic Convergence Rate) *Let Assumption 6.7 and Assumption 6.8 hold. Consider Algorithm 6.2 for solving (6.1). Let* (x^*, y^*, λ^*) *be a KKT* *point of (6.1) and let $r > 0$ be any constant satisfying $r \geq 2\|\lambda^*\|$. Then, for all $\ell \geq 1$, there hold*

$$\left|H(x^{(\ell)}, y^{(\ell)}) - H_*\right| \leq \frac{1}{2\ell}\left(\|x^0 - x^*\|_P^2 + \|y^0 - y^*\|_{\widehat{Q}}^2 + \frac{1}{\rho}(\|\lambda^0\| + r)^2\right),$$

$$\|Ax^{(\ell)} + By^{(\ell)} - c\| \leq \frac{1}{r\ell}\left(\|x^0 - x^*\|_P^2 + \|y^0 - y^*\|_{\widehat{Q}}^2 + \frac{1}{\rho}(\|\lambda^0\| + r)^2\right).$$

Proof By taking $\lambda = 0$ in (6.26), we have

$$H(x^{(\ell)}, y^{(\ell)}) - H_* \leq \frac{1}{2\ell}\left(\|x^0 - x^*\|_P^2 + \|y^0 - y^*\|_{\widehat{Q}}^2 + \frac{1}{\rho}\|\lambda^0\|^2\right). \tag{6.27}$$

To proceed further, we may use $-A^T\lambda^* \in \partial f(x^*)$ and $-B^T\lambda^* \in \partial g(y^*)$ to obtain

$$f(x^{(\ell)}) \geq f(x^*) - \langle\lambda^*, A(x^{(\ell)} - x^*)\rangle,$$

$$g(x^{(\ell)}) \geq g(y^*) - \langle\lambda^*, B(y^{(\ell)} - y^*)\rangle.$$

Adding these two inequalities gives

$$\begin{aligned} H(x^{(\ell)}, y^{(\ell)}) - H_* &\geq -\langle\lambda^*, Ax^{(\ell)} + By^{(\ell)} - c\rangle \\ &\geq -\|\lambda^*\|\|Ax^{(\ell)} + By^{(\ell)} - c\|. \end{aligned} \tag{6.28}$$

Combining (6.28) with (6.26), we have

$$\begin{aligned} &-\|\lambda^*\|\|Ax^{(\ell)} + By^{(\ell)} - c\| + \langle\lambda, Ax^{(\ell)} + By^{(\ell)} - c\rangle \\ &\quad \leq \frac{1}{2\ell}\left(\|x^0 - x^*\|_P^2 + \|y^0 - y^*\|_{\widehat{Q}}^2 + \frac{1}{\rho}(\|\lambda^0\| + \|\lambda\|)^2\right). \end{aligned} \tag{6.29}$$

We may assume $Ax^{(\ell)} + By^{(\ell)} - c \neq 0$ since otherwise we can see from (6.27) and (6.28) that the results in Theorem 6.14 hold trivially. Now, we take

$$\lambda = r\frac{Ax^{(\ell)} + By^{(\ell)} - c}{\|Ax^{(\ell)} + By^{(\ell)} - c\|}$$

with $r \geq 2\|\lambda^*\|$. Note that $\|\lambda\| = r$. We thus obtain from (6.29) that

$$\begin{aligned} &(r - \|\lambda^*\|)\|Ax^{(\ell)} + By^{(\ell)} - c\| \\ &\quad \leq \frac{1}{2\ell}\left(\|x^0 - x^*\|_P^2 + \|y^0 - y^*\|_{\widehat{Q}}^2 + \frac{1}{\rho}(\|\lambda^0\| + r)^2\right). \end{aligned} \tag{6.30}$$

Since $r - \|\lambda^*\| \geq r/2$, we therefore obtain the estimate on $\|Ax^{(\ell)} + By^{(\ell)} - c\|$. Note also that $r - \|\lambda^*\| \geq \|\lambda^*\|$. We can obtain from (6.28) and (6.30) that

$$H(x^{(\ell)}, y^{(\ell)}) - H_* \geq -\frac{1}{2\ell}\left(\|x^0 - x^*\|_P^2 + \|y^0 - y^*\|_{\widehat{Q}}^2 + \frac{1}{\rho}(\|\lambda^0\| + r)^2\right)$$

This together with (6.27) then implies the estimate on $|H(x^{(\ell)}, y^{(\ell)}) - H_*|$. The proof is complete. □

6.4 Non-ergodic Convergence Rate

In this section, we will derive sublinear non-ergodic convergence rate results for Algorithm 6.2. From Theorem 6.12, it follows that $\|u^{k+1} - u^k\|_G^2 \to 0$ as $k \to \infty$. The following simple fact further shows that understanding $\|u^{k+1} - u^k\|_G^2$ is important for Algorithm 6.2.

Lemma 6.15 *If* $\|u^{k+1} - u^k\|_G^2 = 0$ *for some* k, *then* $(x^{k+1}, y^{k+1}, \lambda^{k+1})$ *must be a KKT point of (6.1).*

Proof We observe that $\|u^{k+1} - u^k\|_G^2 = 0$ implies that

$$P\Delta x^{k+1} = 0, \quad \widehat{Q}\Delta y^{k+1} = 0, \quad \Delta\lambda^{k+1} = 0.$$

Since $\widehat{Q} = Q + \rho B^T B$ with Q positive semi-definite and $\Delta\lambda^{k+1} = \rho(Ax^{k+1} + By^{k+1} - c)$, we also have

$$B\Delta y^{k+1} = 0, \quad Q\Delta y^{k+1} = 0, \quad Ax^{k+1} + By^{k+1} - c = 0.$$

Thus, it follows from (6.12) immediately that

$$-A^T\lambda^{k+1} \in \partial f(x^{k+1}), \quad -B^T\lambda^{k+1} \in \partial g(y^{k+1}), \quad Ax^{k+1} + By^{k+1} = c$$

which shows that $(x^{k+1}, y^{k+1}, \lambda^{k+1})$ is a KKT point. □

According to Lemma 6.15, it is natural to consider $\|u^{k+1} - u^k\|_G^2$ as an optimality measure for Algorithm 6.2. We may use $\|u^{k+1} - u^k\|_G^2$ to design a stopping rule, namely, given a small tolerance $\eta > 0$, we may terminate the iteration as long as

$$\|u^{k+1} - u^k\|_G^2 \leq \eta$$

is satisfied for the first time. In order to understand the complexity of this stopping rule, it is necessary to derive a decay estimate for $\|u^{k+1} - u^k\|_G^2$. We will follow

[100, 117] to show that $\|u^{k+1} - u^k\|_G^2$ actually converges to 0 with the rate $o(1/k)$ which is based on its monotonicity as stated below.

Lemma 6.16 *For Algorithm 6.2, the sequence $\{\|u^{k+1} - u^k\|_G^2\}$ is monotonically decreasing.*

Proof By using (6.12), $\lambda^{k+1} = \lambda^k + \rho(Ax^{k+1} + By^{k+1} - c)$ and the monotonicity of ∂f and ∂g, we can obtain

$$\begin{aligned}
0 \leq & \left\langle -A^T(\Delta\lambda^{k+1} - \rho B\Delta y^{k+1} + \rho B\Delta y^k) - P\Delta x^{k+1} + P\Delta x^k, \Delta x^{k+1}\right\rangle \\
& + \left\langle -B^T\Delta\lambda^{k+1} - Q\Delta y^{k+1} + Q\Delta y^k, \Delta y^{k+1}\right\rangle \\
= & - \langle \Delta\lambda^{k+1}, A\Delta x^{k+1} + B\Delta y^{k+1}\rangle + \rho\langle B(\Delta y^{k+1} - \Delta y^k), A\Delta x^{k+1}\rangle \\
& - \langle P(\Delta x^{k+1} - \Delta x^k), \Delta x^{k+1}\rangle - \langle Q(\Delta y^{k+1} - \Delta y^k), \Delta y^{k+1}\rangle.
\end{aligned}$$

Note that

$$A\Delta x^{k+1} + B\Delta y^{k+1} = \frac{1}{\rho}(\Delta\lambda^{k+1} - \Delta\lambda^k).$$

We therefore have

$$\begin{aligned}
0 \leq & - \frac{1}{\rho}\langle \Delta\lambda^{k+1}, \Delta\lambda^{k+1} - \Delta\lambda^k\rangle - \rho\langle B(\Delta y^{k+1} - \Delta y^k), B\Delta y^{k+1}\rangle \\
& + \langle B(\Delta y^{k+1} - \Delta y^k), \Delta\lambda^{k+1} - \Delta\lambda^k\rangle \\
& - \langle P(\Delta x^{k+1} - \Delta x^k), \Delta x^{k+1}\rangle - \langle Q(\Delta y^{k+1} - \Delta y^k), \Delta y^{k+1}\rangle.
\end{aligned}$$

By the polarization identity, we then have

$$\begin{aligned}
0 \leq & \frac{1}{2\rho}\left(\|\Delta\lambda^k\|^2 - \|\Delta\lambda^{k+1}\|^2 - \|\Delta\lambda^k - \Delta\lambda^{k+1}\|^2\right) \\
& + \frac{\rho}{2}\left(\|B\Delta y^k\|^2 - \|B\Delta y^{k+1}\|^2 - \|B(\Delta y^k - \Delta y^{k+1})\|^2\right) \\
& + \frac{1}{2}\left(\|\Delta x^k\|_P^2 - \|\Delta x^{k+1}\|_P^2 - \|\Delta x^k - \Delta x^{k+1}\|_P^2\right) \\
& + \frac{1}{2}\left(\|\Delta y^k\|_Q^2 - \|\Delta y^{k+1}\|_Q^2 - \|\Delta y^k - \Delta y^{k+1}\|_Q^2\right) \\
& + \langle B(\Delta y^{k+1} - \Delta y^k), \Delta\lambda^{k+1} - \Delta\lambda^k\rangle.
\end{aligned}$$

With the help of the matrix G and $\Delta u^k := (\Delta x^k; \Delta y^k; \Delta\lambda^k)$, we obtain

$$\|\Delta u^{k+1}\|_G^2 - \|\Delta u^k\|_G^2 \le -\|\Delta x^k - \Delta x^{k+1}\|_P^2 - \|\Delta y^k - \Delta y^{k+1}\|_Q^2$$
$$- \rho \left\| B(\Delta y^{k+1} - \Delta y^k) - \frac{1}{\rho}(\Delta\lambda^{k+1} - \Delta\lambda^k) \right\|^2$$

which completes the proof. □

Proposition 6.17 *Let Assumptions 6.7 and 6.8 hold. For Algorithm 6.2, there holds*

$$\|u^{k+1} - u^k\|_G^2 \le \frac{\|u^0 - u^*\|_G^2 + \|y^1 - y^0\|_Q^2}{k} \tag{6.31}$$

for all $k \ge 1$ and $\|u^{k+1} - u^k\|_G^2 = o(1/k)$ as $k \to \infty$.

Proof From Lemma 6.11 and Theorem 6.10, (i) it follows that

$$\begin{aligned}
&\sum_{j=1}^{k} \|\Delta u^{j+1}\|_G^2 \\
&\le \sum_{j=1}^{k} ((\|u^j - u^*\|_G^2 + \|\Delta y^j\|_Q^2) - (\|u^{j+1} - u^*\|_G^2 + \|\Delta y^{j+1}\|_Q^2)) \\
&\le \|u^1 - u^*\|_G^2 + \|\Delta y^1\|_Q^2 \le \|u^0 - u^*\|_G^2 + \|y^1 - y^0\|_Q^2
\end{aligned} \tag{6.32}$$

for all $k \ge 1$. By Lemma 6.16, $\{\|\Delta u^{j+1}\|_G^2\}$ is monotonically decreasing. Thus

$$k\|\Delta u^{k+1}\|_G^2 \le \sum_{j=1}^{k} \|\Delta u^{j+1}\|_G^2 \le \|u^0 - u^*\|_G^2 + \|y^1 - y^0\|_Q^2$$

which shows (6.31). Furthermore, by the monotonicity of $\{\|\Delta u^{j+1}\|_G^2\}$ and (6.32), we also have

$$\frac{k}{2}\|\Delta u^{k+1}\|_G^2 \le \sum_{j=[k/2]}^{k} \|\Delta u^{j+1}\|_G^2 \to 0 \tag{6.33}$$

as $k \to \infty$. Thus, $k\|\Delta u^{k+1}\|_G^2 = o(1)$, and hence, $\|\Delta u^{k+1}\|_G^2 = o(1/k)$ as $k \to \infty$. □

Now, we are ready to prove the non-ergodic convergence rate result for Algorithm 6.2 in terms of objective error and constraint error [66, 117].

Theorem 6.18 (Non-ergodic Convergence Rate) *Let Assumptions 6.7 and 6.8 hold. Consider Algorithm 6.2 for solving (6.1) with positive semi-definite matrices P and Q. Then*

$$\left|H(x^k, y^k) - H_*\right| = o\left(\frac{1}{\sqrt{k}}\right) \quad \text{and} \quad \|Ax^k + By^k - c\| = o\left(\frac{1}{\sqrt{k}}\right)$$

as $k \to \infty$.

Proof Since

$$\rho(Ax^k + By^k - c) = \Delta\lambda^k \quad \text{and} \quad \|\Delta\lambda^k\|^2 \le \rho\|\Delta u^k\|_G^2 \tag{6.34}$$

we may use Proposition 6.17 to obtain the estimate $\|Ax^k + By^k - c\| = o(1/\sqrt{k})$ as $k \to \infty$.

In the following, we will focus on deriving the estimate of $|H(x^k, y^k) - H_*|$. Let $\bar{u} := (\bar{x}, \bar{y}, \bar{\lambda})$ be a KKT point of (6.1). By using (6.18), we have

$$\begin{aligned} H(x^k, y^k) - H_* &\le -\langle\bar{\lambda}, Ax^k + By^k - c\rangle + \frac{1}{2}\left(\|u^{k-1} - \bar{u}\|_G^2 - \|u^k - \bar{u}\|_G^2\right) \\ &= -\frac{1}{\rho}\langle\bar{\lambda}, \Delta\lambda^k\rangle - \langle u^{k-1} - \bar{u}, G\Delta u^k\rangle - \frac{1}{2}\|\Delta u^k\|_G^2 \\ &\le \frac{\|\bar{\lambda}\|}{\rho}\|\Delta\lambda^k\| + \|u^{k-1} - \bar{u}\|_G\|\Delta u^k\|_G. \end{aligned} \tag{6.35}$$

By virtue of the monotonicity of $\{\|u^k - \bar{u}\|_G^2\}$ given in Theorem 6.10, we then obtain

$$\begin{aligned} H(x^k, y^k) - H_* &\le \frac{\|\bar{\lambda}\|}{\rho}\|\Delta\lambda^k\| + \|u^0 - \bar{u}\|_G\|\Delta u^k\|_G \\ &\le \left(\|u^0 - \bar{u}\|_G + \frac{\|\bar{\lambda}\|}{\sqrt{\rho}}\right)\|\Delta u^k\|_G. \end{aligned}$$

On the other hand, by using (6.19), we have

$$\begin{aligned} H(x^k, y^k) - H_* &\ge -\langle\bar{\lambda}, Ax^k + By^k - c\rangle = -\frac{1}{\rho}\langle\bar{\lambda}, \Delta\lambda^k\rangle \\ &\ge -\frac{\|\bar{\lambda}\|}{\rho}\|\Delta\lambda^k\| \ge -\frac{\|\bar{\lambda}\|}{\sqrt{\rho}}\|\Delta u^k\|_G. \end{aligned}$$

Therefore

$$\left|H(x^k, y^k) - H_*\right| \le \left(\|u^0 - \bar{u}\|_G + \frac{\|\bar{\lambda}\|}{\sqrt{\rho}}\right)\|\Delta u^k\|_G. \tag{6.36}$$

Now, we can use Proposition 6.17 to conclude the proof. □

6.5 Linear Convergence

According to Theorem 6.12, for the sequence $\{u^k := (x^k; y^k; \lambda^k)\}$ defined by Algorithm 6.2, if it is bounded, it always converges to a KKT point u^* in the G-semi-norm, more precisely, $\|u^k - u^*\|_G^2 + \|\Delta y^k\|_Q^2 \to 0$ as $k \to \infty$. In this section, we will show that this convergence can be linear, i.e., there exists a constant $\delta > 0$ such that

$$\|u^k - u^*\|_G^2 + \|\Delta y^k\|_Q^2 \geq (1+\delta)\left(\|u^{k+1} - u^*\|_G^2 + \|\Delta y^{k+1}\|_Q^2\right) \tag{6.37}$$

for all k provided further conditions on f, g, A, and B are imposed.

In order to obtain (6.37), we will use Lemma 6.11. We will show that, under certain conditions, the left-hand side of (6.22) can be bounded from below by $\delta(\|u^{k+1} - u^*\|_G^2 + \|\Delta y^{k+1}\|_Q^2)$ for some universal constant $\delta > 0$. The following result from linear algebra plays an important role.

Lemma 6.19 *Let M be an $m \times n$ matrix. Then, there exists a positive constant C_0 such that*

$$\|\lambda\| \leq C_0 \|M^T \lambda\|$$

for all $\lambda \in \mathcal{R}(M)$, where $\mathcal{R}(M)$ denotes the range space of M.

Proof Let $r = \text{rank}(M)$. Without loss of generality, we may assume the first r rows of M are linearly independent; we denote by M_r the matrix formed by the first r rows of M. Since any other rows of M are linear combinations of the first r rows, there must exists an $(m-r) \times r$ matrix L such that

$$M = \begin{bmatrix} I_r \\ L \end{bmatrix} M_r,$$

where I_r denotes the $r \times r$ identity matrix. For any $\lambda \in \mathcal{R}(M)$, we can write $\lambda = M\mu$ for some $\mu \in \mathbb{R}^n$. Thus

$$\lambda = \begin{bmatrix} I_r \\ L \end{bmatrix} M_r \mu = \begin{bmatrix} I_r \\ L \end{bmatrix} \lambda_r, \tag{6.38}$$

where $\lambda_r := M_r \mu$. Then

$$\begin{aligned} \|\lambda\|^2 = \lambda^T \lambda = \lambda_r^T [I_r \;\; L^T] \begin{bmatrix} I_r \\ L \end{bmatrix} \lambda_r &= \lambda_r^T (I_r + L^T L) \lambda_r \\ &\leq \lambda_{\max}(I_r + L^T L) \|\lambda_r\|^2. \end{aligned} \tag{6.39}$$

On the other hand, we have from (6.38) that

$$M^T\lambda = M_r^T[I_r\ L^T]\begin{bmatrix} I_r \\ L \end{bmatrix}\lambda_r = M_r^T(I_r + L^T L)\lambda_r = E^T\lambda_r,$$

where $E := (I_r + L^T L)M_r$. Note that E has full row rank. Thus, $\lambda_{\min}(EE^T) > 0$ and

$$\|M^T\lambda\|^2 \geq \lambda_{\min}(EE^T)\|\lambda_r\|^2. \tag{6.40}$$

Combining (6.39) and (6.40), we therefore complete the proof. □

Now, we are ready to prove the following linear convergence result of Algorithm 6.2 due to [69].

Theorem 6.20 *Let Assumptions 6.7 and 6.8 hold. Consider Algorithm 6.2. Let* $u^* = (x^*; y^*; \lambda^*)$ *be a KKT point of (6.1). Then, there exists* $\delta > 0$ *such that*

$$\|u^k - u^*\|_G^2 + \|\Delta y^k\|_Q^2 \geq (1+\delta)\left(\|u^{k+1} - u^*\|_G^2 + \|\Delta y^{k+1}\|_Q^2\right)$$

for all $k \geq 0$*, provided any one of the following conditions holds:*

(i) $P = 0$, $\mu_g > 0$, *B has full row rank, and ∇g is Lipschitz continuous.*
(ii) $\mu_g > 0$, *A has full column rank, B has full row rank, and ∇g is Lipschitz continuous.*
(iii) $\mu_f > 0$, $\mu_g > 0$, *B has full row rank, and ∇g is Lipschitz continuous.*
(iv) $\lambda^0 \in \mathcal{R}([A\ B])$, $\mu_f > 0$, $\mu_g > 0$, *and both ∇f and ∇g are Lipschitz continuous.*
(v) $\lambda^0 \in \mathcal{R}([A\ B])$, $\mu_g > 0$, *A has full column rank, and both ∇f and ∇g are Lipschitz continuous.*

Proof Let r_{k+1} denote the left-hand side of (6.22), i.e.,

$$r_{k+1} := 2\mu_f\|x^{k+1} - x^*\|^2 + 2\mu_g\|y^{k+1} - y^*\|^2 + 2\mu_g\|\Delta y^{k+1}\|^2 + \frac{1}{2}\|u^{k+1} - u^k\|_G^2.$$

It suffices to show that

$$\|u^{k+1} - u^*\|_G^2 + \|\Delta y^{k+1}\|_Q^2 \leq C r_{k+1},$$

where here and in the following C is used to denote a generic constant independent of k.

(i) When $P = 0$, we have

$$\|u^{k+1} - u^*\|_G^2 = \|y^{k+1} - y^*\|_{\widehat{Q}}^2 + \frac{1}{\rho}\|\lambda^{k+1} - \lambda^*\|^2.$$

Since $\mu_g > 0$, the term $\|y^{k+1} - y^*\|^2$ is already contained in r_{k+1}. Therefore, it suffices to show that $\|\lambda^{k+1} - \lambda^*\|^2$ can be bounded by r_{k+1}. Since g is differentiable, we have

$$-B^T\lambda^* = \nabla g(y^*) \quad \text{and} \quad -B^T\lambda^{k+1} - Q\Delta y^{k+1} = \nabla g(y^{k+1}).$$

Therefore

$$B^T(\lambda^{k+1} - \lambda^*) = \nabla g(y^*) - \nabla g(y^{k+1}) - Q\Delta y^{k+1}.$$

By using the Lipschitz continuity of ∇g, we then obtain

$$\begin{aligned}\|B^T(\lambda^{k+1} - \lambda^*)\|^2 &\le C\left(\|y^{k+1} - y^*\|^2 + \|\Delta y^{k+1}\|_Q^2\right)\\ &\le C\left(\|y^{k+1} - y^*\|^2 + \|\Delta y^{k+1}\|_{\widehat{Q}}^2\right).\end{aligned}$$

Since B has full row rank, we have

$$\|\lambda^{k+1} - \lambda^*\|^2 \le C\|B^T(\lambda^{k+1} - \lambda^*)\|^2.$$

Therefore

$$\|\lambda^{k+1} - \lambda^*\|^2 \le C\left(\|y^{k+1} - y^*\|^2 + \|\Delta y^{k+1}\|_{\widehat{Q}}^2\right). \tag{6.41}$$

Note that $\|\Delta y^{k+1}\|_{\widehat{Q}}^2$ is also contained in r_{k+1}. This implies that $\|\lambda^{k+1} - \lambda^*\|^2 \le Cr_{k+1}$.

(ii) Since $\mu_g > 0$, the term $\|y^{k+1} - y^*\|^2$ is contained in r_{k+1}. The control of $\|\lambda^{k+1} - \lambda^*\|^2$ by r_{k+1} has been shown in (i). We only need to estimate $\|x^{k+1} - x^*\|^2$. Note that

$$A(x^{k+1} - x^*) = -B(y^{k+1} - y^*) + \frac{1}{\rho}\Delta\lambda^{k+1}.$$

We have

$$\|A(x^{k+1} - x^*)\|^2 \le C\left(\|y^{k+1} - y^*\|^2 + \|\Delta\lambda^{k+1}\|^2\right).$$

Since A has full column rank, we have

$$\|x^{k+1} - x^*\|^2 \le C\|A(x^{k+1} - x^*)\|^2.$$

Thus

$$\|x^{k+1} - x^*\|^2 \le C\left(\|y^{k+1} - y^*\|^2 + \|\Delta\lambda^{k+1}\|^2\right). \tag{6.42}$$

Therefore, $\|x^{k+1} - x^*\|^2 \le Cr_{k+1}$. Consequently, $\|u^{k+1} - u^*\|_G^2 \le Cr_{k+1}$.

(iii) Because $\mu_f > 0$ and $\mu_g > 0$, the terms $\|x^{k+1} - x^*\|^2$ and $\|y^{k+1} - y^*\|^2$ are contained in r_{k+1}. Under the given conditions, it has been shown in (i) that $\|\lambda^{k+1} - \lambda^*\|^2$ can be bounded by r_{k+1}. Therefore, $\|u^{k+1} - u^*\|_G^2 \le Cr_{k+1}$.

(iv) Since both f and g are differentiable, from the KKT conditions on (x^*, y^*, λ^*), we have

$$-A^T\lambda^* = \nabla f(x^*), \quad -B^T\lambda^* = \nabla g(y^*),$$

and from the optimality conditions on x^{k+1} and y^{k+1}, we have

$$\begin{aligned} -A^T\lambda^{k+1} - \rho A^T B(y^k - y^{k+1}) - P\Delta x^{k+1} &= \nabla f(x^{k+1}), \\ -B^T\lambda^{k+1} - Q\Delta y^{k+1} &= \nabla g(y^{k+1}). \end{aligned}$$

Thus

$$A^T(\lambda^{k+1} - \lambda^*) = \left(\nabla f(x^*) - \nabla f(x^{k+1})\right) + \rho A^T B(y^{k+1} - y^k) - P\Delta x^{k+1},$$

$$B^T(\lambda^{k+1} - \lambda^*) = \left(\nabla g(x^*) - \nabla g(x^{k+1})\right) - Q\Delta y^{k+1}.$$

Let $M = [A\ B]$. Then, by using the Lipschitz continuity of ∇f and ∇g, we have

$$\begin{aligned} &\|M^T(\lambda^{k+1} - \lambda^*)\|^2 \\ &\le C\left(\|\nabla f(x^*) - \nabla f(x^{k+1})\|^2 + \|\nabla g(y^*) - \nabla g(y^{k+1})\|^2\right. \\ &\quad \left. + \|A^T B\Delta y^{k+1}\|^2 + \|P\Delta x^{k+1}\|^2 + \|Q\Delta y^{k+1}\|^2\right) \\ &\le C\left(\|x^{k+1} - x^*\|^2 + \|y^{k+1} - y^*\|^2 + \|B\Delta y^{k+1}\|^2 + \|\Delta y^{k+1}\|_Q^2\right. \\ &\quad \left. + \|\Delta x^{k+1}\|_P^2\right) \\ &\le C\left(\|x^{k+1} - x^*\|^2 + \|y^{k+1} - y^*\|^2 + \|\Delta y^{k+1}\|_{\hat{Q}}^2 + \|\Delta x^{k+1}\|_P^2\right). \end{aligned}$$

Recall that

$$\lambda^{k+1} = \lambda^k + \rho A(x^{k+1} - x^*) + \rho B(y^{k+1} - y^*) = \lambda^k + M\begin{pmatrix} \rho(x^{k+1} - x^*) \\ \rho(y^{k+1} - y^*) \end{pmatrix}.$$

We may use $\lambda^0 \in \mathcal{R}(M)$ and an induction argument to show that $\lambda^k \in \mathcal{R}(M)$ for all k. As the limit point of $\{\lambda^k\}$, we also have $\lambda^* \in \mathcal{R}(M)$. Thus, $\lambda^{k+1}-\lambda^* \in \mathcal{R}(M)$. Now, we may use Lemma 6.19 to obtain

$$\|\lambda^{k+1} - \lambda^*\|^2 \le C\|M^T(\lambda^{k+1} - \lambda^*)\|^2.$$

Consequently

$$\|\lambda^{k+1} - \lambda^*\|^2 \le C\Big(\|x^{k+1} - x^*\|^2 + \|y^{k+1} - y^*\|^2 + \|\Delta y^{k+1}\|_{\widehat{Q}}^2 + \|\Delta x^{k+1}\|_{\widehat{P}}^2\Big). \tag{6.43}$$

Since $\mu_f > 0$ and $\mu_g > 0$, the terms $\|x^{k+1} - x^*\|^2$ and $\|y^{k+1} - y^*\|^2$ are contained in r_{k+1}. Therefore, by using (6.43), we can conclude that $\|u^{k+1} - u^*\|_G^2 \le Cr_{k+1}$.

(v) Since A has full column rank and ∇g is Lipschitz continuous, we have (6.42). Note that $\|\Delta\lambda^{k+1}\|^2$ is contained in r_{k+1}. Since $\mu_g > 0$, the term $\|y^{k+1} - y^*\|^2$ is also contained in r_{k+1}. Thus, it follows from (6.42) that $\|x^{k+1} - x^*\|^2 \le Cr_{k+1}$. Under the given conditions, we also have (6.43), and hence, $\|\lambda^{k+1} - \lambda^*\|^2 \le Cr_{k+1}$. Therefore, $\|u^{k+1} - u^*\|_G^2 \le Cr_{k+1}$.

□

The linear convergence results provided in Theorem 6.20 require at least one of f and g should be strongly convex which however is not satisfied in many practical applications. In Sect. 8.5.2, we will provide further linear convergence results under certain metric subregularity conditions which cover more examples of practical interest.

6.6 Accelerated Alternating Direction Method of Multipliers

In Theorem 6.18, we have shown that for the proximal ADMM given in Algorithm 6.2 the non-ergodic convergence rate in terms of objective error is $o(1/\sqrt{k})$ under general conditions on f and g. In order to obtain ADMM with faster convergence rates, many efforts have been made to incorporate Nesterov's acceleration strategy into the design of algorithms; see, for instance, [55, 91, 173]. In these work, either a faster convergence rate does not obtain, or strong convexity is required to obtain faster convergence rates. Accelerated ADMM with the faster non-ergodic convergence rate $O(1/k)$ in terms of objective error is eventually obtained in [135] under general conditions on f and g. In this section, we will follow [135]. The accelerated proximal ADMM takes the following form.

Algorithm 6.21 *Take initial guess* x^0, y^0, *and* λ^0 *and set* $x^{-1} = x^0$, $y^{-1} = y^0$. *Take* $\rho > 0$ *and* $0 < \tau < 1$. *Set* $\theta_k = \frac{1}{1+k(1-\tau)}$ *for* $k \geq -1$. *For* $k = 0, 1, \cdots$ *do*

$$
\begin{aligned}
\bar{x}^k &= x^k + \frac{\theta_k(1-\theta_{k-1})}{\theta_{k-1}}(x^k - x^{k-1}),\\
\bar{y}^k &= y^k + \frac{\theta_k(1-\theta_{k-1})}{\theta_{k-1}}(y^k - y^{k-1}),\\
x^{k+1} &= \arg\min_x \left\{ f(x) + \langle \lambda^k, Ax\rangle + \frac{\rho}{2\theta_k}\|Ax + B\bar{y}^k - c\|^2 + \frac{1}{2\theta_k}\|x - \bar{x}^k\|_P^2 \right\},\\
y^{k+1} &= \arg\min_y \left\{ g(y) + \langle \lambda^k, By\rangle + \frac{\rho}{2\theta_k}\|Ax^{k+1} + By - c\|^2 + \frac{1}{2\theta_k}\|y - \bar{y}^k\|_Q^2 \right\},\\
\lambda^{k+1} &= \lambda^k + \tau\rho(Ax^{k+1} + By^{k+1} - c),
\end{aligned}
$$

where P *and* Q *are positive semi-definite matrices.*

The analysis of Algorithm 6.21 relies on the following elementary result.

Lemma 6.22 *Let* $\{r_k\}$ *be a sequence such that*

$$
\left\| (1 + (n-1)(1-\tau))\, r_n + \tau \sum_{k=1}^{n-1} r_k \right\| \leq C, \quad n = 1, 2, \cdots, \tag{6.44}
$$

where $0 < \tau < 1$. *Then,* $\|\sum_{k=1}^{n} r_k\| < C/\tau$ *for all* $n = 1, 2, \cdots$.

Proof Let

$$
s_n := \sum_{k=1}^{n} r_k \quad \text{and} \quad \eta_n := (1 + (n-1)(1-\tau))\, r_n + \tau \sum_{k=1}^{n-1} r_k.
$$

Then, $\|\eta_n\| \leq C$ and

$$
r_n = \frac{\eta_n - \tau s_{n-1}}{1 + (n-1)(1-\tau)}.
$$

We will show that $\|s_n\| \leq C/\tau$ for all $n \geq 1$ by induction. From (6.44) with $n = 1$, it follows that $\|s_1\| = \|r_1\| \leq C \leq C/\tau$. Next, we assume that $\|s_n\| \leq C/\tau$ for some $n \geq 1$. Then

$$
s_{n+1} = r_{n+1} + s_n = \frac{\eta_{n+1} - \tau s_n}{1 + n(1-\tau)} + s_n = \frac{\eta_{n+1} + (n+1)(1-\tau)s_n}{1 + n(1-\tau)}.
$$

Therefore, by the induction hypothesis, we have

$$\|s_{n+1}\| \leq \frac{\|\eta_{n+1}\| + (n+1)(1-\tau)\|s_n\|}{1+n(1-\tau)} \leq \frac{C + (n+1)(1-\tau)C/\tau}{1+n(1-\tau)} = \frac{C}{\tau}.$$

The proof is thus complete. □

Now, we are ready to derive the convergence rate of Algorithm 6.21 which is faster than the one on Algorithm 6.2 given in Theorem 6.18 under general conditions.

Theorem 6.23 *Let Assumptions 6.7 and 6.8 hold. Then, for Algorithm 6.21, there hold*

$$\|Ax^k + By^k - c\| \leq \frac{C_1}{1+(k-1)(1-\tau)}, \tag{6.45}$$

$$|H(x^k, y^k) - H(x^*, y^*)| \leq \frac{C_1\|\lambda^*\| + C_2}{1+(k-1)(1-\tau)} \tag{6.46}$$

for all $k \geq 1$, *where*

$$C_1 := \frac{2\|\lambda^0 - \lambda^*\| + 2\sqrt{\|\lambda^0 - \lambda^*\|^2 + \rho\|x^0 - x^*\|_P^2 + \rho\|y^0 - y^*\|_{\widehat{Q}}^2}}{\rho},$$

$$C_2 := \frac{\|\lambda^0 - \lambda^*\|^2}{2\rho} + \frac{1}{2}\|x^0 - x^*\|_P^2 + \frac{1}{2}\|y^* - y^0\|_{\widehat{Q}}^2.$$

Proof By the definition of x^{k+1} and y^{k+1}, we have

$$0 \in \partial f(x^{k+1}) + A^T\lambda^k + \frac{\rho}{\theta_k}A^T(Ax^{k+1} + B\bar{y}^k - c) + \frac{1}{\theta_k}P(x^{k+1} - \bar{x}^k),$$

$$0 \in \partial g(y^{k+1}) + B^T\lambda^k + \frac{\rho}{\theta_k}B^T(Ax^{k+1} + By^{k+1} - c) + \frac{1}{\theta_k}Q(y^{k+1} - \bar{y}^k).$$

Let

$$\bar{\lambda}^{k+1} = \lambda^k + \frac{\rho}{\theta_k}(Ax^{k+1} + B\bar{y}^k - c).$$

Then

$$-A^T\bar{\lambda}^{k+1} - \frac{1}{\theta_k}P(x^{k+1} - \bar{x}^k) \in \partial f(x^{k+1}),$$

$$-B^T\bar{\lambda}^{k+1} - \frac{1}{\theta_k}\widehat{Q}(y^{k+1} - \bar{y}^k) \in \partial g(y^{k+1}).$$

By using the convexity of f and g, we then have

$$f(x) - f(x^{k+1}) \geq -\langle \bar{\lambda}^{k+1}, A(x - x^{k+1}) \rangle - \frac{1}{\theta_k} \langle P(x^{k+1} - \bar{x}^k), x - x^{k+1} \rangle,$$

$$g(y) - g(y^{k+1}) \geq -\langle \bar{\lambda}^{k+1}, B(y - y^{k+1}) \rangle - \frac{1}{\theta_k} \langle \widehat{Q}(y^{k+1} - \bar{y}^k), y - y^{k+1} \rangle.$$

By adding these two inequalities, it follows that

$$\begin{aligned} H(x^{k+1}, y^{k+1}) - H(x, y) \leq\ & \langle \bar{\lambda}^{k+1}, A(x - x^{k+1}) + B(y - y^{k+1}) \rangle \\ & + \frac{1}{\theta_k} \langle P(x^{k+1} - \bar{x}^k), x - x^{k+1} \rangle \\ & + \frac{1}{\theta_k} \langle \widehat{Q}(y^{k+1} - \bar{y}^k), y - y^{k+1} \rangle \end{aligned}$$

for all $x \in \mathbb{R}^n$ and $y \in \mathbb{R}^m$. By taking $(x, y) = (x^k, y^k)$ and $(x, y) = (x^*, y^*)$, respectively, we obtain

$$\begin{aligned} H(x^{k+1}, y^{k+1}) - H(x^k, y^k) \leq\ & \langle \bar{\lambda}^{k+1}, A(x^k - x^{k+1}) + B(y^k - y^{k+1}) \rangle \\ & + \frac{1}{\theta_k} \langle P(x^{k+1} - \bar{x}^k), x^k - x^{k+1} \rangle \\ & + \frac{1}{\theta_k} \langle \widehat{Q}(y^{k+1} - \bar{y}^k), y^k - y^{k+1} \rangle \end{aligned} \tag{6.47}$$

and

$$\begin{aligned} H(x^{k+1}, y^{k+1}) - H(x^*, y^*) \leq\ & \langle \bar{\lambda}^{k+1}, A(x^* - x^{k+1}) + B(y^* - y^{k+1}) \rangle \\ & + \frac{1}{\theta_k} \langle P(x^{k+1} - \bar{x}^k), x^* - x^{k+1} \rangle \\ & + \frac{1}{\theta_k} \langle \widehat{Q}(y^{k+1} - \bar{y}^k), y^* - y^{k+1} \rangle. \end{aligned} \tag{6.48}$$

Now, we multiply (6.47) and (6.48) by $1 - \theta_k$ and θ_k, respectively, and then add them to obtain

$$\begin{aligned} & H(x^{k+1}, y^{k+1}) - (1 - \theta_k) H(x^k, y^k) - \theta_k H(x^*, y^*) \\ & \leq \langle \bar{\lambda}^{k+1}, \theta_k (Ax^* + By^*) + (1 - \theta_k)(Ax^k + By^k) - (Ax^{k+1} + By^{k+1}) \rangle \\ & \quad + \frac{1}{\theta_k} \langle P(x^{k+1} - \bar{x}^k), \theta_k x^* + (1 - \theta_k) x^k - x^{k+1} \rangle \end{aligned}$$

$$+\frac{1}{\theta_k}\langle \widehat{Q}(y^{k+1}-\bar{y}^k), \theta_k y^* + (1-\theta_k)y^k - y^{k+1}\rangle$$
$$= \langle \lambda^*, (1-\theta_k)(Ax^k + By^k - c) - (Ax^{k+1} + By^{k+1} - c)\rangle + I_1 + I_2 + I_3,$$

where

$$I_1 = \langle \bar{\lambda}^{k+1} - \lambda^*, (1-\theta_k)(Ax^k + By^k - c) - (Ax^{k+1} + By^{k+1} - c)\rangle,$$
$$I_2 = \frac{1}{\theta_k}\langle P(x^{k+1} - \bar{x}^k), \theta_k x^* + (1-\theta_k)x^k - x^{k+1}\rangle,$$
$$I_3 = \frac{1}{\theta_k}\langle \widehat{Q}(y^{k+1} - \bar{y}^k), \theta_k y^* + (1-\theta_k)y^k - y^{k+1}\rangle.$$

Regrouping the terms and letting

$$a_k := H(x^k, y^k) - H(x^*, y^*) + \langle \lambda^*, Ax^k + By^k - c\rangle,$$

we have

$$a_{k+1} - (1-\theta_k)a_k \le I_1 + I_2 + I_3. \tag{6.49}$$

We first treat the terms I_2 and I_3. By the polarization identity, we can write

$$I_2 = \frac{1}{2\theta_k}\Big(\|\theta_k x^* + (1-\theta_k)x^k - \bar{x}^k\|_P^2 - \|\theta_k x^* + (1-\theta_k)x^k - x^{k+1}\|_P^2$$
$$- \|x^{k+1} - \bar{x}^k\|_P^2\Big),$$
$$I_3 = \frac{1}{2\theta_k}\Big(\|\theta_k y^* + (1-\theta_k)y^k - \bar{y}^k\|_{\widehat{Q}}^2 - \|\theta_k y^* + (1-\theta_k)y^k - y^{k+1}\|_{\widehat{Q}}^2$$
$$- \|y^{k+1} - \bar{y}^k\|_{\widehat{Q}}^2\Big).$$

Let

$$u^k = \frac{x^k}{\theta_{k-1}} - \frac{1-\theta_{k-1}}{\theta_{k-1}}x^{k-1}, \quad v^k = \frac{y^k}{\theta_{k-1}} - \frac{1-\theta_{k-1}}{\theta_{k-1}}y^{k-1}.$$

Then, by using the definition of $\bar{x}^k$ and $\bar{y}^k$, it is easy to check that

$$\frac{\bar{x}^k}{\theta_k} - \frac{1-\theta_k}{\theta_k}x^k = u^k, \qquad \frac{\bar{y}^k}{\theta_k} - \frac{1-\theta_k}{\theta_k}y^k = v^k.$$

Therefore

$$I_2 = \frac{\theta_k}{2}\left(\|x^* - u^k\|_P^2 - \|x^* - u^{k+1}\|_P^2\right) - \frac{1}{2\theta_k}\|x^{k+1} - \bar{x}^k\|_P^2, \tag{6.50}$$

$$I_3 = \frac{\theta_k}{2}\left(\|y^* - v^k\|_{\widehat{Q}}^2 - \|y^* - v^{k+1}\|_{\widehat{Q}}^2\right) - \frac{1}{2\theta_k}\|y^{k+1} - \bar{y}^k\|_{\widehat{Q}}^2. \tag{6.51}$$

Next, we consider I_1. We first use the facts $\frac{1}{\theta_k} - 1 = \frac{1}{\theta_{k-1}} - \tau$ and $\lambda^{k+1} - \lambda^k = \tau\rho(Ax^{k+1} + By^{k+1} - c)$ to write

$$\begin{aligned}
&(1-\theta_k)(Ax^k + By^k - c) - (Ax^{k+1} + By^{k+1} - c)\\
&= \theta_k\left[\left(\frac{1}{\theta_k} - 1\right)(Ax^k + By^k - c) - \frac{1}{\theta_k}(Ax^{k+1} + By^{k+1} - c)\right]\\
&= \theta_k\left[\left(\frac{1}{\theta_{k-1}} - \tau\right)(Ax^k + By^k - c)\right.\\
&\quad \left. - \left(\frac{1}{\theta_k} - \tau\right)(Ax^{k+1} + By^{k+1} - c) - \frac{\lambda^{k+1} - \lambda^k}{\rho}\right]\\
&= \theta_k(z^k - z^{k+1}),
\end{aligned}$$

where

$$z^k = \left(\frac{1}{\theta_{k-1}} - \tau\right)(Ax^k + By^k - c) + \frac{\lambda^k}{\rho}.$$

Consequently

$$\begin{aligned}
I_1 &= \theta_k\langle\bar{\lambda}^{k+1} - \lambda^*, z^k - z^{k+1}\rangle = \rho\theta_k\left\langle\frac{\bar{\lambda}^{k+1}}{\rho} - \frac{\lambda^*}{\rho}, z^k - z^{k+1}\right\rangle\\
&= \frac{\rho\theta_k}{2}\left(\left\|z^k - \frac{\lambda^*}{\rho}\right\|^2 - \left\|z^{k+1} - \frac{\lambda^*}{\rho}\right\|^2 - \left\|z^k - \frac{\bar{\lambda}^{k+1}}{\rho}\right\|^2 + \left\|z^{k+1} - \frac{\bar{\lambda}^{k+1}}{\rho}\right\|^2\right).
\end{aligned}$$

Note that

$$\begin{aligned}
z^{k+1} - \frac{\bar{\lambda}^{k+1}}{\rho} &= \left(\frac{1}{\theta_k} - \tau\right)(Ax^{k+1} + By^{k+1} - c) + \frac{\lambda^{k+1}}{\rho}\\
&\quad - \left(\frac{\lambda^k}{\rho} + \frac{1}{\theta_k}(Ax^{k+1} + B\bar{y}^k - c)\right)\\
&= \frac{1}{\theta_k}B(y^{k+1} - \bar{y}^k) - \tau(Ax^{k+1} + By^{k+1} - c) + \frac{\lambda^{k+1} - \lambda^k}{\rho}\\
&= \frac{1}{\theta_k}B(y^{k+1} - \bar{y}^k).
\end{aligned}$$

Therefore

$$I_1 = \frac{\rho\theta_k}{2}\left(\left\|z^k - \frac{\lambda^*}{\rho}\right\|^2 - \left\|z^{k+1} - \frac{\lambda^*}{\rho}\right\|^2 - \left\|z^k - \frac{\bar{\lambda}^{k+1}}{\rho}\right\|^2\right) + \frac{\rho}{2\theta_k}\|B(y^{k+1} - \bar{y}^k)\|^2. \tag{6.52}$$

Combining (6.50), (6.51), and (6.52) with (6.49) and noting that $\rho\|B(y^{k+1} - \bar{y}^k)\|^2 \le \|y^{k+1} - \bar{y}^k\|^2_{\widehat{Q}}$, we can obtain

$$\begin{aligned} a_{k+1} - (1-\theta_k)a_k \le{}& \frac{\rho\theta_k}{2}\left(\left\|z^k - \frac{\lambda^*}{\rho}\right\|^2 - \left\|z^{k+1} - \frac{\lambda^*}{\rho}\right\|^2\right) \\ &+ \frac{\theta_k}{2}\left(\|x^* - u^k\|^2_P - \|x^* - u^{k+1}\|^2_P\right) \\ &+ \frac{\theta_k}{2}\left(\|y^* - v^k\|^2_{\widehat{Q}} - \|y^* - v^{k+1}\|^2_{\widehat{Q}}\right). \end{aligned}$$

Multiplying both sides by $2/\theta_k$ and using again the fact $\frac{1-\theta_k}{\theta_k} = \frac{1}{\theta_{k-1}} - \tau$, we obtain

$$\begin{aligned} \frac{2}{\theta_k}a_{k+1} - \frac{2}{\theta_{k-1}}a_k + 2\tau a_k \le{}& \left(\rho\left\|z^k - \frac{\lambda^*}{\rho}\right\|^2 + \|x^* - u^k\|^2_P + \|y^* - v^k\|^2_{\widehat{Q}}\right) \\ &- \left(\rho\left\|z^{k+1} - \frac{\lambda^*}{\rho}\right\|^2 + \|x^* - u^{k+1}\|^2_P + \|y^* - v^{k+1}\|^2_{\widehat{Q}}\right). \end{aligned}$$

Now, we can sum this inequality over k from $k = 0$ to $k = n-1$ and use $\theta_{-1} = 1/\tau$ to obtain

$$\begin{aligned} \frac{2}{\theta_{n-1}}a_n + 2\tau\sum_{k=1}^{n-1}a_k + \rho\left\|z^n - \frac{\lambda^*}{\rho}\right\|^2 &\le \rho\left\|z^0 - \frac{\lambda^*}{\rho}\right\|^2 + \|x^* - u^0\|^2_P + \|y^* - v^0\|^2_{\widehat{Q}} \\ &= \frac{\|\lambda^0 - \lambda^*\|^2}{\rho} + \|x^0 - x^*\|^2_P + \|y^* - y^0\|^2_{\widehat{Q}} \\ &=: C_0. \end{aligned}$$

Recall from (6.19) that $a_k \ge 0$ for all k. Consequently

$$\frac{2}{\theta_{n-1}}a_n + \rho\left\|z^n - \frac{\lambda^*}{\rho}\right\|^2 \le C_0. \tag{6.53}$$

As both terms on the left-hand side of (6.53) are nonnegative, we must have

$$a_n \leq \frac{C_0}{2}\theta_{n-1} \quad \text{and} \quad \left\| z^n - \frac{\lambda^*}{\rho} \right\| \leq \sqrt{\frac{C_0}{\rho}}. \tag{6.54}$$

From the second inequality in (6.54), we can obtain that

$$\left\| z^n - \frac{\lambda^0}{\rho} \right\| \leq \frac{\|\lambda^0 - \lambda^*\|}{\rho} + \sqrt{\frac{C_0}{\rho}}. \tag{6.55}$$

Recall the definition of z^n, we have

$$\begin{aligned} z^n - \frac{\lambda^0}{\rho} &= \left(\frac{1}{\theta_{n-1}} - \tau\right)(Ax^n + By^n - c) + \frac{1}{\rho}(\lambda^n - \lambda^0) \\ &= \left(\frac{1}{\theta_{n-1}} - \tau\right)(Ax^n + By^n - c) + \frac{1}{\rho}\sum_{k=0}^{n-1}(\lambda^{k+1} - \lambda^k) \\ &= \left(\frac{1}{\theta_{n-1}} - \tau\right)(Ax^n + By^n - c) + \tau\sum_{k=0}^{n-1}(Ax^{k+1} + By^{k+1} - c) \\ &= \frac{1}{\theta_{n-1}}(Ax^n + By^n - c) + \tau\sum_{k=1}^{n-1}(Ax^k + By^k - c). \end{aligned}$$

Thus, it follows from (6.55) and the definition of θ_{n-1} that

$$\begin{aligned} &\left\| (1 + (n-1)(1-\tau))(Ax^n + By^n - c) + \tau\sum_{k=1}^{n-1}(Ax^k + By^k - c) \right\| \\ &\leq \frac{\|\lambda^0 - \lambda^*\|}{\rho} + \sqrt{\frac{C_0}{\rho}}. \end{aligned}$$

Now, we can use Lemma 6.22 with $r_k = Ax^k + By^k - c$ to conclude that

$$\tau \left\| \sum_{k=1}^{n-1}(Ax^k + By^k - c) \right\| \leq \frac{\|\lambda^0 - \lambda^*\|}{\rho} + \sqrt{\frac{C_0}{\rho}}.$$

Consequently

$$(1 + (n-1)(1-\tau)) \left\| Ax^n + By^n - c \right\| \leq \frac{2\|\lambda^0 - \lambda^*\|}{\rho} + 2\sqrt{\frac{C_0}{\rho}}$$

which shows (6.45). From the first inequality in (6.54) and $a_n \geq 0$, it follows that

$$\begin{aligned}\left|H(x^n, y^n) - H(x^*, y^*)\right| &\leq \left|\langle \lambda^*, Ax^n + By^n - c\rangle\right| + \frac{C_0}{2}\theta_{n-1}\\ &\leq \|\lambda^*\|\|Ax^n + By^n - c\| + \frac{C_0}{2}\theta_{n-1}\end{aligned}$$

which together with (6.45) and the definition of θ_{n-1} shows (6.46).

□

Problems

6.1 Consider the minimization problem:

$$\min\{\|Dx\|_1 : Ax = b\},$$

where $A \in \mathbb{R}^{m\times n}$, $D \in \mathbb{R}^{p\times n}$, and $b \in \mathbb{R}^m$.

(i) Assume that $A^TA + D^TD$ is invertible. Use ADMM to derive for any $t > 0$ the following iterative algorithm:

$$\begin{cases} x^{k+1} = (A^TA + D^TD)^{-1}\left(A^T(b - \lambda^k/t) + D^T(z^k + \mu^k/t)\right), \\ z^{k+1} = S_{1/t}(Dx^{k+1} - \mu^k/t), \\ \lambda^{k+1} = \lambda^k + t(Ax^{k+1} - b), \\ \mu^{k+1} = \mu^k + t(z^{k+1} - Dx^{k+1}). \end{cases}$$

(ii) Apply the above algorithm to solve the problem:

$$\min\left\{\sum_{i=1}^{n-1}|x_{i+1} - x_i| : Ax = b\right\},$$

write a code for implementation, and test it by taking $A = \texttt{randn(m, n)}$ with $m = 600$ and $n = 1000$ and taking $b = Ax^*$ where $x^* \in \mathbb{R}^n$ is a piecewise constant signal chosen by yourself with at least 10 discontinuous points. When applying the algorithm to reconstruct x^* from b, use $z = 0$, $\lambda = 0$, $\mu = 0$ and $t = 1$. Run your code 3, 5, 10, and 20 iterations and plot the computational results.

6.2 Let A be an $m \times n$ matrix with $m \gg n$ and $b \in \mathbb{R}^m$. Consider the Huber regression:

$$\min_{x\in\mathbb{R}^n} H_\mu(Ax - b),$$

where H_μ is the Huber function with parameter $\mu > 0$; see Problem 4.6. Assume A has full column rank. By introducing $z = Ax - b$ and using ADMM to derive the following algorithm for solving the Huber regression problem

$$\begin{aligned}
x^{k+1} &= (A^T A)^{-1} A^T \left(z^k + b - \lambda^k/\rho\right), \\
v^{k+1} &= Ax^{k+1} - b + \lambda^k/\rho, \\
z^{k+1} &= v^{k+1} - \frac{v^{k+1}}{\max\{\rho|v^{k+1}|, 1 + \rho\mu\}}, \\
\lambda^{k+1} &= \lambda^k + \rho(Ax^{k+1} - z^{k+1} - b).
\end{aligned}$$

Write a code to implement this algorithm.

6.3 For the convex separable minimization problem (6.1) with equality constraint, consider the method [83, 102]:

$$\begin{aligned}
x^{k+1} &\in \arg\min_{x\in\mathbb{R}^n} \left\{ f(x) + \frac{\rho}{2}\left\| Ax + By^k - c + \frac{1}{\rho}\lambda^k \right\|^2 + \frac{1}{2}\|x - x^k\|_P^2 \right\}, \\
y^{k+1} &\in \arg\min_{y\in\mathbb{R}^m} \left\{ g(y) + \frac{\rho}{2}\left\| Ax^{k+1} + By - c + \frac{1}{\rho}\lambda^k \right\|^2 + \frac{1}{2}\|y - y^k\|_Q^2 \right\}, \\
\lambda^{k+1} &= \lambda^k + \gamma\rho(Ax^{k+1} + By^{k+1} - c).
\end{aligned}$$

which is a variant of Algorithm 6.2 and a relaxation factor $\gamma > 0$ is introduced for updating the Lagrangian multipliers. The following items provide a convergence analysis for $0 < \gamma < (1 + \sqrt{5})/2$, and thus, the Lagrangian multipliers can be updated by a larger step size than the one in Algorithm 6.2:

(i) Let $\widehat{Q} := Q + \rho B^T B$ and $\tilde{\lambda}^{k+1} := \lambda^k + \rho(Ax^{k+1} + By^{k+1} - c)$. Adapt the proof of Proposition 6.9 to show for any $(x, y, \lambda) \in \mathbb{R}^n \times \mathbb{R}^m \times \mathbb{R}^p$ that

$$\begin{aligned}
0 \le\ & H(x, y) - H(x^{k+1}, y^{k+1}) + \langle \tilde{\lambda}^{k+1}, Ax + By - c\rangle \\
& - \langle \lambda, Ax^{k+1} + By^{k+1} - c\rangle \\
& + \frac{1}{2}\left(\|x^k - x\|_P^2 + \|y^k - y\|_{\widehat{Q}}^2 + \frac{1}{\gamma\rho}\|\lambda^k - \lambda\|^2 \right) \\
& - \frac{1}{2}\left(\|x^{k+1} - x\|_P^2 + \|y^{k+1} - y\|_{\widehat{Q}}^2 + \frac{1}{\gamma\rho}\|\lambda^{k+1} - \lambda\|^2 \right) \\
& - \frac{1}{2}\|x^{k+1} - x^k\|_P^2 - \frac{1}{2}\|y^{k+1} - y^k\|_{\widehat{Q}}^2 - \frac{1}{2}(2-\gamma)\rho\|Ax^{k+1} + By^{k+1} - c\|^2 \\
& + \rho\langle Ax^{k+1} + By^{k+1} - c, B(y^{k+1} - y^k)\rangle
\end{aligned}$$

for all integers $k \ge 0$.

(ii) Using the first order optimality of y^k and y^{k+1} together with the monotonicity of ∂g to derive

$$\begin{aligned}\rho\langle Ax^{k+1} + By^{k+1} - c, B(y^{k+1} - y^k)\rangle \\ &\leq (1-\gamma)\rho\langle B(y^{k+1} - y^k), Ax^k + By^k - c\rangle \\ &\quad - \frac{1}{2}\|y^{k+1} - y^k\|_Q^2 + \frac{1}{2}\|y^k - y^{k-1}\|_Q^2\end{aligned}$$

for all integers $k \geq 1$.

(iii) Assume $0 < \gamma < (1+\sqrt{5})/2$ and choose μ to satisfy $(1-\gamma)^2 < \mu < 2-\gamma$. Show that

$$\begin{aligned}0 \leq\; & H(x, y) - H(x^{k+1}, y^{k+1}) + \langle \tilde{\lambda}^{k+1}, Ax + By - c\rangle \\ & - \langle \lambda, Ax^{k+1} + By^{k+1} - c\rangle \\ & + \frac{1}{2}(E_k - E_{k+1}) - \frac{1}{2}\|x^{k+1} - x^k\|_P^2 - \frac{1}{2}\|y^{k+1} - y^k\|_Q^2 \\ & - \frac{\mu - (1-\gamma)^2}{2\mu}\rho\|B(y^{k+1} - y^k)\|^2 - \frac{2-\gamma-\mu}{2\gamma^2\rho}\|\lambda^{k+1} - \lambda^k\|^2\end{aligned}$$

for all integers $k \geq 1$, where

$$\begin{aligned}E_k :=\; & \|x^k - x\|_P^2 + \|y^k - y\|_{\widehat{Q}}^2 + \frac{1}{\gamma\rho}\|\lambda^k - \lambda\|^2 + \|y^k - y^{k-1}\|_Q^2 \\ & + \frac{\mu}{\gamma^2\rho}\|\lambda^k - \lambda^{k-1}\|^2.\end{aligned}$$

(iv) Assume (6.1) has a KKT point and $0 < \gamma < (1+\sqrt{5})/2$. Adapt the argument in the proof of Theorem 6.12 to show that

$$H(x^k, y^k) \to H_* \quad \text{and} \quad Ax^k + By^k - c \to 0$$

as $k \to \infty$; moreover, if $\{(x^k, y^k, \lambda^k)\}$ is bounded, show that any of its cluster point is a KKT point of (6.1).

(v) Under the same conditions in (iv), derive a similar ergodic convergence rate as in Theorem 6.14.

6.4 Consider the convex separable minimization problem (6.1) with $f(x) := f_1(x) + f_2(x)$ and $g(y) := g_1(y) + g_2(y)$, where $f_1 : \mathbb{R}^n \to (-\infty, \infty]$ and $g_1 : \mathbb{R}^m \to (-\infty, \infty]$ are proper, lower semicontinuous, convex functions, $f_2 : \mathbb{R}^n \to \mathbb{R}$ and $g_2 : \mathbb{R}^m \to \mathbb{R}$ are continuous differentiable convex functions, and there exist constants L_{f_2} and L_{g_2} such that

$$\|\nabla f_2(x) - \nabla f_2(x')\| \leq L_{f_2}\|x - x'\|, \quad \|\nabla g_2(y) - \nabla g_2(y')\| \leq L_{g_2}\|y - y'\|$$

for all $x, x' \in \mathbb{R}^n$ and $y, y' \in \mathbb{R}^m$. Consider the following algorithm:

$$x^{k+1} \in \arg\min_{x\in\mathbb{R}^n} \left\{ f_1(x) + \langle \nabla f_2(x^k), x\rangle + \frac{\rho}{2}\left\| Ax + By^k - c + \frac{1}{\rho}\lambda^k \right\|^2 \right.$$
$$\left. + \frac{1}{2}\|x - x^k\|_P^2 \right\},$$
$$y^{k+1} \in \arg\min_{y\in\mathbb{R}^m} \left\{ g_1(y) + \langle \nabla g_2(y^k), y\rangle + \frac{\rho}{2}\left\| Ax^{k+1} + By - c + \frac{1}{\rho}\lambda^k \right\|^2 \right.$$
$$\left. + \frac{1}{2}\|y - y^k\|_Q^2 \right\},$$
$$\lambda^{k+1} = \lambda^k + \rho(Ax^{k+1} + By^{k+1} - c)$$

which can be obtained from Algorithm 6.2 with f_2 and g_2 replaced by their linearizations at x^k and y^k, respectively. Assume that (6.1) has a KKT point and P and Q satisfy $P \succ L_{f_2} I$ and $Q \succ L_{g_2} I$. Adapt the arguments in Sects. 6.2 and 6.3 to show that $\{(x^k, y^k)\}$ converges to an optimal solution of (6.1) with $H(x^k, y^k) \to H_*$ as $k \to \infty$. Moreover, for

$$x^{(\ell)} := \frac{1}{\ell}\sum_{k=1}^{\ell} x^k, \quad y^{(\ell)} := \frac{1}{\ell}\sum_{k=1}^{\ell} y^k, \quad \ell \geq 1,$$

show that there holds the ergodic convergence rate

$$|H(x^{(\ell)}, y^{(\ell)}) - H_*| \leq \frac{C}{\ell}, \quad \|Ax^{(\ell)} + By^{(\ell)} - c\| \leq \frac{C}{\ell}$$

for all integers $\ell \geq 1$, where C is a constant independent of ℓ.

6.5 Let X, Y, Z be (infinite dimensional) Hilbert spaces, let $f : Y \to (-\infty, \infty]$ be a proper, lower semicontinuous, strongly convex function with $\mu_f > 0$, and let $A : X \to Z$ and $L : X \to Y$ be bounded linear operators satisfying

$$\|Ax\|^2 + \|Lx\|^2 \geq \alpha\|x\|^2, \quad \forall x \in X$$

for some constant $\alpha > 0$. Given $b \in Z$, consider the minimization problem:

$$\min\{f(Lx) : Ax = b\} \tag{6.56}$$

which arises from linear inverse problems with $f(Lx)$ being used to detect the feature of sought solutions. Assume that the feasible set $\mathcal{F} := \{x : Ax = b\} \cap \operatorname{dom}(f \circ L)$ is non-empty.

(i) Show that (6.56) has a unique solution $x^\dagger$.

(ii) Write (6.56) equivalently as

$$\min\{f(y) : Ax = b \text{ and } Lx - y = 0\}$$

and use ADMM to derive the following algorithm:

$$\begin{aligned}
x^{k+1} &= \arg\min_{x\in X} \left\{\langle \lambda^k, Ax\rangle + \langle \mu^k, Lx\rangle + \frac{\rho}{2}(\|Ax - b\|^2 + \|Lx - y^k\|^2)\right\},\\
y^{k+1} &= \arg\min_{y\in Y} \left\{f(y) - \langle \mu^k, y\rangle + \frac{\rho}{2}\|Lx^{k+1} - y\|^2\right\},\\
\lambda^{k+1} &= \lambda^k + \rho(Ax^{k+1} - b),\\
\mu^{k+1} &= \mu^k + \rho(Lx^{k+1} - y^{k+1}),
\end{aligned}$$

where $\rho > 0$ is a given number. The following items outline a convergence analysis without relying on the existence of a KKT point for (6.56); see [113, 114].

(iii) Adapt the argument in the proof of Lemma 6.16 to show that

$$E_k := \|Ax^k - b\|^2 + \|Lx^k - y^k\|^2 + \|y^k - y^{k-1}\|^2$$

is monotonically decreasing and there is a constant C such that

$$\sum_{k=n}^{\infty}(\|y^{k+1}-y^k\|^2+\|x^{k+1}-x^k\|^2+\|A(x^{k+1}-x^k)\|^2+\|L(x^{k+1}-x^k)\|^2)\le C E_n$$

for all integers $n \ge 1$.

(iv) For any $\hat{x} \in \mathcal{F}$, set $\hat{y} := L\hat{x}$. Note that $\mu^k \in \partial f(y^k)$ for $k \ge 1$, we may consider the Bregman distances $a_k := D_f^{\mu^k}(\hat{y}, y^k)$ and $b_k := D_f^{\mu^k}(y^{k+1}, y^k)$. Show that

$$\begin{aligned}
a_{k+1} - a_k + b_k = &-\rho\left(\|Ax^{k+1} - b\|^2 + \|Lx^{k+1} - y^{k+1}\|^2\right)\\
&- \rho\langle (y^k - y^{k+1}) - (y^{k-1} - y^k), L(\hat{x} - x^{k+1})\rangle
\end{aligned}$$

for all integers $k \ge 1$.

(v) Based on (iii) and (iv), show that $\{y^k\}$ and $\{x^k\}$ are bounded and

$$\sum_{k=1}^{\infty}\left(E_k + D_f^{\mu^k}(y^{k+1}, y^k)\right) < \infty.$$

(vi) Use (iv) and (v) to demonstrate the convergence of $\{a_k := D_f^{\mu^k}(\hat{y}, y^k)\}$ by showing that it is a Cauchy sequence.

(vii) Prove that $\{y^k\}$ is a Cauchy sequence by showing that

$$\lim_{l\to\infty} \sup_{k\geq l} |\langle \mu^k - \mu^l, y^k - \hat{y}\rangle| = 0.$$

(viii) Finally, show that $x^k \to x^\dagger$, $y^k \to Lx^\dagger$ and $f(y^k) \to f(Lx^\dagger)$ as $k \to \infty$.

Chapter 7
Primal-Dual Splitting Algorithms

In this chapter, we study convex-concave bilinear saddle point problems. We begin by analyzing the primal-dual hybrid gradient method, providing a convergence analysis under specific conditions and carefully chosen step sizes. Next, we examine the primal-dual splitting method introduced by Chambolle and Pock, establishing its convergence under constant step sizes. Finally, we explore a primal-dual splitting method with adaptive step sizes, which eliminates the need for prior knowledge of the matrix norm involved in the problem.

7.1 Primal-Dual Hybrid Gradient Method

Consider the minimization problem:

$$\min_{x\in\mathbb{R}^n} \{H(x) := f(Ax) + g(x)\}, \tag{7.1}$$

where A is a matrix of size $m\times n$ and $f : \mathbb{R}^m \to (-\infty, \infty]$ and $g : \mathbb{R}^n \to (-\infty, \infty]$ are proper, lower semicontinuous, convex functions. Let f^* denote the Legendre-Fenchel conjugate of f. Then

$$f(Ax) = \sup_{\lambda\in\mathbb{R}^m} \left\{\langle Ax, \lambda\rangle - f^*(\lambda)\right\}.$$

Consequently

$$\min_{x\in\mathbb{R}^n} \{f(Ax) + g(x)\} = \min_{x\in\mathbb{R}^n} \sup_{\lambda\in\mathbb{R}^m} \Phi(x, \lambda), \tag{7.2}$$

Q. Jin, *Lectures on Nonsmooth Optimization*, Texts in Applied Mathematics 82,
https://doi.org/10.1007/978-3-031-91417-1_7

where x is the primal variable, λ is the dual variable, and

$$\Phi(x, \lambda) = \langle Ax, \lambda\rangle - f^*(\lambda) + g(x). \tag{7.3}$$

The primal-dual hybrid gradient (PDHG) method, which can be viewed as a special instance of the Uzawa method [5], was proposed in [238] for solving (7.1). It is an iterative method based on the formulation (7.2); each iteration consists of two steps:

(i) First, fixed $x = x_c$ and perform one step of the proximal point algorithm to maximize the function $\lambda \to \Phi(x_c, \lambda)$ to get λ_+.
(ii) Next, fixed $\lambda = \lambda_+$ and perform one step of the proximal point algorithm to minimize the function $x \to \Phi(x, \lambda_+)$ to get x_+.

Repeating this loop leads to the algorithm:

$$\lambda^{k+1} = \arg\max_{\lambda\in\mathbb{R}^m} \left\{\Phi(x^k, \lambda) - \frac{1}{2\tau_k}\|\lambda - \lambda^k\|^2\right\},$$

$$x^{k+1} = \arg\min_{x\in\mathbb{R}^n} \left\{\Phi(x, \lambda^{k+1}) + \frac{1}{2\theta_k}\|x - x^k\|^2\right\}$$

with suitable stepsizes $\tau_k > 0$ and $\theta_k > 0$. Using the expression of Φ, we obtain

Algorithm 7.1 (Primal-Dual Hybrid Gradient Method) *Take an initial guess $x_0 \in \mathrm{dom}(g)$ and $\lambda_0 \in \mathrm{dom}(f^*)$, and define*

$$\lambda^{k+1} = \arg\max_{\lambda\in\mathbb{R}^m} \left\{\langle\lambda, Ax^k\rangle - f^*(\lambda) - \frac{1}{2\tau_k}\|\lambda - \lambda^k\|^2\right\},$$

$$x^{k+1} = \arg\min_{x\in\mathbb{R}^n} \left\{\langle\lambda^{k+1}, Ax\rangle + g(x) + \frac{1}{2\theta_k}\|x - x^k\|^2\right\},$$

where $\tau_k, \theta_k > 0$ are step sizes to be chosen properly.

Note that the definition of λ^{k+1} in Algorithm 7.1 involves f^*. In terms of the proximal mapping of f^*, we can write

$$\lambda^{k+1} = \mathrm{prox}_{\tau_k f^*}(\lambda^k + \tau_k Ax^k).$$

In case $\mathrm{prox}_{\tau_k f^*}$ is difficult to calculate, we may use Moreau's decomposition formula given in Theorem 4.25 to obtain

$$\lambda^{k+1} = \lambda^k + \tau_k Ax^k - \tau_k \mathrm{prox}_{f/\tau_k}(\lambda^k/\tau_k + Ax^k)$$

which transformations the calculation of $\mathrm{prox}_{\tau_k f^*}$ to the calculation of prox_{f/τ_k}.

The convergence of the primal-dual hybrid gradient method has been considered in [36, 78, 101]. We will follow the analysis given in [36]. We need the following condition.

Assumption 7.2 *$f : \mathbb{R}^m \to (-\infty, \infty]$ and $g : \mathbb{R}^n \to (-\infty, \infty]$ are proper, lower semicontinuous, convex functions. The function Φ has a saddle point, i.e., there exists $(x^*, \lambda^*) \in \mathbb{R}^n \times \mathbb{R}^m$ such that*

$$\Phi(x^*, \lambda) \le \Phi(x^*, \lambda^*) \le \Phi(x, \lambda^*), \quad \forall x \in \mathbb{R}^n,\ \lambda \in \mathbb{R}^m.$$

Lemma 7.3 *Let Assumption 7.2 hold and consider Algorithm 7.1. Assume that* $\mathrm{dom}(f^*)$ *is a bounded set in $\mathbb{R}^m$. If the step sizes $\{\tau_k\}$ and $\{\theta_k\}$ are chosen such that*

$$\sum_{k=0}^{\infty} \theta_k^2 < \infty, \quad \sum_{k=0}^{\infty} \frac{\theta_k}{\tau_k} < \infty,$$

the sequence $\{x^k\}$ generated by Algorithm 7.1 is bounded and $\sum_{k=0}^{\infty} \|x^{k+1} - x^k\|^2 < \infty$.

Proof Let (x^*, λ^*) be a saddle point of Φ. From the definition of $\{\lambda^{k+1}\}$, we have

$$\frac{\lambda^k - \lambda^{k+1}}{\tau_k} + Ax^k \in \partial f^*(\lambda^{k+1})$$

which implies that

$$f^*(\lambda^*) \ge f^*(\lambda^{k+1}) + \frac{1}{\tau_k}\langle \lambda^k - \lambda^{k+1}, \lambda^* - \lambda^{k+1}\rangle + \langle Ax^k, \lambda^* - \lambda^{k+1}\rangle.$$

Note that

$$2\langle \lambda^k - \lambda^{k+1}, \lambda^* - \lambda^{k+1}\rangle = \|\lambda^k - \lambda^{k+1}\|^2 + \|\lambda^* - \lambda^{k+1}\|^2 - \|\lambda^k - \lambda^*\|^2.$$

Therefore, we obtain

$$\begin{aligned} \frac{\|\lambda^{k+1} - \lambda^*\|^2}{2\tau_k} \le & \frac{\|\lambda^k - \lambda^*\|^2}{2\tau_k} - \frac{\|\lambda^{k+1} - \lambda^k\|^2}{2\tau_k} + f^*(\lambda^*) - f^*(\lambda^{k+1}) \\ & - \langle Ax^k, \lambda^* - \lambda^{k+1}\rangle. \end{aligned} \tag{7.4}$$

Similarly, from the definition of x^{k+1}, we have

$$\frac{x^k - x^{k+1}}{\theta_k} - A^T\lambda^{k+1} \in \partial g(x^{k+1})$$

which implies that

$$g(x^*) \geq g(x^{k+1}) + \frac{1}{\theta_k}\langle x^k - x^{k+1}, x^* - x^{k+1}\rangle - \langle \lambda^{k+1}, A(x^* - x^{k+1})\rangle.$$

Using

$$2\langle x^k - x^{k+1}, x^* - x^{k+1}\rangle = \|x^k - x^{k+1}\|^2 + \|x^* - x^{k+1}\|^2 - \|x^* - x^k\|^2,$$

we therefore have

$$\begin{aligned}\frac{\|x^{k+1} - x^*\|^2}{2\theta_k} \leq \frac{\|x^k - x^*\|^2}{2\theta_k} - \frac{\|x^{k+1} - x^k\|^2}{2\theta_k} + g(x^*) - g(x^{k+1}) \\ + \langle \lambda^{k+1}, A(x^* - x^{k+1})\rangle. \qquad (7.5)\end{aligned}$$

Adding (7.4) and (7.5) gives

$$\begin{aligned}&\frac{\|x^{k+1} - x^*\|^2}{2\theta_k} + \frac{\|\lambda^{k+1} - \lambda^*\|^2}{2\tau_k} \\ &\leq \frac{\|x^k - x^*\|^2}{2\theta_k} + \frac{\|\lambda^k - \lambda^*\|^2}{2\tau_k} - \frac{\|x^{k+1} - x^k\|^2}{2\theta_k} - \frac{\|\lambda^{k+1} - \lambda^k\|^2}{2\tau_k} \\ &\quad + f^*(\lambda^*) - f^*(\lambda^{k+1}) + g(x^*) - g(x^{k+1}) \\ &\quad - \langle Ax^k, \lambda^* - \lambda^{k+1}\rangle + \langle \lambda^{k+1}, A(x^* - x^{k+1})\rangle.\end{aligned}$$

Recall the definition of Φ, we have

$$\begin{aligned}&f^*(\lambda^*) - f^*(\lambda^{k+1}) + g(x^*) - g(x^{k+1}) - \langle Ax^k, \lambda^* - \lambda^{k+1}\rangle \\ &\quad + \langle \lambda^{k+1}, A(x^* - x^{k+1})\rangle \\ &= \Phi(x^*, \lambda^{k+1}) - \Phi(x^{k+1}, \lambda^*) + \langle A(x^{k+1} - x^k), \lambda^* - \lambda^{k+1}\rangle.\end{aligned}$$

Thus

$$\begin{aligned}&\frac{\|x^{k+1} - x^*\|^2}{2\theta_k} + \frac{\|\lambda^{k+1} - \lambda^*\|^2}{2\tau_k} \\ &\leq \frac{\|x^k - x^*\|^2}{2\theta_k} + \frac{\|\lambda^k - \lambda^*\|^2}{2\tau_k} - \frac{\|x^{k+1} - x^k\|^2}{2\theta_k} - \frac{\|\lambda^{k+1} - \lambda^k\|^2}{2\tau_k} \\ &\quad + \Phi(x^*, \lambda^{k+1}) - \Phi(x^{k+1}, \lambda^*) + \langle A(x^{k+1} - x^k), \lambda^* - \lambda^{k+1}\rangle. \qquad (7.6)\end{aligned}$$

Since (x^*, λ^*) is a saddle point of Φ, we have $\Phi(x^*, \lambda^{k+1}) \leq \Phi(x^{k+1}, \lambda^*)$. Therefore

$$\begin{aligned}
&\frac{\|x^{k+1} - x^*\|^2}{2\theta_k} + \frac{\|\lambda^{k+1} - \lambda^*\|^2}{2\tau_k} \\
&\leq \frac{\|x^k - x^*\|^2}{2\theta_k} + \frac{\|\lambda^k - \lambda^*\|^2}{2\tau_k} - \frac{\|x^{k+1} - x^k\|^2}{2\theta_k} - \frac{\|\lambda^{k+1} - \lambda^k\|^2}{2\tau_k} \\
&\quad + \langle A(x^{k+1} - x^k), \lambda^* - \lambda^{k+1} \rangle.
\end{aligned}$$

By the Cauchy-Schwarz inequality, we have

$$\langle A(x^{k+1} - x^k), \lambda^* - \lambda^{k+1} \rangle \leq \frac{t}{2\theta_k}\|x^{k+1} - x^k\|^2 + \frac{\theta_k}{2t}\|A\|^2\|\lambda^{k+1} - \lambda^*\|^2$$

for any $0 < t < 1$. Thus

$$\begin{aligned}
\frac{\|x^{k+1} - x^*\|^2}{2\theta_k} + \frac{\|\lambda^{k+1} - \lambda^*\|^2}{2\tau_k} \leq\ & \frac{\|x^k - x^*\|^2}{2\theta_k} + \frac{\|\lambda^k - \lambda^*\|^2}{2\tau_k} \\
&- \frac{1-t}{2\theta_k}\|x^{k+1} - x^k\|^2 - \frac{\|\lambda^{k+1} - \lambda^k\|^2}{2\tau_k} \\
&+ \frac{\theta_k}{2t}\|A\|^2\|\lambda^* - \lambda^{k+1}\|^2.
\end{aligned}$$

Let D be the diameter of $\mathrm{dom}(f^*)$ which is assumed to be finite. Since $\lambda^k, \lambda^* \in \mathrm{dom}(f^*)$, we have

$$\frac{\|x^{k+1} - x^*\|^2}{2\theta_k} + \frac{1-t}{2\theta_k}\|x^{k+1} - x^k\|^2 \leq \frac{\|x^k - x^*\|^2}{2\theta_k} + \frac{D^2}{2\tau_k} + \frac{\theta_k}{2t}\|A\|^2 D^2.$$

Consequently

$$\|x^{k+1} - x^*\|^2 + (1-t)\|x^{k+1} - x^k\|^2 \leq \|x^k - x^*\|^2 + \frac{D^2\theta_k}{\tau_k} + \frac{\theta_k^2}{t}\|A\|^2 D^2.$$

Summing up for $k = 0, \cdots, n-1$, we obtain

$$\begin{aligned}
&\|x^n - x^*\|^2 + (1-t)\sum_{k=0}^{n-1}\|x^{k+1} - x^k\|^2 \\
&\leq \|x^0 - x^*\|^2 + D^2\sum_{k=0}^{n-1}\frac{\theta_k}{\tau_k} + \frac{\|A\|^2 D^2}{t}\sum_{k=0}^{n-1}\theta_k^2.
\end{aligned}$$

By the conditions on $\{\tau_k\}$ and $\{\theta_k\}$, we obtain the boundedness of $\{x^k\}$ and $\sum_{k=0}^{\infty} \|x^{k+1} - x^k\|^2 < \infty$. □

Next, we will show that along the sequence $\{x^k\}$ defined by Algorithm 7.1, the objective function values $H(x^k)$ converges to the optimal value $H_* := \min\{H(x) : x \in \mathbb{R}^n\}$ as $k \to \infty$ and every limit point of $\{x^k\}$ is an optimal solution of (7.1). This is achieved by interpreting $\{x^k\}$ as an iterative sequence defined by an ε-subgradient method. Let us first give the definition of ε-subgradient.

Definition 7.4 Let $H : \mathbb{R}^n \to (-\infty, \infty]$ be a proper convex function. Given $\varepsilon \geq 0$, a vector $\xi \in \mathbb{R}^n$ is called an ε-subgradient of H at x if

$$H(z) \geq H(x) + \langle \xi, z - x \rangle - \varepsilon, \quad \forall z \in \mathbb{R}^n.$$

We use $\partial_\varepsilon H(x)$ to denote the set of all ε-subgradients of H at x.

We will consider an ε-subgradient method and analyze its convergence. We need the following lemma.

Lemma 7.5 *Let $H : \mathbb{R}^n \to (-\infty, \infty]$ be a proper, lower semicontinuous function such that its sublevels are bounded. Let S denote the set of minimizers of H over $\mathbb{R}^n$ and let $H_* = \min_{x \in \mathbb{R}^n} H$. Then, for any $\varepsilon > 0$, there is $\delta > 0$ such that*

$$\{x : H(x) \leq H_* + \delta\} \subset S + B_\varepsilon(0).$$

Proof Assume the conclusion is not true. Then, there exists $\varepsilon > 0$ such that for any integer $k \geq 1$ there is z^k satisfying $H(z^k) \leq H_* + 1/k$ such that $\operatorname{dist}(z^k, S) \geq \varepsilon$. By the given condition, the set $\{x : H(x) \leq H_* + 1\}$ is bounded, and hence, $\{z^k\}$ is bounded. By taking a subsequence if necessary, we may assume $z^k \to z$ as $k \to \infty$ for some z. By the lower semicontinuity of H, we then have

$$H(z) \leq \liminf_{k \to \infty} H(z^k) \leq \liminf_{k \to \infty} (H_* + 1/k) = H_*$$

Consequently, $H(z) = H_*$, and thus, $z \in S$. This means that $\operatorname{dist}(z^k, S) \to 0$ as $k \to \infty$, contradicting $\operatorname{dist}(z^k, S) \geq \varepsilon > 0$. □

We now prove the following convergence result on an ε-subgradient method for solving convex minimization problems [132].

Proposition 7.6 *Let $H : \mathbb{R}^n \to (-\infty, \infty]$ be a proper, lower semicontinuous, convex function whose sublevel sets are bounded. Let $\{x^k\}$ be a sequence defined by*

$$x^{k+1} = x^k - \theta_k v^k, \quad v^k \in \partial_{\eta_k} H(x^k),$$

where $\{v^k\}$ is bounded and

$$\lim_{k\to\infty} \eta_k = 0, \quad \lim_{k\to\infty} \theta_k = 0, \quad \sum_{k=0}^{\infty} \theta_k = \infty.$$

Then, $H(x^k) \to H_* := \min_{x\in\mathbb{R}^n} H$ as $k \to \infty$, and every limit point of $\{x^k\}$ is a minimizer of H over $\mathbb{R}^n$.

Proof Let S denote the set of minimizers of H. According to Lemma 7.5, for any $\varepsilon > 0$, we can find $\delta > 0$ such that

$$\{x : H(x) \le H_* + \delta\} \subset S + B_\varepsilon(0). \tag{7.7}$$

Moreover, by the given conditions, we can find K such that

$$\theta_k \|v^k\|^2 \le \delta/2, \quad \theta_k \|v^k\| < \varepsilon, \quad \eta_k \le \delta/2, \quad \forall k \ge K. \tag{7.8}$$

Now, let $\hat{x} \in S$ be any element. Then

$$\begin{aligned} \|x^{k+1} - \hat{x}\|^2 &= \|x^k - \hat{x} - \theta_k v^k\|^2 \\ &= \|x^k - \hat{x}\|^2 + \theta_k^2 \|v^k\|^2 - 2\theta_k \langle x^k - \hat{x}, v^k\rangle \\ &= \|x^k - \hat{x}\|^2 + \theta_k \left(\theta_k \|v^k\|^2 - 2\langle x^k - \hat{x}, v^k\rangle\right). \end{aligned} \tag{7.9}$$

Summing Eq. (7.9) gives

$$\sum_{k=K}^{l} \theta_k \left(2\langle x^k - \hat{x}, v^k\rangle - \theta_k \|v^k\|^2\right) \le \|x^K - \hat{x}\|^2 < \infty$$

for all $l \ge K$. Since $\sum_{k=1}^{\infty} \theta_k = \infty$, this implies that there must exist $k_\varepsilon \ge K$ such that

$$2\langle x^{k_\varepsilon} - \hat{x}, v^{k_\varepsilon}\rangle - \theta_{k_\varepsilon} \|v^{k_\varepsilon}\|^2 < \delta/2. \tag{7.10}$$

In view of (7.8) and (7.10), we then obtain

$$\langle v^{k_\varepsilon}, x^{k_\varepsilon} - \hat{x}\rangle < \frac{\delta}{4} + \frac{1}{2}\theta_{k_\varepsilon} \|v^{k_\varepsilon}\|^2 < \frac{\delta}{2}.$$

Consequently, using $v^{k_\varepsilon} \in \partial_{\eta_{k_\varepsilon}} H(x^{k_\varepsilon})$ and (7.8), we have

$$H(\hat{x}) \ge H(x^{k_\varepsilon}) + \langle v^{k_\varepsilon}, \hat{x} - x^{k_\varepsilon}\rangle - \eta_{k_\varepsilon} \ge H(x^{k_\varepsilon}) - \delta \tag{7.11}$$

which implies $H(x^{k_\varepsilon}) \le H_* + \delta$. Therefore, by using (7.7), we obtain $x^{k_\varepsilon} \in S + B_\varepsilon(0)$.

Next, we show by induction that $x^k \subset S + B_{2\varepsilon}(0)$ for all $k \geq k_\varepsilon$. It is trivial for $k = k_\varepsilon$. Assume $x^k \in S + B_{2\varepsilon}(0)$ for some $k \geq k_\varepsilon$. Then, there exists $\hat{x} \in S$ such that $\|x^k - \hat{x}\| < 2\varepsilon$. We will use (7.9) to show $x^{k+1} \in S + B_{2\varepsilon}(0)$ by considering two cases:

(i) If $\theta_k \|v^k\|^2 - 2\langle v^k, x^k - \hat{x}\rangle < 0$, then from (7.9), it follows

$$d(x^{k+1}, S) \leq \|x^{k+1} - \hat{x}\| \leq \|x^k - \hat{x}\| < 2\varepsilon.$$

(ii) If $\theta_k \|v^k\|^2 - 2\langle v^k, x^k - \hat{x}\rangle \geq 0$, then, by using $v^k \in \partial_{\eta_k} H(x^k)$ and (7.8), we can obtain

$$\begin{aligned} H(\hat{x}) &\geq H(x^k) + \langle v^k, \hat{x} - x^k\rangle - \eta_k \\ &\geq H(x^k) - \theta_k \|v^k\|^2 - \eta_k \\ &\geq H(x^k) - \delta \end{aligned}$$

which shows that

$$x^k \in \{x : H(x) \leq H_* + \delta\} \subset S + B_\varepsilon(0).$$

Note that $\|x^{k+1} - x^k\| = \theta_k \|v^k\| < \varepsilon$. Therefore, $x^{k+1} \in S + B_{2\varepsilon}(0)$.

The above result actually shows that $d(x^k, S) \to 0$ as $k \to \infty$. Under the given conditions, S is a bounded, closed, convex set, and $\{x^k\}$ is a bounded sequence. Let x^* be a cluster point of $\{x^k\}$, i.e., there is a subsequence $\{x^{k_l}\}$ of $\{x^k\}$ such that $x^{k_l} \to x^*$ as $l \to 0$. Then, $0 = \lim_{l\to\infty} d(x^{k_l}, S) = d(x^*, S)$ which implies $x^* \in S$, i.e., x^* is a minimizer of H.

Finally, we show $H(x^k) \to H_*$ as $k \to \infty$. Since $H_* \leq H(x^k)$, it suffices to show $H_* \geq a := \limsup_{k\to\infty} H(x^k)$. We may take a subsequence $\{x^{k_l}\}$ of $\{x^k\}$ such that $a = \lim_{l\to\infty} H(x^{k_l})$. Since $\{x^{k_l}\}$ is bounded, by taking a subsequence of $\{x^{k_l}\}$ if necessary, we may assume $x^{k_l} \to x^*$ for some x^*; we have $x^* \in S$. Therefore, since $v^{k_l} \in \partial_{\eta_{k_l}} H(x^{k_l})$ implies

$$H_* = H(x^*) \geq H(x^{k_l}) + \langle v^{k_l}, x^* - x^{k_l}\rangle - \eta_{k_l},$$

we may use the boundedness of $\{v^{k_l}\}$ and $\eta_{k_l} \to 0$ as $l \to \infty$ to conclude $H^* \geq \lim_{l\to\infty} H(x^{k_l}) = a$. The proof is complete. □

Based on Proposition 7.6, we are now ready to prove the following convergence result on Algorithm 7.1.

Theorem 7.7 *Consider the problem (7.1), let Assumption 7.2 hold, and assume the sublevel sets of H are bounded. Assume further that* dom(f^*) *is compact and g is locally Lipschitz in its domain. If the step sizes $\{\tau_k\}$ and $\{\theta_k\}$ are chosen such that*

$$\sum_{k=0}^{\infty} \theta_k = \infty, \quad \sum_{k=0}^{\infty} \theta_k^2 < \infty, \quad \sum_{k=0}^{\infty} \frac{\theta_k}{\tau_k} < \infty, \quad \lim_{k\to\infty} \tau_k = \infty,$$

then for the sequence $\{x^k\}$ generated by Algorithm 7.1 we have $H(x^k) \to H_* := \min_{x\in\mathbb{R}^n} H(x)$ as $k \to \infty$ and every limit point of $\{x^k\}$ is an optimal solution of (7.1).

Proof By the definition of x^{k+1}, we have

$$0 \in A^T\lambda^{k+1} + \partial g(x^{k+1}) + \frac{1}{\theta_k}(x^{k+1} - x^k).$$

Thus, there is $\xi^{k+1} \in \partial g(x^{k+1})$ such that

$$x^{k+1} = x^k - \theta_k(\xi^{k+1} + A^T\lambda^{k+1}). \tag{7.12}$$

Let $v^k :- \xi^{k+1} + A^T\lambda^{k+1}$. According to Proposition 7.6, it sufficcs to show that $\{v^k\}$ is bounded and $v^k \in \partial_{\eta_k} H(x^k)$ for some $\eta_k \geq 0$ with $\eta_k \to 0$ as $k \to \infty$. From Lemma 7.3, it follows that $\{x^k\}$ is bounded and

$$x^{k+1} - x^k \to 0 \quad \text{as } k \to \infty. \tag{7.13}$$

Since g is locally Lipschitz in its domain, there exists a constant M such that

$$\|\xi\| \leq M, \qquad \forall \xi \in \partial g(x^k),\ k = 0, 1, \cdots. \tag{7.14}$$

and

$$g(x^{k+1}) - g(x^k) \to 0 \quad \text{as } k \to \infty. \tag{7.15}$$

From (7.14), we have in particular that $\|\xi^{k+1}\| \leq M$. Since $\lambda^{k+1} \in \text{dom}(f^*)$ and $\text{dom}(f^*)$ is bounded, $\{\lambda^{k+1}\}$ is also bounded. Therefore, $\{v^k\}$ is bounded.

It remains to show $v^k \in \partial_{\eta_k} H(x^k)$ with $0 \leq \eta_k \to 0$ as $k \to \infty$. Since $\xi^{k+1} \in \partial g(x^{k+1})$, we have for all $z \in \mathbb{R}^n$ that

$$g(z) \geq g(x^{k+1}) + \langle \xi^{k+1}, z - x^{k+1}\rangle = g(x^k) + \langle \xi^{k+1}, z - x^k\rangle - \delta_k,$$

where

$$\delta_k := g(x^k) - g(x^{k+1}) - \langle \xi^{k+1}, x^k - x^{k+1}\rangle \geq 0,$$

Thus, $\xi^{k+1} \in \partial_{\delta_k} g(x^k)$. From (7.13), (7.14), and (7.15), it follows that $\delta_k \to 0$ as $k \to \infty$. We next consider the term $A^T\lambda^{k+1}$. By the Fenchel-Young inequality, we have for any $z \in \mathbb{R}^n$ that

$$f(Az) \geq \langle \lambda^{k+1}, Az\rangle - f^*(\lambda^{k+1}) = f(Ax^k) + \langle A^T\lambda^{k+1}, z - x^k\rangle - \varepsilon_k,$$

where

$$\varepsilon_k := f^*(\lambda^{k+1}) + f(Ax^k) - \langle \lambda^{k+1}, Ax^k \rangle \geq 0.$$

This shows that $A^T\lambda^{k+1} \in \partial_{\varepsilon_k}(f \circ A)(x^k)$. We now show that $\varepsilon_k \to 0$ as $k \to \infty$. By using the definition of λ^{k+1}, we can derive that

$$\langle \lambda^{k+1}, Ax^k \rangle - f^*(\lambda^{k+1}) - \frac{\|\lambda^{k+1} - \lambda^k\|^2}{2\tau_k} \geq \langle \lambda, Ax^k \rangle - f^*(\lambda) - \frac{\|\lambda - \lambda^k\|^2}{2\tau_k}$$

for all $\lambda \in \mathbb{R}^m$. This implies that

$$\begin{aligned} f^*(\lambda^{k+1}) - \langle \lambda^{k+1}, Ax^k \rangle &\leq f^*(\lambda) - \langle \lambda, Ax^k \rangle + \frac{1}{2\tau_k}\|\lambda - \lambda^k\|^2 \\ &\leq f^*(\lambda) - \langle \lambda, Ax^k \rangle + \frac{D^2}{2\tau_k} \end{aligned}$$

for all $\lambda \in \text{dom}(f^*)$, where $D = \text{diam}(\text{dom}(f^*))$. Consequently

$$f^*(\lambda^{k+1}) - \langle \lambda^{k+1}, Ax^k \rangle \leq \inf_{\lambda \in \mathbb{R}^m} \left\{ f^*(\lambda) - \langle \lambda, Ax^k \rangle \right\} + \frac{D^2}{2\tau_k} = -f(Ax^k) + \frac{D^2}{2\tau_k}.$$

Thus, $0 \leq \varepsilon_k \leq \frac{D^2}{2\tau_k}$. Using $\tau_k \to \infty$, we obtain $\varepsilon_k \to 0$. The above argument thus shows that $v^k \in \partial_{\eta_k} H(x^k)$ with $\eta_k := \delta_k + \varepsilon_k \to 0$ as $k \to \infty$. The proof is therefore complete. □

Example 7.8 Consider the one-dimensional total variation denoising problem discussed in Example 5.50 which can be stated as

$$\min_{x \in \mathbb{R}^n} \left\{ \frac{1}{2}\|x - b\|_2^2 + \alpha \|Dx\|_1 \right\}. \tag{7.16}$$

where $D : \mathbb{R}^n \to \mathbb{R}^{n-1}$ is the discrete gradient operator. In order to apply Algorithm 7.1, we take $f(z) = \alpha\|z\|_1$ and $g(x) = \frac{1}{2}\|x - b\|_2^2$. Note that

$$f^*(\lambda) = \sup_{z \in \mathbb{R}^{n-1}} \{\langle z, \lambda \rangle - \alpha\|z\|_1\} = \begin{cases} 0, & \text{if } \|\lambda\|_\infty \leq \alpha \\ +\infty, & \text{if } \|\lambda\|_\infty > \alpha. \end{cases}$$

which in particular shows that $\text{dom}(f^*)$ is bounded. Thus, Algorithm 7.1 becomes

$$\begin{aligned} \lambda^{k+1} &= \arg\max_{\|\lambda\|_\infty \leq \alpha} \left\{ \langle \lambda, Dx^k \rangle - \frac{1}{2\tau_k}\|\lambda - \lambda^k\|^2 \right\}, \\ x^{k+1} &= \arg\min_{x \in \mathbb{R}^n} \left\{ \langle \lambda^{k+1}, Dx \rangle + \frac{1}{2}\|x - b\|^2 + \frac{1}{2\theta_k}\|x - x^k\|^2 \right\}. \end{aligned}$$

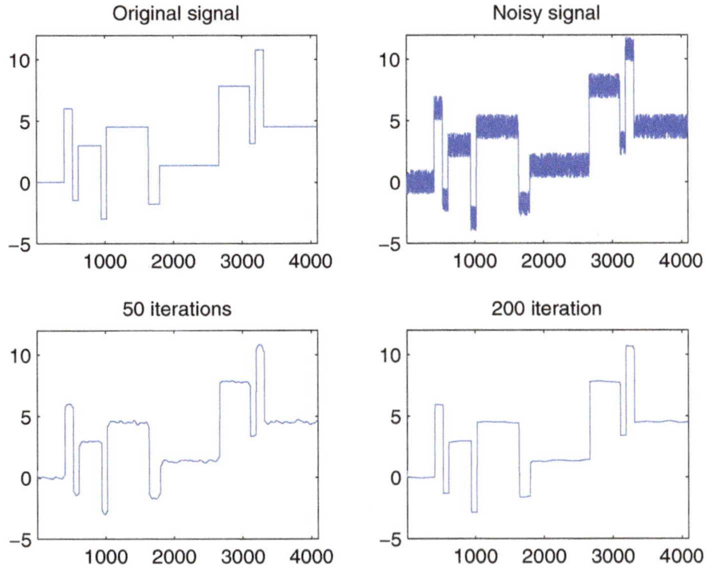

Fig. 7.1 Reconstruct a piecewise constant signal by the one-dimensional total variation denoising model (7.16) which is solved by the method (7.17). (**a**) Original signal. (**b**) Noisy signal. (**c**) 50 iterations. (**d**) 200 iteration

After simplification, it leads to the iteration scheme:

$$\begin{aligned} \lambda^{k+1} &= \max\left\{\min\left\{\lambda^k + \tau_k D x^k, \alpha\right\}, -\alpha\right\}, \\ x^{k+1} &= \frac{1}{1+\theta_k}\left(x^k + \theta_k(b - D^T\lambda^{k+1})\right). \end{aligned} \tag{7.17}$$

In order to illustrate the performance of the method (7.17), we take a piecewise constant signal x^* of size 4000 as shown in Fig. 7.1a, and we then add Gaussian noise in $N(0, 0.1)$ to x^* to produce a noisy signal b which is shown in Fig. 7.1b. Now, we apply the method (7.17) with the initial guess $\lambda^0 = 0$, $x^0 = 0$, the regularization parameter $\alpha = 20$, and the step sizes

$$\tau_k = 0.1 + 0.1k, \qquad \theta_k = \frac{1}{1 + 0.1k}.$$

to reconstruct the true signal. Clearly, $\{\tau_k\}$ and $\{\theta_k\}$ satisfy the requirements in Theorem 7.7, and thus, the corresponding method is convergent. In Fig. 7.1c and d, we plot the computational results after 50 and 200 iterations, which demonstrates the usefulness of the method.

Example 7.9 Let $b \in \mathbb{R}^{m\times n}$ be an image corrupted by Gaussian noise. We want to reconstruct a clean image by solving the two-dimensional total variation denoising problem:

$$\min_{x\in\mathbb{R}^{m\times n}} \left\{\frac{\alpha}{2}\|x-b\|_F^2 + |x|_{TV,a}\right\}, \tag{7.18}$$

where, for any matrix $x \in \mathbb{R}^{m\times n}$, $|x|_{TV,a}$ denotes its anisotropic total variation (see Example 3.29) and $\|x\|_F$ denotes its Frobenius norm which is induced by the inner product

$$\langle x, y\rangle_F = \sum_{i=1}^{m}\sum_{j=1}^{n} x_{i,j}y_{i,j}, \quad x, y \in \mathbb{R}^{m\times n}.$$

By introducing the linear operator $D : \mathbb{R}^{m\times n} \to \mathbb{R}^{m\times n} \times \mathbb{R}^{m\times n}$ and the function $f : \mathbb{R}^{m\times n} \times \mathbb{R}^{m\times n} \to \mathbb{R}$ as defined in Example 3.29, we can rewrite (7.18) in the form:

$$\min_{x\in\mathbb{R}^{m\times n}} \left\{f(Dx) + \frac{\alpha}{2}\|x-b\|_F^2\right\}.$$

When Algorithm 7.1 is used to solve the problem, it gives

$$\begin{aligned} \lambda^{k+1} &= \arg\max_{\lambda\in\mathbb{R}^{m\times n}\times\mathbb{R}^{m\times n}} \left\{\langle\lambda, Dx^k\rangle_F - f^*(\lambda) - \frac{1}{2\tau_k}\|\lambda-\lambda^k\|_F^2\right\}, \\ x^{k+1} &= \arg\min_{x\in\mathbb{R}^{m\times n}} \left\{\langle\lambda^{k+1}, Dx\rangle_F + \frac{\alpha}{2}\|x-b\|_F^2 + \frac{1}{2\theta_k}\|x-x^k\|_F^2\right\}. \end{aligned} \tag{7.19}$$

The Legendre-Fenchel conjugate of f is given by

$$\begin{aligned} f^*(\xi,\eta) &= \sup_{(u,v)\in\mathbb{R}^{m\times n}\times\mathbb{R}^{m\times n}} \left\{\langle(\xi,\eta),(u,v)\rangle_F - f(u,v)\right\}. \\ &= \sum_{i=1}^{m}\sum_{j=1}^{n} \sup_{u_{i,j},v_{i,j}\in\mathbb{R}} \left\{\xi_{i,j}u_{i,j} + \eta_{i,j}v_{i,j} - \big(|u_{i,j}| + |v_{i,j}|\big)\right\} \\ &= \begin{cases} 0 & \text{if } \max\{|\xi_{i,j}|, |\eta_{i,j}|\} \le 1 \text{ for all } i = 1,\cdots,m,\ j = 1,\cdots,n \\ \infty & \text{otherwsie.} \end{cases} \end{aligned}$$

Let

$$\Omega = \Big\{(\xi,\eta) \in \mathbb{R}^{m\times n} \times \mathbb{R}^{m\times n} : \max\{|\xi_{i,j}|, |\eta_{i,j}|\} \le 1 \text{ for } i = 1,\cdots,m \text{ and } j = 1,\cdots,n\Big\}.$$

Then, $f^*(\xi, \eta) = \delta_\Omega(\xi, \eta)$. Combining this with (7.19) yields

$$\begin{aligned} \lambda^{k+1} &= P_\Omega(\lambda^k + \tau_k D x^k), \\ x^{k+1} &= \frac{1}{1+\alpha\theta_k}\left(x^k + \theta_k(\alpha b - D^T \lambda^{k+1})\right), \end{aligned} \tag{7.20}$$

where D^T is determined by $\langle D^T\lambda, x\rangle_F = \langle \lambda, Dx\rangle_F$ whose formula is given in Example 3.29 and the projection $P_\Omega : \mathbb{R}^{m\times n} \times \mathbb{R}^{m\times n} \to \Omega$ is given by

$$P_\Omega(\phi, \psi) = (\xi, \eta),$$

where $\xi, \eta \in \mathbb{R}^{m\times n}$ are given by

$$\xi_{i,j} = \min\left\{\max\left\{\phi_{i,j}, -1\right\}, 1\right\}, \quad \eta_{i,j} = \min\left\{\max\left\{\psi_{i,j}, -1\right\}, 1\right\}.$$

For the numerical simulation, we take an image of size 512×512 and add Gaussian noise, given by $60 * (\texttt{randn}(m, n) - 0.5)$ to obtain the noisy image b. We then reconstruct the image using (7.18) with a regularization parameter $\alpha = 0.02$. The problem (7.18) is solved by the method (7.20), where the step sizes are set as $\tau_k = 2 + 0.1k$ and $\theta_k = \frac{5}{1+0.01k}$. The algorithm runs for 100 iterations. The reconstructed image is shown in Fig. 7.2.

Example 7.10 When an image is contaminated by pepper and salt noise, the model given in Example 7.9 does not work well. We need to find a better model to deal with such situation. We consider the minimization problem:

$$\min_{x\in\mathbb{R}^{m\times n}} \left\{\alpha\|x-b\|_{1,1} + |x|_{TV,a}\right\} \tag{7.21}$$

Fig. 7.2 Reconstruct the images corrupted by Gaussian noise via the total variation denoising model (7.18) which is solved by the PDHG method (7.20)

where

$$\|x\|_{1,1} = \sum_{i=1}^{m}\sum_{j=1}^{n} |x_{i,j}|.$$

When applying the PDHG method, we need to modify (7.20). The updating formula for λ^{k+1} is the same. The updating formula for x^{k+1} becomes

$$\begin{aligned} x^{k+1} &= \arg\min_{x\in\mathbb{R}^{m\times n}} \left\{\langle\lambda^{k+1}, Dx\rangle_F + \alpha\|x-b\|_{1,1} + \frac{1}{2\theta_k}\|x-x^k\|_F^2\right\} \\ &= \arg\min_{x\in\mathbb{R}^{m\times n}} \left\{\alpha\|x-b\|_{1,1} + \frac{1}{2\theta_k}\|x-(x^k-\theta_k D^T\lambda^{k+1})\|^2\right\} \\ &= b + S_{\alpha\theta_k}(x^k - b - \theta_k D^T\lambda^{k+1}). \end{aligned}$$

Thus, we obtain the algorithm:

$$\begin{aligned} \lambda^{k+1} &= P_\Omega(\lambda^k + \tau_k D x^k) \\ x^{k+1} &= b + S_{\alpha\theta_k}(x^k - b - \theta_k D^T\lambda^{k+1}). \end{aligned} \tag{7.22}$$

For the numerical simulation, we consider an image of size 512×512 and introduce salt-and-pepper noise. The resulting noisy image is displayed in the middle of Fig. 7.3. To recover a clean image, we employ the model (7.21) with a regularization parameter $\alpha = 0.8$. The reconstruction is performed using the algorithm (7.22), initialized with $\lambda^0 = 0$ and $x^0 = b$. The step sizes are set as $\tau_k = 2 + 0.001k$ and $\theta_k = 2/\tau_k$. After 1000 iterations, the reconstructed image is shown on the right of Fig. 7.3.

Fig. 7.3 Reconstruct an image corrupted by salt-and-pepper noise via the model (7.21) solved by the PDHG method (7.22)

7.2 Primal-Dual Algorithm with Extrapolation

In this section, we continue the study of the convex minimization problem (7.1). In [50], Chambolle and Pock proposed a class of first-order primal-dual algorithms by introducing an extrapolation step into each iteration step in Algorithm 7.1. The algorithm takes the following form.

Algorithm 7.11 *Choose* $\tau, \sigma > 0$, $\theta \geq 0$, $(x^0, \lambda^0) \in \mathbb{R}^n \times \mathbb{R}^m$ *and set* $x^{-1} = x^0$. *For* $k \geq 0$ *do*

$$\begin{aligned}
\bar{x}^k &= x^k + \theta(x^k - x^{k-1}), \\
\lambda^{k+1} &= \arg\max_{\lambda \in \mathbb{R}^m} \left\{ \langle A\bar{x}^k, \lambda \rangle - f^*(\lambda) - \frac{1}{2\sigma}\|\lambda - \lambda^k\|^2 \right\}, \\
x^{k+1} &= \arg\min_{x \in \mathbb{R}^n} \left\{ g(x) + \langle Ax, \lambda^{k+1} \rangle + \frac{1}{2\tau}\|x - x^k\|^2 \right\}.
\end{aligned}$$

Note that Algorithm 7.11 with $\theta = 0$ becomes Algorithm 7.1 with constant stepsize σ and τ. Algorithm 7.11 allows nonzero θ for extrapolation which may lead to algorithms with faster convergence speed. Algorithm 7.11 with $\theta = 1$ is in fact equivalent to a proximal ADMM applying to the dual problem of (7.1). Recall that the dual problem of (7.1) is

$$\min_{\lambda \in \mathbb{R}^m} \left\{ f^*(\lambda) + g^*(-A^T\lambda) \right\} \tag{7.23}$$

which can be equivalently written as

$$\min_{\lambda \in \mathbb{R}^m} \left\{ f^*(\lambda) + g^*(\mu) : A^T\lambda + \mu = 0 \right\}. \tag{7.24}$$

An application of the proximal ADMM given in Algorithm 6.2 gives the following algorithm:

$$\begin{aligned}
\lambda^{k+1} &= \arg\min_{\lambda \in \mathbb{R}^m} \left\{ f^*(\lambda) - \langle x^k, A^T\lambda \rangle + \frac{\tau}{2}\|A^T\lambda + \mu^k\|^2 + \frac{1}{2}\|\lambda - \lambda^k\|_P^2 \right\}, \\
\mu^{k+1} &= \arg\min_{\mu \in \mathbb{R}^n} \left\{ g^*(\mu) - \langle x^k, \mu \rangle + \frac{\tau}{2}\|\mu + A^T\lambda^{k+1}\|^2 \right\}, \\
x^{k+1} &= x^k - \tau(A^T\lambda^{k+1} + \mu^{k+1}),
\end{aligned} \tag{7.25}$$

where

$$P := \frac{1}{\sigma}I - \tau AA^T,$$

and $\sigma > 0$ is chosen with $\sigma\tau\|A\|^2 < 1$ so that P is positive definite. With this choice of P, we show that Algorithm 7.11 with $\theta = 1$ is equivalent to (7.25). Indeed, by the definition of μ^{k+1} and the Moreau decomposition formula given in Theorem 4.25 (ii) we have

$$\begin{aligned}\mu^{k+1} &= \operatorname{prox}_{g^*/\tau}\left(\frac{1}{\tau}x^k - A^T\lambda^{k+1}\right)\\ &= \frac{1}{\tau}\left(x^k - \tau A^T\lambda^{k+1}\right) - \frac{1}{\tau}\operatorname{prox}_{\tau g}(x^k - \tau A^T\lambda^{k+1}).\end{aligned}$$

Thus, by using the third equation in (7.25), it is easily seen that

$$x^{k+1} = \operatorname{prox}_{\tau g}(x^k - \tau A^T\lambda^{k+1}).$$

With the choice of P, we also have

$$\lambda^{k+1} = \arg\min_{\lambda\in\mathbb{R}^m}\left\{f^*(\lambda) - \langle\lambda, A(x^k - \tau(\mu^k + A^T\lambda^k))\rangle + \frac{1}{2\sigma}\|\lambda - \lambda^k\|^2\right\}.$$

Let $\bar{x}^k = x^k - \tau(\mu^k + A^T\lambda^k)$. Then

$$\lambda^{k+1} = \arg\min_{\lambda\in\mathbb{R}^m}\left\{f^*(\lambda) - \langle\lambda, A\bar{x}^k\rangle + \frac{1}{2\sigma}\|\lambda - \lambda^k\|^2\right\}.$$

By using the third equation in (7.25), i.e., $x^k = x^{k-1} - \tau(A^T\lambda^k + \mu^k)$, we can see that

$$\bar{x}^k = x^k - (x^{k-1} - x^k) = 2x^k - x^{k-1}.$$

The above argument shows that if $\{\lambda^k, \mu^k, x^k\}$ is defined by (7.25), then $\{\bar{x}^k, \lambda^k, x^k\}$ satisfies the equations in Algorithm 7.11 with $\theta = 1$.

Conversely, if $\{\bar{x}^k, \lambda^k, x^k\}$ is defined by Algorithm 7.11 with $\theta = 1$, by setting $\mu^k = \frac{1}{\tau}(x^k - \bar{x}^k) + A^T\lambda^k$, we can reverse the above argument to show that $(\lambda^k, \mu^k, x^k\}$ satisfies (7.25).

Due to the above connections, we may apply the convergence theory for ADMM developed in Chap. 6 to derive convergence results on Algorithm 7.11 with $\theta = 1$. Let $\{\lambda^k, \mu^k, x^k\}$ be the sequence defined by (7.25). According to Theorem 6.10 (i), the quantity

$$\|\lambda^k - \hat{\lambda}\|_P^2 + \tau\|\mu^k - \hat{\mu}\|^2 + \frac{1}{\tau}\|x^k - \hat{x}\|$$

is monotonically decreasing, and thus, $\{\lambda^k, \mu^k, x^k\}$ is a bounded sequence. Consequently, we may use Theorem 6.12 to conclude that there is a KKT point (λ^*, μ^*, x^*) of (7.24) such that

$$\|\lambda^k - \lambda^*\|_P^2 + \tau\|\mu^k - \mu^*\|^2 + \frac{1}{\tau}\|x^k - x^*\| \to 0 \quad \text{as } k \to \infty. \tag{7.26}$$

Note that if $(\hat{\lambda}, \hat{\mu}, \hat{x})$ is a KKT point of (7.24), then $(\hat{x}, \hat{\lambda})$ is a saddle point of Φ; conversely, if $(\hat{x}, \hat{\lambda})$ is a saddle point of Φ, then, by setting $\hat{\mu} = -A^T\hat{\lambda}$, $(\hat{\lambda}, \hat{\mu}, \hat{x})$ is a KKT point of (7.24). Therefore, we may use (7.26) to conclude the following convergence result on Algorithm 7.11 with $\theta = 1$ which has been proved in [50].

Theorem 7.12 *Let Assumption 7.2 hold and consider Algorithm 7.11 with $\theta = 1$, where σ and τ are chosen so that $\tau\sigma\|A\|^2 < 1$. Then, there exists a saddle point (x^*, λ^*) of Φ such that $(x^k, \lambda^k) \to (x^*, \lambda^*)$ as $k \to \infty$.*

Theorem 7.12 provides a convergence result on Algorithm 7.11 with $\theta = 1$ and $\tau\sigma\|A\|^2 < 1$. It is natural to ask if Algorithm 7.11 converges for other values of θ, σ, and τ. In the following, we will follow [12] to show that Algorithm 7.11 converges if

$$\theta > 1/2 \quad \text{and} \quad \tau\sigma\|A\|^2 < 4/(1+2\theta). \tag{7.27}$$

To see this, let $(\hat{x}, \hat{\lambda})$ be any saddle point of Φ and let $\{(x^k, \lambda^k)\}$ be the sequence generated by Algorithm 7.11. We introduce

$$\begin{aligned} F_k &:= f^*(\lambda^k) - f^*(\hat{\lambda}) - \langle A\hat{x}, \lambda^k - \hat{\lambda}\rangle, \\ G_k &:= g(x^k) - g(\hat{x}) + \langle A^T\hat{\lambda}, x^k - \hat{x}\rangle. \end{aligned} \tag{7.28}$$

Since $(\hat{x}, \hat{\lambda})$ is a saddle point of Φ, we must have $F_k \geq 0$ and $G_k \geq 0$. Now, we consider the quantity:

$$\begin{aligned} &\theta(G_{k+1} - G_k) + G_k + F_{k+1} \\ &= \left(\theta - \frac{1}{2}\right)(G_{k+1} - G_k) + G_{k+1} + \frac{1}{2}(G_k - G_{k+1}) + F_{k+1} \end{aligned} \tag{7.29}$$

According to the definition of x^{k+1} and λ^{k+1}, there hold the first-order optimality conditions:

$$\begin{aligned} -\frac{1}{\tau}\Delta x^{k+1} - A^T\lambda^{k+1} &\in \partial g(x^{k+1}), \\ -\frac{1}{\sigma}\Delta\lambda^{k+1} + A(x^k + \theta\Delta x^k) &\in \partial f^*(\lambda^{k+1}), \end{aligned} \tag{7.30}$$

where we use the notation

$$\Delta x^k := x^k - x^{k-1} \quad \text{and} \quad \Delta\lambda^k := \lambda^k - \lambda^{k-1}.$$

By using the first equation in (7.30) and the convexity of g, we have

$$\begin{aligned}
G_{k+1} &= g(x^{k+1}) - g(\hat{x}) + \langle A^T\hat{\lambda}, x^{k+1} - \hat{x}\rangle \\
&\le \left\langle -\frac{1}{\tau}\Delta x^{k+1} - A^T\lambda^{k+1}, x^{k+1} - \hat{x}\right\rangle + \langle A^T\hat{\lambda}, x^{k+1} - \hat{x}\rangle \\
&= -\frac{1}{\tau}\langle \Delta x^{k+1}, x^{k+1} - \hat{x}\rangle + \langle \hat{\lambda} - \lambda^{k+1}, A(x^{k+1} - \hat{x})\rangle,
\end{aligned}$$

$$\begin{aligned}
G_{k+1} - G_k &= g(x^{k+1}) - g(x^k) + \langle A^T\hat{\lambda}, \Delta x^{k+1}\rangle \\
&\le \left\langle -\frac{1}{\tau}\Delta x^{k+1} - A^T\lambda^{k+1}, \Delta x^{k+1}\right\rangle + \langle A^T\hat{\lambda}, \Delta x^{k+1}\rangle \\
&= -\frac{1}{\tau}\|\Delta x^{k+1}\|^2 + \langle \hat{\lambda} - \lambda^{k+1}, A\Delta x^{k+1}\rangle
\end{aligned}$$

and

$$\begin{aligned}
G_k - G_{k+1} &= g(x^k) - g(x^{k+1}) - \langle A^T\hat{\lambda}, \Delta x^{k+1}\rangle \\
&\le \left\langle -\frac{1}{\tau}\Delta x^k - A^T\lambda^k, -\Delta x^{k+1}\right\rangle - \langle A^T\hat{\lambda}, \Delta x^{k+1}\rangle \\
&= \frac{1}{\tau}\langle \Delta x^k, \Delta x^{k+1}\rangle + \langle \lambda^k - \hat{\lambda}, A\Delta x^{k+1}\rangle.
\end{aligned}$$

By using the second equation in (7.30) and the convexity of f^*, we also have

$$\begin{aligned}
F_{k+1} &= f^*(\lambda^{k+1}) - f^*(\hat{\lambda}) - \langle A\hat{x}, \lambda^{k+1} - \hat{\lambda}\rangle \\
&\le \left\langle -\frac{1}{\sigma}\Delta\lambda^{k+1} + A(x^k + \theta\Delta x^k), \lambda^{k+1} - \hat{\lambda}\right\rangle - \langle A\hat{x}, \lambda^{k+1} - \hat{\lambda}\rangle \\
&= -\frac{1}{\sigma}\langle \Delta\lambda^{k+1}, \lambda^{k+1} - \hat{\lambda}\rangle + \langle A(x^k - \hat{x}), \lambda^{k+1} - \hat{\lambda}\rangle + \theta\langle A\Delta x^k, \lambda^{k+1} - \hat{\lambda}\rangle.
\end{aligned}$$

Combining these inequalities with (7.29) and using $\theta - 1/2 > 0$, we can obtain

$$\begin{aligned}
&\theta(G_{k+1} - G_k) + G_k + F_{k+1} \\
&\le \left(\theta - \frac{1}{2}\right)\left(-\frac{1}{\tau}\|\Delta x^{k+1}\|^2 + \langle \hat{\lambda} - \lambda^{k+1}, A\Delta x^{k+1}\rangle\right)
\end{aligned}$$

$$
\begin{aligned}
&-\frac{1}{\tau}\langle\Delta x^{k+1}, x^{k+1}-\hat{x}\rangle+\langle\hat{\lambda}-\lambda^{k+1}, A(x^{k+1}-\hat{x})\rangle \\
&+\frac{1}{2}\left(\frac{1}{\tau}\langle\Delta x^{k}, \Delta x^{k+1}\rangle+\langle\lambda^{k}-\hat{\lambda}, A\Delta x^{k+1}\rangle\right) \\
&-\frac{1}{\sigma}\langle\Delta\lambda^{k+1}, \lambda^{k+1}-\hat{\lambda}\rangle+\langle A(x^{k}-\hat{x}), \lambda^{k+1}-\hat{\lambda}\rangle+\theta\langle A\Delta x^{k}, \lambda^{k+1}-\hat{\lambda}\rangle \\
=&-\frac{\theta}{\tau}\|\Delta x^{k+1}\|^2+\theta\langle\hat{\lambda}-\lambda^{k+1}, A\Delta x^{k+1}\rangle+\theta\langle\lambda^{k+1}-\hat{\lambda}, A\Delta x^{k}\rangle \\
&+\frac{1}{2\tau}\|\Delta x^{k+1}\|^2-\frac{1}{\tau}\langle\Delta x^{k+1}, x^{k+1}-\hat{x}\rangle+\frac{1}{2\tau}\langle\Delta x^{k}, \Delta x^{k+1}\rangle \\
&-\frac{1}{2}\langle\Delta\lambda^{k+1}, A\Delta x^{k+1}\rangle-\frac{1}{\sigma}\langle\Delta\lambda^{k+1}, \lambda^{k+1}-\hat{\lambda}\rangle.
\end{aligned}
$$

Note that

$$
\frac{1}{2\tau}\|\Delta x^{k+1}\|^2-\frac{1}{\tau}\langle\Delta x^{k+1}, x^{k+1}-\hat{x}\rangle=\frac{1}{2\tau}\left(\|x^{k}-\hat{x}\|^2-\|x^{k+1}-\hat{x}\|^2\right)
$$

and

$$
-\frac{1}{\sigma}\langle\Delta\lambda^{k+1}, \lambda^{k+1}-\hat{\lambda}\rangle=\frac{1}{2\sigma}\left(\|\lambda^{k}-\hat{\lambda}\|^2-\|\lambda^{k+1}-\hat{\lambda}\|^2-\|\Delta\lambda^{k+1}\|^2\right).
$$

We therefore obtain

$$
\begin{aligned}
&\theta(G_{k+1}-G_k)+G_k+F_{k+1}+\frac{1}{2\tau}\left(\|x^{k+1}-\hat{x}\|^2-\|x^{k}-\hat{x}\|^2\right) \\
&\le-\frac{\theta}{\tau}\|\Delta x^{k+1}\|^2+\theta\langle\hat{\lambda}-\lambda^{k+1}, A\Delta x^{k+1}\rangle+\theta\langle\lambda^{k+1}-\hat{\lambda}, A\Delta x^{k}\rangle \\
&\quad+\frac{1}{2\tau}\langle\Delta x^{k}, \Delta x^{k+1}\rangle-\frac{1}{2}\langle\Delta\lambda^{k+1}, A\Delta x^{k+1}\rangle \\
&\quad+\frac{1}{2\sigma}\left(\|\lambda^{k}-\hat{\lambda}\|^2-\|\lambda^{k+1}-\hat{\lambda}\|^2-\|\Delta\lambda^{k+1}\|^2\right).
\end{aligned}
$$

By observing that

$$
\begin{aligned}
&\theta\langle\hat{\lambda}-\lambda^{k+1}, A\Delta x^{k+1}\rangle-\frac{1}{2\sigma}\|\lambda^{k+1}-\hat{\lambda}\|^2 \\
&=-\frac{1}{2\sigma}\|\lambda^{k+1}-\hat{\lambda}+\sigma\theta A\Delta x^{k+1}\|^2+\frac{\sigma\theta^2}{2}\|A\Delta x^{k+1}\|^2
\end{aligned}
$$

and

$$
\begin{aligned}
&\theta\langle\lambda^{k+1}-\hat{\lambda}, A\Delta x^k\rangle+\frac{1}{2\sigma}\|\lambda^k-\hat{\lambda}\|^2 \\
&=\theta\langle\Delta\lambda^{k+1}, A\Delta x^k\rangle+\theta\langle\lambda^k-\hat{\lambda}, A\Delta x^k\rangle+\frac{1}{2\sigma}\|\lambda^k-\hat{\lambda}\|^2 \\
&=\theta\langle\Delta\lambda^{k+1}, A\Delta x^k\rangle+\frac{1}{2\sigma}\|\lambda^k-\hat{\lambda}+\sigma\theta A\Delta x^k\|^2-\frac{\sigma\theta^2}{2}\|A\Delta x^k\|^2,
\end{aligned}
$$

we further obtain

$$
\begin{aligned}
&\theta(G_{k+1}-G_k)+G_k+F_{k+1}+\frac{1}{2\tau}\left(\|x^{k+1}-\hat{x}\|^2-\|x^k-\hat{x}\|^2\right) \\
&+\frac{1}{2\sigma}\left(\|\lambda^{k+1}-\hat{\lambda}+\sigma\theta A\Delta x^{k+1}\|^2-\|\lambda^k-\hat{\lambda}+\sigma\theta A\Delta x^k\|^2\right) \\
&\leq-\frac{\theta}{\tau}\|\Delta x^{k+1}\|^2+\frac{\sigma\theta^2}{2}\|A\Delta x^{k+1}\|^2-\frac{\sigma\theta^2}{2}\|A\Delta x^k\|^2+\theta\langle\Delta\lambda^{k+1}, A\Delta x^k\rangle \\
&\quad-\frac{1}{2\sigma}\|\Delta\lambda^{k+1}\|^2+\frac{1}{2\tau}\langle\Delta x^k, \Delta x^{k+1}\rangle-\frac{1}{2}\langle\Delta\lambda^{k+1}, A\Delta x^{k+1}\rangle.
\end{aligned}
$$

By completing square, we have

$$
\begin{aligned}
&\theta\langle\Delta\lambda^{k+1}, A\Delta x^k\rangle-\frac{1}{2}\langle\Delta\lambda^{k+1}, A\Delta x^{k+1}\rangle-\frac{1}{2\sigma}\|\Delta\lambda^{k+1}\|^2 \\
&=-\frac{1}{2\sigma}\left\|\Delta\lambda^{k+1}+\sigma\left(\frac{1}{2}A\Delta x^{k+1}-\theta A\Delta x^k\right)\right\|^2+\frac{\sigma}{2}\left\|\frac{1}{2}A\Delta x^{k+1}-\theta A\Delta x^k\right\|^2 \\
&=-\frac{1}{2\sigma}\left\|\Delta\lambda^{k+1}+\sigma\left(\frac{1}{2}A\Delta x^{k+1}-\theta A\Delta x^k\right)\right\|^2+\frac{\sigma}{8}\|A\Delta x^{k+1}\|^2 \\
&\quad+\frac{\sigma\theta^2}{2}\|A\Delta x^k\|^2-\frac{\sigma\theta}{2}\langle A\Delta x^{k+1}, A\Delta x^k\rangle.
\end{aligned}
$$

Consequently

$$
\begin{aligned}
&\theta(G_{k+1}-G_k)+G_k+F_{k+1}+\frac{1}{2\tau}\left(\|x^{k+1}-\hat{x}\|^2-\|x^k-\hat{x}\|^2\right) \\
&+\frac{1}{2\sigma}\left(\|\lambda^{k+1}-\hat{\lambda}+\sigma\theta A\Delta x^{k+1}\|^2-\|\lambda^k-\hat{\lambda}+\sigma\theta A\Delta x^k\|^2\right) \\
&\leq-\frac{1}{2\sigma}\left\|\Delta\lambda^{k+1}+\sigma\left(\frac{1}{2}A\Delta x^{k+1}-\theta A\Delta x^k\right)\right\|^2-\frac{\theta}{\tau}\|\Delta x^{k+1}\|^2 \\
&\quad+\frac{\sigma(4\theta^2+1)}{8}\|A\Delta x^{k+1}\|^2-\frac{\sigma\theta}{2}\langle A\Delta x^{k+1}, A\Delta x^k\rangle+\frac{1}{2\tau}\langle\Delta x^k, \Delta x^{k+1}\rangle.
\end{aligned}
$$

By adding both sides, the term

$$\left(\frac{\theta}{2\tau}\|\Delta x^{k+1}\|^2 - \frac{\sigma(4\theta^2+1)}{16}\|A\Delta x^{k+1}\|^2\right) \\ - \left(\frac{\theta}{2\tau}\|\Delta x^{k}\|^2 - \frac{\sigma(4\theta^2+1)}{16}\|A\Delta x^{k}\|^2\right)$$

and set

$$V_k := \theta G_k + \frac{1}{2\tau}\|x^k - \hat{x}\|^2 + \frac{1}{2\sigma}\|\lambda^k - \hat{\lambda} + \sigma\theta A\Delta x^k\|^2 \\ + \frac{\theta}{2\tau}\|\Delta x^k\|^2 - \frac{\sigma(4\theta^2+1)}{16}\|A\Delta x^k\|^2, \tag{7.31}$$

we then obtain

$$\begin{aligned}
&V_{k+1} - V_k + G_k + F_{k+1} \\
&\le -\frac{1}{2\sigma}\left\|\Delta\lambda^{k+1} + \sigma\left(\frac{1}{2}A\Delta x^{k+1} - \theta A\Delta x^k\right)\right\|^2 - \frac{\theta}{2\tau}\|\Delta x^{k+1}\|^2 - \frac{\theta}{2\tau}\|\Delta x^k\|^2 \\
&\quad + \frac{\sigma(4\theta^2+1)}{16}\|A\Delta x^{k+1}\|^2 + \frac{\sigma(4\theta^2+1)}{16}\|A\Delta x^k\|^2 \\
&\quad - \frac{\sigma\theta}{2}\langle A\Delta x^{k+1}, A\Delta x^k\rangle + \frac{1}{2\tau}\langle \Delta x^k, \Delta x^{k+1}\rangle.
\end{aligned}$$

Note that

$$\begin{aligned}
&\frac{\sigma(4\theta^2+1)}{16}\|A\Delta x^{k+1}\|^2 + \frac{\sigma(4\theta^2+1)}{16}\|A\Delta x^k\|^2 - \frac{\sigma\theta}{2}\langle A\Delta x^{k+1}, A\Delta x^k\rangle \\
&\quad = \frac{\sigma(4\theta^2+1)}{16}\left\|A\Delta x^{k+1} - \frac{4\theta}{4\theta^2+1}A\Delta x^k\right\|^2 + \frac{\sigma(4\theta^2-1)^2}{16(4\theta^2+1)}\|A\Delta x^k\|^2 \\
&\quad \le \frac{\sigma\|A\|^2(4\theta^2+1)}{16}\left\|\Delta x^{k+1} - \frac{4\theta}{4\theta^2+1}\Delta x^k\right\|^2 + \frac{\sigma\|A\|^2(4\theta^2-1)^2}{16(4\theta^2+1)}\|\Delta x^k\|^2 \\
&\quad = \frac{\sigma\|A\|^2(4\theta^2+1)}{16}\|\Delta x^{k+1}\|^2 - \frac{\sigma\theta\|A\|^2}{2}\langle \Delta x^{k+1}, \Delta x^k\rangle \\
&\qquad + \frac{\sigma\|A\|^2(4\theta^2+1)}{16}\|\Delta x^k\|^2.
\end{aligned}$$

Therefore

$$
\begin{aligned}
&V_{k+1} - V_k + G_k + F_{k+1} \\
&\leq -\frac{1}{2\sigma}\left\|\Delta\lambda^{k+1} + \sigma\left(\frac{1}{2}A\Delta x^{k+1} - \theta A\Delta x^k\right)\right\|^2 + \frac{1-\tau\sigma\theta\|A\|^2}{2\tau}\langle\Delta x^{k+1}, \Delta x^k\rangle \\
&\quad - \frac{8\theta - \tau\sigma\|A\|^2(4\theta^2+1)}{16\tau}\|\Delta x^{k+1}\|^2 - \frac{8\theta - \tau\sigma\|A\|^2(4\theta^2+1)}{16\tau}\|\Delta x^k\|^2.
\end{aligned}
$$

By the given condition (7.27) on θ, σ, and τ, we have

$$
8\theta - \tau\sigma\|A\|^2(4\theta^2+1) \geq 8\theta - \frac{4(4\theta^2+1)}{1+2\theta} = \frac{8\theta - 4}{1+2\theta} > 0,
$$

Thus, we can continue to complete the square to obtain the following result.

Lemma 7.13 *Let Assumption 7.2 hold and consider Algorithm 7.11 with θ, τ, and σ satisfying (7.27). Let $(\hat{x}, \hat{\lambda})$ be any saddle point of Φ. Define F_k and G_k by (7.28) and define V_k by (7.31). Then, there holds*

$$
\begin{aligned}
&V_{k+1} - V_k + G_k + F_{k+1} \\
&\leq -\frac{1}{2\sigma}\left\|\Delta\lambda^{k+1} + \sigma\left(\frac{1}{2}A\Delta x^{k+1} - \theta A\Delta x^k\right)\right\|^2 \\
&\quad - \frac{8\theta - \tau\sigma\|A\|^2(4\theta^2+1)}{16\tau}\left\|\Delta x^{k+1} - \frac{4(1-\tau\sigma\theta\|A\|^2)}{8\theta - \tau\sigma\|A\|^2(4\theta^2+1)}\Delta x^k\right\|^2 \\
&\quad - \frac{(4\theta^2-1)(4-\tau\sigma\|A\|^2(2\theta+1))(4-\tau\sigma\|A\|^2(2\theta-1))}{16\tau(8\theta - \tau\sigma\|A\|^2(4\theta^2+1))}\|\Delta x^k\|^2
\end{aligned}
$$

for all $k \geq 1$.

Based on Lemma 7.13, we are now ready to prove the following convergence result [12].

Theorem 7.14 *Let Assumption 7.2 hold and consider Algorithm 7.11 with θ, τ, and σ satisfying (7.27). Then, $\{(x^k, \lambda^k)\}$ converges to a saddle point of Φ.*

Proof Let $(\hat{x}, \hat{\lambda})$ be any saddle point of Φ. Define F_k, G_k, and V_k by (7.28) and (7.31), respectively. By using the inequality in Lemma 7.13 and noting that all coefficients in front of the terms on the right-hand side are positive, we can conclude that $\{V_k\}$ is monotonically decreasing, $F_k \to 0$, $G_k \to 0$, and

$$
\|\Delta x^k\| \to 0 \quad \text{and} \quad \left\|\Delta\lambda^{k+1} + \sigma\left(\frac{1}{2}A\Delta x^{k+1} - \theta A\Delta x^k\right)\right\| \to 0
$$

and consequently

$$G_k \to 0, \quad \|\Delta x^k\| \to 0 \quad \text{and} \quad \|\Delta\lambda^k\| \to 0 \quad \text{as } k \to \infty. \tag{7.32}$$

Note that $V_k \geq 0$. The monotonicity of V_k then concludes that $\lim_{k\to\infty} V_k$ exists. Thus, we may use (7.32) and the definition of V_k to derive that

$$\lim_{k\to\infty} \left\{ \frac{1}{2\tau}\|x^k - \hat{x}\|^2 + \frac{1}{2\sigma}\|\lambda^k - \hat{\lambda}\|^2 \right\} \text{ exists.} \tag{7.33}$$

According to (7.33), the sequence $\{(x^k, \lambda^k)\}$ is bounded, and hence, it has a convergent subsequence $\{(x^{k_l}, \lambda^{k_l})\}$ with $(x^{k_l}, \lambda^{k_l}) \to (x^*, \lambda^*)$ as $l \to \infty$. By virtue of (7.30), we have

$$-\frac{1}{\tau}\Delta x^{k_l} - A^T\lambda^{k_l} \in \partial g(x^{k_l}), \quad -\frac{1}{\sigma}\Delta\lambda^{k_l} + A(x^{k_l-1} + \theta\Delta x^{k_l-1}) \in \partial f^*(\lambda^{k_l}).$$

Taking $l \to \infty$ and using (7.32) and the closedness of subdifferential, we can obtain $-A^T\lambda^* \in \partial g(x^*)$ and $Ax^* \in \partial f^*(\lambda^*)$ which imply that (x^*, λ^*) is a saddle point of Φ. Therefore, (7.33) holds with $(\hat{x}, \hat{\lambda})$ replaced by (x^*, λ^*). Since

$$\frac{1}{2\tau}\|x^{k_l} - x^*\|^2 + \frac{1}{2\sigma}\|\lambda^{k_l} - \lambda^*\|^2 \to 0 \quad \text{as } l \to \infty,$$

we therefore have

$$\lim_{k\to\infty} \left\{ \frac{1}{2\tau}\|x^k - x^*\|^2 + \frac{1}{2\sigma}\|\lambda^k - \lambda^*\|^2 \right\} = 0.$$

The proof is complete. □

Remark 7.15 The convergence result in Theorem 7.14 for the special case that $\theta = 1$ and $\tau\sigma\|A\|^2 < 4/3$ has been established in [138].

7.3 Primal-Dual Algorithm with Adaptive Step Sizes

According to Theorem 7.12, the convergence of Algorithm 7.11 with $\theta = 1$ requires the constant step sizes σ and τ to satisfy $\sigma\tau\|A\|^2 < 1$, and hence, the norm $\|A\|$ of A needs to be computed. For small-size matrices or sparse matrices, it may not be an issue to obtain $\|A\|$ easily. However, for large-scale dense matrices, computation of $\|A\|$ can be very expensive. Furthermore, even if the norm $\|A\|$ can be estimated, the obtained bound may be too large which can allow only small step size and hence slow down the convergence speed.

To modify Algorithm 7.11 into a convergent method without using the knowledge of $\|A\|$, we may consider the iteration scheme with θ, σ, and τ varying at each step, that is, we will consider the iteration method:

$$\begin{aligned} \bar{x}^k &= x^k + \theta_k(x^k - x^{k-1}), \\ \lambda^{k+1} &= \arg\max_{\lambda\in\mathbb{R}^m} \left\{ \langle A\bar{x}^k, \lambda\rangle - f^*(\lambda) - \frac{1}{2\beta_k\tau_k}\|\lambda - \lambda^k\|^2 \right\}, \\ x^{k+1} &= \arg\min_{x\in\mathbb{R}^n} \left\{ g(x) + \langle Ax, \lambda^{k+1}\rangle + \frac{1}{2\tau_k}\|x - x^k\|^2 \right\}. \end{aligned} \tag{7.34}$$

Let us discuss how to make (7.34) into a convergent method by choosing θ_k, β_k, and τ_k independent of $\|A\|$.

Let $(\hat{x}, \hat{\lambda})$ be any saddle point of Φ and let $k \geq 1$. From the definition of x^k, we can derive as in the proof of Lemma 7.3 that

$$g(x) - g(x^k) \geq \frac{1}{\tau_{k-1}}\langle x^{k-1} - x^k, x - x^k\rangle - \langle A^T\lambda^k, x - x^k\rangle, \quad \forall x.$$

By taking $x = x^{k-1}$ and $x = x^{k+1}$, respectively, we obtain

$$\begin{aligned} g(x^{k-1}) - g(x^k) &\geq \frac{1}{\tau_{k-1}}\langle x^{k-1} - x^k, x^{k-1} - x^k\rangle - \langle A^T\lambda^k, x^{k-1} - x^k\rangle, \\ g(x^{k+1}) - g(x^k) &\geq \frac{1}{\tau_{k-1}}\langle x^{k-1} - x^k, x^{k+1} - x^k\rangle - \langle A^T\lambda^k, x^{k+1} - x^k\rangle. \end{aligned}$$

By multiplying the first inequality by θ_k, then adding with the second inequality, and using the definition of $\bar{x}^k$, we have

$$\begin{aligned} & g(x^{k+1}) - (1+\theta_k)g(x^k) + \theta_k g(x^{k-1}) \\ & \geq \frac{1}{\tau_{k-1}}\langle x^{k-1} - x^k, x^{k+1} - \bar{x}^k\rangle - \langle A^T\lambda^k, x^{k+1} - \bar{x}^k\rangle. \end{aligned}$$

Since $\theta_k(x^{k-1} - x^k) = x^k - \bar{x}^k$, we further have

$$\begin{aligned} & g(x^{k+1}) - (1+\theta_k)g(x^k) + \theta_k g(x^{k-1}) \\ & \geq \frac{1}{\theta_k\tau_{k-1}}\langle x^k - \bar{x}^k, x^{k+1} - \bar{x}^k\rangle - \langle A^T\lambda^k, x^{k+1} - \bar{x}^k\rangle. \end{aligned} \tag{7.35}$$

Next, we use the definition of x^{k+1} and λ^{k+1} to obtain

$$\begin{aligned}
g(\hat{x}) - g(x^{k+1}) &\geq \frac{1}{\tau_k}\langle x^k - x^{k+1}, \hat{x} - x^{k+1}\rangle - \langle A^T\lambda^{k+1}, \hat{x} - x^{k+1}\rangle \\
&\quad + \mu_g\|\hat{x} - x^{k+1}\|^2, \\
f^*(\hat{\lambda}) - f^*(\lambda^{k+1}) &\geq \frac{1}{\beta_k\tau_k}\langle \lambda^k - \lambda^{k+1}, \hat{\lambda} - \lambda^{k+1}\rangle + \langle A\bar{x}^k, \hat{\lambda} - \lambda^{k+1}\rangle,
\end{aligned}$$

where μ_g denotes the modulus of convexity of g. Adding these two inequalities with (7.35) gives

$$\begin{aligned}
& f^*(\hat{\lambda}) - f^*(\lambda^{k+1}) + g(\hat{x}) - (1+\theta_k)g(x^k) + \theta_k g(x^{k-1}) \\
& \geq \frac{1}{\theta_k\tau_{k-1}}\langle x^k - \bar{x}^k, x^{k+1} - \bar{x}^k\rangle + \frac{1}{\tau_k}\langle x^k - x^{k+1}, \hat{x} - x^{k+1}\rangle \\
& \quad + \frac{1}{\beta_k\tau_k}\langle \lambda^k - \lambda^{k+1}, \hat{\lambda} - \lambda^{k+1}\rangle - \langle A^T\lambda^k, x^{k+1} - \bar{x}^k\rangle \\
& \quad - \langle A^T\lambda^{k+1}, \hat{x} - x^{k+1}\rangle + \langle A\bar{x}^k, \hat{\lambda} - \lambda^{k+1}\rangle + \mu_g\|\hat{x} - x^{k+1}\|^2.
\end{aligned}$$

By writing

$$\langle A\bar{x}^k, \hat{\lambda} - \lambda^{k+1}\rangle = \langle A\hat{x}, \hat{\lambda} - \lambda^{k+1}\rangle + \langle \bar{x}^k - \hat{x}, A^T(\hat{\lambda} - \lambda^{k+1})\rangle,$$

we then obtain

$$\begin{aligned}
& f^*(\hat{\lambda}) - f^*(\lambda^{k+1}) + g(\hat{x}) - (1+\theta_k)g(x^k) + \theta_k g(x^{k-1}) \\
& \geq \frac{1}{\theta_k\tau_{k-1}}\langle x^k - \bar{x}^k, x^{k+1} - \bar{x}^k\rangle + \frac{1}{\tau_k}\langle x^k - x^{k+1}, \hat{x} - x^{k+1}\rangle \\
& \quad + \frac{1}{\beta_k\tau_k}\langle \lambda^k - \lambda^{k+1}, \hat{\lambda} - \lambda^{k+1}\rangle - \langle A\hat{x}, \hat{\lambda} - \lambda^{k+1}\rangle + \langle \bar{x}^k - \hat{x}, A^T\hat{\lambda}\rangle \\
& \quad + \langle A^T(\lambda^{k+1} - \lambda^k), x^{k+1} - \bar{x}^k\rangle + \mu_g\|\hat{x} - x^{k+1}\|^2. \qquad (7.36)
\end{aligned}$$

Let

$$\begin{aligned}
D_{f^*}(\lambda, \hat{\lambda}) &:= f^*(\lambda) - f^*(\hat{\lambda}) - \langle A\hat{x}, \lambda - \hat{\lambda}\rangle, \\
D_g(x, \hat{x}) &:= g(x) - g(\hat{x}) - \langle A^T\hat{\lambda}, x - \hat{x}\rangle
\end{aligned}$$

for any $x \in \mathbb{R}^n$ and $\lambda \in \mathbb{R}^m$. Since $(\hat{x}, \hat{\lambda})$ is a saddle point of Φ, we have $D_{f^*}(\lambda, \hat{\lambda}) \geq 0$ and $D_g(x, \hat{x}) \geq 0$. By regrouping the terms in (7.36), it follows that

$$\begin{aligned}
&\tau_k \left(-D_{f^*}(\lambda^{k+1}, \hat{\lambda}) - (1+\theta_k) D_g(x^k, \hat{x}) + \theta_k D_g(x^{k-1}, \hat{x})\right) \\
&\geq \frac{\tau_k}{\theta_k \tau_{k-1}} \langle x^k - \bar{x}^k, x^{k+1} - \bar{x}^k \rangle + \langle x^k - x^{k+1}, \hat{x} - x^{k+1} \rangle \\
&\quad + \frac{1}{\beta_k} \langle \lambda^k - \lambda^{k+1}, \hat{\lambda} - \lambda^{k+1} \rangle + \tau_k \langle A^T(\lambda^{k+1} - \lambda^k), x^{k+1} - \bar{x}^k \rangle \\
&\quad + \mu_g \tau_k \|\hat{x} - x^{k+1}\|^2 \\
&= \frac{\tau_k}{2\theta_k \tau_{k-1}} \left(\|x^k - \bar{x}^k\|^2 - \|x^k - x^{k+1}\|^2 + \|x^{k+1} - \bar{x}^k\|^2\right) \\
&\quad + \frac{1}{2} \left(\|\hat{x} - x^{k+1}\|^2 - \|\hat{x} - x^k\|^2 + \|x^{k+1} - x^k\|^2\right) \\
&\quad + \frac{1}{2\beta_k} \left(\|\hat{\lambda} - \lambda^{k+1}\|^2 - \|\hat{\lambda} - \lambda^k\|^2 + \|\lambda^{k+1} - \lambda^k\|^2\right) \\
&\quad + \tau_k \langle A^T(\lambda^{k+1} - \lambda^k), x^{k+1} - \bar{x}^k \rangle + \mu_g \tau_k \|\hat{x} - x^{k+1}\|^2. \qquad (7.37)
\end{aligned}$$

In order to effectively cancel terms, it is natural to choose θ_k such that $\tau_k/(\theta_k \tau_{k-1}) = 1$, i.e., $\theta_k = \tau_k/\tau_{k-1}$. Next, by the Cauchy-Schwarz inequality, we have

$$\begin{aligned}
\tau_k \langle A^T(\lambda^{k+1} - \lambda^k), x^{k+1} - \bar{x}^k \rangle &\geq -\tau_k \|A^T(\lambda^{k+1} - \lambda^k)\| \|x^{k+1} - \bar{x}^k\| \\
&\geq -\frac{1}{2}\|x^{k+1} - \bar{x}^k\|^2 - \frac{\tau_k^2}{2}\|A^T(\lambda^{k+1} - \lambda^k)\|^2.
\end{aligned}$$

Combining this with (7.37) and using the choice of θ_k, we obtain

$$\begin{aligned}
&\tau_k \left(-D_{f^*}(\lambda^{k+1}, \hat{\lambda}) - (1+\theta_k) D_g(x^k, \hat{x}) + \theta_k D_g(x^{k-1}, \hat{x})\right) \\
&\geq \frac{1 + 2\mu_g \tau_k}{2} \|\hat{x} - x^{k+1}\|^2 - \frac{1}{2}\|\hat{x} - x^k\|^2 + \frac{1}{2\beta_k}\left(\|\hat{\lambda} - \lambda^{k+1}\|^2 - \|\hat{\lambda} - \lambda^k\|^2\right) \\
&\quad + \frac{1}{2}\|x^k - \bar{x}^k\|^2 + \frac{1}{2\beta_k}\|\lambda^{k+1} - \lambda^k\|^2 - \frac{\tau_k^2}{2}\|A^T(\lambda^{k+1} - \lambda^k)\|^2. \qquad (7.38)
\end{aligned}$$

The last two terms on the right-hand side of (7.38) suggest that we may consider choosing τ_k such that

$$\tau_k^2 \|A^T(\lambda^{k+1} - \lambda^k)\|^2 \leq \frac{\delta^2}{\beta_k}\|\lambda^{k+1} - \lambda^k\|^2 \qquad (7.39)$$

for some $\delta \in (0, 1]$. However, we cannot use this inequality to determine τ_k directly because the definition of λ^{k+1} depends on τ_k. Nevertheless, we may choose τ_k by a backtracking line search procedure. Namely, choosing suitable $\tilde{\tau}_k > 0$ and $q \in (0, 1)$, for each integer $j \geq 0$ we set $\tau_k := \tilde{\tau}_k q^j$ and $\theta_k = \tau_k/\tau_{k-1}$, we then use $\bar{x}^k := x^k + \theta_k(x^k - x^{k-1})$ to calculate λ^{k+1} via the second equation in (7.34). Once the first integer $j \geq 0$ is determined to guarantee (7.39), we output the corresponding value of τ_k. By using (7.39), we can obtain from (7.38) that

$$\begin{aligned}
&\frac{1 + 2\mu_g \tau_k}{2}\|\hat{x} - x^{k+1}\|^2 + \frac{\|\hat{\lambda} - \lambda^{k+1}\|^2}{2\beta_k} + \frac{\|x^k - \bar{x}^k\|^2}{2} + \frac{1 - \delta^2}{2\beta_k}\|\lambda^{k+1} - \lambda^k\|^2 \\
&\leq \frac{\|\hat{x} - x^k\|^2}{2} + \frac{\|\hat{\lambda} - \lambda^k\|^2}{2\beta_k} \\
&\quad - \tau_k\left(D_{f^*}(\lambda^{k+1}, \hat{\lambda}) + (1 + \theta_k)D_g(x^k, \hat{x}) - \theta_k D_g(x^{k-1}, \hat{x})\right). \qquad (7.40)
\end{aligned}$$

The first two terms on the left-hand side can be written as

$$(1 + 2\mu_g \tau_k)\left(\frac{\|\hat{x} - x^{k+1}\|^2}{2} + \frac{\|\hat{\lambda} - \lambda^{k+1}\|^2}{2\beta_k(1 + 2\mu_g \tau_k)}\right)$$

Comparing with the first two terms on the right-hand side, in order to obtain a nice recursive relation, we may set $\beta_{k+1} = \beta_k(1 + 2\mu_g \tau_k)$. With this choice of β_{k+1}, we can obtain from (7.40) that

$$\begin{aligned}
&\frac{\beta_{k+1}}{\beta_k}\left(\frac{\|\hat{x} - x^{k+1}\|^2}{2} + \frac{\|\hat{\lambda} - \lambda^{k+1}\|^2}{2\beta_{k+1}}\right) + \frac{\|x^k - \bar{x}^k\|^2}{2} + \frac{1 - \delta^2}{2\beta_k}\|\lambda^{k+1} - \lambda^k\|^2 \\
&\leq \frac{\|\hat{x} - x^k\|^2}{2} + \frac{\|\hat{\lambda} - \lambda^k\|^2}{2\beta_k} \\
&\quad - \tau_k\left(D_{f^*}(\lambda^{k+1}, \hat{\lambda}) + (1 + \theta_k)D_g(x^k, \hat{x}) - \theta_k D_g(x^{k-1}, \hat{x})\right). \qquad (7.41)
\end{aligned}$$

Multiplying (7.41) by $2\beta_k$ and dropping the term involving $D_{f^*}(\lambda^{k+1}, \hat{\lambda})$ which is nonnegative, we can obtain

$$\begin{aligned}
&\beta_{k+1}\|\hat{x} - x^{k+1}\|^2 + \|\hat{\lambda} - \lambda^{k+1}\|^2 + \beta_k\|x^k - \bar{x}^k\|^2 + (1 - \delta^2)\|\lambda^{k+1} - \lambda^k\|^2 \\
&\leq \beta_k\|\hat{x} - x^k\|^2 + \|\hat{\lambda} - \lambda^k\|^2 \\
&\quad - 2\beta_k\tau_k(1 + \theta_k)D_g(x^k, \hat{x}) + 2\beta_k\tau_k\theta_k D_g(x^{k-1}, \hat{x}).
\end{aligned}$$

In order to proceed further, we need $\beta_k \tau_k \theta_k \leq \beta_{k-1}\tau_{k-1}(1+\theta_{k-1})$, which, by recalling $\theta_k = \tau_k/\tau_{k-1}$, is equivalent to

$$\tau_k^2 \leq \frac{\beta_{k-1}}{\beta_k}\tau_{k-1}^2(1+\theta_{k-1}).$$

Since τ_k is defined via $\tilde{\tau}_k$ and $\tau_k \leq \tilde{\tau}_k$, we can achieve this by requiring

$$\tilde{\tau}_k \leq \tau_{k-1}\sqrt{\frac{\beta_{k-1}}{\beta_k}(1+\theta_{k-1})}.$$

Once this is assumed, we consequently have

$$\begin{aligned}
&\beta_{k+1}\|\hat{x}-x^{k+1}\|^2 + \|\hat{\lambda}-\lambda^{k+1}\|^2 + 2\beta_k\tau_k(1+\theta_k)D_g(x^k,\hat{x}) + \beta_k\|x^k-\bar{x}^k\|^2 \\
&\quad + (1-\delta^2)\|\lambda^{k+1}-\lambda^k\|^2 \\
&\leq \beta_k\|\hat{x}-x^k\|^2 + \|\hat{\lambda}-\lambda^k\|^2 + 2\beta_{k-1}\tau_{k-1}(1+\theta_{k-1})D_g(x^{k-1},\hat{x}).
\end{aligned} \tag{7.42}$$

Putting together all the above analysis on the choices of τ_k, θ_k, and β_k leads to the following primal-dual algorithm with line search which was proposed in [151].

Algorithm 7.16 (Primal-Dual Algorithm with Backtracking Line Search) *Choose $x^0 \in \mathrm{dom}(g)$ and $\lambda^0 \in \mathrm{dom}(f^*)$, $q \in (0,1)$, $\delta \in (0,1]$, $\beta_0 > 0$, $\tau_{-1} > 0$, $\theta_{-1} > 0$ and set $x^{-1} = x^0$, $\beta_{-1} = \beta_0$. For $k = 0, 1, \cdots$ do*

(i) *Take*

$$\tilde{\tau}_k \in \left[\tau_{k-1}\sqrt{\frac{\beta_{k-1}}{\beta_k}},\ \tau_{k-1}\sqrt{\frac{\beta_{k-1}}{\beta_k}(1+\theta_{k-1})}\right]. \tag{7.43}$$

(ii) *Determine the smallest integer $j \geq 0$ such that for $\tau_k = \tilde{\tau}_k q^j$ the inequality*

$$\sqrt{\beta_k}\tau_k\|A^T(\lambda^{k+1}-\lambda^k)\| \leq \delta\|\lambda^{k+1}-\lambda^k\| \tag{7.44}$$

holds, where λ^{k+1} is calculated through

$$\begin{aligned}
\theta_k &= \frac{\tau_k}{\tau_{k-1}}; \\
\bar{x}^k &= x^k + \theta_k(x^k - x^{k-1}); \\
\lambda^{k+1} &= \arg\min_{\lambda\in\mathbb{R}^m}\left\{f^*(\lambda) - \langle A\bar{x}^k, \lambda\rangle + \frac{1}{2\beta_k\tau_k}\|\lambda-\lambda^k\|^2\right\}.
\end{aligned}$$

(iii) *Compute*

$$x^{k+1} = \arg\min_{x\in\mathbb{R}^n} \left\{ g(x) + \langle A^T\lambda^{k+1}, x\rangle + \frac{1}{2\tau_k}\|x - x^k\|^2 \right\};$$

$$\beta_{k+1} = \beta_k(1 + 2\mu_g\tau_k).$$

In Algorithm 7.16, the step size τ_k is determined by a backtracking line search procedure. Note that for any $\tau \le \delta/(\sqrt{\beta_k}\|A\|)$ there holds

$$\sqrt{\beta_k}\tau\|A^T(\lambda^{k+1} - \lambda^k)\| \le \delta\|\lambda^{k+1} - \lambda^k\|$$

and $\tilde{\tau}_k q^j \to 0$ as $j \to \infty$, the backtracking line search procedure must terminate in finite many steps, and thus, τ_k is well-defined. In the formulation of Algorithm 7.16, besides those requirements on τ_k, θ_k, and β_k, we also impose a lower bound restriction on $\tilde{\tau}_k$; see (7.43). This additional requirement is used to guarantee a useful lower bound on τ_k. In fact, we have the following result.

Lemma 7.17 *For $\{\tau_k\}$ and $\{\theta_k\}$ defined by Algorithm 7.16, there hold*

$$\sqrt{\beta_k}\tau_k \ge \min\left\{\sqrt{\beta_{-1}}\tau_{-1}, \frac{q\delta}{\|A\|}\right\} \quad \text{and} \quad \theta_k \le \max\left\{\frac{\sqrt{5}+1}{2}, \theta_{-1}\right\} \tag{7.45}$$

for all $k \ge -1$.

Proof We can prove (7.45) by induction. For the first inequality in (7.45), it is clearly true for $k = -1$. For $k \ge 0$, we have $\tau_k = \tilde{\tau}_k q^j$ for some $j \ge 0$, where $\tilde{\tau}_k$ satisfies (7.43). If $j = 0$, then

$$\tau_k = \tilde{\tau}_k \ge \tau_{k-1}\sqrt{\beta_{k-1}/\beta_k}$$

which implies that

$$\sqrt{\beta_k}\tau_k \ge \sqrt{\beta_{k-1}}\tau_{k-1} \ge \min\left\{\sqrt{\beta_{-1}}\tau_{-1}, \frac{q\delta}{\|A\|}\right\}$$

by the induction hypothesis. If $j \ge 1$, then τ_k/q does not make (7.44) to be true. Thus, $\tau_k/q > \delta/(\sqrt{\beta_k}\|A\|)$ which implies that $\sqrt{\beta_k}\tau_k > q\delta/\|A\|$. We therefore obtain the first inequality in (7.45). Next, we prove the second inequality in (7.45). Note that for $k \ge 0$ we have

$$\tau_k \le \tilde{\tau}_k \le \tau_{k-1}\sqrt{\frac{\beta_{k-1}}{\beta_k}(1 + \theta_{k-1})}.$$

Since $\beta_k \geq \beta_{k-1}$ and $\theta_k = \tau_k/\tau_{k-1}$, we thus obtain $\theta_k \leq \sqrt{1+\theta_{k-1}}$. Based on this relation, we can use an induction argument to show that if $\theta_{-1} \leq (\sqrt{5}+1)/2$, then $\theta_k \leq (\sqrt{5}+1)/2$ for all $k \geq -1$, while if $\theta_{-1} \geq (\sqrt{5}+1)/2$, then $\theta_k \leq \theta_{-1}$ for all $k \geq -1$. We thus complete the proof of the second inequality in (7.45). □

For Algorithm 7.16, we have the following convergence result [151].

Theorem 7.18 *Let Assumption 7.2 hold and consider the sequence* $\{(x^k, \lambda^k)\}$ *generated by Algorithm 7.16.*

(i) *When* $\mu_g = 0$*, we take* $\delta \in (0, 1)$*. If* $\{\tau_k\}$ *is bounded, then the whole sequence* $\{(x^k, \lambda^k)\}$ *converges to a saddle point* (x^*, λ^*) *of* Φ*.*
(ii) *When* $\mu_g > 0$*, we may take* $\delta \in (0, 1]$*. Then,* $\|x^k - x^*\| = O(1/k)$ *as* $k \to \infty$*, where* x^* *denotes the unique minimizer of (7.1).*

Proof Let $(\hat{x}, \hat{\lambda})$ be any saddle point of Φ and let

$$a_k := \beta_k\|\hat{x} - x^k\|^2 + \|\hat{\lambda} - \lambda^k\|^2 + 2\beta_{k-1}\tau_{k-1}(1+\theta_{k-1})D_g(x^{k-1}, \hat{x}). \tag{7.46}$$

It then follows from (7.42) that

$$a_{k+1} + \beta_k\|x^k - \bar{x}^k\|^2 + (1-\delta^2)\|\lambda^{k+1} - \lambda^k\|^2 \leq a_k. \tag{7.47}$$

(i) We first consider the case that $\mu_g = 0$ and $\delta \in (0, 1)$. For this case, we have $\beta_k = \beta_0$ for all k. By summing Eq. (7.37), we have

$$a_{k+1} + \beta_0 \sum_{j=1}^{k} \|x^j - \bar{x}^j\|^2 + (1-\delta^2)\sum_{j=1}^{k} \|\lambda^{j+1} - \lambda^j\|^2 \leq a_1 < \infty$$

for all $k \geq 1$. This together with $\delta \in (0, 1)$ implies that $\{(x^k, \lambda^k)\}$ is bounded, $\|x^k - \bar{x}^k\| \to 0$, $\|\lambda^{k+1} - \lambda^k\| \to 0$ as $k \to \infty$. According to Lemma 7.17, $\{\tau_k\}$ has a positive lower bound. We thus have

$$\frac{\lambda^k - \lambda^{k-1}}{\tau_{k-1}} \to 0 \quad \text{and} \quad \frac{x^k - x^{k-1}}{\tau_{k-1}} = \frac{\bar{x}^k - x^k}{\tau_k} \to 0$$

as $k \to \infty$. Let (x^*, λ^*) be a cluster point of $\{(x^k, \lambda^k)\}$. Then, there is a subsequence $\{(x^{k_i}, \lambda^{k_i})\}$ such that $x^{k_i} \to x^*$ and $\lambda^{k_i} \to \lambda^*$ as $i \to \infty$. By the definition of x^{k_i} and λ^{k_i}, we have

$$\frac{x^{k_i-1} - x^{k_i}}{\tau_{k_i-1}} - A^T\lambda^{k_i} \in \partial g(x^{k_i}), \qquad \frac{\lambda^{k_i-1} - \lambda^{k_i}}{\beta_0\tau_{k_i-1}} + A\bar{x}^{k_i} \in \partial f^*(\lambda^{k_i}).$$

Letting $i \to \infty$ and using the closedness of subdifferential, we obtain $-A^T\lambda^* \in \partial g(x^*)$ and $Ax^* \in \partial f^*(\lambda^*)$ which implies that (x^*, λ^*) is a saddle point of Φ.

Next, we show that the whole sequence $\{(x^k, \lambda^k)\}$ converges to (x^*, λ^*). By the lower semicontinuity of g, we first have $g(x^*) \le \liminf_{i\to\infty} g(x^{k_i})$. On the other hand, by the definition of x^{k_i}, we have

$$g(x^*) - g(x^{k_i}) \ge \frac{1}{\tau_{k_i-1}}\langle x^{k_i-1} - x^{k_i}, x^* - x^{k_i}\rangle - \langle A^T\lambda^{k_i}, x^* - x^{k_i}\rangle$$

which implies that $\limsup_{i\to\infty} g(x^{k_i}) \le g(x^*)$. Thus, $g(x^{k_i}) \to g(x^*)$ which implies that $D_g(x^{k_i}, x^*) \to 0$ as $i \to \infty$. Consider a_k defined by (7.46) with $(\hat{x}, \hat{\lambda}) = (x^*, \lambda^*)$. By the boundedness of $\{\tau_k\}$ and $\{\theta_k\}$, we have $a_{k_i+1} \to 0$ as $i \to \infty$. According to (7.47), the sequence $\{a_k\}$ is monotonically decreasing. Consequently, $a_k \to 0$ which in particular implies $(x^k, \lambda^k) \to (x^*, \lambda^*)$ as $k \to \infty$.

(ii) Next, we consider the case that $\mu_g > 0$ and $\delta \in (0, 1]$. It follows from (7.47) that $a_k \le a_1 < \infty$ which in particular implies

$$\beta_k\|\hat{x} - x^k\|^2 \le a_1 < \infty \tag{7.48}$$

for all $k \ge 1$. We need to estimate β_k from below. According to Lemma 7.17, there is a positive constant c_0 such that $\sqrt{\beta_k}\tau_k \ge c_0$. Consequently

$$\beta_{k+1} = \beta_k(1 + 2\mu_g\tau_k) \ge \beta_k + 2c_0\mu_g\sqrt{\beta_k}.$$

By induction, it is easy to show that there is a positive constant c_1 such that $\beta_k \ge c_1 k^2$. Therefore, we can obtain from (7.48) that

$$\|\hat{x} - x^k\|^2 \le \frac{a_1}{\beta_k} \le \frac{a_1}{c_1 k^2}.$$

The proof is thus complete. □

Problems

7.1 Consider the minimization problem:

$$\min_{x\in\mathbb{R}^n} \|Ax - b\|_1,$$

where A is a real matrix of size $m \times n$ with $m \gg n$ and $b \in \mathbb{R}^m$.

(i) Use the primal-dual hybrid gradient method, i.e., Algorithm 7.1, to derive the following algorithm:

$$\lambda^{k+1} = \max\left\{\min\left\{\lambda^k + \tau_k(Ax^k - b), 1\right\}, -1\right\},$$

$$x^{k+1} = x^k - \theta_k A^T \lambda^{k+1},$$

where $\tau_k > 0$ and $\theta_k > 0$ are step sizes.

(ii) Write a code to implement the above algorithm and test it in the following way: use $A = $ `randn(m, n)` to generate a matrix of size $m \times n$ with $m = 1000$ and $n = 400$, pick $x^* \in \mathbb{R}^n$ such that

$$x_i^* = 5t_i^2(1 - t_i) + \sin^2(2\pi t_i), \qquad i = 1, \cdots, n,$$

where $t_i = (i-1)/(n-1)$, next compute Ax^* and generate b by adding 100 outliers using the following Matlab commands:

```
noise = zeros(m,1);
ind = randperm(m,100);
noise1 = randn(m,1)-0.5;
noise(ind) = 100*sign(noise1(ind));
b = Ax* + noise;
```

When applying the algorithm, take $x^0 = 0$, $\lambda^0 = 0$, $\tau_k = 1 + 0.01k$, and $\theta_k = 0.1/(1 + k)$. Run your code 50, 150, and 250 iterations. Plot the computational results.

7.2 Consider the lasso problem:

$$\min_{x \in \mathbb{R}^n} \left\{\frac{1}{2}\|Ax - b\|_2^2 + \alpha\|x\|_1\right\},$$

where $\alpha > 0$, $b \in \mathbb{R}^m$ and A is an $m \times n$ matrix with $m \ll n$.

(i) Use the PDHG method, i.e., Algorithm 7.1, to derive the following algorithm:

$$\lambda^{k+1} = \min\left\{\max\left\{\lambda^k + \tau_k x^k, -\alpha\right\}, \alpha\right\},$$

$$x^{k+1} = \left(I + \theta_k A^T A\right)^{-1}\left(x^k + \theta_k A^T b - \theta_k \lambda^{k+1}\right),$$

where $\tau_k > 0$ and $\theta_k > 0$ are step sizes.

(ii) Write a code to implement the above algorithm and test it in the following way: use $A = $ `randn(m, n)` to generate a matrix of size $m \times n$ with $m = 600$ and

$n = 800$, pick $x^* \in \mathbb{R}^n$ with $\|x\|_0 = 200$; in Matlab, you may produce an x^* by performing the following commands:

```
x_exact = zeros(n,1);
ind = randperm(n,200);
x0 = sign(randn(n,1));
x_exact(ind) = x0(ind);
```

next compute Ax^* and add noise $\varepsilon = 0.1*\texttt{randn(m, 1)}$ to obtain b. When using the algorithm, take $\lambda^0 = 0$, $x^0 = 0$, $\alpha = 1$, $\tau_k = 1 + k$, and $\theta_k = 1/(1 + k)$. Run your code 10, 20, and 30 iterations. Plot the computational results.

7.3 Write a code to implement the primal-dual algorithm with adaptive step sizes, i.e., Algorithm 7.16, for solving the lasso problem given in Problem 7.2.

Chapter 8
Error Bound Conditions and Linear Convergence

In previous chapters, we leveraged the strong convexity of objective functions to establish convergence rates, particularly linear convergence, for various algorithms. However, strong convexity is often too restrictive in practical applications. To address this limitation, we introduce weaker conditions that still ensure linear convergence. Specifically, we present various error bound conditions, explore their equivalence in the convex setting, and demonstrate how they can be used to establish linear convergence for certain algorithms.

8.1 Motivation of Error Bound Conditions

Let $f : \mathbb{R}^n \to (-\infty, \infty]$ be a proper, lower semicontinuous, convex function with

$$S := \arg\min_{x\in\mathbb{R}^n} f(x) \neq \emptyset.$$

In Sect. 4.3, we have considered the proximal point algorithm:

$$x^{k+1} = \text{prox}_{tf}(x^k) \tag{8.1}$$

with constant step size $t > 0$ and shown that x^k converges to a minimizer of f with $f(x^k) - f_* = O(1/k)$, where $f_* := \min_{x\in\mathbb{R}^n} f(x)$.

In this section, we will motivate various error bound conditions guaranteeing linear convergence by considering the proximal point algorithm as a prototypical example. Let $z \in S$ be any element. According to (4.12), we have

$$\|x^{k+1} - z\|^2 - \|x^k - z\|^2 \le 2t(f_* - f(x^{k+1})). \tag{8.2}$$

Q. Jin, *Lectures on Nonsmooth Optimization*, Texts in Applied Mathematics 82,
https://doi.org/10.1007/978-3-031-91417-1_8

If f is strongly convex, then there exists $\mu > 0$ such that $f(x^{k+1}) - f_* \geq \mu\|x^{k+1} - x^*\|^2$, where x^* is the unique element in S. Consequently, it follows from (8.2) that

$$\|x^{k+1} - x^*\|^2 - \|x^k - x^*\|^2 \leq -2t\mu\|x^{k+1} - x^*\|^2$$

which implies that

$$\|x^{k+1} - x^*\| \leq \frac{1}{\sqrt{1 + 2t\mu}}\|x^k - x^*\|$$

and thus x^k converges to x^* linearly. However, strong convexity is usually too restrictive to be satisfied in many practical applications. We need to seek weaker conditions which still guarantee linear convergence. We will consider the proximal point algorithm from different perspectives to derive different error bound conditions to guarantee linear convergence.

8.1.1 Quadratic Growth Condition

Let us start from (8.2) which implies that

$$d^2(x^{k+1}, S) \leq \|x^{k+1} - z\|^2 \leq \|x^k - z\|^2 - 2t(f(x^{k+1}) - f_*)$$

for all $z \in S$. Consequently

$$d^2(x^{k+1}, S) \leq d^2(x^k, S) - 2t(f(x^{k+1}) - f_*).$$

Let $\gamma_{k+1} := (f(x^{k+1}) - f_*)/d^2(x^{k+1}, S)$. Then

$$(1 + 2t\gamma_{k+1})d^2(x^{k+1}, S) \leq d^2(x^k, S).$$

If $\gamma_{k+1} \geq \gamma > 0$ for all k, then

$$d^2(x^{k+1}, S) \leq \frac{1}{1 + 2t\gamma}d^2(x^k, S)$$

and hence $d(x^k, S) \to 0$ linearly as $k \to \infty$. This motivates the following error bound condition.

Definition 8.1 f is said to satisfy a *quadratic growth condition* at $\bar{x} \in S$ if there exist $r > 0$ and $\gamma > 0$ such that

$$\gamma d^2(x, S) \leq f(x) - f_*$$

for all $x \in B_r(\bar{x})$.

It is easy to check any proper, lower semicontinuous, strongly convex function $f : \mathbb{R}^n \to (-\infty, \infty]$ must satisfy a quadratic growth condition. Indeed, from Lemma 5.5, it follows that f is coercive, and thus, by Corollary 1.15 and Proposition 2.28, it has a unique minimizer x^*, i.e., $S = \{x^*\}$. Since $0 \in \partial f(x^*)$, we may use Proposition 3.12 to obtain

$$f(x) - f(x^*) \geq \mu\|x - x^*\|^2 = \mu d^2(x, S), \quad \forall x \in \mathbb{R}^n,$$

where $\mu > 0$ denotes the modulus of convexity of f. However, the converse is not true in general. To see this, consider the function $f(x) := \|x\|_1$ on $\mathbb{R}^n$ which is not strongly convex, it satisfies a quadratic growth condition at its unique minimizer $x^* = 0$ because

$$f(x) - f(x^*) = \|x\|_1 \geq \|x\|^2 = d^2(x, S)$$

for any x satisfying $\|x\| \leq 1$, where $S = \{0\}$. This demonstrates that quadratic growth condition is indeed weaker than strong convexity. We will provide more examples of functions that satisfy a quadratic growth condition but not necessarily strongly convex.

8.1.2 *Subdifferential Error Bound Condition*

Recall that the proximal mapping is firmly nonexpansive, i.e.,

$$\|\text{prox}_{tf}(x) - \text{prox}_{tf}(y)\|^2 + \|(I - \text{prox}_{tf})(x) - (I - \text{prox}_{tf})(y)\|^2 \leq \|x - y\|^2$$

for any $x, y \in \mathbb{R}^n$; see Lemma 4.18. Taking $x = x^k$, $y = x^* \in S$ and noting that $x^* = \text{prox}_{tf}(x^*)$, we then obtain

$$\|x^{k+1} - x^*\|^2 + \|x^k - x^{k+1}\|^2 \leq \|x^k - x^*\|^2, \quad \forall x^* \in S$$

and hence

$$d^2(x^{k+1}, S) + \|x^k - x^{k+1}\|^2 \leq d^2(x^k, S).$$

By the definition of x^{k+1}, we have $(x^k - x^{k+1})/t \in \partial f(x^{k+1})$ which implies that

$$\|x^k - x^{k+1}\|^2 \geq t^2 d^2(0, \partial f(x^{k+1})).$$

Therefore

$$d^2(x^{k+1}, S) + t^2 d^2(0, \partial f(x^{k+1})) \leq d^2(x^k, S).$$

If there is a constant $\gamma > 0$ such that $d(0, \partial f(x^{k+1})) \geq \gamma d(x^{k+1}, S)$ for all k, then

$$(1 + \gamma^2 t^2) d^2(x^{k+1}, S) \leq d^2(x^k, S)$$

and thus $d(x^k, S) \to 0$ linearly as $k \to \infty$. This motivates the following subdifferential error bound condition of f.

Definition 8.2 f is said to satisfy a *subdifferential error bound condition* at $\bar{x} \in S$ if there exist constants $r > 0$ and $\gamma > 0$ such that

$$d(x, S) \leq \gamma d(0, \partial f(x))$$

for all $x \in B_r(\bar{x})$.

Let us illustrate the subdifferential error bound condition by considering strongly convex functions. Let $f : \mathbb{R}^n \to (-\infty, \infty]$ be a proper, lower semicontinuous, strongly convex function with modulus of convexity $\mu > 0$. It is known that f has a unique minimizer x^*, i.e., $S = \{x^*\}$. For any $x \in \text{dom}(\partial f)$, we take an arbitrary $\xi \in \partial f(x)$. Since $0 \in \partial f(x^*)$, it then follows from Proposition 3.12 that

$$2\mu \|x - x^*\|^2 \leq \langle \xi, x - x^* \rangle \leq \|\xi\| \|x - x^*\|$$

which implies $2\mu\|x - x^*\| \leq \|\xi\|$. Since $\xi \in \partial f(x)$ is arbitrary, we thus obtain $2\mu d(x, S) \leq d(0, \partial f(x))$ which shows the subdifferential error bound condition for f.

The subdifferential error bound condition is weaker than strong convexity. To see this, consider again the function $f(x) = \|x\|_1$ on $\mathbb{R}^n$. We have $S = \{0\}$ and, according to Example 3.4, $\xi \in \partial f(x)$ implies $\langle \xi, x \rangle = \|x\|_1$. Thus, for any $x \in \mathbb{R}^n$ satisfying $\|x\| \leq 1$, we have

$$d^2(x, S) = \|x\|^2 \leq \|x\|_1 = \langle \xi, x \rangle \leq \|\xi\| \|x\| = \|\xi\| d(x, S)$$

which shows $d(x, S) \leq \|\xi\|$ for all $\xi \in \partial f(x)$. Therefore

$$d(x, S) \leq d(0, \partial f(x)), \quad \forall x \in B_1(0) \subset \mathbb{R}^n$$

which demonstrates the subdifferential error bound condition for f at its minimizer $x^* = 0$.

In Sect. 8.3, we will generalize the subdifferential error bound condition and introduce the concept of metric subregularity for set-valued mappings (or multifunctions). In that context, subdifferential error bound condition given in Definition 8.2 can be viewed as a special instance of the metric subregularity of ∂f at $(\bar{x}, 0)$.

8.1.3 *Kurdyka-Łojasiewicz Condition*

Next, we consider the proximal point algorithm (8.1) from its very first definition. We have

$$x^{k+1} = \arg\min_{x\in\mathbb{R}^n} \left\{ f(x) + \frac{1}{2t}\|x - x^k\|^2 \right\}$$

which implies that

$$f(x^{k+1}) \leq f(x^k) - \frac{1}{2t}\|x^k - x^{k+1}\|^2. \tag{8.3}$$

Since $(x^k - x^{k+1})/t \in \partial f(x^{k+1})$, we therefore obtain

$$f(x^{k+1}) \leq f(x^k) - \frac{t}{2}d^2(0, \partial f(x^{k+1})).$$

If there is $\gamma > 0$ such that $d^2(0, \partial f(x^{k+1}))/(f(x^{k+1}) - f_*) \geq \gamma$ for all k, then

$$f(x^{k+1}) \leq f(x^k) - \frac{1}{2}\gamma t(f(x^{k+1}) - f_*)$$

which implies that

$$f(x^{k+1}) - f_* \leq \frac{2}{2+\gamma t}(f(x^k) - f_*) \tag{8.4}$$

i.e., $f(x^k) \to f_*$ linearly as $k \to \infty$. This motivates the following special version of the Kurdyka-Łojasiewicz condition.

Definition 8.3 f is said to satisfy the *Kurdyka-Łojasiewicz condition* at a point $\bar{x} \in S$ with exponent $1/2$ if there exist $r > 0$ and $\gamma > 0$ such that

$$\gamma(f(x) - f_*) \leq d^2(0, \partial f(x))$$

for all $x \in B_r(\bar{x})$.

Equation (8.4) gives the linear convergence of $f(x^k)$ to f_*. Based on this, we can derive the linear convergence of x^k. To see this, let $x^* \in S$ be such that $x^k \to x^*$. By using (8.3), we have

$$\|x^k - x^*\| = \lim_{l\to\infty}\|x^k - x^l\| \leq \limsup_{l\to\infty}\sum_{j=k}^{l-1}\|x^{j+1} - x^j\|$$

$$
\begin{aligned}
&= \sum_{j=k}^{\infty} \|x^j - x^{j+1}\| \le \sum_{j=k}^{\infty} \sqrt{2t(f(x^j) - f(x^{j+1})} \\
&\le \sum_{j=k}^{\infty} \sqrt{2t(f(x^j) - f_*)}.
\end{aligned}
$$

Assume f satisfies the Kurdyka-Łojasiewicz condition at x^* specified in Definition 8.3. We may choose an integer k_0 such that $x^k \in B_r(x^*)$, and hence, the estimate (8.4) holds for all $k \ge k_0$. Let $q := 2/(2 + \gamma t)$. Consequently

$$
\|x^k - x^*\| \le \sqrt{2t(f(x^{k_0}) - f_*)} \sum_{j=k}^{\infty} q^{(j-k_0)/2} \le \tilde{C} q^{(k-k_0)/2} \sqrt{f(x^0) - f_*},
$$

where $\tilde{C}$ is a constant independent of k. This shows the linear convergence of x^k to x^*.

When $f : \mathbb{R}^n \to \mathbb{R}$ is a strongly convex differentiable function with the modulus of convexity $\mu > 0$, we have

$$
f(y) \ge f(x) + \langle \nabla f(x), y - x \rangle + \mu \|y - x\|^2, \quad \forall x, y \in \mathbb{R}^n.
$$

For any fixed x, the right-hand side is a quadratic function of y which achieves its minimum $f(x) - \frac{1}{4\mu}\|\nabla f(x)\|^2$ at the point $y = x - \frac{1}{2\mu}\nabla f(x)$. Therefore

$$
f(y) \ge f(x) - \frac{1}{4\mu}\|\nabla f(x)\|^2, \quad \forall x, y \in \mathbb{R}^n
$$

This in particular implies

$$
f(x) - f_* \le \frac{1}{4\mu}\|\nabla f(x)\|^2, \quad \forall x \in \mathbb{R}^n
$$

which is called the Polyak-Łojasiewicz inequality [146, 181]. It is a special case of the Kurdyka-Łojasiewicz inequality given in Definition 8.3. In Chap. 9, we will introduce the general formulation of the Kurdyka-Łojasiewicz condition with a thorough study and investigate its deep impact on the analysis of various algorithms for convex as well as non-convex optimization problems.

8.1.4 Proximal Error Bound Condition

We next consider the proximal point algorithm (8.1) by rewriting it as

$$x^{k+1} = x^k - t\mathcal{P}_t(x^k), \quad \mathcal{P}_t(x^k) := \frac{1}{t}(x^k - \mathrm{prox}_{tf}(x^k)).$$

By the definition of x^{k+1}, we have $\mathcal{P}_t(x^k) := (x^k - x^{k+1})/t \in \partial f(x^{k+1})$. Thus, by the convexity of f, we can obtain

$$\begin{aligned} f(x) &\geq f(x^{k+1}) + \langle \mathcal{P}_t(x^k), x - x^{k+1}\rangle \\ &= f(x^{k+1}) + \langle \mathcal{P}_t(x^k), x - x^k\rangle + t\|\mathcal{P}_t(x^k)\|^2 \end{aligned} \tag{8.5}$$

for all $x \in \mathbb{R}^n$. Consequently, by using the Cauchy-Schwarz inequality, we have for any $x \in S$ that

$$f(x^{k+1}) - f_* \leq \|\mathcal{P}_t(x^k)\| \|x^k - x\| - t\|\mathcal{P}_t(x^k)\|^2$$

and hence

$$f(x^{k+1}) - f_* \leq \|\mathcal{P}_t(x^k)\| d(x^k, S) - t\|\mathcal{P}_t(x^k)\|^2.$$

Let $\gamma_k := d(x^k, S)/\|\mathcal{P}_t(x^k)\|$. Then, we further have

$$f(x^{k+1}) - f_* \leq (\gamma_k - t)\|\mathcal{P}_t(x^k)\|^2.$$

Because $f(x^{k+1}) \geq f_*$, this implies that $\gamma_k \geq t$. By taking $x = x^k$ in (8.5), we have $f(x^k) - f(x^{k+1}) \geq t\|\mathcal{P}_t(x^k)\|^2$. Therefore

$$f(x^{k+1}) - f_* \leq \left(\frac{\gamma_k}{t} - 1\right)(f(x^k) - f(x^{k+1})$$

which implies

$$f(x^{k+1}) - f_* \leq \left(1 - \frac{t}{\gamma_k}\right)(f(x^k) - f_*).$$

Thus, if γ_k is bounded above for all k, then $f(x^k)$ converges to f_* linearly, consequently x^k converges linearly. This motivates the following proximal error bound condition.

Definition 8.4 f is said to satisfy a *proximal error bound condition* at $\bar{x} \in S$ with respect to a fixed parameter $t > 0$ if there exist $r > 0$ and $\gamma > 0$ such that

$$d(x, S) \leq \gamma t^{-1} \|x - \text{prox}_{tf}(x)\|$$

for all $x \in B_r(\bar{x})$.

The definition of proximal error bound condition involves a fixed parameter $t > 0$. Actually, we can take $t = 1$ because of the following monotonicity result [108, 167, 214].

Lemma 8.5 *Let $f : \mathbb{R}^n \to (-\infty, \infty]$ be a proper, lower semicontinuous, convex function. Then, for any $x, v \in \mathbb{R}^n$ and $0 < t_1 \leq t_2$, there hold*

$$\|x - \text{prox}_{t_1 f}(x + t_1 v)\| \leq \|x - \text{prox}_{t_2 f}(x + t_2 v)\|$$

and

$$\frac{1}{t_1}\|x - \text{prox}_{t_1 f}(x + t_1 v)\| \geq \frac{1}{t_2}\|x - \text{prox}_{t_2 f}(x + t_2 v)\|.$$

Proof Let $z_t := \text{prox}_{tf}(x + tv)$ for any $t > 0$. Then, $(x + tv - z_t)/t \in \partial f(z_t)$. By using the monotonicity of ∂f, we have

$$\begin{aligned}
0 &\leq \left\langle \frac{x + t_1 v - z_{t_1}}{t_1} - \frac{x + t_2 v - z_{t_2}}{t_2}, z_{t_1} - z_{t_2} \right\rangle \\
&= \left\langle \frac{x - z_{t_1}}{t_1} - \frac{x - z_{t_2}}{t_2}, -(x - z_{t_1}) + (x - z_{t_2}) \right\rangle \\
&= -\frac{1}{t_1}\|x - z_{t_1}\|^2 - \frac{1}{t_2}\|x - z_{t_2}\|^2 + \left(\frac{1}{t_1} + \frac{1}{t_2}\right) \langle x - z_{t_1}, x - z_{t_2} \rangle \\
&\leq -\frac{1}{t_1}\|x - z_{t_1}\|^2 - \frac{1}{t_2}\|x - z_{t_2}\|^2 + \left(\frac{1}{t_1} + \frac{1}{t_2}\right) \|x - z_{t_1}\| \|x - z_{t_2}\| \\
&= -\left(\|x - z_{t_1}\| - \|x - z_{t_2}\|\right) \left(\frac{1}{t_1}\|x - z_{t_1}\| - \frac{1}{t_2}\|x - z_{t_2}\|\right).
\end{aligned}$$

Since $0 < t_1 \leq t_2$, this implies that $\|x - z_{t_1}\| \leq \|x - z_{t_2}\|$ and $\|x - z_{t_1}\|/t_1 \geq \|x - z_{t_2}\|/t_2$, completing the proof. □

8.1.5 Luo-Tseng Error Bound Condition

When $f(x) = \varphi(x) + h(x)$, where $h : \mathbb{R}^n \to (-\infty, \infty]$ is a proper, lower semicontinuous, convex function and $\varphi : \mathbb{R}^n \to \mathbb{R}$ is a convex continuous differentiable function with L-Lipschitz continuous gradient ∇h, one may formulate the following Luo-Tseng error bound condition [150].

Definition 8.6 $f = \varphi + h$ is said to satisfy a *Luo-Tseng error bound condition* at $\bar{x} \in S$ if there exist $r > 0$ and $\gamma > 0$ such that

$$d(x, S) \leq \gamma \|x - \mathrm{prox}_{\varphi}(x - \nabla h(x))\|$$

for all $x \in B_r(\bar{x})$.

This error bound condition can be motivated from the consideration on the linear convergence of the proximal gradient method:

$$x^{k+1} = \mathrm{prox}_{t\varphi}(x^k - t\nabla h(x^k)). \tag{8.6}$$

The convergence and sublinear convergence rate of (8.6) have been established in Theorem 4.35 under suitable choice of the step size t. We now seek conditions to guarantee a linear convergence of (8.6). To this end, we may rewrite (8.6) as

$$x^{k+1} = x^k - t\mathcal{G}_t(x^k)$$

to see the parallel with the classical gradient method for smooth minimization problems, where

$$\mathcal{G}_t(x) := \frac{1}{t}\left(x - \mathrm{prox}_{t\varphi}(x - t\nabla h(x))\right)$$

which is called the prox-gradient mapping. When $0 < t \leq 1/L$, by using Lemma 4.34, it is easy to derive

$$f(x^k) - f(x^{k+1}) \geq \frac{1}{2L}\|\mathcal{G}_t(x^k)\|^2 \tag{8.7}$$

and

$$f(x^{k+1}) - f_* \leq \langle \mathcal{G}_t(x^k), x^k - \bar{x}\rangle - \frac{1}{2L}\|\mathcal{G}_t(x^k)\|^2 \tag{8.8}$$

for any $\bar{x} \in S$. Note that $\mathcal{G}_t(x^k) = 0$ for some k implies $x^k \in S$ and $x^l = x^k$ for all $l \geq k$. Thus, we may assume $\mathcal{G}_t(x^k) \neq 0$ for all k. From (8.8), it follows

$$f(x^{k+1}) - f_* \leq \|\mathcal{G}_t(x^k)\|\|x^k - \bar{x}\| - \frac{1}{2L}\|\mathcal{G}_t(x^k)\|^2.$$

Since $\bar{x} \in S$ is arbitrary, we have

$$f(x^{k+1}) - f_* \leq \|\mathcal{G}_t(x^k)\| d(x^k, S) - \frac{1}{2L}\|\mathcal{G}_t(x^k)\|^2.$$

Let $\gamma_k := d(x^k, S)/\|\mathcal{G}_t(x^k)\|$. Then, by using (8.7), we further have

$$f(x^{k+1}) - f_* \leq \|\mathcal{G}_t(x^k)\|^2 \left(\gamma_k - \frac{1}{2L}\right) \leq (2L\gamma_k - 1)\left(f(x^k) - f(x^{k+1})\right)$$

which implies that

$$f(x^{k+1}) - f_* \leq \left(1 - \frac{1}{2L\gamma_k}\right)\left(f(x^k) - f_*\right). \tag{8.9}$$

Therefore, if γ_k is bounded above for all integers k, we can obtain a linear convergence in terms of the objective function values. According to Theorem 4.35, $x^k \to \bar{x}$ as $k \to \infty$ for some $\bar{x} \in S$. In case the Luo-Tseng error bound condition given in Definition 8.6 holds for f at $\bar{x}$, we may use Lemma 8.5 to conclude that for a given $0 < t \leq 1/L$ there is a constant $\eta_t > 0$ such that

$$d(x, S) \leq \eta_t \|\mathcal{G}_t(x)\|, \quad \forall x \in B_r(\bar{x}).$$

Thus, there exists an integer k_0 such that $x^k \in B_r(\bar{x})$ and hence $\gamma_k \leq \eta_t$ for all $k \geq k_0$. Consequently, it follows from (8.9) that $f(x^k) - f_* \to 0$ linearly as $k \to \infty$. By using the same argument in Sect. 8.1.3, one can further deduce that $\|x^k - \bar{x}\| \to 0$ linearly as $k \to \infty$.

Note that when $h \equiv 0$, Luo-Tseng error bound condition reduces to the proximal error bound condition. Therefore, it can be viewed as an extension of the proximal error bound condition. The following result shows that these two conditions are actually equivalent.

Theorem 8.7 *Let $f = \varphi + h$, where $h : \mathbb{R}^n \to (-\infty, \infty]$ is proper, lower semicontinuous, convex and $\varphi : \mathbb{R}^n \to \mathbb{R}$ is convex continuous differentiable with L-Lipschitz gradient. Let $\bar{x} \in S := \arg\min_{x\in\mathbb{R}^n} f(x)$ which is assumed to be non-empty. Then, the following two statements are equivalent:*

(i) *There exist $r > 0$ and $\gamma > 0$ such that*

$$d(x, S) \leq \gamma \|x - \operatorname{prox}_f(x)\|, \quad \forall x \in B_r(\bar{x}).$$

(ii) *There exist $r > 0$ and $\tilde{\gamma} > 0$ such that*

$$d(x, S) \leq \tilde{\gamma} \|x - \operatorname{prox}_\varphi(x - \nabla h(x))\|, \quad \forall x \in B_r(\bar{x}).$$

Proof Let $x \in B_r(\bar{x})$, and for any $t > 0$, we set $z_t := \mathcal{G}_t(x)$. Then, $x - t\nabla h(x) \in x - tz_t + t\partial\varphi(x - tz_t)$. Since $\partial f = \partial\varphi + \nabla h$, we have

$$x - t\nabla h(x) + t\nabla h(x - tz_t) \in (I + t\partial f)(x - tz_t)$$

and thus

$$x - tz_t = \mathrm{prox}_{tf}(x - t\nabla h(x) + t\nabla h(x - tz_t)).$$

Consequently, by using the nonexpansiveness of the proximal mapping and the L-Lipschitz continuity of ∇h, we can obtain

$$\begin{aligned}\|x - tz_t - \mathrm{prox}_{tf}(x)\| &= \|\mathrm{prox}_{tf}(x - t\nabla h(x) + t\nabla h(x - tz_t)) - \mathrm{prox}_{tf}(x)\| \\ &\leq t\|\nabla h(x) - \nabla h(x - tz_t)\| \leq Lt^2\|z_t\|\end{aligned}$$

which implies

$$(1 - Lt)\|z_t\| \leq \frac{1}{t}\|x - \mathrm{prox}_{tf}(x)\| \leq (1 + Lt)\|z_t\|. \tag{8.10}$$

If (i) holds, then by using (8.10) with $t = 1$, we have

$$d(x, S) \leq \gamma\|x - \mathrm{prox}_f(x)\| \leq (1 + L)\gamma\|x - \mathrm{prox}_\varphi(x - \nabla h(x))\|$$

which shows that (ii) holds with $\tilde{\gamma} = (1 + L)\gamma$.

If (ii) holds, then by using (8.10) with $t = \min\{1/(2L), 1\}$ and Lemma 8.5, we can obtain

$$\begin{aligned}d(x, S) &\leq \tilde{\gamma}\|x - \mathrm{prox}_\varphi(x - \nabla h(x))\| \leq \tilde{\gamma}\|z_t\| \\ &\leq \frac{2\tilde{\gamma}}{t}\|x - \mathrm{prox}_{tf}(x)\| \\ &\leq \frac{2\tilde{\gamma}}{t}\|x - \mathrm{prox}_f(x)\|\end{aligned}$$

which shows that (i) holds with $\gamma = 2\tilde{\gamma}/t$. □

8.2 Equivalence of Error Bound Conditions

In the above section, we have introduced various error bound conditions that guarantee linear convergence of the proximal point algorithm. It is natural to ask what are the relationships among these conditions. In the following, we will show that all these conditions are actually equivalent in the convex setting.

We start from the following result [74] which demonstrates that the subdifferential error bound condition and the proximal error bound condition are equivalent.

Theorem 8.8 *Let $f : \mathbb{R}^n \to (-\infty, \infty]$ be a proper, lower semicontinuous, convex function with $S := \arg\min_{x \in \mathbb{R}^n} f(x) \neq \emptyset$. Let $\bar{x} \in S$ and let $r > 0$ be a fixed number. Then, the following two conditions are equivalent:*

(i) *There exists a constant $\gamma > 0$ such that*

$$d(x, S) \leq \gamma d(0, \partial f(x)), \quad \forall x \in B_r(\bar{x}).$$

(ii) *There exists a constant $\tilde{\gamma} > 0$ such that*

$$d(x, S) \leq \tilde{\gamma} \|x - \text{prox}_f(x)\|, \quad \forall x \in B_r(\bar{x}).$$

Proof (i) $\Longrightarrow$ (ii). Let $x \in B_r(\bar{x})$ and let $z := \text{prox}_f(x)$. We may use the nonexpansiveness of prox_f and $\bar{x} = \text{prox}_f(\bar{x})$ to obtain $\|z - \bar{x}\| \leq \|x - \bar{x}\|$ which implies that $z \in B_r(\bar{x})$. By the definition of z, we also have $x - z \in \partial f(z)$. Therefore, we may use (i) to obtain

$$d(x, S) \leq \|x - z\| + d(z, S) \leq \|x - z\| + \gamma d(0, \partial f(z)) \leq (\gamma + 1)\|x - z\|$$

which shows that (ii) holds with $\tilde{\gamma} = \gamma + 1$.

(ii) $\Longrightarrow$ (i). Let $x \in B_r(\bar{x})$. For any $v \in \partial f(x)$, we have $x + v \in x + \partial f(x)$ which implies $x = \text{prox}_f(x + v)$. Therefore, by (ii) and the nonexpansiveness of prox_f, we have

$$d(x, S) \leq \tilde{\gamma} \|x - \text{prox}_f(x)\| = \tilde{\gamma} \|\text{prox}_f(x + v) - \text{prox}_f(x)\| \leq \tilde{\gamma} \|v\|$$

which shows that $d(x, S) \leq \tilde{\gamma} d(0, \partial f(x))$, i.e., (i) holds with $\gamma = \tilde{\gamma}$. □

Next, we will prove a result which in particular implies the equivalence between the quadratic growth condition and the Kurdyka-Lojasiewicz condition with exponent $1/2$.

Theorem 8.9 *Let $f : \mathbb{R}^n \to (-\infty, \infty]$ be a proper, lower semicontinuous, convex function with $S := \arg\min_{x \in \mathbb{R}^n} f(x) \neq \emptyset$. Let $\bar{x} \in S$ and $f_* := f(\bar{x})$.*

(i) *If there exist $r > 0$, $\eta \in (0, \infty]$ and a strictly increasing concave function $\varphi \in C[0, \eta) \cap C^1(0, \eta)$ with $\varphi(0) = 0$ such that*

$$\varphi'(f(x) - f_*) d(0, \partial f(x)) \geq 1$$

for all $x \in B_r(\bar{x}) \cap \{f_* < f < f_* + \eta\}$, then

$$d(x, S) \leq 2\varphi(f(x) - f_*)$$

for all $x \in B_r(\bar{x}) \cap \{f_* < f < f_* + \eta\}$.

(ii) *If there exist* $r > 0$, $\eta \in (0, \infty]$ *and a strictly increasing concave function* $\varphi \in C[0, \eta) \cap C^1(0, \eta)$ *with* $\varphi(0) = 0$ *and* $s\varphi'(s) \geq \varphi(s)$ *for all* $s \in (0, \eta)$ *such that*

$$d(x, S) \leq \varphi(f(x) - f_*)$$

for all $x \in B_r(\bar{x}) \cap \{f_* < f < f_* + \eta\}$*, then*

$$\varphi'(f(x) - f_*)d(0, \partial f(x)) \geq 1$$

for all $x \in B_r(\bar{x}) \cap \{f_* < f < f_* + \eta\}$.

Proof

(i) Given any $x \in B_r(\bar{x}) \cap \{f_* < f < f_* + \eta\}$ and consider the sequence $\{x^k\}$ defined by the proximal point algorithm

$$x^k = \arg\min_{z \in \mathbb{R}^n} \left\{ f(z) + \frac{1}{2\tau}\|z - x^{k-1}\|^2 \right\}, \quad k = 1, 2, \cdots$$

with $x^0 := x$, where $\tau > 0$ is a constant. According to Theorem 4.29 and its proof, we have

- $\|x^{k+1} - \bar{x}\| \leq \|x^k - \bar{x}\|$ for all $k \geq 0$. This implies that $x^k \in B_r(\bar{x})$ for all $k \geq 0$.
- $f(x^{k+1}) \leq f(x^k)$ for all k. Consequently, $x^k \in \{f_* \leq f < f_* + \eta\}$ for all $k \geq 0$.
- $f(x^k) \to f_*$ and $x^k \to x^*$ as $k \to \infty$ for some $x^* \in S$.

We now show that for all integers $k \geq 1$ there holds

$$2\|x^{k+1} - x^k\| \leq \|x^k - x^{k-1}\| + 2\Big[\varphi(f(x^k) - f_*) - \varphi(f(x^{k+1}) - f_*)\Big]. \tag{8.11}$$

If $f(x^k) = f_*$, then by the definition of x^{k+1}, we have

$$f(x^{k+1}) + \frac{1}{2\tau}\|x^{k+1} - x^k\|^2 \leq f(x^k) \tag{8.12}$$

which implies $f(x^{k+1}) = f_*$ and $x^{k+1} = x^k$ and hence (8.11) holds trivially. So, we may assume $f(x^k) > f_*$. By the definition of x^k, we have $(x^{k-1} - x^k)/\tau \in \partial f(x^k)$. Thus, by the given condition, we can obtain

$$\varphi'(f(x^k) - f_*) \geq \frac{1}{d(0, \partial f(x^k))} \geq \frac{\tau}{\|x^k - x^{k-1}\|}. \tag{8.13}$$

It follows from the concavity of φ, (8.12) and (8.13) that

$$\begin{aligned}\varphi(f(x^{k+1}) - f_*) &\leq \varphi(f(x^k) - f_*) + \varphi'(f(x^k) - f_*)(f(x^{k+1}) - f(x^k)) \\ &\leq \varphi(f(x^k) - f_*) - \frac{\|x^{k+1} - x^k\|^2}{2\|x^k - x^{k-1}\|}.\end{aligned}$$

Therefore

$$\begin{aligned}2\|x^{k+1} - x^k\| &\leq \|x^k - x^{k-1}\| + \frac{\|x^{k+1} - x^k\|^2}{\|x^k - x^{k-1}\|} \\ &\leq \|x^k - x^{k-1}\| + 2\left[\varphi(f(x^k) - f_*) - \varphi(f(x^{k+1}) - f_*)\right]\end{aligned}$$

which shows (8.11) again.

For any integer $N > 1$, by summing (8.11) over k from 1 to N, we can obtain

$$\begin{aligned}2\sum_{k=1}^{N} \|x^{k+1} - x^k\| &\leq \sum_{k=1}^{N} \|x^k - x^{k-1}\| \\ &\quad + 2\left[\varphi(f(x^1) - f_*) - \varphi(f(x^{N+1}) - f_*)\right] \\ &\leq \sum_{k=1}^{N} \|x^k - x^{k-1}\| + 2\varphi(f(x^0) - f_*).\end{aligned}$$

Consequently

$$\|x^{N+1} - x^0\| \leq \sum_{k=0}^{N} \|x^{k+1} - x^k\| \leq 2\|x^1 - x^0\| + 2\varphi(f(x^0) - f_*).$$

Letting $N \to \infty$ and using $x^0 = x$, we have

$$d(x, S) \leq \|x^* - x\| = \lim_{N\to\infty} \|x^{N+1} - x^0\| \leq 2\|x^1 - x^0\| + 2\varphi(f(x) - f_*).$$

By the definition of x^1, we have $\|x^1 - x^0\|^2 \leq 2\tau(f(x^0) - f(x^1)) \leq 2\tau(f(x) - f_*)$. Therefore

$$d(x, S) \leq \sqrt{8\tau(f(x) - f_*)} + 2\varphi(f(x) - f_*)$$

which is valid for all $\tau > 0$. Letting $\tau \to 0$ then shows the result.

(ii) Let $y \in S$ be any element. For any $x \in B_r(\bar{x}) \cap \{f_* < f < f_* + \eta\}$ with $\partial f(x) \neq \emptyset$, we take any $\xi \in \partial f(x)$. By the convexity of f, we have

$$f_* = f(y) \geq f(x) + \langle \xi, y - x \rangle$$

which implies

$$f(x) - f_* \leq \langle \xi, x - y \rangle \leq \|\xi\| \|x - y\|.$$

Since $y \in S$ and $\xi \in \partial f(x)$ are arbitrary, we can obtain

$$f(x) - f_* \leq d(0, \partial f(x)) d(x, S).$$

By the given condition, we then have

$$f(x) - f_* \leq \varphi(f(x) - f_*) d(0, \partial f(x)).$$

Since $s\varphi'(s) \geq \varphi(s)$ for $s \in (0, r)$, we can further obtain

$$\varphi'(f(x) - f_*) d(0, \partial f(x)) \geq \frac{\varphi(f(x) - f_*) d(0, \partial f(x))}{f(x) - f_*} \geq 1$$

which shows the desired result.

□

Theorem 8.9 was proved in [35] in which part (i) is established by using the subgradient flow generated by ∂f, and hence, the argument relies heavily on the theory of differential inclusion. The argument provided here is taken from [118] which is based on the property of the proximal point algorithm and thus is elementary and more accessible.

Finally, we show the equivalence between the subdifferential error bound condition and the quadratic growth condition. The following result is due to [3, 4].

Theorem 8.10 *Let $f : \mathbb{R}^n \to (-\infty, \infty]$ be a proper, lower semicontinuous, convex function with $S := \arg\min_{x \in \mathbb{R}^n} f(x) \neq \emptyset$. Let $\bar{x} \in S$ and $f_* := f(\bar{x})$. Then, the following two conditions are equivalent:*

(i) *There exist constants $\kappa > 0$ and $r > 0$ such that*

$$d(x, S) \leq \kappa d(0, \partial f(x)), \quad \forall x \in B_r(\bar{x}). \tag{8.14}$$

(ii) *There exist constants $c > 0$ and $\rho > 0$ such that*

$$c d^2(x, S) \leq f(x) - f_*, \quad \forall x \in B_\rho(\bar{x}). \tag{8.15}$$

Proof (i) $\Longrightarrow$ (ii). Under (8.14), we show that (8.15) holds for any $0 < c < 1/(4\kappa)$ and $\rho = 2r/3$. If not, then there exist $z \in B_{2r/3}(\bar{x})$ and $0 < c < 1/(4\kappa)$ such that

$$cd^2(z, S) > f(z) - f_*.$$

By the Ekeland's variational principle given in Theorem 8.11 below with

$$\varepsilon := cd^2(z, S) \quad \text{and} \quad \lambda := \frac{1}{2}d(z, S)$$

we can find $y \in \mathbb{R}^n$ satisfying $\|y - z\| \leq \lambda$ such that

$$f(x) + 4c\lambda\|x - y\| \geq f(y), \quad \forall x \in \mathbb{R}^n. \tag{8.16}$$

This y must be in $B_r(\bar{x})$. Indeed, since $\bar{x} \in S$, we have $\lambda \leq \|z - \bar{x}\|/2$ and thus

$$\|y - \bar{x}\| \leq \|y - z\| + \|z - \bar{x}\| \leq \lambda + \|z - \bar{x}\| \leq \frac{3}{2}\|z - \bar{x}\| < r.$$

Moreover, from (8.16), it follows that the convex function $g(x) := f(x) + 4c\lambda\|x - y\|$ achieves its minimum at y, and thus, by Theorem 3.14 of Moreau-Rockafellar, we have

$$0 \in \partial g(y) = \partial f(y) + 4c\lambda\partial(\|\cdot - y\|)(y).$$

By using Example 3.3, we then obtain $0 \in \partial f(y) + 4c\lambda\mathbb{B}$, where $\mathbb{B}$ denotes the unit ball in $\mathbb{R}^n$ centered at the origin. This implies the existence of $\xi \in \partial f(y)$ such that $\|\xi\| \leq 4c\lambda$. Consequently

$$d\,(0, \partial f(y)) \leq \|\xi\| \leq 4c\lambda.$$

By the definition of λ and the triangle inequality, we have

$$2\lambda = d(z, S) \leq \|z - y\| + d(y, S) \leq \lambda + d(y, S)$$

which implies $0 < \lambda \leq d(y, S)$. Therefore

$$d\,(0, \partial f(y)) \leq 4cd\,(y, S)$$

which is a contradiction to (8.14), considering that $4c < 1/\kappa$ and $y \in B_r(\bar{x})$.

(ii) $\Longrightarrow$ (i). Let $y \in S$ be any elements. Then, for any $x \in B_\rho(\bar{x})$ with $\partial f(x) \neq \emptyset$ and any $\xi \in \partial f(x)$, there holds

$$f(x) - f_* = f(x) - f(y) \leq \langle \xi, x - y \rangle.$$

Therefore, it follows from (8.15) that

$$cd^2(x, S) \leq \langle \xi, x - y \rangle \leq \|\xi\| \|x - y\|.$$

By the arbitrariness of $\xi \in \partial f(x)$ and $y \in S$, we thus have

$$cd^2(x, S) \leq d(0, \partial f(x)) d(x, S)$$

which implies $cd(x, S) \leq d(0, \partial f(x))$ for all $x \in B_\rho(\bar{x})$. □

Theorem 8.11 (Ekeland's Variational Principle) *Let (X, d) be a complete metric space and $f : X \to (-\infty, \infty]$ a proper lower semicontinuous function bounded from below. Let $\varepsilon > 0$ and $u \in X$ be such that*

$$f(u) \leq \inf_{x \in X} f(x) + \varepsilon. \tag{8.17}$$

Then, for any $\lambda > 0$, there exists $v \in X$ such that

(i) $d(u, v) \leq \lambda$.
(ii) $f(v) + \frac{\varepsilon}{\lambda} d(u, v) \leq f(u)$.
(iii) $f(x) + \frac{\varepsilon}{\lambda} d(x, v) \geq f(v)$ *for all* $x \in X$.

Ekeland's variational principle was first proved in [76]. Since then, it has played far reaching influence on the development of nonsmooth analysis [39, 197, 231]. Ekeland's variational principle has the following geometric interpretation: given $u \in X$ satisfying (8.17), if the function $x \to f(u) - (\varepsilon/\lambda) d(x, u)$ does not support f from below, we can find a point v in the set

$$S := \{x \in X : f(x) \leq f(u) - (\varepsilon/\lambda) d(x, u)\}$$

such that the function $x \to f(v) - (\varepsilon/\lambda) d(x, v)$ supports f from below.

Proof We first construct a sequence $\{v^k\} \subset X$ as follows. Set $v^0 := u$. Assuming $v^k \in X$ is constructed, we set

$$S_k := \left\{ x \in X : f(x) + \frac{\varepsilon}{\lambda} d(x, v^k) \leq f(v^k) \right\}$$

and construct $v^{k+1} \in S_k$ by considering two cases: If $f(v^k) = \inf_{S_k} f$, we take $v^{k+1} := v^k$. If $f(v^k) > \inf_{S_k} f$, we take $v^{k+1} \in S_k$ to be any element such that

$$f(v^{k+1}) \leq \inf_{S_k} f + \frac{1}{2} \left(f(v^k) - \inf_{S_k} f \right) \tag{8.18}$$

whose choice is clearly possible; note that

$$f(v^{k+1}) \le \frac{1}{2}\left(f(v^k) + \inf_{S_k} f\right) \le f(v^k).$$

Therefore, for the sequence $\{v^k\}$ chosen above, $\{f(v^k)\}$ is monotonically decreasing. Since f is bounded from below, $\{f(v^k)\}$ is convergent.

Next, we show that $\{v^k\}$ is a convergent sequence. To see this, we may use $v^{k+1} \in S_k$ to obtain

$$d(v^{k+1}, v^k) \le \frac{\lambda}{\varepsilon}\left(f(v^k) - f(v^{k+1})\right).$$

By the triangle inequality, we then have for any integers $0 \le k < l$ that

$$d(v^k, v^l) \le \sum_{j=k}^{l-1} d(v^j, v^{j+1}) \le \frac{\lambda}{\varepsilon}\left(f(v^k) - f(v^l)\right). \tag{8.19}$$

By the convergence of $\{f(v^k)\}$, we can conclude from the above inequality that $d(v^k, v^l) \to 0$ as $k, l \to \infty$, i.e., $\{v^k\}$ is a Cauchy sequence. Since X is complete, there exists $v \in X$ such that $v^k \to v$ as $k \to \infty$.

By taking $l \to \infty$ in (8.19) and using the lower semicontinuity of f, we can obtain

$$f(v) + \frac{\varepsilon}{\lambda} d(v^k, v) \le f(v^k)$$

which shows that $v \in S_k$ for all k, i.e., $v \in \bigcap_{k=0}^{\infty} S_k$. We claim that

$$\bigcap_{k=0}^{\infty} S_k = \{v\}. \tag{8.20}$$

Indeed, if $x \in \bigcap_{k=0}^{\infty} S_k$, we have $x \in S_{k+1} \cap S_k$ and hence

$$\frac{\varepsilon}{\lambda} d(x, v^{k+1}) \le f(v^{k+1}) - f(x) \le f(v^{k+1}) - \inf_{S_k} f$$

for all k. By using (8.18), we further have

$$\frac{\varepsilon}{\lambda} d(x, v^{k+1}) \le f(v^k) - f(v^{k+1}).$$

Letting $k \to \infty$ and using the convergence of $\{f(v^k)\}$, it follows $d(x, v) \le 0$ and thus $x = v$.

We now show that v satisfies the requirements. First, taking $k = 0$ in (8.19) gives

$$f(v^l) + \frac{\varepsilon}{\lambda} d(u, v^l) \leq f(u).$$

Letting $l \to \infty$ and using the lower semicontinuity of f, we obtain

$$f(v) + \frac{\varepsilon}{\lambda} d(u, v) \leq f(u)$$

which shows (ii) and, moreover, by using (8.17), we have

$$\frac{\varepsilon}{\lambda} d(u, v) \leq f(u) - f(v) \leq f(u) - \inf_X f \leq \varepsilon$$

which implies $d(u, v) \leq \lambda$ and thus (i) is proved. Finally, we show (iii). To this end, we first show that $\{S_k\}$ is nested. Note that $x \in S_{k+1}$ implies

$$f(x) + \frac{\varepsilon}{\lambda} d(x, v^{k+1}) \leq f(v^{k+1})$$

and hence, by the triangle inequality and $v^{k+1} \in S_k$, we have

$$\begin{aligned} f(x) + \frac{\varepsilon}{\lambda} d(x, v^k) &\leq f(x) + \frac{\varepsilon}{\lambda} d(x, v^{k+1}) + \frac{\varepsilon}{\lambda} d(v^{k+1}, v^k) \\ &\leq f(v^{k+1}) + \frac{\varepsilon}{\lambda} d(v^{k+1}, v^k) \leq f(v^k) \end{aligned}$$

which means $x \in S_k$. This shows that $S_{k+1} \subset S_k$ for all k, i.e., $\{S_k\}$ is nested. Now, for any $x \in X$ with $x \neq v$, we have from (8.20) that $x \notin S_{k_0}$ for some $k_0 \geq 0$. By the nested property of $\{S_k\}$, we have $x \notin S_k$ for all $k \geq k_0$. Consequently

$$f(x) + \frac{\varepsilon}{\lambda} d(x, v^k) > f(v^k), \quad \forall k \geq k_0.$$

Letting $k \to \infty$ and using the lower semicontinuity of f, we finally obtain $f(x) + \frac{\varepsilon}{\lambda} d(x, v) \geq f(v)$ which shows (iii). □

8.3 Metric Subregularity

In Sect. 8.1.2, we have introduced the subdifferential error bound condition for convex functions to derive linear convergence of the proximal point algorithm. In this section, we will generalize this condition to multifunctions or set-valued maps and introduce the concept of metric subregularity.

A multifunction F from $\mathbb{R}^m$ to $\mathbb{R}^n$ is a map sending every $x \in \mathbb{R}^m$ to a subset $F(x)$ of $\mathbb{R}^n$; we write $F : \mathbb{R}^m \rightrightarrows \mathbb{R}^n$ to indicate F is a multifunction. In contrast, we will write $F : \mathbb{R}^m \to \mathbb{R}^n$ when F is a single-valued map. The domain and range of a multifunction $F : \mathbb{R}^m \rightrightarrows \mathbb{R}^n$ are defined by

$$\begin{aligned}\mathrm{dom}(F) &:= \{x \in \mathbb{R}^m : F(x) \neq \emptyset\},\\ \mathrm{Ran}(F) &:= \{y \in \mathbb{R}^n : y \in F(x) \text{ for some } x \in \mathbb{R}^m\}\end{aligned}$$

respectively. The set

$$\mathrm{Gr}(F) := \{(x, y) \in \mathbb{R}^m \times \mathbb{R}^n : y \in F(x)\}$$

is called the graph of F. The inverse of F is a multifunction $F^{-1} : \mathbb{R}^n \rightrightarrows \mathbb{R}^m$ defined by

$$F^{-1}(y) := \{x \in \mathbb{R}^m : y \in F(x)\}, \quad y \in \mathbb{R}^n.$$

By definition, it is easy to see that $\mathrm{dom}(F^{-1}) = \mathrm{Ran}(F)$, $\mathrm{Ran}(F^{-1}) = \mathrm{dom}(F)$, and

$$\mathrm{Gr}(F^{-1}) = \{(y, x) \in \mathbb{R}^n \times \mathbb{R}^m : (x, y) \in \mathrm{Gr}(F)\}.$$

Example 8.12 Let $f : \mathbb{R}^n \to (-\infty, \infty]$ be a proper convex function. Then, its subdifferential ∂f is a multifunction from $\mathbb{R}^n$ to $\mathbb{R}^n$. If f is lower semicontinuous, it is easy to see that the graph of ∂f is closed. Let f^* denote the Legendre-Fenchel conjugate of f. When f is lower semicontinuous, we may use Theorem 4.9 to conclude

$$x \in (\partial f)^{-1}(\xi) \iff \xi \in \partial f(x) \iff x \in \partial f^*(\xi), \quad \forall x, \xi \in \mathbb{R}^n$$

which demonstrates that $(\partial f)^{-1} = \partial f^*$.

Given a proper convex function $f : \mathbb{R}^n \to (-\infty, \infty]$, it is known that $\bar{x} \in \mathbb{R}^n$ is a minimizer of f if and only if

$$0 \in \partial f(\bar{x}).$$

Thus, finding a minimizer of f is equivalent to solving an inclusion problem related to the multifunction $\partial f : \mathbb{R}^n \rightrightarrows \mathbb{R}^n$. Various iterative algorithms have been developed in the previous chapters to determine a minimizer of f. In order to guarantee a linear convergence of the proximal point algorithm, we have introduced in Sect. 8.1.2 the subdifferential error bound condition of f at a minimizer $\bar{x}$ which states that there exist constants $\kappa \geq 0$ and $r > 0$ such that

$$d(x, (\partial f)^{-1}(0)) \leq \kappa d(0, \partial f(x)), \quad \forall x \in B_r(\bar{x}). \tag{8.21}$$

More generally, given a multifunction $F : \mathbb{R}^m \rightrightarrows \mathbb{R}^n$ and a point $\bar{y} \in \mathbb{R}^n$, we may consider the inclusion problem of determining an $\bar{x} \in \mathbb{R}^m$ such that

$$\bar{y} \in F(\bar{x}).$$

In order to understand the continuous dependence of $\bar{x}$ on $\bar{y}$ and to facilitate the analysis of potential algorithms for solving this inclusion problem, we may consider generalizing (8.21) and introduce the concept of metric subregularity as follows [70, 71, 109, 127].

Definition 8.13 A multifunction $F : \mathbb{R}^m \rightrightarrows \mathbb{R}^n$ is called *metrically subregular* at $(\bar{x}, \bar{y}) \in \mathrm{Gr}(F)$ with a constant $\kappa \geq 0$ if there is an open neighborhood U of $\bar{x}$ such that

$$d(x, F^{-1}(\bar{y})) \leq \kappa d(\bar{y}, F(x)), \quad \forall x \in U. \tag{8.22}$$

If U can be taken as the whole space $\mathbb{R}^m$, then F is called *globally metrically subregular* at $(\bar{x}, \bar{y})$.

Global metric subregularity requires (8.22) to hold for all $x \in \mathbb{R}^m$ which is too restrictive to include many examples. On the other hand, the metric subregularity requires (8.22) to hold in a neighborhood U of $\bar{x}$ which could be too small to accommodate the iterates generated by an algorithm. However, in algorithmic analysis, the boundedness of the generated sequence is typically available. This observation motivates the introduction of bounded metric subregularity [237], which provides a more flexible condition suited for analyzing iterative methods.

Definition 8.14 A multifunction $F : \mathbb{R}^m \rightrightarrows \mathbb{R}^n$ is called *bounded metrically subregular* at $(\bar{x}, \bar{y}) \in \mathrm{Gr}(F)$ if for any number $r > 0$ there is a constant $\kappa > 0$ such that

$$d(x, F^{-1}(\bar{y})) \leq \kappa d(\bar{y}, F(x)), \quad \forall x \in B_r(\bar{x}).$$

Example 8.15 Consider the convex function $f(x) = \|x\|$. Its subdifferential ∂f is bounded metrically subregular at any $(\bar{x}, \bar{\xi}) \in \mathrm{Gr}(\partial f)$. To see this, recall that

$$\partial f(x) = \begin{cases} x/\|x\|, & \text{if } x \neq 0, \\ \overline{B}_1(0), & \text{if } x = 0. \end{cases}$$

Thus, $\|\bar{\xi}\| \leq 1$. If $\|\bar{\xi}\| < 1$, then $(\partial f)^{-1}(\bar{\xi}) = \{0\}$ and thus $d\left(x, (\partial f)^{-1}(\bar{\xi})\right) = \|x\|$ for any x, while for $x \neq 0$, we have

$$d\left(\bar{\xi}, \partial f(x)\right) = \left\| \bar{\xi} - \frac{x}{\|x\|} \right\| \geq 1 - \|\bar{\xi}\| > 0.$$

Therefore, for any $M > 0$ and any x with $\|x\| \le M$, there holds

$$d\left(x, (\partial f)^{-1}(\bar{\xi})\right) \le M \le \frac{M}{1 - \|\bar{\xi}\|} d\left(\bar{\xi}, \partial f(x)\right).$$

If $\|\bar{\xi}\| = 1$, then $(\partial f)^{-1}(\bar{\xi}) = \{\alpha\bar{\xi} : \alpha \ge 0\}$ and thus $\|x\|\bar{\xi} \in (\partial f)^{-1}(\bar{\xi})$ for any x. Therefore, for any $x \ne 0$ with $\|x\| \le M$, we have

$$d\left(x, (\partial f)^{-1}(\bar{\xi})\right) \le \left\|x - \|x\|\bar{\xi}\right\| = \|x\| \left\| \frac{x}{\|x\|} - \bar{\xi} \right\| \le M d\left(\bar{\xi}, \partial f(x)\right)$$

which holds trivially for $x = 0$. This shows the bounded metric subregularity of ∂f at any $(\bar{x}, \bar{\xi}) \in \mathrm{Gr}(\partial f)$. □

Example 8.16 Consider $g(x) = \delta_{\bar{B}_r(0)}(x)$, the indicator function of the closed ball of radius r with center at the origin. It is easy to see that its subdifferential is given by

$$\partial g(x) = \begin{cases} \{0\} & \text{if } \|x\| < r, \\ \{\alpha x : \alpha \ge 0\} & \text{if } \|x\| = r, \\ \emptyset & \text{if } \|x\| > r. \end{cases}$$

We will show that ∂g is globally metrically subregular at any point $(\bar{x}, \bar{\xi}) \in \mathrm{Gr}(\partial g)$. To see this, we will consider two cases: $\bar{\xi} = 0$ and $\bar{\xi} \ne 0$.

If $\bar{\xi} = 0$, we must have

$$d(x, (\partial g)^{-1}(\bar{\xi})) \le d(\bar{\xi}, \partial g(x)), \quad \forall x \in \mathbb{R}^n.$$

This is trivial for $\|x\| > r$ because $\partial g(x) = \emptyset$ and thus $d(\bar{\xi}, \partial g(x)) = \infty$. If $\|x\| \le r$, by noting that $(\partial g)^{-1}(\bar{\xi}) = \bar{B}_r(0)$ we have $d(x, (\partial g)^{-1}(\bar{\xi})) = 0$ and the assertion holds again.

In the following, we assume $\bar{\xi} \ne 0$. Then, $\|\bar{x}\| = r$ and $\bar{\xi} = \alpha\bar{x}$ for some $\alpha > 0$. We claim

$$d(x, (\partial g)^{-1}(\bar{\xi})) \le \frac{2}{\alpha} d(\bar{\xi}, \partial g(x)), \quad \forall x \in \mathbb{R}^n. \tag{8.23}$$

We only need to consider $x \in \bar{B}_r(0)$. Note that $(\partial g)^{-1}(\bar{\xi}) = \{\bar{x}\}$. We have

$$d^2(x, (\partial g)^{-1}(\bar{\xi})) = \|x - \bar{x}\|^2 = r^2 + \|x\|^2 - 2\langle x, \bar{x}\rangle.$$

If $\|x\| < r$, we have

$$d^2(x, (\partial g)^{-1}(\bar{\xi})) \le 4r^2 \quad \text{and} \quad d^2(\bar{\xi}, \partial g(x)) = \|\bar{\xi}\|^2 = \alpha^2 r^2$$

and hence (8.23) holds. If $\|x\| = r$, then $d^2(x, (\partial g)^{-1}(\bar{\xi})) = 2r^2 - 2\langle x, \bar{x}\rangle$ and

$$d^2(\bar{\xi}, \partial g(x)) = \inf_{\beta \geq 0} \|\alpha\bar{x} - \beta x\|^2 = \alpha^2 \inf_{t \geq 0} \|\bar{x} - tx\|^2$$
$$= \alpha^2 \inf_{t \geq 0} \left(r^2 + r^2t^2 - 2t\langle x, \bar{x}\rangle\right).$$

If $\langle x, \bar{x}\rangle \leq 0$, then, by noting that $d^2(x, (\partial g)^{-1}(\bar{\xi})) \leq 2r^2 + 2\|x\|\|\bar{x}\| = 4r^2$,

$$d^2(\bar{\xi}, \partial g(x)) \geq \alpha^2 r^2 \geq \frac{\alpha^2}{4} d^2(x, (\partial g)^{-1}(\bar{\xi})).$$

If $\langle x, \bar{x}\rangle > 0$, then

$$d^2(\bar{\xi}, \partial g(x)) = \alpha^2 \left(r^2 - \frac{\langle x, \bar{x}\rangle^2}{r^2}\right) = \frac{\alpha^2}{r^2}(r^2 + \langle x, \bar{x}\rangle)(r^2 - \langle x, \bar{x}\rangle)$$
$$\geq \alpha^2(r^2 - \langle x, \bar{x}\rangle) = \frac{\alpha^2}{2} d^2(x, (\partial g)^{-1}(\bar{\xi})).$$

Therefore, (8.23) always holds. □

Verifying the metric subregularity of a given multifunction is usually challenging. In the following, we will provide a class of bounded metrically subregular multifunctions by considering the polyhedral multifunctions.

Definition 8.17

(i) A multifunction $F : \mathbb{R}^m \rightrightarrows \mathbb{R}^n$ is called *polyhedral convex* if its graph is a polyhedral convex set in $\mathbb{R}^m \times \mathbb{R}^n$.
(ii) A multifunction $F : \mathbb{R}^m \rightrightarrows \mathbb{R}^n$ is called *polyhedral* if its graph $\mathrm{Gr}(F)$ is the union of finitely many sets that are polyhedral convex in $\mathbb{R}^m \times \mathbb{R}^n$.

Let $F : \mathbb{R}^m \rightrightarrows \mathbb{R}^n$ be a multifunction. By definition, it is easy to see that if F is polyhedral convex (or polyhedral), so is its inverse F^{-1}. Moreover, we have the following result.

Theorem 8.18 *Let Φ and Ψ be polyhedral multifunctions from $\mathbb{R}^m$ to $\mathbb{R}^n$. Then, $\Phi + \Psi$ is also polyhedral.*

Proof Since Φ and Ψ are polyhedral, there exist polyhedral convex sets $G_i, i = 1, \cdots, p$ and $E_j, j = 1, \cdots, q$ such that

$$\mathrm{Gr}(\Phi) = \bigcup_{i=1}^{p} G_i \quad \text{and} \quad \mathrm{Gr}(\Psi) = \bigcup_{j=1}^{q} E_j.$$

Therefore

$$\mathrm{Gr}(\Phi+\Psi) = \{(x, y+z) \in \mathbb{R}^m \times \mathbb{R}^n : (x, y) \in \mathrm{Gr}(\Phi) \text{ and } (x, z) \in \mathrm{Gr}(\Psi)\}$$
$$= \bigcup_{i=1}^{p} \bigcup_{j=1}^{q} \Gamma_{ij},$$

where

$$\Gamma_{ij} := \{(x, y+z) \in \mathbb{R}^m \times \mathbb{R}^n : (x, y) \in G_i \text{ and } (x, z) \in E_j\}.$$

It suffices to show that each Γ_{ij} is a polyhedral convex set. By introducing the linear maps

$$T_1 : \mathbb{R}^m \times \mathbb{R}^n \times \mathbb{R}^n \to \mathbb{R}^m \times \mathbb{R}^n \times \mathbb{R}^m \times \mathbb{R}^n, \quad (x, y, z) \to (x, y, x, z)$$

and

$$T_2 : \mathbb{R}^m \times \mathbb{R}^n \times \mathbb{R}^m \times \mathbb{R}^n \to \mathbb{R}^m \times \mathbb{R}^n, \quad (x, y, w, z) \to (x, y+z),$$

we can write $\Gamma_{ij} = T_2((G_i \times E_j) \cap T_1(\mathbb{R}^m \times \mathbb{R}^n \times \mathbb{R}^n))$. Since polyhedral convex property is preserved under linear maps (see Theorem 8.64), each Γ_{ij} must be a polyhedral convex set. □

In the following, we will show that every polyhedral multifunction is bounded metrically subregular. We first consider the polyhedral convex multifunctions and investigate their Lipschitz continuity via Hoffmann's error bound given in Theorem 5.16. We need the notion of Hausdorff distance between sets. Given two sets C and D in $\mathbb{R}^n$, their Hausdorff distance is defined by

$$h(C, D) := \max\left\{\sup_{x\in C} d(x, D), \sup_{x\in D} d(x, C)\right\}.$$

We will use the Hausdorff distance to measure the continuity of multifunctions.

Definition 8.19 A multifunction $F : \mathbb{R}^m \rightrightarrows \mathbb{R}^n$ is said to be *Lipschitz continuous* relative to a set $D \subset \mathrm{dom}(F)$ if there is a constant $\kappa \geq 0$ such that

$$h(F(x'), F(x)) \leq \kappa \|x' - x\|, \quad \forall x, x' \in D.$$

Theorem 8.20 *Any polyhedral convex multifunction $F : \mathbb{R}^m \rightrightarrows \mathbb{R}^n$ is Lipschitz continuous relative to its domain.*

Proof Since $F : \mathbb{R}^m \rightrightarrows \mathbb{R}^n$ is polyhedral convex, by definition, its graph $\mathrm{Gr}(F)$ is a polyhedral convex set, i.e., there exist matrices $A \in \mathbb{R}^{p\times m}$, $B \in \mathbb{R}^{p\times n}$ and a vector $c \in \mathbb{R}^p$ such that

$$\mathrm{Gr}(F) = \left\{(x, y) \in \mathbb{R}^m \times \mathbb{R}^n : Ax + By \le c\right\}.$$

Consequently, for any $x \in \mathrm{dom}(F)$, we have

$$F(x) = \{y \in \mathbb{R}^n : By \le c - Ax\}.$$

Now, let $x, x' \in \mathrm{dom}(F)$ be any two points. For any $y \in F(x)$, we have from Theorem 5.16 that

$$d(y, F(x')) \le \kappa\|(Ax' + By - c)_+\|,$$

where $\kappa \ge 0$ is a constant determined completely by B. Note that $y \in F(x)$ implies $Ax + By - c \le 0$, we have

$$Ax' + By - c = (Ax + By - c) + A(x' - x) \le A(x' - x).$$

Therefore

$$d(y, F(x')) \le \kappa\|[A(x' - x)]_+\| \le \kappa\|A(x' - x)\|$$

and hence

$$\sup_{y\in F(x)} d(y, F(x')) \le \kappa\|A(x' - x)\| \le c\|x' - x\|$$

with a constant $c \ge 0$ determined completely by A and B. By symmetry, we thus obtain $h(F(x'), F(x)) \le c\|x' - x\|$ for all $x, x' \in \mathrm{dom}(F)$. The proof is complete. □

Based on the Lipschitz continuity of polyhedral convex multifunctions, we now show that every polyhedral multifunction is outer Lipschitz continuous defined below.

Definition 8.21 A multifunction $F : \mathbb{R}^m \rightrightarrows \mathbb{R}^n$ is called *outer Lipschitz continuous* at $\bar{x}$ relative to a set $D \subset \mathrm{dom}(F)$ with $\bar{x} \in D$ if there is a constant $\kappa \ge 0$ and a neighborhood V of $\bar{x}$ such that

$$\sup_{y\in F(x)} d(y, F(\bar{x})) \le \kappa\|x - \bar{x}\|, \quad \forall x \in V \cap D.$$

If F is outer Lipschitz continuous at every point $x \in D$ relative to D with the same constant κ, then F is said to be outer Lipschitz continuous relative to D.

Theorem 8.22 *Any polyhedral multifunction $F : \mathbb{R}^m \rightrightarrows \mathbb{R}^n$ is outer Lipschitz continuous relative to its domain.*

Proof We will show that there is a constant $\kappa > 0$ such that for any $\bar{x} \in \mathrm{dom}(F)$ there is a neighborhood V of $\bar{x}$ such that

$$\sup_{y \in F(x)} d(y, F(\bar{x})) \leq \kappa \|x - \bar{x}\|, \quad \forall x \in V \cap \mathrm{dom}(F). \tag{8.24}$$

Since F is polyhedral, there exist $k \geq 1$ polyhedral convex sets $G_1, \cdots, G_k$ in $\mathbb{R}^m \times \mathbb{R}^n$ such that

$$\mathrm{Gr}(F) = \bigcup_{i=1}^{k} G_i.$$

For each $i = 1, \cdots, k$, we define the multifunction $F_i : \mathbb{R}^m \rightrightarrows \mathbb{R}^n$ by

$$F_i(x) := \left\{ y \in \mathbb{R}^n : (x, y) \in G_i \right\}, \quad \forall x \in \mathbb{R}^m.$$

Note that $\mathrm{Gr}(F_i) = G_i$ and G_i is polyhedral convex, the multifunction F_i is polyhedral convex. It follows from Theorem 8.20 that each F_i is Lipschitz continuous, i.e., there exists a constant $\kappa_i > 0$ such that

$$h(F_i(x), F_i(x')) \leq \kappa_i \|x - x'\|, \quad \forall x, x' \in \mathrm{dom}(F_i).$$

Given $\bar{x} \in \mathrm{dom}(F)$, we define

$$I := \left\{ i : 1 \leq i \leq k \text{ and there is } y \in \mathbb{R}^n \text{ such that } (\bar{x}, y) \in G_i \right\}.$$

Clearly, $I \neq \emptyset$ and, for each $i \notin I$, the set $\{\bar{x}\} \times \mathbb{R}^n$ and G_i are disjoint. Since both $\{\bar{x}\} \times \mathbb{R}^n$ and G_i are polyhedral convex sets, by Proposition 5.33, they have positive distance, and hence, there is a neighborhood V_i of $\bar{x}$ such that $(V_i \times \mathbb{R}^n) \cap G_i = \emptyset$. Let $V := \bigcap_{i \notin I} V_i$. Then, V is still a neighborhood of $\bar{x}$ and

$$(V \times \mathbb{R}^n) \cap \left(\bigcup_{i \notin I} G_i \right) = \emptyset.$$

Consequently

$$(V \times \mathbb{R}^n) \cap \mathrm{Gr}(F) \subset \bigcup_{i \in I} G_i.$$

For $x \in V \cap \mathrm{dom}(F)$, let $y \in F(x)$ be any element. We have

$$(x, y) \in (V \times \mathbb{R}^n) \cap \mathrm{Gr}(F) \subset \bigcup_{i \in I} G_i.$$

Thus, there is $i \in I$ such that $(x, y) \in G_i$ and hence $y \in F_i(x)$. Therefore, by noting that $F(\bar{x}) = \bigcup_{j \in I} F_j(\bar{x})$, we obtain

$$d(y, F(\bar{x})) \leq d(y, F_i(\bar{x})) \leq h(F_i(x), F_i(\bar{x})) \leq \kappa_i \|x - \bar{x}\| \leq \kappa \|x - \bar{x}\|,$$

where $\kappa := \max_{1 \leq i \leq k} \kappa_i$. Because $y \in F(x)$ is arbitrary, we thus obtain (8.24). □

Now, we are ready to show that polyhedral multifunctions are bounded metrically subregular [190].

Theorem 8.23 *Let $F : \mathbb{R}^m \rightrightarrows \mathbb{R}^n$ be a polyhedral multifunction. Then, there is a constant $\kappa > 0$ such that for any $y \in \mathbb{R}^n$ there is a number $\varepsilon > 0$ such that*

$$d(x, F^{-1}(y)) \leq \kappa\, d(y, F(x)), \quad \forall x \text{ satisfying } d(y, F(x)) < \varepsilon. \tag{8.25}$$

Consequently, for any $r > 0$ and any $y \in \mathbb{R}^n$, there is a number $\eta > 0$ such that

$$d(x, F^{-1}(y)) \leq \eta d(y, F(x)), \quad \forall x \text{ satisfying } d(x, F^{-1}(y)) < r. \tag{8.26}$$

Proof Since $F : \mathbb{R}^n \rightrightarrows \mathbb{R}^m$ is polyhedral, so is its inverse $F^{-1} : \mathbb{R}^m \rightrightarrows \mathbb{R}^n$. According to Theorem 8.22, F^{-1} is outer Lipschitz continuous at every point in its domain, i.e., there is a constant $\kappa > 0$ such that for any $y \in \mathbb{R}^m$ there is a number $\varepsilon > 0$ such that

$$\sup_{x \in F^{-1}(y')} d(x, F^{-1}(y)) \leq \kappa \|y' - y\|, \quad \forall y' \in B_\varepsilon(y) \cap \mathrm{dom}(F^{-1}).$$

For any $x \in \mathbb{R}^m$ satisfying $d(y, F(x)) < \varepsilon$, by the closedness of $F(x)$, we can find $\bar{y} \in F(x)$ such that $\|y - \bar{y}\| = d(y, F(x)) < \varepsilon$. Consequently

$$d(x, F^{-1}(y)) \leq \kappa \|\bar{y} - y\| = \kappa d(y, F(x)).$$

This shows (8.25).

Next, we show (8.26). Let $x \in \mathbb{R}^m$ be any point satisfying $d(x, F^{-1}(y)) < r$. If $d(y, F(x)) \geq \varepsilon$, then we have

$$d(x, F^{-1}(y)) \leq r \leq \frac{r}{\varepsilon} d(y, F(x)).$$

If $d(y, F(x)) < \varepsilon$, then we may apply (8.25) to conclude $d(x, F^{-1}(y)) \leq \kappa d(y, F(x))$. The proof is therefore complete. □

8.4 Semi-algebraic Functions

In the previous section, we introduced the concept of bounded metric subregularity and demonstrated that every polyhedral multifunction possesses this property. This regularity condition is particularly useful for establishing the linear convergence of algorithms. However, many multifunctions fail to satisfy this condition. To address such cases, we consider weaker regularity conditions that may still provide meaningful convergence rate estimates, albeit weaker than linear convergence. This leads us to introduce the following regularity condition for a multifunction $F : \mathbb{R}^m \rightrightarrows \mathbb{R}^n$ at a point $(\bar{x}, \bar{y}) \in \mathrm{Gr}(F)$: for any $r > 0$ there exist constants $\alpha \in (0, 1]$ and $\kappa \geq 0$ such that

$$d(x, F^{-1}(\bar{y})) \leq \kappa \left[d(\bar{y}, F(x))\right]^{\alpha}, \quad \forall x \in B_r(\bar{x}).$$

We refer to this condition as bounded Hölder metric subregularity of F at $(\bar{x}, \bar{y})$. In this section, we will focus on semi-algebraic multifunctions and show that they are bounded Hölder metrically subregular. Furthermore, we will also show that every proper, lower semicontinuous, semi-algebraic function $f : \mathbb{R}^n \to (-\infty, \infty]$ with $S := \arg\min_{x \in \mathbb{R}^n} f(x) \neq \emptyset$ satisfies the so-called bounded Hölder growth condition, i.e., for any $\bar{x} \in S$ and $r > 0$, there exist constants $\alpha \in (0, 1]$ and $c \geq 0$ such that

$$d(x, S) \leq c \left(f(x) - f(\bar{x})\right)^{\alpha}, \quad \forall x \in B_r(\bar{x}).$$

To establish these results, we first provide a brief overview of semi-algebraic geometry. For further details, we refer the reader to [62, 95], among others.

Definition 8.24 A subset A of $\mathbb{R}^n$ is called *semi-algebraic* if it is a union of finite many sets of the form

$$\left\{x \in \mathbb{R}^n : f(x) = 0 \text{ and } f_i(x) < 0 \text{ for } i = 1, \cdots, p\right\},$$

where $f, f_i : \mathbb{R}^n \to \mathbb{R}$ with $i = 1, \cdots, p$ are real polynomials.

By definition, it is straightforward to see the following facts about semi-algebraic sets:

(i) Every polyhedral set is semi-algebraic.
(ii) If A and B are semi-algebraic sets in $\mathbb{R}^n$, so are $A \cap B$, $A \cup B$, and $A \setminus B$, i.e., the class of semi-algebraic sets is stable under the operation of finite union, finite intersection, and complementation.
(iii) If $A \subset \mathbb{R}^m$ and $B \subset \mathbb{R}^n$ are semi-algebraic, then $A \times B$ is a semi-algebraic set in $\mathbb{R}^{m+n}$.
(iv) Every semi-algebraic set in $\mathbb{R}$ is a finite union of points and intervals.

Example 8.25 For any $\bar{x} \in \mathbb{R}^n$ and $r > 0$, the open ball $B_r(\bar{x})$ and the closed ball $\bar{B}_r(\bar{x})$ are semi-algebraic because

$$B_r(\bar{x}) = \{x \in \mathbb{R}^n : \|x - \bar{x}\|^2 < r^2\} \text{ and } \bar{B}_r(\bar{x}) = B_r(\bar{x}) \cup \{x \in \mathbb{R}^n : \|x - \bar{x}\|^2 = r^2\}.$$

Example 8.26 Let $A \subset \mathbb{R}^m \times \mathbb{R}^n$ be a semi-algebraic set and define the set

$$B := \{(y, x) \in \mathbb{R}^n \times \mathbb{R}^m : (x, y) \in A\}.$$

Since A can be written as a finite union of sets of the form

$$\{(x, y) \in \mathbb{R}^m \times \mathbb{R}^n : f(x, y) = 0 \text{ and } f_i(x, y) < 0 \text{ for } i = 1, \cdots, p\}$$

for real polynomials f and f_i, $i = 1, \cdots, p$, the set B is also a finite union of the sets of the form:

$$\{(y, x) \in \mathbb{R}^n \times \mathbb{R}^m : f(x, y) = 0 \text{ and } f_i(x, y) < 0 \text{ for } i = 1, \cdots, p\}.$$

Therefore, B is also semi-algebraic.

Next, we state the famous Tarski-Seidenberg theorem [201, 208] which is one of the fundamental results in semi-algebraic geometry; its proof is not included here; the interested readers may consult [62], for instance.

Theorem 8.27 (Tarski-Seidenberg Theorem) *Let S be a semi-algebraic set in $\mathbb{R}^{m+n}$ and let $\pi : \mathbb{R}^{m+n} \to \mathbb{R}^m$ be the canonical projection:*

$$\pi(x_1, \cdots, x_m, x_{m+1}, \cdots, x_{m+n}) = (x_1, \cdots, x_m)$$

Then, $\pi(S)$ is a semi-algebraic set in $\mathbb{R}^m$.

Tarski-Seidenberg theorem has the following direct consequence which will be used frequently.

Lemma 8.28 *Let $S \subset \mathbb{R}^m \times \mathbb{R}^n$, $U \subset \mathbb{R}^m$, and $V \subset \mathbb{R}^n$ be semi-algebraic sets. Then, both*

$$A := \{x \in U : \exists y \in V \text{ such that } (x, y) \in S\}$$

and

$$B := \{x \in U : (x, y) \in S \text{ for all } y \in V\}$$

are semi-algebraic sets in $\mathbb{R}^m$.

Proof Let $\pi : \mathbb{R}^m \times \mathbb{R}^n \to \mathbb{R}^m$ be the canonical projection. We have $A = \pi(S \cap (U \times V))$. Since $S \cap (U \times V)$ is semi-algebraic, it follows from the Tarski-Seidenberg theorem that A is semi-algebraic. Note also that

$$U \setminus B = \{x \in U : \exists y \in V \text{ such that } (x, y) \in S^c\}.$$

Since S^c is semi-algebraic, by the first part, we can conclude that $U \setminus B$ and hence B is semi-algebraic. □

Lemma 8.29 *The closure, interior, and boundary of semi-algebraic sets are semi-algebraic.*

Proof Let $A \subset \mathbb{R}^n$ be a semi-algebraic set. Then

$$\overline{A} = \{x \in \mathbb{R}^n : \forall t > 0\, \exists y \in A \text{ such that } \|y - x\|^2 < t\}.$$

Since the set

$$S := \{(x, t, y) \in \mathbb{R}^n \times (0, \infty) \times A : \|x - y\|^2 < t\}$$

is semi-algebraic, by using Lemma 8.28 twice, we can conclude that

$$T := \{(x, t) \in \mathbb{R}^n \times (0, \infty) : \exists y \in A \text{ such that } (x, t, y) \in S\}$$

and

$$\overline{A} = \{x \in \mathbb{R}^n : (x, t) \in T \text{ for all } t > 0\}$$

are semi-algebraic.

Note that the interior of A is given by $\text{int}(A) = \mathbb{R}^n \setminus \overline{A^c}$, it is semi-algebraic. Since the boundary of A is given by $\overline{A} \cap \overline{A^c}$, it is also semi-algebraic. □

Definition 8.30 A function $f : \mathbb{R}^n \to (-\infty, \infty]$ is called *semi-algebraic* if its graph

$$\text{Gr}(f) := \left\{(x, f(x)) \in \mathbb{R}^{n+1} : x \in \text{dom}(f)\right\}$$

is a semi-algebraic subset of $\mathbb{R}^{n+1}$. Given a subset $A \subset \mathbb{R}^n$, a function $f : A \to (-\infty, \infty]$ is called *semi-algebraic* if its extension $f_A : \mathbb{R} \to (-\infty, \infty]$ defined by

$$f_A(x) = \begin{cases} f(x) & \text{if } x \in A, \\ +\infty & \text{otherwise} \end{cases}$$

is semi-algebraic.

According to the Tarski-Seidenberg theorem, it is straightforward to see that the domain and range of a semi-algebraic function are semi-algebraic sets. Furthermore, by definition, it is clear that real polynomial functions and indicator functions of semi-algebraic sets are semi-algebraic.

Example 8.31 The function $x \to \|x\|$ on $\mathbb{R}^n$ is semi-algebraic because its graph

$$\Gamma = \{(x, t) \in \mathbb{R}^n \times \mathbb{R} : t = \|x\|\} = \{(x, t) \in \mathbb{R}^n \times \mathbb{R} : \|x\|^2 - t^2 = 0 \text{ and } t \geq 0\}$$

is clearly a semi-algebraic set.

Example 8.32 For the function $x \to \|x\|_\infty$ on $\mathbb{R}^n$, its graph is

$$\Gamma = \left\{(x, t) : t = \max_{1 \leq i \leq n} |x_i|\right\} = \bigcup_{i=1}^{n} \{(x, t) : |x_i| = t \text{ and } |x_j| \leq t \text{ for } j \neq i\}$$

which is clearly a semi-algebraic set in $\mathbb{R}^{n+1}$. Thus, $\|x\|_\infty$ is semi-algebraic.

Lemma 8.33 *A function $f : \mathbb{R}^n \to [-\infty, \infty]$ is semi-algebraic if and only if the set $A := \{(x, t) \in \mathbb{R}^n \times \mathbb{R} : t \leq f(x)\}$ is semi-algebraic.*

Proof If f is semi-algebraic, then its graph $\mathrm{Gr}(f)$ is semi-algebraic and hence the set

$$\begin{aligned} B &:= \{(x, \lambda, t) \in \mathbb{R}^n \times \mathbb{R} \times \mathbb{R} : \lambda = f(x),\ t \leq \lambda\} \\ &= (\mathrm{Gr}(f) \times \mathbb{R}) \cap \{(x, \lambda, t) \in \mathbb{R}^n \times \mathbb{R} \times \mathbb{R} :\ t \leq \lambda\} \end{aligned}$$

is semi-algebraic. Let $\pi : \mathbb{R}^n \times \mathbb{R} \times \mathbb{R} \to \mathbb{R}^n \times \mathbb{R}$ be the projection defined by $\pi(x, \lambda, t) = (x, t)$. Then, $A = \pi(B)$. It thus follows from the Tarski-Seidenberg theorem that A is semi-algebraic.

Conversely, if A is a semi-algebraic set, then the set

$$\begin{aligned} C &:= \{(x, \lambda, t) \in \mathbb{R}^n \times \mathbb{R} \times \mathbb{R} : \lambda \leq f(x),\ t < \lambda\} \\ &= (A \times \mathbb{R}) \cap \{(x, \lambda, t) \in \mathbb{R}^n \times \mathbb{R} \times \mathbb{R} : t < \lambda\} \end{aligned}$$

is semi-algebraic. By the Tarski-Seidenberg theorem, the set $\pi(C) = \{(x, t) \in \mathbb{R}^n \times \mathbb{R} : t < f(x)\}$ is semi-algebraic and so is $\mathrm{Gr}(f) = A \setminus \pi(C)$. Consequently, f is semi-algebraic. □

For semi-algebraic functions in $\mathbb{R}$, we have the following polynomially bounded result.

Proposition 8.34 *If $f : A \subset \mathbb{R} \to \mathbb{R}$ is semi-algebraic, then there exist constants $B, C > 0$ and a rational exponent $q \geq 1$ such that*

$$|f(x)| \leq C|x|^q, \quad \forall x \in A \text{ and } |x| > B.$$

Proof Since f is semi-algebraic, its graph can be expressed as $\mathrm{Gr}(f) = \cup_{i=1}^m G_i$, where each G_i is a non-empty set of the form

$$G_i := \{(x, y) \in \mathbb{R}^2 : p_i(x, y) = 0,\ q_{i,1}(x, y) < 0,\ \cdots,\ q_{i,k_i}(x, y) < 0\}$$

with real polynomials p_i and $q_{i,j}$, $j = 1, \cdots, k_i$. We can write

$$p_i(x, y) = a_{i,0}(x)y^{d_i} + a_{i,1}(x)y^{d_i-1} + \cdots + a_{i,d_i-1}(x)y + a_{i,d_i}(x),$$

where $a_{i,l}$ are all real polynomials in x with $a_{i,0} \neq 0$. We claim that $d_i \geq 1$. Suppose, for contradiction, that $d_i = 0$. Then, for any $(x_0, y_0) \in G_i$, we have $p_i(x_0, y_0) = a_{i,0}(x_0) = 0$. By the continuity of $q_{i,j}$ for $j = 1, \cdots, k_i$, we can find $\varepsilon > 0$ such that $\{x_0\} \times (y_0 - \varepsilon, y_0 + \varepsilon) \in G_i$. However, this contradicts the fact that $\mathrm{Gr}(f)$ is the graph of a function, which must assign a unique y-value for each x. Hence, we must have $d_i \geq 1$.

Now, consider the real polynomial $p(x, y) := p_1(x, y)p_2(x, y) \cdots p_m(x, y)$ which can be expressed as

$$p(x, y) = a_0(x)y^d + a_1(x)y^{d-1} + \cdots + a_{d-1}(x)y + a_d(x).$$

According to the property of p_i shown above, we have $d \geq 1$ and $a_0 \neq 0$. As a nonzero polynomial in a single variable, there exists a constant $B > 0$ such that $a_0(x) \neq 0$ for all x satisfying $|x| > B$. We claim that $p(x, y) \neq 0$ for all $(x, y) \in \mathbb{R}^2$ satisfying

$$|x| > B \quad \text{and} \quad |y| > \max_{i=1,\cdots,d} \left(d \left| \frac{a_i(x)}{a_0(x)} \right| \right)^{1/i}.$$

This follows directly from the inequality

$$|p(x, y)| \geq |a_0(x)||y|^d - \sum_{i=1}^d |a_i(x)||y|^{d-i} > |a_0(x)||y|^d - \sum_{i=1}^d \frac{1}{d}|a_0(x)||y|^d = 0.$$

Since $p(x, f(x)) = 0$ for all $x \in A$. we can use the above claim to conclude that

$$|f(x)| \leq \max_{i=1,\cdots,d} \left(d \left| \frac{a_i(x)}{a_0(x)} \right| \right)^{1/i}, \qquad \forall x \in A \text{ and } |x| > B.$$

Since a_i are polynomials, from the above equation, it follows that there exist a constant $C > 0$ and a rational exponent $q \geq 1$ such that $|f(x)| \leq C|x|^q$ for all $x \in A$ satisfying $|x| > B$. □

According to Proposition 8.34, we can immediately conclude that the exponential function e^x is not semi-algebraic on $\mathbb{R}$.

The definition of semi-algebraicity for functions can be extended for maps and even for multifunctions which is defined below.

Definition 8.35 A multifunction $F : \mathbb{R}^m \rightrightarrows \mathbb{R}^n$ is called *semi-algebraic* if its graph $\mathrm{Gr}(F)$ is a semi-algebraic set in $\mathbb{R}^m \times \mathbb{R}^n$.

Example 8.36 If a multifunction $F : \mathbb{R}^m \rightrightarrows \mathbb{R}^n$ is semi-algebraic, then $\mathrm{Gr}(F)$ is a semi-algebraic set in $\mathbb{R}^m \times \mathbb{R}^n$. By Example 8.26, $\mathrm{Gr}(F^{-1})$ is a semi-algebraic set in $\mathbb{R}^n \times \mathbb{R}^m$, and thus, the inverse multifunction $F^{-1} : \mathbb{R}^n \rightrightarrows \mathbb{R}^m$ of F is also semi-algebraic.

Lemma 8.37 *Images and inverse images of semi-algebraic sets under semi-algebraic multifunctions are semi-algebraic.*

Proof Let $F : \mathbb{R}^m \rightrightarrows \mathbb{R}^n$ be a semi-algebraic multifunction and let $A \subset \mathbb{R}^m$ and $B \subset \mathbb{R}^n$ be semi-algebraic sets. Then

$$F(A) = \{y \in \mathbb{R}^n : \exists x \in A \text{ such that } (x, y) \in \mathrm{Gr}(F)\},$$

$$F^{-1}(B) = \{x \in \mathbb{R}^m : \exists y \in B \text{ such that } (x, y) \in \mathrm{Gr}(F)\}.$$

thus, it follows from Lemma 8.28 that both $F(A)$ and $F^{-1}(B)$ are semi-algebraic sets. □

Lemma 8.38 *The sum, product, and composition of semi-algebraic maps are semi-algebraic.*

Proof We only prove the result for composition, and the results for sum and product can be proved similarly. Let $F : \mathbb{R}^n \to \mathbb{R}^m$ and $H : \mathbb{R}^p \to \mathbb{R}^n$ be semi-algebraic maps. Then, the graph of $F \circ H$ is

$$\begin{aligned}\mathrm{Gr}(F \circ H) &= \{(x, y) \in \mathbb{R}^p \times \mathbb{R}^m : y = F(H(x))\} \\ &= \{(x, y) \in \mathbb{R}^p \times \mathbb{R}^m : \exists z \in \mathbb{R}^n \text{ with } y = F(z) \text{ and } z = H(x)\} \\ &= \pi\{(x, y, z) \in \mathbb{R}^p \times \mathbb{R}^m \times \mathbb{R}^n : (y, z) \in \mathrm{Gr}(F),\ (x, z) \in \mathrm{Gr}(H)\},\end{aligned}$$

where $\pi : \mathbb{R}^p \times \mathbb{R}^m \times \mathbb{R}^n \to \mathbb{R}^p \times \mathbb{R}^m$ is the canonical projection. Since $\mathrm{Gr}(F)$ and $\mathrm{Gr}(H)$ are semi-algebraic, by Tarski-Seidenberg theorem, $\mathrm{Gr}(F \circ H)$ is semi-algebraic. □

Example 8.39 The ℓ^1-norm $\|x\|_1$ on $\mathbb{R}^n$ is a semi-algebraic function. Indeed, since $\|x\|_1 = \sum_{i=1}^n |x_i|$, by using Lemma 8.38, it suffices to show the absolute value function $t \to |t|$ on $\mathbb{R}$ is semi-algebraic which can be easily verified because its graph

$$\begin{aligned}\mathcal{G} &= \{(t, s) : s = |t|\} \\ &= \{(t, s) : t > 0, s = t\} \cup \{(t, s) : t < 0, s = -t\} \cup \{(t, s) : t = 0, s = 0\}\end{aligned}$$

is clearly a semi-algebraic set in $\mathbb{R}^2$.

Example 8.40 Let $f : \mathbb{R}^n \to (-\infty, \infty]$ be a proper convex semi-algebraic function. Consider its subdifferential ∂f which is a multifunction from $\mathbb{R}^n$ to $\mathbb{R}^n$. The graph of ∂f is

$$\begin{aligned}\mathrm{Gr}(\partial f) &= \{(x, \xi) \in \mathbb{R}^n \times \mathbb{R}^n : f(y) \geq f(x) + \langle \xi, y - x\rangle \text{ for all } y \in \mathbb{R}^n\} \\ &= \{(x, \xi) \in \mathbb{R}^n \times \mathbb{R}^n : (x, \xi, y) \in S \text{ for all } y \in \mathbb{R}^n\},\end{aligned}$$

where

$$S := \{(x, \xi, y) \in \mathbb{R}^n \times \mathbb{R}^n \times \mathbb{R}^n : f(y) - f(x) - \langle \xi, y - x\rangle \geq 0\}.$$

Since f is semi-algebraic, by Lemma 8.38, the function

$$\varphi(x, \xi, y) := f(y) - f(x) - \langle \xi, y - x\rangle$$

is also semi-algebraic, and hence, $S = \varphi^{-1}([0, \infty))$ is semi-algebraic by Lemma 8.37. Thus, it follows from Lemma 8.28 that $\mathrm{Gr}(\partial f)$ is a semi-algebraic set, and hence, ∂f is a semi-algebraic multifunction.

If in addition f is lower semicontinuous, then $\partial f^* = (\partial f)^{-1}$, where f^* is the Legendre-Fenchel conjugate of f. Thus, from the above fact and Example 8.36, it follows that ∂f^* is also semi-algebraic.

Example 8.41 Let $g : \mathbb{R}^n \to (-\infty, \infty]$ be a semi-algebraic function that is bounded below and let $F : \mathbb{R}^m \rightrightarrows \mathbb{R}^n$ be a semi-algebraic multifunction. Then, the function $\varphi : \mathbb{R}^m \to (-\infty, \infty]$ defined by

$$\varphi(x) := \begin{cases} \inf\{g(y) : y \in F(x)\} & \text{if } F(x) \neq \emptyset, \\ +\infty & \text{if } F(x) = \emptyset \end{cases}$$

is semi-algebraic. To see this, according to Lemma 8.33, it suffices to show the set

$$B := \{(x, t) \in \mathbb{R}^m \times \mathbb{R} : t > \varphi(x)\}$$

is semi-algebraic. By the definition of φ, we have

$$\begin{aligned}B &= \{(x, t) \in \mathbb{R}^m \times \mathbb{R} : \exists y \in F(x) \text{ such that } t > g(y)\} \\ &= \{(x, t) \in \mathbb{R}^m \times \mathbb{R} : \exists y \in \mathbb{R}^n \text{ such that } (x, t, y) \in S\},\end{aligned}$$

where

$$S := \{(x, t, y) \in \mathbb{R}^m \times \mathbb{R} \times \mathbb{R}^n : (x, y) \in \mathrm{Gr}(F) \text{ and } t - g(y) > 0\}.$$

By the given conditions on g and F, it is easy to see that S is semi-algebraic. Therefore, we may use Lemma 8.28 to conclude the semi-algebraicity of B.

Example 8.42 Let $A \subset \mathbb{R}^n$ be a semi-algebraic set and define the function

$$\varphi(x) := d(x, A), \quad x \in \mathbb{R}^n.$$

By introducing the multifunction $F : \mathbb{R}^n \rightrightarrows \mathbb{R}^n$ defined by $F(x) := A - x$ for each $x \in \mathbb{R}^n$, we can see that

$$\varphi(x) = \inf\{\|y\| : y \in F(x)\}.$$

Note that

$$\mathrm{Gr}(F) = \{(x, y) \in \mathbb{R}^n \times \mathbb{R}^n : x + y \in A\} = \psi^{-1}(A),$$

where $\psi : \mathbb{R}^n \times \mathbb{R}^n \to \mathbb{R}^n$ given by $\psi(x, y) = x + y$ is semi-algebraic. Thus, it follows from Lemma 8.37 that $\mathrm{Gr}(F)$ is semi-algebraic and so is F. Since $\|y\|$ is semi-algebraic (see Example 8.31), we can use Example 8.41 to conclude that φ is a semi-algebraic function.

We next prove the following important result on semi-algebraic functions which is a refined version of Łojasiewicz inequality [62, 147].

Proposition 8.43 *Let K be a compact semi-algebraic set in $\mathbb{R}^n$ and let $\phi : K \to [0, \infty)$ and $\psi : K \to [0, \infty]$ be semi-algebraic functions. If ϕ is continuous, ψ is lower semicontinuous and*

$$\{x \in K : \psi(x) = 0\} \subset \{x \in K : \phi(x) = 0\}, \tag{8.27}$$

then there exist constants $c > 0$ and $\alpha \in (0, 1]$ such that $\phi(x) \leq c[\psi(x)]^\alpha$ for all $x \in K$.

Proof We define the multifunction $F : (0, \infty) \rightrightarrows \mathbb{R}^n$ by

$$F(t) := \{x \in K : \phi(x) = 1/t\}, \quad \forall t > 0.$$

Since

$$\mathrm{Gr}(F) = \{(t, x) : t > 0,\ x \in K \text{ and } t\phi(x) = 1\},$$

K is semi-algebraic, and $t\phi(x)$ is semi-algebraic by Lemma 8.38, we can see that $\mathrm{Gr}(F)$ is a semi-algebraic set and thus F is a semi-algebraic multifunction. Furthermore, since ϕ is continuous and K is compact, we conclude that $F(t)$ is a compact set for each $t > 0$. Next, we define the function:

$$\theta(t) := \begin{cases} \inf\{\psi(x) : x \in F(t)\} & \text{if } F(t) \neq \emptyset, \\ +\infty & \text{if } F(t) = \emptyset. \end{cases}$$

According to Example 8.41, θ is a semi-algebraic function. We claim that $\theta(t) > 0$ for all $t > 0$. To see this, suppose there is $t > 0$ such that $\theta(t) = 0$. Then, there exists a sequence $\{x_k\} \subset F(t)$ such that $\psi(x_k) \to 0$. Since $F(t)$ is compact, we can extract a subsequence of $\{x_k\}$ if necessary such that $x_k \to \bar{x} \in F(t)$. By the lower semicontinuity of ψ, we obtain

$$0 \leq \psi(\bar{x}) \leq \liminf \psi(x_k) = 0$$

which implies $\psi(\bar{x}) = 0$ and hence $\phi(\bar{x}) = 0$ by (8.27). This contradicts the fact $\phi(\bar{x}) = 1/t > 0$.

Now, we introduce the function $\eta : (0, \infty) \to [0, \infty)$ defined by

$$\eta(t) := 1/\theta(t), \quad t > 0.$$

Clearly, η is a semi-algebraic function. By Proposition 8.34, there exist constants $B, C > 0$ and a rational exponent $q \geq 1$ such that

$$0 \leq \eta(t) \leq Ct^q, \quad \forall t > B.$$

To complete the proof, we write $K = K_0 \cup K_1 \cup K_2$, where

$$\begin{aligned} K_0 &:= \{x \in K : \phi(x) = 0\}, \\ K_1 &:= \{x \in K : 0 < \phi(x) < 1/B\}, \\ K_2 &:= \{x \in K : \phi(x) \geq 1/B\} \end{aligned}$$

For any $x \in K_1$, setting $t = 1/\phi(x)$ ensures $x \in F(t)$ and $t > B$. Consequently

$$\psi(x) \geq \theta(t) = \frac{1}{\eta(t)} \geq \frac{1}{Ct^q} = \frac{1}{C}\phi(x)^q.$$

This inequality also holds trivially for $x \in K_0$, leading to

$$\phi(x) \leq C^{1/q}[\psi(x)]^{1/q}, \quad \forall x \in K_0 \cup K_1. \tag{8.28}$$

On the other hand, let

$$D := \min\left\{\psi(x)/\phi(x)^q : x \in K_2\right\}.$$

By using the lower semicontinuity of ψ, the continuity of ϕ, the definition of K_2, and (8.27), we can conclude $D > 0$. Thus, $\psi(x) \geq D\phi(x)^q$ for all $x \in K_2$. Combining this with (8.28), we thus complete the proof. □

By virtue of Proposition 8.43, we are now ready to prove a Hölder growth error bound for semi-algebraic functions.

Theorem 8.44 *Let $f : \mathbb{R}^n \to (-\infty, \infty]$ be a proper, lower semicontinuous, semi-algebraic function and let $\gamma \in \mathbb{R}$ be such that*

$$S := \{x \in \mathbb{R}^n : f(x) \leq \gamma\} \neq \emptyset.$$

Then, for any $\bar{x} \in \mathbb{R}^n$ and $r > 0$, there exist constants $c > 0$ and $\alpha \in (0, 1]$ such that

$$d(x, S) \leq c[f(x) - \gamma]_+^\alpha, \quad \forall x \in B_r(\bar{x}).$$

Proof Let

$$\phi(x) := d(x, S) \quad \text{and} \quad \psi(x) := [f(x) - \gamma]_+.$$

Then, ϕ is continuous and ψ is lower semicontinuous on $\mathbb{R}^n$. By the semi-algebraicity of f, we can see that both ϕ and ψ are semi-algebraic. Moreover, $\psi(x) = 0$ implies $\phi(x) = 0$. Thus, we may apply Proposition 8.43 to ϕ and ψ on the compact semi-algebraic set $\bar{B}_r(\bar{x})$ to conclude the proof. □

We can also use Proposition 8.43 to prove the following bounded Hölder metric subregularity result for semi-algebraic multifunctions with closed graphs [134].

Theorem 8.45 *Let $F : \mathbb{R}^m \rightrightarrows \mathbb{R}^n$ be a semi-algebraic multifunction with closed graph. Then, F is bounded Hölder metrically subregular at any point $(\bar{x}, \bar{y}) \in \mathrm{Gr}(F)$, i.e., for any $r > 0$, there exist constants $\kappa > 0$ and $\alpha \in (0, 1]$ such that*

$$d(x, F^{-1}(\bar{y})) \leq \kappa \, [d(\bar{y}, F(x))]^\alpha, \quad \forall x \in B_r(\bar{x}).$$

Proof Let

$$\phi(x) := d(x, F^{-1}(\bar{y})) \quad \text{and} \quad \psi(x) := d(\bar{y}, F(x)), \quad x \in \mathbb{R}^m.$$

According to Examples 8.41 and 8.42, both ϕ and ψ are nonnegative semi-algebraic functions and ϕ is continuous. Since the graph of F is closed, $F(x)$ is closed for each x. Thus, $\psi(x) = 0$ implies $\bar{y} \in F(x)$ and hence $x \in F^{-1}(\bar{y})$; consequently, $\phi(x) = 0$. We next show that ψ is lower semicontinuous. Let $\{x^k\} \subset \mathbb{R}^n$ be a sequence with $x^k \to x \in \mathbb{R}^m$ and let $a := \liminf_{k\to\infty} \psi(x^k)$. If $a = \infty$, we automatically have $\psi(x) \leq a$. So, we may assume $a < \infty$. We can find a subsequence $\{x^{k_l}\}$ of $\{x^k\}$ such that $a = \lim_{l\to\infty} \psi(x^{k_l})$. Since $F(x^{k_l})$ is closed, there exists $y^{k_l} \in F(x^{k_l})$ such that $\psi(x^{k_l}) = \|\bar{y} - y^{k_l}\|$. Thus, $\lim_{l\to\infty} \|\bar{y} - y^{k_l}\| = a < \infty$ which means $\{y^{k_l}\}$ is bounded. Therefore, by taking a subsequence if necessary, we may assume $y^{k_l} \to y$ for some $y \in \mathbb{R}^n$. Since $x^{k_l} \to x$ and $y^{k_l} \in F(x^{k_l})$, by the closedness of the graph of F, we can conclude $(x, y) \in \mathrm{Gr}(F)$, i.e., $y \in F(x)$. Hence

$$a = \|\bar{y} - y\| \geq d(\bar{y}, F(x)) = \psi(x).$$

Thus, we always have $\psi(x) \leq \liminf_{k\to\infty} \psi(x^k)$. Now, we may use Proposition 8.43 to conclude that for any $r > 0$ there exist $\kappa > 0$ and $\alpha \in (0, 1]$ such that

$$\phi(x) \leq \kappa\, [\psi(x)]^\alpha\,, \qquad \forall x \in B_r(\bar{x})$$

which shows the desired result. □

8.5 Linear Convergence and Beyond

In this section, we demonstrate how to derive (linear) convergence rates under error bound conditions by considering the proximal gradient method, the alternating direction method of multipliers, and the primal-dual splitting algorithm. We need the following calculus result [6].

Lemma 8.46 *Let $\{\Delta_k\}$ be a sequence of nonnegative numbers satisfying*

$$\Delta_k^\theta \leq C(\Delta_{k-1} - \Delta_k) \tag{8.29}$$

for all $k \geq 1$, where $C > 0$ and $\theta > 1$ are constants. Then, there is a constant $\tilde{C} > 0$ such that

$$\Delta_k \leq \tilde{C}(1+k)^{-\frac{1}{\theta-1}}$$

for all $k \geq 0$.

Proof According to (8.29), $\{\Delta_k\}$ is monotonically decreasing. If $\Delta_{k_0} = 0$ for some k_0, then $\Delta_k = 0$ for all $k \geq k_0$ and the result holds trivially. Therefore, we may assume $\Delta_k > 0$ for all $k \geq 0$.

We first show that there is a constant $\mu_0 > 0$ such that

$$\Delta_k^{1-\theta} - \Delta_{k-1}^{1-\theta} \geq \mu_0 \tag{8.30}$$

for all $k \geq 1$. To see this, we rewrite (8.29) as $1 \leq C(\Delta_{k-1} - \Delta_k)\Delta_k^{-\theta}$. If $\Delta_k > \Delta_{k-1}/2$, then

$$1 \leq 2^\theta C(\Delta_{k-1} - \Delta_k)\Delta_{k-1}^{-\theta} \leq 2^\theta C \int_{\Delta_k}^{\Delta_{k-1}} s^{-\theta} ds = \frac{2^\theta C}{\theta - 1}\left(\Delta_k^{1-\theta} - \Delta_{k-1}^{1-\theta}\right)$$

which implies (8.30) with $\mu_0 = (\theta - 1)/(2^\theta C)$. If $\Delta_k \le \Delta_{k-1}/2$, then, by using $1 - \theta < 0$, we have $\Delta_k^{1-\theta} \ge 2^{\theta-1}\Delta_{k-1}^{1-\theta}$. Thus

$$\Delta_k^{1-\theta} - \Delta_{k-1}^{1-\theta} \ge (2^{\theta-1} - 1)\Delta_{k-1}^{1-\theta} \ge (2^{\theta-1} - 1)\Delta_0^{1-\theta}$$

and we again obtain (8.30).

From (8.30), it follows that

$$\Delta_k^{1-\theta} - \Delta_0^{1-\theta} = \sum_{j=1}^{k}(\Delta_j^{1-\theta} - \Delta_{j-1}^{1-\theta}) \ge \mu_0 k.$$

Consequently, $\Delta_k^{1-\theta} \ge \Delta_0^{1-\theta} + \mu_0 k$. Since $1 - \theta < 0$, we must have

$$\Delta_k \le (\Delta_0^{1-\theta} + \mu_0 k)^{-\frac{1}{\theta-1}}$$

for all k. This shows the desired estimate. □

8.5.1 Proximal Gradient Method

Consider the convex minimization problem:

$$\min_{x\in\mathbb{R}^n} \{F(x) := f(x) + g(x)\}. \tag{8.31}$$

Under the assumption that $g : \mathbb{R}^n \to (-\infty, \infty]$ is proper, lower semicontinuous, convex and $f : \mathbb{R}^n \to \mathbb{R}$ is convex differentiable with globally Lipschitz continuous gradient ∇f, we have considered solving (8.31) by the proximal gradient method in Sect. 4.4 and established convergence results under constant step size. The global Lipschitz continuity of ∇f is a severe condition in some applications. How to weaken this condition attracts a lot of research work [15, 22, 23, 196]. In this subsection, we reconsider the proximal gradient method under the following weaker conditions.

Assumption 8.47

(i) $f, g : \mathbb{R}^n \to (-\infty, \infty]$ *are proper, lower semicontinuous, convex functions with* $\mathscr{D} := \mathrm{int}(\mathrm{dom}(f)) \cap \mathrm{dom}(g) \ne \emptyset$. *Moreover,* $S := \arg\min_{x\in\mathbb{R}^n}\{F(x) := f(x) + g(x)\} \ne \emptyset$.

(ii) *f is differentiable on $\mathscr{D}$ and ∇f is locally Lipschitz continuous at every $x \in \mathscr{D}$, i.e., there exist $r_x > 0$ and $L_x > 0$ such that*

$$\|\nabla f(y) - \nabla f(z)\| \le L_x \|y - z\|$$

for all $y, z \in \mathscr{D} \cap B_{r_x}(x)$.

(iii) *For any $x \in \mathscr{D}$, there holds $\{z \in \mathbb{R}^n : F(z) \le F(x)\} \subset \mathscr{D}$.*

To implement the proximal gradient method, the choice of the step size is crucial. Under Assumption 8.47, the gradient ∇f is not necessarily globally defined and no Lipschitz constant is available to guide our choice of step size. Instead, we consider choosing the step size by a backtracking line search.

Algorithm 8.48 *Take an initial guess $x^0 \in \mathscr{D} = \mathrm{int}(\mathrm{dom}(f)) \cap \mathrm{dom}(g)$. Pick $\sigma > 0$ and $0 < \rho < 1$. Set $t_{-1} := \sigma$. For $k = 0, 1, \cdots$ do the following:*

(i) *Choose t_k to be the largest number $t \in \{t_{k-1}\rho^j : j = 0, 1, \cdots\}$ such that*

$$f(J_t(x^k)) \le f(x^k) + \langle \nabla f(x^k), J_t(x^k) - x^k \rangle + \frac{1}{2t}\|J_t(x^k) - x^k\|^2,$$

where $J_t(x^k) := \mathrm{prox}_{tg}(x^k - t\nabla f(x^k))$ for any $t > 0$.

(ii) *Set $x^{k+1} := J_{t_k}(x^k)$.*

The line search for choosing t_k in Algorithm 8.48 is called the Beck-Teboulle backtracking line search [20].

Under Assumption 8.47, we will show that Algorithm 8.48 is well-defined, and the produced sequence $\{x^k\}$ converges to an optimal solution of (8.31). Furthermore, this convergence is linear when a quadratic growth condition holds. We need the following lemma.

Lemma 8.49 *If $x \in \mathrm{dom}(g)$, then for any $v \in \mathbb{R}^n$ there holds $\|\mathrm{prox}_{tg}(x + tv) - x\| \to 0$ as $t \to 0+$.*

Proof Let $z_t := \mathrm{prox}_{tg}(x+tv)$. According to Lemma 8.5, $\|z_t - x\|$ is monotonically increasing with respect to $t > 0$. Let $\{t_k\}$ be a sequence of positive numbers with $t_k \to 0$ as $k \to \infty$. Since the monotonicity of $\{\|z_{t_k} - x\|\}$ implies the boundedness of $\{z_{t_k}\}$, by taking a subsequence if necessary, we may assume $z_{t_k} \to \bar{x}$ as $k \to \infty$ for some $\bar{x} \in \mathbb{R}^n$. By the lower semicontinuity of g, we have $g(\bar{x}) \le \liminf_{k\to\infty} g(z_{t_k}) < \infty$ which implies that $g(z_{t_k}) > g(\bar{x}) - 1$ for large k. Recall that

$$z_{t_k} = \mathrm{prox}_{t_k g}(x + t_k v) = \arg\min_{y\in\mathbb{R}^n} \left\{ g(y) + \frac{1}{2t_k}\|y - (x + t_k v)\|^2 \right\},$$

we have

$$g(z_{t_k}) + \frac{1}{2t_k}\|z_{t_k} - (x + t_k v)\|^2 \le g(x) + \frac{1}{2t_k}\|x - (x + t_k v)\|^2 = g(x) + \frac{t_k}{2}\|v\|^2.$$

Therefore

$$\begin{aligned}\|z_{t_k} - x - t_k v\|^2 &\le 2t_k \left(g(x) - g(z_{t_k})\right) + t_k^2 \|v\|^2 \\ &\le 2t_k \left(g(x) - g(\bar{x}) + 1\right) + t_k^2 \|v\|^2.\end{aligned}$$

Consequently, $\|z_{t_k} - x - t_k v\|^2 \to 0$ and hence $\|z_{t_k} - x\| \to 0$ as $k \to \infty$. This together with the monotonicity of $\|z_t - x\|$ thus implies $\|z_t - x\| \to 0$ as $t \to 0+$. □

Based on Lemma 8.49, we now show that Algorithm 8.48 is well-defined. It suffices to show that if $x^k \in \mathscr{D}$ then t_k exists and $x^{k+1} \in \mathscr{D}$. To see this, by the local Lipschitz continuity of f at x^k, we can find $r_k > 0$ and $L_k > 0$ such that $B_{r_k}(x^k) \subset \operatorname{int}(\operatorname{dom}(f))$ and

$$\|\nabla f(x) - \nabla f(\tilde{x})\| \le L_k \|x - \tilde{x}\|, \quad \forall x, \tilde{x} \in \mathscr{D} \cap B_{r_k}(x^k). \tag{8.32}$$

By using Lemma 8.49, we have $\|J_t(x^k) - x^k\| \to 0$ as $t \to 0$ which implies that $J_t(x^k) \in \mathscr{D} \cap B_{r_k}(x^k)$ for sufficiently small $t > 0$ because $J_t(x^k) \in \operatorname{dom}(g)$ always holds. Therefore, by using (8.32) and the descent lemma, we have

$$\begin{aligned}f(J_t(x^k)) - f(x^k) - \langle \nabla f(x^k), J_t(x^k) - x^k \rangle &\le \frac{L_k}{2} \|J_t(x^k) - x^k\|^2 \\ &\le \frac{1}{2t} \|J_t(x^k) - x^k\|^2\end{aligned}$$

for $t = t_{k-1}\rho^j$ with sufficiently large j. This shows the existence of t_k.

Next, we show that $x^{k+1} \in \mathscr{D}$. According to (iii) in Assumption 8.47, it suffices to show $F(x^{k+1}) \le F(x^k)$. By the definition of x^{k+1}, we have

$$\frac{x^k - x^{k+1}}{t_k} - \nabla f(x^k) \in \partial g(x^{k+1}).$$

By the convexity of g, we then obtain

$$g(x) - g(x^{k+1}) \ge \left\langle \frac{x^k - x^{k+1}}{t_k} - \nabla f(x^k), x - x^{k+1} \right\rangle, \quad \forall x \in \mathbb{R}^n.$$

By the definition of t_k and x^{k+1} and the convexity of f, we also have

$$\begin{aligned}f(x^{k+1}) &\le f(x^k) + \langle \nabla f(x^k), x^{k+1} - x^k \rangle + \frac{1}{2t_k} \|x^{k+1} - x^k\|^2 \\ &\le f(x) + \langle \nabla f(x^k), x^{k+1} - x \rangle + \frac{1}{2t_k} \|x^{k+1} - x^k\|^2.\end{aligned}$$

Combining the above two equations gives

$$\begin{aligned}F(x) = f(x) + g(x) &\geq f(x^{k+1}) + g(x^{k+1}) + \frac{1}{t_k}\langle x^k - x^{k+1}, x - x^{k+1}\rangle \\ &\quad - \frac{1}{2t_k}\|x^{k+1} - x^k\|^2 \\ &= F(x^{k+1}) - \frac{1}{2t_k}\left(\|x^k - x\|^2 - \|x^{k+1} - x\|^2\right)\end{aligned}$$

and hence

$$F(x^{k+1}) \leq F(x) + \frac{1}{2t_k}\left(\|x^k - x\|^2 - \|x^{k+1} - x\|^2\right), \quad \forall x \in \mathbb{R}^n. \tag{8.33}$$

Taking $x = x^k$ gives

$$F(x^{k+1}) \leq F(x^k) - \frac{1}{2t_k}\|x^{k+1} - x^k\|^2 \tag{8.34}$$

which in particular implies $F(x^{k+1}) \leq F(x^k)$ and thus Algorithm 8.48 is well-defined.

Lemma 8.50 *Let Assumption 8.47 hold. Then, Algorithm 8.48 is well-defined and (8.33) holds. Moreover, there is a constant $\gamma > 0$ such that $t_k \geq \gamma$ for all k.*

Proof According to the above analysis, it remains only to show the existence of a constant $\gamma > 0$ such that $t_k \geq \gamma$ for all k.

If there is $\bar{k}$ such that $x^{\bar{k}+1} = x^{\bar{k}}$, then $x^{\bar{k}} = J_{t_{\bar{k}}}(x^{\bar{k}})$ which means $x^{\bar{k}}$ is an optimal solution of (8.31). Consequently, by the definition of the algorithm, we can see $t_k = t_{\bar{k}}$ and $x^k = x^{\bar{k}}$ for all $k \geq \bar{k}$ and the result holds. Thus, in the following, we may assume $x^{k+1} \neq x^k$ for all k. Assume the result is not true. Since $\{t_k\}$ is monotonically decreasing, we have $\lim_{k\to\infty} t_k = 0$. By using (8.33) with $x \in S$, it is easy to deduce that $\{x^k\}$ is bounded. Thus, we may find a subsequence $\{k_l\}$ of positive integers such that $t_{k_l} < t_{k_l - 1}$ for all l and $x^{k_l} \to \bar{x}$ as $l \to \infty$ for some $\bar{x} \in \mathbb{R}^n$. By the lower semicontinuity of F and the monotonicity of $\{F(x^k)\}$, we have $F(\bar{x}) \leq \liminf_{k\to\infty} F(x^k) \leq F(x^0)$. Since $x^0 \in \mathscr{D}$, so is $\bar{x}$ by (iii) in Assumption 8.47. Consequently, by (ii) in Assumption 8.47, there exist $\bar{r} > 0$ and $\bar{L} > 0$ such that $B_{\bar{r}}(\bar{x}) \subset \operatorname{int}(\operatorname{dom}(f))$ and

$$\|\nabla f(y) - \nabla f(z)\| \leq \bar{L}\|y - z\|, \quad \forall y, z \in \mathscr{D} \cap B_{\bar{r}}(\bar{x}). \tag{8.35}$$

Let $\hat{t}_{k_l} := t_{k_l}/\rho$. Then, $t_{k_l} < \hat{t}_{k_l} \le t_{k_l-1}$. Thus, by setting $\hat{x}^{k_l} := J_{\hat{t}_{k_l}}(x^{k_l})$, we have from the definition of t_{k_l} that

$$f(\hat{x}^{k_l}) > f(x^{k_l}) + \langle \nabla f(x^{k_l}), \hat{x}^{k_l} - x^{k_l} \rangle + \frac{1}{2\hat{t}_{k_l}} \|\hat{x}^{k_l} - x^{k_l}\|^2. \tag{8.36}$$

By using Lemma 8.5, we have

$$\|\hat{x}^{k_l} - x^{k_l}\| \le \frac{\hat{t}_{k_l}}{t_{k_l}} \|J_{t_{k_l}}(x^{k_l}) - x^{k_l}\| = \frac{1}{\rho}\|x^{k_l+1} - x^{k_l}\|.$$

By using (8.34), the monotonicity of $\{F(x^k)\}$ and $t_k \le t_{-1} = \sigma$, it is easy to deduce $\|x^{k+1} - x^k\| \to 0$ as $k \to \infty$. Therefore, $\|\hat{x}^{k_l} - x^{k_l}\| \to 0$, and hence, $\hat{x}^{k_l} \to \bar{x}$ as $l \to \infty$. This implies that $\hat{x}^{k_l} \in \mathcal{D} \cap B_{\bar{r}}(\bar{x})$ for sufficiently large l. Consequently, it follows from (8.36) that

$$\begin{aligned} f(\hat{x}^{k_l}) &> f(\hat{x}^{k_l}) + \langle \nabla f(\hat{x}^{k_l}), x^{k_l} - \hat{x}^{k_l} \rangle + \langle \nabla f(x^{k_l}), \hat{x}^{k_l} - x^{k_l} \rangle \\ &\quad + \frac{1}{2\hat{t}_{k_l}} \|\hat{x}^{k_l} - x^{k_l}\|^2 \end{aligned}$$

which together with (8.35) shows that

$$\begin{aligned} \frac{1}{2\hat{t}_{k_l}} \|\hat{x}^{k_l} - x^{k_l}\|^2 &\le \langle \nabla f(\hat{x}^{k_l}) - \nabla f(x^{k_l}), \hat{x}^{k_l} - x^{k_l} \rangle \\ &\le \|\nabla f(\hat{x}^{k_l}) - \nabla f(x^{k_l})\| \|\hat{x}^{k_l} - x^{k_l}\| \\ &\le \bar{L}\|\hat{x}^{k_l} - x^{k_l}\|^2. \end{aligned}$$

We note that $\hat{x}^{k_l} \ne x^{k_l}$ for all l; if not, then $\hat{x}^{k_l} = x^{k_l}$ for some l implies that x^{k_l} is an optimal solution and hence $x^{k_l+1} = x^{k_l}$ which is a contradiction. Thus, we can deduce that $1/(2\hat{t}_{k_l}) \le \bar{L}$, i.e., $t_{k_l} \ge \rho/(2\bar{L})$ for all l, contradicting the assumption $t_{k_l} \to 0$ as $l \to \infty$. □

Now, we are ready to show the following convergence rate result for Algorithm 8.48 under bounded Hölder growth conditions [23].

Theorem 8.51 *Let Assumption 8.47 hold and consider the sequence $\{x^k\}$ generated by Algorithm 8.48. Then, $\{x^k\}$ converges to some point $x^* \in S$. If, in addition, F satisfies the growth error bound at x^* in the sense that there exist $r > 0$, $\kappa > 0$ and $\alpha \ge 2$ such that*

$$F(x) - F(x^*) \ge \kappa d(x, S)^\alpha, \quad \forall x \in B_r(x^*), \tag{8.37}$$

then there holds the following convergence rate:

(i) *If* $\alpha = 2$, *then there exists an integer* $k_0 \geq 0$ *such that*

$$\|x^{k+1} - x^*\| \leq \frac{1}{\sqrt{1 + \gamma\kappa/2}} \|x^k - x^*\|, \tag{8.38}$$

$$F(x^{k+1}) - F(x^*) \leq \frac{\sqrt{1 + \gamma\kappa/2} + 1}{2\sqrt{1 + \gamma\kappa/2}} \left(F(x^k) - F(x^*)\right) \tag{8.39}$$

for all $k \geq k_0$, *where* $\gamma > 0$ *is the number determined in Lemma 8.50.*

(ii) *If* $\alpha > 2$, *then there exist an integer* $k_0 \geq 0$ *and a constant* C *such that*

$$\|x^k - x^*\| \leq C(1 + k)^{-\frac{1}{\alpha - 2}} \quad \text{and} \quad F(x^k) - F(x^*) \leq C(1 + k)^{-\frac{2}{\alpha - 2}} \tag{8.40}$$

for all $k \geq k_0$.

Proof According to (8.33) and Lemma 8.50, we have

$$\|x^{k+1} - x\|^2 - \|x^k - x\|^2 \leq 2\gamma \left(F(x) - F(x^{k+1})\right), \quad \forall x \in S. \tag{8.41}$$

Thus, we may use the same argument in the proof of Theorem 4.35 to show that x^k converges to some $x^* \in S$.

Next, we consider deriving the convergence rates under the condition (8.37). Since $x^k \to x^*$, there exists an integer $k_0 \geq 0$ such that $x^k \in B_r(x^*)$ for all $k \geq k_0$. It then follows from (8.41) and (8.37) that

$$\begin{aligned} \|x^{k+1} - x^*\|^2 - \|x^k - x^*\|^2 &\leq -2\gamma \left(F(x^{k+1}) - F(x^*)\right) \\ &\leq -2\gamma\kappa d(x^{k+1}, S)^\alpha \end{aligned} \tag{8.42}$$

for all $k \geq k_0$. Let $x \in S$ be any element. Note that (8.41) implies that $\{\|x^k - x\|\}$ is monotonically decreasing. We have

$$\begin{aligned} \|x^{k+1} - x^*\| &\leq \|x^{k+1} - x\| + \|x^* - x\| \\ &= \|x^{k+1} - x\| + \lim_{l \to \infty} \|x^l - x\| \\ &\leq 2\|x^{k+1} - x\|. \end{aligned}$$

Since $x \in S$ is arbitrary, this implies that $\|x^{k+1} - x^*\| \leq 2d(x^{k+1}, S)$. Therefore, it follows from (8.42) that

$$\|x^{k+1} - x^*\|^2 - \|x^k - x^*\|^2 \leq -2^{1-\alpha}\gamma\kappa \|x^{k+1} - x^*\|^\alpha \tag{8.43}$$

(i) If $\alpha = 2$, from (8.43), we immediately obtain (8.38). To derive (8.39), we will use (8.33) with $x = x^*$ and (8.34). We need to bound $\|x^k - x^*\|^2 - \|x^{k+1} - x^*\|^2$ in terms of $\|x^k - x^{k+1}\|^2$. Note that

$$\begin{aligned}&\|x^k - x^*\|^2 - \|x^{k+1} - x^*\|^2 \\ &= \left(\|x^k - x^*\| - \|x^{k+1} - x^*\|\right)\left(\|x^k - x^*\| + \|x^{k+1} - x^*\|\right).\end{aligned}$$

Let $\eta > 1$ be a number to be determined. By using (8.38), we have

$$\begin{aligned}\|x^k - x^*\| + \|x^{k+1} - x^*\| &= \eta\|x^k - x^*\| + (1-\eta)\|x^k - x^*\| + \|x^{k+1} - x^*\| \\ &\le \eta\|x^k - x^*\| \\ &\quad + \left((1-\eta)\sqrt{1+\gamma\kappa/2} + 1\right)\|x^{k+1} - x^*\|.\end{aligned}$$

By choosing $\eta > 1$ such that $(1-\eta)\sqrt{1+\gamma\kappa/2} + 1 = -\eta$, i.e.,

$$\eta = (\sqrt{1+\gamma\kappa/2} + 1)/(\sqrt{1+\gamma\kappa/2} - 1),$$

we can obtain

$$\|x^k - x^*\| + \|x^{k+1} - x^*\| \le \eta\left(\|x^k - x^*\| - \|x^{k+1} - x^*\|\right).$$

Consequently

$$\begin{aligned}\|x^k - x^*\|^2 - \|x^{k+1} - x^*\|^2 &\le \eta\left(\|x^k - x^*\| - \|x^{k+1} - x^*\|\right)^2 \\ &\le \eta\|x^k - x^{k+1}\|^2.\end{aligned}$$

Now, we may use (8.33) and (8.34) to obtain

$$\begin{aligned}F(x^{k+1}) - F(x^*) &\le \frac{1}{2t_k}\left(\|x^k - x^*\|^2 - \|x^{k+1} - x^*\|^2\right) \\ &\le \frac{\eta}{2t_k}\|x^k - x^{k+1}\|^2 \\ &\le \eta\left(F(x^k) - F(x^{k+1})\right)\end{aligned}$$

which implies that

$$F(x^{k+1}) - F(x^*) \le \frac{\eta}{1+\eta}\left(F(x^k) - F(x^*)\right)$$

and hence (8.39) follows.

(ii) If $\alpha > 2$, we may use (8.43) and Lemma 8.46 to obtain

$$\|x^k - x^*\|^2 \leq C(1+k)^{-\frac{2}{\alpha-2}}$$

which implies the first estimate in (8.40). By virtue of (8.41), we then obtain

$$F(x^{k+1}) - F(x^*) \leq \frac{1}{2\gamma}\|x^k - x^*\|^2 \leq C(1+k)^{-\frac{2}{\alpha-2}}$$

which show the second estimate in (8.40).

□

8.5.2 Alternating Direction Method of Multipliers

In Chap. 6, we have considered solving the convex minimization problem (6.1) by the proximal alternating direction of method of multipliers, i.e., Algorithm 6.2, and obtained various convergence results. In particular, we have established in Theorem 6.20 some linear convergence results on Algorithm 6.2. However, these results require strong convexity on f or g and full row rank conditions on the coefficient matrix A or B. In this section, we will establish some new linear convergence results on Algorithm 6.2 without using strong convexity.

Let $\{u^k := (x^k, y^k, \lambda^k)\}$ be a sequence generated by Algorithm 6.2. With the same notation Δx^k, Δy^k, $\Delta\lambda^k$, and Δu^k as defined in Chap. 6, it has been shown in Lemma 6.11 that

$$\|u^{k+1} - u^*\|_G^2 + \|\Delta y^{k+1}\|_Q^2 \leq \|u^k - u^*\|_G^2 + \|\Delta y^k\|_Q^2 - \|\Delta u^{k+1}\|_G^2 \tag{8.44}$$

for any KKT point $u^* := (x^*, y^*, \lambda^*)$ of (6.1), where G is the matrix defined in (6.38). If $\{u^k\}$ is bounded, it has been proved in Theorem 6.12 that there is a KKT point $u^\dagger := (x^\dagger, y^\dagger, \lambda^\dagger)$ of (6.1) such that $\|u^k - u^\dagger\|_G^2 \to 0$ as $k \to \infty$. Recall that $u^* := (x^*, y^*, \lambda^*) \in \mathbb{R}^n \times \mathbb{R}^m \times \mathbb{R}^p$ is a KKT point of (6.1) if

$$-A^T\lambda^* \in \partial f(x^*), \quad -B^T\lambda^* \in \partial g(y^*) \quad \text{and} \quad Ax^* + By^* = c.$$

If we introduce the multifuncton $F : \mathbb{R}^n \times \mathbb{R}^m \times \mathbb{R}^p \rightrightarrows \mathbb{R}^n \times \mathbb{R}^m \times \mathbb{R}^p$ defined by

$$F(u) := \begin{pmatrix} \partial f(x) + A^T\lambda \\ \partial g(y) + B^T\lambda \\ Ax + By - c \end{pmatrix}, \quad \forall u = \begin{pmatrix} x \\ y \\ \lambda \end{pmatrix} \in \mathbb{R}^n \times \mathbb{R}^m \times \mathbb{R}^p, \tag{8.45}$$

then u^* is a KKT point means $0 \in F(u^*)$ or, equivalently, $u^* \in F^{-1}(0)$. In the following, we will derive the convergence rates of Algorithm 6.2 under bounded metric subregularity conditions on F at $(u^\dagger, 0) \in \mathrm{Gr}(F)$; see [117]; the part on linear convergence strengthens the corresponding result in [226].

Theorem 8.52 *Let Assumptions 6.7 and 6.8 hold. Consider the sequence* $\{u^k := (x^k, y^k, \lambda^k)\}$ *generated by Algorithm 6.2. Assume* $\{u^k\}$ *is bounded and let* $u^\dagger$ *be the KKT point of (6.1) such that* $\|u^k - u^\dagger\|_G^2 \to 0$ *as* $k \to \infty$. *Let* R *be a number such that* $\|u^k - u^\dagger\| \le R$ *for all* k *and assume that there exist* $\kappa > 0$ *and* $\alpha \in (0, 1]$ *such that*

$$d(u, F^{-1}(0)) \le \kappa [d(0, F(u))]^\alpha, \quad \forall u \in B_R(u^\dagger). \tag{8.46}$$

(i) *If* $\alpha = 1$, *then there exists a constant* $0 < q < 1$ *such that*

$$\|u^{k+1} - u^\dagger\|_G^2 + \|\Delta y^{k+1}\|_Q^2 \le q^2 \left(\|u^k - u^\dagger\|_G^2 + \|\Delta y^k\|_Q^2\right) \tag{8.47}$$

for all $k \ge 0$ *and consequently there exist* $C > 0$ *and* $0 < q < 1$ *such that*

$$\begin{aligned} \|u^k - u^\dagger\|_G, \ \|\Delta u^k\|_G &\le C q^k, \\ \|Ax^k + By^k - c\| &\le C q^k, \\ |H(x^k, y^k) - H_*| &\le C q^k \end{aligned} \tag{8.48}$$

for all $k > 0$.

(ii) *If* $\alpha \in (0, 1)$ *then there is a constant* C *such that*

$$\|u^k - u^\dagger\|_G^2 + \|\Delta y^k\|_Q^2 \le C(k+1)^{-\frac{\alpha}{1-\alpha}} \tag{8.49}$$

and consequently

$$\begin{aligned} \|u^k - u^\dagger\|_G, \ \|\Delta u^k\|_G &\le C(k+1)^{-\frac{1}{2(1-\alpha)}}, \\ \|Ax^k + By^k - c\| &\le C(k+1)^{-\frac{1}{2(1-\alpha)}}, \\ |H(x^k, y^k) - H_*| &\le C(k+1)^{-\frac{1}{2(1-\alpha)}} \end{aligned} \tag{8.50}$$

for all $k \ge 0$.

Proof By using (8.44) with $u^* = u^\dagger$, we can obtain

$$\begin{aligned} \|u^{k+1} - u^\dagger\|_G^2 + \|\Delta y^{k+1}\|_Q^2 &\le \|u^k - u^\dagger\|_G^2 + \|\Delta y^k\|_Q^2 - \|\Delta u^{k+1}\|_G^2 \\ &= \|u^k - u^\dagger\|_G^2 + \|\Delta y^k\|_Q^2 - \eta \|\Delta u^{k+1}\|_G^2 \\ &\quad - (1-\eta)\|\Delta u^{k+1}\|_G^2, \end{aligned} \tag{8.51}$$

where $\eta \in (0, 1)$ is any number. According to (6.12)

$$\begin{pmatrix} \rho A^T B \Delta y^{k+1} - P \Delta x^{k+1} \\ -Q \Delta y^{k+1} \\ A x^{k+1} + B y^{k+1} - c \end{pmatrix} \in F(u^{k+1}).$$

Thus, by using $\Delta \lambda^{k+1} = \rho(Ax^{k+1} + By^{k+1} - c)$, we can obtain

$$\begin{aligned} d^2(0, F(u^{k+1})) &\leq \|\rho A^T B \Delta y^{k+1} - P \Delta x^{k+1}\|^2 + \| - Q \Delta y^{k+1}\|^2 \\ &\quad + \|Ax^{k+1} + By^{k+1} - c\|^2 \\ &\leq 2\|P \Delta x^{k+1}\|^2 + 2\rho^2 \|A\|^2 \|B \Delta y^{k+1}\|^2 \\ &\quad + \|Q \Delta y^{k+1}\|^2 + \frac{1}{\rho^2} \|\Delta \lambda^{k+1}\|^2 \\ &\leq \gamma \|\Delta u^{k+1}\|_G^2, \end{aligned} \tag{8.52}$$

where

$$\gamma := \max \left\{ 2\|P\|, 2\rho\|A\|^2, \|Q\|, \frac{1}{\rho} \right\}.$$

Combining this with (8.51) gives

$$\begin{aligned} \|u^{k+1} - u^\dagger\|_G^2 + \|\Delta y^{k+1}\|_Q^2 &\leq \|u^k - u^\dagger\|_G^2 + \|\Delta y^k\|_Q^2 - \eta \|\Delta u^{k+1}\|_G^2 \\ &\quad - \frac{1 - \eta}{\gamma} d^2(0, F(u^{k+1})). \end{aligned}$$

Since $\|u^k - u^\dagger\| \leq R$ for all k and F satisfies (8.46), one can see that

$$d(u^{k+1}, F^{-1}(0)) \leq \kappa [d(0, F(u^{k+1}))]^\alpha, \quad \forall k \geq 0.$$

Consequently

$$\begin{aligned} \|u^{k+1} - u^\dagger\|_G^2 + \|\Delta y^{k+1}\|_Q^2 &\leq \|u^k - u^\dagger\|_G^2 + \|\Delta y^k\|_Q^2 - \eta \|\Delta u^{k+1}\|_G^2 \\ &\quad - \frac{1 - \eta}{\gamma \kappa^{2/\alpha}} [d(u^{k+1}, F^{-1}(0))]^{2/\alpha}. \end{aligned}$$

For any $u = (x, y, \lambda) \in \mathbb{R}^n \times \mathbb{R}^m \times \mathbb{R}^p$, let

$$d_G(u, F^{-1}(0)) := \inf_{\bar{u} \in F^{-1}(0)} \|u - \bar{u}\|_G$$

which measures the "distance" from u to $F^{-1}(0)$ under the semi-norm $\|\cdot\|_G$. It is easy to see that

$$d_G^2(u, F^{-1}(0)) \leq \|G\| d^2(u, F^{-1}(0)),$$

where $\|G\|$ denotes the norm of the operator G. Then, we have

$$\begin{aligned}\|u^{k+1} - u^\dagger\|_G^2 + \|\Delta y^{k+1}\|_Q^2 \leq \|u^k - u^\dagger\|_G^2 + \|\Delta y^k\|_Q^2 - \eta\|\Delta u^{k+1}\|_G^2 \\ - \frac{1-\eta}{\gamma(\kappa^2\|G\|)^{1/\alpha}}[d_G(u^{k+1}, F^{-1}(0))]^{2/\alpha}.\end{aligned}$$

Now, let $\bar{u} \in F^{-1}(0)$ be any point. Then

$$\|u^{k+1} - u^\dagger\|_G \leq \|u^{k+1} - \bar{u}\|_G + \|u^\dagger - \bar{u}\|_G.$$

Since $\|u^k - u^\dagger\|_G \to 0$, we have $\|u^k - \bar{u}\|_G \to \|u^\dagger - \bar{u}\|_G$ as $k \to \infty$. According to Theorem 6.10, $\{\|u^k - \bar{u}\|_G^2\}$ is monotonically decreasing. Thus

$$\|u^{k+1} - u^\dagger\|_G \leq \|u^{k+1} - \bar{u}\|_G + \lim_{l\to\infty} \|u^l - \bar{u}\|_G \leq 2\|u^{k+1} - \bar{u}\|_G.$$

Since $\bar{u} \in F^{-1}(0)$ is arbitrary, we thus have

$$\|u^{k+1} - u^\dagger\|_G \leq 2d_G(u^{k+1}, F^{-1}(0)).$$

Therefore

$$\begin{aligned}\|u^{k+1} - u^\dagger\|_G^2 + \|\Delta y^{k+1}\|_Q^2 \leq \|u^k - u^\dagger\|_G^2 + \|\Delta y^k\|_Q^2 - \eta\|\Delta u^{k+1}\|_G^2 \\ - \frac{1-\eta}{\gamma(4\kappa^2\|G\|)^{1/\alpha}}\|u^{k+1} - u^\dagger\|_G^{2/\alpha}.\end{aligned}$$

By using the fact $\|\Delta u^k\|_G \to 0$ established in Proposition 6.17, we can find a constant $C > 0$ such that

$$\|\Delta u^{k+1}\|_G^2 \geq C\|\Delta u^{k+1}\|_G^{2/\alpha}.$$

Note that $\|\Delta u^{k+1}\|_G^2 \geq \|\Delta y^{k+1}\|_Q^2$. Thus

$$\begin{aligned}\|u^{k+1} - u^\dagger\|_G^2 + \|\Delta y^{k+1}\|_Q^2 \leq \|u^k - u^\dagger\|_G^2 + \|\Delta y^k\|_Q^2 - C\eta\|\Delta y^{k+1}\|_Q^{2/\alpha} \\ - \frac{1-\eta}{\gamma(4\kappa^2\|G\|)^{1/\alpha}}\|u^{k+1} - u^\dagger\|_G^{2/\alpha}.\end{aligned}$$

Choose η such that

$$\eta = \frac{1}{1 + C\gamma(4\kappa^2\|G\|)^{1/\alpha}}.$$

Then

$$\begin{aligned}
&\|u^{k+1} - u^\dagger\|_G^2 + \|\Delta y^{k+1}\|_Q^2 \\
&\le \|u^k - u^\dagger\|_G^2 + \|\Delta y^k\|_Q^2 - C\eta\left(\|\Delta y^{k+1}\|_Q^{2/\alpha} + \|u^{k+1} - u^\dagger\|_G^{2/\alpha}\right).
\end{aligned}$$

Using the inequality $(a+b)^p \le 2^{p-1}(a^p + b^p)$ for $a, b \ge 0$ and $p \ge 1$, we then obtain

$$\begin{aligned}
&\|u^{k+1} - u^\dagger\|_G^2 + \|\Delta y^{k+1}\|_Q^2 \\
&\le \|u^k - u^\dagger\|_G^2 + \|\Delta y^k\|_Q^2 - 2^{1-1/\alpha}C\eta\left(\|u^{k+1} - u^\dagger\|_G^2 + \|\Delta y^{k+1}\|_Q^2\right)^{1/\alpha}.
\end{aligned} \tag{8.53}$$

(i) If $\alpha = 1$, then we obtain the linear convergence:

$$(1 + C\eta)\left(\|u^{k+1} - u^\dagger\|_G^2 + \|\Delta y^{k+1}\|_Q^2\right) \le \|u^k - u^\dagger\|_G^2 + \|\Delta y^k\|_Q^2$$

which is (8.47) with $q = 1/\sqrt{1 + C\eta}$. By using Lemma 6.11 and (8.47), we immediately obtain the first estimate in (8.48). By using (6.34) and (6.36) we then obtain the last two estimates in (8.48).

(ii) If $\alpha \in (0, 1)$, we may use (8.53) and Lemma 8.46 to obtain (8.49). To derive the first estimate in (8.50), we may use Lemma 6.11 to obtain

$$\sum_{j=l}^{k} \|\Delta u^j\|_G^2 \le \|u^l - u^\dagger\|_G^2 + \|\Delta y^l\|_Q^2$$

for all integers $1 \le l < k$. By using the monotonicity of $\{\|\Delta u^j\|_G^2\}$ shown in Lemma 6.16 and the estimate (8.49), we have

$$(k - l + 1)\|\Delta u^k\|_G^2 \le C(l + 1)^{-\frac{\alpha}{1-\alpha}}.$$

Taking $l = [k/2]$, the largest integers $\le k/2$, gives

$$\|\Delta u^k\|_G^2 \le C(k + 1)^{-\frac{\alpha}{1-\alpha} - 1} = C(k + 1)^{-\frac{1}{1-\alpha}}$$

with a possibly different generic constant C. This shows the first estimate in (8.50). Based on this, we can use (6.34) and (6.36) to obtain the last two estimates in (8.50). The proof is therefore complete.

□

Remark 8.53 In Theorem 8.23, we have shown that every polyhedral multifunction is bounded metrically subregular. Therefore, if ∂f and ∂g are polyhedral multifunctions, then the multifunction F defined by (8.45) is also polyhedral, and thus, (8.46) with $\alpha = 1$ holds.

On the other hand, if f and g are semi-algebraic convex functions, then the multifunction F satisfies (8.46) for some $\alpha \in (0, 1]$. Indeed, the semi-algebraicity of f and g implies that their subdifferentials ∂f and ∂g are semi-algebraic multifunctions with closed graph; consequently, F is semi-algebraic with closed graph. According to Theorem 8.45, F is bounded Hölder metrically subregular at any point $(\bar{u}, \bar{\xi})$ on its graph, which in particular implies (8.46).

8.5.3 *Primal-Dual Splitting Method*

In Sect. 7.2, we have considered the primal-dual splitting algorithm of Chambolle and Pock, i.e., Algorithm 7.11 with $\theta = 1$, for solving the convex minimization problem (7.1) which can be written as the saddle point problem

$$\min_{x \in \mathbb{R}^n} \sup_{\lambda \in \mathbb{R}^m} \left\{ \Phi(x, \lambda) := \langle Ax, \lambda \rangle - f^*(\lambda) + g(x) \right\},$$

where A is a matrix of size $m \times n$, $f : \mathbb{R}^m \to (-\infty, \infty]$ and $g : \mathbb{R}^n \to (-\infty, \infty]$ are proper, lower semicontinuous, convex functions, and f^* denotes the Legendre-Fenchel conjugate of f. Algorithm 7.11 with $\theta = 1$ can be written equivalently as

$$\begin{aligned}
x^{k+1} &= \operatorname{prox}_{\tau g}(x^k - \tau A^T \lambda^k), \\
\bar{x}^{k+1} &= 2x^{k+1} - x^k, \\
\lambda^{k+1} &= \operatorname{prox}_{\sigma f^*}(\lambda^k + \sigma A \bar{x}^{k+1}).
\end{aligned} \tag{8.54}$$

Assuming that Φ has a saddle point, the convergence of the algorithm has been established in Theorem 7.12 when the step sizes $\tau > 0$ and $\sigma > 0$ are chosen such that $\tau\sigma \|A\|^2 < 1$. In this subsection, we will follow an idea in [133] to show the linear convergence of the algorithm under a metric subregularity condition.

According to the definition of proximal mapping, we can obtain from (8.54) that

$$\frac{1}{\tau}(x^k - x^{k+1}) - A^T(\lambda^k - \lambda^{k+1}) \in A^T\lambda^{k+1} + \partial g(x^{k+1}),$$
$$\frac{1}{\sigma}(\lambda^k - \lambda^{k+1}) - A(x^k - x^{k+1}) \in -Ax^{k+1} + \partial f^*(\lambda^{k+1}).$$

By introducing the multifunction $F : \mathbb{R}^n \times \mathbb{R}^m \rightrightarrows \mathbb{R}^n \times \mathbb{R}^m$ and the linear mapping $S : \mathbb{R}^n \times \mathbb{R}^m \to \mathbb{R}^n \times \mathbb{R}^m$ defined by

$$F(z) := (A^T\lambda + \partial g(x), -Ax + \partial f^*(\lambda)), \quad Sz := (x/\tau - A^T\lambda, \lambda/\sigma - Ax) \tag{8.55}$$

for any $z := (x, \lambda) \in \mathbb{R}^n \times \mathbb{R}^m$, we can obtain for $z^k := (x^k, \lambda^k)$ that

$$S(z^k - z^{k+1}) \in F(z^{k+1}). \tag{8.56}$$

By using the property of ∂g and ∂f^*, it is straightforward to show that the graph of F is closed. Furthermore, by the monotonicity of ∂g and ∂f^*, one can see that F is also monotone. Indeed, let $z_i := (x_i, \lambda_i)$, $i = 1, 2$, be any two elements in $\mathbb{R}^n \times \mathbb{R}^m$ with $F(z_i) \neq \emptyset$. Then, for any $\theta_i := (\xi_i, \eta_i) \in F(z_i)$, we have

$$\xi_i - A^T\lambda_i \in \partial g(x_i) \quad \text{and} \quad \eta_i + Ax_i \in \partial f^*(\lambda_i).$$

Therefore, by the monotonicity of ∂g and ∂f^*, we can obtain

$$\begin{aligned} 0 &\le \langle (\xi_1 - A^T\lambda_1) - (\xi_2 - A^T\lambda_2), x_1 - x_2\rangle + \langle (\eta_1 + Ax_1) - (\eta_2 + Ax_2), \lambda_1 - \lambda_2\rangle \\ &= \langle \xi_1 - \xi_2, x_1 - x_2\rangle + \langle \eta_1 - \eta_2, \lambda_1 - \lambda_2\rangle \\ &= \langle \theta_1 - \theta_2, z_1 - z_2\rangle \end{aligned}$$

which shows the monotonicity of F.

Let $\mathcal{Z}^*$ denote the set of all saddle points of Φ which is assumed to be non-empty. Let $\hat{z} := (\hat{x}, \hat{\lambda})$ be any saddle point of Φ. Then, $0 \in F(\hat{z})$. This together with (8.56) and the monotonicity of F shows that

$$0 \le \langle S(z^k - z^{k+1}), z^{k+1} - \hat{z}\rangle. \tag{8.57}$$

Under the assumption $\tau\sigma\|A\|^2 < 1$, it is easy to see that the linear mapping S is positive definite self-adjoint. Thus, there exist constants $0 < \alpha \le \beta$ such that

$$\alpha\|z\|^2 \le \langle Sz, z\rangle \le \beta\|z\|^2, \quad \forall z \in \mathbb{R}^n \times \mathbb{R}^m. \tag{8.58}$$

Let $\|z\|_S^2 := \langle Sz, z\rangle$. Then, it follows from (8.57) and the polarization identity that

$$\|z^{k+1} - \hat{z}\|_S^2 \le \|z^k - \hat{z}\|_S^2 - \|z^{k+1} - z^k\|_S^2. \tag{8.59}$$

Based on (8.59), we can now prove the following convergence result on the method (8.54).

Theorem 8.54 *Let $f : \mathbb{R}^m \to (-\infty, \infty]$ and $g : \mathbb{R}^n \to (-\infty, \infty]$ be proper, lower semicontinuous, convex functions and let A be a matrix of size $m \times n$. Assume that Φ has a saddle point. Consider the algorithm (8.54) with $\tau\sigma\|A\|^2 < 1$. Then, $z^k := (x^k, \lambda^k)$ converges to some saddle point $z^* := (x^*, \lambda^*)$ of Φ as $k \to \infty$. If, in addition, the multifunction F is Hölder metrically subregular at $(z^*, 0)$, i.e., there exist $r > 0$, $\alpha \in (0, 1]$ and $\kappa > 0$ such that*

$$d(z, F^{-1}(0)) \le \kappa\, [d(0, F(z))]^\alpha, \quad \forall z \in B_r(z^*), \tag{8.60}$$

then there exists an integer k_0 such that for $\alpha = 1$ there holds

$$\|z^{k+1} - z^*\|_S^2 \le \frac{4\kappa^2\beta^2}{1 + 4\kappa^2\beta^2}\|z^k - z^*\|_S^2, \quad \forall k \ge k_0 \tag{8.61}$$

while for $0 < \alpha < 1$ there is a constant $C > 0$ such that

$$\|z^k - z^*\|_S^2 \le C(1 + k)^{-\frac{\alpha}{1-\alpha}}, \quad \forall k \ge k_0. \tag{8.62}$$

Proof According to (8.59), we have

$$\|z^{k+1} - \hat{z}\|_S^2 \le \|z^k - \hat{z}\|_S^2, \quad \forall \hat{z} \in \mathcal{Z}^* \tag{8.63}$$

and

$$\|z^{k+1} - z^k\|_S^2 \to 0 \quad \text{as } k \to \infty.$$

By virtue of (8.58), the sequence $\{z^k\}$ is bounded and $z^{k+1} - z^k \to 0$ as $k \to \infty$. Thus, there is a subsequence $\{z^{k_l}\}$ of $\{z^k\}$ such that $z^{k_l} \to z^*$ as $l \to \infty$ for some $z^* := (x^*, \lambda^*) \in \mathbb{R}^n \times \mathbb{R}^m$. We also have $z^{k_l+1} \to z^*$. By setting $k = k_l$ in (8.56) and noting $S(z^{k_l+1} - z^{k_l}) \to 0$ as $l \to \infty$, we may use the closedness of the graph of F to conclude $0 \in F(z^*)$ which means $z^* \in \mathcal{Z}^*$. Consequently, it follows from (8.63) that $\|z^k - z^*\|_S^2$ is monotonically decreasing with respect to k. Since $\|z^{k_l} - z^*\|_S^2 \to 0$ as $l \to \infty$, we must have $\|z^k - z^*\|_S^2 \to 0$ as $k \to \infty$ which shows the convergence.

Next, we derive the convergence rate of $\{z^k\}$ under the condition (8.60). Since $z^k \to z^*$ as $k \to \infty$, there exists an integer k_0 such that $z^k \in B_r(z^*)$ for all $k \ge k_0$. Note that $F^{-1}(0) = \mathcal{Z}^*$, we may use (8.56) and (8.58) to obtain for $k \ge k_0$ that

$$d(z^{k+1}, \mathcal{Z}^*) \le \kappa[d(0, F(z^{k+1}))]^\alpha \le \kappa\|S(z^{k+1} - z^k)\|^\alpha \le \kappa\beta^{\alpha/2}\|z^{k+1} - z^k\|_S^\alpha.$$

Combining this with (8.59) with $\hat{z} = z^*$ gives

$$\|z^{k+1} - z^*\|_S^2 \le \|z^k - z^*\|_S^2 - \frac{1}{\kappa^{2/\alpha}\beta}[d(z^{k+1}, \mathcal{Z}^*)]^{2/\alpha}, \quad \forall k \ge k_0.$$

Let $\hat{z} \in \mathcal{Z}^*$ be any element. We have from (8.63) that

$$\begin{aligned}\|z^{k+1} - z^*\|_S &\le \|z^{k+1} - \hat{z}\|_S + \|z^* - \hat{z}\|_S = \|z^{k+1} - \hat{z}\|_S + \lim_{l\to\infty} \|z^l - \hat{z}\|_S \\ &\le 2\|z^{k+1} - \hat{z}\|_S \le 2\beta^{1/2}\|z^{k+1} - \hat{z}\|\end{aligned}$$

which implies $\|z^{k+1} - z^*\|_S \le 2\beta^{1/2} d(z^{k+1}, \mathcal{Z}^*)$. Therefore

$$\|z^{k+1} - z^*\|_S^2 \le \|z^k - z^*\|_S^2 - \frac{1}{(2\kappa)^{2/\alpha}\beta^{1+1/\alpha}}\|z^{k+1} - z^*\|_S^{2/\alpha}.$$

Consequently, for $\alpha = 1$ we immediately obtain (8.61), while for $0 < \alpha < 1$, we may use Lemma 8.46 to obtain (8.62). □

Remark 8.55 The first part of Theorem 8.54 provides a new proof of Theorem 7.12. For the second part on convergence rates, it requires the Hölder metric subregularity of F at $(z^*, 0)$. This can be guaranteed under various conditions on f and g:

(i) First, consider the case that g is μ-strongly convex with $\mu > 0$ and f is convex differentiable with L-Lipschitz continuous gradient for some $L > 0$. According to Theorem 5.9, f^* is $1/L$-strongly convex. Note that for any $z = (x, \lambda) \in \mathbb{R}^n \times \mathbb{R}^m$ and any $(\xi, \eta) \in F(z)$, we have

$$\xi - A^T\lambda \in \partial g(x) \quad \text{and} \quad \eta + Ax \in \partial f^*(\lambda).$$

Note also that $\hat{z} = (\hat{x}, \hat{\lambda}) \in \mathcal{Z}^* = F^{-1}(0)$ implies $-A^T\hat{\lambda} \in \partial g(\hat{x})$ and $A\hat{x} \in \partial f^*(\hat{\lambda})$. Therefore, it follows from Proposition 3.12 that

$$\mu\|x - \hat{x}\|^2 \le \langle \xi - A^T(\lambda - \hat{\lambda}), x - \hat{x}\rangle, \ \frac{1}{L}\|\lambda - \hat{\lambda}\|^2 \le \langle \eta + A(x - \hat{x}), \lambda - \hat{\lambda}\rangle$$

and hence

$$\begin{aligned}&\mu\|x - \hat{x}\|^2 + \frac{1}{L}\|\lambda - \hat{\lambda}\|^2 \\ &\le \langle \xi, x - \hat{x}\rangle + \langle \eta, \lambda - \hat{\lambda}\rangle \le \|\xi\|\|x - \hat{x}\| + \|\eta\|\|\lambda - \hat{\lambda}\| \\ &\le \left(\frac{1}{\mu}\|\xi\|^2 + L\|\eta\|^2\right)^{1/2}\left(\mu\|x - \hat{x}\|^2 + \frac{1}{L}\|\lambda - \hat{\lambda}\|^2\right)^{1/2}.\end{aligned}$$

Therefore

$$\mu\|x - \hat{x}\|^2 + \frac{1}{L}\|\lambda - \hat{\lambda}\|^2 \le \frac{1}{\mu}\|\xi\|^2 + L\|\eta\|^2$$

which implies

$$\|x - \hat{x}\|^2 + \|\lambda - \hat{\lambda}\|^2 \le \frac{\max\{1/\mu, L\}}{\min\{\mu, 1/L\}}\left(\|\xi\|^2 + \|\eta\|^2\right).$$

Since $(\hat{x}, \hat{\lambda}) \in F^{-1}(0)$ and $(\xi, \eta) \in F(z)$ are arbitrary, we thus obtain

$$d(z, F^{-1}(0)) \le \kappa d(0, F(z)), \quad \forall z \in \mathbb{R}^n \times \mathbb{R}^m,$$

where $\kappa := \max\{1/\mu, L\}$. This shows that F is globally metrically subregular at $(\hat{z}, 0)$ for any $\hat{z} \in \mathcal{Z}^*$.

(ii) Consider next the case that both f and g are proper, lower semicontinuous, semi-algebraic, convex functions. According to Example 8.40, ∂g and ∂f^* are semi-algebraic. Thus, by definition, F is semi-algebraic. By Theorem 8.45, F is therefore bounded Hölder metrically subregular at any point in its graph. If both ∂f and ∂g are polyhedral, then $\partial f^* = (\partial f)^{-1}$ is polyhedral and so is F. By Theorem 8.23, F is bounded metrically subregular at any point in its graph.

We conclude this subsection by extending the primal-dual splitting algorithm to solve the convex minimization problem:

$$\min_{x \in \mathbb{R}^n} \{f(Ax) + g(x) + h(x)\},$$

where f and g are as above and $h : \mathbb{R}^n \to \mathbb{R}$ is a convex differentiable function with an L-Lipschitz continuous gradient ∇h. Note that this minimization problem is equivalent to the saddle point problem:

$$\min_{x \in \mathbb{R}^n} \sup_{\lambda \in \mathbb{R}^m} \left\{\langle Ax, \lambda\rangle - f^*(\lambda) + g(x) + h(x)\right\}. \tag{8.64}$$

Note that $\hat{z} = (\hat{x}, \hat{\lambda})$ is a saddle point of (8.64) if and only if $0 \in \tilde{F}(\hat{z})$, where $\tilde{F} : \mathbb{R}^n \times \mathbb{R}^m \rightrightarrows \mathbb{R}^n \times \mathbb{R}^m$ is defined by

$$\tilde{F}(z) := F(z) + (\nabla h(x), 0), \quad z = (x, \lambda) \in \mathbb{R}^n \times \mathbb{R}^m$$

with F being defined in (8.55). If we apply the primal-dual splitting algorithm (8.54) to solve it directly, we may produce an iterative sequence $\{z^k := (x^k, \lambda^k)\}$ with x^{k+1} being calculated from x^k and λ^k by solving the minimization problem:

$$\begin{aligned} x^{k+1} &= \text{prox}_{\tau(g+h)}(x^k - \tau A^T \lambda^k) \\ &= \arg\min_{x\in\mathbb{R}^n} \left\{ g(x) + h(x) + \frac{1}{2\tau}\|x - (x^k - \tau A^T \lambda^k)\|^2 \right\}. \end{aligned}$$

Due to the appearance of h, solving this problem can be difficult even if the proximal mapping of g is easy to calculate. To resolve this issue, we may decouple g and h by replacing $h(x)$ by its linearization $h(x^k) + \langle \nabla h(x^k), x - x^k\rangle$ at x^k and define x^{k+1} by the new updating rule:

$$\begin{aligned} x^{k+1} &= \arg\min_{x\in\mathbb{R}^n} \left\{ g(x) + h(x^k) + \langle \nabla h(x^k), x - x^k\rangle + \frac{1}{2\tau}\|x - (x^k - \tau A^T \lambda^k)\|^2 \right\} \\ &= \arg\min_{x\in\mathbb{R}^n} \left\{ g(x) + \frac{1}{2\tau}\|x - (x^k - \tau A^T \lambda^k - \tau \nabla h(x^k))\|^2 \right\}. \end{aligned}$$

Using this formula to replace the updating formula of x^{k+1} in (8.54), we can obtain the following primal-dual splitting algorithm [60, 219]:

$$\begin{aligned} x^{k+1} &= \text{prox}_{\tau g}(x^k - \tau(A^T \lambda^k + \nabla h(x^k))), \\ \bar{x}^{k+1} &= 2x^{k+1} - x^k, \\ \lambda^{k+1} &= \text{prox}_{\sigma f^*}(\lambda^k + \sigma A\bar{x}^{k+1}) \end{aligned} \tag{8.65}$$

for solving (8.64). With F and S defined as in (8.55), we can see that the sequence $\{z^k := (x^k, \lambda^k)\}$ defined by (8.65) satisfies

$$S(z^k - z^{k+1}) - (\nabla h(x^k), 0) \in F(z^{k+1}). \tag{8.66}$$

We may use the same argument in the above to analyze the algorithm (8.65). First, by using (8.66) and the monotonicity of F, we can obtain

$$0 \le \langle S(z^k - z^{k+1}), z^{k+1} - \hat{z}\rangle + \langle \nabla h(x^k) - \nabla h(\hat{x}), \hat{x} - x^{k+1}\rangle$$

for any saddle point $\hat{z} := (\hat{x}, \hat{\lambda})$ of (8.64). By the given property on h, we may use Proposition 2.34 to obtain

$$\begin{aligned} &\langle \nabla h(x^k) - \nabla h(\hat{x}), \hat{x} - x^{k+1}\rangle \\ &= \langle \nabla h(x^k) - \nabla h(\hat{x}), \hat{x} - x^k\rangle + \langle \nabla h(x^k) - \nabla h(\hat{x}), x^k - x^{k+1}\rangle \end{aligned}$$

$$\leq -\frac{1}{L}\|\nabla h(x^k) - \nabla h(\hat{x})\|^2 + \|\nabla h(x^k) - \nabla h(\hat{x})\|\|x^k - x^{k+1}\|$$
$$\leq \frac{L}{4}\|x^k - x^{k+1}\|^2.$$

Therefore

$$0 \leq \langle S(z^k - z^{k+1}), z^{k+1} - \hat{z}\rangle + \frac{L}{4}\|x^k - x^{k+1}\|^2$$
$$= \frac{1}{2}\|z^k - \frac{1}{2}\hat{z}\|_S^2 - \|z^{k+1} - \frac{1}{2}\hat{z}\|_S^2 - \|z^{k+1} - z^k\|_{\tilde{S}}^2, \tag{8.67}$$

where $\tilde{S} : \mathbb{R}^n \times \mathbb{R}^m \to \mathbb{R}^n \times \mathbb{R}^m$ is the linear self-adjoint operator defined by

$$\tilde{S}z = Sz - ((L/2)x, 0), \quad \forall z = (x, \lambda) \in \mathbb{R}^n \times \mathbb{R}^m.$$

It is easy to check if σ and τ are chosen such that $\sigma\tau\|A\|^2 < 1 - L\tau/2$ then $\tilde{S}$ is positive definite. Based on (8.67), we can follow the same argument in the proof of Theorem 8.54 to conclude that $z^k = (x^k, \lambda^k)$ converges to a saddle point $z^* = (x^*, \lambda^*)$ of (8.64). If in addition $\tilde{F}$ is Hölder metrically subregular at $(z^*, 0)$, we can derive the same convergence rate for z^k as in Theorem 8.54.

8.6 Appendix: Polyhedral Convex Sets

In this appendix, we will collect some basic facts on polyhedral convex sets; for more information, please refer to [191]. Recall that a set D in $\mathbb{R}^n$ is called polyhedral convex if there exist a matrix A of size $m \times n$ and a vector $b \in \mathbb{R}^m$ such that $D = \{x \in \mathbb{R}^n : Ax \leq b\}$. In case $b = 0$, then D is a polyhedral convex cone. In the following, we will show that the image of polyhedral convex sets under affine transforms are polyhedral convex and the Minkowski sum of polyhedral convex sets are again polyhedral convex. We start with the following result on polyhedral convex sets under the canonical projections.

Theorem 8.56 *The canonical projections of polyhedral convex sets are still polyhedral convex.*

Proof Let $D = \{x \in \mathbb{R}^n : Ax \leq b\}$ be a polyhedral convex set, where $A = (a_{ij})$ is an $m \times n$ matrix and $b \in \mathbb{R}^m$. It suffices to show that under the canonical projection $\pi : \mathbb{R}^n \to \mathbb{R}^{n-1}$ given by $\pi(x_1, \cdots, x_{n-1}, x_n) = (x_1, \cdots, x_{n-1})$, the set $\pi(D)$ is polyhedral convex. We use the Fourier-Motzkin elimination. Let

$$I_+ := \{i : a_{in} > 0\}, \quad I_0 := \{i : a_{in} = 0\}, \quad I_- := \{i : a_{in} < 0\}$$

and let a_i denote the i-th column of A^T. We need to eliminate the variable x_n in the inequalities $\langle a_i, x\rangle \le b_i$ for $i \in I_+ \cup I_-$. Taking $i \in I_+$ and $l \in I_-$, multiplying $\langle a_l, x\rangle \le b_l$ by a_{in} and $\langle a_i, x\rangle \le b_i$ by $-a_{ln}$, and then adding them gives

$$a_{in}\langle a_l, x\rangle - a_{ln}\langle a_i, x\rangle \le a_{in}b_l - a_{ln}b_i. \tag{8.68}$$

Note that Eq. (8.68) does not contain the variable x_n. Now, we define D' to be the set consisting of those $x' = (x_1, \cdots, x_{n-1}) \in \mathbb{R}^{n-1}$ satisfying (8.68) for all $(i, l) \in I_+ \times I_-$ and

$$\sum_{j=1}^{n-1} a_{ij}x_i \le b_i, \quad \forall i \in I_0.$$

Clearly, D' is a polyhedral convex set. It remains only to show $\pi(D) = D'$. For any $x' \in \pi(D)$, there exists an x_n such that $(x', x_n) \in D$. Thus, by the above elimination procedure, we have $x' \in D'$. Consequently, $\pi(D) \subset D'$. To see $D' \subset \pi(D)$, let $x' \in D'$ be any element. We need to find an x_n such that $(x', x_n) \in D$ which equivalently requires

$$\sum_{j=1}^{n-1} a_{ij}x_j + a_{in}x_n \le b_i, \quad \forall i \in I_+ \quad \text{and} \quad \sum_{j=1}^{n-1} a_{lj}x_j + a_{ln}x_n \le b_l, \quad \forall l \in I_-,$$

i.e.,

$$\max_{l\in I_-}\left\{\frac{1}{a_{ln}}\left(b_l - \sum_{j=1}^{n-1} a_{lj}x_j\right)\right\} \le x_n \le \min_{i\in I_+}\left\{\frac{1}{a_{in}}\left(b_i - \sum_{j=1}^{n-1} a_{ij}x_j\right)\right\}.$$

Because x' satisfies (8.68) for all $(i, l) \in I_+ \times I_-$, such an x_n does exist and therefore $D' \subset \pi(D)$. □

Definition 8.57 A *finitely generated cone* in $\mathbb{R}^n$ is a set of the form $\{B\lambda : \lambda \ge 0\}$, where B is an $n \times p$ matrix.

Theorem 8.58 (Weyl's Theorem) *Any finitely generated cone is a polyhedral convex cone.*

Proof Let $C = \{B\lambda : \lambda \ge 0\}$ be a finitely generated cone in $\mathbb{R}^n$, where B is an $n \times p$ matrix. Then

$$\begin{aligned} C &= \{x \in \mathbb{R}^n : x = B\lambda \text{ for some } \lambda \in \mathbb{R}^p \text{ with } \lambda \ge 0\} \\ &= \{x \in \mathbb{R}^n : \exists \lambda \in \mathbb{R}^p \text{ such that } x - B\lambda \le 0,\ -x + B\lambda \le 0,\ -\lambda \le 0\} \end{aligned}$$

which means that C is the canonical projection of the polyhedral convex cone

$$C' = \{(x, \lambda) \in \mathbb{R}^n \times \mathbb{R}^p : x - B\lambda \leq 0, -x + B\lambda \leq 0, -\lambda \leq 0\}$$

into $\mathbb{R}^n$. Therefore, it follows from Theorem 8.56 that C is a polyhedral convex cone. □

According to Theorem 8.58, we can see immediately that any finitely generated cone is a closed set. This provides an alternative proof of Lemma 5.31.

As a direct consequence of Weyl's theorem, we can obtain the following useful Farkas lemma.

Theorem 8.59 (Farkas Lemma) *Given $B \in \mathbb{R}^{m\times n}$ and $b \in \mathbb{R}^m$. Then exactly one of the following can occur:*

(i) *There exists $\lambda \in \mathbb{R}^m$ with $\lambda \geq 0$ such that $B\lambda = b$.*
(ii) *There exists $a \in \mathbb{R}^n$ such that $B^T a \leq 0$ and $\langle a, b\rangle > 0$.*

Proof These two cases cannot happen simultaneously since otherwise we would have

$$0 < \langle a, b\rangle = \langle a, B\lambda\rangle = \langle B^T a, \lambda\rangle \leq 0.$$

Next, we show that one of them must occur. Let $C = \{B\lambda : \lambda \geq 0\}$. By Theorem 8.58. we have $C = \{x : Ax \leq 0\}$ for some matrix A. If (i) does not happen, then $b \notin C$, and hence, A^T must have a column a such that $\langle a, b\rangle > 0$. Since each column b_j of B is in C, we have $\langle a, b_j\rangle \leq 0$, and thus, $B^T a \leq 0$. This shows that (ii) does occur. □

Next, we will show the converse of Weyl's theorem, i.e., every polyhedral convex cone is a finitely generated cone. To this end, we introduce the polar or dual of cones.

Definition 8.60 The *polar or dual* of a cone $C \subset \mathbb{R}^n$ is the set

$$C^* := \{\xi \in \mathbb{R}^n : \langle \xi, x\rangle \leq 0 \text{ for all } x \in C\}.$$

Given two cones $C, D \subset \mathbb{R}^n$, it is easy to see that $C \subset D$ implies $D^* \subset C^*$. Furthermore

$$C \subset C^{**} \quad \text{and} \quad C^* = C^{***}. \tag{8.69}$$

Indeed, for any $x \in C$, we have by the definition of C^* that $\langle \xi, x\rangle \leq 0$ for all $\xi \in C^*$ and thus $x \in C^{**}$; this shows $C \subset C^{**}$. Applying this established result to C^* gives $C^* \subset C^{***}$. Conversely, $\xi \in C^{***}$ implies $\langle \xi, x\rangle \leq 0$ for all $x \in C^{**}$. Since $C \subset C^{**}$, we have $\langle \xi, x\rangle \leq 0$ for all $x \in C$ and thus $\xi \in C^*$. This shows $C^{***} \subset C^*$ and thus $C^* = C^{***}$.

For polyhedral convex cones and finitely generated cones, there hold more precise statements related to their polar cones.

Lemma 8.61 *Let A and B be matrices of size $m \times n$ and $n \times p$, respectively.*

(i) *If $C = \{B\lambda : \lambda \geq 0\}$, then $C^* = \{x : B^T x \leq 0\}$ and $C^{**} = C$.*
(ii) *If $D = \{x : Ax \leq 0\}$, then $D^* = \{A^T\lambda : \lambda \geq 0\}$ and $D^{**} = D$.*

Proof

(i) Note that

$$\begin{aligned} x \in C^* &\iff \langle x, B\lambda\rangle \leq 0 \text{ for all } \lambda \geq 0 \\ &\iff \langle B^T x, \lambda\rangle \leq 0 \text{ for all } \lambda \geq 0 \\ &\iff B^T x \leq 0. \end{aligned}$$

Therefore, $C^* = \{x : B^T x \leq 0\}$.

Next, we show $C^{**} = C$. By (8.69), we have $C \subset C^{**}$. To show $C^{**} \subset C$, let $b \notin C$ be any element. By Farkas lemma, there is a such that $\langle a, b\rangle > 0$ and $B^T a \leq 0$. Note that $a \in C^*$ and thus $b \notin C^{**}$.

(ii) Let $C = \{A^T\lambda : \lambda \geq 0\}$. By (i), we have $C^* = D$ and hence $D^* = C^{**} = C$. Consequently, $D^{**} = C^* = D$.

□

Theorem 8.62 (Minkowski Theorem) *Any polyhedral convex cone is a finitely generated cone.*

Proof If D is a polyhedral convex cone, then, by (ii) in Lemma 8.61, $D^{**} = D$ and D^* is a finitely generated cone. By Weyl's theorem, D^* is also a polyhedral convex cone. By using again (ii) in Lemma 8.61, D^{**} is a finitely generated cone. This completes the proof as $D = D^{**}$. □

Combining Theorems 8.58 and 8.62, it follows that a set $D \subset \mathbb{R}^n$ is a polyhedral convex cone if and only if it is a finitely generated cone. The following result extends this equivalent statement for general polyhedral convex sets.

Theorem 8.63 (Minkowski-Weyl Duality) *A set $D \subset \mathbb{R}^n$ is a polyhedral convex set if and only if*

$$D = \left\{ B\lambda + C\mu : \lambda, \mu \geq 0 \text{ and } \sum_i \lambda_i = 1 \right\} \tag{8.70}$$

for some suitable matrices B and C.

Proof Let $D \subset \mathbb{R}^n$ be a polyhedral convex set and write $D = \{x : Ax \leq b\}$, where A is an $m \times n$ matrix and $b \in \mathbb{R}^m$. Consider the set:

$$Q := \left\{ \begin{pmatrix} x \\ t \end{pmatrix} : Ax \leq tb,\ t \geq 0 \right\} = \left\{ \begin{pmatrix} x \\ t \end{pmatrix} : \begin{pmatrix} A & -b \\ 0 & -1 \end{pmatrix} \begin{pmatrix} x \\ t \end{pmatrix} \leq 0 \right\}$$

in $\mathbb{R}^n \times \mathbb{R}$ which is called the *homogenization* of D. Clearly, Q is a polyhedral convex cone and

$$x \in D \iff \begin{pmatrix} x \\ 1 \end{pmatrix} \in Q. \tag{8.71}$$

According to Theorem 8.62, Q is a finitely generated cone, i.e., there is a matrix M of size $(n+1) \times q$ such that $Q = \{M\eta : \eta \geq 0\}$. Let β^T be the last row of M. Then, by the definition of Q, $\langle \beta, \eta \rangle \geq 0$ for all $\eta \geq 0$ which implies $\beta \geq 0$. By reordering and scaling the columns of M, which does not change the set Q, we may assume $\beta^T = (\mathbf{1}^T, \mathbf{0}^T)$, where $\mathbf{1}$ is a vector with all entries equal 1, similarly for $\mathbf{0}$. Decomposing M and splitting η accordingly as

$$M = \begin{pmatrix} B & C \\ \mathbf{1}^T & \mathbf{0}^T \end{pmatrix} \quad \text{and} \quad \eta = \begin{pmatrix} \lambda \\ \mu \end{pmatrix}$$

gives

$$Q = \left\{ \begin{pmatrix} B\lambda + C\mu \\ \sum_i \lambda_i \end{pmatrix} : \lambda, \mu \geq 0 \right\}.$$

Based on (8.71), we consequently obtain

$$D = \left\{ B\lambda + C\mu : \lambda, \mu \geq 0 \text{ and } \sum_i \lambda_i = 1 \right\}.$$

Conversely, let D be the set defined by (8.70), and we will show D is a polyhedral convex set. Consider the set:

$$P := \left\{ \begin{pmatrix} B\lambda + C\mu \\ \sum_i \lambda_i \end{pmatrix} : \lambda, \mu \geq 0 \right\} = \left\{ \begin{pmatrix} B & C \\ \mathbf{1}^T & \mathbf{0}^T \end{pmatrix} \begin{pmatrix} \lambda \\ \mu \end{pmatrix} : \lambda, \mu \geq 0 \right\}$$

which is clearly a finitely generated cone in $\mathbb{R}^{n+1}$. By virtue of Theorem 8.58, P is a polyhedral convex cone, i.e., there is a matrix A' such that

$$P = \left\{ \begin{pmatrix} x \\ t \end{pmatrix} : A' \begin{pmatrix} x \\ t \end{pmatrix} \leq 0 \right\}.$$

Write $A' = (A \ \ -b)$, where $-b$ denotes the last column of A', then

$$P = \left\{ \begin{pmatrix} x \\ t \end{pmatrix} : Ax - tb \leq 0 \right\}.$$

Note that $x \in D$ if and only if $\binom{x}{1} \in P$. Consequently, $D = \{x : Ax \leq b\}$, i.e., D is a polyhedral convex set. □

Theorem 8.64

(i) *Images of polyhedral convex sets under affine transforms are polyhedral convex.*
(ii) *Minkowski sums of polyhedral convex sets are polyhedral convex.*

Proof

(i) Let D be a polyhedral convex set. By Theorem 8.63, we can write D in the form (8.70). Let T be an affine transform defined by $Tx := Mx + \alpha$ with a suitable matrix M and a vector α. Then

$$T(D) = \left\{MB\lambda + MC\mu + \alpha : \lambda, \mu \geq 0 \text{ and } \sum_i \lambda_i = 1\right\}$$
$$= \left\{(MB + A)\lambda + MC\mu : \lambda, \mu \geq 0 \text{ and } \sum_i \lambda_i = 1\right\},$$

where A is a matrix whose every column is α. This together with Theorem 8.63 shows that $T(D)$ is a polyhedral convex set.

(ii) Let C and D be polyhedral convex sets in $\mathbb{R}^n$. Define $Q := \{(x, y) : x \in C, y \in D\}$. Clearly, Q is a polyhedral convex set. Note that $C + D = T(Q)$, where $T : \mathbb{R}^n \times \mathbb{R}^n \to \mathbb{R}^n$ is the linear transform defined by $T(x, y) := x + y$. Thus, by (i), $C + D$ is polyhedral convex.

□

Problems

8.1 Let $f : \mathbb{R}^n \to \mathbb{R}$ be a differentiable convex function with a non-empty set of minimizers S and set $f_* := \min_{x\in\mathbb{R}^n} f(x)$. We say f is restricted strongly convex with constant $\nu > 0$ if it satisfies the restricted secant inequality

$$\langle \nabla f(x), x - P_S(x)\rangle \geq \nu d^2(x, S), \quad \forall x \in \mathbb{R}^n,$$

where $P_S(x)$ denotes the Euclidean projection of x onto S. The notion of restricted strong convexity was introduced in [130] and was further studied and used in [199, 232], for instance. Show that if f is restricted strongly convex with $\nu > 0$, then f satisfies the quadratic growth condition:

$$d^2(x, S) \leq \gamma\,(f(x) - f_*), \quad \forall x \in \mathbb{R}^n \tag{8.72}$$

with $\gamma := 2/\nu$. Conversely, if f satisfies the quadratic growth condition (8.72) with $\gamma > 0$, show that f must be restricted strongly convex with constant $\nu := 1/\gamma$.

8.2 Let $A \in \mathbb{R}_+^{m\times n}$ be an $m \times n$ matrix with nonnegative entries and nonzero rows and let $b \in \mathbb{R}_{++}^m$ be a vector with positive components. Consider the minimization problem:

$$\min_{x\in\mathbb{R}^n} \left\{ F(x) := h(Ax) + \delta_{\mathbb{R}_+^n}(x) \right\} \tag{8.73}$$

where $h : \mathbb{R}^m \to (-\infty, \infty]$ is the function defined by

$$h(y) = \begin{cases} \sum_{i=1}^m \left(b_i \log \dfrac{b_i}{y_i} - b_i + y_i \right) & \text{if } y \in \mathbb{R}_{++}^m, \\ +\infty & \text{otherwise.} \end{cases}$$

This problem arises from recovering a signal x from the measurement b of Ax corrupted by Poisson noise; see, for instance, [24, 64, 218].

(i) Show that (8.73) has an optimal solution.
(ii) Let S denote the set of all optimal solutions of (8.73) and let $\bar{x} \in S$. Show that

$$S = \{u \in \mathbb{R}^n : Au = A\bar{x},\ \langle \nabla f(\bar{x}), u\rangle = 0 \text{ and } u \in \mathbb{R}_+^n\}.$$

(iii) Given $\bar{x} \in S$. Use the Hoffman's error bound result to show that F satisfies the quadratic growth condition at $\bar{x}$ in the sense that for any $R > 0$ there is a constant $\nu > 0$ such that

$$F(x) - F(\bar{x}) \geq \nu d^2(x, S), \quad \forall x \in B_R(\bar{x}).$$

8.3 Let $\Phi : \mathbb{R}^m \rightrightarrows \mathbb{R}^n$ be a multifunction with closed values and let $D \subset \operatorname{dom}(\Phi)$ be a non-empty set. Show that Φ is Lipschitz continuous relative to D with constant κ if and only if

$$d(x, \Phi(y)) \leq \kappa d(y, \Phi^{-1}(x) \cap D)$$

for all $x \in \mathbb{R}^n$ and $y \in D$.

8.4 A multifunction $F : \mathbb{R}^m \rightrightarrows \mathbb{R}^n$ is called calm at $(\bar{x}, \bar{y}) \in \operatorname{Gr}(F)$ if there exist $\kappa > 0$ and $r > 0$ such that

$$d(y, F(\bar{x})) \leq \kappa d(\bar{x}, F^{-1}(y)), \quad \forall y \in B_r(\bar{y}).$$

It is clear that F is calm at $(\bar{x}, \bar{y}) \in \mathrm{Gr}(F)$ if and only if F^{-1} is metrically subregular at $(\bar{y}, \bar{x}) \in \mathrm{Gr}(F^{-1})$. Given a multifunction $F : \mathbb{R}^m \rightrightarrows \mathbb{R}^n$ and $(\bar{x}, \bar{y}) \in \mathrm{Gr}(F)$, show that the following statements are equivalent:

(i) F is calm at $(\bar{x}, \bar{y})$.
(ii) There exist $\kappa_1 > 0$ and $r_1 > 0$ such that

$$d(y, F(\bar{x})) \leq \kappa_1 d(\bar{x}, F^{-1}(y) \cap B_{r_1}(\bar{x})), \quad \forall y \in B_{r_1}(\bar{y}).$$

(iii) There exist $\kappa_2 > 0$ and $r_2 > 0$ such that

$$F(x) \cap B_{r_2}(\bar{y}) \subset F(\bar{x}) + \kappa_2 \|x - \bar{x}\| B_1(0), \quad \forall x \in B_{r_2}(\bar{x}).$$

(iv) There exist $\kappa_3 > 0$ and $r_3 > 0$ such that

$$F(x) \cap B_{r_3}(\bar{y}) \subset F(\bar{x}) + \kappa_3 \|x - \bar{x}\| B_1(0), \quad \forall x \in \mathbb{R}^m.$$

Here, $B_1(0)$ denotes the unit ball in $\mathbb{R}^n$ with center at the origin.

8.5 Let $f : \mathbb{R}^n \to (-\infty, \infty]$ be a convex function. Set $\Gamma := \mathbb{R}^n \times \mathbb{R} \times \mathbb{R}^n \times \{-1\}$ and let $T : \mathbb{R}^n \times \mathbb{R} \times \mathbb{R}^n \times \mathbb{R} \to \mathbb{R}^n \times \mathbb{R}^n$ be the linear map defined by $T(x, a, \xi, b) = (x, \xi)$. Show that

$$\mathrm{Gr}(\partial f) = G := T\left(\mathrm{Gr}(\delta_{\mathrm{epi}(f)}) \cap \Gamma\right),$$

where $\delta_{\mathrm{epi}(f)}$ denotes the indicator function of the epigraph $\mathrm{epi}(f)$ of f.

8.6 Let C be a non-empty polyhedral convex set in $\mathbb{R}^n$ and let δ_C denote the indicator function of C. Show that $\partial \delta_C$ is a polyhedral multifunction.

8.7 Let $f : \mathbb{R}^n \to (-\infty, \infty]$ be a convex polyhedral function. Show that ∂f is a polyhedral multifunction.

8.8 Show the following properties of semi-algebraic sets:

(i) If A and B are semi-algebraic sets in $\mathbb{R}^n$, so are $A \cap B$, $A \cup B$, and $A \setminus B$.
(ii) If $A \subset \mathbb{R}^m$ and $B \subset \mathbb{R}^n$ are semi-algebraic, then $A \times B \subset \mathbb{R}^{m \times n}$ is also semi-algebraic.
(iii) Every semi-algebraic set in $\mathbb{R}$ is a finite union of points and intervals.

8.9 Let U be an open semi-algebraic set in $\mathbb{R}^n$ and let $f : U \to \mathbb{R}$ be a semi-algebraic function. Show that, if f admits a partial derivative $\partial f / \partial x_i$ on U, then this derivative is semi-algebraic.

8.10 In Sect. 8.5.3, we developed the algorithm (8.65) for solving the saddle point problem (8.64). By interchanging the roles of the primal and dual variables, from (8.65), we obtain the following algorithm:

$$\begin{aligned}
\lambda^{k+1} &= \text{prox}_{\sigma f^*}(\lambda^k + \sigma A x^k),\\
\bar{\lambda}^{k+1} &= 2\lambda^{k+1} - \lambda^k,\\
x^{k+1} &= \text{prox}_{\tau g}(x^k - \tau(A^T \bar{\lambda}^{k+1} + \nabla h(x^k)))
\end{aligned}$$

which also can solve (8.64). Adapt the argument from Sect. 8.5.3 to show that the sequence $z^k := (x^k, \lambda^k)$ converges to a saddle point $z^* = (x^*, \lambda^*)$ of (8.64). If the multifunction $\tilde{F} : \mathbb{R}^n \times \mathbb{R}^m \rightrightarrows \mathbb{R}^n \times \mathbb{R}^m$ defined by

$$\tilde{F}(z) := (A^T\lambda + \nabla h(x) + \partial g(x), -Ax + \partial f^*(\lambda)), \quad z = (x, \lambda) \in \mathbb{R}^n \times \mathbb{R}^m$$

is Hölder metric subregular at $(z^*, 0) \in \text{Gr}(\tilde{F})$, i.e., there exist $r > 0$, $\alpha \in (0, 1]$ and $\kappa > 0$ such that

$$d(z, \tilde{F}^{-1}(0)) \leq \kappa \left[d(0, \tilde{F}(z))\right]^\alpha, \quad \forall z \in B_r(z^*),$$

show that for $\alpha = 1$ there exist constants $C > 0$ and $0 < q < 1$ such that

$$\|z^{k+1} - z^*\|^2 \leq C q^k, \quad \forall k \geq 0$$

while for $0 < \alpha < 1$ there is a constant $C > 0$ such that

$$\|z^k - z^*\|^2 \leq C(1 + k)^{-\frac{\alpha}{1-\alpha}}, \quad \forall k \geq 0.$$

Chapter 9
Optimization with Kurdyka-Łojasiewicz Property

In Sect. 8.1.3, we introduced the Kurdyka-Łojasiewicz (KL) condition for convex functions. In this chapter, we extend the KL condition to non-convex functions and apply it to analyze algorithms for solving non smooth optimization problems where the objective functions may lack convexity. Additionally, we present numerous examples of functions that satisfy the KL condition. In particular, we demonstrate that semi-algebraic functions belong to the class of KL functions.

9.1 Limiting Subdifferential

Recall that a function $f : \mathbb{R}^n \to \mathbb{R}$ is differentiable at a point $x \in \mathbb{R}^n$ if there is a vector $\xi \in \mathbb{R}^n$, called the gradient of f at x, such that

$$\lim_{y \neq x, y \to x} \frac{f(y) - f(x) - \langle \xi, y - x \rangle}{\|y - x\|} = 0.$$

At a point where f is not differentiable, no gradient is available. In order to formulate the KL conditions for non-smooth functions, we need to introduce an object as a replacement of the gradient. This has been done for convex functions in Chap. 3 where convex subgradient and subdifferential have been introduced. It is natural to expect that such an object exists for non-smooth non-convex functions which not only plays the role of a gradient but also enjoys some properties of subdifferentials for convex functions. This leads us to introduce the following definition [156, 194].

Q. Jin, *Lectures on Nonsmooth Optimization*, Texts in Applied Mathematics 82,
https://doi.org/10.1007/978-3-031-91417-1_9

Definition 9.1 Let $f : \mathbb{R}^n \to (-\infty, \infty]$ be a proper function, and let $x \in \text{dom}(f)$.

(i) A vector $\xi \in \mathbb{R}^n$ is called a *Fréchet subgradient* of f at x if

$$\liminf_{y \neq x, y \to x} \frac{f(y) - f(x) - \langle \xi, y - x \rangle}{\|y - x\|} \geq 0.$$

The set of all Fréchet subgradients of f at x is denoted by $\hat{\partial} f(x)$ which is called the *Fréchet subdifferential* of f at x. When $x \notin \text{dom}(f)$, we set $\hat{\partial} f(x) = \emptyset$.

(ii) A vector $\xi \in \mathbb{R}^n$ is called a *limiting subgradient* or simply *subgradient* of f at $x \in \mathbb{R}^n$ if there exist sequences $\{x^k\}$ and $\{\xi^k\}$ in $\mathbb{R}^n$ with the properties

$$\xi^k \in \hat{\partial} f(x^k), \quad x^k \to x, \quad f(x^k) \to f(x) \quad \text{and} \quad \xi^k \to \xi$$

as $k \to \infty$. The set of all subgradients of f at x, denoted by $\partial f(x)$, is called the *limiting subdifferential* or simply the *subdifferential* of f at x. We set

$$\text{dom}(\partial f) := \{x \in \mathbb{R}^n : \partial f(x) \neq \emptyset\}.$$

According to the definition of Fréchet subgradient and the property of lim inf, it is easily seen that $\xi \in \hat{\partial} f(x)$ if and only if for any $\varepsilon > 0$ there is a neighborhood U of x such that

$$f(y) - f(x) - \langle \xi, y - x \rangle \geq -\varepsilon \|y - x\|, \quad \forall y \in U.$$

We will use this equivalent statement frequently in the following.

Example 9.2 If f is differentiable at x, then $\hat{\partial} f(x) = \{\nabla f(x)\}$. Indeed, by the differentiability of f, we immediately see that $\nabla f(x) \in \hat{\partial} f(x)$. Conversely, let $\xi \in \hat{\partial} f(x)$. Then, by the differentiability of f at x, for any $\varepsilon > 0$, we have for y sufficiently close to x that

$$-\varepsilon \|y - x\| \leq f(y) - f(x) - \langle \xi, y - x \rangle = o(\|y - x\|) + \langle \nabla f(x) - \xi, y - x \rangle.$$

Therefore there exists a number $r > 0$ such that

$$\langle \nabla f(x) - \xi, y - x \rangle \geq -2\varepsilon \|y - x\|, \quad \forall y \in B_r(x).$$

Consequently $\langle \nabla f(x) - \xi, u \rangle \geq -2\varepsilon$ for all vectors u satisfying $\|u\| = 1$. Since $\varepsilon > 0$ is arbitrary, we have $\langle \nabla f(x) - \xi, u \rangle \geq 0$ for all u with $\|u\| = 1$ which in turn implies $\nabla f(x) - \xi = 0$. Thus $\hat{\partial} f(x) = \{\nabla f(x)\}$. Therefore, if f is continuously differentiable around x, then $\partial f(x) = \hat{\partial} f(x) = \{\nabla f(x)\}$. □

Example 9.3 Consider the function $f(x) = |x|^q$ on $\mathbb{R}$ with $0 < q < 1$. Note that f is continuous differentiable everywhere expect $x = 0$. Thus

$$\partial f(x) = \hat{\partial} f(x) = \{q\text{sign}(x)|x|^{q-1}\}, \quad x \neq 0.$$

Next we determine $\hat{\partial} f(0)$. Since $0 < q < 1$, for any $\xi \in \mathbb{R}$, there holds $f(x) - f(0) - \xi(x-0) = |x|^q - \xi x \geq 0$ for sufficiently small x and thus $\xi \in \hat{\partial} f(0)$. Since $\xi \in \mathbb{R}$ is arbitrary, we have $\hat{\partial} f(0) = \mathbb{R}$ and hence $\partial f(0) = \mathbb{R}$. □

Lemma 9.4 *The subdifferential ∂f is closed in the sense that if $\{x^k\}$ and $\{\xi^k\}$ are such that*

$$x^k \to x, \quad \xi^k \to \xi, \quad f(x^k) \to f(x) \quad \textit{and} \quad \xi^k \in \partial f(x^k),$$

then $\xi \in \partial f(x)$.

Proof Since $\xi^k \in \partial f(x^k)$, by definition we can find $\hat{x}^k$ and $\hat{\xi}^k \in \hat{\partial} f(\hat{x}^k)$ such that

$$\|\hat{x}^k - x^k\| + \|\hat{\xi}^k - \xi^k\| + |f(\hat{x}^k) - f(x^k)| \leq \frac{1}{k}.$$

Note that

$$\begin{aligned}\|\hat{x}^k - x\| + \|\hat{\xi}^k - \xi\| + |f(\hat{x}^k) - f(x)| \leq \|\hat{x}^k - x^k\| &+ \|\hat{\xi}^k - \xi^k\| \\ &+ |f(\hat{x}^k) - f(x^k)| \\ &+ \|x^k - x\| + \|\xi^k - \xi\| \\ &+ |f(x^k) - f(x)|.\end{aligned}$$

We have $\hat{x}^k \to x$, $\hat{\xi}^k \to \xi$ and $f(\hat{x}^k) \to f(x)$. Moreover, $\hat{\xi}^k \in \hat{\partial} f(\hat{x}^k)$. By definition, we therefore have $\xi \in \partial f(x)$. □

In Definition 9.1, we use the same notation $\partial f(x)$ for limiting subdifferential as we have used for the subdifferential of convex functions; see Definition 3.1. This does not cause any confusion because for a proper lower semicontinuous convex function $f : \mathbb{R}^n \to (-\infty, \infty]$, its subdifferential defined in Definition 3.1 is identically equal to the one defined in Definition 9.1. In fact, for any $\xi \in \partial f(x)$ defined by Definition 3.1, we have

$$f(y) - f(x) - \langle \xi, y - x\rangle \geq 0 \quad \text{for all } y \in \mathbb{R}^n$$

which immediately implies $\xi \in \hat{\partial} f(x)$. Conversely, let $\xi \in \hat{\partial} f(x)$. Then for any $\varepsilon > 0$, there is a number $r > 0$ such that

$$f(z) - f(x) - \langle \xi, z - x\rangle \geq -\varepsilon\|z - x\|, \quad \forall z \in B_r(x).$$

For any $y \in \mathbb{R}^n$, we have $z := x + t(y - x) \in B_r(x)$ for sufficiently small $t > 0$. Thus, by the convexity of f, we have for sufficiently small $t > 0$ that

$$\begin{aligned} -\varepsilon t\|y - x\| &\leq f(x + t(y - x)) - f(x) - t\langle \xi, y - x\rangle \\ &\leq tf(y) - tf(x) - t\langle \xi, y - x\rangle \end{aligned}$$

which implies

$$f(y) - f(x) - \langle \xi, y - x\rangle \geq -\varepsilon\|y - x\|, \quad \forall y \in \mathbb{R}^n.$$

Since $\varepsilon > 0$ can be arbitrarily small, we must have $f(y) - f(x) - \langle \xi, y - x\rangle \geq 0$ for all $y \in \mathbb{R}^n$. Therefore $\xi \in \partial f(x)$ in the sense of Definition 3.1.

Recall that the lower semi-continuity of f implies that the subdifferential ∂f in the sense of Definition 3.1 is closed by taking limit. Thus, we may use the above fact and Lemma 9.4 to conclude that the two subdifferentials defined in Definitions 3.1 and 9.1 are identically equal for proper lower semicontinuous convex functions.

Lemma 9.5 *Let $f : \mathbb{R}^n \to (-\infty, \infty]$ be a proper function. If x is a local minimizer of f, then $0 \in \partial f(x)$.*

Proof Since x is a local minimizer, we have $f(y) \geq f(x)$ and thus $f(y) - f(x) - \langle 0, y - x\rangle \geq 0$ for all y in a neighborhood of x. By the definition of Fréchet subdifferential, $0 \in \hat{\partial} f(x)$ and consequently $0 \in \hat{\partial} f(x) \subset \partial f(x)$. □

It is known that the converse of Lemma 9.5 is true for convex functions. However, the converse is not necessarily true for non-convex functions.

Definition 9.6 Let $f : \mathbb{R}^n \to (-\infty, \infty]$ be a proper function. A point $x \in \mathbb{R}^n$ with $0 \in \partial f(x)$ is called a *limiting critical point* or simply *critical point* of f.

Example 9.7 Consider the counting function $f(x) = \|x\|_0$ on $\mathbb{R}^n$. It is clear that every point in $\mathbb{R}^n$ is a local minimizer of f and thus every point is a critical point. We now determine the subdifferential of f by showing that

$$\partial f(x) = \hat{\partial} f(x) = \{\xi \in \mathbb{R}^n : \xi_i = 0 \text{ for } i \in \text{supp}(x)\}$$

for any $x \in \mathbb{R}^n$, where supp(x) denotes the support of x, i.e.,

$$\text{supp}(x) := \{i \in \{1, \cdots, n\} : x_i \neq 0\}.$$

To see this, we first show that if $\xi \in \hat{\partial} f(x)$, then $\xi_i = 0$ for $i \in \text{supp}(x)$. By the definition of $\hat{\partial} f(x)$, we have

$$0 \leq \liminf_{y \neq x, y \to x} \frac{\|y\|_0 - \|x\|_0 - \langle \xi, y - x\rangle}{\|y - x\|}.$$

For any fixed $i \in \operatorname{supp}(x)$, we take $y \in \mathbb{R}^n$ such that $y_j = 0$ for $j \notin \operatorname{supp}(x)$, $y_j = x_j$ for $j \in \operatorname{supp}(x) \setminus \{i\}$ and $y_i = x_i + t$. Then $\|y\|_0 = \|x\|_0$ for sufficiently small t. Consequently $0 \le \liminf_{t\to 0} t\xi_i/|t|$ which implies $\xi_i = 0$.

Conversely, for any $\xi \in \mathbb{R}^n$ with $\xi_i = 0$ for $i \in \operatorname{supp}(x)$, we show $\xi \in \hat{\partial} f(x)$ by proving that for any y sufficiently close to x, there holds

$$\|y\|_0 - \|x\|_0 - \langle \xi, y - x\rangle \ge 0. \tag{9.1}$$

We can find a number $r > 0$ such that $\operatorname{supp}(x) \subset \operatorname{supp}(y)$ and $\langle \xi, y - x\rangle \le 1$ for all $y \in B_r(x)$. Let $y \in B_r(x)$ be any element. If $\operatorname{supp}(y) = \operatorname{supp}(x)$, then $\|y\|_0 = \|x\|_0$ and $\langle \xi, y - x\rangle = 0$. Thus (9.1) holds trivially. If $\operatorname{supp}(y) \ne \operatorname{supp}(x)$, then $\|y\|_0 - \|x\|_0 \ge 1$ and thus $\|y\|_0 - \|x\|_0 - \langle \xi, y - x\rangle \ge 1 - \langle \xi, y - x\rangle \ge 0$. This shows (9.1) again.

We thus obtain the expression of $\hat{\partial} f(x)$ for any $x \in \mathbb{R}^n$. By this characterization, it is readily to see that $\partial f(x) = \hat{\partial} f(x)$. □

In the following, we will give some calculus rules for subdifferentials.

Proposition 9.8 *Let $H : \mathbb{R}^{n_1} \times \cdots \times \mathbb{R}^{n_\ell} \to \mathbb{R}$ be continuous differentiable. For each $i = 1, \cdots, \ell$, let $f_i : \mathbb{R}^{n_i} \to (-\infty, \infty]$ be proper lower semicontinuous. Consider the function*

$$\Phi(x_1, \cdots, x_\ell) := H(x_1, \cdots, x_\ell) + \sum_{i-1}^{\ell} f_i(x_i).$$

Then

$$\partial\Phi(x_1, \cdots, x_\ell) = \prod_{i=1}^{\ell} (\nabla_i H(x_1, \cdots, x_\ell) + \partial f_i(x_i)) \tag{9.2}$$

for all $(x_1, \cdots, x_\ell) \in \mathbb{R}^{n_1} \times \cdots \times \mathbb{R}^{n_\ell}$, where $\nabla_i H$ denotes the partial gradient of H with respect to the variable x_i.

Proof Let $x := (x_1, \cdots, x_\ell)$ be fixed. We first show that the set on the left-hand side of (9.2) is contained in the set on the right-hand side. To see this, let $\xi := (\xi_1, \cdots, \xi_\ell) \in \partial\Phi(x_1, \cdots, x_\ell)$ with $\xi_i \in \mathbb{R}^{n_i}$ for $i = 1, \cdots, \ell$. By definition, there exist sequences $\{x^k := (x_1^k, \cdots, x_\ell^k)\}$ and $\{\xi^k := (\xi_1^k, \cdots, \xi_\ell^k)\}$ such that

$$\xi^k \in \hat{\partial}\Phi(x^k), \quad x^k \to x, \quad \Phi(x^k) \to \Phi(x) \quad \text{and} \quad \xi^k \to \xi \tag{9.3}$$

as $k \to \infty$. Since $\xi^k \in \hat{\partial}\Phi(x^k)$, for any $\varepsilon > 0$, there is a neighborhood U of x^k such that

$$\Phi(y) - \Phi(x^k) - \langle \xi^k, y - x^k\rangle \ge -\varepsilon\|y - x^k\|, \quad \forall y \in U.$$

For each $i \in \{1, \cdots, \ell\}$, then, by the differentiability of H, we have for y_i sufficient close to x_i^k that

$$\begin{aligned} -\varepsilon\|y_i - x_i^k\| &\le \Phi(x_1^k, \cdots, x_{i-1}^k, y_i, x_{i+1}^k, \cdots, x_\ell^k) - \Phi(x^k) - \langle \xi_i^k, y_i - x_i^k\rangle \\ &= H(x_1^k, \cdots, x_{i-1}^k, y_i, x_{i+1}^k, \cdots, x_\ell^k) - H(x^k) \\ &\quad + f_i(y_i) - f_i(x_i^k) - \langle \xi_i^k, y_i - x_i^k\rangle \\ &= o(\|y_i - x_i^k\|) + f_i(y_i) - f_i(x_i^k) - \langle \xi_i^k - \nabla_i H(x^k), y_i - x_i^k\rangle. \end{aligned}$$

Therefore, there is a number $r > 0$ such that

$$f_i(y_i) - f_i(x_i^k) - \langle \xi_i^k - \nabla_i H(x^k), y_i - x_i^k\rangle \ge -2\varepsilon\|y_i - x_i^k\|, \quad \forall y_i \in B_r(x_i^k).$$

Since $\varepsilon > 0$ can be arbitrarily small, we thus have $\xi_i^k - \nabla_i H(x^k) \in \hat{\partial} f_i(x_i^k)$. According to (9.3) and the continuous differentiability of H, we have

$$x_i^k \to x_i, \quad \xi_i^k - \nabla_i H(x^k) \to \xi_i - \nabla_i H(x) \quad \text{and} \quad \sum_{j=1}^{\ell} f_j(x_j^k) \to \sum_{j=1}^{\ell} f_j(x_j)$$

as $k \to \infty$. Since each f_j is lower semicontinuous at x_j, we also have

$$f_j(x_j) \le \liminf_{k\to\infty} f_j(x_j^k), \quad j = 1, \cdots, \ell.$$

Consequently

$$\begin{aligned} \limsup_{k\to\infty} f_i(x_i^k) &= \limsup_{k\to\infty} \left(\sum_{j=1}^{\ell} f_j(x_j^k) - \sum_{j\ne i} f_j(x_j^k) \right) \\ &= \sum_{j=1}^{\ell} f_j(x_j) - \liminf_{k\to\infty} \sum_{j\ne i} f_j(x_j^k) \\ &\le \sum_{j=1}^{\ell} f_j(x_j) - \sum_{j\ne i} \liminf_{k\to\infty} f_j(x_j^k) \\ &\le \sum_{j=1}^{\ell} f_j(x_j) - \sum_{j\ne i} f_j(x_j) = f_i(x_i). \end{aligned}$$

Therefore $f_i(x_i^k) \to f_i(x_i)$ as $k \to \infty$ for each i. By definition, we then have $\xi_i - \nabla_i H(x) \in \partial f_i(x_i)$, i.e., $\xi_i \in \nabla_i H(x) + \partial f_i(x_i)$ for each i which means that ξ is contained in the set on the right-hand side of (9.2).

Conversely, let $\xi = (\xi_1, \cdots, \xi_\ell)$ be an element in the set on the right-hand side of (9.2). Then $\xi_i - \nabla_i H(x) \in \partial f_i(x_i)$ for each i. Thus, there exist $\{x_i^k\}$ and $\{\eta_i^k\}$ such that

$$\eta_i^k \in \hat{\partial} f_i(x_i^k), \quad x_i^k \to x_i, \quad f_i(x_i^k) \to f_i(x_i) \quad \text{and} \quad \eta_i^k \to \xi_i - \nabla_i H(x) \tag{9.4}$$

as $k \to \infty$. Since $\eta_i^k \in \hat{\partial} f_i(x_i^k)$, for any $\varepsilon > 0$, there is a number $r > 0$ such that

$$f_i(y_i) - f_i(x_i^k) - \langle \eta_i^k, y_i - x_i^k \rangle \geq -\varepsilon \| y_i - x_i^k \|$$

for all $y_i \in B_r(x_i^k)$ and $i = 1, \cdots, \ell$. Let $x^k := (x_1^k, \cdots, x_\ell^k)$ and $\eta^k := (\eta_1^k, \cdots, \eta_\ell^k)$. Then, by the differentiability of H and by shrinking r if necessary, we have

$$\begin{aligned}\Phi(y) - \Phi(x^k) - \langle \eta^k + \nabla H(x^k), y - x^k \rangle &= H(y) - H(x^k) - \langle \nabla H(x^k), y - x^k \rangle \\ &\quad + \sum_{i=1}^{\ell} \Big(f_i(y_i) - f_i(x_i^k) \\ &\qquad - \langle \eta_i^k, y_i - x_i^k \rangle \Big) \\ &\geq o(\|y - x^k\|) - \varepsilon \sum_{i=1}^{\ell} \|y_i - x_i^k\| \\ &\geq o(\|y - x^k\|) - \ell^{1/2} \varepsilon \|y - x^k\| \\ &\geq -2\ell^{1/2} \varepsilon \|y - x^k\| \end{aligned}$$

for all $y := (y_1, \cdots, y_\ell) \in B_r(x_1^k) \times \cdots \times B_r(x_\ell^k)$. Since $\varepsilon > 0$ is arbitrary, this shows $\eta^k + \nabla H(x^k) \in \hat{\partial} \Phi(x^k)$. By using the continuous differentiability of H and (9.4), we have $x^k \to x$, $\Phi(x^k) \to \Phi(x)$, and $\eta^k + \nabla H(x^k) \to \xi$ as $k \to \infty$. Therefore $\xi \in \partial \Phi(x)$. The proof is complete. □

Proposition 9.9 *For each $i = 1, \cdots, \ell$, let $f_i : \mathbb{R}^n \to (-\infty, \infty]$ be proper lower semicontinuous. Define*

$$f(x) := \min\{f_i(x) : i = 1, \cdots, \ell\}, \quad x \in \mathbb{R}^n,$$

and let $I(x) := \{i \in \{1, \cdots, \ell\} : f_i(x) = f(x)\}$. Then

$$\partial f(x) \subset \bigcup_{i \in I(x)} \partial f_i(x).$$

Proof Let $\xi \in \partial f(x)$. Then there exist $\{x^k\}$ and $\{\xi^k\}$ such that $\xi^k \in \hat{\partial} f(x^k)$, $x^k \to x$, $f(x^k) \to f(x)$, and $\xi^k \to \xi$ as $k \to \infty$. As $\xi^k \in \hat{\partial} f(x^k)$, for any $\varepsilon > 0$, there is a number $r > 0$ such that

$$f(y) - f(x^k) - \langle \xi^k, y - x^k \rangle \geq -\varepsilon \|y - x^k\|, \quad \forall y \in B_r(x^k).$$

For each $i \notin I(x)$, from the lower semi-continuity of f_i, it follows that $f(x) < f_i(x) \leq \liminf_{k\to\infty} f_i(x^k)$ which together with $f(x^k) \to f(x)$ shows that $f_i(x^k) > f(x^k)$ and thus $I(x^k) \subset I(x)$ for large k. Therefore for each large k, we can take $i_k \in I(x^k) \subset I(x)$ such that $f_{i_k}(x^k) = f(x^k)$. Due to the finiteness of the set $I(x)$, some index from $I(x)$ must repeatedly appear infinitely many times among $\{i_k\}$. Thus, by taking a subsequence if necessary, there must have an $i \in I(x)$ such that $f_i(x^k) = f(x^k)$ for all k. Consequently,

$$f_i(y) - f_i(x^k) - \langle \xi^k, y - x^k \rangle \geq f(y) - f(x^k) - \langle \xi^k, y - x^k \rangle \geq -\varepsilon \|y - x^k\|$$

for all $y \in B_r(x^k)$. Since $\varepsilon > 0$ is arbitrary, we have $\xi^k \in \hat{\partial} f_i(x^k)$ for all k. Moreover, noting that $x^k \to x$, $\xi^k \to \xi$ and $f_i(x^k) = f(x^k) \to f(x) = f_i(x)$ as $k \to \infty$, we can conclude that $\xi \in \partial f_i(x)$. □

Next we will generalize Proposition 9.9 to the so-called inf-projection. Given a proper function $F : \mathbb{R}^n \times \mathbb{R}^q \to (-\infty, \infty]$, we define

$$f(x) := \inf_{y\in\mathbb{R}^q} F(x, y) \quad \text{and} \quad Y(x) := \arg\min_{y\in\mathbb{R}^q} F(x, y). \tag{9.5}$$

The function f is called the inf-projection of F. We will investigate the property of f via the property of F. The following condition on F is required.

Definition 9.10 A proper function $F(x, y)$ on $\mathbb{R}^n \times \mathbb{R}^q$ is said to be *level-bounded in y locally uniformly in x* if for each $\bar{x} \in \mathbb{R}^n$ and $\alpha \in \mathbb{R}$ there is a neighborhood U of $\bar{x}$ such that the set

$$\{(x, y) : x \in U \text{ and } F(x, y) \leq \alpha\}$$

is bounded in $\mathbb{R}^n \times \mathbb{R}^q$.

Proposition 9.11 *Let $F : \mathbb{R}^n \times \mathbb{R}^q \to (-\infty, \infty]$ be a function that is proper, lower semicontinuous, and level-bounded in $y \in \mathbb{R}^q$ locally uniformly in $x \in \mathbb{R}^n$. Then,*

(i) *The inf-projection f is proper and lower semicontinuous, and, for each $\bar{x} \in$ dom(f), the set $Y(\bar{x})$ is non-empty and compact.*
(ii) *Let $\bar{x} \in$ dom(f) be any fixed point. Then for any $\delta > 0$, there exists $\varepsilon > 0$ such that*

$$Y(x) \subset \{y \in \mathbb{R}^q : d(y, Y(\bar{x})) < \delta\}$$

whenever $x \in B_\varepsilon(\bar{x})$ and $|f(x) - f(\bar{x})| < \varepsilon$.

(iii) *For any* $\bar{x} \in \mathbb{R}^n$, *there holds*

$$\partial f(\bar{x}) \subset \bigcup_{y \in Y(\bar{x})} \{\xi \in \mathbb{R}^n : (\xi, 0) \in \partial F(\bar{x}, y)\}. \tag{9.6}$$

Proof

(i) We first show f is proper. Let $\bar{x} \in \mathbb{R}^n$. If $\{y : F(\bar{x}, y) \leq \alpha\} = \emptyset$ for any α, then $f(\bar{x}) = +\infty$. If there is a number α such that $\{y : F(\bar{x}, y) \leq \alpha\} \neq \emptyset$, then it is compact by the conditions on F. Thus, it follows from Theorem 1.12 that $f(\bar{x}) = F(\bar{x}, \bar{y})$ for some $\bar{y} \in \mathbb{R}^q$ which means $f(\bar{x})$ is finite and $Y(\bar{x}) \neq \emptyset$. By the properness of F, the latter case always occurs at some point $\bar{x}$. Thus f is proper and $Y(\bar{x}) \neq \emptyset$ at each $\bar{x} \in \text{dom}(f)$. To see the compactness of $Y(\bar{x})$ for $\bar{x} \in \text{dom}(f)$, let $\{y^k\} \subset Y(\bar{x})$ be any sequence. Then $F(\bar{x}, y^k) = f(\bar{x}) < \infty$, and, by the condition on F, we can conclude that $\{y^k\}$ is bounded. Thus $\{y^k\}$ has a convergent subsequence $\{y^{k_l}\}$ with a limit $\bar{y}$. By the lower semi-continuity of F, we then obtain

$$f(\bar{x}) \leq F(\bar{x}, \bar{y}) \leq \liminf_{l \to \infty} F(\bar{x}, y^{k_l}) = f(\bar{x})$$

and thus $F(\bar{x}, \bar{y}) = f(\bar{x})$ which means $\bar{y} \in Y(\bar{x})$. Therefore, $Y(\bar{x})$ is compact.

Next we show the lower semi-continuity of f. Let $\{x^k\} \subset \mathbb{R}^n$ be a sequence converging to $\bar{x}$. In case $f(\bar{x}) = \infty$, we must have $f(x^k) \to \infty$ as $k \to \infty$. If this is not true, then, by taking a subsequence of $\{x^k\}$ if necessary, we can find a number α such that $f(x^k) \leq \alpha$ for all k. Thus, there exists a sequence $\{y^k\}$ such that $F(x^k, y^k) = f(x^k) \leq \alpha$. By the level-boundedness of F in y locally uniformly in x, $\{y^k\}$ must be a bounded sequence. Thus, by taking a subsequence, we have $y^k \to \bar{y}$ as $k \to \infty$ for some $\bar{y}$. Now we may use the definition of f and the lower semi-continuity of F to derive that

$$f(\bar{x}) \leq F(\bar{x}, \bar{y}) \leq \liminf_{k \to \infty} F(x^k, y^k) = \liminf_{k \to \infty} f(x^k) \leq \alpha < \infty$$

which is a contradiction. So in this case we have $f(\bar{x}) = \lim_{k \to \infty} f(x^k)$. We next consider the case $f(\bar{x}) < \infty$. If $\liminf_{k \to \infty} f(x^k) = \infty$, nothing needs to prove. So we may assume $\alpha := \liminf_{k \to \infty} f(x^k)$ is finite. Thus, by taking a subsequence if necessary, we have $f(x^k) \leq \alpha + 1$ for all k. Now we can repeat the above argument to deduce that $f(\bar{x}) \leq \liminf_{k \to \infty} f(x^k)$.

(ii) We use a contradiction argument. If the assertion is not true, then there exist $\delta > 0$, a sequences $\{x^k\} \subset \mathbb{R}^n$ satisfying

$$x^k \in B_{1/k}(\bar{x}) \quad \text{and} \quad |f(x^k) - f(\bar{x})| < 1/k$$

and a sequence $\{y^k\} \subset \mathbb{R}^q$ such that

$$y^k \in Y(x^k) \quad \text{and} \quad d(y^k, Y(\bar{x})) \geq \delta$$

for all k. As $F(x^k, y^k) = f(x^k) \leq f(\bar{x}) + 1$, it follows from the level-boundedness of F in y locally uniformly in x that $\{y^k\}$ is a bounded sequence and, hence, by taking a subsequence if necessary, $y^k \to \bar{y}$ as $k \to \infty$ for some $\bar{y}$. Since $x^k \to \bar{x}$ and $f(x^k) \to f(\bar{x})$, we may use the lower semi-continuity of F to obtain

$$f(\bar{x}) \leq F(\bar{x}, \bar{y}) \leq \liminf_{k\to\infty} F(x^k, y^k) = \liminf_{k\to\infty} f(x^k) = f(\bar{x}).$$

This implies $F(\bar{x}, \bar{y}) = f(\bar{x})$ and so $\bar{y} \in Y(\bar{x})$. Consequently $d(y^k, Y(\bar{x})) \to 0$ as $k \to \infty$ which is a contradiction.

(iii) Let $\xi \in \partial f(\bar{x})$. Then there exist $\{x^k\}$ and $\{\xi^k\}$ such that $\xi^k \in \hat{\partial} f(x^k)$, $x^k \to \bar{x}$, $f(x^k) \to f(\bar{x})$, and $\xi^k \to \xi$ as $k \to \infty$. Since $\xi^k \in \hat{\partial} f(x^k)$, for any $\varepsilon > 0$, there is a number $r > 0$ such that

$$f(x) - f(x^k) - \langle \xi^k, x - x^k \rangle \geq -\varepsilon \|x - x^k\|, \quad \forall x \in B_r(x^k).$$

Note that $f(x) \leq F(x, y)$ for all $y \in \mathbb{R}^q$ and, since $f(x^k)$ is finite, by (i) we have $f(x^k) = F(x^k, y^k)$ for some $y^k \in Y(x^k)$. Therefore,

$$\begin{aligned} &F(x, y) - F(x^k, y^k) - \langle (\xi^k, 0), (x, y) - (x^k, y^k) \rangle \\ &\geq f(x) - f(x^k) - \langle \xi^k, x - x^k \rangle \geq -\varepsilon \|x - x^k\| \geq -\varepsilon \|(x, y) - (x^k, y^k)\| \end{aligned}$$

for all $(x, y) \in B_r(x^k) \times \mathbb{R}^q$. Since $\varepsilon > 0$ is arbitrary, we therefore have $(\xi^k, 0) \in \hat{\partial} F(x^k, y^k)$. As $x^k \to \bar{x}$ and $f(x^k) \to f(\bar{x})$, it follows from (ii) that $d(y^k, Y(\bar{x})) \to 0$ as $k \to \infty$. Since $Y(\bar{x})$ is compact, $\{y^k\}$ must be bounded. Thus, by taking a subsequence if necessary, we have $y^k \to \bar{y}$ for some $\bar{y} \in Y(\bar{x})$. Note that $F(x^k, y^k) = f(x^k) \to f(\bar{x}) = F(\bar{x}, \bar{y})$ and $(\xi^k, 0) \to (\xi, 0)$, by the definition of the limiting subdifferential, we have $(\xi, 0) \in \partial F(\bar{x}, \bar{y})$ for some $\bar{y} \in Y(\bar{x})$.

□

9.2 Kurdyka-Łojasiewicz Condition

In [146, 147] Łojasiewicz showed that for any real analytic function f, there is a $\theta \in (0, 1)$ such that

$$\frac{|f(x) - f(x^*)|^\theta}{\|\nabla f(x)\|}$$

remains bounded around any critical point x^* of f. In [129], Kurdyka generalized the result to C^1 functions whose graphs belong to an o-minimal structure.

With the preparation in Sect. 9.1, we now formulate the Kurdyka-Łojasiewicz condition for non-smooth non-convex functions which was introduced in [31, 32]. In order to make the exposition easier, we first introduce the concept of desingularization functions.

Definition 9.12 A function $\varphi : [0, \eta) \to \mathbb{R}$ is called a *desingularization function* if $\varphi \in C[0, \eta) \cap C^1(0, \eta)$, $\varphi(0) = 0$ and $\varphi'(s) > 0$ for all $s \in (0, \eta)$. The set of all desingularization functions on $[0, \eta)$ is denoted by $\mathcal{K}(0, \eta)$.

Definition 9.13 (KL Condition) A proper function $f : \mathbb{R}^n \to (-\infty, \infty]$ is said to satisfy the *Kurdyka-Łojasiewicz (KL) condition* at $x^* \in \text{dom}(f)$ if there exist $\eta \in (0, \infty]$, a ball $B_r(x^*)$ centered at x^* and a concave desingularization function $\varphi \in \mathcal{K}(0, \eta)$ such that the Kurdyka-Łojasiewicz inequality

$$\varphi'(f(x) - f(x^*))d(0, \partial f(x)) \geq 1$$

holds for all $x \in B_r(x^*) \cap \{f(x^*) < f < f(x^*) + \eta\}$.

Definition 9.14 (KL Function) A proper function $f : \mathbb{R}^n \to (-\infty, \infty]$ satisfying the KL condition at each point of $\text{dom}(f)$ is called a *KL function*.

Definition 9.15 (KL Exponent) A proper function $f : \mathbb{R}^n \to (-\infty, \infty]$ is said to satisfy the KL condition at $x^* \in \text{dom}(f)$ with exponent $\alpha \in [0, 1)$ if f satisfies the KL condition at x^* with $\varphi(s) = cs^{1-\alpha}$ for some $c > 0$, i.e., there exist $c, r > 0$ and $\eta \in (0, \infty]$ such that

$$d(0, \partial f(x)) \geq c(f(x) - f(x^*))^\alpha$$

for all $x \in B_r(x^*) \cap \{f(x^*) < f < f(x^*) + \eta\}$. f is called a KL function with exponent α if f is a KL function and has exponent α at any point in $\text{dom}(f)$.

As we will see in Sect. 9.3, the KL condition plays an important role in the convergence analysis of algorithms for non-smooth non-convex optimization problems, and the exponent of KL functions is closely related to the rate of convergence. From the definition, it is easy to see that a KL function with exponent $\alpha \in [0, 1)$ must be a KL function with exponent β for any $\beta \in [\alpha, 1)$.

Lemma 9.16 *A proper lower semicontinuous function* $f : \mathbb{R}^n \to (-\infty, \infty]$ *satisfies the KL condition with exponent* α *for any* $\alpha \in [0, 1)$ *at any noncritical point* $\bar{x} \in \text{dom}(f)$.

Proof Let $\bar{x} \in \text{dom}(f)$ be a noncritical point. We first claim that there must exist a constant $c > 0$ such that

$$\|x - \bar{x}\| + |f(x) - f(\bar{x})| \leq c \implies d(0, \partial f(x)) \geq c. \tag{9.7}$$

Suppose this is not true, then there exists a sequence $\{x^k\} \subset \mathbb{R}^n$ such that

$$\|x^k - \bar{x}\| + |f(x^k) - f(\bar{x})| \le \frac{1}{k} \quad \text{and} \quad d(0, \partial f(x^k)) \le \frac{1}{k}.$$

Thus $x^k \to \bar{x}$, $f(x^k) \to f(\bar{x})$, and we can find $\xi^k \in \partial f(x^k)$ with $\|\xi^k\| \le 1/k \to 0$. Therefore, by the closedness of ∂f given in Lemma 9.4, we have $0 \in \partial f(\bar{x})$ which is a contradiction. By virtue of (9.7), now we take $\varphi(s) = s/c$ and obtain

$$\varphi'(f(x) - f(\bar{x}))d(0, \partial f(x)) = \frac{1}{c}d(0, \partial f(x)) \ge 1$$

whenever $x \in B_{c/2}(\bar{x}) \cap \{f(\bar{x}) < f < f(\bar{x}) + c/2\}$. Thus f satisfies the KL condition at $\bar{x}$ with exponent 0 and hence with exponent α for any $\alpha \in [0, 1)$. □

From Lemma 9.16, it follows that for a proper lower semicontinuous function $f : \mathbb{R}^n \to (-\infty, \infty]$, we need only to check the KL condition at critical points which in fact is the most important in applications.

Example 9.17 Let $f : \mathbb{R}^n \to (-\infty, \infty]$ be a proper, lower semicontinuous, convex, KL function with exponent $\alpha \in [0, 1)$. Then, for any $\lambda > 0$, its Moreau envelop $M_\lambda f$ must be a KL function with exponent $\max\{\alpha, 1/2\}$. To see this, according to Lemma 9.16, it suffices to verify the KL condition at the minimizers of $M_{\lambda f}$. Let $\bar{x}$ be a minimizer of $M_{\lambda f}$. Then, by Theorem 4.23, $\bar{x}$ is also a minimizer of f and $f(\bar{x}) = M_{\lambda f}(\bar{x})$. By the KL property of f at $\bar{x}$, there exist $C > 0$, $r > 0$, and $\eta \in (0, \infty]$ such that

$$f(x) - f(\bar{x}) \le Cd(0, \partial f(x))^{1/\alpha}, \quad \forall x \in B_r(\bar{x}) \cap \{f(\bar{x}) \le f < f(\bar{x}) + \eta\}. \tag{9.8}$$

Let $x \in B_r(\bar{x}) \cap \{M_{\lambda f}(\bar{x}) \le M_{\lambda f} < M_{\lambda f}(\bar{x}) + \eta\}$ be any point. By noting that

$$M_{\lambda f}(x) = f(\text{prox}_{\lambda f}(x)) + \frac{1}{2\lambda}\|\text{prox}_{\lambda f}(x) - x\|^2$$

and $\nabla M_{\lambda f}(x) = (x - \text{prox}_{\lambda f}(x))/\lambda$, we have $f(\bar{x}) \le f(\text{prox}_{\lambda f}(x)) \le M_{\lambda f}(x) < f(\bar{x}) + \eta$ and

$$M_{\lambda f}(x) - M_{\lambda f}(\bar{x}) = f(\text{prox}_{\lambda f}(x)) - f(\bar{x}) + \frac{\lambda}{2}\|\nabla M_{\lambda f}(x)\|^2.$$

Since $\|\text{prox}_{\lambda f}(x) - \bar{x}\| = \|\text{prox}_{\lambda f}(x) - \text{prox}_{\lambda f}(\bar{x})\| \le \|x - \bar{x}\| < r$, we may use (9.8) to conclude

$$M_{\lambda f}(x) - M_{\lambda f}(\bar{x}) \le Cd(0, \partial f(\text{prox}_{\lambda f}(x)))^{1/\alpha} + \frac{\lambda}{2}\|\nabla M_{\lambda f}(x)\|^2.$$

By the definition of $\text{prox}_{\lambda f}(x)$, we have

$$\nabla M_{\lambda f}(x) = \frac{x - \text{prox}_{\lambda f}(x)}{\lambda} \in \partial f(\text{prox}_{\lambda f}(x)).$$

Therefore

$$M_{\lambda f}(x) - M_{\lambda f}(\bar{x}) \leq C\|\nabla M_{\lambda f}(x)\|^{1/\alpha} + \frac{\lambda}{2}\|\nabla M_{\lambda f}(x)\|^2$$

which implies the desired conclusion.

Example 9.18 Let $f : \mathbb{R}^n \to (-\infty, \infty]$ be a proper, lower semicontinuous, KL function with exponent $\alpha \in [1/2, 1)$. For any $\gamma > 0$, consider the function $F : \mathbb{R}^n \times \mathbb{R}^n \to (-\infty, \infty]$ defined by

$$F(x, y) := f(x) + \frac{\gamma}{2}\|x - y\|^2.$$

We will show that F is also a KL function with exponent α. According to Lemma 9.16, we only need to verify the KL condition of F at its critical points. By Proposition 9.8, we have

$$\partial F(x, y) = \begin{pmatrix} \partial f(x) + \gamma(x - y) \\ \gamma(y - x) \end{pmatrix}.$$

Therefore

$$d(0, \partial F(x, y)) = \sqrt{\inf_{\xi \in \partial f(x)} \|\xi + \gamma(x - y)\|^2 + \|\gamma(y - x)\|^2}.$$

By the triangle inequality and the Cauchy-Schwarz inequality, we can see

$$\|\xi\|^2 \leq (\|\xi + \gamma(x - y)\| + \|\gamma(x - y)\|)^2 \leq 3\|\xi + \gamma(x - y)\|^2 + \frac{3}{2}\|\gamma(x - y)\|^2.$$

Thus

$$\begin{aligned} d(0, \partial F(x, y)) &\geq \sqrt{\frac{1}{3}\inf_{\xi \in \partial f(x)} \|\xi\|^2 + \frac{1}{2}\|\gamma(x - y)\|^2} \\ &\geq \frac{1}{\sqrt{6}}\inf_{\xi \in \partial f(x)} \|\xi\| + \frac{1}{2}\|\gamma(x - y)\| \\ &= \frac{1}{\sqrt{6}} d(0, \partial f(x)) + \frac{\gamma}{2}\|x - y\|. \end{aligned}$$

Let $(\bar{x}, \bar{y})$ be any fixed critical point of F. Then $\bar{y} = \bar{x}$. Since f satisfies the KL condition at $\bar{x}$ with exponent α, there exist $c_0 > 0$, $r \in (0, 1/2]$, and $\eta \in (0, \infty]$ such that

$$f(x) - f(\bar{x}) \leq c_0 \, [d(0, \partial f(x))]^{1/\alpha}$$

for all $x \in B_r(\bar{x}) \cap \{f < f(\bar{x}) + \eta\}$. Let $B_r(\bar{x}, \bar{x})$ denote the open ball in $\mathbb{R}^n \times \mathbb{R}^n$ with center $(\bar{x}, \bar{x})$ and radius r. Then for any $(x, y) \in B_r(\bar{x}, \bar{x}) \cap \{F < F(\bar{x}, \bar{x}) + \eta\}$, we have $x \in B_r(\bar{x})$ and $f(x) \leq F(x, y) < F(\bar{x}, \bar{x}) + \eta = f(\bar{x}) + \eta$. Consequently, we can find a constant $c_1 > 0$ such that

$$\begin{aligned}
[d(0, \partial F(x, y))]^{1/\alpha} &\geq c_1 \left(c_0 \, [d(0, \partial f(x))]^{1/\alpha} + \frac{\gamma}{2} \|x - y\|^{1/\alpha} \right) \\
&\geq c_1 \left(f(x) - f(\bar{x}) + \frac{\gamma}{2} \|x - y\|^{1/\alpha} \right).
\end{aligned}$$

Since $1/\alpha \leq 2$ and $\|x - y\| \leq \|x - \bar{x}\| + \|y - \bar{x}\| < 2r \leq 1$, we finally obtain

$$[d(0, \partial F(x, y))]^{1/\alpha} \geq c_1 \left(f(x) - f(\bar{x}) + \frac{\gamma}{2} \|x - y\|^2 \right) = c_1 \left(F(x, y) - F(\bar{x}, \bar{x}) \right)$$

for all $(x, y) \in B_r(\bar{x}, \bar{x}) \cap \{F < F(\bar{x}, \bar{x}) + \eta\}$ and the KL condition of F at $(\bar{x}, \bar{x})$ is verified.

The following example shows that any locally strongly convex functions are KL functions with exponent $1/2$.

Example 9.19 A function $f : \mathbb{R}^n \to (-\infty, \infty]$ is said to be locally strongly convex if for any $\bar{x} \in \mathbb{R}^n$ there is a ball $B_r(\bar{x})$ centered at $\bar{x}$ and a constant $c > 0$ such that

$$f(ty + (1 - t)x) + ct(1 - t)\|y - x\|^2 \leq tf(y) + (1 - t)f(x)$$

for all $x, y \in B_r(\bar{x}) \cap \mathrm{dom}(f)$ and $0 \leq t \leq 1$. Then for any $x \in B_r(\bar{x})$ with $\partial f(x) \neq \emptyset$, we have

$$\begin{aligned}
tf(\bar{x}) + (1 - t)f(x) &\geq f(x + t(\bar{x} - x)) + ct(1 - t)\|x - \bar{x}\|^2 \\
&\geq f(x) + \langle \xi, t(\bar{x} - x) \rangle + ct(1 - t)\|\bar{x} - x\|^2
\end{aligned}$$

for any $\xi \in \partial f(x)$ and $0 < t \leq 1$. Regrouping the terms gives

$$f(\bar{x}) - f(x) \geq \langle \xi, \bar{x} - x \rangle + c(1 - t)\|\bar{x} - x\|^2.$$

By taking $t \to 0$ and using the Cauchy-Schwarz inequality, it follows that

$$f(\bar{x}) - f(x) \geq \langle \xi, \bar{x} - x \rangle + c\|\bar{x} - x\|^2 \geq c\|\bar{x} - x\|^2 - \|\xi\| \|\bar{x} - x\| \geq -\frac{1}{4c}\|\xi\|^2$$

that $\Omega \subset \bigcup_{i=1}^{\ell} B_{r_{x_i}}(x_i)$. Now we may choose $r > 0$ such that

$$\{z : d(z, \Omega) < r\} \subset \bigcup_{i=1}^{\ell} B(x_i, r_{x_i})$$

and set

$$\eta := \min\{\eta_{x_1}, \cdots, \eta_{x_\ell}\}, \quad \varphi = \sum_{i=1}^{\ell} \varphi_{x_i}.$$

Then $\varphi \in \mathcal{K}(0, \eta)$ is a concave function. By the choices of r and η, for any $z \in \{d(z, \Omega) < r\} \cap \{\mu < f < \mu + \eta\}$, there must exist $j \in \{1, \cdots, \ell\}$ such that

$$z \in B_{r_{x_j}}(x_j) \cap \{\mu < f < \mu + \eta_{x_j}\}.$$

Thus

$$\begin{aligned} \varphi'(f(z) - \mu)d(0, \partial f(z)) &= \sum_{i=1}^{\ell} \varphi'_{x_i}(f(z) - \mu)d(0, \partial f(z)) \\ &\geq \varphi'_{x_j}(f(z) - \mu)d(0, \partial f(z)) \\ &\geq 1. \end{aligned}$$

The proof is thus complete. □

For the KL property of inf-projection of KL functions, we have the following result [230].

Proposition 9.23 *Let $F : \mathbb{R}^n \times \mathbb{R}^q \to (-\infty, \infty]$ be a proper lower semicontinuous function that is level-bounded in $y \in \mathbb{R}^q$ locally uniformly in $x \in \mathbb{R}^n$. Let $f(x)$ and $Y(x)$ be defined in (9.5) for $x \in \mathbb{R}^n$. Let $\bar{x} \in \mathrm{dom}(f)$. If F satisfies the KL condition with exponent $\alpha \in [0, 1)$ at every point in $\{\bar{x}\} \times Y(\bar{x})$, then f is proper, lower semicontinuous, and satisfies the KL condition at $\bar{x}$ with exponent α.*

Proof It follows from Proposition 9.11 (i) that f is proper, lower semicontinuous and $Y(\bar{x}) \neq \emptyset$ is compact. Since F satisfies the KL condition with exponent $\alpha \in [0, 1)$ at every point in the compact set $\Omega := \{\bar{x}\} \times Y(\bar{x})$ on which $F \equiv f(\bar{x})$, we may use Lemma 9.22 to find $c, r > 0$ and $\eta \in (0, \infty]$ such that

$$F(x, y) - f(\bar{x}) \leq c d(0, \partial F(x, y))^{1/\alpha} \tag{9.13}$$

for all (x, y) satisfying $d((x, y), \Omega) < r$, and $F(x, y) < f(\bar{x}) + \eta$. According to Proposition 9.11 (ii) we can find $0 < \varepsilon \leq \min\{\eta, r/2\}$ such that

$$Y(x) \subset \{y \in \mathbb{R}^q : d(y, Y(\bar{x})) < r/2\} \tag{9.14}$$

whenever $x \in B_\varepsilon(\bar{x})$ and $|f(x) - f(\bar{x})| < \varepsilon$. Now we claim that

$$f(x) - f(\bar{x}) \le c d(0, \partial f(x))^{1/\alpha} \tag{9.15}$$

for $x \in B_\varepsilon(\bar{x}) \cap \{f < f(\bar{x}) + \varepsilon\}$. Indeed, for any $\xi \in \partial f(x)$, by Proposition 9.11, (iii) we have $(\xi, 0) \in \partial F(x, y)$ for some $y \in Y(x)$. Note from (9.14) that

$$d((x, y), \Omega) \le \|x - \bar{x}\| + d(y, Y(\bar{x})) < \varepsilon + r/2 \le r.$$

Since $F(x, y) = f(x) < f(\bar{x}) + \varepsilon$, it follows from (9.13) and $(\xi, 0) \in \partial F(x, y)$ that

$$f(x) - f(\bar{x}) = F(x, y) - f(\bar{x}) \le c d(0, \partial F(x, y))^{1/\alpha} \le c\|\xi\|^{1/\alpha}.$$

Since $\xi \in \partial f(x)$ is arbitrary, we therefore obtain (9.15). □

In general, the verification of the KL condition for a function at its critical points is highly nontrivial and presents a lot of challenges. The following theorem, due to [31, 32], provides many examples of KL functions from the class of semi-algebraic functions.

Theorem 9.24 *If $f : \mathbb{R}^n \to (-\infty, \infty]$ is a proper, lower semicontinuous, semi-algebraic function, then for any $\bar{x} \in \mathrm{dom}(f)$ and $r > 0$, there exist $c > 0$, $\alpha \in [0, 1)$, and $\eta \in (0, \infty]$ such that*

$$d(0, \partial f(x)) \ge c(f(x) - f(\bar{x}))^\alpha$$

for all $x \in B_r(\bar{x}) \cap \{f(\bar{x}) < f < f(\bar{x}) + \eta\}$.

We defer the proof of Theorem 9.24 to Sect. 9.5, since the argument is intricate and involves other tools to be developed. Moreover, we will show in Sect. 9.5 an even more general result which states that any definable function in an o-minimal structure satisfies the KL condition.

Example 9.25 Consider the ℓ^0-norm $\|\cdot\|_0$ on $\mathbb{R}^n$, where, for each $x \in \mathbb{R}^n$, $\|x\|_0$ denotes the number of nonzero entries of x, i.e.,

$$\|x\|_0 := \sum_{i=1}^n \mathrm{sign}(|x_i|).$$

This function is important in compressive sensing and sparsity recovery. To show this function is semi-algebraic, it suffices to show that the single variable function $t \to \mathrm{sign}(|t|)$ is semi-algebraic. Note that the graph of this function is

$$\begin{aligned} \mathcal{G} &= \{(t, s) \in \mathbb{R}^2 : s = \mathrm{sign}(|t|)\} \\ &= \{(t, s) : t > 0, s = 1\} \cup \{(t, s) : t < 0, s = 1\} \cup \{(t, s) : t = 0, s = 0\} \end{aligned}$$

which, by definition, is clearly a semi-algebraic set in $\mathbb{R}^2$. Furthermore, one can verify that $\|\cdot\|_0$ is lower semicontinuous. To see this, let $\{x^k\} \subset \mathbb{R}^n$ be a sequence that converges to $\bar{x} \in \mathbb{R}^n$ as $k \to \infty$. Then $x_i^k \to \bar{x}_i$ as $k \to \infty$ for every $i = 1, \cdots, n$. If $\bar{x}_i \neq 0$ for some i, then $x_i^k \neq 0$ for large k. Therefore,

$$\{i \in \{1, \cdots, n\} : x_i \neq 0\} \subset \{i \in \{1, \cdots, n\} : x_i^k \neq 0\}$$

for large k. Consequently,

$$\|\bar{x}\|_0 = \sum_{i=1}^{n} \text{sign}(|\bar{x}_i|) \leq \sum_{i=1}^{n} \text{sign}(|x_i^k|) = \|x^k\|_0$$

for large k and hence $\|\bar{x}\|_0 \leq \liminf_{k\to\infty} \|x^k\|_0$. According to Theorem 9.24, $\|\cdot\|_0$ is therefore a KL function on $\mathbb{R}^n$.

9.3 A General Convergence Framework

In this section, we will provide an abstract convergence theorem for algorithms designed for solving optimization problems. The methodology encoded in the framework is applicable for dealing with a wider range of optimization methods. Assume there is a minimization problem. By some algorithm, we produce an iterative sequence $\{x^k\} \subset \mathbb{R}^n$. With a proper choice of a function $\mathcal{M} : \mathbb{R}^n \to (-\infty, \infty]$, which will be called a merit function, we assume that the following three conditions hold:

(H1). (Sufficient decrease condition) There exists a constant $a > 0$ such that

$$\mathcal{M}(x^{k+1}) + a\|x^{k+1} - x^k\|^2 \leq \mathcal{M}(x^k)$$

for all integers $k \geq 0$.

(H2). (Relative error condition) There exists a constant $b > 0$ such that, for each integer $k \geq 0$, there exists $w^{k+1} \in \partial\mathcal{M}(x^{k+1})$ with

$$\|w^{k+1}\| \leq b\|x^{k+1} - x^k\|.$$

(H3). (Continuity condition) There exists a subsequence $\{x^{k_j}\}$ of $\{x^k\}$ and an element x^* such that

$$x^{k_j} \to x^* \quad \text{and} \quad \mathcal{M}(x^{k_j}) \to \mathcal{M}(x^*), \qquad \text{as } j \to \infty.$$

The condition (H1) models a descent property of the optimization methods and is called a sufficient-decrease condition. The condition (H2) motivates from the well-known fact that most algorithms in optimization are generated by an infinite sequence of subproblems which involve exact or inexact minimization processes. This condition precisely reflects relative inexact optimality conditions for such minimization subproblems. The condition (H3) does not require f to be continuous. For many algorithms that generate $\{x^k\}$, their nature forces $\{x^k\}$ to be bounded and to comply with (H3).

The following theorem, due to [8], provides a global convergence result on $\{x^k\}$ under the conditions (H1), (H2), (H3), and the KL property of the merit function $\mathcal{M}$.

Theorem 9.26 (Global Convergence) *Let* $\mathcal{M} : \mathbb{R}^n \to (-\infty, \infty]$ *be a proper lower semicontinuous function and let* $\{x^k\}$ *be a sequence satisfying (H1), (H2) and (H3). If* $\mathcal{M}$ *satisfies the KL property at the cluster point* x^* *specified in (H3), then* $\{x^k\}$ *has a finite length with* $x^k \to x^*$ *as* $k \to \infty$ *and* x^* *is a critical point of* $\mathcal{M}$.

Proof Since $\mathcal{M}$ satisfies the KL condition at x^*, we can find $\eta \in (0, \infty]$, $r > 0$ and a concave function $\varphi \in \mathcal{K}(0, \eta)$ such that

$$\varphi'(\mathcal{M}(x) - \mathcal{M}(x^*))d(0, \partial\mathcal{M}(x)) \geq 1$$

for all $x \in B_r(x^*) \cap \{\mathcal{M}(x^*) < \mathcal{M} < \mathcal{M}(x^*) + \eta\}$. According to (H1), $\{\mathcal{M}(x^k)\}$ is monotonically decreasing. By (H3), we have $x^{k_j} \to x^*$ and $\mathcal{M}(x^{k_j}) \to \mathcal{M}(x^*)$ as $j \to \infty$ for a subsequence $\{x^{k_j}\}$ of $\{x^k\}$. Therefore, $\mathcal{M}(x^k) \geq \mathcal{M}(x^*)$ and $\mathcal{M}(x^k) \to \mathcal{M}(x^*)$ as $k \to \infty$. Thus, we can find k_N such that

$$\|x^{k_N} - x^*\| + 2\sqrt{\frac{\mathcal{M}(x^{k_N}) - \mathcal{M}(x^*)}{a}} + \frac{b}{a}\varphi(\mathcal{M}(x^{k_N}) - \mathcal{M}(x^*)) < r \tag{9.16}$$

and

$$\mathcal{M}(x^*) \leq \mathcal{M}(x^k) < \mathcal{M}(x^*) + \eta \quad \text{for all } k \geq k_N. \tag{9.17}$$

We will use an induction argument to show that $x^k \in B_r(x^*)$ for all $k \geq k_N$. According to (9.16), this is trivial for $k = k_N$. Now we assume that $x^k \in B_r(x^*)$ for all $k_N \leq k \leq j$ for some $j \geq k_N$. We will show that $x^{j+1} \in B_r(x^*)$. We will achieve this by showing that

$$\begin{aligned} &2\|x^{k+1} - x^k\| \\ &\leq \|x^k - x^{k-1}\| + \frac{b}{a}\left[\varphi(\mathcal{M}(x^k) - \mathcal{M}(x^*)) - \varphi(\mathcal{M}(x^{k+1}) - \mathcal{M}(x^*))\right] \end{aligned} \tag{9.18}$$

for all $k_N \leq k \leq j$. If $\mathcal{M}(x^k) = \mathcal{M}(x^*)$, then it follows from (H1) that $\mathcal{M}(x^{k+1}) = \mathcal{M}(x^*)$ and $x^{k+1} = x^k$ which demonstrate that (9.18) holds true. So we may assume

$\mathcal{M}(x^k) > \mathcal{M}(x^*)$. Now we can use the KL property and (H2) to obtain

$$\varphi'(\mathcal{M}(x^k) - \mathcal{M}(x^*)) \geq \frac{1}{\operatorname{dist}(0, \partial\mathcal{M}(x^k))} \geq \frac{1}{\|w^k\|} \geq \frac{1}{b\|x^k - x^{k-1}\|}. \tag{9.19}$$

Since φ is concave, we have

$$\varphi(s) \leq \varphi(t) + \varphi'(t)(s-t) \quad \text{for all } s, t \in [0, \eta).$$

This in particular implies that

$$\begin{aligned}
&\varphi(\mathcal{M}(x^{k+1}) - \mathcal{M}(x^*)) \\
&\leq \varphi(\mathcal{M}(x^k) - \mathcal{M}(x^*)) + \varphi'(\mathcal{M}(x^k) - \mathcal{M}(x^*))(\mathcal{M}(x^{k+1}) - \mathcal{M}(x^k)).
\end{aligned}$$

Therefore, by (H1), $\varphi' > 0$ and (9.19), we can obtain

$$\begin{aligned}
\varphi(\mathcal{M}(x^k) - \mathcal{M}(x^*)) &- \varphi(\mathcal{M}(x^{k+1}) - \mathcal{M}(x^*)) \\
&\geq \varphi'(\mathcal{M}(x^k) - \mathcal{M}(x^*))(\mathcal{M}(x^k) - \mathcal{M}(x^{k+1})) \\
&\geq \varphi'(\mathcal{M}(x^k) - \mathcal{M}(x^*))a\|x^{k+1} - x^k\|^2 \\
&\geq \frac{a\|x^{k+1} - x^k\|^2}{b\|x^k - x^{k-1}\|}.
\end{aligned}$$

Consequently, by using the inequality $2ab \leq a^2 + b^2$, it follows that

$$\begin{aligned}
&2\|x^{k+1} - x^k\| \\
&\leq 2\sqrt{\frac{b}{a}\|x^k - x^{k-1}\| \left[\varphi(\mathcal{M}(x^k) - \mathcal{M}(x^*)) - \varphi(\mathcal{M}(x^{k+1}) - \mathcal{M}(x^*))\right]} \\
&\leq \|x^k - x^{k-1}\| + \frac{b}{a}\left[\varphi(\mathcal{M}(x^k) - \mathcal{M}(x^*)) - \varphi(\mathcal{M}(x^{k+1}) - \mathcal{M}(x^*))\right]
\end{aligned}$$

which shows (9.18).

Now we sum (9.18) over k from $k = k_N + 1$ to $k = j$ to obtain

$$\begin{aligned}
2\sum_{k=k_N+1}^{j} \|x^{k+1} - x^k\| \leq &\sum_{k=k_N+1}^{j} \|x^k - x^{k-1}\| \\
&+ \frac{b}{a}\left[\varphi(\mathcal{M}(x^{k_N+1}) - \mathcal{M}(x^*)) - \varphi(\mathcal{M}(x^{j+1}) - \mathcal{M}(x^*))\right]
\end{aligned}$$

$$\leq \sum_{k=k_N+1}^{j} \|x^k - x^{k-1}\| + \frac{b}{a}\varphi(\mathcal{M}(x^{k_N+1}) - \mathcal{M}(x^*))$$

$$\leq \sum_{k=k_N+1}^{j} \|x^k - x^{k-1}\| + \frac{b}{a}\varphi(\mathcal{M}(x^{k_N}) - \mathcal{M}(x^*)).$$

This implies that

$$\sum_{k=k_N+1}^{j} \|x^{k+1} - x^k\| + \|x^{j+1} - x^j\|$$
$$\leq \|x^{k_N+1} - x^{k_N}\| + \frac{b}{a}\varphi(\mathcal{M}(x^{k_N}) - \mathcal{M}(x^*)). \tag{9.20}$$

Therefore

$$\|x^{j+1} - x^*\| \leq \sum_{k=k_N+1}^{j} \|x^{k+1} - x^k\| + \|x^{k_N+1} - x^{k_N}\| + \|x^{k_N} - x^*\|$$
$$\leq \|x^{k_N} - x^*\| + 2\|x^{k_N+1} - x^{k_N}\| + \frac{b}{a}\varphi(\mathcal{M}(x^{k_N}) - \mathcal{M}(x^*)).$$

By using (H1) with $k = k_N$, we have

$$\|x^{k_N+1} - x^{k_N}\| \leq \sqrt{\frac{\mathcal{M}(x^{k_N}) - \mathcal{M}(x^{k_N+1})}{a}} \leq \sqrt{\frac{\mathcal{M}(x^{k_N}) - \mathcal{M}(x^*)}{a}}.$$

Consequently, by virtue of (9.16), we then obtain

$$\|x^{j+1} - x^*\| \leq \|x^{k_N} - x^*\| + 2\sqrt{\frac{\mathcal{M}(x^{k_N}) - \mathcal{M}(x^*)}{a}} + \frac{b}{a}\varphi(\mathcal{M}(x^{k_N}) - \mathcal{M}(x^*)) < r$$

which implies that $x^{j+1} \in B_r(x^*)$.

Since $x^k \in B_r(x^*)$ for all $k \geq k_N$, we can see that (9.20) holds for all $j \geq k_N$. In particular

$$\sum_{k=k_N+1}^{\infty} \|x^{k+1} - x^k\| \leq \|x^{k_N+1} - x^{k_N}\| + \frac{b}{a}\varphi(\mathcal{M}(x^{k_N}) - \mathcal{M}(x^*)) < \infty$$

which shows that $\{x^k\}$ has a finite length. Since a sequence with finite length property is a Cauchy sequence, thus $\{x^k\}$ is convergent. Since, by (H3), $\{x^k\}$ has a subsequence converging to x^*, we must have $x^k \to x^*$ as $k \to \infty$.

By using (H1) and (H2), we have $w^k \in \partial\mathcal{M}(x^k)$ with $w^k \to 0$ as $k \to \infty$. Since $x^k \to x^*$ and $\mathcal{M}(x^k) \to \mathcal{M}(x^*)$ as $k \to \infty$, we may use the closedness of subdifferential shown in Lemma 9.4 to conclude that $0 \in \partial\mathcal{M}(x^*)$. Thus, x^* is a critical point of $\mathcal{M}$. □

Next we investigate the convergence rates for the sequence $\{x^k\}$ satisfying (H1), (H2), and (H3) with the merit function $\mathcal{M}$ satisfying the KL condition with an exponent $\alpha \in [0, 1)$. We have the following result [6].

Theorem 9.27 (Convergence Rates) *Let all the conditions in Theorem 9.26 hold. Assume further that the function $\mathcal{M}$ satisfies the KL condition at x^* with an exponent $\alpha \in [0, 1)$. Then the following convergence rate result holds:*

(i) *If $\alpha = 0$, then $\{x^k\}$ converges in finite many steps.*
(ii) *If $\alpha \in (0, 1/2]$, there exist constants $C > 0$ and $q \in (0, 1)$ such that*

$$\|x^k - x^*\| \le Cq^k, \quad \forall k \ge 0.$$

(iii) *If $\alpha \in (1/2, 1)$, then there exist a constant $C > 0$ such that*

$$\|x^k - x^*\| \le C(1+k)^{-\frac{1-\alpha}{2\alpha-1}}, \quad \forall k \ge 0.$$

Proof Since $\mathcal{M}$ satisfies the KL condition at x^* with an exponent $\alpha \in [0, 1)$, there exist $c, r > 0$ and $\eta \in (0, \infty]$ such that

$$(\mathcal{M}(x) - \mathcal{M}(x^*))^\alpha \le cd(0, \partial\mathcal{M}(x))$$

whenever $x \in B_r(x^*) \cap \{\mathcal{M}(x^*) < \mathcal{M} < \mathcal{M}(x^*) + \eta\}$. Let $\varphi(s) := \tilde{c}s^{1-\alpha}$ with $\tilde{c} = \frac{1}{c(1-\alpha)}$. According to Theorem 9.26, by dropping finite many terms from $\{x^k\}$, we may assume that $x^k \in B_r(x^*)$ and $\mathcal{M}(x^*) \le \mathcal{M}(x^k) < \mathcal{M}(x^*) + \eta$ for all $k \ge 1$ and hence (9.18) holds for all k. Let $l > j \ge 1$ be any two integers. We can sum (9.18) over k from $k = j$ to $k = l$ to obtain

$$\begin{aligned} 2\sum_{k=j}^{l} \|x^{k+1} - x^k\| &\le \sum_{k=j}^{l} \|x^k - x^{k-1}\| \\ &\quad + \frac{b}{a}\left[\varphi(\mathcal{M}(x^j) - \mathcal{M}(x^*)) - \varphi(\mathcal{M}(x^{l+1}) - \mathcal{M}(x^*))\right] \\ &\le \sum_{k=j}^{l} \|x^k - x^{k-1}\| + \frac{b}{a}\varphi(\mathcal{M}(x^j) - \mathcal{M}(x^*)). \end{aligned}$$

This implies that

$$\sum_{k=j}^{l} \|x^{k+1} - x^k\| + \|x^{l+1} - x^l\| \leq \|x^j - x^{j-1}\| + \frac{b}{a}\varphi(\mathcal{M}(x^j) - \mathcal{M}(x^*)).$$

By taking $l \to \infty$, we have

$$\sum_{k=j}^{\infty} \|x^{k+1} - x^k\| \leq \|x^j - x^{j-1}\| + \frac{b}{a}\varphi(\mathcal{M}(x^j) - \mathcal{M}(x^*)). \tag{9.21}$$

Let

$$\Delta_j := \sum_{k=j}^{\infty} \|x^{k+1} - x^k\|.$$

Then $\|x^j - x^{j-1}\| = \Delta_{j-1} - \Delta_j$ and $\Delta_j \to 0$ as $j \to \infty$ by the finite length property of $\{x^k\}$. Therefore, it follows from (9.21) that

$$\begin{aligned}\Delta_j &\leq \Delta_{j-1} - \Delta_j + \frac{b}{a}\varphi(\mathcal{M}(x^j) - \mathcal{M}(x^*)) \\ &= \Delta_{j-1} - \Delta_j + \frac{b\tilde{c}}{a}(\mathcal{M}(x^j) - \mathcal{M}(x^*))^{1-\alpha}.\end{aligned} \tag{9.22}$$

By using the KL condition of $\mathcal{M}$ at x^* and (H2), we have

$$\begin{aligned}(\mathcal{M}(x^j) - \mathcal{M}(x^*))^{1-\alpha} \leq c^{\frac{1-\alpha}{\alpha}}\|w^j\|^{\frac{1-\alpha}{\alpha}} &\leq (bc)^{\frac{1-\alpha}{\alpha}}\|x^j - x^{j-1}\|^{\frac{1-\alpha}{\alpha}} \\ &= (bc)^{\frac{1-\alpha}{\alpha}}(\Delta_{j-1} - \Delta_j)^{\frac{1-\alpha}{\alpha}}.\end{aligned}$$

Combining this with (9.22) gives

$$\Delta_j \leq \Delta_{j-1} - \Delta_j + C_1(\Delta_{j-1} - \Delta_j)^{\frac{1-\alpha}{\alpha}}, \tag{9.23}$$

where $C_1 := ((bc^{1-2\alpha})^{\frac{1}{\alpha}}/[(1-\alpha)a]$.

Case 1. $\alpha = 0$ If there is an integer k such that $x^k = x^*$, then it follows from (H1) that

$$\mathcal{M}(x^{k+1}) + a\|x^{k+1} - x^k\|^2 \leq \mathcal{M}(x^k) = \mathcal{M}(x^*).$$

Since $\mathcal{M}(x^{k+1}) \geq \mathcal{M}(x^*)$, we must have $\mathcal{M}(x^{k+1}) = \mathcal{M}(x^*)$ and $x^{k+1} = x^k = x^*$. Repeating this argument, we have $x^j = x^*$ and $\mathcal{M}(x^j) = \mathcal{M}(x^*)$ for all $j \geq k$. Therefore, if $\{x^k\}$ does not converge in finite many steps, then $x^k \neq x^*$ and

$\mathcal{M}(x^k) > \mathcal{M}(x^*)$ for all k. So we may use the KL condition of $\mathcal{M}$ at x^* and (H2) to conclude that

$$1 \leq cd(0, \partial\mathcal{M}(x^k)) \leq bc\|x^k - x^{k-1}\| \quad \text{for all } k.$$

However, by using (H1), we also have

$$a\|x^k - x^{k-1}\|^2 \leq \mathcal{M}(x^{k-1}) - \mathcal{M}(x^k) \to 0 \quad \text{as } k \to \infty.$$

We therefore obtain a contradiction.

Case 2. $\alpha \in (0, 1/2]$ Now we have $\frac{1-\alpha}{\alpha} \geq 1$. Since $\Delta_j \to 0$ as $j \to \infty$, we can find a constant $C_2 > 0$ such that

$$C_1(\Delta_{j-1} - \Delta_j)^{\frac{1-\alpha}{\alpha}} \leq C_2(\Delta_{j-1} - \Delta_j)$$

for all j. It then follows from (9.23) that

$$\Delta_j \leq (1 + C_2)(\Delta_{j-1} - \Delta_j)$$

which implies that

$$\Delta_j \leq q\Delta_{j-1} \quad \text{with } q := \frac{1 + C_2}{2 + C_2} < 1.$$

Thus $\Delta_j \leq \Delta_0 q^j$ and hence $\|x^j - x^*\| = \lim_{k\to\infty} \|x^j - x^k\| \leq \Delta_j \leq \Delta_0 q^j$ for all j.

Case 3. $\alpha \in (1/2, 1)$ Now $\frac{1-\alpha}{\alpha} < 1$. Since $\Delta_j \to 0$ as $j \to \infty$, we can find a constant $C_3 > 0$ such that

$$\Delta_{j-1} - \Delta_j \leq C_3(\Delta_{j-1} - \Delta_j)^{\frac{1-\alpha}{\alpha}}$$

for all j. It then follows from (9.23) that

$$\Delta_j \leq (C_1 + C_3)(\Delta_{j-1} - \Delta_j)^{\frac{1-\alpha}{\alpha}}$$

or equivalently

$$\Delta_j^{\frac{\alpha}{1-\alpha}} \leq C_4(\Delta_{j-1} - \Delta_j) \quad \text{for all } j$$

for some constant $C_4 > 0$. Since $\alpha/(1-\alpha) > 1$, we may use Lemma 8.46 to conclude that $\Delta_j \le C_5(1+j)^{-\frac{1-\alpha}{2\alpha-1}}$. for some constant C_5. Consequently

$$\|x^j - x^*\| \le \Delta_j \le C_5(1+j)^{-\frac{1-\alpha}{2\alpha-1}}$$

for all j. The proof is thus complete. □

9.4 Analysis of Algorithms

In the previous section, we have provided a general framework on convergence and convergence rates of an iterative sequence admitting certain properties with a merit function satisfying the KL condition. In this section, we will provide applications for various methods including the forward-backward splitting method, the proximal alternating linearized minimization method (PALM), the Douglas-Rachford splitting algorithm, and alternating minimization. We will frequently use the following result which can be viewed as a generalization of Lemma 4.34 in convex setting to non-convex setting.

Lemma 9.28 *Let $h : \mathbb{R}^n \to \mathbb{R}$ be a continuous differentiable function with L-Lipschitz continuous gradient ∇h, and let $\sigma : \mathbb{R}^n \to (-\infty, \infty]$ be a proper and lower semicontinuous function. For any $t > 0$ and $v, w \in \mathbb{R}^n$, we define $u^+ \in \mathbb{R}^n$ by*

$$u^+ \in \arg\min_{x\in\mathbb{R}^n} \left\{\sigma(x) + \frac{1}{2t}\|x - (v - t\nabla h(w))\|^2\right\}$$

which is assumed to exist. Then for any $\eta > 0$ and $u \in \mathbb{R}^n$, there holds

$$\begin{aligned} h(u^+) + \sigma(u^+) \le h(u) + \sigma(u) + \frac{L+\eta}{2}\|u^+ - u\|^2 + \frac{L^2}{2\eta}\|u - w\|^2 \\ + \frac{1}{2t}\|u - v\|^2 - \frac{1}{2t}\|u^+ - v\|^2. \end{aligned} \tag{9.24}$$

If $u = v = w$, then there holds

$$h(u^+) + \sigma(u^+) \le h(u) + \sigma(u) - \frac{1}{2}\left(\frac{1}{t} - L\right)\|u^+ - u\|^2. \tag{9.25}$$

Proof From the definition of u^+, it follows that

$$\sigma(u^+) + \frac{1}{2t}\|u^+ - v + t\nabla h(w)\|^2 \le \sigma(u) + \frac{1}{2t}\|u - v + t\nabla h(w)\|^2.$$

By expanding the terms, we have

$$\sigma(u^+)+\frac{1}{2t}\|u^+-v\|^2+\langle\nabla h(w),u^+-v\rangle\leq\sigma(u)+\frac{1}{2t}\|u-v\|^2+\langle\nabla h(w),u-v\rangle,$$

that is

$$\sigma(u^+)\leq\sigma(u)+\langle\nabla h(w),u-u^+\rangle+\frac{1}{2t}\|u-v\|^2-\frac{1}{2t}\|u^+-v\|^2.$$

Since h is continuous differentiable with L-Lipschitz continuous gradient, we may use the descent lemma, i.e., Lemma 2.30 to obtain

$$h(u^+)\leq h(u)+\langle\nabla h(u),u^+-u\rangle+\frac{L}{2}\|u^+-u\|^2.$$

Adding the above two equations gives

$$\begin{aligned}h(u^+)+\sigma(u^+)\leq h(u)+\sigma(u)+\langle\nabla h(u)-\nabla h(w),u^+-u\rangle+\frac{L}{2}\|u^+-u\|^2\\+\frac{1}{2t}\|u-v\|^2-\frac{1}{2t}\|u^+-v\|^2.\end{aligned}\tag{9.26}$$

By virtue of the Cauchy-Schwarz inequality and the L-Lipschitz continuity of ∇h, we have for any $\eta>0$ that

$$\begin{aligned}\langle\nabla h(u)-\nabla h(w),u^+-u\rangle&\leq\frac{\eta}{2}\|u^+-u\|^2+\frac{1}{2\eta}\|\nabla h(u)-\nabla h(w)\|^2\\&\leq\frac{\eta}{2}\|u^+-u\|^2+\frac{L^2}{2\eta}\|u-w\|^2.\end{aligned}$$

Combining this with (9.26), we thus obtain (9.24). When $u=v=w$, from (9.26), we immediately obtain (9.25). □

9.4.1 Forward-Backward Splitting Method

The forward-backward splitting method, also called the proximal gradient method, has been studied in Chap. 4 for convex minimization problems whose objective functions are the sums of two convex functions; one of them is continuous differentiable with Lipschitz continuous gradient. In this section, we will consider the forward-backward splitting method for solving the non-convex minimization problem

$$\min_{x\in\mathbb{R}^n}\{F(x):=f(x)+g(x)\},\tag{9.27}$$

where $f : \mathbb{R}^n \to \mathbb{R}$ is a continuous differentiable function with L-Lipschitz continuous gradient ∇f and $g : \mathbb{R}^n \to (-\infty, \infty]$ is a proper lower semicontinuous function. We assume that (9.27) has a solution. The forward-backward splitting method for solving (9.27) can be described as follows.

Algorithm 9.29 *Take an initial guess $x^0 \in \mathbb{R}^n$. For $k = 0, 1, \cdots$, determine*

$$x^{k+1} \in \arg\min_{x \in \mathbb{R}^n} \left\{ g(x) + \frac{1}{2t_k} \|x - (x^k - t_k \nabla f(x^k))\|^2 \right\}$$

with suitably chosen step size $t_k > 0$.

In the formulation of Algorithm 9.29, we assume x^{k+1} is well-defined at each iteration step. This can be guaranteed under certain conditions on g; for instance, each x^{k+1} is well-defined if g is bounded from below by a linear function.

We will use the framework in Sect. 9.3 to analyze Algorithm 9.29. We assume that the step size $\{t_k\}$ satisfies $0 < \underline{t} \le t_k \le \bar{t} < 1/L$ for all k. By using (9.25) in Lemma 9.28 with $h = f$, $\sigma = g$, $u = x^k$ and $u^+ = x^{k+1}$, we can obtain

$$\begin{aligned} F(x^{k+1}) &\le F(x^k) - \frac{1}{2}\left(\frac{1}{t_k} - L\right) \|x^{k+1} - x^k\|^2 \\ &\le F(x^k) - \frac{1}{2}\left(\frac{1}{\bar{t}} - L\right) \|x^{k+1} - x^k\|^2 \end{aligned}$$

which shows (H1). By the definition of x^{k+1}, Lemma 9.5 and Proposition 9.8, we have

$$0 \in \partial g(x^{k+1}) + \frac{1}{t_k}(x^{k+1} - x^k) + \nabla f(x^k)$$

Therefore

$$\begin{aligned} w^{k+1} &:= -\frac{1}{t_k}(x^{k+1} - x^k) + \left(\nabla f(x^{k+1}) - \nabla f(x^k)\right) \\ &\in \nabla f(x^{k+1}) + \partial g(x^{k+1}) = \partial F(x^{k+1}). \end{aligned}$$

Note that

$$\|w^{k+1}\| \le \frac{1}{t_k}\|x^{k+1} - x^k\| + \|\nabla f(x^{k+1}) - \nabla f(x^k)\| \le \left(\frac{1}{\underline{t}} + L\right) \|x^{k+1} - x^k\|.$$

This shows that (H2) holds. Now we can apply Theorems 9.26 and 9.27 to obtain the following convergence result.

Theorem 9.30 *Let $F = f + g : \mathbb{R}^n \to (-\infty, \infty]$ be a proper, lower semicontinuous, KL function, where $f : \mathbb{R}^n \to \mathbb{R}$ is continuous differentiable with*

L-Lipschitz continuous gradient and g is proper and lower semicontinuous. Let $\{x^k\}$ be a sequence generated by Algorithm 9.29 with $0 < \underline{t} \le t_k \le \bar{t} < 1/L$. If $\{x^k\}$ is bounded, then it has a finite length and converges to a critical point of F. If F is a KL function with exponent $\alpha \in [0, 1)$, then $\{x^k\}$ enjoys the convergence rate given in Theorem 9.27.

Proof In order to apply Theorems 9.26 and 9.27, we need to verify (H1), (H2) and (H3). We have verified (H1) and (H2). It remains only to verify (H3). Since $\{x^k\}$ is assumed bounded, it has a subsequence $\{x^{k_j}\}$ such that $x^{k_j} \to x^*$ as $j \to \infty$. By the lower semi-continuity of g, we have

$$g(x^*) \le \liminf_{j\to\infty} g(x^{k_j}).$$

Thus, by the continuity of f, we have

$$F(x^*) = f(x^*) + g(x^*) \le \liminf_{j\to\infty}\left(f(x^{k_j}) + g(x^{k_j})\right) = \liminf_{j\to\infty} F(x^{k_j})$$

According to (H1), $\{F(x^k)\}$ is monotonically decreasing. Thus $F(x^k) \ge F(x^*) > -\infty$ for all k. Consequently, it follows from (H1) that $\|x^{k+1} - x^k\| \to 0$ as $k \to \infty$. This in particular implies that $x^{k_j} - x^{k_j-1} \to 0$ and $x^{k_j-1} \to x^*$ as $j \to \infty$. Now, we use the minimizing property of x^{k_j} to obtain

$$\begin{aligned} &g(x^{k_j}) + \frac{1}{2t_{k_j-1}}\left\|x^{k_j} - x^{k_j-1} + t_{k_j-1}\nabla f(x^{k_j-1})\right\|^2 \\ &\le g(x^*) + \frac{1}{2t_{k_j-1}}\left\|x^* - x^{k_j-1} + t_{k_j-1}\nabla f(x^{k_j-1})\right\|^2 \end{aligned}$$

which can be equivalently written as

$$\begin{aligned} g(x^{k_j}) \le g(x^*) &+ \langle \nabla f(x^{k_j-1}), x^* - x^{k_j}\rangle + \frac{1}{2t_{k_j-1}}\|x^* - x^{k_j-1}\|^2 \\ &- \frac{1}{2t_{k_j-1}}\|x^{k_j} - x^{k_j-1}\|^2. \end{aligned}$$

By taking $j \to \infty$ and using the condition $0 < \underline{t} \le t_k \le \bar{t}$, we then obtain

$$\limsup_{j\to\infty} g(x^{k_j}) \le g(x^*).$$

Therefore $\lim_{j\to\infty} g(x^{k_j}) = g(x^*)$. This together with the continuity of f shows that $F(x^{k_j}) \to F(x^*)$ as $j \to \infty$. Therefore (H3) holds. We thus complete the proof by invoking Theorems 9.26 and 9.27. □

Remark 9.31 Theorem 9.30 requires $\{x^k\}$ to be bounded. This can be guaranteed if the sublevel set $\{x \in \mathbb{R}^n : F(x) \leq F(x^0)\}$ of F is bounded as $\{F(x^k)\}$ is monotonically decreasing.

Example 9.32 Let A be a nonzero $m \times n$ matrix and $b \in \mathbb{R}^m$. Consider the minimization problem

$$\min_{x\in\mathbb{R}^n} \left\{\lambda\|x\|_0 + \frac{1}{2}\|Ax - b\|^2\right\} \tag{9.28}$$

which is used to determine sparse solutions, where $\lambda > 0$ is a regularization parameter and $\|x\|_0$ denotes the ℓ^0-norm of x which counts the number of nonzero entries in x. According to Example 9.25, the objective function in (9.28) is lower semicontinuous and semi-algebraic and hence is a KL function with exponent α for some $\alpha \in [0, 1)$. In order to apply the forward-backward splitting method to solve (9.28), we need to determine a solution of the minimization problem

$$\min \left\{\lambda\|x\|_0 + \frac{1}{2}\|x - v\|^2\right\} \tag{9.29}$$

for any $\lambda > 0$ and $v \in \mathbb{R}^n$. Since the objective function in (9.29) is separable, we only need to consider the one-dimensional case. We have the following result.

Lemma 9.33 *Given $\lambda > 0$ and $a \in \mathbb{R}$, the one-dimensional minimization problem*

$$\min_{t\in\mathbb{R}} \left\{\lambda \text{sign}(|t|) + \frac{1}{2}(t - a)^2\right\}$$

has an optimal solution given by

$$h_\lambda(a) = a\text{sign}\{(a^2 - 2\lambda)_+\} = \begin{cases} a & \text{if} |a| > \sqrt{2\lambda} \\ 0 & \text{if} |a| \leq \sqrt{2\lambda} \end{cases}$$

which is called the one-dimensional hard thresholding of a with threshold $\lambda > 0$.

Proof Let $\varphi(t) := \lambda\text{sign}(|t|) + \frac{1}{2}(t - a)^2$, and let $\hat{t}$ be a minimizer of φ over $\mathbb{R}$. If $\hat{t} \neq 0$, then $\hat{t}$ is a critical point of the function $t \to \lambda + \frac{1}{2}(t - a)^2$ and hence $\hat{t} = a$. This shows that either $\hat{t} = 0$ or $\hat{t} = a$. Further, if $\hat{t} \neq 0$, then $\hat{t} = a$ and its minimality implies that $\varphi(a) = \varphi(\hat{t}) \leq \varphi(0)$, i.e., $\lambda \leq a^2/2$ which forces $|a| \geq \sqrt{2\lambda}$. Therefore, if $|a| < \sqrt{2\lambda}$, then $\hat{t} = 0$; indeed, this can be verified directly as for any $t \neq 0$, there holds

$$\varphi(0) - \varphi(t) = \frac{1}{2}a^2 - \lambda - \frac{1}{2}(t - a)^2 < -\frac{1}{2}(t - a)^2 \leq 0.$$

If $|a| > \sqrt{2\lambda}$, then $\varphi(a) = \lambda < \frac{1}{2}a^2 = \varphi(0)$. Thus $\hat{t} \neq 0$, and as shown above, we must have $\hat{t} = a$.

For the case $|a| = \sqrt{2\lambda}$, note that $\varphi(a) = \lambda = \frac{1}{2}a^2 = \varphi(0)$ and for any $t \neq 0$, there holds

$$\varphi(t) - \varphi(0) = \lambda + \frac{1}{2}(t-a)^2 - \frac{1}{2}a^2 = \frac{1}{2}(t-a)^2 \geq 0.$$

Therefore, both $\hat{t} = 0$ and $\hat{t} = a$ are minimizers of φ over $\mathbb{R}$; by taking $\hat{t} = 0$, we obtain the desired conclusion. □

Based on Lemma 9.33, the minimization problem (9.29) has an optimal solution given by

$$H_\lambda(v) = (h_\lambda(v_1), \cdots, h_\lambda(v_n))$$

which is call the *hard thresholding* of $v := (v_1, \cdots, v_n)$ with threshold λ. Applying the forward-backward splitting method, i.e., Algorithm 9.29 with $f(x) = \frac{1}{2}\|Ax - b\|^2$ and $g(x) = \lambda\|x\|_0$, to solve (9.28) gives the iteration scheme

$$x^{k+1} = H_{\lambda t_k}(x^k - t_k A^T(Ax^k - b)). \tag{9.30}$$

Note that

$$\|\nabla f(x) - \nabla f(y)\| = \|A^T A(x-y)\| \leq \|A\|^2\|x-y\|, \quad \forall x, y \in \mathbb{R}^n.$$

Therefore, if the step sizes t_k are chosen such that

$$0 < \underline{t} \leq t_k \leq \bar{t} < 1/\|A\|^2,$$

then we may use Theorem 9.30 to conclude that for the sequence $\{x^k\}$ generated by (9.30), if bounded, converges to a critical point of $\lambda\|x\|_0 + \frac{1}{2}\|Ax - b\|^2$ which, according to Proposition 9.8 and Example 9.7, is a point $x^* \in \mathbb{R}^n$ such that $[A^T(Ax^* - b)]_i = 0$ whenever $x_i^* \neq 0$.

9.4.2 Proximal Alternating Linearized Minimization Method

In this section, we consider the minimization problem

$$\min\left\{F(x) := H(x) + \sum_{i=1}^{s} f_i(x_i) : x \in \mathbb{R}^n\right\}, \tag{9.31}$$

where $x = (x_1, \cdots, x_s)$ with $x_i \in \mathbb{R}^{n_i}$ for $i = 1, \cdots, s$ and $n = \sum_{i=1}^{s} n_i$. Here we assume that $H : \mathbb{R}^n \to \mathbb{R}$ is a continuous differentiable function and each $f_i : \mathbb{R}^{n_i} \to (-\infty, \infty]$ is a proper lower semicontinuous function. We will use $\nabla_i H$ to denote the partial gradient of H with respect to x_i. This model is rich enough to cover many applications including compressed sensing, machine learning, matrix factorization, blind decomposition, and image reconstruction.We will assume that H and $f_i, i = 1, \cdots, s$, satisfy the following conditions.

Assumption 9.34

(i) $H : \mathbb{R}^{n_1} \times \cdots \times \mathbb{R}^{n_s} \to (-\infty, \infty)$ *is continuous differentiable and each* $f_i : \mathbb{R}^{n_i} \to (-\infty, \infty]$ *is proper, lower semicontinuous. All these functions* H *and* f_i, $i = 1, \cdots, s$ *are bounded from below.*

(ii) *For each* $i = 1, \cdots, s$ *and each fixed* $(x_1, \cdots, x_{i-1}, x_{i+1}, \cdots, x_s) \in \mathbb{R}^{n_1} \times \cdots \times \mathbb{R}^{n_{i-1}} \times \mathbb{R}^{n_{i+1}} \times \cdots \times \mathbb{R}^{n_s}$, *there exists a number*

$$L_i(x_1, \cdots, x_{i-1}, x_{i+1}, \cdots, x_s) > 0$$

such that

$$\begin{aligned}&\|\nabla_i H(x_1, \cdots, x_{i-1}, u, x_{i+1}, \cdots, x_s) - \nabla_i H(x_1, \cdots, x_{i-1}, v, x_{i+1}, \cdots, x_s)\|\\ &\le L_i(x_1, \cdots, x_{i-1}, x_{i+1}, \cdots, x_s)\|u - v\|\end{aligned}$$

for all $u, v \in \mathbb{R}^{n_i}$.

(iii) *For each* $i = 1, \cdots, s$ *and each bounded subset* Ω *in* $\prod_{j=1}^{i-1} \mathbb{R}^{n_j} \times \prod_{j=i+1}^{s} \mathbb{R}^{n_j}$, *there exist* $L_i^+ \ge L_i^- > 0$ *such that*

$$L_i^- \le L_i(x_1, \cdots, x_{i-1}, x_{i+1}, \cdots, x_s) \le L_i^+$$

for all $(x_1, \cdots, x_{i-1}, x_{i+1}, \cdots, x_s) \in \Omega$.

(iv) ∇H *is Lipschitz continuous on bounded subsets of* $\mathbb{R}^{n_1} \times \cdots \times \mathbb{R}^{n_s}$, *that is, for each bounded subset* $B_1 \times \cdots \times B_s \subset \mathbb{R}^{n_1} \times \cdots \times \mathbb{R}^{n_s}$, *there exists* $M > 0$ *such that*

$$\|\nabla H(x_1, \cdots x_s) - \nabla H(y_1, \cdots y_s)\| \le M\|(x_1, \cdots, x_s) - (y_1, \cdots, y_s)\|$$

for all $(x_1, \cdots, x_s), (y_1, \cdots, y_s) \in B_1 \times \cdots \times B_s$.

The standard method for solving (9.31) is the so-called *Gauss-Seidel iteration scheme*, also known as *alternating minimization method* or *block coordinate descent method* in modern era. Starting with some given initial point $x^0 := (x_1^0, \cdots, x_s^0)$, this method generates a sequence $\{x^k := (x_1^k, \cdots, x_s^k)\}$ iteratively by the scheme

$$\begin{cases} x_1^{k+1} \in \arg\min_{x_1 \in \mathbb{R}^{n_1}} F(x, x_2^k, x_3^k, \cdots, x_s^k), \\ x_2^{k+1} \in \arg\min_{x_2 \in \mathbb{R}^{n_2}} F(x_1^{k+1}, x_2, x_3^k, \cdots, x_s^k), \\ \vdots \qquad\qquad\qquad \vdots \\ x_s^{k+1} \in \arg\min_{x_s \in \mathbb{R}^{n_s}} F(x_1^{k+1}, \cdots, x_{s-1}^{k+1}, x_s). \end{cases} \tag{9.32}$$

The alternating minimization method has been studied extensively in the literature [10, 14, 16, 213]. One of the key assumptions necessary to proving convergence is that F must be strongly convex of one argument, while the other arguments are fixed in order to guarantee that each x_i^{k+1} is uniquely defined. This technical condition can be removed if we modify the alternating minimization method by adding proximal terms, which leads to the proximal alternating minimization method [7]

$$\begin{cases} x_1^{k+1} \in \arg\min_{x_1 \in \mathbb{R}^{n_1}} \left\{ f_1(x_1) + H(x_1, x_2^k, \cdots, x_s^k) + \frac{\gamma_1}{2}\|x_1 - x_1^k\|^2 \right\}, \\ x_2^{k+1} \in \arg\min_{x_2 \in \mathbb{R}^{n_2}} \left\{ f_2(x_2) + H(x_1^{k+1}, x_2, x_3^k, \cdots, x_s^k) + \frac{\gamma_2}{2}\|x_2 - x_2^k\|^2 \right\} \\ \vdots \qquad\qquad\qquad \vdots \\ x_s^{k+1} \in \arg\min_{x_s \in \mathbb{R}^{n_s}} \left\{ f_s(x_s) + H(x_1^{k+1}, \cdots, x_{s-1}^{k+1}, x_s) + \frac{\gamma_s}{2}\|x_s - x_s^k\|^2 \right\}, \end{cases} \tag{9.33}$$

where γ_i, $i = 1, \cdots, s$, are positive numbers. Under very general conditions, convergence of subsequence has been proved. This method, however, suffers from the drawback that each step requires exact minimization of a non-convex and non-smooth problem which usually can only be solved inexactly. For instance, the determination of x_i^{k+1} requires to solve the minimization problem

$$x_i^{k+1} \in \arg\min_{x_i \in \mathbb{R}^{n_i}} \left\{ f_i(x_i) + H(x_{<i}^{k+1}, x_i, x_{>i}^k) + \frac{\gamma_i}{2}\|x_i - x_i^k\|^2 \right\}, \tag{9.34}$$

where for $x = (x_1, \cdots, x_s)$ we used the convention

$$x_{<i} := (x_1, \cdots, x_{i-1}), \qquad x_{>i} := (x_{i+1}, \cdots, x_s); \tag{9.35}$$

the notation $x_{\le i}$ and $x_{\ge i}$ can be understood similarly. Even if f_i can be a simple function, the appearance of $H(x_{<i}^{k+1}, x_i, x_{>i}^k)$ in (9.34) makes the determination of x_i^{k+1} difficult. To alleviate this drawback, one may consider to replace

$H(x^{k+1}_{<i}, x_i, x^k_{>i})$ by its linearization at x_i^k which is given by

$$H(x^{k+1}_{<i}, x^k_{\geq i}) + \langle \nabla_i H(x^{k+1}_{<i}, x^k_{\geq i}), x_i - x_i^k \rangle.$$

This leads to the following proximal alternating linearized minimization method [34].

Algorithm 9.35 (Proximal Alternating Linearized Minimization) *Pick an initial guess* $x^0 = (x_1^0, \cdots, x_s^0) \in \mathbb{R}^{n_1} \times \cdots \times \mathbb{R}^{n_s}$ *and* $\gamma_i > 1$ *for* $i = 1, \cdots, s$. *For* $k = 0, 1, \ldots$, *define* $x^{k+1} := (x_1^{k+1}, \cdots, x_s^{k+1})$ *by successively calculating*

$$\gamma_i^k := \gamma_i L_i(x^{k+1}_{<i}, x^k_{>i}),$$

$$x_i^{k+1} \in \arg\min_{x_i \in \mathbb{R}^{n_i}} \left\{ f_i(x_i) + \langle \nabla_i H(x^{k+1}_{<i}, x^k_{\geq i}), x_i - x_i^k \rangle + \frac{\gamma_i^k}{2} \|x_i - x_i^k\|^2 \right\}$$

for $i = 1, \cdots, s$.

We will use Theorems 9.26 and 9.27 to produce convergence results concerning Algorithm 9.35. Therefore, we need to verify (H1), (H2), and (H3) with $\mathcal{M} := F$.

Lemma 9.36 *Let Assumptions 9.34 hold, and let* $\{x^k := (x_1^k, \cdots, x_s^k)\}$ *be a sequence generated by Algorithm 9.35. Then*

$$F(x^{k+1}) \leq F(x^k) - \frac{1}{2} \sum_{i=1}^{s} (\gamma_i - 1) L_i^k \|x_i^{k+1} - x_i^k\|^2, \tag{9.36}$$

where $L_i^k = L_i(x^{k+1}_{<i}, x^k_{>i})$. *If* $\{x^k\}$ *is bounded, then (H1) holds, i.e., there is a positive constant* $\rho > 0$ *such that*

$$F(x^{k+1}) \leq F(x^k) - \rho \|x^{k+1} - x^k\|^2 \tag{9.37}$$

for all $k \geq 0$.

Proof For $i = 1, \cdots, s$, by the definition of x_i^{k+1}, we have

$$x_i^{k+1} \in \arg\min_{x_i \in \mathbb{R}^{n_i}} \left\{ f_i(x_i) + \frac{\gamma_i^k}{2} \left\| x_i - \left(x_i^k - \frac{1}{\gamma_i^k} \nabla_i H(x^{k+1}_{<i}, x^k_{\geq i}) \right) \right\|^2 \right\}.$$

Thus, based on (ii) in Assumption 9.34, we may use (9.25) in Lemma 9.28 with $\sigma = f_i$, $h = H(x^{k+1}_{<i}, \cdot, x^k_{>i})$, $t = 1/\gamma_i^k$, $u = x_i^k$ and $u^+ = x_i^{k+1}$ to obtain

$$f_i(x_i^{k+1}) + H(x^{k+1}_{\leq i}, x^k_{>i}) \leq f_i(x_i^k) + H(x^{k+1}_{<i}, x^k_{\geq i}) - \frac{\gamma_i^k - L_i^k}{2} \|x_i^{k+1} - x_i^k\|^2.$$

Summing this inequality over i from 1 to s gives

$$\begin{aligned}&\sum_{i=1}^{s} f_i(x_i^{k+1}) + \sum_{i=1}^{s} H(x_{\le i}^{k+1}, x_{>i}^{k})\\ &\le \sum_{i=1}^{s} f_i(x_i^{k}) + \sum_{i=1}^{s} H(x_{\le i-1}^{k+1}, x_{>i-1}^{k}) - \frac{1}{2}\sum_{i=1}^{s}(\gamma_i^k - L_i^k)\|x_i^{k+1} - x_i^k\|^2.\end{aligned}$$

After cancellation and using $\gamma_i^k = \gamma_i L_i^k$, we can obtain

$$\sum_{i=1}^{s} f_i(x_i^{k+1}) + H(x^{k+1}) \le \sum_{i=1}^{s} f_i(x_i^{k}) + H(x^k) - \frac{1}{2}\sum_{i=1}^{s}(\gamma_i - 1)L_i^k\|x_i^{k+1} - x_i^k\|^2$$

This shows (9.36).

If $\{x^k\}$ is a bounded sequence, we may use (iii) in Assumption 9.34 to find constants $L_i^- > 0$ such that $L_i^k \ge L_i^-$ for all k and $i = 1, \cdots s$. Thus, by using (9.36) and $\gamma_i > 1$, we may conclude (9.37) with $\rho = \frac{1}{2}\min\{(\gamma_i - 1)L_i^- : i = 1, \cdots, s\}$. □

Lemma 9.37 *Let Assumption 9.34 hold, and let $\{x^k\}$ be a sequence generated by Algorithm 9.35. If $\{x^k\}$ is bounded, then (H2) holds. To be more precisely, for each integer $k \ge 0$, define*

$$\xi_i^{k+1} := \gamma_i^k(x_i^k - x_i^{k+1}) + \nabla_i H(x^{k+1}) - \nabla_i H(x_{<i}^{k+1}, x_{\ge i}^{k})$$

for $i = 1, \cdots, s$. Then $\xi^{k+1} := (\xi_1^{k+1}, \cdots, \xi_s^{k+1}) \in \partial F(x^{k+1})$ and there exists $C > 0$ such that

$$\|\xi^{k+1}\| \le C\|x^{k+1} - x^k\|$$

for all $k \ge 0$.

Proof By the minimizing property of x_i^{k+1}, Lemma 9.5 and Proposition 9.8, we have

$$0 \in \partial f_i(x_i^{k+1}) + \nabla_i H(x_{<i}^{k+1}, x_{\ge i}^{k}) + \gamma_i^k(x_i^{k+1} - x_i^k)$$

which implies that

$$\begin{aligned}\xi_i^{k+1} &:= \gamma_i^k(x_i^k - x_i^{k+1}) + \nabla_i H(x^{k+1}) - \nabla_i H(x_{<i}^{k+1}, x_{\ge i}^{k})\\ &\in \partial f_i(x_i^{k+1}) + \nabla_i H(x^{k+1}).\end{aligned}$$

Thus, it follows from Proposition 9.8 that $\xi^{k+1} := (\xi_1^{k+1}, \cdots, \xi_s^{k+1}) \in \partial F(x^{k+1})$.

We next estimate $\|\xi^{k+1}\|$. Since $\{x^k\}$ is bounded, according to (iii) in Assumption 9.34, we can find constants L_i^+ such that $L_i^k \leq L_i^+$ for all $k \geq 0$ and $i = 1, \cdots, s$. Furthermore, according to (iv) in Assumption 9.34, we can find a constant M independent of k such that

$$\|\nabla_i H(x^{k+1}) - \nabla_i H(x_{<i}^{k+1}, x_{\geq i}^k)\| \leq M\|x_{\geq i}^{k+1} - x_{\geq i}^k\| \leq M\|x^{k+1} - x^k\|.$$

Thus

$$\begin{aligned}
\|\xi_i^{k+1}\| &\leq \gamma_i^k \|x_i^k - x_i^{k+1}\| + \|\nabla_i H(x^{k+1}) - \nabla_i H(x_{<i}^{k+1}, x_{\geq i}^k)\| \\
&\leq \gamma_i L_i^k \|x_i^k - x_i^{k+1}\| + M\|x^{k+1} - x^k\| \\
&\leq (\gamma_i L_i^+ + M)\|x^{k+1} - x^k\|.
\end{aligned}$$

Therefore

$$\|\xi^{k+1}\| \leq \sum_{i=1}^s \|\xi_i^{k+1}\| \leq \left(\sum_{i=1}^s \gamma_i L_i^+ + sM\right) \|x^{k+1} - x^k\|.$$

The proof is complete. □

Lemma 9.38 *Let Assumption 9.34 hold, and let $\{x^k\}$ be a sequence generated by Algorithm 9.35. If $\{x^k\}$ is bounded, then (H3) holds, i.e., there exist $x^* \in \mathbb{R}^n$ and a subsequence $\{x^{k_l}\}$ of $\{x^k\}$ such that $x^{k_l} \to x^*$ and $F(x^{k_l}) \to F(x^*)$ as $l \to \infty$.*

Proof Since $\{x^k\}$ is bounded, it has a subsequence $\{x^{k_l}\}$ that converges to a point x^*, i.e.,

$$\lim_{l\to\infty} \|x_i^{k_l} - x_i^*\| = 0, \quad i = 1, \cdots, s \tag{9.38}$$

Since each f_i is lower semicontinuous, we have

$$f_i(x_i^*) \leq \liminf_{l\to\infty} f(x_i^{k_l}), \quad i = 1, \cdots, s. \tag{9.39}$$

According to Lemma 9.36, $\{F(x^k)\}$ is monotonically decreasing. Since (i) in Assumption 9.34 implies that F is bounded from below, $\lim_{k\to\infty} F(x^k)$ must exist and is finite. Therefore, by using Lemma 9.36 again, we can conclude that $\|x^k - x^{k-1}\| \to 0$ as $k \to \infty$, i.e.,

$$\lim_{k\to\infty} \|x_i^k - x_i^{k-1}\| \to 0, \quad i = 1, \cdots, s \tag{9.40}$$

It then follows from (9.38) that

$$\lim_{l\to\infty} \|x_i^{k_l-1} - x^*\| = 0, \quad i = 1, \cdots, s. \tag{9.41}$$

By using the minimizing property of $x_i^{k_l}$, we have

$$f_i(x_i^{k_l}) + \langle \nabla_i H(x_{<i}^{k_l}, x_{\geq i}^{k_l-1}), x_i^{k_l} - x_i^{k_l-1}\rangle + \frac{\gamma_i^{k_l-1}}{2}\|x_i^{k_l} - x_i^{k_l-1}\|^2$$
$$\leq f_i(x_i^*) + \langle \nabla_i H(x_{<i}^{k_l}, x_{\geq i}^{k_l-1}), x_i^* - x_i^{k_l-1}\rangle + \frac{\gamma_i^{k_l-1}}{2}\|x_i^* - x_i^{k_l-1}\|^2. \tag{9.42}$$

As $\{x^k\}$ is bounded, we may use (iii) in Assumption 9.34 to assert that $\gamma_i^{k_l-1} \leq \gamma_i L_i^+$ for some constants L_i^+ for $i = 1, \cdots, s$. Thus, by virtue of the continuity of ∇H, (9.38), (9.40) and (9.41), it follows from (9.42) that

$$\limsup_{l\to\infty} f_i(x_i^{k_l}) \leq f_i(x_i^*), \quad i = 1, \cdots, s. \tag{9.43}$$

The combination of (9.39) and (9.43) gives

$$\lim_{l\to\infty} f_i(x_i^{k_l}) = f_i(x_i^*) \quad i - 1, \cdots, s.$$

Since H is continuous, we also have $\lim_{l\to\infty} H(x^{k_l}) = H(x^*)$. Therefore

$$\lim_{l\to\infty} F(x^{k_l}) = \lim_{l\to\infty} \left\{ H(x^{k_l}) + \sum_{i=1}^{s} f_i(x_i^{k_l}) \right\} = H(x^*) + \sum_{i=1}^{s} f_i(x_i^*) = F(x^*)$$

which shows (H3). □

Due to Lemmas 9.36–9.38, we can apply Theorems 9.26 and 9.27 directly to obtain the following convergence result on Algorithm 9.35.

Theorem 9.39 *Let Assumption 9.34 hold, let F be a KL function, and let $\{x^k\}$ be a sequence generated by Algorithm 9.35. If $\{x^k\}$ is bounded, then it has a finite length and converges to a critical point x^* of the function F. If in addition F is a KL function with an exponent $\alpha \in [0, 1)$, then $\{x^k\}$ converges to x^* with the rate given in Theorem 9.27.*

In the above result, $\{x^k\}$ is required to be bounded. This can be guaranteed if the sublevel set $\{F \leq F(x^0)\}$ is bounded because Lemma 9.36 implies that $\{x^k\}$ always stays in this set.

9.4.3 Douglas-Rachford Splitting Algorithm

In Chap. 4, we have considered the Douglas-Rachford splitting algorithm for convex composite minimization problems in which the objective function is a sum of two convex functions. In this section, we will extend the Douglas-Rachford splitting algorithm to solve non-convex optimization problems. We consider the problem of the form

$$\min_{x\in\mathbb{R}^n} \{f(x)+g(x)\}, \tag{9.44}$$

where f and g are proper functions on $\mathbb{R}^n$ which are not necessarily convex. We will make the following assumption

Assumption 9.40

(i) $f:\mathbb{R}^n\to(-\infty,\infty)$ *is differentiable and* ∇f *is* L*-Lipschitz. Moreover, there exists* $\sigma\in\mathbb{R}$ *such that* $f(x)+\frac{\sigma}{2}\|x\|^2$ *is convex.*
(ii) $g:\mathbb{R}^n\to(-\infty,\infty]$ *is proper, lower semicontinuous and*

$$\arg\min_{y\in\mathbb{R}^n}\left\{g(y)+\frac{1}{2\gamma}\|y-x\|^2\right\}$$

is nonempty for any $x\in\mathbb{R}^n$ *and any* $\gamma>0$.
(iii) *the minimization problem (9.44) has a solution.*

The number σ in Assumption 9.40 can be in $[-L,L]$. Indeed, since $f(x)+\frac{\sigma}{2}\|x\|^2$ is convex, we have from Proposition 2.31 that

$$\langle(\nabla f(x)+\sigma x)-(\nabla f(y)+\sigma y),x-y\rangle\geq 0$$

which together with the L-Lipschitz continuity of ∇f implies that

$$-\sigma\|x-y\|^2\leq\langle\nabla f(x)-\nabla f(y),x-y\rangle\leq L\|x-y\|^2$$

for any $x,y\in\mathbb{R}^n$. Therefore $\sigma\geq -L$. On the other hand, by using the L-Lipschitz continuity of ∇f, we have

$$\begin{aligned}
&\left\langle\nabla\left(f+\frac{L}{2}\|\cdot\|^2\right)(x)-\nabla\left(f+\frac{L}{2}\|\cdot\|^2\right)(y),x-y\right\rangle\\
&=\langle\nabla f(x)-\nabla f(y),x-y\rangle+L\|x-y\|^2\\
&\geq-\|\nabla f(x)-\nabla f(y)\|\|x-y\|+L\|x-y\|^2\\
&\geq 0
\end{aligned}$$

which, by using again Proposition 2.31, implies that $f(x) + \frac{L}{2}\|x\|^2$ is a convex function.

The Douglas-Rachford splitting algorithm for solving (9.44) essentially takes the same form as in Algorithm 4.56 in Chap. 4, i.e.,

Algorithm 9.41 *Take an initial guess* z^0 *and a step size* $\gamma > 0$. *For* $k = 0, 1, \cdots$ *do*

$$
\begin{aligned}
x^k &\in \arg\min_{x\in\mathbb{R}^n} \left\{ f(x) + \frac{1}{2\gamma}\|x - z^k\|^2 \right\}, \\
y^k &\in \arg\min_{y\in\mathbb{R}^n} \left\{ g(y) + \frac{1}{2\gamma}\|y - (2x^k - z^k)\|^2 \right\}, \\
z^{k+1} &= z^k + (y^k - x^k).
\end{aligned}
$$

For convex problems, the convergence of Douglas-Rachford splitting algorithm has been established for any $\gamma > 0$. For non convex problems, however, the convergence analysis requires γ to be suitably small; for instance, according to (i) in Assumption 9.40, we need $0 < \gamma < 1/\sigma$ in order to guarantee each x^k to be well-defined; further restrictions on γ will be given below. The analysis on Algorithm 9.41 is based on the framework in Sect. 9.3 with a proper choice of a merit function $\mathcal{M}$. To motivate such a function, we first use the definition of x^k to obtain

$$
0 = \nabla f(x^k) + \frac{1}{\gamma}(x^k - z^k) \tag{9.45}
$$

which implies that $z^k = x^k + \gamma\nabla f(x^k)$. Thus, we may use the definition of y^k to further obtain

$$
\begin{aligned}
y^k &\in \arg\min_{y\in\mathbb{R}^n} \left\{ g(y) + \frac{1}{2\gamma}\|y - (x^k - \gamma\nabla f(x^k))\|^2 \right\} \\
&= \arg\min_{y\in\mathbb{R}^n} \left\{ g(y) + f(x^k) + \langle \nabla f(x^k), y - x^k\rangle + \frac{1}{2\gamma}\|y - x^k\|^2 \right\}
\end{aligned} \tag{9.46}
$$

which suggests us to consider the quantity

$$
\begin{aligned}
&f(x^k) + g(y^k) + \langle \nabla f(x^k), y^k - x^k\rangle + \frac{1}{2\gamma}\|y^k - x^k\|^2 \\
&= f(x^k) + g(y^k) + \frac{1}{\gamma}\langle z^k - x^k, y^k - x^k\rangle + \frac{1}{2\gamma}\|x^k - y^k\|^2.
\end{aligned} \tag{9.47}
$$

This motivates us to consider the function

$$
\mathcal{M}(x, y, z) := f(x) + g(y) + \frac{1}{\gamma}\langle x - z, x - y\rangle + \frac{1}{2\gamma}\|x - y\|^2 \tag{9.48}
$$

which was introduced in [177, 210] by the so-called Douglas-Rachford envelope.

In the following, we will give a convergence analysis on Algorithm 9.41 by following the arguments in [136, 210]

Lemma 9.42 *Let Assumption 9.40 hold, and consider Algorithm 9.41. Assume that* $\gamma > 0$ *is chosen such that*

$$c_0 := \frac{1}{2\gamma} - \frac{\sigma}{2} - \gamma L^2 > 0. \tag{9.49}$$

Then

$$\mathcal{M}(x^{k+1}, y^{k+1}, z^{k+1}) + c_0\|x^{k+1} - x^k\|^2 \leq \mathcal{M}(x^k, y^k, z^k) \tag{9.50}$$

for all $k \geq 0$.

Proof According to (9.46) and (9.47), we have

$$\begin{aligned}
&\mathcal{M}(x^{k+1}, y^{k+1}, z^{k+1}) \\
&= \min_{y\in\mathbb{R}^n} \left\{ f(x^{k+1}) + g(y) + \langle \nabla f(x^{k+1}), y - x^{k+1}\rangle + \frac{1}{2\gamma}\|y - x^{k+1}\|^2 \right\} \\
&\leq f(x^{k+1}) + g(y^k) + \langle \nabla f(x^{k+1}), y^k - x^{k+1}\rangle + \frac{1}{2\gamma}\|y^k - x^{k+1}\|^2.
\end{aligned}$$

Let

$$\rho(x^k, x^{k+1}) := f(x^k) - f(x^{k+1}) - \langle \nabla f(x^{k+1}), x^k - x^{k+1}\rangle.$$

Then

$$\begin{aligned}
&\mathcal{M}(x^{k+1}, y^{k+1}, z^{k+1}) \\
&\leq f(x^k) - \rho(x^k, x^{k+1}) + g(y^k) + \langle \nabla f(x^{k+1}), y^k - x^k\rangle + \frac{1}{2\gamma}\|y^k - x^{k+1}\|^2 \\
&= \mathcal{M}(x^k, y^k, z^k) + \frac{1}{2\gamma}(\|y^k - x^{k+1}\|^2 - \|y^k - x^k\|^2) - \rho(x^k, x^{k+1}) \\
&\quad + \langle \nabla f(x^{k+1}) - \nabla f(x^k), y^k - x^k\rangle \\
&= \mathcal{M}(x^k, y^k, z^k) + \frac{1}{2\gamma}\|x^k - x^{k+1}\|^2 + \frac{1}{\gamma}\langle y^k - x^k, x^k - x^{k+1}\rangle \\
&\quad + \langle \nabla f(x^{k+1}) - \nabla f(x^k), y^k - x^k\rangle - \rho(x^k, x^{k+1}).
\end{aligned}$$

By virtue of (9.45), it follows that

$$x^k - y^k = z^k - z^{k+1} = x^k - x^{k+1} + \gamma\left(\nabla f(x^k) - \nabla f(x^{k+1})\right).$$

Therefore

$$\begin{aligned}
&\mathcal{M}(x^k, y^k, z^k) - \mathcal{M}(x^{k+1}, y^{k+1}, z^{k+1}) \\
&\geq \rho(x^k, x^{k+1}) - \frac{1}{2\gamma}\|x^k - x^{k+1}\|^2 - \frac{1}{\gamma}\langle y^k - x^k, x^k - x^{k+1}\rangle \\
&\quad - \langle \nabla f(x^{k+1}) - \nabla f(x^k), y^k - x^k\rangle \\
&= \rho(x^k, x^{k+1}) + \frac{1}{2\gamma}\|x^k - x^{k+1}\|^2 - \gamma\|\nabla f(x^k) - \nabla f(x^{k+1})\|^2. \qquad (9.51)
\end{aligned}$$

Since $f(x) + \frac{\sigma}{2}\|x\|^2$ is convex, we may use Proposition 2.31 to derive that

$$\rho(x^k, x^{k+1}) \geq -\frac{\sigma}{2}\|x^k - x^{k+1}\|^2.$$

Therefore, by using the L-Lipschitz continuity of ∇f, we further obtain

$$\mathcal{M}(x^k, y^k, z^k) - \mathcal{M}(x^{k+1}, y^{k+1}, z^{k+1}) \geq \left(\frac{1}{2\gamma} - \frac{\sigma}{2} - \gamma L^2\right)\|x^k - x^{k+1}\|^2$$

which completes the proof. □

Remark 9.43

(i) In Lemma 9.42, a larger step size γ is possible if f is convex. In fact, we can show that the result in Lemma 9.42 still holds for any $0 < \gamma < 1/L$. To see this, we first use Proposition 2.34 to obtain

$$\rho(x^k, x^{k+1}) \geq \frac{1}{2L}\|\nabla f(x^k) - \nabla f(x^{k+1})\|^2.$$

Therefore, it follows from (9.51) that

$$\begin{aligned}
&\mathcal{M}(x^k, y^k, z^k) - \mathcal{M}(x^{k+1}, y^{k+1}, z^{k+1}) \\
&\geq \frac{1}{2\gamma}\|x^k - x^{k+1}\|^2 + \left(\frac{1}{2L} - \gamma\right)\|\nabla f(x^k) - \nabla f(x^{k+1})\|^2.
\end{aligned}$$

For $\gamma \le 1/(2L)$, nothing needs to be proved. So we may assume $1/(2L) < \gamma < 1/L$. By the L-Lipschitz continuity of ∇f, we then obtain

$$\mathcal{M}(x^k, y^k, z^k) - \mathcal{M}(x^{k+1}, y^{k+1}, z^{k+1}) \ge \left(\frac{1}{2\gamma} + \frac{L}{2} - \gamma L^2\right) \|x^k - x^{k+1}\|^2$$

which confirms the claim by noting that $\frac{1}{2\gamma} - \frac{L}{2} - \gamma L^2 > 0$ for $0 < \gamma < 1/L$.

(ii) By a delicate analysis, a tight upper bound on the step size was obtained in [210] in which Algorithm 9.41 was even considered with the equation $z^{k+1} = z^k + y^k - x^k$ replaced by the equation

$$z^{k+1} = z^k + \lambda(y^k - x^k)$$

for $0 < \lambda < 4$; the case $\lambda = 2$ has been considered in [139].

Proposition 9.44 *Let the conditions in Lemma 9.42 hold. Let $\{(x^k, y^k, z^k)\}$ be a sequence generated by Algorithm 9.41. If $\{(x^k, y^k, z^k)\}$ has a cluster point (x^*, y^*, z^*), then*

$$x^k - y^k \to 0 \quad \text{and} \quad \mathcal{M}(x^k, y^k, z^k) \to \mathcal{M}(x^*, y^*, z^*) \quad \text{as } k \to \infty.$$

Moreover $x^* = y^*$ and $0 \in \nabla f(x^*) + \partial g(x^*)$.

Proof As (x^*, y^*, z^*) is a cluster point of $\{(x^k, y^k, z^k)\}$, it has a subsequence $\{(x^{k_l}, y^{k_l}, z^{k_l})\}$ such that $(x^{k_l}, y^{k_l}, z^{k_l}) \to (x^*, y^*, z^*)$ as $l \to \infty$. By summing (9.50) in Lemma 9.42 over k from $k = 0$ to $k = k_l - 1$, we can obtain

$$c_0 \sum_{k=0}^{k_l - 1} \|x^{k+1} - x^k\|^2 \le \mathcal{M}(x^0, y^0, z^0) - \mathcal{M}(x^{k_l}, y^{k_l}, z^{k_l}).$$

Since $\mathcal{M}$ is lower semicontinuous, we may take $\limsup_{l\to\infty}$ to obtain

$$c_0 \sum_{k=0}^{\infty} \|x^{k+1} - x^k\|^2 \le \mathcal{M}(x^0, y^0, z^0) - \mathcal{M}(x^*, y^*, z^*).$$

As $\mathcal{M}$ is proper, we have $\mathcal{M}(x^*, y^*, z^*) > -\infty$. Therefore

$$\sum_{k=0}^{\infty} \|x^{k+1} - x^k\|^2 < \infty.$$

This in particular implies that $\|x^{k+1} - x^k\| \to 0$ as $k \to \infty$. Recall Eq. (9.45). It then follows from the L-Lipschitz continuity of ∇f that

$$\begin{aligned}\|z^{k+1} - z^k\| &= \|x^{k+1} - x^k + \gamma(\nabla f(x^{k+1}) - \nabla f(x^k))\| \\ &\leq (1 + \gamma L)\|x^{k+1} - x^k\|. \end{aligned} \tag{9.52}$$

This together with the equation $y^k - x^k = z^{k+1} - z^k$ enables us to further obtain $\|z^{k+1} - z^k\| \to 0$ and $\|y^k - x^k\| \to 0$ as $k \to \infty$. Consequently, $y^{k_l} - x^{k_l} \to 0$ as $l \to \infty$ which implies that $y^* = x^*$. By using the minimizing property of y^{k_l}, we have

$$g(y^{k_l}) + \frac{1}{2\gamma}\|y^{k_l} - 2x^{k_l} + z^{k_l}\|^2 \leq g(y^*) + \frac{1}{2\gamma}\|y^* - 2x^{k_l} + z^{k_l}\|^2$$

from which we can deduce that $\limsup_{l\to\infty} g(y^{k_l}) \leq g(y^*)$. By the lower semi-continuity of g, we also have $g(y^*) \leq \liminf_{l\to\infty} g(y^{k_l})$. Therefore $g(y^{k_l}) \to g(y^*)$ as $l \to \infty$ and hence $\mathcal{M}(x^{k_l}, y^{k_l}, z^{k_l}) \to \mathcal{M}(x^*, y^*, z^*)$ as $l \to \infty$. Since $\mathcal{M}(x^k, y^k, z^k)$ is monotonically decreasing, we thus have $\mathcal{M}(x^k, y^k, z^k) \to \mathcal{M}(x^*, y^*, z^*)$ as $k \to \infty$. Recall that

$$0 = \nabla f(x^{k_l}) + \frac{1}{\gamma}(x^{k_l} - z^{k_l}) \quad \text{and} \quad 0 \in \partial g(y^{k_l}) + \frac{1}{\gamma}(y^{k_l} - 2x^{k_l} + z^{k_l}).$$

By the continuity of ∇f and the closedness of ∂g, we may take $l \to \infty$ to derive that $0 = \nabla f(x^*) + \frac{1}{\gamma}(x^* - z^*)$ and

$$0 \in \partial g(y^*) + \frac{1}{\gamma}(y^* - 2x^* + z^*) = \partial g(x^*) + \frac{1}{\gamma}(z^* - x^*).$$

Therefore $0 \in \nabla f(x^*) + \partial g(x^*)$. □

Lemma 9.45 *Let the conditions in Lemma 9.42 hold. Let $\{(x^k, y^k, z^k)\}$ be a sequence generated by Algorithm 9.41. Then there is a constant $c_1 > 0$ such that*

$$d(0, \partial\mathcal{M}(x^k, y^k, z^k)) \leq c_1\|x^{k+1} - x^k\|$$

for all $k \geq 1$.

Proof According to Proposition 9.8 and the definition of $\mathcal{M}$, we have

$$\partial\mathcal{M}(x^k, y^k, z^k)) = \begin{pmatrix} \nabla f(x^k) + \frac{1}{\gamma}(x^k - z^k) + \frac{2}{\gamma}(x^k - y^k) \\ \partial g(y^k) + \frac{1}{\gamma}(y^k - 2x^k + z^k) \\ \frac{1}{\gamma}(y^k - x^k) \end{pmatrix}.$$

Recall that $\nabla f(x^k) + \frac{1}{\gamma}(x^k - z^k) = 0$. By the definition of y^k, we also have $-\frac{1}{\gamma}(y^k - 2x^k + z^k) \in \partial g(y^k)$. Therefore,

$$\begin{pmatrix} \frac{2}{\gamma}(x^k - y^k) \\ 0 \\ \frac{1}{\gamma}(y^k - x^k) \end{pmatrix} \in \partial \mathcal{M}(x^k, y^k, z^k).$$

Consequently, by virtue of $y^k - x^k = z^{k+1} - z^k$, we have

$$d(0, \partial \mathcal{M}(x^k, y^k, z^k)) \leq \frac{3}{\gamma}\|y^k - x^k\| = \frac{3}{\gamma}\|z^{k+1} - z^k\|.$$

This together with (9.52) then completes the proof. □

Theorem 9.46 *Let Assumption 9.40 hold, and let $\{(x^k, y^k, z^k)\}$ be a sequence generated by Algorithm 9.41 in which the step size γ is chosen to satisfy (9.49). Assume that $\{(x^k, y^k, z^k)\}$ has a cluster point (x^*, y^*, z^*). If the function $\mathcal{M}$ defined by (9.48) is a KL function, then $\{(x^k, y^k, z^k)\}$ has a finite length and the whole sequence converges to (x^*, y^*, z^*). If in addition $\mathcal{M}$ is a KL function with an exponent $\alpha \in [0, 1)$, then $\{(x^k, y^k, z^k)\}$ enjoys the convergence rate as given in Theorem 9.27.*

Proof Since $\mathcal{M}$ satisfies the KL condition at (x^*, y^*, z^*), we can find $\eta \in (0, \infty]$, a neighborhood U of (x^*, y^*, z^*), and a concave function $\varphi \in \mathcal{K}(0, \eta)$ such that

$$\varphi'(\mathcal{M}(x, y, z) - \mathcal{M}_*)d(0, \partial \mathcal{M}(x, y, z)) \geq 1$$

whenever $(x, y, z) \in U \cap \{\mathcal{M}_* < \mathcal{M} < \mathcal{M}_* + \eta\}$, where $\mathcal{M}_* := \mathcal{M}(x^*, y^*, z^*)$. We may take $r > 0$ such that the set

$$\left\{(x, y, z) \in \mathbb{R}^n \times \mathbb{R}^n \times \mathbb{R}^n : \|x - x^*\| < 2r,\ \|y - y^*\| < 3r,\ \|z - z^*\| < 2r\right\}$$

is contained in U.

Since (x^*, y^*, z^*) is a cluster point of $\{(x^k, y^k, z^k)\}$, there is a subsequence $\{(x^{k_l}, y^{k_l}, z^{k_l})\}$ such that $(x^{k_l}, y^{k_l}, z^{k_l}) \to (x^*, y^*, z^*)$ as $l \to \infty$. Moreover, it follows from Proposition 9.44 that $y^* = x^*$ and

$$x^k - y^k \to 0 \quad \text{and} \quad \mathcal{M}(x^k, y^k, z^k) \to \mathcal{M}_* \quad \text{as } k \to \infty.$$

Thus, we can find k_N such that

$$\|x^{k_N} - x^*\| < r,\ \ \|z^{k_N} - z^*\| < r,\ \ \varphi(\mathcal{M}(x^{k_N}, y^{k_N}, z^{k_N}) - \mathcal{M}_*) < \frac{c_0 r}{(1 + \gamma L)c_1}$$

and

$$\|y^k - x^k\| < r, \quad \mathcal{M}_* \le \mathcal{M}(x^k, y^k, z^k) < \mathcal{M}_* + \eta$$

for all $k \ge k_N$. We will show by induction that $(x^k, y^k, z^k) \in U$ for all $k \ge k_N$ which trivially holds for $k = k_N$ according to the choice of k_N. Assume that $(x^k, y^k, z^k) \in U$ for all $k_N \le k \le j$ for some $j \ge k_N$. By using the KL condition of $\mathcal{M}$, Lemmas 9.42 and 9.45, we may follow the proof of Theorem 9.26 to derive that

$$\begin{aligned}
&\varphi(\mathcal{M}(x^k, y^k, z^k) - \mathcal{M}_*) - \varphi(\mathcal{M}(x^{k+1}, y^{k+1}, z^{k+1}) - \mathcal{M}_*) \\
&\ge \frac{c_0}{c_1}\|x^k - x^{k+1}\|
\end{aligned} \tag{9.53}$$

for all $k_N \le k \le j$. Consequently

$$\sum_{k=k_N}^{j} \|x^{k+1} - x^k\| \le \frac{c_1}{c_0}\varphi(\mathcal{M}(x^{k_N}, y^{k_N}, z^{k_N}) - \mathcal{M}_*). \tag{9.54}$$

Therefore

$$\begin{aligned}
\|x^{j+1} - x^*\| &\le \|x^{k_N} - x^*\| + \sum_{k=k_N}^{j} \|x^{k+1} - x^k\| \\
&\le \|x^{k_N} - x^*\| + \frac{c_1}{c_0}\varphi(\mathcal{M}(x^{k_N}, y^{k_N}, z^{k_N}) - \mathcal{M}_*) \\
&< 2r
\end{aligned}$$

By virtue of (9.52), we also have

$$\begin{aligned}
\|z^{j+1} - z^*\| &\le \|z^{k_N} - z^*\| + (1 + \gamma L)\sum_{k=k_N}^{j} \|x^{k+1} - x^k\| \\
&\le \|z^{k_N} - z^*\| + \frac{(1 + \gamma L)c_1}{c_0}\varphi(\mathcal{M}(x^{k_N}, y^{k_N}, z^{k_N}) - \mathcal{M}_*) \\
&< 2r.
\end{aligned}$$

In view of the fact $y^* = x^*$, we can obtain

$$\|y^{j+1} - y^*\| \le \|y^{j+1} - x^{j+1}\| + \|x^{j+1} - x^*\| < 3r.$$

Thus $(x^{j+1}, y^{j+1}, z^{j+1}) \in U$.

Since $(x^k, y^k, z^k) \in U$ for all $k \geq k_N$, (9.54) holds for all $j \geq k_N$ which shows that $\{x^k\}$ has a finite length. From (9.52), it then follows that $\{z^k\}$ has a finite length. By using $y^k - x^k = z^{k+1} - z^k$, we have

$$\begin{aligned}\|y^{k+1} - y^k\| &\leq \|y^{k+1} - x^{k+1}\| + \|y^k - x^k\| + \|x^{k+1} - x^k\| \\ &= \|z^{k+2} - z^{k+1}\| + \|z^{k+1} - z^k\| + \|x^{k+1} - x^k\|\end{aligned}$$

which demonstrates that $\{y^k\}$ has a finite length. Therefore, the whole sequence $\{(x^k, y^k, z^k)\}$ converges to (x^*, y^*, z^*).

As (9.53) holds for all $k \geq k_N$, we may follow the same argument in the proof of Theorem 9.27 with minor modification to obtain convergence rates when $\mathcal{M}$ is a KL function with an exponent $\alpha \in [0, 1)$. □

Proposition 9.44 and Theorem 9.46 require the sequence $\{(x^k, y^k, z^k)\}$ generated by Algorithm 9.41 to have a cluster point which can be guaranteed if the sequence itself is bounded. The following result provides a sufficient condition to ensure the boundedness of $\{(x^k, y^k, z^k)\}$.

Lemma 9.47 *Let Assumption 9.40 hold, and let $\{(x^k, y^k, z^k)\}$ be a sequence generated by Algorithm 9.41 in which the step size γ is chosen to satisfy (9.49). If the sublevel set*

$$\{x \in \mathbb{R}^n : f(x) + g(x) \leq \mathcal{M}(x^0, y^0, z^0)\}$$

of $f + g$ is bounded, then $\{(x^k, y^k, z^k)\}$ is a bounded sequence.

Proof According to Lemma 9.42, we have $\mathcal{M}(x^k, y^k, z^k) \leq \mathcal{M}(x^0, y^0, z^0)$ for all k. Note that

$$\begin{aligned}\mathcal{M}(x^k, y^k, z^k) &= f(x^k) + \langle \nabla f(x^k), y^k - x^k\rangle + g(y^k) + \frac{1}{2\gamma}\|x^k - y^k\|^2 \\ &= f(y^k) + g(y^k) - \Big[f(y^k) - f(x^k) - \langle \nabla f(x^k), y^k - x^k\rangle\Big] \\ &\quad + \frac{1}{2\gamma}\|x^k - y^k\|^2 \\ &\geq f(y^k) + g(y^k) + \frac{1 - \gamma L}{2\gamma}\|x^k - y^k\|^2.\end{aligned}$$

We therefore have

$$f(y^k) + g(y^k) + \frac{1 - \gamma L}{2\gamma}\|x^k - y^k\|^2 \leq \mathcal{M}(x^0, y^0, z^0) \tag{9.55}$$

for all k. Since γ satisfies (9.49) and $\sigma \geq -L$, we have $\gamma < 1/L$. It then follows from (9.55) that $f(y^k) + g(y^k) \leq \mathcal{M}(x^0, y^0, z^0)$ and hence $\{y^k\}$ is bounded by the given condition. According to Assumption 9.40 (iii), $f + g$ is bounded from below. Thus from (9.55), we can conclude that $\|x^k - y^k\|$ is bounded and so is $\{x^k\}$. Finally, by using $z^k = x^k + \gamma \nabla f(x^k)$, we obtain the boundedness of $\{z^k\}$. □

9.4.4 Alternating Minimization for Fourier Phase Retrieval

Given a vector $x = (x_1, \cdots, x_n)^T \in \mathbb{C}^n$ consisting n complex entries, its discrete Fourier transform is defined by $\hat{x} = (\hat{x}_1, \cdots, \hat{x}_n)^T \in \mathbb{C}^n$, where

$$\hat{x}_k := \sum_{j=1}^{n} x_j e^{-\mathrm{i}\frac{2\pi jk}{n}}, \quad k = 1, \cdots, n$$

and i is the imaginary unit; we denote $\hat{x}$ by $\mathcal{F}(x)$. By virtue of the identity

$$\sum_{l=1}^{n} e^{\mathrm{i}\frac{2\pi(k-j)l}{n}} = \begin{cases} n & \text{if } j \equiv k \bmod n, \\ 0 & \text{otherwise} \end{cases}$$

it is easy to see that $\|\mathcal{F}(x)\|_2^2 = n\|x\|_2^2$ for all $x \in \mathbb{C}^n$. Thus $\mathcal{F} : \mathbb{C}^n \to \mathbb{C}^n$ is invertible. It is easy to check that for any $x \in \mathbb{C}^n$, the j-th component of $\mathcal{F}^{-1}(x)$ is given by

$$[\mathcal{F}^{-1}(x)]_j = \frac{1}{n}\sum_{k=1}^{n} x_k e^{\mathrm{i}\frac{2\pi jk}{n}}, \quad j = 1, \cdots, n.$$

Fourier phase retrieval is to reconstruct a signal $x^* \in \mathbb{R}^n$ from a measurement c of the modulus of its Fourier transform $|\mathcal{F}(x^*)|$, where, for a complex vector $z = (z_1, \cdots, z_n)^T \in \mathbb{C}^n$, we use the convention $|z| := (|z_1|, \cdots, |z_n|)^T$. It seems "natural" to reconstruct x^* by considering the nonlinear least square problem

$$\min_{x\in\mathbb{R}^n} \frac{1}{2} \||\mathcal{F}(x)| - c\|_2^2 . \tag{9.56}$$

This model, however, cannot capture the feature of the sought signal well. Prior knowledge on the sought signal should be used. Let $g : \mathbb{R}^n \to (-\infty, \infty]$ be a prior function which is proper, lower semicontinuous, and bounded from below. We consider the regularized nonlinear least square problem

$$\min_{x\in\mathbb{R}^n} \left\{\frac{1}{2n} \||\mathcal{F}(x)| - c\|_2^2 + g(x)\right\} . \tag{9.57}$$

This minimization problem can be reformulated in an alternative way for which we may use the alternating minimization solver. To this end, let

$$Z_c := \{z \in \mathbb{C}^n : |\mathcal{F}(z)| = c\}.$$

We have the following observation [178]

Lemma 9.48 *For any $x \in \mathbb{R}^n$, there holds*

$$\min_{z \in Z_c} \|x - z\|_2^2 = \frac{1}{n} \||\mathcal{F}(x)| - c\|_2^2.$$

Proof Recall that $\|x - z\|_2^2 = \|\mathcal{F}(x) - \mathcal{F}(z)\|_2^2/n$. Therefore,

$$z^* \in \arg\min_{z \in Z_c} \|x - z\|_2^2$$

if and only if $z^* = \mathcal{F}^{-1}(\omega^*)$ and

$$\omega^* \in \arg\min_{|\omega|=c} \|\mathcal{F}(x) - \omega\|_2^2.$$

Let $\omega^* = (\omega_1^*, \cdots, \omega_n^*)^T$. Then

$$\omega_j^* \in \arg\min_{|\omega_j|=c_j} \left|[\mathcal{F}(x)]_j - \omega_j\right|,$$

where $[\mathcal{F}(x)]_j$ denotes the j-th component of $\mathcal{F}(x)$. Consequently,

$$\omega_j^* = \begin{cases} c_j \frac{[\mathcal{F}(x)]_j}{|[\mathcal{F}(x)]_j|} & \text{if } [\mathcal{F}(x)]_j \neq 0, \\ c_j e^{\mathrm{i}\theta_j} & \text{if } [\mathcal{F}(x)]_j = 0, \end{cases}$$

where θ_j can be any angle. Hence

$$\begin{aligned} \min_{z \in Z_c} \|x - z\|_2^2 &= \|x - z^*\|_2^2 = \|x - \mathcal{F}^{-1}(\omega^*)\|_2^2 = \frac{1}{n}\|\mathcal{F}(x) - w^*\|_2^2 \\ &= \frac{1}{n}\sum_{j=1}^n |[\mathcal{F}(x)]_j - \omega_j^*|^2. \end{aligned}$$

Note that if $[\mathcal{F}(x)]_j \neq 0$, then

$$|[\mathcal{F}(x)]_j - \omega_j^*| = \left|[\mathcal{F}(x)]_j - c_j \frac{[\mathcal{F}(x)]_j}{|[\mathcal{F}(x)]_j|}\right| = \left||[\mathcal{F}(x)]_j| - c_j\right|;$$

and if $[\mathcal{F}(x)]_j = 0$, then

$$|[\mathcal{F}(x)]_j - \omega_j^*| = |\omega_j^*| = c_j = \big||[\mathcal{F}(x)]_j| - c_j\big|.$$

Thus

$$\min_{z\in Z_c} \|x - z\|_2^2 = \frac{1}{n}\sum_{j=1}^{n} \big||[\mathcal{F}(x)]_j| - c_j\big|^2 = \frac{1}{n}\||\mathcal{F}(x)| - c\|_2^2.$$

The proof is complete. □

Lemma 9.48 shows that (9.57) is equivalent to the problem

$$\min_{x\in\mathbb{R}^n, z\in Z_c} \left\{\frac{1}{2}\|x - z\|_2^2 + g(x)\right\} \tag{9.58}$$

based on which we may formulate the alternating minimization algorithm that is an extension of the Fineup method [82].

Algorithm 9.49 *Take an initial guess $x_0 \in \mathbb{R}^n$. For $k = 0, 1, \cdots$ do*

$$z^{k+1} \in \arg\min_{z\in Z_c} \left\{\frac{1}{2}\|x^k - z\|_2^2\right\},$$

$$x^{k+1} \in \arg\min_{x\in\mathbb{R}^n} \left\{\frac{1}{2}\|x - z^{k+1}\|_2^2 + g(x)\right\}.$$

Note that z^{k+1} is a projection of x^k onto the set Z_c. According to the proof of Lemma 9.48, z^{k+1} can be computed by $z^{k+1} = \mathcal{F}^{-1}(\omega^k)$, where

$$\omega_j^k = \begin{cases} c_j \frac{[\mathcal{F}(x^k)]_j}{|[\mathcal{F}(x^k)]_j|} & \text{if } [\mathcal{F}(x^k)]_j \neq 0, \\ c_j e^{\mathrm{i}\theta_j^k} & \text{if } [\mathcal{F}(x^k)]_j = 0 \end{cases}$$

with θ_j^k being an arbitrary angle.

By writing $z \in \mathbb{C}^n$ as $z = w_1 + \mathrm{i}w_2$ in terms of its real and imaginary parts, where $w_1 = \Re(z)$ and $w_2 = \Im(z)$, and introducing

$$\tilde{Z}_c = \left\{(w_1, w_2) \in \mathbb{R}^n \times \mathbb{R}^n : z = w_1 + \mathrm{i}w_2 \in Z_c\right\},$$

we can write (9.58) equivalently as

$$\min_{x\in\mathbb{R}^n, (w_1,w_2)\in\tilde{Z}_c} \left\{\frac{1}{2}\|x - w_1\|_2^2 + \frac{1}{2}\|w_2\|_2^2 + g(x)\right\},$$

and Algorithm 9.49 can be restated as

$$\begin{aligned} (w_1^{k+1}, w_2^{k+1}) &\in \arg \min_{(w_1,w_2)\in\tilde{Z}_c} \left\{\frac{1}{2}\|x^k - w_1\|_2^2 + \frac{1}{2}\|w_2\|_2^2\right\}, \\ x^{k+1} &\in \arg\min_{x\in\mathbb{R}^n} \left\{\frac{1}{2}\|x - w_1^{k+1}\|_2^2 + g(x)\right\}. \end{aligned} \tag{9.59}$$

In the following, we will provide a convergence theory for the alternating minimization algorithm for the Fourier phase retrieval. Let

$$\tilde{K}(x, w_1, w_2) := \frac{1}{2}\|x - w_1\|_2^2 + \frac{1}{2}\|w_2\|_2^2 + g(x) + \delta_{\tilde{Z}_c}(w_1, w_2).$$

We have the following convergence result [178].

Theorem 9.50 *Let $g : \mathbb{R}^n \to (-\infty, \infty]$ be a proper, lower semicontinuous, convex function bounded from below, and let $\{(x^k, w_1^k, w_2^k)\}$ be generated by the alternating minimization algorithm (9.59). Then*

(i) *$\{\tilde{K}(x^k, w_1^k, w_2^k)\}$ is monotonically decreasing and*

$$\sum_{k=1}^{\infty}\|x^k - x^{k-1}\|^2 < \infty.$$

(ii) *$\{(x^k, w_1^k, w_2^k)\}$ is a bounded sequence in $\mathbb{R}^n \times \mathbb{R}^n \times \mathbb{R}^n$, and any accumulation point of $\{(x^k, w_1^k, w_2^k)\}$ is a critical point of $\tilde{K}$.*

(iii) *Let Ω be the set of all accumulation points of $\{(x^k, w_1^k, w_2^k)\}$. Then Ω is a compact set and $\tilde{K}$ is constant on Ω.*

(iv) *If, in addition, g is semi-algebraic, then*

$$\sum_{k=1}^{\infty}\|x^k - x^{k-1}\| < \infty$$

and thus the whole sequence $\{x^k\}$ is convergent.

Proof

(i) Since x^k is a minimizer of the function

$$x \to \tilde{K}(x, w_1^k, w_2^k) = \frac{1}{2}\|x - w_1^k\|^2 + \frac{1}{2}\|w_2^k\|^2 + g(x) + \delta_{\tilde{Z}_c}(w_1^k, w_2^k)$$

which is strongly convex, we thus have

$$\tilde{K}(x^k, w_1^k, w_2^k) + \frac{1}{2}\|x^k - x^{k-1}\|^2 \le \tilde{K}(x^{k-1}, w_1^k, w_2^k).$$

By the definition of (w_1^k, w_2^k), we also have

$$\tilde{K}(x^{k-1}, w_1^k, w_2^k) \leq \tilde{K}(x^{k-1}, w_1^{k-1}, w_2^{k-1}).$$

Consequently

$$\tilde{K}(x^k, w_1^k, w_2^k) + \frac{1}{2}\|x^k - x^{k-1}\|^2 \leq \tilde{K}(x^{k-1}, w_1^{k-1}, w_2^{k-1}). \tag{9.60}$$

This implies that $\{\tilde{K}(x^k, w_1^k, w_2^k)\}$ is monotonically decreasing and

$$\sum_{k=2}^{\infty} \|x^k - x^{k-1}\|^2 \leq \tilde{K}(x^1, w_1^1, w_2^1) < \infty.$$

(ii) By definition, $\tilde{Z}_c$ is a compact subset of $\mathbb{R}^n \times \mathbb{R}^n$. Since $\{(w_1^k, w_2^k)\} \subset \tilde{Z}_c$, it is a bounded sequence. From the monotonicity of $\tilde{K}(x^k, w_1^k, w_2^k)$, we also have

$$\frac{1}{2}\|x^k - w_1^k\|^2 + g(x^k) \leq \tilde{K}(x^k, w_1^k, w_2^k) \leq \tilde{K}(x^1, w_1^1, w_2^1) < \infty.$$

Thus, by virtue of the boundedness of $\{w_1^k\}$ and the boundedness from below of g, $\{x^k\}$ is also a bounded sequence.

Let (x^*, w_1^*, w_2^*) be any accumulation point of $\{(x^k, w_1^k, w_2^k)\}$. We show that (x^*, w_1^*, w_2^*) is a critical point of $\tilde{K}$, i.e., $0 \in \partial\tilde{K}(x^*, w_1^*, w_2^*)$. Note that

$$\partial\tilde{K}(x, w_1, w_2) = ((x - w_1) + \partial g(x)) \times ((w_1 - x, w_2) + \partial\delta_{\tilde{Z}_c}(w_1, w_2)).$$

We only need to show that

$$w_1^* - x^* \in \partial g(x^*), \quad (x^* - w_1^*, -w_2^*) \in \partial\delta_{\tilde{Z}_c}(w_1^*, w_2^*). \tag{9.61}$$

As an accumulation point, we can find a subsequence $\{(x^{k_j}, w_1^{k_j}, w_2^{k_j})\}$ of $\{(x^k, w_1^k, w_2^k)\}$ such that

$$(x^{k_j}, w_1^{k_j}, w_2^{k_j}) \to (x^*, w_1^*, w_2^*) \quad \text{as } j \to \infty. \tag{9.62}$$

According to (i), we have $x^k - x^{k-1} \to 0$ as $k \to \infty$. Thus $x^{k_j-1} \to x^*$ as $j \to \infty$. By the first-order optimality condition of $(x^{k_j}, w_1^{k_j}, w_2^{k_j})$, we have

$$w_1^{k_j} - x^{k_j} \in \partial g(x^{k_j}), \quad (x^{k_j-1} - w_1^{k_j}, -w_2^{k_j}) \in \partial\delta_{\tilde{Z}_c}(w_1^{k_j}, w_2^{k_j}).$$

By taking $j \to \infty$, we will be able to obtain (9.61) by the closedness of subdifferential if we can show that

$$g(x^{k_j}) \to g(x^*) \quad \text{and} \quad \delta_{\tilde{Z}_c}(w_1^{k_j}, w_2^{k_j}) \to \delta_{\tilde{Z}_c}(w_1^*, w_2^*) \tag{9.63}$$

as $j \to \infty$. The second one in (9.63) holds trivially, since $(w_1^{k_j}, w_2^{k_j}) \in \tilde{Z}_c$ implies $(w_1^*, w_2^*) \in \tilde{Z}_c$ and hence $\delta_{\tilde{Z}_c}(w_1^{k_j}, w_2^{k_j}) = \delta_{\tilde{Z}_c}(w_1^*, w_2^*) = 0$. To show the first one in (9.63), we may use the minimality of x^{k_j} to obtain

$$g(x^{k_j}) + \frac{1}{2}\|x^{k_j} - w_1^{k_j}\|^2 \le g(x^*) + \frac{1}{2}\|x^* - w_1^{k_j}\|^2.$$

Therefore

$$\begin{aligned}\limsup_{j\to\infty} g(x^{k_j}) &\le g(x^*) + \lim_{j\to\infty}\left(\frac{1}{2}\|x^* - w_1^{k_j}\|^2 - \frac{1}{2}\|x^{k_j} - w_1^{k_j}\|^2\right)\\ &= g(x^*).\end{aligned}$$

By using the lower semi-continuity of g, we also have

$$g(x^*) \le \liminf_{j\to\infty} g(x^{k_j}).$$

Therefore, we obtain the first equation in (9.63).

(iii) By definition Ω is a closed set. Since $\{(x^k, w_1^k, w_2^k)\}$ is bounded, Ω must be bounded. Consequently, Ω is compact. Next we show that $\tilde{K}$ is constant on Ω. Let

$$\mu := \lim_{k\to\infty} \tilde{K}(x^k, w_1^k, w_2^k)$$

whose existence is guaranteed by (i). It suffices to show that $\tilde{K}(x^*, w_1^*, w_2^*) = \mu$ for all $(x^*, w_1^*, w_2^*) \in \Omega$. For $(x^*, w_1^*, w_2^*) \in \Omega$, we can find a subsequence $\{(x^{k_j}, w_1^{k_j}, w_2^{k_j})\}$ of $\{(x^k, w_1^k, w_2^k)\}$ such that (9.62) holds. We may use the same argument in the proof of (ii) to conclude that $g(x^{k_j}) \to g(x^*)$ as $j \to \infty$. Thus, by noting that $(w_1^{k_j}, w_2^{k_j}), (w_1^*, w_2^*) \in \tilde{Z}_c$, we have

$$\begin{aligned}\mu &= \lim_{j\to\infty} \tilde{K}(x^{k_j}, w_1^{k_j}, w_2^{k_j}) = \lim_{j\to\infty}\left(\frac{1}{2}\|x^{k_j} - w_1^{k_j}\|^2 + \frac{1}{2}\|w_2^{k_j}\|^2 + g(x^{k_j})\right)\\ &= \frac{1}{2}\|x^* - w_1^*\|^2 + \frac{1}{2}\|w_2^*\|^2 + g(x^*)\\ &= K(x^*, w_1^*, w_2^*).\end{aligned}$$

(iv) If there is some k such that $\tilde{K}(x^k, w_1^k, w_2^k) = \mu$, then by (i) we have $\tilde{K}(x^n, w_1^n, w_2^n) = \mu$ and hence $x^n = x^k$ for all $n \geq k$ which shows the desired result. Therefore, we may assume $\tilde{K}(x^k, w_1^k, w_2^k) > \mu$ for all k.

We need to use the uniformized KL property. Note that $\tilde{Z}_c$ is a semi-algebraic set, its indicator function $\delta_{\tilde{Z}_c}$ is semi-algebraic. Since g is semi-algebraic, the function $\tilde{K}(x, w_1, w_2)$, as the sum of g, $\delta_{\tilde{Z}_c}$ and the quadratic function $\frac{1}{2}\|x - w_1\|^2 + \frac{1}{2}\|w_2\|^2$, is therefore semi-algebraic. Thus $\tilde{K}$ is a KL function. According to (iii), Ω is a compact set and $\tilde{K}$ is a constant μ on Ω. Therefore, by the uniformized KL property, we can find $\rho > 0$, $\eta > 0$ and a desingularization function $\varphi \in \mathcal{K}(0, \eta)$ such that

$$\varphi'(\tilde{K}(x, w_1, w_2) - \mu)d(0, \partial\tilde{K}(x, w_1, w_2)) \geq 1$$

for all

$$(x, w_1, w_2) \in (\Omega + B_\rho(0)) \cap \{\mu < \tilde{K}(x, w_1, w_2) < \mu + \eta\}.$$

By the definition of Ω, we have $d((x^k, w_1^k, w_2^k), \Omega) \to 0$ as $k \to \infty$. Therefore, we can find an integer k_0 such that

$$(x^k, w_1^k, w_2^k) \in \Omega + B_\rho(0) \quad \text{and} \quad \mu < \tilde{K}(x^k, w_1^k, w_2^k) < \mu + \eta$$

for all $k \geq k_0$. Consequently,

$$\varphi'(\tilde{K}(x^k, w_1^k, w_2^k) - \mu)d(0, \partial\tilde{K}(x^k, w_1^k, w_2^k)) \geq 1, \quad \forall k \geq k_0.$$

By the concavity of φ and (9.60), we have

$$\begin{aligned}
&\varphi(\tilde{K}(x^k, w_1^k, w_2^k) - \mu) - \varphi(\tilde{K}(x^{k+1}, w_1^{k+1}, w_2^{k+1}) - \mu) \\
&\geq \varphi'(\tilde{K}(x^k, w_1^k, w_2^k) - \mu)\left(\tilde{K}(x^k, w_1^k, w_2^k) - \tilde{K}(x^{k+1}, w_1^{k+1}, w_2^{k+1})\right) \\
&\geq \frac{1}{2}\|x^{k+1} - x^k\|^2 \varphi'(\tilde{K}(x^k, w_1^k, w_2^k) - \mu) \\
&\geq \frac{\|x^{k+1} - x^k\|^2}{2d(0, \partial\tilde{K}(x^k, w_1^k, w_2^k))}.
\end{aligned}$$

We need to estimate $d(0, \partial\tilde{K}(x^k, w_1^k, w_2^k))$. By the first-order optimality condition for (x^k, w_1^k, w_2^k), we have

$$w_1^k - x^k \in \partial g(x^k), \quad (x^{k-1} - w_1^k, -w_2^k) \in \partial\delta_{\tilde{Z}_c}(w_1^k, w_2^k).$$

According to the formula for $\partial \tilde{K}(x^k, w_1^k, w_2^k)$, it follows that

$$(0, x^{k-1} - x^k, 0) \in \partial \tilde{K}(x^k, w_1^k, w_2^k).$$

Therefore

$$d(0, \partial \tilde{K}(x^k, w_1^k, w_2^k)) \leq \|x^k - x^{k-1}\|.$$

Consequently

$$\varphi(\tilde{K}(x^k, w_1^k, w_2^k) - \mu) - \varphi(\tilde{K}(x^{k+1}, w_1^{k+1}, w_2^{k+1}) - \mu) \geq \frac{\|x^{k+1} - x^k\|^2}{2\|x^k - x^{k-1}\|}$$

which implies that

$$\begin{aligned} &2\|x^{k+1} - x^k\| \\ &\leq 2\sqrt{2\|x^k - x^{k-1}\| \left(\varphi(\tilde{K}(x^k, w_1^k, w_2^k) - \mu) - \varphi(\tilde{K}(x^{k+1}, w_1^{k+1}, w_2^{k+1}) - \mu)\right)} \\ &\leq \|x^k - x^{k-1}\| + 2\left(\varphi(\tilde{K}(x^k, w_1^k, w_2^k) - \mu) - \varphi(\tilde{K}(x^{k+1}, w_1^{k+1}, w_2^{k+1}) - \mu)\right) \end{aligned}$$

for all $k > k_0$. Summing over k from $k_0 + 1$ to an arbitrarily large integer n, we can obtain

$$\begin{aligned} 2\sum_{k=k_0+1}^{n} \|x^{k+1} - x^k\| \leq& \sum_{k=k_0+1}^{n} \|x^k - x^{k-1}\| \\ &+ 2\sum_{k=k_0+1}^{n} \left(\varphi(\tilde{K}(x^k, w_1^k, w_2^k) - \mu) - \varphi(\tilde{K}(x^{k+1}, w_1^{k+1}, w_2^{k+1}) - \mu)\right) \\ \leq& \sum_{k=k_0+1}^{n} \|x^k - x^{k-1}\| + 2\varphi(\tilde{K}(x^{k_0+1}, w_1^{k_0+1}, w_2^{k_0+1}) - \mu). \end{aligned}$$

This implies that

$$\sum_{k=k_0+1}^{n} \|x^{k+1} - x^k\| \leq \|x^{k_0+1} - x^{k_0}\| + 2\varphi(\tilde{K}(x^{k_0+1}, w_1^{k_0+1}, w_2^{k_0+1}) - \mu).$$

Since this is true for any n, we therefore obtain the desired result. □

9.5 O-Minimal Structures, Definable Functions, and KL Property

In this section, we will prove Theorem 9.24 which in particular implies that every proper, lower semicontinuous, semi-algebraic function is a KL function. Actually we will go far beyond this by showing that every proper, lower semicontinuous, definable function in an arbitrary o-minimal structure is a KL function.

9.5.1 *Differentiable Manifolds*

We need a little preparation on differential geometry; for more information, one may refer to [38, 123], for instance.

A differentiable manifold is a topological space that locally resembles Euclidean space and has a structure that allows for differentiation. Formally, a differentiable manifold is defined as follows:

Definition 9.51 Let $l \geq 1$ be an integer. An n-dimensional C^l-*manifold* is a pair $(\mathcal{M}, \mathcal{A})$, where $\mathcal{M}$ is a second-countable, Hausdorff topological space, and $\mathcal{A} = \{(U_\alpha, \varphi_\alpha)\}_{\alpha\in I}$ is an atlas of $\mathcal{M}$, that is,

- $\bigcup_{\alpha\in I} U_\alpha = \mathcal{M}$;
- For each $\alpha \in I$, U_α is an open set of $\mathcal{M}$, $\varphi_\alpha(U_\alpha)$ is an open set of $\mathbb{R}^n$ and $\varphi_\alpha : U_\alpha \to \varphi(U_\alpha)$ is a homeomorphism; each $(U_\alpha, \varphi_\alpha)$ is called a coordinate chart;
- For any $\alpha, \beta \in I$, the transition map

$$\varphi_\beta \circ \varphi_\alpha^{-1} : \varphi_\alpha(U_\alpha \cap U_\beta) \to \varphi_\beta(U_\alpha \cap U_\beta)$$

  is C^l.

Given a n-dimensional C^l-manifold $\mathcal{M}$ with $l \geq 1$. A function $f : \mathcal{M} \to \mathbb{R}$ is called C^l if for each coordinate chart $(U_\alpha, \varphi_\alpha)$, the function $f \circ \varphi_\alpha^{-1} : \varphi(U_\alpha) \to \mathbb{R}$ is C^l. We use $C^l(\mathcal{M})$ to denote the collection of all C^l functions on $\mathcal{M}$. Given $p \in \mathcal{M}$ and let (U, φ) be a coordinate chart with $p \in U$. For each $i = 1, \cdots, n$, we define the coordinate basis vector $\partial_i|_p$ by

$$\partial_i|_p f := \frac{\partial}{\partial x_i}(f \circ \varphi^{-1})\bigg|_{\varphi(p)}, \quad \forall f \in C^l(\mathcal{M}).$$

The tangent space $T_p\mathcal{M}$ of $\mathcal{M}$ at p is then defined as

$$T_p\mathcal{M} := \operatorname{span}\{\partial_1|_p, \cdots, \partial_n|_p\}.$$

Every element in $T_p\mathcal{M}$ is called a tangent vector of $\mathcal{M}$ at p. A C^l vector field v on $\mathcal{M}$ means $v|_p \in T_p\mathcal{M}$ for each $p \in \mathcal{M}$ and for any coordinate chart (U, φ), with corresponding coordinate basis vectors $\partial_1, \cdots, \partial_n$, there holds

$$v = \sum_{i=1}^{n} v_i \partial_i,$$

where every v_i is a C^l function on U.

Given $p \in \mathcal{M}$ and $v \in T_p\mathcal{M}$, we can find a C^l-curve $\gamma : (-\varepsilon, \varepsilon) \to \mathcal{M}$ with $\gamma(0) = p$ such that

$$\frac{d}{dt} f(\gamma(t))\Big|_{t=0} = v(f), \quad \forall f \in C^l(\mathcal{M}).$$

We then denote $\gamma'(0) = v$. Indeed, by taking a coordinate chart (U, φ) with $p \in U$ and writing $v = \sum_{i=1}^{n} v^i \partial_i|_p$, then we can find a curve $\tilde{\gamma} : (-\varepsilon, \varepsilon) \to \mathbb{R}^n$ such that $\tilde{\gamma}(0) = \varphi(p)$ and $\frac{d\tilde{\gamma}(0)}{dt} = (v^1, \cdots, v^n)$. Let $\gamma := \varphi^{-1} \circ \tilde{\gamma}$ which is well-defined by shrinking $\varepsilon > 0$ if necessary. Then for any $f \in C^l(\mathcal{M})$, we have

$$\begin{aligned}\frac{d}{dt} f(\gamma(t))\Big|_{t=0} &= \frac{d}{dt}[(f \circ \varphi^{-1})(\tilde{\gamma})]\Big|_{t=0} = \sum_{i=1}^{n} \frac{\partial}{\partial x_i}(f \circ \varphi^{-1})\Big|_{\varphi(p)} \frac{d\tilde{\gamma}^i(0)}{dt} \\ &= \sum_{i=1}^{n} v^i \partial_i|_p f = v(f).\end{aligned}$$

Definition 9.52 A subset $\mathcal{M}$ of $\mathbb{R}^n$ is called a k-dimensional C^l-*submanifold* with $l \geq 1$ if for every $p \in \mathcal{M}$ there is an open set $U \subset \mathbb{R}^n$ containing p, an open set $V \subset \mathbb{R}^n$ and a C^l-diffeomorphism $F : U \to V$ such that

$$F(\mathcal{M} \cap U)) = \{x = (x^1, \cdots, x^n) \in V : x^{k+1} = \cdots = x^n = 0\}.$$

Example 9.53 For a C^l function f with $l \geq 1$ defined on an open set $O \subset \mathbb{R}^n$, consider its graph

$$\mathrm{Gr}(f) := \{(x, y) \in \mathbb{R}^{n+1} : x \in O \text{ and } y = f(x)\}.$$

For each $p = (x, f(x)) \in \mathrm{Gr}(f)$, we take $U = V = O \times \mathbb{R}$ and define $F : U \to V$ by $F(x, y) = (x, y - f(x))$. Then F is a C^l diffeomorphism and

$$F(\mathrm{Gr}(f) \cap U) = \{(x, 0) : x \in O\} = \{(x, y) \in V : x \in O \text{ and } y = 0\}.$$

Therefore $\mathrm{Gr}(f)$ is an n-dimensional C^l-submanifold of $\mathbb{R}^{n+1}$.

Definition 9.54 Given an integer $l \geq 1$, a C^l-*Riemannian manifold* of dimension n is a pair $(\mathcal{M}, g)$, where $\mathcal{M}$ is an n-dimensional C^l-manifold, and g is a smoothly varying inner product of class C^l on its tangent spaces. The C^l-smoothness of g means that for each coordinate chart (U, φ), with corresponding coordinate basis vectors $\partial_1, \cdots, \partial_n$, the functions $g_{ij} := g(\partial_i, \partial_j)$ are C^l-smooth. We refer to g as a Riemannian metric on $\mathcal{M}$.

Let $(\mathcal{M}, g)$ be a Riemmanian manifold. Given a smooth scalar function $f : \mathcal{M} \to \mathbb{R}$, the gradient of f, denoted by ∇f, is the unique vector field on $\mathcal{M}$ such that

$$g_p(\nabla f, v) := v(f), \quad \forall v \in T_p\mathcal{M}$$

for each $p \in \mathcal{M}$, where $g_p(\cdot, \cdot)$ denotes the inner product induced by the Riemannian metric g at p. Let (U, φ) be any coordinate chart, with coordinate vector basis $\partial_1, \cdots, \partial_n$, we write $\nabla f = \sum_{j=1}^n a^j \partial_j$. By definition, we have

$$\partial_i f = g(\nabla f, \partial_i) = \sum_{j=1}^n a^j g(\partial_i, \partial_j) = \sum_{j=1}^n a^j g_{ij}.$$

Let $(g^{ij}) := (g_{ij})^{-1}$. Then, solving for a^j, we get $a^j = \sum_{i=1}^n g^{ij} \partial_i f$ and thus

$$\nabla f = \sum_{i,j=1}^n g^{ij} \partial_i f \partial_j.$$

Furthermore, for any $v \in T_p\mathcal{M}$, let $\gamma : (-\varepsilon, \varepsilon) \to \mathcal{M}$ be a C^l curve with $\gamma(0) = p$ and $\gamma'(0) = v$. Then for any $f \in C^l(\mathcal{M})$, there holds

$$\left.\frac{d}{dt} f(\gamma(t))\right|_{t=0} = v(f) = g_p(\nabla f, v).$$

9.5.2 O-Minimal Structure and Definable Functions

We have considered semi-algebraic geometry in Sect. 8.4 from which one can observe that many properties of semi-algebraic sets can be deduced from a number of basic axioms. Based on this observation, the o-minimal structures were introduced and considered in [180, 215] to generalize the semi-algebraic structure. Let us first recall the definition of o-minimal structures [129, 216, 217].

Definition 9.55 (O-minimal Structure) An *o-minimal structure* on the real field $(\mathbb{R}, +, \cdot)$ is a sequence $\mathcal{O} = (\mathcal{O}_n)_n$ of collections $\mathcal{O}_n$, $n \in \mathbb{N}$, of subsets of $\mathbb{R}^n$ satisfying the following axioms:

(i) For every $n \in \mathbb{N}$, the collection $\mathcal{O}_n$ is closed under Boolean operations, i.e., finite intersections, unions, and complements.
(ii) If $A \in \mathcal{O}_n$ and $B \in \mathcal{O}_m$, then $A \times B \in \mathcal{O}_{n+m}$.
(iii) For any set $A \in \mathcal{O}_{n+1}$, there holds $\pi(A) \in \mathcal{O}_n$, where $\pi : \mathbb{R}^{n+1} \to \mathbb{R}^n$ denotes the canonical projection onto $\mathbb{R}^n$,
(iv) $\mathcal{O}_n$ contains the family of algebraic subsets of $\mathbb{R}^n$, that is, every set of the form $\{x \in \mathbb{R}^n : p(x) = 0\}$, where $p : \mathbb{R}^n \to \mathbb{R}$ is a polynomial function.
(v) The elements of $\mathcal{O}_1$ are exactly the finite unions of open intervals and points.

Every set in $\mathcal{O}_n$ is called a *definable* subset of $\mathbb{R}^n$.

Remark 9.56 Let $\mathcal{SA}_n$ denote the collection of all semi-algebraic sets in $\mathbb{R}^n$. According to the properties of semi-algebraic sets, $\mathcal{SA} := \{\mathcal{SA}_n\}$ is an o-minimal structure on $(\mathbb{R}, +, \cdot)$. Actually $\mathcal{SA}$ is the smallest o-minimal structure over $(\mathbb{R}, +, \cdot)$. To see this, let $\mathcal{O} = (\mathcal{O}_n)_n$ be any o-minimal structure. Then for any real polynomial $f : \mathbb{R}^n \to \mathbb{R}$, we have

$$\{x \in \mathbb{R}^n : f(x) < 0\} = \pi(A),$$

where

$$\begin{aligned} A &:= \{(x, t) \in \mathbb{R}^{n+1} : t - f(x) = 0\} \cap \{(x, t) \in \mathbb{R}^{n+1} : t < 0\} \\ &= \{(x, t) \in \mathbb{R}^{n+1} : t - f(x) = 0\} \cap (\mathbb{R}^n \times (-\infty, 0)). \end{aligned}$$

According to (v), we have $(-\infty, 0) \in \mathcal{O}_1$ and hence $\mathbb{R}^n \times (-\infty, 0) \in \mathcal{O}_{n+1}$ by (ii). By (iv) and (i), we then have $A \in \mathcal{O}_{n+1}$ and therefore, by (iii), $\{x \in \mathbb{R}^n : f(x) < 0\} = \pi(A) \in \mathcal{O}_n$. By using this established fact, we can conclude from (i) and (iv) that

$$\{x \in \mathbb{R}^n : f(x) = 0,\ f_i(x) < 0,\ i = 1, \cdots, p\} \in \mathcal{O}_n$$

for any real polynomials f and $f_i, i = 1, \cdots, p$. Thus $\mathcal{SA}_n \subset \mathcal{O}_n$.

Definition 9.57 Given an o-minimal structure $\mathcal{O} = (\mathcal{O}_n)_n$ over the real field $(\mathbb{R}, +, \cdot)$ and a set $A \subset \mathbb{R}^n$, a mapping $F : A \to \mathbb{R}^m$ is called *definable* in $\mathcal{O}$ if its graph

$$Gr(F) := \{(x, y) \in A \times \mathbb{R}^m : y = F(x)\}$$

is a definable set in $\mathbb{R}^n \times \mathbb{R}^m$, i.e., $Gr(F) \in \mathcal{O}_{n+m}$.

Based on the axioms of o-minimal structures, we can use the similar argument in the proof of Lemma 8.28 to obtain the following result.

Lemma 9.58 *Let $S \subset \mathbb{R}^m \times \mathbb{R}^n$, $U \subset \mathbb{R}^m$, and $V \subset \mathbb{R}^n$ are definable sets in an o-minimal structure. Then both*

$$A := \{u \in U : \exists v \in V \text{ such that } (u, v) \in S\}$$

and

$$B := \{u \in U : (u, v) \in S \text{ for all } v \in V\}$$

are definable sets of $\mathbb{R}^m$ in the same o-minimal structure.

According to the definition of definable sets and definable maps, we may use Lemma 9.58 and the same arguments presented in Sect. 8.4 to derive the following facts:

- Images and inverse images of definable sets under definable maps are definable.
- The interior and closure of definable sets are definable.
- The sum and composition of definable maps are definable.
- A function $f : \mathbb{R}^n \to [-\infty, \infty]$ is definable if and only if the set $\{(x, t) \in \mathbb{R}^n \times \mathbb{R} : t \leq f(x)\}$ is definable.
- If $A \subset \mathbb{R}^m$ is a definable set, then $x \to d(x, A)$ is a definable function on $\mathbb{R}^m$. More generally, if $g : A \subset \mathbb{R}^m \to (-\infty, \infty]$ is a definable function that is bounded from below and $F : A \to \mathbb{R}^n$ is a definable mapping, then the function $\varphi : \mathbb{R}^n \to (-\infty, \infty]$ defined by

$$\varphi(y) := \begin{cases} \inf\{g(x) : x \in F^{-1}(y)\} & \text{if } F^{-1}(y) \neq \emptyset, \\ +\infty & \text{if } F^{-1}(y) = \emptyset \end{cases}$$

  is also definable.

Example 9.59 Let D_f denote the set of discontinuity points of a definable function $f : \mathbb{R}^n \to \mathbb{R}$. Then D_f is definable. To see this, we note that

$$\begin{aligned} D_f &= \{x \in \mathbb{R}^n : f \text{ is not continuous at } x\} \\ &= \{x \in \mathbb{R}^n : \exists \varepsilon > 0, \forall \delta > 0, \exists z \in \mathbb{R}^n \text{ with } \|z - x\| < \delta \\ &\qquad \& |f(z) - f(x)| \geq \varepsilon\}. \end{aligned}$$

Since the set

$$\begin{aligned} S := \{&(x, \varepsilon, \delta, z) \in \mathbb{R}^n \times (0, \infty) \times (0, \infty) \times \mathbb{R}^n : \|x - z\| < \delta \\ &\& |f(x) - f(z)| \geq \varepsilon\} \end{aligned}$$

is definable, by repeated using Lemma 9.58 three times, we can conclude that D_f is also definable.

For single variable definable functions $f : \mathbb{R} \to \mathbb{R}$, we can say more about the set D_f. Actually we have the following result.

Lemma 9.60 *If $f : (a, b) \to \mathbb{R}$ is definable, then D_f is finite.*

Proof Assume D_f is infinite. Since any definable set in $\mathbb{R}$ is a finite union of open intervals and points, D_f must contain an open interval I.

We first claim that, for any interval I' contained in I, $f(I')$ contains an interval. If not, since $f(I')$ is definable in $\mathbb{R}$, it must be finite, say $f(I') = \{u_1, \cdots, u_k\}$. Then $I' \subset \cup_{i=1}^k f^{-1}(u_i)$ and each $f^{-1}(u_i)$ is definable. Thus, there must exist $1 \le i \le k$ such that $f^{-1}(u_i)$ contains an interval (α, β) contained in I'. Since $f(x) = u_i$ for all $x \in (\alpha, \beta) \subset I' \subset D_f$, f is continuous on (α, β), contradicting $(\alpha, \beta) \subset D_f$.

By the above claim, $f(I)$ contains an open interval. Thus, we may find $a_1 < b_1$ with $b_1 - a_1 \le 1$ such that $(a_1, b_1) \subset f(I)$. Note that for any $x \in (a_1, b_1)$, there is $t \in I$ such that $x = f(t)$ and hence $t \in I \cap f^{-1}(a_1, b_1)$. As different x corresponds to different t, $I \cap f^{-1}(a_1, b_1)$ must be infinite. Since $I \cap f^{-1}(a_1, b_1)$ is definable, it must contain an open interval $I_1 := (\alpha_1, \beta_1) \subset I$ such that $f(I_1) \subset (a_1, b_1)$. By applying the above claim again, $f(I_1)$ contains an open interval. Thus, we may find $a_2 < b_2$ with $b_2 - a_2 \le 1/2$ such that $(a_2, b_2) \subset f(I_1)$. Since $I_1 \cap f^{-1}(a_2, b_2)$ is infinite, by using the definability of $I_1 \cap f^{-1}(a_2, b_2)$, we can conclude it must contain an open interval $I_2 := (\alpha_2, \beta_2) \subset I_1$ such that $\alpha_1 < \alpha_2 < \beta_2 < \beta_1$ and $f(I_2) \subset (a_2, b_2)$. Repeating the above procedure, we can find a sequence of open intervals $\{I_n := (\alpha_n, \beta_n)\}$ such that

$$\alpha_1 < \alpha_2 < \cdots < \alpha_n < \cdots < \beta_n < \cdots < \beta_2 < \beta_1$$

and each $f(I_n)$ is contained in an interval of length $1/n$. This implies that f is continuous at each point $x \in \cap_{n=1}^{\infty} I_n \neq \emptyset$ which is contained in D_f. We thus obtain a contradiction. □

Lemma 9.61 (Monotonicity Lemma) *Let $f : (a, b) \to \mathbb{R}$ be a definable function and $l \in \mathbb{N}$. Then there is a finite number of points $a = t_0 < t_1 < \cdots < t_k < t_{k+1} = b$ such that, on every interval (t_i, t_{i+1}), f is C^l and either strictly monotone or constant.*

Proof From Lemma 9.60, it follows that D_f is finite. Thus $(a, b) \setminus D_f$ is a finite union of open intervals, and f is continuous on every such a subinterval. Therefore, it suffices to show the result by assuming f is continuous on (a, b).

For each $x \in (a, b)$, we claim that

$$f'_+(x) := \lim_{y \searrow x} \frac{f(y) - f(x)}{y - x} \in [-\infty, \infty]$$

must exist. If not, then there exists a number $r \in \mathbb{R}$ such that

$$\liminf_{y \searrow x} \frac{f(y) - f(x)}{y - x} < r < \limsup_{y \searrow x} \frac{f(y) - f(x)}{y - x}.$$

This implies that for any $\varepsilon > 0$, there exist $z_1, z_2 \in (x, x + \varepsilon)$ such that

$$f(z_1) - f(x) < r(z_1 - x), \quad f(z_2) - f(x) > r(z_2 - x).$$

Therefore $(x, x + \varepsilon) \cap I_{x,r}^+ \neq \emptyset$ and $(x, x + \varepsilon) \cap I_{x,r}^- \neq \emptyset$, where

$$I_{x,r}^+ := \{z \in (x, b) : f(z) - f(x) > r(z - x)\},$$
$$I_{x,r}^- := \{z \in (x, b) : f(z) - f(x) \le r(z - x)\}.$$

Clearly both $I_{x,r}^+$ and $I_{x,r}^-$ are definable sets in $\mathbb{R}$ with $I_{x,r}^+ \cap I_{x,r}^- = \emptyset$ and $I_{x,r}^+ \cup I_{x,r}^- = (x, b)$. Thus there must exist $\varepsilon > 0$ such that one of them contains the interval $(x, x + \varepsilon)$ while the other one does not have any overlap with it. We thus derive a contradiction.

By a similar argument, we can show that

$$f'_-(x) := \lim_{y \nearrow x} \frac{f(y) - f(x)}{y - x} \in [-\infty, \infty]$$

must exist for each $x \in (a, b)$.

Note that both f_+ and f_- are definable functions on (a, b). We next show that the four sets

$$\{f'_+ = +\infty\}, \quad \{f'_+ = -\infty\}, \quad \{f'_- = +\infty\} \quad \text{and} \quad \{f'_- = -\infty\}$$

are finite. We only consider the set $\{f'_+ = +\infty\}$, since other sets can be considered similarly. Note that

$$\{f'_+ = +\infty\} = \{x \in (a, b) : \forall A > 0, \exists \delta > 0, \forall y \in (x, x + \delta) \cap (a, b)$$
$$\text{there holds } f(y) - f(x) > A(y - x)\}.$$

By Lemma 9.58, it is definable. If $\{f'_+ = +\infty\}$ is infinite, as a definable set in $\mathbb{R}$, it must contain an open interval I. Take $a, b \in I$ with $a < b$, and consider the function

$$g(x) := f(x) - \frac{f(b) - f(a)}{b - a}(x - a), \quad x \in I$$

Clearly $g(b) = g(a)$ and $g'_+ = f'_+ - (f(b) - f(a))/(b - a) = +\infty$ on I which implies $g(b) > g(a)$. We thus obtain a contradiction.

Therefore

$$(a, b) \setminus (\{f'_+ = +\infty\} \cup \{f'_+ = -\infty\} \cup \{f'_- = +\infty\} \cup \{f'_- = -\infty\})$$

is a finite union of open intervals. Thus, we may assume on (a, b) the function f is continuous and both f'_+ and f'_- are finite. By applying Lemma 9.60 to f, f'_+ and f'_- we can even assume that f, f'_+, and f_- are continuous on (a, b).

We now claim that $f'_+ = f'_-$ on (a, b). If not, there exists $x_0 \in (a, b)$ such that $f_+(x_0) \neq f_-(x_0)$; we may assume $f'_-(x_0) < f'_+(x_0)$. Then, by the continuity of f'_+ and f'_-, we can find a number $r \in \mathbb{R}$ and an open interval $J \subset (a, b)$ containing x_0 such that $f'_- < r < f'_+$ on J. Thus

$$(f(x) - rx)'_+ > 0 \quad \text{and} \quad (f(x) - rx)'_- < 0 \text{ on } J.$$

Therefore, we may use Lemma 9.62 below to conclude $f(x) - rx$ is strictly increasing and strictly decreasing on J simultaneously which is a contradiction.

Thus f' exists, and both f and f' are continuous on (a, b). We may write

$$\begin{aligned}(a, b) = \{x \in (a, b) : f'(x) > 0\} \cup \{x \in (a, b) : f'(x) = 0\} \\ \cup \{x \in (a, b) : f'(x) < 0\}\end{aligned}$$

each set on the right-hand side is definable and hence is a finite union of open intervals and points. Hence, there must exist $a = t_0 < t_1 < \cdots < t_{n-1} < t_n = b$ such that on each subinterval (t_i, t_{i+1}) either $f' > 0$ or $f' = 0$ or $f' < 0$. The proof is complete. □

Lemma 9.62 *If f is continuous and $f'_+ > 0$ (or $f'_- < 0$) on an open interval I, then f is strictly increasing (or strictly decreasing) on I.*

Proof We assume $f'_+ > 0$ on I and show f is strictly increasing. If this is not true, then there exist $\alpha, \beta \in I$ with $\alpha < \beta$ such that $f(\alpha) \geq f(\beta)$. Since $f'_+(\alpha) > 0$, we can find γ with $\alpha < \gamma < \beta$ such that $f(\alpha) < f(\gamma)$. By the continuity of f, there exists $\tilde{\gamma} \in (\alpha, \beta)$ such that f achieves its maximum on $[\alpha, \beta]$ at $\tilde{\gamma}$. Consequently $f'_+(\tilde{\gamma}) \leq 0$ which is a contradiction. □

Given a set $A \subset \mathbb{R}^n$. By the property of closure, for any $x \in \overline{A}$, there exists a sequence $\{x_k\} \subset A$ such that $x_k \to x$ as $k \to \infty$. When A is a definable set, we have the following much stronger result whose proof can be found in [61, Theorem 3.2].

Lemma 9.63 (Curve Selection Lemma) *Let A be a definable subset of $\mathbb{R}^n$, and let x be an element of $\overline{A}$. Then for all $l \in \mathbb{N}$, there exist $\varepsilon > 0$ and a C^l definable curve $c : [0, \varepsilon) \to \mathbb{R}^n$ such that $c(0) = x$ and $c(t) \in A$ for all $t \in (0, \varepsilon)$.*

We now prove the following KL inequality for C^1 definable functions on bounded C^1 definable submanifolds of $\mathbb{R}^n$; see [129].

Theorem 9.64 *Let U be a non-empty bounded definable C^1 submanifold of $\mathbb{R}^n$, and let $f : U \to (0, \infty)$ be a C^1 definable function with $\inf_U f = 0$. Then there exist $\rho > 0$ and a strictly increasing C^1 definable function $\psi : [0, \rho) \to \mathbb{R}$ with $\psi(0) = 0$ such that*

$$\psi'(f(x))\|\nabla f(x)\| = \|\nabla(\psi \circ f)(x)\| \geq 1 \tag{9.64}$$

for all $x \in U \cap \{0 < f \leq \rho\}$.

Proof For $t > 0$, let

$$f^{-1}(t) := \{x \in U : f(x) = t\}$$

be the level set of f, which is definable. Since $\inf_U f = 0$, we have $f^{-1}(t) \neq \emptyset$ for all sufficiently small $t > 0$. Define the function

$$\varphi(t) := \inf\left\{\|\nabla f(x)\| : x \in f^{-1}(t)\right\}$$

which is definable and finite on an interval $(0, \varepsilon]$ for some $\varepsilon > 0$. We will show, by shrinking ε if necessary, that

$$\varphi(t) > 0, \quad \forall t \in (0, \varepsilon]. \tag{9.65}$$

Assume that (9.65) holds. Then, for each $t \in (0, \varepsilon]$, there exists $x \in f^{-1}(t)$ such that

$$\|\nabla f(x)\| \leq 2\varphi(t) = 2\varphi(f(x)).$$

Thus $f^{-1}(t) \cap \Delta \neq \emptyset$ for all $t \in (0, \varepsilon]$, where

$$\Delta := \{x \in U : 0 < f(x) \leq \varepsilon, \|\nabla f(x)\| \leq 2\varphi(f(x))\}$$

is a definable set. Now, let $\{t_k\}$ be a sequence such that $0 < t_k \to 0$ as $k \to \infty$ and pick $x_k \in f^{-1}(t_k) \cap \Delta$ for each k. Then

$$(x_k, t_k) \in \Gamma_\Delta(f),$$

where $\Gamma_\Delta(f)$ denotes the graph of f restricted to Δ. Since $\{x_k\} \subset U$ is bounded, it has an accumulation point $b \in \overline{U}$. So $(b, 0)$ is an accumulation point of $\{(x_k, t_k)\}$. Consequently

$$(b, 0) \in \overline{\Gamma_\Delta(f) \setminus \{(b, 0)\}},$$

By Lemma 9.63, there exists a C^1 definable curve $\tilde{\eta} : [0, \delta] \to \mathbb{R}^{n+1}$ for some $\delta > 0$ such that $\tilde{\eta}(0) = (b, 0)$ and $\tilde{\eta}((0, \delta]) \subset \Gamma_\Delta(f)$. Writing

$$\tilde{\eta}(s) = (\eta(s), f(\eta(s))), \quad s \in [0, \delta],$$

we define $g(s) := f(\eta(s))$ which satisfies $g(0) = 0$ and $g(s) > 0$ for $s \in (0, \delta]$. By Lemma 9.61, possibly after shrinking δ, the function g is strictly increasing on $(0, \delta]$ and maps onto $(0, \rho]$ for some $0 < \rho \leq \varepsilon$; moreover $g'(s) > 0$ for $s \in (0, \delta]$. Thus

$$\tilde{\psi}(t) := g^{-1}(t), \quad t \in (0, \rho]$$

is a well-defined C^1 function mapping $(0, \rho]$ onto $(0, \delta]$. Let $B > 0$ be such that $\|\eta'(s)\| \leq B$ for $s \in [0, \delta]$. We now show that (9.64) holds with $\psi := 2B\tilde{\psi}$. Indeed, for any $x \in U$ satisfying $0 < f(x) \leq \rho$, we have

$$\|\nabla(\psi \circ f)(x)\| = 2B\tilde{\psi}'(f(x))\|\nabla f(x)\| \geq 2B\tilde{\psi}'(f(x))\varphi(f(x))$$

by the definition of φ. Since $0 < t := f(x) \leq \rho$, we have $s := \tilde{\psi}(t) \in (0, \delta]$ and thus $f(x) = t = g(s) = f(\eta(s))$. Since $\eta(s) \in \Delta$, we have

$$\varphi(f(x)) = \varphi(f(\eta(s))) \geq \frac{1}{2}\|\nabla f(\eta(s))\|.$$

From $g(s) = f(\eta(s))$, we obtain

$$g'(s) = \langle \nabla f(\eta(s)), \eta'(s) \rangle \leq \|\eta'(s)\|\|\nabla f(\eta(s))\| \leq B\|\nabla f(\eta(s))\|.$$

Thus $\varphi(f(x)) \geq \frac{1}{2B}g'(s)$ which leads to

$$\|\nabla(\psi \circ f)(x)\| \geq \tilde{\psi}'(f(x))g'(s) = \tilde{\psi}'(t)g'(s).$$

Since $t = g(\tilde{\psi}(t))$ for any $t \in (0, \rho]$, we have $1 = g'(\tilde{\psi}(t))\tilde{\psi}'(t) = g'(s)\tilde{\psi}'(t)$. Therefore $\|\nabla(\psi \circ f)(x)\| \geq 1$ for all $x \in U \cap \{0 < f \leq \rho\}$ and (9.64) follows.

It remains only to show (9.65). Assume it is not true, then there exists $\{t_k\}$ with $0 < t_k \to 0$ as $k \to \infty$ such that $\varphi(t_k) = 0$ for all k. By the definition of $\varphi(t_k)$, we can find $x_k \in f^{-1}(t_k)$ such that $\|\nabla f(x_k)\| < t_k = f(x_k)$. Define

$$\Sigma := \{x \in U : \|\nabla f(x)\| < f(x)\}$$

which is a definable set. Then $\{x_k\} \subset \Sigma$. Let $\Gamma_\Sigma(f)$ be the graph of f restricted to Σ. Then $(x_k, t_k) = (x_k, f(x_k)) \in \Gamma_\Sigma(f)$ for all k. Since $\{x_k\} \subset U$ is bounded, it has an accumulation point a. Then $(a, 0)$ is an accumulation point of $\{(x_k, t_k)\}$ and thus

$$(a, 0) \in \overline{\Gamma_\Sigma(f) \setminus \{(a, 0)\}}.$$

Since $\Gamma_\Sigma(f)$ is definable, by Lemma 9.63, there is a C^1 definable curve $\tilde{\gamma} : [0, \varepsilon'] \to \mathbb{R}^{n+1}$ for some $\varepsilon' > 0$ such that $\tilde{\gamma}(0) = (a, 0)$ and $\tilde{\gamma}((0, \varepsilon']) \subset \Gamma_\Sigma(f)$. Write

$$\tilde{\gamma}(s) = (\gamma(s), f(\gamma(s))).$$

Then $\gamma(s) \in \Sigma$ and the function $h(s) := f(\gamma(s))$ is definable with $h(0) = 0$ and $h(s) > 0$ for $s \in (0, \varepsilon']$. By Lemma 9.61, h is strictly increasing on $(0, \varepsilon']$ by shrinking ε' if necessary. Note that $h'(s) = \langle \nabla f(\gamma(s)), \gamma'(s) \rangle$, we have

$$|h'(s)| \leq \|\nabla f(\gamma(s))\| \|\gamma'(s)\| \leq A \|\nabla f(\gamma(s))\| < A f(\gamma(s)) = A h(s), \tag{9.66}$$

where A is a number such that $\|\gamma'(s)\| \leq A$ for all $s \in [0, \varepsilon']$. This implies that $h'(s) \to 0$ as $s \to 0$. Since $h'(s)$ is definable, by Lemma 9.61 and the property of h, we must have $h'(s) > 0$ and $h'(s)$ is strictly increasing on $(0, \varepsilon']$ by shrinking ε'. Consequently, by the same derivation for (9.66), we have

$$0 < h'(s) \leq A h(s), \quad \forall s \in (0, \varepsilon'].$$

By the mean value theorem and the strict monotonicity of $h'(s)$, we have for any $s \in (0, \varepsilon')$ that

$$h(s) = h(s) - h(0) = s h'(\bar{s}) \leq s h'(s),$$

where $\bar{s} \in (0, s)$. Therefore $0 < h'(s) \leq A s h'(s)$ and hence $1 \leq A s$ for all $s \in (0, \varepsilon']$ which is a contradiction. □

9.5.3 *Stratification*

Based on Theorem 9.64, we next show that any proper, lower semi-continuous, definable function $f : \mathbb{R}^n \to (-\infty, \infty]$ possesses the KL property. Since f is not necessarily differentiable, its graph may not be a smooth submanifold of $\mathbb{R}^{n+1}$. However, it may be possible to decompose its graph into finite many pieces, each of which is a smooth submanifold. Projecting these pieces onto $\mathbb{R}^n$ gives a partition of $\mathrm{dom}(f)$. Restricting f onto each set in this partition, we may use Theorem 9.64

together with other auxiliary results to demonstrate the KL property of f. To describe the partition more precisely, let us introduce the stratification.

Definition 9.65

(i) Let $X \subset \mathbb{R}^n$ be a nonempty set, and let $p \geq 1$ be an integer. A collection of subsets $\mathcal{X} := \{X_i\}_{i\in I}$ of X is called a C^p *stratification* of X if it is a locally finite partition of X, each X_i is a C^p submanifold of $\mathbb{R}^n$ and, for each $i \neq j$

$$\overline{X_i} \cap X_j \neq \emptyset \implies X_j \subset \overline{X_i} \setminus X_i.$$

The submanifolds X_i are called *strata* of $\mathcal{X}$.

(ii) Given a finite collection $\{A_1, \cdots, A_q\}$ of subsets of X, a stratification $\mathcal{X} = \{X_i\}_{i\in I}$ of X is called *compatible* with $\{A_1, \cdots, A_q\}$ if each A_i is a locally finite union of strata X_j.

Example 9.66 Consider the set

$$X := \{(x, y) : x^3 - y^2 \geq 0\}$$

in $\mathbb{R}^2$. It is easy to see that the four sets

$$X_1 := \{(x, y) : x^3 - y^2 > 0\}, \qquad X_2 := \{(x, y) : x^3 = y^2, y > 0\},$$
$$X_3 := \{(x, y) : x^3 = y^2, y < 0\}, \quad X_4 := \{(0, 0)\}$$

form a stratification of X.

Let $\mathcal{X} = \{X_i\}_{i\in I}$ be a stratification of a set $X \subset \mathbb{R}^n$. Consider a point $x \in \overline{X_i} \cap X_j$ for $i, j \in I$ with $i \neq j$. We may find a sequence $\{x^k\} \subset X_i$ such that $x^k \to x$ as $k \to \infty$. It is natural to consider the relation between the tangent spaces $T_{x^k} X_i$ and $T_x X_j$. To this end, we need a distance between two vector subspaces in $\mathbb{R}^n$.

Definition 9.67 Let V and W be two vector subspaces of $\mathbb{R}^n$. Their distance is defined by

$$d(V, W) := \max\left\{ \sup_{\substack{v\in V\\ \|v\|=1}} \inf_{\substack{w\in W\\ \|w\|=1}} \|v - w\|, \ \sup_{\substack{w\in W\\ \|w\|=1}} \inf_{\substack{v\in V\\ \|v\|=1}} \|v - w\| \right\}.$$

A sequence $\{V_k\}$ of subspaces of $\mathbb{R}^n$ is said to converge to a subspace V of $\mathbb{R}^n$, denoted by $V_k \to V$, if $d(V_k, V) \to 0$ as $k \to \infty$.

Let $\mathcal{V}$ denote the collection of all vector subspaces of $\mathbb{R}^n$. It is easy to check that d defined in Definition 9.67 is indeed a distance function on $\mathcal{V}$, and thus $(\mathcal{V}, d)$ is a metric space.

Lemma 9.68 *Any sequence $\{V_k\}$ of subspaces of $\mathbb{R}^n$ has a subsequence $\{V_{k_l}\}$ such that $d(V_{k_l}, V) \to 0$ as $l \to \infty$ for some subspace V of $\mathbb{R}^n$.*

Proof By taking a subsequence if necessary, we may find an integer $0 \le d \le n$ such that every V_k is a d-dimensional subspace of $\mathbb{R}^n$. For each V_k, we choose an orthonormal basis $\{v_{k,i} : i = 1, \cdots, d\}$. Since $\|v_{k,i}\| = 1$, we may find a subsequence $\{k_l\}$ of positive integers such that $v_{k_l,i} \to v_i$ as $l \to \infty$ for each $i = 1, \cdots, d$, where v_i are vectors in $\mathbb{R}^n$. Clearly, $\{v_i : i = 1, \cdots, d\}$ are orthonormal. Define $V := \text{span}\{v_1, \cdots, v_d\}$. We now show that

$$d(V_{k_l}, V) \to 0 \quad \text{as } l \to \infty.$$

To see this, for any $x \in V_{k_l}$ with $\|x\| = 1$, we can write $x = \sum_{i=1}^d \alpha_i v_{k_l,i}$ with $\sum_{i=1}^d |\alpha_i|^2 = 1$. Define $\tilde{x} := \sum_{i=1}^d \alpha_i v_i$. Then $\tilde{x} \in V$ satisfies $\|\tilde{x}\| = 1$ and, by the triangle inequality and the Cauchy-Schwarz inequality, we have

$$\inf_{\substack{v \in V \\ \|v\|=1}} \|x - v\| \le \|x - \tilde{x}\| \le \sum_{i=1}^d |\alpha_i| \|v_{k_l,i} - v_i\| \le \left(\sum_{i=1}^d \|v_{k_l,i} - v_i\|^2 \right)^{1/2}$$

which implies

$$\sup_{\substack{x \in V_{k_l} \\ \|x\|=1}} \inf_{\substack{v \in V \\ \|v\|=1}} \|x - v\| \le \left(\sum_{i=1}^d \|v_{k_l,i} - v_i\|^2 \right)^{1/2}.$$

By a similar argument, we also have

$$\sup_{\substack{v \in V \\ \|v\|=1}} \inf_{\substack{x \in V_{k_l} \\ \|x\|=1}} \|x - v\| \le \left(\sum_{i=1}^d \|v_{k_l,i} - v_i\|^2 \right)^{1/2}.$$

Therefore,

$$d(V_{k_l}, V) \le \left(\sum_{i=1}^d \|v_{k_l,i} - v_i\|^2 \right)^{1/2} \to 0$$

as $l \to \infty$. □

Lemma 9.69 *Let $\{V_k\}$ and V be subspaces of $\mathbb{R}^n$ with $d(V_k, V) \to 0$ as $k \to \infty$. If $\xi_k \in \mathbb{R}^n$ is normal to V_k for each k and $\xi_k \to \xi$ as $k \to \infty$ for some $\xi \in \mathbb{R}^n$, then ξ is normal to V.*

Proof Let $v \in V$ be any unit vector. Since $d(V_k, V) \to 0$, we have

$$\inf_{\substack{\tilde{v} \in V_k \\ \|\tilde{v}\|=1}} \|v - \tilde{v}\| \le d(V_k, V) \to 0$$

as $k \to \infty$. Thus, there is $v_k \in V_k$ with $\|v_k\| = 1$ for each k such that $v_k \to v$ as $k \to \infty$. Since ξ_k is normal to V_k, we have $\langle \xi_k, v_k \rangle = 0$. Consequently $\langle \xi, v \rangle = \lim_{k\to\infty} \langle \xi_k, v_k \rangle = 0$. Due to the arbitrariness of v, this shows that ξ is normal to V. □

Definition 9.70 Let X be a non-empty set of $\mathbb{R}^n$ and $p \ge 1$ an integer. A C^p stratification $\mathcal{X} = (X_i)_{i\in I}$ of X is called *Whitney* if for each $x \in \overline{X_i} \cap X_j$ with $i \ne j$ and for each sequence $\{x_k\} \subset X_i$, we have

$$x_k \to x,\ T_{x_k} X_i \to \mathcal{T} \implies T_x X_j \subset \mathcal{T},$$

where $\mathcal{T}$ is a vector subspace of $\mathbb{R}^n$.

We have the following existence result on Whitney stratification for the graphs of lower semicontinuous definable functions; one may refer to [205, II.1.17] for a proof; see also [32, Lemma 8].

Proposition 9.71 *Let $\mathcal{O}$ be an o-minimal structure on $(\mathbb{R}, +, \cdot)$, let $f : \mathbb{R}^n \to (-\infty, \infty]$ be a proper, lower semi-continuous, definable function and let $\mathcal{B} := \{B_1, \cdots, B_q\}$ be a collection of definable subsets of $\mathbb{R}^n$. Then for any $p \ge 1$, there exists a C^p-Whitney stratification $\{S_1, \cdots, S_l\}$ of $Gr(f)$ such that*

(i) *each S_i is definable in $\mathcal{O}$;*
(ii) *the stratification $\{S_1, \cdots, S_l\}$ is non-vertical in the sense that $e_{n+1} \notin T_u S_i$ for all $i \in \{1, \cdots, l\}$ and $u \in S_i$, where $e_{n+1} = (0, \cdots, 0, 1) \in \mathbb{R}^{n+1}$;*
(iii) *the collection $\{X_1, \cdots, X_l\}$, with*

$$X_i = \pi(S_i), \quad i = 1, \cdots, l,$$

is a C^p-Whitney stratification of dom(f) *compatible with $\mathcal{B}$, where $\pi : \mathbb{R}^{n+1} \to \mathbb{R}^n$ is the canonical projection defined by $\pi(x_1, \cdots, x_n, x_{n+1}) = (x_1, \cdots, x_n)$.*

We have the following projection formula onto tangent spaces for lower semi-continuous functions whose graph admit a non-vertical Whitney stratification [32].

Lemma 9.72 *Let $f : \mathbb{R}^n \to (-\infty, \infty]$ be a proper, lower semicontinuous function. Assume that the graph $Gr(f)$ of f admits a non-vertical C^1 Whitney stratification $\mathcal{S} := \{S_i\}_{i\in I}$. For each $x \in$ dom(f), let S_x be the stratum containing $(x, f(x))$, and* *let $X_x := \pi(S_x)$, where $\pi : \mathbb{R}^{n+1} \to \mathbb{R}^n$ is the canonical projection. There holds*

$$P_{T_x X_x}(\partial f(x)) = \begin{cases} \emptyset & \text{if } \partial f(x) = \emptyset, \\ \{\nabla_R f(x)\} & \text{if } \partial f(x) \neq \emptyset, \end{cases} \tag{9.67}$$

where $\nabla_R f(x)$ denotes the Riemannian gradient of f restricted to X_x.

Proof Let $X_i := \pi(S_i)$ for $i \in I$. The non-verticality of the stratification $\{S_i\}_{i\in I}$ of $\mathrm{Gr}(f)$ implies that each X_i is a C^1-submanifold of $\mathbb{R}^n$ and the restriction of f to X_i is a C^1 function.

We first consider the Fréchet subdifferential $\hat{\partial} f(x)$ and show that if $\hat{\partial} f(x) \neq \emptyset$ then

$$P_{T_x X_x}(\hat{\partial} f(x)) = \{\nabla_R f(x)\}. \tag{9.68}$$

To see this, we take any $\xi \in \hat{\partial} f(x)$. Let $v \in T_x X_x$ be any tangent vector. Then there is a C^1 curve $c : (-\varepsilon, \varepsilon) \to X_x$ with $c(0) = x$ and $c'(0) = v$. Consequently, the function $f \circ c : (-\varepsilon, \varepsilon) \to \mathbb{R}$ is C^1 with

$$\left.\frac{d}{dt} f(c(t))\right|_{t=0} = \langle \nabla_R f(x), v\rangle. \tag{9.69}$$

On the other hand, since $\xi \in \hat{\partial} f(x)$, for any $\eta > 0$, we can find $r > 0$ such that

$$f(y) - f(x) - \langle \xi, y - x\rangle \geq -\eta\|y - x\|, \quad \forall y \in B_r(x).$$

Taking $y = c(t)$ gives

$$f(c(t)) - f(c(0)) - \langle \xi, c(t) - c(0)\rangle \geq -\eta\|c(t) - c(0)\|$$

for small $t > 0$. Dividing the above equation by $t > 0$ and letting $t \to 0$, we can obtain

$$\left.\frac{d}{dt} f(c(t))\right|_{t=0} - \langle \xi, v\rangle \geq -\eta\|v\|.$$

Since $\eta > 0$ can be arbitrary, we must have

$$\left.\frac{d}{dt} f(c(t))\right|_{t=0} \geq \langle \xi, v\rangle. \tag{9.70}$$

Combining (9.69) and (9.70) shows that $\langle \nabla_R f(x), v\rangle \geq \langle \xi, v\rangle$ for all $v \in T_x X_x$ and hence

$$\langle \nabla_R f(x), v\rangle = \langle \xi, v\rangle, \quad \forall v \in T_x X_x. \tag{9.71}$$

Consequently $P_{T_x X_x}(\xi) = \nabla_R f(x)$.

Next we show (9.67). Let $\xi \in \partial f(x)$. Then we have $\{x_k\} \subset \operatorname{dom}(\hat{\partial} f)$, $\xi_k \in \hat{\partial} f(x_k)$ such that $x_k \to x$, $f(x_k) \to f(x)$, and $\xi_k \to \xi$ as $k \to \infty$. Due to the local finiteness of $\mathcal{S}$, by taking a subsequence if necessary, we may find a stratum S_i such that $u_k := (x_k, f(x_k)) \in S_i$ for all k.

If $S_i = S_x$, then we have from (9.68) that $P_{T_x X_x}(\xi_k) = \nabla_R f(x_k)$ for all k. Taking $k \to \infty$, using the continuity of the projection and the continuity of $\nabla_R f$ on X_x, we can obtain $P_{T_x X_x}(\xi) = \nabla_R f(x)$ as desired.

In the following, we assume $S_i \neq S_x$. Since $u_k := (x_k, f(x_k)) \to (x, f(x)) \in S_x$, we have $(x, f(x)) \in \overline{S_i} \cap S_x \neq \emptyset$. Consider the sequence $\{T_{u_k} S_i\}$ of tangent spaces. By using Lemma 9.68, we may assume that $\{T_{u_k} S_i\}$ converges to a vector space $\mathcal{T}$ as $k \to \infty$. Since the stratification is Whitney, we have $T_{(x, f(x))} S_x \subset \mathcal{T}$. Since $\xi_k \in \hat{\partial} f(x_k)$, we may use the same argument for deriving (9.71) to obtain

$$\langle \nabla_R f(x_k), v\rangle = \langle \xi_k, v\rangle, \quad \forall v \in T_{x_k} X_i, \tag{9.72}$$

where $X_i = \pi(S_i)$. Since S_i is a stratum on the graph of f, it is easy to see that every $\tilde{v} \in T_{u_k} S_i$ has the form $\tilde{v} = (v, a)$, where $v \in T_{x_k} X_i$ and $a = \langle \nabla_R f(x_k), v\rangle$. Therefore we can conclude from (9.72) that $\langle (\xi_k, -1), \tilde{v}\rangle = 0$ for all $\tilde{v} \in T_{u_k} S_i$ which means $(\xi_k, -1)$ is normal to $T_{u_k} S_i$. Thus it follows from Lemma 9.69 that $(\xi, -1) = \lim_{k\to\infty}(\xi_k, -1)$ is normal to $\mathcal{T}$ and hence is normal to $T_{(x, f(x))} S_x$. Note that

$$T_{(x, f(x))} S_x \subset Y := T_x X_x + \mathbb{R} e_{n+1}.$$

Thus the projection of $(\xi, -1)$ onto Y is also normal to $T_{(x, f(x))} S_x$. Note also that the projection of $(\xi, -1)$ onto Y is $(P_{T_x X_x}(\xi), -1)$. Therefore $(P_{T_x X_x}(\xi), -1) \in Y$ is normal to $T_{(x, f(x))} S_x$. Since $(\nabla_R f(x), -1) \in Y$ is also normal to $T_{(x, f(x))} S_x$ which is a codimension one subspace of Y, thus these two normal vectors must be the same and hence $\nabla_R f(x) = P_{T_x X_x}(\xi)$. □

9.5.4 *Proof of the KL Property*

With the above preparation, we are now ready to prove that every lower semicontinuous definable function is a KL function.

Theorem 9.73 ([32]) *Let $f : \mathbb{R}^n \to (-\infty, \infty]$ be a proper, lower semicontinuous, definable function. Then for any $\bar{x} \in \operatorname{dom}(f)$ and $r > 0$, there exist $\rho > 0$ and a* *strictly increasing continuous definable function* $\psi : [0, \rho) \to [0, \infty)$*, which is* C^1 *on* $(0, \rho)$ *with* $\psi(0) = 0$*, such that*

$$\psi'(f(x) - f(\bar{x}))d(0, \partial f(x)) \geq 1 \tag{9.73}$$

for all $x \in B_r(\bar{x}) \cap \{f(\bar{x}) < f \leq f(\bar{x}) + \rho\}$*. Furthermore, this* ψ *can be taken to a concave function.*

Proof Let $U := \{x \in B_r(\bar{x}) : f(x) > f(\bar{x})\}$ which is a definable set. According to Proposition 9.71, there is a non-vertical C^1 Whitney definable stratification $\{S_1, \cdots, S_l\}$ of $\mathrm{Gr}(f)$ such that $\mathcal{X} := \{X_i := \pi(S_i) : i = 1, \cdots, l\}$ is a definable stratification of $\mathrm{dom}(f)$ compatible with U. Thus there is an index set $I \subset \{1, \cdots, l\}$ such that $U = \cup_{i \in I} X_i$. Note that each X_i with $i \in I$ is a bounded definable C^1 submanifold of $\mathbb{R}^n$ and the function $f_i := f|_{X_i} - f(\bar{x})$ is a C^1 definable function on X_i. By virtue of Theorem 9.64, there exist $\rho_i > 0$ and a strictly increasing C^1 definable function $\psi_i : [0, \rho_i) \to \mathbb{R}$ with $\psi_i(0) = 0$ such that

$$\psi_i'(f_i(x))\|\nabla f_i(x)\| = \|\nabla(\psi_i \circ f_i)(x)\| \geq 1, \quad \forall x \in X_i \cap \{0 < f_i \leq \rho_i\}.$$

Note that $\nabla f_i(x) = \nabla_R f(x)$ for $x \in X_i$, where $\nabla_R f(x)$ denotes the Riemannian gradient of f restricted to the stratum $X_x \in \mathcal{X}$ containing x. We have

$$\|\nabla_R f(x)\| \geq \frac{1}{\psi_i'(f(x) - f(\bar{x}))}, \quad \forall x \in X_i \cap \{f(\bar{x}) < f \leq f(\bar{x}) + \rho_i\}. \tag{9.74}$$

Let $\rho := \min\{\rho_i : i \in I\}$ and for any $i, j \in I$ with $i \neq j$ consider the definable function $\psi_i'(t) - \psi_j'(t)$ on $(0, \rho)$. By Lemma 9.61, this function has a constant sign on $(0, \rho)$ by shrinking ρ if necessary. Since I is a finite set, there must exist $i_0 \in I$ such that $\psi_{i_0}'(t) \geq \psi_i'(t)$ for $t \in (0, \rho)$ and $i \in I$. Let $\psi := \psi_{i_0}$. Then it follows from (9.74) that

$$\|\nabla_R f(x)\| \geq \frac{1}{\psi'(f(x) - f(\bar{x}))}, \quad \forall x \in B_r(\bar{x}) \cap \{f(\bar{x}) < f \leq f(\bar{x}) + \rho\}.$$

If $\partial f(x) \neq \emptyset$, it then follows from Lemma 9.72 that for any $\xi \in \partial f(x)$, there holds $P_{T_x X_x}(\xi) = \nabla_R f(x)$ and hence $\|\xi\| \geq \|\nabla_R f(x)\|$. Consequently $d(0, \partial f(x)) \geq \|\nabla_R f(x)\|$ which holds trivially when $\partial f(x) = \emptyset$ by the convention $d(0, \emptyset) = \infty$. Therefore

$$d(0, \partial f(x)) \geq \frac{1}{\psi'(f(x) - f(\bar{x}))}$$

for all $x \in B_r(\bar{x}) \cap \{f(\bar{x}) < f \leq f(\bar{x}) + \rho\}$ and (9.73) is established.

We next show that ψ can be taken to a concave function. Since ψ' is a definable function, by applying Lemma 9.61 to ψ' and by shrinking $\rho > 0$ if necessary, we can guarantee either $\psi'' \leq 0$ on $(0, \rho]$ or $\psi'' > 0$ on $(0, \rho]$. If the former holds, then ψ is already a concave function. If the latter happens, then $\psi'(t) < \psi'(\rho)$ for $t \in (0, \rho)$. Consequently (9.73) holds with ψ replaced by the concave function $\tilde{\psi}(t) := \psi'(\rho)t$ because $\tilde{\psi}'(t) = \psi'(\rho) > \psi'(t)$ for $t \in (0, \rho)$. The proof is thus complete. □

Finally, we apply Theorem 9.73 to establish Theorem 9.24 by showing that, for a lower semicontinuous semi-algebraic function, the concave function ψ in Theorem 9.73 can be chosen as

$$\psi(s) = cs^{1-\alpha} \qquad \text{for some } c > 0 \text{ and } \alpha \in [0, 1).$$

To justify this choice, we use the following result from Puiseux series expansion [30, 220].

Lemma 9.74 (Puiseux Lemma) *Let $f : (0, \eta) \to \mathbb{R}$ be a semi-algebraic function. Then there exist integers m, p with $p \geq 1$, a sequence $\{a_i\}_{i \geq m}$ of real numbers, and a real number $\varepsilon \in (0, \eta]$ such that f can be represented as the convergent series*

$$f(t) = \sum_{i=m}^{\infty} a_i t^{i/p}, \quad \forall t \in (0, \varepsilon)$$

which is called the Puiseux expansion of f on $(0, \varepsilon)$.

Proof of Theorem 9.24 Since f is semi-algebraic, we may use the o-minimal structure $\{\mathcal{SA}_n\}$. According to Theorem 9.73, there exist $\rho > 0$ and a strictly increasing, continuous, semi-algebraic function $\psi : [0, \rho) \to [0, \infty)$ with $\psi \in C^1(0, \rho)$ and $\psi(0) = 0$ such that

$$d(0, \partial f(x)) \geq \frac{1}{\psi'(f(x) - f(\bar{x}))}$$

for all $x \in B_r(\bar{x}) \cap \{f(\bar{x}) < f < f(\bar{x}) + \rho\}$. By applying Lemma 9.74 to ψ and using $\psi(0) = 0$, we can obtain the Puiseux expansion

$$\psi(t) = \sum_{i=1}^{\infty} a_i t^{i/p}$$

of ψ on some interval $(0, \eta)$ for some number $0 < \eta \leq \rho$ and integer $p \geq 1$. Consequently

$$\psi'(t) = |\psi'(t)| \leq Dt^{1/p-1}, \quad \forall t \in (0, \eta)$$

for some $D > 0$. Therefore,

$$d(0, \partial f(x)) \geq \frac{1}{D}(f(x) - f(\bar{x}))^{1-1/p}$$

for all $x \in B_r(\bar{x}) \cap \{f(\bar{x}) < f < f(\bar{x}) + \eta\}$. □

Problems

9.1 Let $f : \mathbb{R}^n \to (-\infty, \infty]$ and $g : \mathbb{R}^n \to (-\infty, \infty]$ be two proper functions. Show that

$$\partial f(x) + \partial g(x) \subset \partial(f + g)(x), \quad \forall x \in \mathbb{R}^n.$$

9.2 Let $f : \mathbb{R}^n \to (-\infty, \infty]$ be a proper function which is lower semicontinuous at $\bar{x} \in \operatorname{dom}(f)$ with $f(\bar{x}) > 0$. For any $p > 0$, show that

$$\hat{\partial}(f^p)(\bar{x}) = pf(\bar{x})^{p-1}\hat{\partial}f(\bar{x}) \quad \text{and} \quad \partial(f^p)(\bar{x}) = pf(\bar{x})^{p-1}\partial f(\bar{x}).$$

9.3 Let $f : \mathbb{R}^n \to (-\infty, \infty]$ be a proper function. Let $p > 0$, and let $\bar{x} \in \operatorname{dom}(f)$ with $f(\bar{x}) > 0$. Assume that f is lower semicontinuous at $\bar{x}$ and f^p satisfies the KL condition at $\bar{x}$ with exponent $\alpha \in [0, 1)$. Show that f also satisfies the KL condition at $\bar{x}$ with exponent α.

9.4 Consider the function f in the form $f(x) = g(x) + h(x)$, where $g : \mathbb{R}^n \to (-\infty, \infty]$ is a proper, lower semicontinuous, convex function and $h : \mathbb{R}^n \to \mathbb{R}$ is a continuously differentiable function with a locally Lipschitz continuous gradient. Let S denote the set of critical points of f. We say the *Luo-Tseng error bound condition* holds for f if for any $\gamma > \inf_{\mathbb{R}^n} f$ there exist $c, \varepsilon > 0$ such that

$$d(x, S) \leq c\|\operatorname{prox}_g(x - \nabla h(x)) - x\|$$

whenever $f(x) < \gamma$ and $\|\operatorname{prox}_g(x - \nabla h(x)) - x\| < \varepsilon$.

(i) Assume that $S \neq \emptyset$ and the Luo-Tseng error bound condition holds. Show that for any $\bar{x} \in S$, there is $r > 0$ such that

$$d(x, S) \leq d(0, \partial f(x))$$

for all $x \in B_r(\bar{x}) \cap \{f < f(\bar{x}) + 1\}$.

(ii) If in addition the critical values of f is separated in the sense that, for any $\bar{x} \in S$, there is $\delta > 0$ such that $y \in B(\bar{x}, \delta) \cap S$ implies $f(y) = f(\bar{x})$, show that f is a KL function with exponent $1/2$.

9.5 Let $f : \mathbb{R}^n \to (-\infty, \infty]$ be a proper, lower semicontinuous function and $\bar{x} \in \mathrm{dom}(f)$. Assume that there exist $r > 0$ $\eta > 0$ and a desingularizing function $\varphi \in \mathcal{K}(0, \eta)$ such that

$$\varphi'(f(x) - f(\bar{x}))d(0, \hat{\partial} f(x)) \geq 1 \tag{9.75}$$

for all $x \in B_r(\bar{x}) \cap \{f(\bar{x}) < f < f(\bar{x}) + \eta\}$. Show that there must exist $0 < \tilde{r} \leq r$ and $0 < \tilde{\eta} \leq \eta$ such that

$$d(x, S) \leq 2\varphi(f(x) - f(\bar{x}))$$

for all $x \in B_{\tilde{r}}(\bar{x}) \cap \{f(\bar{x}) \leq f < f(\bar{x}) + \tilde{\eta}\}$, where $S := \{x \in \mathbb{R}^n : f(x) \leq f(\bar{x})\}$.

9.6 For any nonzero $x \in \mathbb{R}^n$, show that

$$\partial \left(\frac{\|\cdot\|_1}{\|\cdot\|_2} \right)(x) = \frac{1}{\|x\|_2} \partial(\|\cdot\|_1)(x) - \frac{\|x\|_1}{\|x\|_2^3} x.$$

9.7 Let $A \in \mathbb{R}^{m \times n}$ with $m \ll n$ and $b \in \mathbb{R}^m$ with $0 \neq b \in \mathrm{Ran}(A)$. The non-convex model

$$\min \left\{ \frac{\|x\|_1}{\|x\|_2} : x \in \mathbb{R}^n \text{ and } Ax = b \right\} \tag{9.76}$$

has been introduced in [185, 229] to determine a sparse solution of $Ax = b$. In [222], the following algorithm has been proposed for solving (9.76):

$$x^{k+1} = \arg\min \left\{ \|x\|_1 - \frac{\|x^k\|_1}{\|x^k\|_2^2} \langle x, x^k \rangle + \frac{\alpha}{2} \|x - x^k\|^2 : Ax = b \right\},$$

where $\alpha > 0$ and the initial guess x^0 satisfies $Ax^0 = b$. If $\{x^k\}$ is bounded, use Theorem 9.26 to show that $\{x^k\}$ converges to a critical point of the function

$$f(x) := \frac{\|x\|_1}{\|x\|_2} + \delta_{A^{-1}(b)}(x)$$

which is a proper, lower semicontinuous, semi-algebraic function.

9.8 Consider the minimization problem

$$\min_{x \in \mathbb{R}^n} \{H(x) := f(x) + g(x)\}, \tag{9.77}$$

where the function $f : \mathbb{R}^n \to \mathbb{R}$ is continuously differentiable with L-Lipschitz continuous gradient ∇f, the function $g : \mathbb{R}^n \to (-\infty, \infty]$ is proper, lower semicontinuous, and there is a constant $\gamma > 0$ such that the function $x \to$

$g(x) + \frac{1}{2\gamma}\|x\|^2$ is bounded below on $\mathbb{R}^n$, and moreover the function H is coercive on $\mathbb{R}^n$. In [170], an inertial proximal algorithm for non-convex optimization, which was termed as `iPiano`, was proposed to solve (9.77) which takes the form

$$x^{k+1} = \arg\min_{x\in\mathbb{R}^n} \left\{ g(x) + \frac{1}{2\alpha}\|x - (x^k - \alpha\nabla f(x^k) + \beta(x^k - x^{k-1}))\|^2 \right\}$$

with the initial guess $x^{-1} = x^0 \in \mathrm{dom}(H)$, where $\alpha > 0$ and $\beta \geq 0$ are suitable parameters. Show that for any $0 < \alpha < \gamma$ and $\beta \geq 0$, the sequence $\{x^k\}$ is well-defined. Assume further that

$$0 \leq \beta < \frac{1}{2} \quad \text{and} \quad 0 < \alpha < \min\left\{\frac{1-2\beta}{L}, \gamma\right\}$$

and consider the function $\Phi(x, y) := H(x) + \delta\|x - y\|^2$ on $\mathbb{R}^n \times \mathbb{R}^n$ with $\delta := \frac{1-\beta}{2\alpha} - \frac{L}{2} > 0$. Show the following result:

(i) For each integer $k \geq 0$, there holds

$$\Phi(x^{k+1}, x^k) + a\|x^k - x^{k-1}\|^2 \leq \Phi(x^k, x^{k-1})$$

with $a := \delta - \beta/(2\alpha) > 0$;

(ii) For each integer $k \geq 0$, there exists $w^{k+1} \in \partial\Phi(x^{k+1}, x^k)$ such that

$$\|w^{k+1}\| \leq \frac{4}{\alpha}\left(\|x^{k+1} - x^k\| + \|x^k - x^{k-1}\|\right).$$

(iii) There is a subsequence $\{x^{k_j}\}$ of $\{x^k\}$ such that $x^{k_j} \to x^*$ and $\Phi(x^{k_j}, x^{k_j-1}) \to H(x^*)$ as $j \to \infty$. Moreover, x^* is a critical point of H;

(iv) If Φ additionally satisfies the KL condition at the point (x^*, x^*), then the sequence $\{x^k\}$ has a finite length, i.e.,

$$\sum_{k=0}^{\infty} \|x^{k+1} - x^k\| < \infty$$

and $x^k \to x^*$ as $k \to \infty$;

(v) Assume that H satisfies the KL condition with exponent $\theta \in [1/2, 1)$.

(a) If $\theta = 1/2$, then $H(x^k) \to H(x^*)$ and $x^k \to x^*$ linearly;

(b) If $\theta \in (1/2, 1)$, then

$$H(x^k) - H(x^*) = O(k^{\frac{1}{1-2\theta}}) \quad \text{and} \quad \|x^k - x^*\| = O(k^{\frac{1-\theta}{1-2\theta}}).$$

9.9 Consider the minimization problem

$$\min\{f(x)+g(x)\} = \min\{f(x)+g(y) : y-x=0\}, \tag{9.78}$$

where $f : \mathbb{R}^n \to \mathbb{R}$ is a continuously differentiable function with L-Lipschitz continuous gradient ∇f and $g : \mathbb{R}^n \to (-\infty, \infty]$ is a proper, lower semicontinuous function bounded below. An application of ADMM yields the iteration scheme

$$\begin{aligned} y^{k+1} &\in \arg\min_{y\in\mathbb{R}^n} \mathscr{L}_\rho(x^k, y, \lambda^k), \\ x^{k+1} &\in \arg\min_{x\in\mathbb{R}^n} \mathscr{L}_\rho(x, y^{k+1}, \lambda^k), \\ \lambda^{k+1} &= \lambda^k + \rho(y^{k+1} - x^{k+1}), \end{aligned} \tag{9.79}$$

where

$$\mathscr{L}_\rho(x, y, \lambda) := f(x) + g(y) + \langle \lambda, y-x\rangle + \frac{\rho}{2}\|y-x\|^2$$

and $\rho > 0$ is a number. Show that, if $\rho > 2L$, then the method (9.79) is well-defined and the following results hold:

(i) For all integers $k \geq 0$, there holds

$$\mathscr{L}_\rho(x^{k+1}, y^{k+1}, \lambda^{k+1}) + a\|x^{k+1} - x^k\|^2 \leq \mathscr{L}_\rho(x^k, y^k, \lambda^k),$$

where $a := \frac{1}{2}(\rho - L) - \frac{L^2}{\rho} > 0$.

(ii) There is a constant $b > 0$ such that

$$d\left(0, \partial\mathscr{L}_\rho(x^{k+1}, y^{k+1}, \lambda^{k+1})\right) \leq b\|x^{k+1} - x^k\|, \quad \forall k \geq 0.$$

(iii) If in addition g is coercive, then the sequence $\{(x^k, y^k, \lambda^k)\}$ is bounded.

(iv) Any cluster point (x^*, y^*, λ^*) of $\{(x^k, y^k, \lambda^k)\}$ is a KKT point of the problem, i.e., $x^* = y^*$, $\lambda^* = \nabla f(x^*)$ and $-\lambda^* \in \partial g(y^*)$.

(v) If f and g are semi-algebraic and g is coercive, then $\{(x^k, y^k, \lambda^k)\}$ must converge to a KKT point of the problem and

$$\sum_{k=0}^{\infty} \|x^{k+1} - x^k\| < \infty.$$

Chapter 10
Semi-smooth Newton Methods

Many optimization problems can be equivalently reduced to solving equations of the form $F(x) = y$, where F is a mapping that may not be Fréchet differentiable, thus preventing the direct use of the classical Newton method. In this chapter, we introduce the semi-smooth Newton method and demonstrate its superlinear convergence under the concept of Newton differentiability of F, a condition that is much weaker than Fréchet differentiability. Newton differentiability can be guaranteed by semi-smoothness, which is defined in terms of Clarke's generalized Jacobian. It turns out that a large class of tame functions are semi-smooth. We also discuss a globalization of the semi-smooth Newton method and present various applications in optimization.

10.1 Newton Differentiability and Semi-smooth Newton Method

Let X and Y be Banach spaces, and let $F : \mathscr{D} \subset X \to Y$ be a mapping with domain $\mathscr{D}$. We consider finding a solution of the nonlinear equation

$$F(x) = 0, \tag{10.1}$$

where F might be non-differentiable. This equation may arise from many applications; in Sect. 10.5, we will present examples of non-smooth optimization problems which lead to such equations.

When F is Fréchet differentiable, Newton method is a prominent algorithm for root finding problems due to its superlinear convergence. Recall that F is Fréchet differentiable at $x \in X$ if there is a bounded linear operator $A : X \to Y$ such that

$$\|F(x + h) - F(x) - Ah\| = o(\|h\|) \tag{10.2}$$

Q. Jin, *Lectures on Nonsmooth Optimization*, Texts in Applied Mathematics 82,
https://doi.org/10.1007/978-3-031-91417-1_10

as $\|h\| \to 0$. We will denote A by $F'(x)$ and call it the Fréchet derivative of F at x. Assuming that F is Fréchet differentiable such that the Fréchet derivative $F'(x)$ is invertible with $F'(x)^{-1}$ being uniformly bounded in a neighborhood of a sought solution x^*, Newton method takes the form

$$x^{k+1} = x^k - F'(x^k)^{-1}F(x^k). \tag{10.3}$$

To see the superlinear convergence of (10.3), we may use $F(x^*) = 0$ to write

$$\begin{aligned}
\|x^{k+1} - x^*\| &= \|x^k - x^* - F'(x^k)^{-1}(F(x^k) - F(x^*))\| \\
&= \|F'(x^k)^{-1}[F(x^k) - F(x^*) - F'(x^k)(x^k - x^*)]\| \\
&\le C\|F(x^k) - F(x^*) - F'(x^k)(x^k - x^*)\|,
\end{aligned}$$

where C is a constant such that $\|F'(x)^{-1}\| \le C$ for all x in a neighborhood of x^*. Let $\mathscr{L}(X, Y)$ denote the Banach space consisting of all bounded linear operators from X to Y. By using the Fréchet differentiability of F at x^* and assuming the mapping $x \to F'(x)$ from X to $\mathscr{L}(X, Y)$ is continuous at x^*, we can further derive that

$$\begin{aligned}
\|x^{k+1} - x^*\| &\le C\|F(x^k) - F(x^*) - F'(x^*)(x^k - x^*)\| \\
&\quad + C\|(F'(x^k) - F'(x^*))(x^k - x^*)\| \\
&= o(\|x^k - x^*\|)
\end{aligned}$$

which demonstrates the superlinear convergence.

We are going to generalize Newton method to treat the equation (10.1) with non-differentiable mapping F with the hope that the superlinear convergence is preserved. Checking the above analysis on Newton method, the key part is to estimate the term $F(x^k) - F(x^*) - F'(x^k)(x^k - x^*)$ which can be written as

$$F(x^* + e^k) - F(x^*) - F'(x^* + e^k)e^k,$$

where $e^k := x^k - x^*$. If we are able to find a mapping $G : X \to \mathscr{L}(X, Y)$ such that $G(x)$ is invertible with $G(x)^{-1}$ uniformly bounded in a neighborhood of x^* and

$$\|F(x^* + e^k) - F(x^*) - G(x^* + e^k)e^k\| = o(\|e^k\|),$$

then we can still obtain superlinear convergence of the method (10.3) with $F'(x^k)^{-1}$ replaced by $G(x^k)^{-1}$. This motivates the following definition of Newton differentiability [54, 128].

Definition 10.1 Let X and Y be Banach spaces. A mapping $F : \mathscr{D} \subset X \to Y$ is said to be *Newton differentiable* at $x \in \mathscr{D}$ if there exists a mapping $G : \mathscr{D} \to \mathscr{L}(X, Y)$ such that

$$\|F(x+h) - F(x) - G(x+h)h\| = o(\|h\|) \tag{10.4}$$

as $\|h\| \to 0$. The mapping G is then called a *Newton derivative* of F at x and is denoted as $D_N F$.

The definition of Newton differentiability looks similar to the definition of Fréchet differentiability. However, they are essentially different. In the definition of Fréchet differentiability of F at x, the single bounded linear operator $F'(x)$, the Fréchet derivative at x, is used which can be determined by considering the Gâteaux derivative

$$F'(x; h) := \lim_{t \to 0} \frac{F(x+th) - F(x)}{t}.$$

The definition of Newton differentiability involves a family of bounded linear operators, and it does not present a constructive procedure to find a Newton derivative. How to determine a Newton derivative is indeed an involved procedure which will be discussed further.

When F is Fréchet differentiable at x, by definition F is also Newton differentiable at x with the Newton derivative given by $D_N F(\tilde{x}) := F'(x)$ for all $\tilde{x} \in X$. When F is Fréchet differentiable in a neighborhood of x and the mapping $u \to F'(u)$ is continuous at x, then $D_N F(\tilde{x}) := F'(\tilde{x})$ provides another Newton derivative of F at x. This shows that Newton derivative of F at a point x may not be unique. Next we provide an example which demonstrates that Newton differentiability does not imply Fréchet differentiability and a mapping may have infinitely many Newton derivatives.

Example 10.2 Consider the function $F(x) = |x|$ on $\mathbb{R}$. Clearly F is not differentiable at $x = 0$. However, we can check that

$$G(x) = \begin{cases} 1 & \text{if } x > 0, \\ \delta & \text{if } x = 0 \\ -1 & \text{if } x < 0 \end{cases}$$

is a Newton derivative of F at $x = 0$, where $\delta \in \mathbb{R}$ is any number. In fact, by direct calculation, one can see that

$$F(0+h) - F(0) - G(0+h)h = |h| - G(h)h = 0$$

for any $h \in \mathbb{R}$.

Now let us return to the consideration of the equation (10.1). If F is Newton differentiable at the sought solution x^* with a Newton derivative $D_N F$ such that $D_N F(x)$ is invertible for all x in a neighborhood of x^*, then, instead of the Newton method (10.3), we may consider the method

$$x^{k+1} = x^k - D_N F(x^k)^{-1} F(x^k) \tag{10.5}$$

which is called a *semi-smooth Newton method*. The next result shows that the semi-smooth Newton method is indeed superlinear convergent.

Theorem 10.3 *Let X and Y be Banach spaces, and let $F : X \to Y$ be a mapping. Consider the equation (10.1) which is assumed to have a solution x^*. Assume that F is Newton differentiable at x^* with a Newton derivative $D_N F$. If there is a neighborhood $\mathscr{N}(x^*)$ of x^* and a constant M such that $D_N F(x)$ is nonsingular with $\|D_N F(x)^{-1}\| \le M$ for all $x \in \mathscr{N}(x^*)$, then the sequence $\{x^k\}$ defined by the semi-smooth Newton method (10.5) converges to x^* superlinearly provided that $\|x^0 - x^*\|$ is sufficiently small.*

Proof By using the Newton differentiability of F at x^*, we may take a small $\rho > 0$ such that $B_\rho(x^*) \subset \mathscr{N}(x^*)$ and

$$\|F(x^* + h) - F(x^*) - D_N F(x^* + h)h\| \le \frac{1}{2M}\|h\| \tag{10.6}$$

for all $\|h\| < \rho$. We first claim that if x_0 is chosen such that $\|x_0 - x^*\| < \rho$, then $x^k \in B_\rho(x^*)$ for all $k \ge 0$. This is clearly true for $k = 0$. Now we assume $x^k \in B_\rho(x^*)$ for some $k \ge 0$. Then, by using (10.5), $F(x^*) = 0$, and the fact that $\|D_N F(x)^{-1}\| \le M$ for all $x \in B_\rho(x^*)$, we have

$$\begin{aligned}\|x^{k+1} - x^*\| &= \|x^k - x^* - D_N F(x^k)^{-1}(F(x^k) - F(x^*))\| \\ &= \|D_N F(x^k)^{-1}(F(x^k) - F(x^*) - D_N F(x^k)(x^k - x^*))\| \\ &\le M\|F(x^k) - F(x^*) - D_N F(x^k)(x^k - x^*)\|.\end{aligned} \tag{10.7}$$

It then follows from (10.6) that

$$\|x^{k+1} - x^*\| \le \frac{1}{2}\|x^k - x^*\| < \rho$$

which demonstrates that $x^{k+1} \in B_\rho(x^*)$, and thus the claim is proved by induction. The above argument in fact shows also that

$$\|x^k - x^*\| \le \frac{1}{2}\|x^{k-1} - x^*\| \le \cdots \le \frac{1}{2^k}\|x^0 - x^*\|$$

for all $k \geq 0$. Therefore $\|x^k - x^*\| \to 0$ as $k \to \infty$. Finally, we may use (10.7) and the Newton differentiability of F at x^* to conclude

$$\|x^{k+1} - x^*\| = o(\|x^k - x^*\|) \quad \text{as } k \to \infty$$

which shows the superlinear convergence. □

Remark 10.4

(i) If $D_N F$ is a Newton derivative of F at x^* with the property

$$\|F(x^* + h) - F(x^*) - D_N F(x^* + h)h\| = O(\|h\|^2)$$

as $\|h\| \to 0$, then from the proof of Theorem 10.3, it is easy to see that

$$\|x^{k+1} - x^*\| = O(\|x^k - x^*\|^2)$$

and thus the semi-smooth Newton method (10.5) exhibits quadratic convergence.

(ii) If $D_N F(x)$ is a Newton derivative of F at x^* such that

$$F(x^* + h) - F(x^*) - D_N F(x^* + h)h = 0 \tag{10.8}$$

for sufficiently small h, then the semi-smooth Newton method (10.5) can produce the exact solution after a finite number of iterations. Indeed, since $x^k \to x^*$ as $k \to \infty$, it follows from (10.8) that there must exist k_0 such that

$$F(x^{k_0}) - F(x^*) - D_N F(x^{k_0})(x^{k_0} - x^*) = 0.$$

Since $F(x^*) = 0$, this implies that

$$x^{k_0} - D_N F(x^{k_0})^{-1} F(x^{k_0}) = x^*.$$

Consequently $x^{k_0+1} = x^*$.

We next provide some calculus rules for Newton derivatives. The following facts can be easily verified:

(i) Let X and Y be Banach spaces. Let $F : X \to Y$ and $G : X \to Y$ be two mappings with Newton derivatives $D_N F$ and $D_N G$, respectively. Then $\alpha D_N F + \beta D_N G$ is a Newton derivative of $\alpha F + \beta G$ for any $\alpha, \beta \in \mathbb{R}$.

(ii) Let $X, Y_i, i = 1, \cdots, n$, be Banach spaces, and for each i, let $F_i : X \to Y_i$ be a mapping with a Newton derivative $D_N F_i$. Then the mapping $F : X \to Y_1 \times \cdots \times Y_n$ defined by $F(x) = (F_1(x), \cdots, F_n(x))$ is also Newton differentiable with Newton derivative

$$D_N F(x) = (D_N F_1(x), \cdots, D_N F_n(x)).$$

For Newton derivatives, there also holds the following chain rule.

Proposition 10.5 *Consider the mappings $F : X \to Y$ and $G : Y \to Z$, where X, Y, and Z are Banach spaces. Assume that F is Newton differentiable at $x \in X$ with a Newton derivative $D_N F$ and G is Newton differentiable at $y := F(x)$ with a Newton derivative $D_N G$. If $D_N F$ and $D_N G$ are uniformly bounded in a neighborhood of x and y, respectively, then the composite $G \circ F$ is Newton differentiable at x with a Newton derivative given by*

$$D_N(G \circ F)(\tilde{x}) := D_N G(F(\tilde{x})) \circ D_N F(\tilde{x})$$

for all $\tilde{x}$ in a neighborhood of x.

Proof By the Newton differentiability of F at x, we have

$$\|F(x+h) - F(x) - D_N F(x+h)h\| = o(\|h\|)$$

as $\|h\| \to 0$. Let $\tau := F(x+h) - F(x)$. Then $F(x+h) = y + \tau$ and, by virtue of the uniformly boundedness of $D_N F$ in a neighborhood of x, we can obtain

$$\|\tau\| \le \|D_N F(x+h)h\| + o(\|h\|) = O(\|h\|).$$

Note that

$$\begin{aligned}
&(G \circ F)(x+h) - (G \circ F)(x) - D_N(G \circ F)(x+h)h \\
&= G(F(x+h)) - G(F(x)) - D_N G(F(x+h)) \circ D_N F(x+h)h \\
&= [G(y+\tau) - G(y) - D_N G(y+\tau)\tau] \\
&\quad + D_N G(y+\tau)\,[F(x+h) - F(x) - D_N F(x+h)h].
\end{aligned}$$

Therefore, by using the Newton differentiability of G at y and the uniformly boundedness of $D_N G$ in a neighborhood of y, we can derive that

$$\begin{aligned}
&\|(G \circ F)(x+h) - (G \circ F)(x) - D_N(G \circ F)(x+h)h\| \\
&\le \|G(y+\tau) - G(y) - D_N G(y+\tau)\tau\| \\
&\quad + \|D_N G(y+\tau)\,[F(x+h) - F(x) - D_N F(x+h)h\| \\
&\le o(\|\tau\|) + \|D_N G(y+\tau)\|\,\|F(x+h) - F(x) - D_N F(x+h)h\| \\
&= o(\|\tau\|) + o(\|h\|) = o(\|h\|)
\end{aligned}$$

as $\|h\| \to 0$. The proof is complete. □

Although Newton differentiability does not imply Fréchet differentiability, it places certain restrictions on the continuity of F as the following result indicates.

Lemma 10.6 *Let X and Y be Banach spaces, and let $F : X \to Y$ be a mapping. The following two assertions are equivalent:*

(i) *F is Newton differentiable at x with a Newton derivative $D_N F$ that is uniformly bounded in a neighborhood of x;*
(ii) *F is Lipschitz continuous at x in the sense that there exist a constant L and a neighborhood $\mathcal{N}(x)$ of x such that*

$$\|F(z) - F(x)\| \le L\|z - x\|, \quad \forall z \in \mathcal{N}(x).$$

Proof (i) $\Rightarrow$ (ii) is trivial. In the following, we focus on proving (ii) $\Rightarrow$ (i). According to the Hanh-Banach extension theorem, for any $z \in X$ with $z \neq x$, there is a bounded linear functional g_z on X such that $\|g_z\| = 1$ and $g_z(z - x) = \|z - x\|$. We define

$$G(z)h := \begin{cases} \dfrac{1}{\|z - x\|} g_z(h)(F(z) - F(x)), & z \neq x, \\ 0, & z = x \end{cases}$$

for all $h \in X$. It is easily seen that $G(z) \in \mathcal{L}(X, Y)$ for each $z \in \mathcal{N}(x)$ with $\|G(z)h\| \le L\|h\|$. Furthermore

$$F(z) - F(x) - G(z)(z - x) = 0$$

for all z. Therefore, G is a Newton derivative of F at x which is uniformly bounded in $\mathcal{N}(x)$. □

10.2 Clarke Generalized Jacobians and Semi-smoothness

In this section, we will provide a systematic way to construct Newton derivatives. Due to Lemma 10.6, we will consider locally Lipschitz continuous mappings. We will rely on the following Rademacher's theorem which may not hold in infinite dimensions. Therefore we will restrict ourselves to consider mappings between two finite dimensional spaces.

Theorem 10.7 (Rademacher's Theorem) *Let $X \subset \mathbb{R}^n$ be an open set, and let $F : X \to \mathbb{R}^m$ be locally Lipschitz continuous. Then F is differentiable almost everywhere in X.*

The proof of Rademacher's theorem can be found in many books on real analysis; see [79] for instance. Let $X \subset \mathbb{R}^n$ be an open set. For a mapping $F : X \to \mathbb{R}^m$, we use $\mathcal{D}_F$ to denote the set of points in X at which F is differentiable; we will also use

$F'(x)$ to denote the Jacobian of F at $x \in \mathscr{D}_F$. If F is locally Lipschitz continuous on X, according to Rademacher's theorem, the set $X \setminus \mathscr{D}_F$ form a set of Lebesgue measure zero. Consequently, any point $x \in X$ is a limiting point of $\mathscr{D}_F$.

Definition 10.8 Let $X \subset \mathbb{R}^n$ be an open set, and let $F : X \to \mathbb{R}^m$ be a locally Lipschitz continuous mapping. For each $x \in X$, the set

$$\partial_B F(x) := \{G \in \mathbb{R}^{m \times n} : \exists \{x^k\} \subset \mathscr{D}_F \text{ such that } x^k \to x \text{ and } F'(x^k) \to G\}$$

is called the *Bouligand-subdifferential* (or *B-subdifferential*) of F at x. The convex hull of $\partial_B F(x)$, i.e.,

$$\partial_C F(x) := \operatorname{conv}(\partial_B F(x))$$

is called the *Clarke's generalized Jacobian* of F at x.

Let $X \subset \mathbb{R}^n$ be an open set and $F : X \to \mathbb{R}^m$ be locally Lipschitz continuous. By definition, it is easy to see that if $x \in \mathscr{D}_F$ then $\{\nabla F(x)\} \subset \partial_B F(x) \subset \partial_C F(x)$ and if F is continuously differentiable in a neighborhood of x, then $\{\nabla F(x)\} = \partial_B F(x) = \partial_C F(x)$. The following result gives some useful properties of $\partial_B F$ and $\partial_C F$; see [57].

Proposition 10.9 *Let $X \subset \mathbb{R}^n$ be an open set, and let $F : X \to \mathbb{R}^m$ be locally Lipschitz continuous. Then the following statements hold:*

(i) *$\partial_B F$ and $\partial_C F$ are locally bounded on X.*
(ii) *$\partial_C F$ is closed at x in the sense that if $x_i \to x$ and $G_i \in \partial_C F(x_i)$ with $G_i \to G$, then $G \in \partial_C F(x)$.*
(iii) *$\partial_B F(x)$ and $\partial_C F(x)$ are non-empty compact subsets in $\mathbb{R}^{m \times n}$ for each $x \in X$.*

Proof

(i) Since F is locally Lipschitz on X, for any $\bar{x} \in X$, we can find an open set $U \subset X$ and a constant L such that

$$\|F(x) - F(y)\| \le L\|x - y\|, \quad \forall x, y \in U. \tag{10.9}$$

For any $x \in U$, if $x \in \mathscr{D}_F$, then for any $h \in \mathbb{R}^n$, we have

$$F'(x)h = \lim_{t \to 0} \frac{F(x + th) - F(x)}{t}.$$

This together with (10.9) then implies that $\|F'(x)h\| \le L\|h\|$ for all $h \in \mathbb{R}^n$. Consequently, $\|F'(x)\| \le L$ for all $x \in U \cap \mathscr{D}_F$.

To see the boundedness of $\partial_B F$ on U, we take any $x \in U$ and $G \in \partial_B F(x)$. By definition, we can find $\{x_i\} \subset U \cap \mathscr{D}_F$ such that $x_i \to x$ and $F'(x_i) \to G$ as $i \to \infty$. Thus

$$\|G\| = \lim_{i \to \infty} \|F'(x_i)\| \le L.$$

To see the boundedness of $\partial_C F$ on U, we note that every element in $\partial_C F(x)$ is a convex combination of elements from $\partial_B F(x)$, and consequently $\|G\| \le L$ for every $G \in \partial_C F(x)$ with $x \in U$.

(ii) We first show that $\partial_B F$ is closed at x in the sense that if $x_i \to x$ and $G_i \in \partial_B F(x_i)$ with $G_i \to G$, then $G \in \partial_B F(x)$. Indeed, for each i, by the definition of $G_i \in \partial_B F(x_i)$, we can find $\bar{x}_i \in \mathscr{D}_F$ such that

$$\|\bar{x}_i - x_i\| \le \frac{1}{i} \quad \text{and} \quad \|F'(\bar{x}_i) - G_i\| \le \frac{1}{i}.$$

This implies that $\bar{x}_i \to x$ and $F'(\bar{x}_i) \to G$. Therefore $G \in \partial_B F(x)$ by definition.

Next we show that $\partial_C F$ is closed at x in the sense that if $x_i \to x$ and $G_i \in \partial_C F(x_i)$ with $G_i \to G$, then $G \in \partial_C F(x)$. As $G_i \in \partial_C F(x_i) = \mathrm{conv}(\partial_B F(x_i))$, by Caratheodory's Theorem, we can write

$$G_i = \sum_{\ell=1}^{mn+1} t_i^{(\ell)} G_i^{(\ell)}, \tag{10.10}$$

where $t_i^{(\ell)} \ge 0$ with $\sum_{\ell=1}^{mn+1} t_i^{(\ell)} = 1$ and $G_i^{(\ell)} \in \partial_B F(x_i)$ for $\ell = 1, \cdots, mn+1$. Clearly, for each ℓ, $\{t_i^{(\ell)}\}$ is bounded. Moreover, from (i) it follows for each ℓ the set $\{G_i^{(\ell)}\}$ is also bounded. Thus, by taking subsequences if necessary, we may assume that $t_i^{(\ell)} \to t^{(\ell)}$ and $G_i^{(\ell)} \to G^{(\ell)}$ as $i \to \infty$ for all $\ell = 1, \cdots, mn+1$. By the closedness of $\partial_B F$, we have $G^{(\ell)} \in \partial_B F(x)$ for all $\ell = 1, \cdots, mn+1$. Note that $t^{(\ell)} \ge 0$ and $\sum_{\ell=1}^{mn+1} t^{(\ell)} = 1$. Moreover, from (10.10), we can obtain

$$G = \sum_{\ell=1}^{mn+1} t^{(\ell)} G^{(\ell)}.$$

Since $G^{(\ell)} \in \partial_B F(x)$ for all ℓ, we must have $G \in \partial_C F(x)$ by definition.

(iii) The boundedness of $\partial_B F(x)$ and $\partial_C F(x)$ follows from (i). The closedness of $\partial_B F(x)$ and $\partial_C F(x)$ follows from (ii). Now we show that $\partial_B F(x)$ and $\partial_C F(x)$ are non-empty. By Redemacher's theorem, we can find $\{x_i\} \subset \mathscr{D}_F$ with $x_i \to x$ as $i \to \infty$. By the locally Lipschitz continuity of F, we know that $\{F'(x_i)\}$ is bounded. Thus, by taking a subsequence of $\{x_i\}$ if necessary, we can guarantee that $F'(x_i) \to G$ for some $G \in \mathbb{R}^{m\times n}$. Thus, $G \in \partial_B F(x)$ by definition which implies $\partial_B F(x) \neq \emptyset$. Consequently, $\partial_C F(x) \neq \emptyset$.

□

In the following, we will provide certain criterion to guarantee that a mapping $F : X \subset \mathbb{R}^n \to \mathbb{R}^m$ is Newton differentiable and a Newton derivative can be constructed by using elements from $\partial_C F(x)$. The concept of semi-smoothness plays an important role. The semi-smoothness was first introduced in [154] for scalar valued functions. It was further generalized in [184] to cover mappings between finite dimensional spaces.

Definition 10.10 Let $X \subset \mathbb{R}^n$ be an open set, and let $F : X \to \mathbb{R}^m$ be locally Lipschitz continuous.

(i) F is called *directionally differentiable* at $x \in X$ if

$$F'(x; h) := \lim_{t\to 0^+} \frac{F(x+th) - F(x)}{t}$$

exists for all $h \in \mathbb{R}^n$.

(ii) F is called *semismooth* at $x \in X$ if F is directionally differentiable at x and, for every $V \in \partial_C F(x+h)$,

$$Vh - F'(x; h) = o(\|h\|) \quad \text{as } h \to 0. \tag{10.11}$$

It is clear that any C^1 mapping is semi-smooth. However, there exist semi-smooth mappings that are not C^1.

Example 10.11 Consider the function $F(x) := \|x\|_2$ on $\mathbb{R}^n$. This function is continuously differentiable at all $x \neq 0$ and is non-differentiable at $x = 0$. It is easy to see that

$$\partial_C F(x) = \partial_B F(x) = \left\{\frac{x}{\|x\|_2}\right\}, \quad x \neq 0$$

and

$$\partial_B F(0) = \{v \in \mathbb{R}^n : \|v\|_2 = 1\}, \qquad \partial_C F(0) = \{v \in \mathbb{R}^n : \|v\|_2 \leq 1\}.$$

Furthermore, F is directionally differentiable at $x = 0$ with $F'(0; h) = \|h\|_2$ for all $h \in \mathbb{R}^n$. Consequently, since any $V \in \partial_C F(h)$ has the form $V = h/\|h\|_2$ for $h \neq 0$, we have

$$Vh - F'(0; h) = \left\langle \frac{h}{\|h\|_2}, h \right\rangle - \|h\|_2 = 0$$

which shows that F is semi-smooth at $x = 0$.

We will show that, for locally Lipschitz continuous mappings, semi-smoothness implies Newton differentiability and a Newton derivative can be constructed by using elements from the Clarke's generalized Jacobians. To this end, we need the following result [204].

Lemma 10.12 *Let $F : \mathbb{R}^n \to \mathbb{R}^m$ be Lipschitz in a neighborhood of a point $x \in \mathbb{R}^n$. If F is directionally differentiable at x, then $F'(x; \cdot)$ is Lipschitz continuous and positively homogeneous and F is Bouligand differentiable at x in the sense that*

$$\lim_{\|h\|\to 0} \frac{F(x+h) - F(x) - F'(x; h)}{\|h\|} = 0.$$

Proof The positive homogeneity of $F(x; \cdot)$ follows easily from the definition. Since F is Lipschitz continuous around x, there is a neighborhood U of x and a constant L such that

$$\|F(y) \quad F(z)\| \leq L\|y - z\|, \quad \forall y, z \in U.$$

Consequently, for any $h, h' \in \mathbb{R}^n$, we have

$$\begin{aligned} \|F'(x; h) - F'(x; h')\| &= \left\| \lim_{t\to 0+} \frac{F(x+th) - F(x+th')}{t} \right\| \\ &\leq \lim_{t\to 0+} \left\| \frac{F(x+th) - F(x+th')}{t} \right\| \\ &\leq L\|h - h'\| \end{aligned}$$

which shows that $F'(x; \cdot)$ is Lipschitz continuous.

We next show the Bouligand differentiability of F at x by a contradiction argument. If F is not Bouligand differentiable at x, then there exist a number $\varepsilon > 0$ and a sequence $\{h_k\} \subset \mathbb{R}^n$ with $\|h_k\| \to 0$ such that

$$\frac{\|F(x+h_k) - F(x) - F'(x; h_k)\|}{\|h_k\|} \geq \varepsilon > 0$$

for all k. Let $t_k := \|h_k\|$ and $v_k := h_k/\|h_k\|$. Then $\|v_k\| = 1$. So, by passing to a subsequence if necessary, we can assume $v_k \to v$ as $k \to \infty$. Then, by the positively homogeneity of $h \to F'(x; h)$ and the directional differentiability of F at x, we have

$$\begin{aligned} \varepsilon &\leq \left\| \frac{F(x+t_k v_k) - F(x)}{t_k} - F'(x; v_k) \right\| \\ &\leq \left\| \frac{F(x+t_k v) - F(x)}{t_k} - F'(x; v) \right\| + \frac{\|F(x+t_k v_k) - F(x+t_k v)\|}{t_k} \end{aligned}$$

$$
\begin{aligned}
&+ \|F'(x; v_k) - F'(x; v)\| \\
&\leq \left\| \frac{F(x + t_k v) - F(x)}{t_k} - F'(x; v) \right\| + 2L\|v_k - v\| \\
&\to 0
\end{aligned}
$$

as $k \to \infty$, which is a contradiction. □

Now we are ready to prove the following important result.

Theorem 10.13 *Let $X \subset \mathbb{R}^n$ be an open set, and let $F : X \to \mathbb{R}^m$ be a locally Lipschitz continuous mapping. Assume that F is semi-smooth at $x \in X$. Define $\mathcal{G}_F : X \to \mathbb{R}^{m \times n}$ such that $\mathcal{G}_F(y) \in \partial_C F(y)$ for each $y \in X$. Then $\mathcal{G}_F$ is a Newton derivative of F at x.*

Proof Since F is semi-smooth at x and hence is directionally differentiable at x, it follows from Lemma 10.12 that

$$
F(x + h) - F(x) - F'(x; h) = o(\|h\|) \quad \text{as } \|h\| \to 0.
$$

By the definition of $\mathcal{G}_F$, we have $\mathcal{G}_F(x + h) \in \partial_C F(x + h)$. Therefore, we may use the definition of semi-smoothness to further obtain

$$
\begin{aligned}
\|F(x + h) - F(x) - \mathcal{G}_F(x + h)h\| &\leq \|F(x + h) - F(x) - F'(x; h)\| \\
&\quad + \|\mathcal{G}_F(x + h)h - F'(x; h)\| \\
&= o(\|h\|)
\end{aligned}
$$

as $\|h\| \to 0$ which completes the proof. □

Theorem 10.13 provides a procedure for constructing a Newton derivative. When a mapping F is locally Lipschitz continuous and semi-smooth, we can choose elements from the Clarke's generalized Jacobians to obtain a Newton derivative. Since the Jacobians of F at differentiable points can be calculated systematically, the B-subdifferential is built from the limits of sequences of Jacobians at differentiable points and the Clarke's generalized Jacobian is the convex hull of the B-subdifferential, we therefore have a constructive procedure for obtaining a Newton derivative.

Note that the definition of semi-smoothness involves all elements from the Clarke's generalized Jacobians which may contain many elements and thus can make it difficult to verify semi-smoothness in general. In the following, we will provide some equivalent statements which make the verification of semi-smoothness somewhat easier.

Theorem 10.14 *Let $X \subset \mathbb{R}^n$ be an open set, let $F : X \to \mathbb{R}^m$ be locally Lipschitz continuous, and let $x \in X$. The following statements are equivalent:*

(i) *F is semi-smooth at x.*
(ii) *F is directionally differentiable at x and*

$$\lim_{x+h \in \mathscr{D}_F,\, \|h\| \to 0} \frac{F'(x+h;\, h) - F'(x;\, h)}{\|h\|} = 0. \tag{10.12}$$

(iii) *F is directionally differentiable at x and for every $V \in \partial_B F(x+h)$,*

$$Vh - F'(x;\, h) = o(\|h\|) \quad \text{as } h \to 0. \tag{10.13}$$

Proof "(i) $\Longrightarrow$(ii)" Since $x + h \in \mathscr{D}_F$, we have $F'(x + h; h) = F'(x + h)h$ and $F'(x + h) \in \partial_C F(x + h)$. Thus (10.12) follows immediately from (10.11).

"(ii) $\Longrightarrow$(iii)" Since F is locally Lipschitz continuous around x, we can find $r > 0$ and $L > 0$ such that $B_r(x) \subset X$ and

$$\|F(y) - F(z)\| \le L\|y - z\|, \quad \forall y, z \in B_r(x). \tag{10.14}$$

Let $\varepsilon > 0$ be an arbitrarily small number. By (10.12), we can find $0 < \delta \le r$ such that

$$\|F'(x + h; h) - F'(x; h)\| \le \varepsilon\|h\| \tag{10.15}$$

for all h such that $\|h\| < \delta$ and $x+h \in \mathscr{D}_F$. Now for any $V \in \partial_B F(x+h)$ with $h \ne 0$ and $\|h\| < \delta/2$, by definition, we can find $\bar{h}$ with $\|\bar{h} - h\| < \min\{\delta/2, \varepsilon\|h\|/L\}$ and $x + \bar{h} \in \mathscr{D}_F$ such that $\|V - F'(x + \bar{h})\| < \varepsilon$. Note that $F'(x + \bar{h}; h) = F'(x + \bar{h})h$, we have

$$\|Vh - F'(x + \bar{h}; h)\| = \|(V - F'(x + \bar{h}))h\| \le \varepsilon\|h\|. \tag{10.16}$$

Consequently, by writing

$$\begin{aligned} Vh - F'(x; h) = {} & \left[Vh - F'(x + \bar{h}; h)\right] + \left[F'(x + \bar{h}; h) - F'(x + \bar{h}; \bar{h})\right] \\ & + \left[F'(x + \bar{h}; \bar{h}) - F'(x; \bar{h})\right] + \left[F'(x; \bar{h}) - F'(x; h)\right] \end{aligned}$$

and noting that $\|\bar{h}\| \le \|h\| + \|\bar{h} - h\| < \delta$, we can use (10.15), (10.16) and the Lipschitz continuity of $F'(x + \bar{h}; \cdot)$ and $F'(x; \cdot)$ shown in Lemma 10.12 to obtain

$$\begin{aligned} \|Vh - F'(x; h)\| \le {} & \|Vh - F'(x + \bar{h}; h)\| + \|F'(x + \bar{h}; h) - F'(x + \bar{h}; \bar{h})\| \\ & + \|F'(x + \bar{h}; \bar{h}) - F'(x; \bar{h})\| + \|F'(x; \bar{h}) - F'(x; h)\| \end{aligned}$$

$$\le \varepsilon\|h\| + \varepsilon\|\bar{h}\| + 2L\|\bar{h} - h\|$$
$$\le \left(4 + \frac{\varepsilon}{L}\right)\varepsilon\|h\|.$$

Since $\varepsilon > 0$ is arbitrarily small, we therefore obtain (10.13).

"(iii) $\Longrightarrow$(i)" According to Caratheodory's theorem, see Theorem 2.5, each $V \in \partial_C F(x+h) = \text{conv}(\partial_B F(x+h))$ can be written as

$$V = \sum_{\ell=1}^{mn+1} \lambda_\ell V_\ell,$$

where $\lambda_l \ge 0$ with $\sum_{\ell=1}^{mn+1} \lambda_\ell = 1$ and $V_\ell \in \partial_B F(x+h)$. Since

$$Vh - F'(x;h) = \sum_{\ell=1}^{mn+1} \lambda_\ell (V_\ell h - F'(x;h)),$$

we can obtain (10.11) from (10.13) immediately. □

10.3 A Globalization of Semi-smooth Newton Method

In Sect. 10.1, we have considered the semi-smooth Newton method and established its superlinear convergence when the initial guess is sufficiently close to the sought solution. In order to make the method more flexible in practical applications, a globalized version should be developed so that the resulting method is convergent for any initial guess and the local superlinear convergence property is preserved.

In this section, we will consider an equation of the form

$$F(x) = 0,$$

where the mapping $F : \mathbb{R}^n \to \mathbb{R}^n$ arises as the gradient $\nabla\varphi(x)$ of a convex differentiable function $\varphi : \mathbb{R}^n \to \mathbb{R}$ whose gradient is locally Lipschitz continuous. We will develop a globally convergent semi-smooth Newton method which is locally superlinear convergent. We start with the following result which demonstrates the symmetric positive semi-definite property of the Clarke generalized Jacobians of any mapping $F : \mathbb{R}^n \to \mathbb{R}^n$ arising from such a way.

Lemma 10.15 *Let $\varphi : \mathbb{R}^n \to \mathbb{R}$ be a convex differentiable function whose gradient $F(x) := \nabla\varphi(x)$ is locally Lipschitz continuous. Then every $V \in \partial_C F(x)$ is symmetric positive semi-definite for any $x \in \mathbb{R}^n$.*

Proof Let $\mathscr{D}_F$ denote the set of points at which F is differentiable. By the definitions of $\partial_C F$ and $\partial_B F$, it suffices to show that for any $x \in \mathscr{D}_F$, the Jacobian

$F'(x)$ is symmetric positive semi-definite. The symmetry of $F'(x)$ is a classical result in analysis; see [171, Section 3.3.4]. To see the positive semi-definiteness of $F'(x)$ for $x \in \mathscr{D}_F$, by the convexity of φ and Proposition 2.31, we have

$$\langle F(x + tv) - F(x), v \rangle \geq 0$$

for any $v \in \mathbb{R}^n$ and $t > 0$. Consequently

$$\langle F'(x)v, v \rangle = \lim_{t \to 0+} \left\langle \frac{F(x + tv) - F(x)}{t}, v \right\rangle \geq 0, \quad \forall v \in \mathbb{R}^n.$$

This completes the proof. □

For the semi-smooth Newton method considered in Sect. 10.1, when x^k is defined, the next iterate is defined by $x^{k+1} = x^k + d^k$, where d^k is the solution of the linear system $V_k d = -F(x^k)$ for some $V_k \in \partial_C F(x^k)$. Although Lemma 10.15 shows that V_k is symmetric positive semi-definite, it still can be singular and if so then d^k is not well-defined. To resolve this issue, one may pick a suitable number $\varepsilon_k > 0$ and consider instead the linear system

$$(V_k + \varepsilon_k I)d = -F(x^k). \tag{10.17}$$

Since $V_k + \varepsilon_k I$ is positive definite, this linear system is unique solvable. In case this linear system is difficult to be solved exactly, and one may resort to iterative algorithms, such as the conjugate gradient method, to solve it approximately. It is hoped that the linear system is solved up to certain accuracy and the obtained d^k is a descent direction for φ at x^k. In order to produce a globally convergent method, one needs to define $x^{k+1} = x^k + \alpha_k d^k$ with the step-size α_k chosen by suitable line search strategy, such as the Armijo rule. Putting all these consideration together leads to the following globalized semi-smooth Newton method [174, 236].

Algorithm 10.16 *Fix parameters $\mu \in (0, 1/2)$, $\delta \in (0, 1)$, $\tau \in (0, 1]$, $\bar{\eta} > 0$, and $\bar{\varepsilon} > 0$. Take an initial guess $x^0 \in \mathbb{R}^n$. For $k = 0, 1, \cdots$ do*

(i) *Calculate*

$$\eta_k := \min\{\bar{\eta}, \|F(x^k)\|^{1+\tau}\} \quad \text{and} \quad \varepsilon_k := \min\{\bar{\varepsilon}, \|F(x^k)\|\};$$

(ii) *Pick $V_k \in \partial_C F(x^k)$ and solve the linear system (10.17) exactly or approximately by the conjugate gradient method to find a search direction d^k such that*

$$\|(V_k + \varepsilon_k I)d^k + F(x^k)\| \leq \eta_k;$$

(iii) *Determine* m_k *by the Armijo rule, i.e.,* m_k *is chosen to be the first nonnegative integer such that*

$$\varphi(x^k + \delta^m d^k) \le \varphi(x^k) + \mu\delta^m \langle F(x^k), d^k\rangle$$

and set $\alpha_k := \delta^{m_k}$*;*

(iv) *Set* $x^{k+1} := x^k + \alpha_k d^k$.

In Algorithm 10.16 at each step where $F(x^k) \neq 0$, we need to solve a linear system (10.17). Let $\tilde{V}_k := V_k + \varepsilon_k I$. If (10.17) is solved exactly, then $d^k = -\tilde{V}_k^{-1} F(x^k)$ and consequently

$$\frac{1}{\lambda_{\max}(\tilde{V}_k)} \le -\frac{\langle F(x^k), d^k\rangle}{\|F(x^k)\|^2} \le \frac{1}{\lambda_{\min}(\tilde{V}_k)}, \tag{10.18}$$

where $\lambda_{\min}(\tilde{V}_k)$ and $\lambda_{\max}(\tilde{V}_k)$ denote the smallest and largest eigenvalues of $\tilde{V}_k$, respectively. In case solving (10.17) exactly is expensive, we may use the following conjugate gradient method to solve it approximately.

Algorithm 10.17 (Conjugate Gradient Method) *Let* A *be an* $n \times n$ *symmetric positive definite matrix and* $b \in \mathbb{R}^n$ *with* $b \neq 0$*. Solve the linear system* $Au = b$ *by the following procedure:*

(i) *Pick the initial guess* $u^0 = 0$*, set* $r^0 = b$*,* $p^0 = r^0$ *and* $j = 0$*;*

(ii) *While* $r^j \neq 0$ *do*

$$\alpha_j = \|r^j\|^2/\langle p^j, Ap^j\rangle;$$
$$u^{j+1} = u^j + \alpha_j p^j;$$
$$r^{j+1} = r^j - \alpha_j Ap^j;$$
$$\beta_{j+1} := \|r^{j+1}\|^2/\|r^j\|^2;$$
$$p^{j+1} := r^{j+1} + \beta_{j+1} p^j;$$
$$j = j + 1;$$

End

The conjugate gradient method is a prominent algorithm for solving linear system; see [169, 212], for instance. It is known that the method converges to the exact solution $u^* := A^{-1}b$ in at most n steps; moreover, there hold the following important properties:

$$\langle r^j, r^i\rangle = 0, \quad \langle p^j, Ap^i\rangle = 0 \quad \text{and} \quad \langle r^j, p^i\rangle = 0 \tag{10.19}$$

for all $j = 1, \cdots, n$ and $i = 0, \cdots, j-1$. By applying Algorithm 10.17 to solve (10.17) approximately, it was observed in [236] that the produced search direction d^k satisfies (10.18). Actually this is the direct consequence of the following result on Algorithm 10.17.

Lemma 10.18 *Consider Algorithm 10.17. For all* $j = 1, \cdots, n$ *there holds*

$$\frac{1}{\lambda_{\max}(A)} \le \frac{\langle u^j, b\rangle}{\|b\|^2} \le \frac{1}{\lambda_{\min}(A)}.$$

Proof For $j \ge 1$, we have from the definition of u^j that

$$\langle u^j, b\rangle = \langle u^{j-1}, b\rangle + \alpha_{j-1}\langle p^{j-1}, b\rangle. \tag{10.20}$$

Recursively using this identity and the initialization $u^0 = 0$, we can obtain

$$\langle u^j, b\rangle = \sum_{i=0}^{j-1} \alpha_i \langle p^i, b\rangle. \tag{10.21}$$

By the definition of p^j, (10.19), the initialization $p^0 = r^0 = b$ and the definition of β_i, we have $\langle p^0, b\rangle = \|r^0\|^2$ and, for $j \ge 1$,

$$\langle p^j, b\rangle = \langle r^j, b\rangle + \beta_j\langle p^{j-1}, b\rangle = \beta_j\langle p^{j-1}, b\rangle = \left(\prod_{i=1}^{j} \beta_i\right)\langle p^0, b\rangle = \|r^j\|^2. \tag{10.22}$$

Thus, from (10.20), $u^0 = 0$, $p^0 = r^0 = b$ and the definition of α_0, it follows for $j \ge 1$ that

$$\langle u^j, b\rangle \ge \langle u^{j-1}, b\rangle \ge \cdots \ge \langle u^1, b\rangle = \alpha_0\langle p^0, b\rangle = \frac{\|r^0\|^2}{\langle r^0, Ar^0\rangle}\|b\|^2 \ge \frac{\|b\|^2}{\lambda_{\max}(A)}.$$

On the other hand, by using (10.21) and (10.22), we have

$$\langle u^j, b\rangle = \sum_{i=0}^{j-1} \alpha_i \|r^i\|^2. \tag{10.23}$$

Letting $e^i := A^{-1}b - u^i$ and using the notation $\|z\|_A^2 := \langle z, Az\rangle$ for any $z \in \mathbb{R}^n$, we can see

$$\begin{aligned}\|e^i\|_A^2 &= \|e^{i+1} + (u^{i+1} - u^i)\|_A^2 = \|e^{i+1} + \alpha_i p^i\|_A^2 \\ &= \|e^{i+1}\|_A^2 + \alpha_i^2\|p^i\|_A^2 + 2\alpha_i\langle Ae^{i+1}, p^i\rangle.\end{aligned}$$

By induction, it is easy to see that $Ae^{i+1} = b - Au^{i+1} = r^{i+1}$. Thus we may use (10.19) to conclude $\langle Ae^{i+1}, p^i\rangle = 0$. Consequently,

$$\|e^i\|_A^2 - \|e^{i+1}\|_A^2 = \alpha_i^2\|p^i\|_A^2 = \alpha_i^2\langle p^i, Ap^i\rangle = \alpha_i\|r^i\|^2$$

by the definition of α_i. Combining this with (10.23) gives

$$\langle u^j, b\rangle = \|e^0\|_A^2 - \|e^j\|_A^2 \le \|e^0\|_A^2 = \|A^{-1}b\|_A^2 = \langle b, A^{-1}b\rangle \le \frac{\|b\|^2}{\lambda_{\min}(A)}.$$

The proof is therefore complete. □

According to (10.18), the direction d^k obtained at each iteration in Algorithm 10.16 is indeed a descent direction for φ at x^k. Therefore, the convergence can be established by a standard argument; see Theorem 10.21. In order to guarantee a local superlinear convergence, we need to impose certain condition on F at the sought solution x^*; in particular, we will assume F is semi-smooth at x^* and all elements in $\partial_B F(x^*)$ are nonsingular. The following two results will be used [112, 174].

Lemma 10.19 *Let $\varphi : \mathbb{R}^n \to \mathbb{R}$ be continuous differentiable with $F := \nabla\varphi$ being semi-smooth at a point $x \in \mathbb{R}^n$. Then for any $V \in \partial_C F(x+h)$, there holds*

$$\varphi(x+h) - \varphi(x) - \langle\nabla\varphi(x), h\rangle - \frac{1}{2}\langle Vh, h\rangle = o(\|h\|^2)$$

as $\|h\| \to 0$.

Proof By using Lemma 10.12 and the semi-smoothness of F at x, we may derive for any $V \in \partial_C F(x+h)$ that

$$\begin{aligned}
&\varphi(x+h) - \varphi(x) - \langle\nabla\varphi(x), h\rangle - \frac{1}{2}\langle Vh, h\rangle \\
&\quad = \int_0^1 \langle F(x+th) - F(x), h\rangle dt - \frac{1}{2}\langle Vh, h\rangle \\
&\quad = \int_0^1 \langle F(x+th) - F(x) - F'(x; th), h\rangle dt \\
&\qquad + \frac{1}{2}\langle F'(x; h) - Vh, h\rangle \\
&\quad = o(\|h\|^2)
\end{aligned}$$

which completes the proof. □

Lemma 10.20 *Let $\varphi : \mathbb{R}^n \to \mathbb{R}$ be a convex differentiable function with locally Lipschitz gradient $F := \nabla\varphi$, let $x \in \mathbb{R}^n$, and assume that all $V \in \partial_B F(x)$ are nonsingular. Then there exist a neighborhood $\mathscr{N}$ of x and a constant $c_0 > 0$ such that*

$$\langle Vh, h\rangle \geq c_0\|h\|^2, \quad \forall h \in \mathbb{R}^n \tag{10.24}$$

for all $y \in \mathscr{N}$ and $V \in \partial_C F(y)$.

Proof According to Lemma 10.15, every element in $\partial_B F(x)$ is symmetric positive semi-definite. Thus, under the given conditions, every element in $\partial_B F(x)$ must be positive define. Since $\partial_C F(x) = \text{conv}(\partial_B F(x))$, every element in $\partial_C F(x)$ is also symmetric positive definite.

Now we use a contradiction argument to show (10.24). If it is not true, then there exist $\{y^k\} \subset \mathbb{R}^n$ with $y^k \to x$, $V_k \in \partial_C F(y^k)$, and $\{h^k\} \subset \mathbb{R}^n$ with $\|h^k\| = 1$ such that

$$\langle V_k h^k, h^k\rangle \to 0 \quad \text{as } k \to \infty. \tag{10.25}$$

Since F is locally Lipschitz continuous, $\{V_k\}$ is bounded. By passing to a subsequence if necessary, we may assume $V_k \to V$ and $h^k \to h$ as $k \to \infty$. Clearly $\|h\| = 1$ and $V \in \partial_C F(x)$ by Proposition 10.9 (ii). Taking $k \to \infty$ in (10.25) gives $\langle Vh, h\rangle = 0$ which contradicts the positive definiteness of V. □

Now we are ready to formulate and prove the global convergence and local superlinear convergence result for Algorithm 10.16.

Theorem 10.21 *Consider the equation $F(x) = 0$, where $F := \nabla\varphi$ and $\varphi : \mathbb{R}^n \to \mathbb{R}$ is a convex differentiable function whose gradient is locally Lipschitz continuous. Let $\{x^k\}$ be the sequence produced by Algorithm 10.16 . If every sublevel set of φ is bounded, then $\{x^k\}$ is bounded and every cluster point of $\{x^k\}$ is a solution of $F(x) = 0$. If, in addition, $F(x) = 0$ has a unique solution x^*, then $x^k \to x^*$ as $k \to \infty$. Moreover, if F is semi-smooth at x^* and all $V \in \partial_B F(x^*)$ are nonsingular, then*

$$\|x^{k+1} - x^*\| = o(\|x^k - x^*\|)$$

as $k \to \infty$, i.e., x^k converges to x^* asymptotically superlinearly.

Proof According to Lemma 10.18, every d^k is a descent direction. Thus, with the step sizes chosen by the Armijo rule, we can guarantee $\varphi(x^{k+1}) \leq \varphi(x^k)$ for all k, and hence $\{x^k\}$ is contained in the sublevel set $\{x \in \mathbb{R}^n : \varphi(x) \leq \varphi(x^0)\}$ which is bounded. Thus $\{x^k\}$ is bounded, and hence $\{x^k\}$ contains a convergent subsequence $\{x^{k_l}\}$ with $x^{k_l} \to \hat{x}$ as $l \to \infty$. Moreover, by the monotonicity of $\{\varphi(x^k)\}$ and the Armijo rule, it is easy to derive that

$$\lim_{k\to\infty} \alpha_k\langle F(x^k), d^k\rangle = 0. \tag{10.26}$$

We claim that $F(\hat{x}) = 0$. If it is not true, then there exist positive numbers $\gamma_1 \geq \gamma_0 > 0$ such that

$$\gamma_0 \leq \|F(x^{k_l})\| \leq \gamma_1, \quad \forall l. \tag{10.27}$$

By the definition of ε_{k_l} and (10.27), we have $\varepsilon_{k_l} \geq \gamma_2 := \min\{\bar{\varepsilon}, \gamma_0\}$ and thus $\lambda_{\min}(\tilde{V}_{k_l}) \geq \varepsilon_{k_l} \geq \gamma_2$. On the other hand, by using $\varepsilon_{k_l} \leq \bar{\varepsilon}$ and the boundedness of V_{k_l}, we have $\lambda_{\max}(\tilde{V}_{k_l}) \leq \gamma_3$ for some constant $\gamma_3 > 0$. Therefore

$$0 < \gamma_2 \leq \lambda_{\min}(\tilde{V}_{k_l}) \leq \lambda_{\max}(\tilde{V}_{k_l}) \leq \gamma_3, \quad \forall l \tag{10.28}$$

Now we will derive contradictions by considering two cases.
Case 1. $m_{k_l} = 0$ for infinite many l. By taking a subsequence if necessary, we may assume $m_{k_l} = 0$ and hence $\alpha_{k_l} = 1$ for all l. It then follows from (10.26) that $\langle F(x^{k_l}), d^{k_l}\rangle \to 0$ as $l \to \infty$. By using (10.18), we have

$$\frac{1}{\lambda_{\max}(\tilde{V}_{k_l})} \leq -\frac{\langle F(x^{k_l}), d^{k_l}\rangle}{\|F(x^{k_l})\|^2} \leq -\frac{\langle F(x^{k_l}), d^{k_l}\rangle}{\gamma_0^2} \to 0$$

as $l \to \infty$, which is a contradiction to (10.28).
Case 2. there exists l_0 such that $m_{k_l} \geq 1$ for all $l \geq l_0$. By the definition of m_{k_l}, we have

$$\varphi(x^{k_l} + \delta^{m_{k_l}-1}d^{k_l}) > \varphi(x^{k_l}) + \mu\delta^{m_{k_l}-1}\langle F(x^{k_l}), d^{k_l}\rangle.$$

By the local Lipschitz continuity of F and the descent lemma, there is a number $L > 0$ such that

$$\varphi(x^{k_l} + \delta^{m_{k_l}-1}d^{k_l}) \leq \varphi(x^{k_l}) + \langle F(x^{k_l}), \delta^{m_{k_l}-1}d^{k_l}\rangle + \frac{L}{2}\|\delta^{m_{k_l}-1}d^{k_l}\|^2.$$

Combining the above two inequalities gives

$$\frac{L}{2}\delta^{2m_{k_l}-2}\|d^{k_l}\|^2 \geq (\mu - 1)\delta^{m_{k_l}-1}\langle F(x^{k_l}), d^{k_l}\rangle$$

which implies

$$\alpha_{k_l} = \delta^{m_{k_l}} \geq \frac{2(1-\mu)\delta}{L}\frac{\langle -F(x^{k_l}), d^{k_l}\rangle}{\|d^{k_l}\|^2}.$$

This together with (10.26) shows that $\langle F(x^{k_l}), d^{k_l}\rangle^2/\|d^{k_l}\|^2 \to 0$ as $l \to \infty$. By using (10.18), we then have

$$\frac{\|F(x^{k_l})\|}{\|d^{k_l}\|} \to 0 \quad \text{as } l \to \infty. \tag{10.29}$$

However, by the requirement on d^{k_l}, the definition of η_{k_l}, (10.27) and (10.28), we have

$$\bar{\eta} \geq \eta_{k_l} \geq \|\tilde{V}_{k_l} d^{k_l} + F(x^{k_l})\| \geq \|\tilde{V}_{k_l} d^{k_l}\| - \|F(x^{k_l})\| \geq \gamma_2 \|d^{k_l}\| - \gamma_1$$

and hence $\|d^{k_l}\| \leq (\bar{\eta} + \gamma_1)/\gamma_2$. Therefore, $\|F(x^{k_l})\|/\|d^{k_l}\| \geq \gamma_0\gamma_2/(\bar{\eta} + \gamma_1) > 0$ which is a contradiction to (10.29).

The above argument actually shows that any cluster point of $\{x^k\}$ is a solution of $F(x) = 0$. When $F(x) = 0$ is assumed to have a unique solution x^*, every convergent subsequence of $\{x^k\}$ must converge to x^* and therefore the whole sequence must converge to x^* as well.

Next we show the local superlinear convergence under the conditions that F is semi-smooth at x^* and every $V \in \partial_B F(x^*)$ is nonsingular. According to Lemma 10.20, we can conclude the existence of a positive number c_0 such that

$$\langle (V_k + \varepsilon_k I)d, d\rangle \geq c_0 \|d\|^2, \quad \forall d \in \mathbb{R}^n. \tag{10.30}$$

and hence a constant C such that

$$\|(V_k + \varepsilon_k I)^{-1}\| \leq C \tag{10.31}$$

for sufficiently large k. Therefore, by using $F(x^*) = 0$, we have

$$\begin{aligned}
\|x^k + d^k - x^*\| &= \|x^k - x^* + (V_k + \varepsilon_k I)^{-1}((V_k + \varepsilon_k I)d^k + F(x^k) - F(x^k))\| \\
&\leq \|x^k - x^* - (V_k + \varepsilon_k I)^{-1} F(x^k)\| + C\|(V_k + \varepsilon_k I)d^k + F(x^k)\| \\
&\leq \|(V_k + \varepsilon_k I)^{-1}(F(x^k) - F(x^*) - (V_k + \varepsilon_k I)(x^k - x^*))\| + C\eta_k \\
&\leq C\left(\|F(x^k) - F(x^*) - V_k(x^k - x^*)\| + \varepsilon_k\|x^k - x^*\| + \eta_k\right).
\end{aligned}$$

By the local Lipschitz continuity of F, we can find a number L such that

$$\|F(x^k)\| = \|F(x^k) - F(x^*)\| \leq L\|x^k - x^*\|.$$

Thus, by the definition of ε_k and η_k, we can conclude

$$\varepsilon_k = O(\|x^k - x^*\|) \quad \text{and} \quad \eta_k = O(\|x^k - x^*\|^{1+\tau}). \tag{10.32}$$

This together with Theorem 10.13 implies

$$\|x^k + d^k - x^*\| = o(\|x^k - x^*\|) \tag{10.33}$$

for sufficiently large k. Consequently,

$$\lim_{k\to\infty} \frac{\|d^k\|}{\|x^k - x^*\|} = 1. \tag{10.34}$$

Thus, by using (10.34), (10.30) and (10.32), we can obtain

$$\begin{aligned}\langle F(x^k), d^k\rangle &= \langle F(x^k) + (V_k + \varepsilon_k I)d^k, d^k\rangle - \langle (V_k + \varepsilon_k I)d^k, d^k\rangle \\ &\le \eta_k \|d_k\| - \langle (V_k + \varepsilon_k I)d^k, d^k\rangle \\ &\le -c_1\|d^k\|^2 \end{aligned} \tag{10.35}$$

for sufficiently large k, where c_1 is a positive constant.

To obtain the superlinear convergence, according to (10.33), it suffices to show $x^{k+1} = x^k + d^k$ for large k which amounts to showing $m_k = 0$ in the Armijo rule for choosing step size; we will follow [80, 174]. Since F is locally Lipschitz continuous, by the descent lemma, (10.33) and (10.34), we have

$$\begin{aligned}\varphi(x^k + d^k) &= \varphi(x^*) + \langle F(x^*), x^k + d^k - x^*\rangle + O(\|x^k + d^k - x^*\|^2) \\ &= \varphi(x^*) + o(\|d^k\|^2). \end{aligned} \tag{10.36}$$

By Lemma 10.19 and $F(x^*) = 0$, we also have

$$\varphi(x^k) = \varphi(x^*) + \frac{1}{2}\langle V(x^k - x^*), x^k - x^*\rangle + o(\|x^k - x^*\|^2)$$

for any $V \in \partial_C F(x^k)$. By the boundedness of V, (10.33), (10.34) and the identity

$$\begin{aligned}\langle V(x^k - x^*), x^k - x^*\rangle = \langle Vd^k, d^k\rangle &+ \langle V(x^k + d^k - x^*), x^k + d^k - x^*\rangle \\ &- 2\langle V(x^k + d^k - x^*), d^k\rangle \end{aligned}$$

we can see that

$$\langle V(x^k - x^*), x^k - x^*\rangle = \langle Vd^k, d^k\rangle + o(\|d^k\|^2).$$

Consequently

$$\varphi(x^k) = \varphi(x^*) + \frac{1}{2}\langle Vd^k, d^k\rangle + o(\|d^k\|^2). \tag{10.37}$$

Combining (10.36) and (10.37) gives

$$\varphi(x^k + d^k) - \varphi(x^k) - \frac{1}{2}\langle F(x^k), d^k\rangle = -\frac{1}{2}\langle F(x^k) + Vd^k, d^k\rangle + o(\|d^k\|^2).$$

Since $V \in \partial_C F(x^k)$, by using Theorem 10.13, $F(x^*) = 0$, (10.33), and (10.34), we have

$$F(x^k) + Vd^k = [F(x^k) - F(x^*) - V(x^k - x^*)] + V(x^k + d^k - x^*) = o(\|d^k\|).$$

Therefore

$$\varphi(x^k + d^k) - \varphi(x^k) - \frac{1}{2}\langle F(x^k), d^k\rangle = o(\|d^k\|^2) \tag{10.38}$$

Based on (10.35) and (10.38), we have for $\mu \in (0, 1/2)$ that

$$\varphi(x^k + d^k) - \varphi(x^k) - \mu\langle F(x^k), d^k\rangle = \left(\frac{1}{2} - \mu\right)\langle F(x^k), d^k\rangle + o(\|d^k\|^2) \le 0$$

for sufficiently large k. The proof is complete. □

10.4 Tame Functions and Semi-smoothness

In this section, we will provide abundant examples of semi-smooth functions by showing that locally Lipschitz continuous tame functions are semi-smooth [33, 110].

Definition 10.22

(i) A subset $A \subset \mathbb{R}^n$ is called tame if for every $r > 0$ the intersection $A \cap [-r, r]^n$ is definable in an o-minimal structure over $(\mathbb{R}, +, \cdot)$.
(ii) A mapping $F : U \subset \mathbb{R}^n \to \mathbb{R}^m$ is called tame if its graph is a tame set in $\mathbb{R}^n \times \mathbb{R}^m$.

We need the following result concerning definable functions of single variable.

Lemma 10.23 *Let $\phi, \psi : (0, \varepsilon) \to \mathbb{R}$ be definable functions continuous at $t = 0$ with $\phi(0) = \psi(0) = 0$. If $\psi'(t) > 0$ for $t > 0$ sufficiently small, then*

$$\lim_{t\to 0+} \frac{\phi(t)}{\psi(t)} = \ell \Longrightarrow \lim_{t\to 0+} \frac{\phi'(t)}{\psi'(t)} = \ell.$$

Proof According to Lemma 9.61, ϕ and ψ are continuous differentiable on $(0, \varepsilon)$ by shrinking ε if necessary. If $\lim_{t\to 0+} \phi(t)/\psi(t) = \ell$, then for any $\delta > 0$, there exists $0 < \varepsilon' \le \varepsilon$ such that $|\phi(t)/\psi(t) - \ell| < \delta$ for $t \in (0, \varepsilon')$ and hence

$$\eta_1(t) := \phi(t) - (\ell - \delta)\psi(t) > 0 \quad \text{and} \quad \eta_2(t) := \phi(t) - (\ell + \delta)\psi(t) < 0$$

on $(0, \varepsilon')$ with $\eta_1(0) = \eta_2(0) = 0$. By using Lemma 9.61 and by shrinking $\varepsilon' > 0$ if necessary, we can conclude that η_1 is strictly increasing and η_2 is strictly decreasing on $(0, \varepsilon')$. Therefore, $\eta_1'(t) \geq 0$ and $\eta_2'(t) \leq 0$ which imply $|\phi'(t)/\psi'(t) - \ell| \leq \delta$ for $t \in (0, \varepsilon')$. The proof is thus complete. □

Theorem 10.24 *Any locally Lipschitz continuous tame mapping is semi-smooth.*

Proof Let $F : U \to \mathbb{R}^m$ be a locally Lipschitz continuous tame function with an open domain $U \subset \mathbb{R}^n$; since semi-smoothness is of local nature, we may assume F is definable. For any fixed $x \in U$ and $v \in \mathbb{R}^n$, the mapping

$$t \to \eta(t) := \frac{F(x + tv) - F(x)}{t}$$

is definable on $(0, \varepsilon)$ for some $\varepsilon > 0$. Applying the monotonicity lemma, i.e., Lemma 9.61, to components of $\eta(t)$ shows that each component of η is monotone on $(0, \varepsilon)$ by shrinking ε if necessary. The local Lipschitz continuity of F implies the boundedness of $\eta(t)$ for small t. Thus, we may use the monotone convergence theorem to conclude the convergence of $\eta(t)$ as $t \to 0+$ and hence the existence of directional derivative $F'(x; v)$ of F at x in the direction v. This shows the directional differentiability of F on U.

To show F is semi-smooth at any given $x \in U$, according to Theorem 10.14, it remains only to show that

$$\|F'(x + h; h) - F'(x; h)\| = o(\|h\|) \quad \text{as } \|h\| \to 0$$

which amounts to showing

$$\max_{v \in \mathbb{S}^{n-1}} \|F'(x + tv; v) - F'(x; v)\| \to 0 \quad \text{as } t \to 0, \tag{10.39}$$

where $\mathbb{S}^{n-1}$ denotes the unit sphere in $\mathbb{R}^n$. We will use a contradiction argument to show (10.39). Assume it is not true, then there exist $\alpha > 0$ and a sequence $\{(t_k, v_k)\} \subset (0, \infty) \times \mathbb{S}^{n-1}$ with $t_k \to 0$ such that

$$\|F'(x + t_k v_k; v_k) - F'(x; v_k)\| \geq \alpha > 0 \quad \text{for all } k.$$

By taking a subsequence of $\{v_k\}$ if necessary, we may assume $v_k \to v$ as $k \to \infty$ for some $v \in \mathbb{S}^{n-1}$. Consider the set

$$A := \{(t, h) \in (0, \infty) \times \mathbb{S}^{n-1} : \|F'(x + th; h) - F'(x; h)\| \geq \alpha\}$$

which is definable with $(t_k, v_k) \in A$ for all k. Thus $(0, v) \in \overline{A}$. By the curve selection lemma, i.e., Lemma 9.63, we can find a C^1 definable curve

$$s \to (t(s), h(s)), \quad s \in (0, \delta), \tag{10.40}$$

contained in A with $(t(s), h(s)) \to (0, v)$ as $s \to 0+$. Since $t(s) > 0$ for $s \in (0, \delta)$ and $t(s) \to 0$ as $s \to 0+$, it follows from Lemma 9.61 that $s \to t(s)$ is strictly increasing on $(0, \delta)$; by using its inverse function, which is definable, we may reparametrize the curve (10.40) as $(t, p(t))$ with $t \in (0, \varepsilon)$ for some $\varepsilon > 0$ which is still definable. Let

$$\xi(t) := t(p(t) - v).$$

Then, $\xi(t)$ is definable and, as $p(t) \to v$, we have $\xi(t)/t \to 0$ as $t \to 0+$. Thus we may apply Lemma 10.23 to each component of $\xi(t)/t$ to conclude $\xi'(t) \to 0$ as $t \to 0+$.

Consider the definable mapping $t \to F(x + tp(t))$ which is differentiable for small $t > 0$ by the monotonicity lemma. We claim that

$$\frac{d}{dt} F(x + tp(t)) = F'(x + tp(t); v + \xi'(t)) \tag{10.41}$$

for small $t > 0$. To see this, for small $\tau > t > 0$, we write

$$\begin{aligned} &\frac{F(x + \tau p(\tau)) - F(x + tp(t))}{\tau - t} - F'(x + tp(t); v + \xi'(t)) \\ &= \frac{F(x + tp(t) + (\tau - t)(v + \xi'(t))) - F(x + tp(t))}{\tau - t} \\ &\quad - F'(x + tp(t); v + \xi'(t)) + I(\tau, t), \end{aligned} \tag{10.42}$$

where

$$I(\tau, t) := \frac{F(x + \tau p(\tau)) - F(x + tp(t) + (\tau - t)(v + \xi'(t)))}{\tau - t}.$$

By the local Lipschitz continuity of F, there is a number L such that

$$\begin{aligned} \|I(\tau, t)\| &\le \frac{L}{\tau - t} \left\| \tau p(\tau) - \big(tp(t) + (\tau - t)(v + \xi'(t))\big) \right\| \\ &= \frac{L}{\tau - t} \|\xi(\tau) - \xi(t) - (\tau - t)\xi'(t)\| \to 0 \end{aligned}$$

as $\tau \to t$. Therefore, by using the directional differebtiability of F, we may obtain (10.41) from (10.42) by taking $\tau \to t$.

We next show that

$$\lim_{t \to 0+} F'(x + tp(t); v + \xi'(t)) = F'(x; v). \tag{10.43}$$

To see this, by using the local Lipschitz continuity of F, we can derive that

$$\frac{1}{t}\|F(x+tp(t)) - F(x+tv)\| \leq \frac{L\|\xi(t)\|}{t} \to 0 \quad \text{as } t \to 0+$$

which implies

$$\lim_{t\to 0+} \frac{F(x+tp(t)) - F(x)}{t} = \lim_{t\to 0+} \frac{F(x+tv) - F(x)}{t} = F'(x; v).$$

Therefore, by applying Lemma 10.23, we can obtain

$$\lim_{t\to 0+} \frac{d}{dt} F(x+tp(t)) = F'(x; v)$$

which together with (10.41) shows (10.43).

Finally, by using the local Lipschitz continuity of F, (10.43), and the facts $p(t) \to v$ and $\xi'(t) \to 0$, we can find a number L such that

$$\begin{aligned}
&\|F'(x+tp(t); p(t)) - F'(x; p(t))\| \\
&\leq \|F'(x; p(t)) - F'(x; v)\| + \|F'(x+tp(t); v+\xi'(t)) - F'(x; v)\| \\
&\quad + \|F'(x+tp(t); p(t)) - F'(x+tp(t); v+\xi'(t))\| \\
&\leq L\|p(t) - v\| + L(\|p(t) - v\| + \|\xi'(t)\|) \\
&\quad + \|F'(x+tp(t); v+\xi'(t)) - F'(x; v)\| \to 0
\end{aligned}$$

as $t \to 0$. This is, however, a contradiction, because $(t, p(t)) \in A$ implies

$$\|F'(x+tp(t); p(t)) - F'(x; p(t))\| \geq \alpha > 0$$

for all small $t > 0$. □

Corollary 10.25 *Let $U \subset \mathbb{R}^n$ be an open semi-algebraic set, and let $F : U \to \mathbb{R}^m$ be a locally Lipchitz continuous semi-algebraic mapping. Then F is semi-smooth near every point in U; moreover, for any $x \in U$, there is a number $\gamma > 0$ such that*

$$\|F(x+h) - F(x) - Vh\| = O(\|h\|^{1+\gamma}) \quad \textit{as } \|h\| \to 0$$

for any $V \in \partial_C F(x+h)$.

Proof Let $x \in U$, and consider the functions

$$\varphi_x(t) := \sup_{\|v\|=1} \left\| \frac{1}{t}(F(x+tv) - F(x)) - F'(x; v) \right\|$$

and

$$\psi_x(t) := \sup_{\|v\|=1} \|F'(x+tv; v) - F'(x; v)\|$$

which are semi-algebraic on $(0, \varepsilon)$ for some $\varepsilon > 0$; see Example 8.41. According to the proof of Lemma 10.12 and Theorem 10.24, we have $\varphi_x(t) = o(1)$ and $\psi_x(t) = o(1)$ as $t \to 0+$. By using Lemma 9.74, we then obtain $\varphi_x(t) = O(t^\gamma)$ and $\psi_x(t) = O(t^\gamma)$ for some $\gamma > 0$. By using $\psi_x(t) = O(t^\gamma)$ and the proof of Theorem 10.14, we have $\|Vh - F'(x; h)\| = O(\|h\|^{1+\gamma})$, which together with $\varphi_x(t) = O(t^\gamma)$ implies the desired estimate. □

10.5 Applications

In this section, we will give applications of the semi-smooth Newton method on typical problems in optimization.

10.5.1 Linear Complementarity Problem

Given a matrix $A \in \mathbb{R}^{n\times n}$ and vectors $f, g \in \mathbb{R}^n$, we consider the linear complementarity problem which aims at finding vectors $y, \lambda \in \mathbb{R}^n$ such that

$$\begin{cases} Ay + \lambda = f, \\ y \le g, \quad \lambda \ge 0, \quad \langle \lambda, y - g \rangle = 0. \end{cases} \tag{10.44}$$

Many problems can be reduced to solving (10.44); these include the linear and quadratic programming problems, the bimatrix game problem, a number of free boundary problems in fluid mechanics, and so on. One may refer to [63] for the details of many concrete examples.

The existence and uniqueness of solutions of the linear complementarity problem (10.44) for any given vectors $f, g \in \mathbb{R}^n$ depend heavily on the property of the matrix A which is defined below.

Definition 10.26 A square matrix is called a P-matrix if all its principal minors are positive.

The concept of P-matrix was first introduced in [81] in 1962. Since then, the importance of P-matrices in optimization, differential equations, statistics, and various areas has been well documented. There are numerous equivalent formulations of P-matrices [26]. We remark that any positive definite matrix is a P-matrix; a P-matrix in general, however, is not necessarily to be positive definite unless it is symmetric.

For the linear complementarity problem (10.44), there holds the following existence and uniqueness result whose proof can be found in the books [26, 63].

Theorem 10.27 *The linear complementarity problem (10.44) has a unique solution for every $f, g \in \mathbb{R}^n$ if and only if $A \in \mathbb{R}^{n\times n}$ is a P-matrix.*

We now consider solving (10.44) by the semi-smooth Newton method. Note that the second line in (10.44) is equivalent to

$$\min\{\lambda, -c(y-g)\} = 0$$

or equivalently

$$\lambda - \max\{0, \lambda + c(y-g)\} = 0$$

for any fixed positive constant c, where min and max operations are understood in componentwise sense. By introducing the map $M : \mathbb{R}^n \to \mathbb{R}^n$ defined by

$$M(u) := \max\{0, u\}, \quad \forall u \in \mathbb{R}^n, \tag{10.45}$$

we can see that finding a solution of (10.44) becomes the root-finding problem

$$F(y, \lambda) = 0,$$

where $F : \mathbb{R}^n \times \mathbb{R}^n \to \mathbb{R}^n \times \mathbb{R}^n$ is given by

$$F(y, \lambda) = \begin{pmatrix} Ay + \lambda - f \\ \lambda - M(\lambda + c(y-g)) \end{pmatrix}. \tag{10.46}$$

Note that F is not smooth. We need to show that F is Newton differentiable and find a Newton derivative for F. We first consider the map M defined by (10.45)

Proposition 10.28 *The map M defined by (10.45) is Newton differentiable and has a Newton derivative given by*

$$\mathcal{G}_M(u) = \operatorname{diag}(G_1(u_1), \ldots, G_n(u_n)), \tag{10.47}$$

where $G_i : \mathbb{R} \to \mathbb{R}$ is given by

$$G_i(x) = \begin{cases} 0 & \text{if } x \le 0, \\ 1 & \text{if } x > 0. \end{cases} \tag{10.48}$$

Proof For any $u \in \mathbb{R}^n$, let

$$\beta := \begin{cases} +\infty & \text{if } u = 0, \\ \min\{|u_i| : u_i \neq 0\} & \text{if } u \neq 0. \end{cases}$$

By straightforward calculation, it is easy to see that

$$M(u+h) - M(u) - \mathcal{G}_M(u+h)h = 0$$

for any $h \in \mathbb{R}^n$ satisfying $\|h\| < \beta$. This in particular shows that M is Newton differentiable and $\mathcal{G}_M$ is a Newton derivative. □

According to Proposition 10.28 and the chain rule given in Proposition 10.5, we can see that the map F defined by (10.46) is Newton differentiable with a Newton derivative given by

$$\mathcal{G}_F(y,\lambda) := \begin{pmatrix} A & I \\ -c\mathcal{G}_M(\lambda + c(y-g)) & I - \mathcal{G}_M(\lambda + c(y-g)) \end{pmatrix},$$

where I denotes the identity matrix on $\mathbb{R}^n$. Thus the semi-smooth Newton method for solving $F(y,\lambda) = 0$ takes the form

$$\begin{pmatrix} y^{k+1} \\ \lambda^{k+1} \end{pmatrix} = \begin{pmatrix} y^k \\ \lambda^k \end{pmatrix} + \begin{pmatrix} \delta y^k \\ \delta \lambda^k \end{pmatrix}, \tag{10.49}$$

where δy^k and $\delta\lambda^k$ are determined by

$$\mathcal{G}_F(y^k, \lambda^k) \begin{pmatrix} \delta y^k \\ \delta \lambda^k \end{pmatrix} = -F(y^k, \lambda^k),$$

i.e.,

$$\begin{aligned} &\begin{pmatrix} A & I \\ -c\mathcal{G}_M(\lambda^k + c(y^k-g)) & I - \mathcal{G}_M(\lambda^k + c(y^k-g)) \end{pmatrix} \begin{pmatrix} \delta y^k \\ \delta \lambda^k \end{pmatrix} \\ &= -\begin{pmatrix} Ay^k + \lambda^k - f \\ \lambda^k - M(\lambda^k + c(y^k - g)) \end{pmatrix}. \end{aligned} \tag{10.50}$$

In order to formulate the semi-smooth Newton method more explicitly, we introduce

$$\mathcal{A}_k := \{i : \lambda_i^k + c(y_i^k - g_i) > 0\}, \quad \mathcal{I}_k := \{i : \lambda_i^k + c(y_i^k - g_i) \le 0\}$$

which are the active and inactive sets associated with (y^k, λ^k), respectively. We may arrange the coordinates in such a way that the active and inactive ones occur in consecutive order; namely, we write

$$y = \begin{pmatrix} y_{\mathcal{I}_k} \\ y_{\mathcal{A}_k} \end{pmatrix}, \quad \lambda = \begin{pmatrix} \lambda_{\mathcal{I}_k} \\ \lambda_{\mathcal{A}_k} \end{pmatrix}.$$

Accordingly, the matrix A has the block matrix representation given by

$$A = \begin{pmatrix} A_{\mathcal{I}_k\mathcal{I}_k} & A_{\mathcal{I}_k\mathcal{A}_k} \\ A_{\mathcal{A}_k\mathcal{I}_k} & A_{\mathcal{A}_k\mathcal{A}_k} \end{pmatrix}.$$

With the above notation, the equation (10.50) can be written as

$$\begin{pmatrix} A_{\mathcal{I}_k\mathcal{I}_k} & A_{\mathcal{I}_k\mathcal{A}_k} & I_{\mathcal{I}_k} & 0 \\ A_{\mathcal{A}_k\mathcal{I}_k} & A_{\mathcal{A}_k\mathcal{A}_k} & 0 & I_{\mathcal{A}_k} \\ 0 & 0 & I_{\mathcal{I}_k} & 0 \\ 0 & -cI_{\mathcal{A}_k} & 0 & 0 \end{pmatrix} \begin{pmatrix} \delta y^k_{\mathcal{I}_k} \\ \delta y^k_{\mathcal{A}_k} \\ \delta\lambda^k_{\mathcal{I}_k} \\ \delta\lambda^k_{\mathcal{A}_k} \end{pmatrix} = - \begin{pmatrix} (Ay^k + \lambda^k - f)_{\mathcal{I}_k} \\ (Ay^k + \lambda^k - f)_{\mathcal{A}_k} \\ \lambda^k_{\mathcal{I}_k} \\ -c(y^k - g)_{\mathcal{A}_k} \end{pmatrix}. \tag{10.51}$$

From (10.51), it follows that

$$A_{\mathcal{I}_k\mathcal{I}_k}\delta y^k_{\mathcal{I}_k} + A_{\mathcal{I}_k\mathcal{A}_k}\delta y^k_{\mathcal{A}_k} + \delta\lambda^k_{\mathcal{I}_k} = -(Ay^k + \lambda^k - f)_{\mathcal{I}_k}, \tag{10.52}$$

$$A_{\mathcal{A}_k\mathcal{I}_k}\delta y^k_{\mathcal{I}_k} + A_{\mathcal{A}_k\mathcal{A}_k}\delta y^k_{\mathcal{A}_k} + \delta\lambda^k_{\mathcal{A}_k} = -(Ay^k + \lambda^k - f)_{\mathcal{A}_k}, \tag{10.53}$$

$$\delta\lambda^k_{\mathcal{I}_k} = -\lambda^k_{\mathcal{I}_k}, \tag{10.54}$$

$$-c\delta y^k_{\mathcal{A}_k} = c(y^k - g)_{\mathcal{A}_k}. \tag{10.55}$$

By using (10.54) and (10.55), it is easy to see that

$$\lambda^{k+1}_{\mathcal{I}_k} = \lambda^k_{\mathcal{I}_k} + \delta\lambda^k_{\mathcal{I}_k} = 0, \tag{10.56}$$

$$y^{k+1}_{\mathcal{A}_k} = y^k_{\mathcal{A}_k} + \delta y^k_{\mathcal{A}_k} = y^k_{\mathcal{A}_k} - (y^k - g)_{\mathcal{A}_k} = g_{\mathcal{A}_k}. \tag{10.57}$$

We next determine $y^{k+1}_{\mathcal{I}_k}$ and $\lambda^{k+1}_{\mathcal{A}_k}$. Observing that the combination of (10.52) and (10.53) is essentially the equation

$$A\delta y^k + \delta\lambda^k = -(Ay^k + \lambda^k - f)$$

which is equivalent to

$$Ay^{k+1} + \lambda^{k+1} = f. \tag{10.58}$$

Consider (10.58) on the inactive set $\mathcal{I}_k$ gives

$$A_{\mathcal{I}_k\mathcal{I}_k}y^{k+1}_{\mathcal{I}_k} + A_{\mathcal{I}_k\mathcal{A}_k}y^{k+1}_{\mathcal{A}_k} + \lambda^{k+1}_{\mathcal{I}_k} = f_{\mathcal{I}_k}$$

which together with (10.56) and (10.57) implies that

$$A_{\mathcal{I}_k\mathcal{I}_k}y^{k+1}_{\mathcal{I}_k} = f_{\mathcal{I}_k} - A_{\mathcal{I}_k\mathcal{A}_k}g_{\mathcal{A}_k}. \tag{10.59}$$

Since A is a P-matrix, as a principal submatrix $A_{\mathcal{I}_k\mathcal{I}_k}$ is invertible and thus we may solve (10.59) to obtain $y^{k+1}_{\mathcal{I}_k}$. Once y^{k+1} is determined, we may consider (10.58) on the active set $\mathcal{A}_k$ to obtain

$$\lambda^{k+1}_{\mathcal{A}_k} = f_{\mathcal{A}_k} - (Ay^{k+1})_{\mathcal{A}_k}. \tag{10.60}$$

We may repeat this procedure until a convergence criterion is satisfied. This leads to the following algorithm.

Algorithm 10.29 *Choose initial guess y^0, λ^0, and take a positive number $c > 0$. For $k = 0, 1, \cdots$, set*

$$\mathcal{A}_k := \{i : \lambda^k_i + c(y^k_i - g_i) > 0\},$$
$$\mathcal{I}_k := \{i : \lambda^k_i + c(y^k_i - g_i) \le 0\},$$

and define y^{k+1} and λ^{k+1} by

$$\lambda^{k+1}_{\mathcal{I}_k} = 0, \quad y^{k+1}_{\mathcal{A}_k} = g_{\mathcal{A}_k},$$
$$A_{\mathcal{I}_k\mathcal{I}_k} y^{k+1}_{\mathcal{I}_k} = f_{\mathcal{I}_k} - A_{\mathcal{I}_k\mathcal{A}_k} g_{\mathcal{A}_k},$$
$$\lambda^{k+1}_{\mathcal{A}_k} = f_{\mathcal{A}_k} - (Ay^{k+1})_{\mathcal{A}_k}$$

until a stopping criterion is satisfied.

It is worth pointing out that Algorithm 10.29 coincides with the primal-dual active set method developed in [25]. This fact was first observed in [105] which demonstrates an amazing connection between the semi-smooth Newton method and the primal-dual active set method.

In order to apply Theorem 10.3 to obtain a convergence result on Algorithm 10.29, we need to show that $\mathcal{G}_F(y, \lambda)^{-1}$ is uniformly bounded at least in a neighborhood of the sought solution.

Given $(y, \lambda) \in \mathbb{R}^n \times \mathbb{R}^n$, we introduce the associated active and inactive sets

$$\mathcal{A} := \{i : \lambda_i + c(y_i - g_i) > 0\}, \quad \mathcal{I} := \{i : \lambda_i + c(y_i - g_i) \le 0\}.$$

As before, we arrange the coordinates such that the active and inactive ones occur in consecutive order. Then $\mathcal{G}_F(y, \lambda)$ has the block matrix representation

$$\mathcal{G}_F(y, \lambda) = \begin{pmatrix} A_{\mathcal{I}\mathcal{I}} & A_{\mathcal{I}\mathcal{A}} & I_{\mathcal{I}} & 0 \\ A_{\mathcal{A}\mathcal{I}} & A_{\mathcal{A}\mathcal{A}} & 0 & I_{\mathcal{A}} \\ 0 & 0 & I_{\mathcal{I}} & 0 \\ 0 & -cI_{\mathcal{A}} & 0 & 0 \end{pmatrix}.$$

By using the same procedure for deriving (10.56), (10.57), (10.59) and (10.60), we can see that $\mathcal{G}_F(y,\lambda)^{-1}$ has the representation

$$\mathcal{G}_F(y,\lambda)^{-1} = \begin{pmatrix} A_{\mathcal{I}\mathcal{I}}^{-1} & 0 & -A_{\mathcal{I}\mathcal{I}} & \frac{1}{c}A_{\mathcal{I}\mathcal{I}}^{-1}A_{\mathcal{I}\mathcal{A}} \\ 0 & 0 & 0 & -\frac{1}{c}I_{\mathcal{A}} \\ 0 & 0 & I_{\mathcal{I}} & 0 \\ -A_{\mathcal{A}\mathcal{I}}A_{\mathcal{I}\mathcal{I}}^{-1} & I_{\mathcal{A}} & A_{\mathcal{A}\mathcal{I}}A_{\mathcal{I}\mathcal{I}}^{-1} & \frac{1}{c}(A_{\mathcal{A}\mathcal{A}} - A_{\mathcal{A}\mathcal{I}}A_{\mathcal{I}\mathcal{I}}^{-1}A_{\mathcal{I}\mathcal{A}}) \end{pmatrix}$$

from which it is easy to see that $\mathcal{G}_F(y,\lambda)^{-1}$ is uniformly bounded on $\mathbb{R}^n \times \mathbb{R}^n$.

Theorem 10.30 *Let A be a P-matrix, and let (y^*,λ^*) be the unique solution of (10.44). Then the sequence (y^k,λ^k) produced by Algorithm 10.29 converges to (y^*,λ^*) in finite many steps provided $\|y^0 - y^*\| + \|\lambda^0 - \lambda^*\|$ is sufficiently small.*

Proof From the proof of Proposition 10.28, it is known that for the map M defined by (10.45), there holds

$$M(u+h) - M(u) - \mathcal{G}_M(u+h)h = 0$$

for sufficiently small $\|h\|$. With the help of this fact, we can easily obtain

$$F(y^*+h,\lambda^*+\tau) - F(y^*,\lambda^*) - \mathcal{G}_F(y^*+h,\lambda^*+\tau)\begin{pmatrix} h \\ \tau \end{pmatrix} = 0$$

if $\|h\| + \|\tau\|$ is sufficiently small. Therefore, we may use Theorem 10.3 and Remark 10.4 (ii) to conclude the proof. □

Next we will show that Algorithm 10.29 has the global convergence property if A is an M-matrix [105, 111].

Definition 10.31 A square matrix A is called an M-matrix if it is nonsingular, $a_{ij} \le 0$ for $i \neq j$, and $A^{-1} \ge 0$. Here, for a matrix B, we use $B \ge 0$ to mean that every component of B is nonnegative.

There exist many equivalent characterizations of M-matrices; see [26, Theorem 6.2.3]. In particular, it has been shown that a matrix $A \in \mathbb{R}^{n\times n}$ is an M-matrix if and only if $a_{ij} \le 0$ for $i \neq j$ and every principal minor of A is positive. According to this fact, it is easy to see that every M-matrix is a P-matrix and any principal submatrix of an M-matrix $A \in \mathbb{R}^{n\times n}$ is also an M-matrix. Consequently, for any subset $\mathcal{I}$ of $\{1,2,\cdots,n\}$, let $\mathcal{A} = \{1,2,\cdots,n\}\setminus\mathcal{I}$, then $A_{\mathcal{I}\mathcal{I}}^{-1} \ge 0$, $A_{\mathcal{I}\mathcal{A}} \le 0$ and thus $A_{\mathcal{I}\mathcal{I}}^{-1}A_{\mathcal{I}\mathcal{A}} \le 0$.

Theorem 10.32 *Let A be an M-matrix, and let (y^*,λ^*) be the unique solution of (10.44). For any initial guess (y^0,λ^0), the sequence $\{(y^k,\lambda^k)\}$ produced by Algorithm 10.29 converges to (y^*,λ^*) in finite many steps. Moreover, $y^* \le y^{k+1} \le y^k$ for all $k \ge 1$ and $y^k \le g$ for all $k \ge 2$.*

Proof
(i) We first show that $y^{k+1} \le y^k$ for $k \ge 1$. According to (10.56) and (10.57), we have

$$\lambda_i^k (y_i^k - g_i) = 0, \quad \forall k \ge 1 \text{ and } i = 1, \cdots, n \tag{10.61}$$

which implies either $\lambda_i^k = 0$ or $y_i^k = g_i$ for each i. Note that for $i \in \mathcal{A}_k$, there holds $\lambda_i^k + c(y_i^k - g_i) > 0$. Therefore, we must have $y_i^k \ge g_i$ for $i \in \mathcal{A}_k$. Since (10.57) gives $y_i^{k+1} = g_i$, we thus $y_i^k \ge y_i^{k+1}$ for $i \in \mathcal{A}_k$. Hence

$$\delta y_{\mathcal{A}_k}^k = y_{\mathcal{A}_k}^{k+1} - y_{\mathcal{A}_k}^k \le 0.$$

It remains to show $\delta y_{\mathcal{I}_k}^k \le 0$. Recall that $Ay^k + \lambda^k = f$ and $Ay^{k+1} + \lambda^{k+1} = f$. Therefore, $A\delta y^k + \delta\lambda^k = 0$ and thus

$$-\delta\lambda_{\mathcal{I}_k}^k = (A\delta y^k)_{\mathcal{I}_k} = A_{\mathcal{I}_k\mathcal{I}_k}\delta y_{\mathcal{I}_k}^k + A_{\mathcal{I}_k\mathcal{A}_k}\delta y_{\mathcal{A}_k}^k.$$

which implies that

$$\delta y_{\mathcal{I}_k}^k = -A_{\mathcal{I}_k\mathcal{I}_k}^{-1}\delta\lambda_{\mathcal{I}_k}^k - A_{\mathcal{I}_k\mathcal{I}_k}^{-1}A_{\mathcal{I}_k\mathcal{A}_k}\delta y_{\mathcal{A}_k}^k. \tag{10.62}$$

By using $\lambda_{\mathcal{I}_k}^k + c(y_{\mathcal{I}_k}^k - g_{\mathcal{I}_k}) \le 0$ and (10.61), it is easy to see that $\lambda_{\mathcal{I}_k}^k \le 0$. Hence, it follows from (10.54) that $\delta\lambda_{\mathcal{I}_k}^k \ge 0$. This together with (10.62), $\delta y_{\mathcal{A}_k}^k \le 0$ and the properties of M-matrices, shows that $\delta y_{\mathcal{I}_k}^k \le 0$.
(ii) We next show that $y^k \le g$ for all $k \ge 2$. According to (i), it suffices to show that $y^2 \le g$. Let i be any index. If $y_i^1 \le g_i$, then by (i) we have $y_i^2 \le y_i^1 \le g_i$. If $y_i^1 > g_i$, then by (10.61), we have $\lambda_i^1 = 0$. Consequently $\lambda_i^1 + c(y_i^1 - g_i) > 0$. This means $i \in \mathcal{A}_1$ and thus $y_i^2 = g_i$ by (10.57).
(iii) We then show that $y^k \ge y^*$ for all $k \ge 1$. Recall that $\lambda^* \ge 0$ and $y^* \le g$. We have $y_{\mathcal{A}_{k-1}}^* \le g_{\mathcal{A}_{k-1}} = y_{\mathcal{A}_{k-1}}^k$ by (10.57). It remains only to show that $y_{\mathcal{I}_{k-1}}^k \ge y_{\mathcal{I}_{k-1}}^*$. Since $Ay^* + \lambda^* = f$ and $Ay^k + \lambda^k = f$, we have $A(y^k - y^*) + \lambda^k - \lambda^* = 0$. Thus

$$\begin{aligned} -(\lambda^k - \lambda^*)_{\mathcal{I}_{k-1}} = (A(y^k - y^*))_{\mathcal{I}_{k-1}} &= A_{\mathcal{I}_{k-1}\mathcal{I}_{k-1}}(y_{\mathcal{I}_{k-1}}^k - y_{\mathcal{I}_{k-1}}^*) \\ &\quad + A_{\mathcal{I}_{k-1}\mathcal{A}_{k-1}}(y_{\mathcal{A}_{k-1}}^k - y_{\mathcal{A}_{k-1}}^*). \end{aligned}$$

According to (10.56) and (10.57), we have $\lambda_{\mathcal{I}_{k-1}}^k = 0$ and $y_{\mathcal{A}_{k-1}}^k = g_{\mathcal{A}_{k-1}}$. Therefore, we can obtain from the above equation that

$$y_{\mathcal{I}_{k-1}}^k - y_{\mathcal{I}_{k-1}}^* = A_{\mathcal{I}_{k-1}\mathcal{I}_{k-1}}^{-1}\lambda_{\mathcal{I}_{k-1}}^* + A_{\mathcal{I}_{k-1}\mathcal{I}_{k-1}}^{-1}A_{\mathcal{I}_{k-1}\mathcal{A}_{k-1}}(y_{\mathcal{A}_{k-1}}^* - g_{\mathcal{A}_{k-1}}).$$

Since $\lambda^*_{\mathcal{I}_{k-1}} \geq 0$ and $y^*_{\mathcal{A}_{k-1}} \leq g_{\mathcal{A}_{k-1}}$, we may use the properties of M-matrices to conclude that $y^k_{\mathcal{I}_{k-1}} \geq y^*_{\mathcal{I}_{k-1}}$.

(iv) Now we claim that there exists an integer k_0 such that $\lambda^k \geq 0$ for all $k \geq k_0$. If this is not true, then there exists a sequence $\{k_j\}$ of positive integers with $k_j \to \infty$ such that each λ^{k_j} has a negative component. By the finite dimensionality and by taking a subsequence of $\{k_j\}$ if necessary, we may find an index i such that $\lambda_i^{k_j} < 0$ for all j. We now fix an integer j and derive a contradiction by showing that $\lambda_i^k = 0$ for all $k > k_j$. Indeed, since $\lambda_i^{k_j} < 0$, by (10.61), we have $y_i^{k_j} = g_i$. Consequently $\lambda_i^{k_j} + c(y_i^{k_j} - g_i) < 0$ which implies that $i \in \mathcal{I}_{k_j}$ and thus $\lambda_i^{k_j+1} = 0$ by (10.56). By virtue of (ii), we then have $\lambda_i^{k_j+1} + c(y_i^{k_j+1} - g_i) \leq 0$ and thus $i \in \mathcal{I}_{k_j+1}$. By repeating this argument, we can conclude $i \in \mathcal{I}_k$ and hence $\lambda_i^k = 0$ for all $k > k_j$.

(v) Finally, we complete the proof. Since $y^* \leq y^k \leq g$ and $y^{k+1} \leq y^k$, there exists $\bar{y}$ such that $\bar{y} = \lim_{k\to\infty} y^k \leq g$. As $\lambda^k = f - Ay^k$, we have for $\bar{\lambda} := f - A\bar{y}$ that $\bar{\lambda} = \lim_{k\to\infty} \lambda^k \geq 0$ by (iv). Note also that $\langle \lambda^k, y^k - g\rangle = 0$, we thus have $\langle \bar{\lambda}, \bar{y} - g\rangle = 0$. Therefore $(\bar{y}, \bar{\lambda})$ is a solution of (10.44). By uniqueness, we have $(\bar{y}, \bar{\lambda}) = (y^*, \lambda^*)$ and thus $(y^k, \lambda^k) \to (y^*, \lambda^*)$ as $k \to \infty$. The finite convergence then can be concluded by Theorem 10.30. □

10.5.2 Elastic-Net Regression

Let A be a matrix of size $m \times n$ and $b \in \mathbb{R}^m$; we consider the elastic-net regression problem of the form

$$\min_{x\in\mathbb{R}^n} \left\{ \frac{1}{2}\|Ax - b\|^2 + \alpha\|x\|_1 + \frac{\beta}{2}\|x\|^2 \right\}, \tag{10.63}$$

where $\alpha > 0$ and $\beta > 0$ are suitably chosen regularization parameters. This problem is different from Lasso in that a quadratic term is added to make the objective function strongly convex, and thus it has a unique minimizer. When seeking for a solution with sparsity, β is usually chosen to be a small positive number.

Many algorithms discussed in the previous chapters, including the proximal gradient method and the alternating direction method of multipliers, can be used to determine the minimizer of (10.63) numerically. In this section, we will consider solving (10.63) by the semi-smooth Newton method [93]. To this end, we will reformulate (10.63) as an equivalent root-finding equation. According to Theorems 3.9, 3.14, and Example 4.15, we have for any $\gamma > 0$ that

$$\begin{aligned} x^* \text{ is a minimizer of (10.63)} &\iff 0 \in A^T(Ax^* - b) + \beta x^* + \alpha\partial(\|\cdot\|_1)(x^*) \\ &\iff -\beta\gamma x^* - \gamma A^T(Ax^* - b)) \in \alpha\gamma\partial(\|\cdot\|_1)(x^*) \end{aligned}$$

$$\iff x^* = \text{prox}_{\alpha\gamma\|\cdot\|_1}(x^* - \gamma(A^T(Ax^* - b) + \beta x^*))$$

$$\iff x^* = S_{\alpha\gamma}(x^* - \gamma(A^T(Ax^* - b) + \beta x^*)),$$

where $S_{\alpha\gamma}$ denotes the soft thresholding operator. Therefore, finding a minimizer of (10.63) is equivalent to solving the root-finding equation

$$F(x) := x - S_{\alpha\gamma}(x - \gamma(A^T(Ax - b) + \beta x)) = 0 \tag{10.64}$$

on $\mathbb{R}^n$. To apply the semi-smooth Newton method to (10.64), we need to show that the mapping F is Newton differentiable with an explicit Newton derivative. Note that the key part of F is the soft-thresholding operator; thus, we first show that the soft-thresholding operator is Newton differentiable.

Proposition 10.33 *The soft-thresholding operator S_τ on $\mathbb{R}^n$ with parameter $\tau > 0$ is Newton differentiable and has a Newton derivative $\mathcal{G}_{S_\tau} : \mathbb{R}^n \to \mathbb{R}^n$ given by*

$$[\mathcal{G}_{S_\tau}(x)h]_i = \begin{cases} h_i & \text{for } |x_i| > \tau, \\ 0 & \text{for } |x_i| \le \tau \end{cases} \tag{10.65}$$

for any $x, h \in \mathbb{R}^n$, where $[\mathcal{G}_{S_\tau}(x)h]_i$ denotes the i-th component of $\mathcal{G}_{S_\tau}(x)h$.

Proof Let $x \in \mathbb{R}^n$. For any $h \in \mathbb{R}^n$, we set

$$D(x, h) := S_\tau(x + h) - S_\tau(x) - G_{S_\tau}(x + h)h.$$

We will show that $D(x, h) = 0$ if $\|h\|_\infty$ is sufficiently small. Indeed, for any index i with $|x_i| = \tau$, as $[S_\tau(x)]_i = 0$ we may check directly that

$$[D(x, h)]_i = [S_\tau(x + h)]_i - [\mathcal{G}_{S_\tau}(x + h)h]_i = 0 \tag{10.66}$$

for all h satisfying $\|h\|_\infty < \tau$. Thus, if $|x_i| = \tau$ for all i, we are done. Therefore, we may assume $\{i : |x_i| \neq \tau\} \neq \emptyset$. By assuming further that

$$\|h\|_\infty < \min\{||x_i| - \tau| : |x_i| \neq \tau\},$$

we can check directly that $[D(x, h)]_i = 0$ for all i with $|x_i| \neq \tau$. The proof is complete. □

By using Propositions 10.5 and 10.33, we can see that the mapping F defined in (10.64) is Newton differentiable with a Newton derivative given by

$$\mathcal{G}_F(x) = I - \mathcal{G}_{S_{\alpha\gamma}}(x - \gamma(A^T(Ax - b) + \beta x))(I - \gamma(\beta I + A^T A)). \tag{10.67}$$

Moreover

$$F(x+h)-F(x)-\mathcal{G}_F(x+h)h=0 \tag{10.68}$$

for sufficiently small h. Consequently, the semi-smooth Newton method for solving (10.64) takes the form $x^{k+1}=x^k+d^k$, where $d^k\in\mathbb{R}^n$ is determined by

$$\mathcal{G}_F(x^k)d^k=-F(x^k),$$

that is

$$\begin{aligned}&\Big[I-\mathcal{G}_{S_{\alpha\gamma}}((1-\beta\gamma)x^k-\gamma A^T(Ax^k-b))(I-\gamma(\beta I+A^TA))\Big]d^k\\&=-\Big[x^k-S_{\alpha\gamma}((1-\beta\gamma)x^k-\gamma A^T(Ax^k-b))\Big].\end{aligned} \tag{10.69}$$

Now we introduce

$$\begin{aligned}\mathcal{A}_k&:=\{i:|(1-\beta\gamma)x_i^k-[\gamma A^T(Ax^k-b)]_i|>\alpha\gamma\},\\ \mathcal{I}_k&:=\{i:|(1-\beta\gamma)x_i^k-[\gamma A^T(Ax^k-b)]_i|\le\alpha\gamma\}\end{aligned} \tag{10.70}$$

which are the active and inactive sets associated with x^k, respectively. We may arrange the coordinates of elements in $\mathbb{R}^n$, so that the active and inactive ones occur in consecutive order. Thus for any $h\in\mathbb{R}^n$, we may write it as

$$h=\begin{pmatrix}h_{\mathcal{A}_k}\\ h_{\mathcal{I}_k}\end{pmatrix}.$$

Accordingly, the matrix $M:=A^TA$ has the block-wise representation

$$M=\begin{pmatrix}M_{\mathcal{A}_k\mathcal{A}_k} & M_{\mathcal{A}_k\mathcal{I}_k}\\ M_{\mathcal{I}_k\mathcal{A}_k} & M_{\mathcal{I}_k\mathcal{I}_k}\end{pmatrix}.$$

and, by Proposition 10.33,

$$\mathcal{G}_{S_{\alpha\gamma}}\Big((1-\beta\gamma)x^k-\gamma A^T(Ax^k-b)\Big)=\begin{pmatrix}I_{\mathcal{A}_k} & 0\\ 0 & 0\end{pmatrix},$$

Therefore, for the Newton derivative $\mathcal{G}_F$ defined by (10.67), we have

$$\mathcal{G}_F(x^k)=\begin{pmatrix}I_{\mathcal{A}_k} & 0\\ 0 & I_{\mathcal{I}_k}\end{pmatrix}-\begin{pmatrix}I_{\mathcal{A}_k} & 0\\ 0 & 0\end{pmatrix}(I-\gamma(\beta I+M))$$

$$= \begin{pmatrix} \beta\gamma I_{\mathcal{A}_k} & 0 \\ 0 & I_{\mathcal{I}_k} \end{pmatrix} + \gamma \begin{pmatrix} I_{\mathcal{A}_k} & 0 \\ 0 & 0 \end{pmatrix} \begin{pmatrix} M_{\mathcal{A}_k\mathcal{A}_k} & M_{\mathcal{A}_k\mathcal{I}_k} \\ M_{\mathcal{I}_k\mathcal{A}_k} & M_{\mathcal{I}_k\mathcal{I}_k} \end{pmatrix}$$

$$= \begin{pmatrix} \beta\gamma I_{\mathcal{A}_k} + \gamma M_{\mathcal{A}_k\mathcal{A}_k} & \gamma M_{\mathcal{A}_k\mathcal{I}_k} \\ 0 & I_{\mathcal{I}_k} \end{pmatrix}. \tag{10.71}$$

Consequently, the semi-smooth Newton method (10.69) becomes

$$\begin{pmatrix} \beta\gamma I_{\mathcal{A}_k} + \gamma M_{\mathcal{A}_k\mathcal{A}_k} & \gamma M_{\mathcal{A}_k\mathcal{I}_k} \\ 0 & I_{\mathcal{I}_k} \end{pmatrix} \begin{pmatrix} d^k_{\mathcal{A}_k} \\ d^k_{\mathcal{I}_k} \end{pmatrix} = - \begin{pmatrix} r^k_{\mathcal{A}_k} \\ r^k_{\mathcal{I}_k} \end{pmatrix}, \tag{10.72}$$

where

$$r^k = x^k - S_{\alpha\gamma}((1-\beta\gamma)x^k - \gamma A^T(Ax^k - b)).$$

According to (10.72), we have

$$d^k_{\mathcal{I}_k} = -r^k_{\mathcal{I}_k}, \quad (\beta I_{\mathcal{A}_k} + M_{\mathcal{A}_k\mathcal{A}_k})d^k_{\mathcal{A}_k} + M_{\mathcal{A}_k\mathcal{I}_k} d^k_{\mathcal{I}_k} = -\frac{1}{\gamma} r^k_{\mathcal{A}_k}, \tag{10.73}$$

Therefore, by noting that $r^k_{\mathcal{I}_k} = x^k_{\mathcal{I}_k}$, we have

$$x^{k+1}_{\mathcal{I}_k} = x^k_{\mathcal{I}_k} + d^k_{\mathcal{I}_k} = 0.$$

Note that $M_{\mathcal{A}_k\mathcal{A}_k}$ is symmetric positive semidefinite, the matrix $\beta I_{\mathcal{A}_k} + M_{\mathcal{A}_k\mathcal{A}_k}$ is invertible. Therefore, we can solve the second equation in (10.73) to find $d^k_{\mathcal{A}_k}$. This leads to the following algorithm for solving the elastic-net regression problem (10.63).

Algorithm 10.34 *Choose an initial guess x^0 and a number $\gamma > 0$. For $k = 0, 1, \cdots$, do the following:*

(i) *Compute the residual*

$$r^k = F(x^k) = x^k - S_{\alpha\gamma}((1-\beta\gamma)x^k - \gamma A^T(Ax^k - b)),$$

(ii) *Determine the active and inactive sets $\mathcal{A}_k$ and $\mathcal{I}_k$ by (10.70).*

(iii) *Determine $d^k_{\mathcal{A}_k}$ by solving*

$$(\beta I_{\mathcal{A}_k} + M_{\mathcal{A}_k\mathcal{A}_k})d^k_{\mathcal{A}_k} = M_{\mathcal{A}_k\mathcal{I}_k} x^k_{\mathcal{I}_k} - \frac{1}{\gamma} r^k_{\mathcal{A}_k}$$

and update x^{k+1} by setting $x^{k+1}_{\mathcal{I}_k} = 0$ and $x^{k+1}_{\mathcal{A}_k} = x^k_{\mathcal{A}_k} + d^k_{\mathcal{A}_k}$.

According to (10.71), it is easy to see that $\mathcal{G}_F(x^k)$ is invertible with

$$\mathcal{G}_F(x^k)^{-1} = \begin{pmatrix} (\beta_\gamma I + M_{\mathcal{A}_k\mathcal{A}_k})^{-1} & -\gamma(\beta_\gamma I + M_{\mathcal{A}_k\mathcal{A}_k})^{-1} M_{\mathcal{A}_k \mathcal{I}_k} \\ 0 & I_{\mathcal{I}_k} \end{pmatrix}$$

which implies that $\|\mathcal{G}_F(x^k)^{-1}\|$ is uniformly bounded for all k. Thus, we may use Theorem 10.3 and its proof to obtain the following convergence result on Algorithm 10.34.

Theorem 10.35 *Consider the elastic-net regression problem (10.63), and let x^* be its unique minimizer. Then the sequence $\{x^k\}$ produced by Algorithm 10.34 converges to x^* in finite many steps provided $\|x^0 - x^*\|$ is sufficiently small.*

10.5.3 Inverse Problem with Uniform Noise

Let A be an $m \times n$ matrix and $b \in \mathbb{R}^m$. We consider the minimization problem

$$\min_{x\in\mathbb{R}^n} \frac{1}{2}\|x\|_2^2 \quad \text{subject to } \|Ax - b\|_\infty \le \eta, \tag{10.74}$$

where $\eta > 0$ denotes the noise level when the data is corrupted by uniform noise. This minimization problem occurs in many applications including inverse problems, image compression, and optimal control. We will use the semi-smooth Newton method to find the solution of (10.74) numerically [223]. To this end, we first use the splitting technique to rewrite (10.74) as

$$\min_{x\in\mathbb{R}^n, r\in\mathbb{R}^m} \frac{1}{2}\|x\|_2^2 \quad \text{subject to } \|r\|_\infty \le \eta, \quad r = Ax - b. \tag{10.75}$$

Let $\Omega := \{r \in \mathbb{R}^m : \|r\|_\infty \le \eta\}$, and let δ_Ω be the indicator function of Ω. Then (10.75) can be equivalently formulated as

$$\min_{x\in\mathbb{R}^n, r\in\mathbb{R}^m} \frac{1}{2}\|x\|_2^2 + \delta_\Omega(r) \quad \text{subject to } r = Ax - b. \tag{10.76}$$

Note that the Lagrangian function corresponding to (10.76) is

$$\mathcal{L}(x, r, \lambda) := \frac{1}{2}\|x\|_2^2 + \delta_\Omega(r) + \langle \lambda, r + b - Ax\rangle. \tag{10.77}$$

According to the theory of Karush-Kuhn-Tucker, (x^*, r^*) is a minimizer of (10.76) if and only if there is $\lambda^* \in \mathbb{R}^m$ such that the following KKT conditions hold:

$$\begin{aligned} 0 &= \nabla_x \mathcal{L}(x^*, r^*, \lambda^*) = x^* - A^T \lambda^*, \\ 0 &\in \partial_r \mathcal{L}(x^*, r^*, \lambda^*) = \lambda^* + \partial \delta_\Omega(r^*), \\ 0 &= Ax^* - b - r^*. \end{aligned} \tag{10.78}$$

According to the second equation in (10.78), we have for any $\beta > 0$ that

$$\begin{aligned} r^* &= (I + \beta \partial \delta_\Omega)^{-1}(r^* - \beta\lambda^*) \\ &= \arg\min_{r \in \mathbb{R}^m} \left\{ \delta_\Omega(r) + \frac{1}{2\beta} \|r - (r^* - \beta\lambda^*)\|^2 \right\} \\ &= P_\Omega(r^* - \beta\lambda^*), \end{aligned}$$

where P_Ω denotes the Euclidean projection of $\mathbb{R}^m$ onto Ω which is given by

$$P_\Omega(z) = \max\{\min\{z, \eta\}, -\eta\}.$$

Consequently, (x^*, r^*, λ^*) can be obtained by solving the equations

$$0 = x^* - A^T\lambda^*, \quad 0 = r^* - P_\Omega(r^* - \beta\lambda^*), \quad 0 = Ax^* - b - r^*. \tag{10.79}$$

From the first and third equations in (10.79), we have $r^* = AA^T\lambda^* - b$. By plugging this into the second equation in (10.79), we can see that λ^* is a root of the equation

$$F(\lambda) := AA^T\lambda - b - P_\Omega(AA^T\lambda - b - \beta\lambda) = 0. \tag{10.80}$$

If we can solve (10.80) for λ^*, we may find x^* and r^* by setting $x^* = A^T\lambda^*$ and $r^* = Ax^* - b$.

In order to use the semi-smooth Newton method to solve (10.80), we need to show the Newton differentiability of the mapping F defined in (10.80) and determine a Newton derivative. According to the definition of F, we need to consider the projection P_Ω. By using the similar argument in the proof of Proposition 10.33, we can obtain the following result.

Proposition 10.36 *The projection mapping P_Ω is Newton differentiable and has a Newton derivative $\mathcal{G}_{P_\Omega}$ given by*

$$(\mathcal{G}_{P_\Omega}(z)h)_i = \begin{cases} h_i & \text{for } |z_i| < \eta, \\ 0 & \text{for } |z_i| \geq \eta. \end{cases} \tag{10.81}$$

Moreover, for any $z \in \mathbb{R}^n$, *there holds*

$$P_\Omega(z+h) - P_\Omega(z) - \mathcal{G}_{P_\Omega}(z+h)h = 0$$

if $\|h\|$ *is sufficiently small.*

By using Proposition 10.36 and the chain rule given in Proposition 10.5, we can see that F defined in (10.80) is Newton differentiable with a Newton derivative $\mathcal{G}_F$ given by

$$\mathcal{G}_F(\lambda) = M - \mathcal{G}_{P_\Omega}(M\lambda - b - \beta\lambda)(M - \beta I),$$

where $M := AA^T$. Moreover

$$F(\lambda + h) - F(\lambda) - \mathcal{G}_F(\lambda + h)h = 0$$

for sufficiently small $\|h\|$. Consequently, the semi-smooth Newton method for solving (10.80) takes the form $\lambda^{k+1} = \lambda^k + d^k$, where d^k is determined by $\mathcal{G}_F(\lambda^k)d^k = -F(\lambda^k)$, that is

$$\begin{aligned} &\Big[M - \mathcal{G}_{P_\Omega}(M\lambda^k - b - \beta\lambda^k)(M - \beta I)\Big] d^k \\ &= -\Big[M\lambda^k - b - P_\Omega(M\lambda^k - b - \beta\lambda^k)\Big]. \end{aligned} \tag{10.82}$$

Now we introduce

$$\begin{aligned} \mathcal{A}_k &:= \{i : |(M\lambda^k - b - \beta\lambda^k)_i| > \eta\}, \\ \mathcal{I}_k &:= \{i : |(M\lambda^k - b - \beta\lambda^k)_i| \le \eta\} \end{aligned} \tag{10.83}$$

which are the active and inactive sets associated with y^k, respectively. We may rearrange the coordinates of elements, so that the active and inactive ones occur in consecutive order. Accordingly, we can represent M as

$$M = \begin{pmatrix} M_{\mathcal{A}_k\mathcal{A}_k} & M_{\mathcal{A}_k\mathcal{I}_k} \\ M_{\mathcal{I}_k\mathcal{A}_k} & M_{\mathcal{I}_k\mathcal{I}_k} \end{pmatrix}.$$

Furthermore, by the definition of $\mathcal{G}_{P_\Omega}$, we can see that $\mathcal{G}_{P_\Omega}(M\lambda^k - f - \beta\lambda^k)$ has the representation

$$\begin{pmatrix} 0 & 0 \\ 0 & I_{\mathcal{I}_k} \end{pmatrix},$$

Therefore $\mathcal{G}_F(\lambda^k)$ has the representation

$$\begin{aligned}\mathcal{G}_F(\lambda^k) &= \begin{pmatrix} M_{\mathcal{A}_k\mathcal{A}_k} & M_{\mathcal{A}_k\mathcal{I}_k} \\ M_{\mathcal{I}_k\mathcal{A}_k} & M_{\mathcal{I}_k\mathcal{I}_k} \end{pmatrix} - \begin{pmatrix} 0 & 0 \\ 0 & I_{\mathcal{I}_k} \end{pmatrix} \left[\begin{pmatrix} M_{\mathcal{A}_k\mathcal{A}_k} & M_{\mathcal{A}_k\mathcal{I}_k} \\ M_{\mathcal{I}_k\mathcal{A}_k} & M_{\mathcal{I}_k\mathcal{I}_k} \end{pmatrix} - \beta \begin{pmatrix} I_{\mathcal{A}_k} & 0 \\ 0 & I_{\mathcal{I}_k} \end{pmatrix} \right] \\ &= \begin{pmatrix} M_{\mathcal{A}_k\mathcal{A}_k} & M_{\mathcal{A}_k\mathcal{I}_k} \\ 0 & \beta I_{\mathcal{I}_k} \end{pmatrix}\end{aligned}$$

This together with (10.82) leads to the following semi-smooth Newton method to solve (10.80) and hence (10.74).

Algorithm 10.37 *Choose an initial guess λ^0 and a number $\beta > 0$. For $k = 0, 1, \cdots$, do the following:*

(i) *Calculate $x^k = A^T\lambda^k$, $r^k = Ax^k - b$, and*

$$h^k := F(\lambda^k) = r^k - P_\Omega(r^k - \beta\lambda^k).$$

(ii) *Determine the active and inactive sets $\mathcal{A}_k$ and $\mathcal{I}_k$ by (10.83).*
(iii) *Calculate the Newton step d^k by solving*

$$\begin{pmatrix} M_{\mathcal{A}_k\mathcal{A}_k} & M_{\mathcal{A}_k\mathcal{I}_k} \\ 0 & \beta I_{\mathcal{I}_k} \end{pmatrix} \begin{pmatrix} d^k_{\mathcal{A}_k} \\ d^k_{\mathcal{I}_k} \end{pmatrix} = - \begin{pmatrix} h^k_{\mathcal{A}_k} \\ h^k_{\mathcal{I}_k} \end{pmatrix}$$

and update $\lambda^{k+1} := \lambda^k + d^k$.

When A is of full row rank, one can easily show that $G_F(\lambda)$ is invertible and $\|\mathcal{G}_F(\lambda)^{-1}\|$ is uniformly bounded. Therefore, we may use Theorem 10.3 to conclude the finite convergence of Algorithm 3.

10.5.4 Semi-smooth Newton-Based Proximal Point Algorithm

When solving a minimization problem using an iterative method, it is common to encounter situations where the computation of iterates requires solving subproblems with a different algorithm. In such cases, the semi-smooth Newton method serves as a powerful tool for this purpose [140, 155, 207, 235].

In this section, we discuss the semi-smooth Newton-based proximal point algorithm. As an illustrative example, we consider its application to the Lasso problem.

$$\min_{x\in\mathbb{R}^n} \left\{ \frac{1}{2}\|Ax - b\|^2 + \alpha\|x\|_1 \right\}, \tag{10.84}$$

where A is an $m \times n$ matrix, $b \in \mathbb{R}^m$ and $\alpha > 0$ is a regularization parameter.

If we apply the semi-smooth Newton method directly to solve (10.84), as done in Sect. 10.5.2, we may encounter the equation (10.72) with $\beta = 0$ for computing the Newton step. However, when $m \ll n$, the coefficient matrix may become singular, making the method ill-defined.

In the following, we will demonstrate how the semi-smooth Newton method can be integrated to resolve subproblems within the proximal point algorithm. Recall that the proximal point algorithm for solving (10.84) requires solving a minimization problem

$$\min_{x \in \mathbb{R}^n} \left\{ \frac{1}{2}\|Ax - b\|^2 + \alpha \|x\|_1 + \frac{1}{2t_k}\|x - x^k\|^2 \right\} \tag{10.85}$$

at each iteration step, where $t_k > 0$ is a step size. However, solving (10.85) exactly is impossible in general. Instead of solving (10.85) exactly, we will solve it approximately by a semi-smooth Newton method and then use the obtained result in the next iteration step. To make things more transparent, assume x^k is available, and let $\hat{x}^{k+1}$ be the exact solution of (10.85), i.e.,

$$\hat{x}^{k+1} = \arg \min_{x \in \mathbb{R}^n} \left\{ \frac{1}{2}\|Ax - b\|^2 + \alpha \|x\|_1 + \frac{1}{2t_k}\|x - x^k\|^2 \right\}. \tag{10.86}$$

By the first-order optimality condition on $\hat{x}^{k+1}$, we have

$$0 \in A^T(A\hat{x}^{k+1} - b) + \alpha \partial(\|\cdot\|_1)(\hat{x}^{k+1}) + \frac{1}{t_k}(\hat{x}^{k+1} - x^k)$$

which implies

$$x^k - t_k A^T(A\hat{x}^{k+1} - b) \in (I + \alpha t_k \partial(\|\cdot\|_1))(\hat{x}^{k+1})$$

and hence

$$\hat{x}^{k+1} = \text{prox}_{\alpha t_k \|\cdot\|_1}\left(x^k - t_k A^T(A\hat{x}^{k+1} - b)\right).$$

Let $\hat{\xi}^{k+1} = A\hat{x}^{k+1}$. Then

$$\hat{x}^{k+1} = \text{prox}_{\alpha t_k \|\cdot\|_1}\left(x^k - t_k A^T(\hat{\xi}^{k+1} - b)\right), \tag{10.87}$$

$$\hat{\xi}^{k+1} = A\, \text{prox}_{\alpha t_k \|\cdot\|_1}\left(x^k - t_k A^T(\hat{\xi}^{k+1} - b)\right). \tag{10.88}$$

This shows that $\hat{\xi}^{k+1}$ is a solution of (10.88). If we can solve (10.88) exactly, then we may use (10.87) to determine $\hat{x}^{k+1}$ exactly. However, solving (10.88) exactly is impossible in general. We may consider solving it approximately by a semi-smooth

Newton method. To this end, let

$$F(\xi) := \xi - A\,\mathrm{prox}_{\alpha t_k \|\cdot\|_1}\left(x^k - t_k A^T(\xi - b)\right).$$

Then (10.88) is equivalent to $F(\xi) = 0$.

Lemma 10.38 *Let*

$$\mathcal{U}(\xi) := \frac{1}{2}\|\xi\|^2 + \frac{1}{2t_k}\|x^k - t_k A^T(\xi - b)\|^2 - \alpha M_{\alpha t_k\|\cdot\|_1}(x^k - t_k A^T(\xi - b)),$$

where $M_{\alpha t_k\|\cdot\|_1}$ denotes the Moreau envelop of $\|\cdot\|_1$ with parameter αt_k. Then $\mathcal{U}$ is a continuous differentiable strongly convex function with $\nabla\mathcal{U}(\xi) = F(\xi)$.

Proof From the properties of Moreau envelop, see Theorem 4.23; it follows that $\mathcal{U}(\xi)$ is continuous differentiable and

$$\begin{aligned}\nabla\mathcal{U}(\xi) &= \xi - A(x^k - t_k A^T(\xi - b)) + \alpha t_k A(\nabla M_{\alpha t_k\|\cdot\|_1})(x^k - t_k A^T(\xi - b)) \\ &= \xi - A(x^k - t_k A^T(\xi - b)) \\ &\quad + A\left(x^k - t_k A^T(\xi - b) - \mathrm{prox}_{\alpha t_k\|\cdot\|_1}(x^k - t_k A^T(\xi - b))\right) \\ &= F(\xi).\end{aligned}$$

Next we show that $\mathcal{U}$ is strongly convex. We can write $\mathcal{U}(\xi) = \frac{1}{2}\|\xi\|^2 + \widetilde{\mathcal{U}}(\xi)$, where

$$\widetilde{\mathcal{U}}(\xi) := \frac{1}{2t_k}\|x^k - t_k A^T(\xi - b)\|^2 - \alpha M_{\alpha t_k\|\cdot\|_1}(x^k - t_k A^T(\xi - b)).$$

It suffices to show $\widetilde{\mathcal{U}}$ is convex. By the Moreau envelop decomposition formula, see Theorem 4.25 (i); we have

$$\widetilde{\mathcal{U}}(\xi) = \alpha M_{\|\cdot\|_1^*/(\alpha t_k)}\left((x^k - t_k A^T(\xi - b))/(\alpha t_k)\right)$$

Since $M_{\|\cdot\|_1^*/(\alpha t_k)}$ is convex, this implies the convexity of $\widetilde{\mathcal{U}}$. □

The above result shows that F is a gradient mapping. Therefore, we may use the globalized semi-smooth Newton method given in Algorithm 10.16 to solve $F(\xi) = 0$ which has the unique solution $\hat{\xi}^{k+1}$. Let ξ^{k+1} denote the output result of the semi-smooth Newton method satisfying

$$\|F(\xi^{k+1})\| = \|\nabla\mathcal{U}(\xi^{k+1})\| < \frac{\varepsilon_k}{2t_k}$$

and define

$$x^{k+1} = \mathrm{prox}_{\alpha t_k \|\cdot\|_1}\left(x^k - t_k A^T(\xi^{k+1} - b)\right).$$

Then by the strong convexity of $\mathcal{U}$ and $\nabla\mathcal{U}(\hat{\xi}^{k+1}) = 0$, we have

$$\|\xi^{k+1} - \hat{\xi}^{k+1}\| \le 2\|\nabla\mathcal{U}(\xi^{k+1}) - \nabla\mathcal{U}(\hat{\xi}^{k+1})\| = 2\|\nabla\mathcal{U}(\xi^{k+1})\| < \frac{\varepsilon_k}{t_k}.$$

Consequently, by the definition of x^{k+1} and $\hat{x}^{k+1}$ and the non-expansiveness of the proximal mapping, we have

$$\|x^{k+1} - \hat{x}^{k+1}\| \le t_k\|A^T(\xi^{k+1} - \hat{\xi}^{k+1})\| \le \|A\|\varepsilon_k.$$

Next we consider the convergence of $\{x^k\}$. We will show that $\{x^k\}$ converges to a solution of (10.84) as $k \to \infty$. Note that $\hat{x}^{k+1}$ is defined by (10.86). Therefore, $\{x^k\}$ can be viewed as a sequence obtained via an inexact proximal point algorithm. Thus, the convergence of $\{x^k\}$ can be obtained correspondingly by the following result; see [192].

Theorem 10.39 *Let $f : \mathbb{R}^n \to (-\infty, \infty]$ be a proper, lower semicontinuous, convex function with $S := \arg\min_{x\in\mathbb{R}^n} f(x) \ne \emptyset$. Given x^0, and define a sequence $\{x^k\}$ such that*

$$\|x^{k+1} - \mathrm{prox}_{t_k f}(x^k)\| < \varepsilon_k, \quad k = 0, 1, \cdots,$$

where $\sum_{k=0}^{\infty} \varepsilon_k < \infty$ and $\{t_k\}$ is a sequence of positive numbers bounded away from zero. Then $\{x^k\}$ converges to some element $\hat{x} \in S$.

Proof Let $\hat{x}^{k+1} := \mathrm{prox}_{t_k f}(x^k)$. By the non-expansiveness of the proximal mapping, we have for any $\bar{x} \in S$ that

$$\|x^{k+1} - \bar{x}\| \le \|x^{k+1} - \hat{x}^{k+1}\| + \|\mathrm{prox}_{t_k f}(x^k) - \mathrm{prox}_{t_k f}(\bar{x})\| \le \varepsilon_k + \|x^k - \bar{x}\|,$$

where we used $\mathrm{prox}_{t_k f}(\bar{x}) = \bar{x}$. Since $\sum_k \varepsilon_k < \infty$, this implies that $\lim_{k\to\infty} \|x^k - \bar{x}\|$ exists; see Lemma 3.20, which in particular shows the boundedness of $\{x^k\}$. Moreover, by the firmly non-expansiveness of the proximal mapping, we have

$$\begin{aligned}
&\|\hat{x}^{k+1} - \bar{x}\|^2 + \|x^k - \hat{x}^{k+1}\|^2 \\
&= \|\mathrm{prox}_{t_k f}(x^k) - \mathrm{prox}_{t_k f}(\bar{x})\|^2 + \|(I - \mathrm{prox}_{t_k f})(x^k) - (I - \mathrm{prox}_{t_k f})(\bar{x})\|^2 \\
&\le \|x^k - \bar{x}\|^2.
\end{aligned}$$

Thus

$$\|x^k - \hat{x}^{k+1}\|^2 \le \|x^k - \bar{x}\|^2 - \|\hat{x}^{k+1} - \bar{x}\|^2. \tag{10.89}$$

Since $\|x^{k+1} - \hat{x}^{k+1}\| < \varepsilon_k \to 0$ as $k \to \infty$, we can conclude $\lim_{k\to\infty} \|\hat{x}^{k+1} - \bar{x}\| = \lim_{k\to\infty} \|x^{k+1} - \bar{x}\|$. Therefore, by taking $k \to \infty$ in 10.89), we obtain

$$\|x^k - \hat{x}^{k+1}\| \to 0 \quad \text{as } k \to \infty.$$

Recall that $\{x^k\}$ is bounded. So there is a subsequence $\{x^{k_l}\}$ of $\{x^k\}$ such that $x^{k_l} \to \hat{x}$ as $l \to \infty$ for some $\hat{x} \in \mathbb{R}^n$. We claim $\hat{x} \in S$. To see this, by the definition of $\hat{x}^{k_l+1}$, we have $\hat{x}^{k_l+1} = \text{prox}_{t_{k_l} f}(x^{k_l})$ and hence

$$(x^{k_l} - \hat{x}^{k_l+1})/t_{k_l} \in \partial f(\hat{x}^{k_l+1}). \tag{10.90}$$

Since $\{t_k\}$ is bounded away from zero, we have $(x^{k_l} - \hat{x}^{k_l+1})/t_{k_l} \to 0$. Because $\hat{x}^{k_l+1} \to \hat{x}$ and f is lower semicontinuous, by letting $l \to \infty$ in (10.90), we then obtain $0 \in \partial f(\hat{x})$, i.e., $\hat{x} \in S$. Finally, since $\lim_{k\to\infty} \|x^k - \hat{x}\|$ exists and $\lim_{l\to\infty} \|x^{k_l} - \hat{x}\| = 0$, we must have $\lim_{k\to\infty} \|x^k - \hat{x}\| = 0$. □

Problems

10.1 Let $f : \mathbb{R}^n \to \mathbb{R}$ be a locally Lipschitz continuous function. This problem devotes to establishing an equivalent characterization of $\partial_C f$.

(i) Given $x, v \in \mathbb{R}^n$, we define

$$f^\circ(x; v); = \limsup_{t\searrow 0, y\to x} \frac{f(y + tv) - f(y)}{t}$$

which is called the Clarke directional derivative of f at x in the direction v. Show that, for each fixed $x \in \mathbb{R}^n$, the function $v \to f^\circ(x; v)$ is finite, sublinear, and globally Lipschitz continuous.

(ii) For each $x \in \mathbb{R}^n$, let

$$A(x) := \{\xi \in \mathbb{R}^n : \langle \xi, v\rangle \le f^\circ(x; v) \text{ for all } v \in \mathbb{R}^n\}.$$

Show that each $A(x)$ is a convex set. Moreover, if $x_k \to x$, and $\xi_k \in A(x_k)$ with $\xi_k \to \xi$, then $\xi \in A(x)$.

(iii) Show that

$$\partial_C f(x) = \{\xi \in \mathbb{R}^n : \langle \xi, v\rangle \le f^\circ(x; v) \text{ for all } v \in \mathbb{R}^n\}, \quad \forall x \in \mathbb{R}^n$$

and

$$f^{\circ}(x; v) = \sup\{\langle \xi, v \rangle : \xi \in \partial_C f(x)\}, \quad \forall x, v \in \mathbb{R}^n.$$

10.2 Let $f : X \to \mathbb{R}$ be convex. Show that $f^{\circ}(x; v) = f'(x; v)$ for all $x, v \in \mathbb{R}^n$ and $\partial_C f(x) = \partial f(x)$ for all $x \in \mathbb{R}^n$, where ∂f denotes the convex subdifferential of f.

10.3 Let $f_i : \mathbb{R}^n \to \mathbb{R}$, $i = 1, \cdots, N$, be continuously differentiable functons. Define

$$f(x) := \max\{f_1(x), \cdots, f_N(x)\}, \quad x \in \mathbb{R}^n$$

and set $I(x) := \{i \in \{1, \cdots, N\} : f_i(x) = f(x)\}$ which is called the active index set of f at x. Show the following results:

(i) $f : \mathbb{R}^n \to \mathbb{R}$ is locally Lipschitz continuous.
(ii) For any $x \in \mathbb{R}^n$, there holds

$$\partial_C f(x) = \operatorname{conv}\{\nabla f_i(x) : i \in I(x)\}.$$

(iii) f is Newton differentiable for all $x \in \mathbb{R}^n$, and every $D_N f$ with $D_N f(x) \in \partial_C f(x)$ for each x is a Newton derivative.

10.4 Let $\Omega \subset \mathbb{R}^n$ be a non-empty, closed, convex set. Consider the Euclidean projection P_Ω from $\mathbb{R}^n$ to Ω.

(i) For any $x \in \mathbb{R}^n$, show that $\partial_C P_\Omega(x) \neq \emptyset$ and any $V(x) \in \partial_C P_\Omega(x)$ is symmetric positive semi-definite with $\|V(x)\| \leq 1$.
(ii) If Ω is a non-empty, closed, convex cone, show that for any $x \in \mathbb{R}^n$ and $V(x) \in \partial_C P_\Omega(x)$, there holds $V(x)x = P_\Omega(x)$.

10.5 Let $f : \mathbb{R}^n \to (-\infty, \infty]$ be a proper, lower semicontinuous, convex function. Let $t > 0$ and $x \in \mathbb{R}^n$. Show that for every matrix $V \in \partial_C \mathrm{prox}_{tf}(x) \subset \mathbb{R}^{n\times n}$, both V and $I - V$ are symmetric and positive semi-definite.

10.6 Let $G : \mathbb{R}^m \to \mathbb{R}^n$ be a map that is continuously differentiable in an open neighborhood of a point $\bar{x} \in \mathbb{R}^m$, and let $\Theta : \mathbb{R}^n \to \mathbb{R}^p$ be a map that is locally Lipschitz continuous in a neighborhood of $G(\bar{x})$. Define $F(x) := \Theta(G(x))$ for $x \in \mathbb{R}^m$. If $G'(\bar{x}) : \mathbb{R}^m \to \mathbb{R}^n$ is surjective, show that

$$\partial_B F(\bar{x}) = \partial_B \Theta(G(\bar{x}))G'(\bar{x}).$$

(Hint: See [53, Lemma 1])

10.7 Let $F : \mathbb{R}^n \to \mathbb{R}^n$ be locally Lipschitz continuous, let $\bar{x} \in \mathbb{R}^n$ be fixed, and assume that all $V \in \partial_B F(\bar{x})$ are nonsingular. Show that there exist a neighborhood

$\mathcal{O}$ of $\bar{x}$ and a constant C such that for all $x \in \mathcal{O}$ every $V \in \partial_B F(x)$ is nonsingular and $\|V^{-1}\| \leq C$.

10.8 Consider the composite minimization problem

$$\min_{x \in \mathbb{R}^n} \{f(x) + g(x)\}, \tag{10.91}$$

where $f : \mathbb{R}^n \to \mathbb{R}$ is a twice continuously differentiable function and $g : \mathbb{R}^n \to (-\infty, \infty]$ is a proper, lower semicontinuous, convex function.

(i) Show that any optimal solution of (10.91), if exists, must satisfy the equation

$$F(x) := x - \text{prox}_{tg}(x - t\nabla f(x)) = 0 \tag{10.92}$$

for any $t > 0$.

(ii) For a fixed $t > 0$, if prox_{tg} is semi-smooth, show that $F : \mathbb{R}^n \to \mathbb{R}^n$ is also semi-smooth.

(iii) Given $x \in \mathbb{R}^n$ and let $\lambda_{\max}(\nabla^2 f(x))$ denote the largest eigenvalue of $\nabla^2 f(x)$. If $0 < t < 1/\lambda_{\max}(\nabla^2 f(x))$, show that

$$\partial_B F(x) = \left\{ I - M(I - t\nabla^2 f(x)) : M \in \partial_B \text{prox}_{tg}(x - t\nabla f(x)) \right\}.$$

(iv) When applying the semi-smooth Newton method (10.5) to solve (10.92), we may choose $D_N F(x) \in \partial_B F(x)$ for each x. To ensure that the method is well defined and exhibits local convergence, it is necessary to guarantee that every element of $\partial_B F(x^*)$ is invertible at the sought solution x^*. Assume that $0 < t < 1/\lambda_{\max}(\nabla^2 f(x^*))$ and that $\nabla^2 f(x^*)$ is positive semi-definite. Show that every element in $\partial_B F(x^*)$ is invertible if and only if

$$v^T \nabla^2 f(x^*) v > 0, \quad \forall 0 \neq v \in \text{Null}(I - M)$$

for every $M \in \partial_B \text{prox}_{tg}(x^* - t\nabla f(x^*))$. (Hint: see [107, Proposition 3.1])

Chapter 11
Stochastic Algorithms

Optimization of finite-sum problems has become fundamental in many fields, particularly in machine learning and data science. These problems, which are often of the form

$$\min_{x\in\mathbb{R}^d}\left\{f(x)=\frac{1}{N}\sum_{i=1}^{N}f_i(x)\right\}, \tag{11.1}$$

involve minimizing the average of a large number of component functions $f_i(x)$, each typically representing the loss associated with a single data point. Such problems arise frequently in empirical risk minimization, where $f_i(x)$ is the loss of a predictive model on the i-th training example. When the number of components N is large, computing the full gradient $\nabla f(x)$ at each iteration of traditional gradient-based methods becomes computationally expensive, motivating the use of stochastic algorithms that offer more efficient alternatives.

Stochastic algorithms exploit randomness to approximate the full gradient by sampling one or a subset of the component functions at each iteration, significantly reducing the per-iteration cost. In this chapter, we will explore several stochastic algorithms for solving large-scale optimization problems, including the stochastic gradient descent (SGD), and various variance-reduced variants such as the stochastic variance-reduced gradient (SVRG) and stochastic average gradient acceleration (SAGA). We will also discuss the randomized block-coordinate descent method and the stochastic primal-dual hybrid gradient method.

11.1 Probability Theory and Martingale Theory

In this section, we collect some important definitions and theorems from probability theory and martingale theory [28].

Q. Jin, *Lectures on Nonsmooth Optimization*, Texts in Applied Mathematics 82,
https://doi.org/10.1007/978-3-031-91417-1_11

When setting up a mathematical model, we need to decide a so-called sample space and then decide on a system of subsets to which we will assign a probability. The natural type of system to consider is a σ-algebra.

Definition 11.1 Given a set Ω, we use $\mathcal{P}(\Omega)$ to denote its power set. A class $\mathcal{F} \subseteq \mathcal{P}(\Omega)$ of subsets of Ω is called a *σ-algebra* on Ω if

(i) $\Omega \in \mathcal{F}$;
(ii) If $A \in \mathcal{F}$, then $A^c := \Omega \setminus A \in \mathcal{F}$;
(iii) If $\{A_k\}_{k=1}^{\infty} \subseteq \mathcal{F}$ then $\cup_{k=1}^{\infty} A_k \in \mathcal{F}$.

The pair $(\Omega, \mathcal{F})$ is called a *measurable space* or an *event space*.

Given a collection $\mathcal{G} \subset \mathcal{P}(\Omega)$, we use $\sigma(\mathcal{G})$ to denote the σ-algebra generated by $\mathcal{G}$ which is the smallest σ-algebra containing $\mathcal{G}$, i.e.,

$$\sigma(\mathcal{G}) := \bigcap \{\mathcal{F} : \mathcal{F} \text{ is a } \sigma\text{-algebra and } \mathcal{G} \subseteq \mathcal{F} \subseteq \mathcal{P}(\Omega)\}.$$

In probability theory, σ-algebra makes up the subsets to which one can assign probabilities in a reasonable manner. Here is the formal definition of the probability measures.

Definition 11.2 Let $(\Omega, \mathcal{F})$ be an event space. A mapping $\mathbb{P} : \mathcal{F} \to [0, \infty)$ is called a *probability measure* if

(i) $\mathbb{P}(\Omega) = 1$;
(ii) $\mathbb{P}(\cup_{k=1}^{\infty} A_k) = \sum_{k=1}^{\infty} \mathbb{P}(A_k)$ if $\{A_k\}_{k=1}^{\infty} \subseteq \mathcal{F}$ is mutually disjoint.

We call the triple $(\Omega, \mathcal{F}, \mathbb{P})$ a *probability space*.

Given an event space $(\Omega, \mathcal{F})$, a mapping $X : \Omega \to [-\infty, \infty]$ is called a (real) *random variable* if it is $\mathcal{F}$-measurable, i.e., $\{X \le a\} := \{\omega \in \Omega : X(\omega) \le a\} \in \mathcal{F}$ for every $a \in \mathbb{R}$. Given a family of random variables $\{X_\alpha\}_{\alpha \in \Lambda}$ on $(\Omega, \mathcal{F})$, we use $\sigma(\{X_\alpha\}_{\alpha \in \Lambda})$ to denote the smallest σ-algebra on which each X_α is measurable and call it the σ-algebra generated by $\{X_\alpha\}_{\alpha \in \Lambda}$.

Definition 11.3 Let $(\Omega, \mathcal{F}, \mathbb{P})$ be a probability space. The *expected value* or *expectation* of a random variable X, denoted as $\mathbb{E}[X]$, is defined by the integral

$$\mathbb{E}[X] := \int_{\Omega} X(\omega) d\, \mathbb{P}(\omega)$$

if the integral exists.

For a random variable X with a finite expectation $\mathbb{E}[X]$, its *variance* is defined by

$$\mathrm{Var}(X) := \mathbb{E}[|X - \mathbb{E}[X]|^2].$$

It is straightforward to check that there holds the variance decomposition formula

$$\mathbb{E}[|X - \mathbb{E}[X]|^2] = \mathbb{E}[X^2] - |\mathbb{E}[X]|^2.$$

Let $(\Omega, \mathcal{F}, \mathbb{P})$ be a probability space. For $p \geq 1$, we use $L^p(\Omega, \mathcal{F}, \mathbb{P})$ to denote the space consisting of all random variables X satisfying $\mathbb{E}[|X|^p] = \int_\Omega |X|^p d\mathbb{P} < \infty$. Let $\mathcal{G}$ be a sub σ-algebra of $\mathcal{F}$. Then a random variable with respect to $\mathcal{F}$ is not necessarily a random variable with respect to $\mathcal{G}$. However, given a $\mathcal{F}$-measurable random variable, we can define a unique $\mathcal{G}$-measurable random variable via the conditional expectation.

Definition 11.4 Given a probability space $(\Omega, \mathcal{F}, \mathbb{P})$ and a σ-algebra $\mathcal{G} \subset \mathcal{F}$. For a random variable X with respect to $\mathcal{F}$ with $X \in L^1(\Omega, \mathcal{F}, \mathbb{P})$, its *conditional expectation* given $\mathcal{G}$, denoted as $\mathbb{E}[X|\mathcal{G}]$ is defined to be a $\mathcal{G}$-measurable random variable such that

$$\int_G X d\mathbb{P} - \int_G \mathbb{E}[X|\mathcal{G}] d\mathbb{P}$$

for all $G \in \mathcal{G}$.

According to the Radon-Nikodym theorem, $\mathbb{E}[X|\mathcal{G}]$ is well-defined and is unique. Given a set $A \in \mathcal{F}$, the conditional probability of A given $\mathcal{G}$, denoted by $\mathbb{P}(A|\mathcal{G})$, is defined by

$$\mathbb{P}(A|\mathcal{G}) := \mathbb{E}[I_A|\mathcal{G}],$$

where I_A denotes the characteristic function of A, i.e., $I_A(\omega) = 1$ if $\omega \in A$ and 0 otherwise.

In the following result, we list some key properties of the conditional expectation whose proofs can be found in standard textbooks.

Proposition 11.5 *Let $(\Omega, \mathcal{F}, \mathbb{P})$ be a probability space, and let $\mathcal{G}$ be a σ-algebra satisfying $\mathcal{G} \subset \mathcal{F}$. Let $X, Y \in L^1(\Omega, \mathcal{F}, \mathbb{P})$. Then the following properties hold almost surely:*

(i) $\mathbb{E}[\alpha X + \beta Y|\mathcal{G}] = \alpha\mathbb{E}[X|\mathcal{G}] + \beta\mathbb{E}[Y|\mathcal{G}]$ *for all constants α and β.*
(ii) $\mathbb{E}[\mathbb{E}[X|\mathcal{G}]] = \mathbb{E}[X]$.
(iii) *If $X \leq Y$, then* $\mathbb{E}[X|\mathcal{G}] \leq \mathbb{E}[Y|\mathcal{G}]$.
(iv) *For any convex function $\phi : \mathbb{R} \to \mathbb{R}$, there holds the Jensen's inequality* $\phi(\mathbb{E}[X|\mathcal{G}]) \leq \mathbb{E}[\phi(X)|\mathcal{G}]$.
(v) *If $XY \in L^1(\Omega, \mathcal{F}, \mathbb{P})$ and X is $\mathcal{G}$-measurable, then* $\mathbb{E}[XY|\mathcal{G}] = X\mathbb{E}[Y|\mathcal{G}]$.
(vi) *If $\mathcal{H}$ is a σ-algebra with $\mathcal{H} \subset \mathcal{G}$, then* $\mathbb{E}[\mathbb{E}[X|\mathcal{G}]|\mathcal{H}] = \mathbb{E}[X|\mathcal{H}]$.

Let $(\Omega, \mathcal{F}, \mathbb{P})$ be a probability space. A (discrete) stochastic process means a sequence of $\mathcal{F}$-measurable random variables $\{X_k\}$. A family of sub σ-algebras $\{\mathcal{F}_k\}$

of $\mathcal{F}$ is called a filtration if $\mathcal{F}_k \subset \mathcal{F}_m$ whenever $k \le m$. A stochastic process $\{X_k\}$ is said to be adapted to the filtration $\{\mathcal{F}_k\}$ if X_k is $\mathcal{F}_k$-measurable for every k.

Given a stochastic process $\{X_k\}$, we may define $\mathcal{F}_k := \sigma(X_0, \cdots, X_k)$ for each k and call $\{\mathcal{F}_k\}$ the natural filtration with respect to $\{X_k\}$. Clearly, $\{X_k\}$ is adapted to its natural filtration.

Definition 11.6 Let $\{M_k\}$ be a stochastic process in $L^1(\Omega, \mathcal{F}, \mathbb{P})$ adapted to a filtration $\{\mathcal{F}_k\}$.

(i) If $M_k = \mathbb{E}[M_{k+1}|\mathcal{F}_k]$ for all k, then $\{M_k\}_{k\ge 0}$ is called a *martingale* with respect to $\{\mathcal{F}_k\}$.
(ii) If $\mathbb{E}[M_{k+1}|\mathcal{F}_k] \ge M_k$ for all k, then $\{M_k\}$ is called a *submartingale* with respect to $\{\mathcal{F}_k\}$.
(iii) If $\mathbb{E}[(M_{k+1}|\mathcal{F}_k] \le M_k$ for all k, then $\{M_k\}$ is called a *supermartingale* with respect to $\{\mathcal{F}_k\}$.

By using Proposition 11.5, it is easily seen that if $\{M_k\}$ is a submartingale with respect to a filtration $\{\mathcal{F}_k\}$, then

$$\mathbb{E}[M_k|\mathcal{F}_l] \ge M_l, \quad \forall l \le k \tag{11.2}$$

and if $\varphi : \mathbb{R} \to \mathbb{R}$ is a nondecreasing convex function, then $\{\varphi(M_k)\}$ is also a submartingale. If $\{M_k\}$ is a martingale with respect to $\{\mathcal{F}_k\}$, then the sequence $\{X_k := M_k - M_{k-1}\}$ has the property

$$\mathbb{E}[X_k|\mathcal{F}_{k-1}] = \mathbb{E}[M_k|\mathcal{F}_{k-1}] - \mathbb{E}[M_{k-1}|\mathcal{F}_{k-1}] = M_{k-1} - M_{k-1} = 0.$$

In general, given a random process $\{X_k\}$ in $L^1(\Omega, \mathcal{F}, \mathbb{P})$ adapted to a filtration $\{\mathcal{F}_k\}$, it is called a *martingale difference sequence* if $\mathbb{E}[X_k|\mathcal{F}_{k-1}] = 0$ for every k.

We next prove the martingale convergence theorem. To this end, we need first to introduce the number of upcrossings of a stochastic process across a given interval.

Definition 11.7 (Upcrossings) Let $\{M_k\}_{k=0}^{\infty}$ be a stochastic process on a probability space $(\Omega, \mathcal{F}, \mathbb{P})$. Given real numbers $a < b$ and an integer $N \ge 1$, we define $U_N[a, b](\omega)$, for each $\omega \in \Omega$, to be the largest nonnegative integer n such that there exist integers

$$0 \le s_1 < t_1 < s_2 < t_2 < \cdots < s_n < t_n \le N$$

for which

$$M_{s_i}(\omega) < a < b < M_{t_i}(\omega), \quad 1 \le i \le n.$$

We call $U_N[a, b]$ the number of *upcrossings* of $\{M_k\}$ across $[a, b]$ up to N.

Clearly, for each $\omega \in \Omega$, $U_N[a,b](\omega)$ is nondecreasing in N. Let

$$U_\infty[a,b](\omega) := \lim_{N\to\infty} U_N[a,b](\omega).$$

Then $U_\infty[a,b](\omega)$ is either a nonnegative integer or infinity.

We claim the sequence $\{M_k(\omega)\}$ converges if and only if $U_\infty[a,b](\omega)$ is finite for all $a < b$.

To see this, we first assume that $\{M_k(\omega)\}$ is convergent and let $l := \lim_k M_k(\omega)$. If $l > a$, then there exists N such that $M_k(\omega) > a$ for all $k \geq N$ which means that no upcrossing can happen after the time N. Consequently, $U_\infty[a,b](\omega) \leq N$ is finite. If $l \leq a$, then $l < b$. We can infer that $M_k(\omega) < b$ for all $k \geq N$ for some N, again this gives $U_\infty[a,b](\omega) \leq N$.

Conversely, suppose that the sequence $\{M_k(\omega)\}$ does not converge, which means

$$l := \liminf_{k\to\infty} M_k(\omega) < \limsup_{k\to\infty} M_k(\omega) =: u.$$

Then we can pick a, b such that $l < a < b < u$. Thus for any integer N, we can find integers $p > q > N$ such that $M_p(\omega) > b$ and $M_q(\omega) < a$. Consequently, we can find infinitely many integers

$$1 \leq s_1 < t_1 < s_2 < t_2 < \cdots < s_n < t_n < \cdots$$

such that $M_{s_i}(\omega) < a < b < M_{t_i}(\omega)$ for all i. Therefore $U_\infty[a,b](\omega) = \infty$.

According to the above claim, we have

$$\begin{aligned} \{M_k \text{ diverges}\} &= \bigcup_{a<b,a,b\in\mathbb{R}} \{U_\infty[a,b] = \infty\} \\ &= \bigcup_{a<b,a,b\in\mathbb{Q}} \{U_\infty[a,b] = \infty\}. \end{aligned} \tag{11.3}$$

Thus it is important to estimate the probability of the events $\{U_\infty[a,b] = \infty\}$ for $a, b \in \mathbb{Q}$. This can be achieved by Doob's upcrossing lemma. In what follows, for any $x \in \mathbb{R}$, we set $x^+ := \max\{x, 0\}$ and $x^- := -\min\{x, 0\}$.

Lemma 11.8 (Doob's Upcrossing Lemma) *Let $\{M_k\}_{k=0}^\infty$ be a supermartingale, and let $U_N[a,b]$ be the number of upcrossings of $\{M_k\}$ across the interval $[a,b]$ until time N, with $a < b$. Then*

$$(b-a)\mathbb{E}[U_N[a,b]] \leq \mathbb{E}[(M_N - a)^-].$$

Proof Define a sequence $\{C_k\}_k$ of random variables as follows:

$$C_0(\omega) = \begin{cases} 1, & \text{if } M_0(\omega) < a, \\ 0, & \text{Otherwise.} \end{cases}$$

and, for $k \geq 1$,

$$C_k(\omega) = \begin{cases} 1, & \text{if } C_{k-1}(\omega) = 1 \text{ and } M_{k-1}(\omega) \leq b, \\ 1, & \text{if } C_{k-1}(\omega) = 0 \text{ and } M_{k-1}(\omega) < a, \\ 0, & \text{Otherwise.} \end{cases}$$

Clearly, C_k is $\mathcal{F}_{k-1}$-measurable for each $k \geq 1$. Now we define

$$Y_k := \sum_{i=1}^{k} C_i(M_i - M_{i-1}), \quad k \geq 1.$$

and claim that

$$Y_N \geq (b-a)U_N[a,b] - (M_N - a)^-. \tag{11.4}$$

To prove this claim, let $m := U_N[a,b]$, we set $t_0 = 0$, and for $i = 1, 2, \cdots$, define

$$s_i = \inf\{k : k \geq t_{i-1} \text{ and } M_k < a\},$$
$$t_i = \inf\{k : k > s_i \text{ and } M_k > b\}.$$

Then $t_m \leq N$ and $t_{m+1} > N$. Furthermore, by the definition of $\{C_k\}$, we have $C_k = 0$ for $1 \leq k \leq s_1$ and

$$C_k = \begin{cases} 1 & \text{if } s_i + 1 \leq k \leq t_i, \\ 0 & \text{if } t_i + 1 \leq k \leq s_{i+1}, \end{cases} \qquad i = 1, 2, \cdots.$$

Let

$$R_N := \sum_{l=t_m+1}^{N} C_l(M_l - M_{l-1}).$$

Then we have

$$\begin{aligned} Y_N &= \sum_{i=1}^{m} \sum_{l=s_i+1}^{t_i} C_l(M_l - M_{l-1}) + R_N \\ &= \sum_{i=1}^{m} (M_{t_i} - M_{s_i}) + R_N \\ &\geq (b-a)m + R_N \end{aligned} \tag{11.5}$$

We need to consider the term R_N. If $s_{m+1} \geq N$, then $R_N = 0$ as $C_k = 0$ for all $t_m + 1 \leq k \leq N$; if $s_{m+1} < N$, then

$$\begin{aligned} R_N &= \sum_{k=s_{m+1}+1}^{N} C_k(M_k - M_{k-1}) = \sum_{k=s_{m+1}+1}^{N} (M_k - M_{k-1}) \\ &= M_N - M_{s_{m+1}} \geq M_N - a. \end{aligned}$$

Therefore

$$R_N \geq \min\{M_N - a, 0\} = -(M_N - a)^-.$$

Combining this with (11.5), we obtain the claim (11.4).

By the construction of Y_k, the supermartingale property of $\{M_k\}$, and the fact that C_i is $\mathcal{F}_{i-1}$-measurable, it can be seen that

$$\mathbb{E}[Y_1|\mathcal{F}_0] = C_1\mathbb{E}[M_1 - M_0|\mathcal{F}_0] = C_1 \left(\mathbb{E}[M_1|\mathcal{F}_0] - M_0\right) \leq 0$$

and, for $k \geq 2$,

$$\begin{aligned} \mathbb{E}[Y_k|\mathcal{F}_{k-1}] &= \mathbb{E}\left[Y_{k-1} + C_k(M_k - M_{k-1})|\mathcal{F}_{k-1}\right] \\ &= Y_{k-1} + C_k \left(\mathbb{E}[M_k|\mathcal{F}_{k-1}] - M_{k-1}\right) \\ &\leq Y_{k-1}. \end{aligned}$$

By further using the properties of conditional expectation, we can obtain $\mathbb{E}[Y_k|\mathcal{F}_1] \leq Y_1$ and thus

$$\mathbb{E}[Y_N] = \mathbb{E}\left[\mathbb{E}[Y_N|\mathcal{F}_1]\right] \leq \mathbb{E}[Y_1] = \mathbb{E}\left[\mathbb{E}[Y_1|\mathcal{F}_0]\right] \leq 0.$$

This together with (11.4) shows the desired inequality. □

Based on Doob's upcrossing lemma, we now can prove the celebrated result of Doob's martingale convergence theorem.

Theorem 11.9 (Supermartingale Convergence Theorem) *If $\{M_k\}_{k\geq 0}$ is a supermartingale that is bounded in L^1, i.e., $\sup_k \mathbb{E}[|M_k|] < \infty$, then M_k converges almost surely to a random variable M_∞ with $\mathbb{E}[|M_\infty|] < \infty$.*

Proof Let $a, b \in \mathbb{Q}$ be any two numbers satisfying $a < b$. By Doob's upcrossing lemma, i.e., Lemma 11.8, we have

$$(b - a)\mathbb{E}[U_N[a, b]] \leq \mathbb{E}[(M_N - a)^-] \leq \mathbb{E}[|M_N|] + |a| \leq \sup_k \mathbb{E}[|M_k|] + |a|.$$

Since $U_N[a, b]$ is nondecreasing in N and has point-wise limit $U_\infty[a, b]$, by the monotone convergence theorem, we have

$$\mathbb{E}\left[U_\infty[a, b]\right] = \lim_{N\to\infty} \mathbb{E}[U_N[a, b]] \le \frac{1}{b-a}\left(\sup_k \mathbb{E}[|M_k|] + |a|\right) < \infty.$$

Hence $\mathbb{P}(U_\infty[a, b] = \infty) = 0$. By using (11.3) and the countability of $\mathbb{Q}$, we can conclude that $\mathbb{P}(\{M_k\} \text{ diverges}) = 0$, which means there is a random variable M_∞ such that $M_k \to M_\infty$ almost surely. By Fatou's Lemma, we have

$$\mathbb{E}[|M_\infty|] = \mathbb{E}\left[\lim_{k\to\infty} |M_k|\right] \le \liminf_{k\to\infty} \mathbb{E}[|M_k|] \le \sup_k \mathbb{E}[|M_k|] < \infty.$$

The proof is thus complete. □

To investigate iterative method which demonstrates random characteristics, we need the following important consequence of martingale convergence theorem, due to [89]; see also [182, Lemma 9].

Lemma 11.10 *Let $\{X_k\}$ and $\{U_k\}$ be two sequences of nonnegative random variables with $\mathbb{E}[X_0] < \infty$ and*

$$\mathbb{E}[X_{k+1}|\mathcal{F}_k] \le (1+\delta_k)X_k - U_k + \epsilon_k, \tag{11.6}$$

where $\{\mathcal{F}_k := \sigma(X_0, \cdots, X_k)\}$ is the natural filtration and $\{\delta_k\}$ and $\{\epsilon_k\}$ are nonnegative deterministic sequences with

$$\sum_{k=0}^{\infty} \delta_k < \infty \qquad \text{and} \qquad \sum_{k=0}^{\infty} \epsilon_k < \infty. \tag{11.7}$$

Then $\sum_{k=0}^{\infty} U_k < \infty$ almost surely and X_k converges almost surely to a random variable $X_\infty \ge 0$ with $\mathbb{E}[X_\infty] < \infty$.

Proof Since $\sum_{k=1}^{\infty} \delta_k < \infty$, by using the fact that $\log(1+t) \le t$ for $t \ge 0$, we have

$$\log \prod_{k=0}^{\infty}(1+\delta_k) = \sum_{k=0}^{\infty} \log(1+\delta_k) \le \sum_{k=0}^{\infty} \delta_k < \infty. \tag{11.8}$$

This implies that $\prod_{k=0}^{\infty}(1+\delta_k) < \infty$. Thus, it makes sense to define

$$\tilde{X}_k := \prod_{i=k}^{\infty}(1+\delta_i)X_k + \sum_{i=k}^{\infty} \epsilon_i \prod_{j=i+1}^{\infty}(1+\delta_j), \quad \tilde{U}_k := \prod_{i=k+1}^{\infty}(1+\delta_i)U_k \tag{11.9}$$

for every $k \geq 0$. By the given conditions, we have $\mathbb{E}(\tilde{X}_0) < \infty$ and

$$\begin{aligned}
\mathbb{E}[\tilde{X}_{k+1}|\mathcal{F}_k] &= \prod_{i=k+1}^{\infty}(1+\delta_i)\mathbb{E}[X_{k+1}|\mathcal{F}_k] + \sum_{i=k+1}^{\infty}\epsilon_i \prod_{j=i+1}^{\infty}(1+\delta_j) \\
&\leq \prod_{i=k+1}^{\infty}(1+\delta_i)[(1+\delta_k)X_k - U_k + \epsilon_k] + \sum_{i=k+1}^{\infty}\epsilon_i \prod_{j=i+1}^{\infty}(1+\delta_j) \\
&= \prod_{i=k}^{\infty}(1+\delta_i)X_k - \tilde{U}_k + \epsilon_k \prod_{i=k+1}^{\infty}(1+\delta_i) + \sum_{i=k+1}^{\infty}\epsilon_i \prod_{j=i+1}^{\infty}(1+\delta_j) \\
&= \prod_{i=k}^{\infty}(1+\delta_i)X_k + \sum_{i=k}^{\infty}\epsilon_i \prod_{j=i+1}^{\infty}(1+\delta_j) - \tilde{U}_k \\
&= \tilde{X}_k - \tilde{U}_k.
\end{aligned} \tag{11.10}$$

Taking the full expectation gives

$$\mathbb{E}[\tilde{U}_k] \leq \mathbb{E}[\tilde{X}_k] - \mathbb{E}[\tilde{X}_{k+1}], \quad \forall k \geq 0$$

which implies

$$\sum_{k=0}^{\infty}\mathbb{E}[\tilde{U}_k] \leq \mathbb{E}[\tilde{X}_0] < \infty.$$

Since $0 \leq U_k \leq \tilde{U}_k$ for all k, we have $\sum_{k=0}^{\infty}\mathbb{E}[U_k] < \infty$ and thus $\sum_{k=0}^{\infty}U_k < \infty$ almost surely.

From (11.10) and $\tilde{U}_k \geq 0$, we also have $\mathbb{E}[\tilde{X}_{k+1}|\mathcal{F}_k] \leq \tilde{X}_k$ for all k. Therefore, $\{\tilde{X}_k\}$ is a supermartingale with respect to $\{\mathcal{F}_k\}$. By Theorem 11.9, there exists a random variable $\tilde{X}_\infty \geq 0$ with $\mathbb{E}[\tilde{X}_\infty] < \infty$ such that $\tilde{X}_k \to \tilde{X}_\infty$ almost surely as $k \to \infty$. Note that

$$0 \leq \log\prod_{i=k}^{\infty}(1+\delta_i) = \sum_{i=k}^{\infty}\log(1+\delta_i) \leq \sum_{i=k}^{\infty}\delta_i \to 0 \qquad \text{as } k \to \infty$$

which implies $\prod_{i=k}^{\infty}(1+\delta_i) \to 1$. By (11.9), we have

$$X_k = \left[\prod_{i=k}^{\infty}(1+\delta_i)\right]^{-1}\left[\tilde{X}_k - \sum_{i=k}^{\infty}\epsilon_i \prod_{j=i+1}^{\infty}(1+\delta_j)\right].$$

Consequently, $X_k \to X_\infty := \tilde{X}_\infty$ as $k \to \infty$ almost surely. □

Remark 11.11 The result in Lemma 11.10 was extended in [189] to allow $\{\delta_k\}$ and $\{\epsilon_k\}$ to be stochastic processes that are summable almost surely. This extended result is known as the Robbins-Siegmund theorem.

11.2 Stochastic (Sub)gradient Methods

We first give two examples to demonstrate how the minimization problems of the form (11.1) arise.

Example 11.12 Consider solving the linear system $Ax = b$, where A is an $N \times d$ matrix and $b \in \mathbb{R}^N$. It can be equivalently written as

$$\min_{x\in\mathbb{R}^d}\left\{f(x) := \frac{1}{2N}\|Ax - b\|^2\right\}.$$

Let a_i denote the i-th column of A^T and b_i the i-th component of b. Then f has the form (11.1) with $f_i(x) = \frac{1}{2}(\langle a_i, x\rangle - b_i)^2$.

Example 11.13 Consider the classification problem in machine learning. Given the training data (a_j, y_j), $j = 1, \cdots, N$, with $a_j \in \mathbb{R}^n$ and $y_j \in \{-1, +1\}$, we want to find a function

$$p(a; x) = \frac{1}{1 + \exp(\langle a, x\rangle)}$$

parametrized by a vector $x \in \mathbb{R}^n$ which is chosen so that

$$p(a_j, x) \approx 1 \text{ when } y_j = +1 \quad \text{and} \quad p(a_j, x) \approx 0 \text{ when } y_j = -1.$$

Once this function is determined, we may use it to classify new input data a by calculating $p(a, x)$. The optimal value of x can be found by considering the minimization problem

$$\min_{x\in\mathbb{R}^n}\left\{L(x) := -\frac{1}{N}\left(\sum_{j:y_j=-1}\log(1 - p(a_j, x)) + \sum_{j:y_j=+1}\log p(a_j, x)\right)\right\}$$

which is called the *logistic regression*. The objective function L can be written as

$$L(x) = -\frac{1}{N}\left\{\sum_{j:y_j=-1}\langle a_j, x\rangle - \sum_{j=1}^{N}\log\left(1 + \exp(\langle a_j, x\rangle)\right)\right\}$$

$$= \frac{1}{N} \sum_{j=1}^{N} \left[u_j \langle a_j, x \rangle + \log \left(1 + \exp(\langle a_j, x \rangle) \right) \right],$$

where $u_j = 0$ when $y_j = +1$ and $u_j = -1$ when $y_j = -1$. This problem clearly takes the form (11.1).

If we apply the gradient method to solve (11.1), at each iteration step, we need to compute the full gradient

$$\nabla f = \frac{1}{N} \sum_{i=1}^{N} \nabla f_i$$

which requires the computation of ∇f_i for all $i = 1, \cdots, N$. When N is very large, the computation can become prohibitively expensive. Therefore, it is essential to develop stochastic algorithms that efficiently handle large-scale optimization problems by using only randomly selected partial information.

The following result is useful for analyzing stochastic algorithms we will introduce in the remaining part of this chapter.

Lemma 11.14 *Let $f, h : \mathbb{R}^d \to (-\infty, \infty]$ be proper, lower semicontinuous, convex functions. Let $\mu_f \geq 0$ and $\mu_g \geq 0$ denote the modulus of convexity of f and g, respectively. Let $\gamma > 0$. Then for any $u, v, w \in \mathbb{R}^d$, $\xi \in \partial f(v)$ and $z = \text{prox}_{\gamma h}(w)$ there holds*

$$\begin{aligned} f(z) + h(z) \leq f(u) + h(u) + \left(\frac{1}{2\gamma} - \mu_f \right) \|v - u\|^2 - \left(\frac{1}{2\gamma} + \mu_h \right) \|z - u\|^2 \\ - \frac{1}{2\gamma} \|z - v\|^2 + \frac{1}{\gamma} \langle w - v + \gamma\xi, z - u \rangle + D_f^{\xi}(z, v), \end{aligned} \tag{11.11}$$

where $D_f^{\xi}(z, v) := f(z) - f(v) - \langle \xi, z - v \rangle$ is the Bregman distance induced by f at v in the direction ξ. If f is continuous differentiable and ∇f is L-Lipschitz continuous, there holds

$$\begin{aligned} f(z) + h(z) \leq f(u) + h(u) + \left(\frac{1}{2\gamma} - \mu_f \right) \|v - u\|^2 - \left(\frac{1}{2\gamma} + \mu_h \right) \|z - u\|^2 \\ - \left(\frac{1}{2\gamma} - \frac{L}{2} \right) \|z - v\|^2 + \frac{1}{\gamma} \langle w - v + \gamma\xi, z - u \rangle. \end{aligned} \tag{11.12}$$

Proof By using the convexity of f and $\xi \in \partial f(v)$, we have

$$\begin{aligned} f(u) &\geq f(v) + \langle \xi, u - v \rangle + \mu_f \|v - u\|^2 \\ &= f(z) + \langle \xi, u - z \rangle + \mu_f \|v - u\|^2 - D_f^{\xi}(z, v). \end{aligned}$$

Since $z = \text{prox}_{\gamma h}(w)$, we have $(w - z)/\gamma \in \partial h(z)$. Thus, by the convexity of h, we can obtain

$$h(u) \geq h(z) + \frac{1}{\gamma} \langle w - z, u - z \rangle + \mu_h \|z - u\|^2.$$

Adding the above two equations gives

$$\begin{aligned} f(u) + h(u) \geq f(z) + h(z) - \frac{1}{\gamma}\langle v - z, z - u\rangle - D_f^{\xi}(z, v) + \mu_f \|v - u\|^2 \\ + \mu_h \|z - u\|^2 - \frac{1}{\gamma}\langle w - v + \gamma\xi, z - u\rangle. \end{aligned}$$

Noting that

$$\frac{1}{\gamma}\langle v - z, z - u\rangle = \frac{1}{2\gamma}\left(\|v - u\|^2 - \|z - u\|^2 - \|z - v\|^2\right),$$

we then obtain

$$\begin{aligned} f(u) + h(u) \geq f(z) + h(z) - \left(\frac{1}{2\gamma} - \mu_f\right)\|v - u\|^2 + \left(\frac{1}{2\gamma} + \mu_h\right)\|z - u\|^2 \\ + \frac{1}{2\gamma}\|z - v\|^2 - \frac{1}{\gamma}\langle w - v + \gamma\xi, z - u\rangle - D_f^{\xi}(z, v) \end{aligned}$$

which shows (11.11). When f is continuous differentiable and ∇f is L-Lipschitz continuous, we have from the descent lemma that $D_f^{\xi}(z, v) \leq \frac{L}{2}\|z - v\|^2$. Combining this with (11.11) shows (11.12). □

11.2.1 Proximal Stochastic Subgradient Method

Consider the minimization problem

$$\min_{x \in \mathbb{R}^n} \left\{ \Phi(x) := \sum_{i=1}^{N} \Phi_i(x) \right\}, \tag{11.13}$$

where, for each $i \in \{1, \cdots, N\}$, $\Phi_i(x) = f_i(x) + h_i(x)$ with $f_i, h_i : \mathbb{R}^n \to (-\infty, \infty]$ being proper, lower semicontinuous, convex functions. We make the following assumption.

Assumption 11.15

(i) *For each* $i \in \{1, \cdots, N\}$ *the functions* $f_i, h_i : \mathbb{R}^n \to (-\infty, \infty]$ *are proper, lower semicontinuous, and convex. Moreover, both* f_i *and* h_i *are Lipschitz continuous on* $\mathrm{dom}(f_i)$.
(ii) *The problem (11.13) has an optimal solution, i.e.,* $S := \arg\min_{x \in \mathbb{R}^n} \Phi(x) \neq \emptyset$. *We set* $\Phi_* := \min_{x \in \mathbb{R}^n} \Phi(x)$.

In Sect. 3.3, we introduced the incremental subgradient method for solving problems of the form (11.13). In that method, the component functions are incorporated into the algorithm sequentially in a cyclic manner to update the iterative solutions. It is known that the order in which these component functions are used significantly affects the convergence speed. Unfortunately, determining an optimal order manually is challenging.

In this section, we present a stochastic variant of the incremental subgradient method, which uses a randomly sampled component function at each iteration step. To solve (11.13), we consider the following proximal stochastic subgradient method.

Algorithm 11.16 (Proximal Stochastic Subgradient Method) *Pick an initial guess* $x^0 \in \cap_{i=1}^N \mathrm{dom}(h_i)$ *and step sizes* $\{\gamma_k\}$. *For* $k = 0, 1, \cdots$ *do the following*

(i) *Pick* $i_k \in \{1, \cdots, N\}$ *randomly via a uniform distribution;*
(ii) *Pick* $\xi^k \in \partial f_{i_k}(x^k)$ *and calculate*

$$w^k := x^k - \gamma_k \xi^k; \qquad x^{k+1} := \mathrm{prox}_{\gamma_k h_{i_k}}(w^k).$$

In this algorithm, at each iteration, the iterate x^{k+1} inherits randomness from the choice of i_k and thus depends on all previous indices i_l for $0 \leq l \leq k$. We use $\mathbb{E}_{i_k}$ to denote the expectation taken with respect to i_k, conditioned on all prior indices i_l for $0 \leq l < k$. The following convergence result for Algorithm 11.16 is established in [27].

Theorem 11.17 *Let Assumption 11.15 hold and consider Algorithm 11.16, where the step sizes* $\{\gamma_k\}$ *are chosen such that*

$$\gamma_k > 0, \quad \sum_{k=0}^{\infty} \gamma_k = \infty \quad \text{and} \quad \sum_{k=1}^{\infty} \gamma_k^2 < \infty.$$

Then, almost surely, there is an optimal solution x^* *of (11.13) such that* $\{x^k\}$ *converges to* x^* *as* $k \to \infty$.

Proof Let $\bar{x}$ be any optimal solution of (11.13). By using (11.11) in Lemma 11.14 with $f = f_{i_k}$, $h = h_{i_k}$, $u = \bar{x}$, $v = x^k$, $w = w^k$, $z = x^{k+1}$, $\xi = \xi^k$, and $\gamma = \gamma_k$, we can obtain

$$\Phi_{i_k}(x^{k+1}) \le \Phi_{i_k}(\bar{x}) + \frac{1}{2\gamma_k}\left(\|x^k - \bar{x}\|^2 - \|x^{k+1} - \bar{x}\|^2 - \|x^{k+1} - x^k\|^2\right) + D_{f_{i_k}}^{\xi^k}(x^{k+1}, x^k),$$

where we take $\mu_f = \mu_h = 0$. According to Assumption 11.15, there is a number L such that $\|\xi^k\| \le L$,

$$\begin{aligned} D_{f_{i_k}}^{\xi^k}(x^{k+1}, x^k) &\le |f_{i_k}(x^{k+1}) - f_{i_k}(x^k)| + |\langle \xi^k, x^{k+1} - x^k\rangle| \\ &\le (L + \|\xi^k\|)\|x^{k+1} - x^k\| \le 2L\|x^{k+1} - x^k\| \end{aligned}$$

and

$$\begin{aligned} \Phi_{i_k}(x^k) - \Phi_{i_k}(x^{k+1}) &\le |f_{i_k}(x^k) - f_{i_k}(x^{k+1})| + |h_{i_k}(x^k) - h_{i_k}(x^{k+1})| \\ &\le 2L\|x^{k+1} - x^k\|. \end{aligned}$$

Therefore

$$\Phi_{i_k}(x^k) \le \Phi_{i_k}(\bar{x}) + \frac{1}{2\gamma_k}\left(\|x^k - \bar{x}\|^2 - \|x^{k+1} - \bar{x}\|^2 - \|x^{k+1} - x^k\|^2\right) + 4L\|x^{k+1} - x^k\|.$$

Since $4L\|x^{k+1} - x^k\| \le \frac{1}{2\gamma_k}\|x^{k+1} - x^k\|^2 + 8L^2\gamma_k$, we thus obtain

$$\Phi_{i_k}(x^k) \le \Phi_{i_k}(\bar{x}) + \frac{1}{2\gamma_k}\left(\|x^k - \bar{x}\|^2 - \|x^{k+1} - \bar{x}\|^2\right) + 8L^2\gamma_k,$$

or equivalently

$$\|x^{k+1} - \bar{x}\|^2 \le \|x^k - \bar{x}\|^2 - 2\gamma_k(\Phi_{i_k}(x^k) - \Phi_{i_k}(\bar{x})) + 16L^2\gamma_k^2.$$

Taking the expectation with respect to i_k gives

$$\begin{aligned} \mathbb{E}_{i_k}[\|x^{k+1} - \bar{x}\|^2] &\le \|x^k - \bar{x}\|^2 - \frac{2\gamma_k}{N}\sum_{i=1}^{N}(\Phi_i(x^k) - \Phi_i(\bar{x})) + 16L^2\gamma_k^2 \\ &= \|x^k - \bar{x}\|^2 - \frac{2\gamma_k}{N}(\Phi(x^k) - \Phi(\bar{x})) + 16L^2\gamma_k^2. \end{aligned}$$

Since $\Phi(x^k)-\Phi(\bar{x}) \geq 0$ and $\sum_{k=1}^{\infty} \gamma_k^2 < \infty$, we may use Lemma 11.10 to conclude there is an event $\Omega_{\bar{x}}$ of probability one such that, along any sample path in $\Omega_{\bar{x}}$,

$$\sum_{k=0}^{\infty} \gamma_k(\Phi(x^k) - \Phi_*) < \infty \tag{11.14}$$

and the sequence $\{\|x^k - \bar{x}\|\}$ converges.

Now we show the almost sure convergence of $\{x^k\}$. By the separability of the Euclidean space, we can find a countable subset $C \subset S$ such that C is dense in S. Let

$$\widetilde{\Omega} := \bigcap_{x \in C} \Omega_x.$$

Since C is countable and $\mathbb{P}(\Omega_x) = 1$ for each $x \in C$, we have $\mathbb{P}(\widetilde{\Omega}) = 1$. Along each sample path in $\widetilde{\Omega}$, the sequence $\{\|x^k - x\|\}$ is convergent for any $x \in C$. In particular, $\{x^k\}$ is bounded. Note that (11.14) holds and thus by using $\sum_{k=0}^{\infty} \gamma_k = \infty$, we can conclude $\liminf_{k\to\infty} \Phi(x^k) = \Phi_*$. Consequently, there is a subsequence $\{x^{k_l}\}$ of $\{x^k\}$ such that $\Phi(x^{k_l}) \to \Phi_*$ and $x^{k_l} \to x^*$ for some $x^* \in \mathbb{R}^d$. By the lower semi-continuity of Φ, we can see that $\Phi(x^*) = \Phi_*$, i.e., x^* is an optimal solution of (11.13). To show $\|x^k - x^*\| \to 0$, for any $\varepsilon > 0$, we can find $x_\varepsilon \in C$ such that $\|x_\varepsilon - x^*\| < \varepsilon$. Since $\{\|x^k - x_\varepsilon\|\}$ converges and $\|x^{k_l} - x_\varepsilon\| \to \|x^* - x_\varepsilon\|$, we have $\|x^k - x_\varepsilon\| \to \|x^* - x_\varepsilon\|$. Therefore,

$$\limsup_{k\to\infty} \|x^k - x^*\| \leq \lim_{k\to\infty} \|x^k - x_\varepsilon\| + \|x_\varepsilon - x^*\| = 2\|x_\varepsilon - x^*\| < 2\varepsilon.$$

Since $\varepsilon > 0$ is arbitrary, we must have $\|x^k - x^*\| \to 0$ as $k \to \infty$. □

11.2.2 Proximal Stochastic Gradient Method

In Sect. 11.2.1, we examined Algorithm 11.16 for solving the minimization problem (11.13). Although no differentiability is assumed for any of the component functions, the requirement of Lipschitz continuity may be too restrictive for many practical applications. For instance, consider Example 11.12, in which the component functions are given by $f_i(x) = (\langle a_i, x\rangle - b_i)^2$, which unfortunately are not Lipschitz continuous on $\mathbb{R}^d$. However, these component functions are continuously differentiable with Lipschitz continuous gradients. This observation motivates us to consider problems of the form (11.13) where each f_i is continuously differentiable. By leveraging the differentiability structure, we can establish a more robust convergence theory that applies to a broader range of applications not covered by the theory in Sect. 11.2.1.

Specifically, in this section, we consider the minimization problem

$$\min \left\{ F(x) := \frac{1}{N} \sum_{i=1}^{N} f_i(x) + h(x) \right\}, \tag{11.15}$$

where each f_i is convex, continuously differentiable, and $h : \mathbb{R}^d \to (-\infty, \infty]$ is proper, lower semicontinuous, and convex. Under this formulation, Algorithm 11.13 becomes the proximal stochastic gradient method as stated below.

Algorithm 11.18 (Proximal SGD) *Pick an initial guess $x^0 \in \mathrm{dom}(h)$ and step sizes $\{\gamma_k\}$. For $k = 0, 1, \cdots$, do the following*

(i) *Pick $i_k \in \{1, \cdots, N\}$ randomly via a uniform distribution;*
(ii) *Calculate $x^{k+1} := \mathrm{prox}_{\gamma_k h}(x^k - \gamma_k \nabla f_{i_k}(x^k))$.*

When $h = 0$, Algorithm 11.18 reduces to the classic stochastic gradient descent (SGD) method, which can be written as

$$x^{k+1} = x^k - \gamma_k \nabla f_{i_k}(x^k),$$

where i_k is a randomly sampled data point. This approach was originally proposed in [188] by Robbins and Monro and has since become foundational in machine learning, as it provides an efficient way to approximate the true gradient while only requiring a single data sample per iteration.

Introducing a nontrivial h in Algorithm 11.18 allows for the incorporation of a regularization term, which can further enhance the method's flexibility. The regularization term can be particularly useful in handling constraints or encouraging specific properties, like sparsity, in the solution. As a result, Algorithm 11.18 becomes more adaptable to complex objectives common in modern data science, including high-dimensional statistical learning or cases where structured regularization is key for interpretability and model performance.

To rigorously analyze the convergence properties of Algorithm 11.18, we will establish results under the following standard conditions. These conditions involve assumptions on the smoothness of f, and the nature of the regularization term h, each of which plays a role in determining the convergence rate and stability of the algorithm.

Assumption 11.19

(i) *Every f_i is a continuously differentiable convex function, and there is a constant L such that each ∇f_i is L-Lipschitz continuous, i.e.,*

$$\|\nabla f_i(x) - \nabla f_i(y)\| \leq L\|x - y\|, \quad \forall x, y \in \mathbb{R}^d.$$

(ii) *$h : \mathbb{R}^d \to (-\infty, \infty]$ is proper, lower semicontinuous, and convex.*
(iii) *The problem (11.15) has an optimal solution. We set $F_* := \min_{x \in \mathbb{R}^d} F(x)$.*

Based on Assumption 11.19, we will perform the convergence analysis of the proximal SGD method. In the following, we will use

$$f(x) := \frac{1}{N}\sum_{i=1}^{N} f_i(x).$$

According to (i) in Assumption 11.19, it is easy to see that f is convex, continuous differentiable and its gradient ∇f is L-Lipschitz continuous.

Let $\bar{x}$ be any solution of (11.15). For the proximal SGD with $0 < \gamma_k \leq 1/L$, we can use (11.12) in Lemma 11.14 with $u = \bar{x}$, $v = x^k$, $w = x^k - \gamma_k \nabla f_{i_k}(x^k)$, $\xi = \nabla f(x^k)$, $\gamma = \gamma_k$, and $z = x^{k+1}$ to obtain

$$F(x^{k+1}) \leq F_* + \left(\frac{1}{2\gamma_k} - \mu_f\right)\|x^k - \bar{x}\|^2 - \left(\frac{1}{2\gamma} + \mu_h\right)\|x^{k+1} - \bar{x}\|^2 \\ - \left\langle \nabla f_{i_k}(x^k) - \nabla f(x^k), x^{k+1} - x \right\rangle.$$

Setting $\mu_f = \mu_h = 0$ and multiplying the both sides by $2\gamma_k$ gives

$$\begin{aligned} &2\gamma_k\left(F(x^{k+1}) - F_*\right) \\ &\leq \|x^k - \bar{x}\|^2 - \|x^{k+1} - \bar{x}\|^2 - 2\gamma_k\left\langle \nabla f_{i_k}(x^k) - \nabla f(x^k), x^{k+1} - \bar{x} \right\rangle \\ &= \|x^k - \bar{x}\|^2 - \|x^{k+1} - \bar{x}\|^2 - 2\gamma_k\left\langle \nabla f_{i_k}(x^k) - \nabla f(x^k), \hat{x}^{k+1} - \bar{x} \right\rangle \\ &\quad - 2\gamma_k\left\langle \nabla f_{i_k}(x^k) - \nabla f(x^k), x^{k+1} - \hat{x}^{k+1} \right\rangle, \end{aligned}$$

where $\hat{x}^{k+1} := \mathrm{prox}_{\gamma_k h}(x^k - \gamma_k \nabla f(x^k))$ which is independent of i_k. By the non-expansiveness of $\mathrm{prox}_{\gamma_k h}$, we have

$$\begin{aligned} \|x^{k+1} - \hat{x}^{k+1}\| &= \|\mathrm{prox}_{\gamma_k h}(x^k - \gamma_k \nabla f_{i_k}(x^k)) - \mathrm{prox}_{\gamma_k h}(x^k - \gamma_k \nabla f(x^k))\| \\ &\leq \gamma_k \|\nabla f_{i_k}(x^k) - \nabla f(x^k)\|. \end{aligned}$$

Therefore

$$2\gamma_k\left(F(x^{k+1}) - F_*\right) \leq \|x^k - \bar{x}\|^2 - \|x^{k+1} - \bar{x}\|^2 + 2\gamma_k^2\|\nabla f_{i_k}(x^k) - \nabla f(x^k)\|^2 \\ - 2\gamma_k\left\langle \nabla f_{i_k}(x^k) - \nabla f(x^k), \hat{x}^{k+1} - \bar{x} \right\rangle.$$

Taking the expectation with respect to i_k gives

$$
\begin{aligned}
2\gamma_k \mathbb{E}_{i_k}\left[F(x^{k+1}) - F_*\right] \leq \|x^k - \bar{x}\|^2 - \mathbb{E}_{i_k}[\|x^{k+1} - \bar{x}\|^2] \\
+ 2\gamma_k^2 \mathbb{E}_{i_k}\left[\|\nabla f_{i_k}(x^k) - \nabla f(x^k)\|^2\right],
\end{aligned}
\tag{11.16}
$$

where we used $\mathbb{E}_{i_k}[\nabla f_{i_k}(x^k)] = \nabla f(x^k)$. By using the variance decomposition formula, we can obtain

$$
\begin{aligned}
&\mathbb{E}_{i_k}\left[\|\nabla f_{i_k}(x^k) - \nabla f(x^k)\|^2\right] \\
&= \mathbb{E}_{i_k}\left[\|\nabla f_{i_k}(x^k) - \nabla f(\bar{x}) - \left(\nabla f(x^k) - \nabla f(\bar{x})\right)\|^2\right] \\
&\leq \mathbb{E}_{i_k}\left[\|\nabla f_{i_k}(x^k) - \nabla f(\bar{x})\|^2\right].
\end{aligned}
$$

By virtue of the inequality $\|a + b\|^2 \leq 2(\|a\|^2 + \|b\|^2)$, we further obtain

$$
\begin{aligned}
&\mathbb{E}_{i_k}\left[\|\nabla f_{i_k}(x^k) - \nabla f(x^k)\|^2\right] \\
&\leq 2\mathbb{E}_{i_k}\left[\|\nabla f_{i_k}(x^k) - \nabla f_{i_k}(\bar{x})\|^2\right] + 2\mathbb{E}_{i_k}\left[\|\nabla f_{i_k}(\bar{x}) - \nabla f(\bar{x})\|^2\right] \\
&= \frac{2}{N}\sum_{i=1}^{N} \|\nabla f_i(x^k) - \nabla f_i(\bar{x})\|^2 + 2\sigma_F,
\end{aligned}
$$

where

$$
\sigma_F := \frac{1}{N}\sum_{i=1}^{N} \|\nabla f_i(\bar{x}) - \nabla f(\bar{x})\|^2 \tag{11.17}
$$

which represents the gradient noise at $\bar{x}$. By using the co-coercivity of convex functions f_i, see Proposition 2.34; we have

$$
\begin{aligned}
&\mathbb{E}_{i_k}\left[\|\nabla f_{i_k}(x^k) - \nabla f(x^k)\|^2\right] \\
&\leq \frac{4L}{N}\sum_{i=1}^{N} \left(f_i(x^k) - f_i(\bar{x}) - \langle \nabla f_i(\bar{x}), x^k - \bar{x}\rangle\right) + 2\sigma_F \\
&= 4L\left(f(x^k) - f(\bar{x}) - \langle \nabla f(\bar{x}), x^k - \bar{x}\rangle\right) + 2\sigma_F.
\end{aligned}
$$

Combining this with Eq. (11.16) gives the following result.

Lemma 11.20 *Let Assumption 11.19 hold. Consider Algorithm 11.18 with* $0 < \gamma_k \le 1/L$. *Let* $\bar{x}$ *be any solution of (11.15). There holds*

$$2\gamma_k \mathbb{E}_{i_k}[F(x^{k+1}) - F_*] \le \|x^k - \bar{x}\|^2 - \mathbb{E}_{i_k}[\|x^{k+1} - \bar{x}\|^2] + 4\sigma_F \gamma_k^2 + 8L\gamma_k^2 \Big(f(x^k) - f(\bar{x}) - \langle \nabla f(\bar{x}), x^k - \bar{x}\rangle\Big)$$

for all integers $k \ge 0$.

Based on the above lemma, we will prove the following convergence result on the proximal SGD method.

Theorem 11.21 *Let Assumption 11.19 hold. Consider Algorithm 11.18 with* $\{\gamma_k\}$ *chosen such that*

$$\gamma_k > 0, \quad \sum_{k=0}^{\infty} \gamma_k = \infty \quad \text{and} \quad \sum_{k=0}^{\infty} \gamma_k^2 < \infty.$$

If $0 < \gamma_0 < 1/(4L)$ *and* $\{\gamma_k\}$ *is nonincreasing, then almost surely* $\{x^k\}$ *converges to a solution of (11.15) and* $F(x^k) \to F_*$ *as* $k \to \infty$. *Moreover, for* $\bar{x}^k$ *defined by*

$$\bar{x}^k := \frac{1}{\sum_{l=0}^{k} \gamma_l} \sum_{l=0}^{k} \gamma_l x^l. \tag{11.18}$$

there holds

$$\mathbb{E}[F(\bar{x}^k) - F_*] \le \frac{\|x^0 - \bar{x}\|^2 + 2\gamma_0(F(x^0) - F_*) + 4\sigma_F \sum_{l=0}^{k-1} \gamma_l^2}{2(1 - 4L\gamma_0)\sum_{l=0}^{k} \gamma_l}$$

for all integers $k \ge 0$, *where* $\bar{x}$ *denotes any solution of (11.15).*

Proof In order to use Lemma 11.20, we need to estimate

$$f(x^k) - f(\bar{x}) - \langle \nabla f(\bar{x}), x^k - \bar{x}\rangle.$$

Since $\bar{x}$ is a solution of (11.15), we have $-\nabla f(\bar{x}) \in \partial h(\bar{x})$ and hence

$$-\langle \nabla f(\bar{x}), x^k - \bar{x}\rangle \le h(x^k) - h(\bar{x})$$

by the convexity of h. Consequently

$$\begin{aligned} f(x^k) - f(\bar{x}) - \langle \nabla f(\bar{x}), x^k - \bar{x}\rangle &\le f(x^k) - f(\bar{x}) + h(x^k) - h(\bar{x}) \\ &= F(x^k) - F_*. \end{aligned}$$

Combining this with the equation in Lemma 11.20 gives

$$2\gamma_k \mathbb{E}_{i_k}[F(x^{k+1}) - F_*] \leq \|x^k - \bar{x}\|^2 - \mathbb{E}_{i_k}[\|x^{k+1} - \bar{x}\|^2] + 4\sigma_F \gamma_k^2 + 8L\gamma_k^2(F(x^k) - F_*).$$

By using $\gamma_{k+1} \leq \gamma_k \leq \gamma_0$, we then have

$$2\gamma_{k+1} \mathbb{E}_{i_k}[F(x^{k+1}) - F_*] \leq \|x^k - \bar{x}\|^2 - \mathbb{E}_{i_k}[\|x^{k+1} - \bar{x}\|^2] + 4\sigma_F \gamma_k^2 + 8L\gamma_0\gamma_k(F(x^k) - F_*). \tag{11.19}$$

We first show the almost sure convergence of $\{x^k\}$. Let

$$X_k := \|x^k - \bar{x}\|^2 + 2\gamma_k(F(x^k) - F_*), \quad k \geq 0.$$

It then follows from (11.19) that

$$\mathbb{E}_{i_k}[X_{k+1}] \leq X_k - (2 - 8L\gamma_0)\gamma_k(F(x^k) - F_*) + 4\sigma_F \gamma_k^2.$$

Thus, we may use Lemma 11.10 to conclude that

$$\sum_{k=0}^{\infty} \gamma_k(F(x^k) - F_*) < \infty \quad \text{and} \quad \lim_{k\to\infty} \left(\|x^k - \bar{x}\|^2 + 2\gamma_k(F(x^k) - F_*)\right) \text{ exists}$$

almost surely. The first equation implies $\gamma_k(F(x^k) - F_*) \to 0$ almost surely, and hence we can conclude from the second equation that $\lim_{k\to\infty} \|x^k - \bar{x}\|$ exists almost surely. Therefore, for any solution $\bar{x}$ of (11.15), the event

$$\left\{\sum_{k=0}^{\infty} \gamma_k(F(x^k) - F_*) < \infty \text{ and } \lim_{k\to\infty} \|x^k - \bar{x}\| \text{ exists}\right\}$$

has probability one. Thus, we may follow the argument in the proof of Theorem 11.17 to conclude that $\{x^k\}$ converges to a solution of (11.15) almost surely.

To show the second part, we take the full expectation on Eq. (11.19) to obtain

$$2\gamma_{k+1} \mathbb{E}[F(x^{k+1}) - F_*] \leq \mathbb{E}[\|x^k - \bar{x}\|^2] - \mathbb{E}[\|x^{k+1} - \bar{x}\|^2] + 4\sigma_F \gamma_k^2 + 8L\gamma_0\gamma_k \mathbb{E}[F(x^k) - F_*].$$

Summing this inequality gives

$$
\begin{aligned}
2\sum_{l=1}^{k}\gamma_l\mathbb{E}[F(x^l)-F_*] \\
&\le \sum_{l=0}^{k-1}\left(\mathbb{E}[\|x^l-\bar{x}\|^2]-\mathbb{E}[\|x^{l+1}-\bar{x}\|^2]\right)+4\sigma_F\sum_{l=0}^{k-1}\gamma_l^2 \\
&\quad +8L\gamma_0\sum_{l=0}^{k-1}\gamma_l\mathbb{E}[F(x^l)-F_*] \\
&\le \|x^0-\bar{x}\|^2+4\sigma_F\sum_{l=0}^{k-1}\gamma_l^2+8L\gamma_0\sum_{l=0}^{k-1}\gamma_l\mathbb{E}[F(x^l)-F_*]
\end{aligned}
$$

which implies that

$$
\begin{aligned}
2(1-4L\gamma_0)\sum_{l=0}^{k}\gamma_l\mathbb{E}[F(x^l)-F_*] \\
&\le \|x^0-\bar{x}\|^2+2\gamma_0(F(x^0)-F_*)+4\sigma_F\sum_{l=0}^{k-1}\gamma_l^2.
\end{aligned}
$$

By using the definition of $\bar{x}^k$ and the convexity of F, we have

$$
\mathbb{E}[F(\bar{x}^k)-F_*]\le\frac{1}{\sum_{l=0}^{k}\gamma_l}\sum_{l=0}^{k}\gamma_l\mathbb{E}[F(x^l)-F_*].
$$

Combining this with the above equation shows the desired result. □

When F is strongly convex, we have the following sublinear convergence rate result on the proximal SGD method.

Theorem 11.22 *Let Assumption 11.19 hold and consider Algorithm 11.18. Assume that F is strongly convex with modulus of convexity $\mu>0$, and let x^* denote the unique solution of (11.15). If the step size is chosen as*

$$
\gamma_k:=\frac{2\mu}{8L^2+\mu^2(k+4)},\tag{11.20}
$$

then

$$\mathbb{E}\left[\|x_k - x^*\|^2\right] \leq \frac{C}{8L^2 + \mu^2(k+4)} \tag{11.21}$$

for all $k \geq 0$, where C is a constant independent of k.

Proof By the strong convexity of F, we have

$$F(x^{k+1}) - F_* \geq \mu\|x^{k+1} - x^*\|^2.$$

Thus, it follows from Lemma 11.20 that

$$\begin{aligned}&(1 + 2\mu\gamma_k)\mathbb{E}_{i_k}[\|x^{k+1} - x^*\|^2] \\ &\leq \|x^k - x^*\|^2 + 4\sigma_F\gamma_k^2 + 8L\gamma_k^2\left(f(x^k) - f(x^*) - \langle\nabla f(x^*), x^k - x^*\rangle\right).\end{aligned}$$

By the L-Lipschitz continuity of ∇f, we then have

$$(1 + 2\mu\gamma_k)\mathbb{E}_{i_k}[\|x^{k+1} - x^*\|^2] \leq (1 + 4L^2\gamma_k^2)\|x^k - x^*\|^2 + 4\sigma_F\gamma_k^2.$$

Taking the full expectation and setting $A_k := \mathbb{E}[\|x^k - x^*\|^2]$, we have

$$(1 + 2\mu\gamma_k)A_{k+1} \leq (1 + 4L^2\gamma_k^2)A_k + 4\sigma_F\gamma_k^2. \tag{11.22}$$

Consequently

$$A_{k+1} \leq \frac{1 + 4L^2\gamma_k^2}{1 + 2\mu\gamma_k}A_k + 4\sigma_F\gamma_k^2.$$

By the choice of γ_k, it is easy to check that

$$\frac{1 + 4L^2\gamma_k^2}{1 + 2\mu\gamma_k} \leq 1 - \mu\gamma_k.$$

Therefore

$$A_{k+1} \leq (1 - \mu\gamma_k)A_k + 4\sigma_F\gamma_k^2. \tag{11.23}$$

Based on this equation, in the following, we use an induction argument to show that (11.21) holds with

$$C := \max\{16\sigma_F, (8L^2 + 4\mu^2)\|x_0 - x^*\|^2\}.$$

This is trivial for $k = 0$. Assume (11.21) is proved for some $k \geq 0$. Then it follows from (11.23) that

$$\begin{aligned} A_{k+1} &\leq \left(1 - \frac{2\mu^2}{8L^2 + \mu^2(k+4)}\right) \frac{C}{8L^2 + \mu^2(k+4)} + \frac{16\sigma_F \mu^2}{(8L^2 + \mu^2(k+4))^2} \\ &= \frac{C}{8L^2 + \mu^2(k+5)} + \Delta_k, \end{aligned}$$

where

$$\begin{aligned} \Delta_k &:= \left(\frac{C}{8L^2 + \mu^2(k+4)} - \frac{C}{8L^2 + \mu^2(k+5)}\right) - \frac{2C\mu^2}{(8L^2 + \mu^2(k+4))^2} \\ &\quad + \frac{16\sigma_F \mu^2}{(8L^2 + \mu^2(k+4))^2} \\ &= \frac{C\mu^2}{(8L^2 + \mu^2(k+4))(8L^2 + \mu^2(k+5))} - \frac{(2C - 16\sigma_F)\mu^2}{(8L^2 + \mu^2(k+4))^2}. \end{aligned}$$

We need to show $\Delta_k \leq 0$ which is equivalent to showing

$$C\mu^2(8L^2 + \mu^2(k+4)) - (2C - 16\sigma_F)\mu^2(8L^2 + \mu^2(k+5)) \leq 0.$$

This follows immediately by noting that the left-hand side can be written as

$$-\mu^2\left[(C - 16\sigma_F)(8L^2 + \mu^2(k+4)) + (2C - 16\sigma_F)\mu^2\right]$$

which is nonpositive as $C \geq 16\sigma_F$. The proof is therefore complete. □

Note that in deriving the above convergence rate results for the proximal SGD method, the term σ_F presents significant challenges and prevents us from achieving similar results as for the deterministic proximal gradient method. In the next section, we will introduce the variance reduction techniques to deal with this issue. On the other hand, if we could guarantee that $\sigma_F = 0$, we would expect improved convergence rates. This observation motivates the introduction of the following definition.

Definition 11.23 We say the problem (11.15) satisfies the *interpolation condition* if there is a solution x^* of (11.15) such that

$$\nabla f_i(x^*) = \nabla f(x^*), \quad i = 1, \cdots, N. \tag{11.24}$$

Here are some comments on interpolation condition. Assume (11.15) satisfies the interpolation condition in the sense of (11.24). Since $0 \in \nabla f(x^*) + \partial h(x^*)$ as x^* is a solution of (11.15), we have from (11.24) that $0 \in \nabla f_i(x^*) + \partial h(x^*)$ which implies x^* is a minimizer of $f_i + h$ over $\mathbb{R}^d$ for every i. Actually, under the interpolation condition, we can show that any solution $\bar{x}$ of (11.15) minimizes $f_i + h$ over $\mathbb{R}^d$ for every i. Indeed, by noting that

$$0 = F(\bar{x}) - F(x^*) = \frac{1}{N}\sum_{i=1}^{N}\big((f_i(\bar{x}) + h(\bar{x})) - (f_i(x^*) + h(x^*))\big)$$

and $f_i(\bar{x}) + h(\bar{x}) \geq f_i(x^*) + h(x^*)$ for every i, we have

$$f_i(\bar{x}) + h(\bar{x}) = f_i(x^*) + h(x^*)$$

and hence $\bar{x}$ is a minimizer of $f_i + h$ over $\mathbb{R}^d$ for every i. Conversely, if h is also continuously differentiable and there exists a solution x^* of (11.15) such that x^* minimizes $f_i + h$ over $\mathbb{R}^d$ for each i, then we have

$$\nabla f(x^*) + \nabla h(x^*) = 0 = \nabla f_i(x^*) + \nabla h(x^*), \quad i = 1, \cdots, N$$

which implies (11.24) and thus the interpolation condition must hold.

For convex minimization problems that satisfy the interpolation condition, the proximal SGD method can achieve significantly faster convergence rates, sometimes approaching those of deterministic methods, even without variance reduction techniques. This improvement arises because, with no gradient noise at the optimal point, the variance of the gradient estimates reduces naturally as the iterates approach the solution. The following result demonstrates this enhanced convergence behavior.

Theorem 11.24 *Let Assumption 11.19 hold and assume that the problem (11.15) satisfies the interpolation condition. Consider Algorithm 11.18 with a constant step-size $\gamma_k = \gamma$ for all k. If $0 < \gamma < 1/(4L)$, then*

$$\mathbb{E}[F(\bar{x}^k) - F^*] \leq \frac{C}{k+1} \tag{11.25}$$

for all $k \geq 0$, where $\bar{x}^k := \frac{1}{k+1}\sum_{l=0}^{k} x^l$ and C is a constant independent of k.

If in addition F is strongly convex with modulus of convexity $\mu > 0$ and $0 < \gamma < \mu/(2L^2)$, then for the unique minimizer x^* of (11.15), there holds

$$\mathbb{E}\left[\|x_k - x^*\|^2\right] \leq q^k \|x_0 - x^*\|^2. \tag{11.26}$$

for all $k \geq 0$, where $q := (1 + 4L^2\gamma^2)/(1 + 2\mu\gamma) < 1$.

Proof By using $\sigma_F = 0$ and following the proof of Theorem 11.21, it is straightforward to obtain (11.25). When F is strongly convex, we have from (11.22) that

$$(1 + 2\mu\gamma)A_{k+1} \leq (1 + 4L^2\gamma^2)A_k$$

which implies $A_{k+1} \leq qA_k$ for all $k \geq 0$. Recursively using this inequality completes the proof of (11.26). □

11.3 Variance-Reduced Stochastic Gradient Methods

In Sect. 11.2.2, we have considered the proximal SGD method, which updates the solution based on the gradient of a randomly selected component function. Despite its simplicity, the proximal SGD method suffers from high variance in the gradient estimates, which can slow down convergence speed. To address this issue, in this section, we will consider various variance-reduction techniques, such as the stochastic variance reduced gradient (SVRG) and the stochastic average gradient acceleration (SAGA). These methods improve convergence rates by reducing the variance of the stochastic gradient estimates while maintaining the computational efficiency of stochastic methods.

To motivate the variance reduction techniques, let us consider the convex minimization problem (11.15) of finite sum. In the proximal SGD method, a randomly selected partial gradient $\nabla f_{i_k}(x^k)$ is used to replace the full gradient of f. However, if the problem (11.15) does not satisfy the interpolation condition, the gradient noise σ_F, as defined in (11.17), remains a positive quantity for any solution x^* of (11.15). Consequently, the variance of the stochastic gradient, i.e., $\mathbb{E}_{i_k}[\|\nabla f_{i_k}(x^k) - \nabla f(x^k)\|^2]$, may stay significant as k increases, causing the proximal SGD iterates to converge slowly.

To accelerate the proximal SGD method, it is necessary to reduce the variance caused by stochastic gradients. One way to achieve this is by using covariates. That is, at the kth iteration step, we take N elements v_i^k, $i = 1, \cdots, N$ to be determined, and express the full gradient $\nabla f(x^k)$ as

$$\nabla f(x^k) = \frac{1}{N}\sum_{i=1}^{N} \nabla f_i(x^k) = \frac{1}{N}\sum_{i=1}^{N}\left(\nabla f_i(x^k) - v_i^k + v_i^k\right).$$

This gives

$$\nabla f(x^k) = \frac{1}{N}\sum_{i=1}^{N}\left(\nabla f_i(x^k) - v_i^k + \frac{1}{N}\sum_{j=1}^{N} v_j^k\right).$$

Thus, if $i_k \in \{1, \cdots, N\}$ is selected randomly via a uniform distribution, then

$$g^k := \nabla f_{i_k}(x^k) - v_{i_k}^k + \frac{1}{N}\sum_{j=1}^N v_j^k \tag{11.27}$$

is an unbiased estimate of $\nabla f(x^k)$, which we can use to update x^{k+1} by $x^{k+1} = x^k - \gamma g^k$ with a suitable step size γ. Utilizing the variance decomposition formula, the variance of g^k can be bounded as

$$\begin{aligned}\mathbb{E}_{i_k}[\|g^k - \nabla f(x^k)\|^2] &= \mathbb{E}_{i_k}\left[\|\nabla f_{i_k}(x^k) - v_{i_k}^k\|^2\right] - \left\|\frac{1}{N}\sum_{j=1}^N \left(\nabla f_j(x^k) - v_j^k\right)\right\|^2 \\ &\le \mathbb{E}_{i_k}\left[\left\|\nabla f_{i_k}(x^k) - v_{i_k}^k\right\|^2\right] = \frac{1}{N}\sum_{i=1}^N \left\|\nabla f_i(x^k) - v_i^k\right\|^2.\end{aligned}$$

This suggests that if each v_i^k is chosen such that v_i^k approaches $\nabla f_i(x^k)$ as k increases, then the variance can be reduced. Various variance reduction techniques can be developed through judicious choices of v_i^k. In the following two subsections, we will focus on the SVRG and SAGA methods. For additional variance reduction techniques, please refer to [40, 73, 92, 168, 198], for instance.

11.3.1 SVRG: Stochastic Variance-Reduced Gradient Method

As mentioned above, it is essential to choose v_i^k carefully so that g^k, given in (11.27), can effectively substitute for the full gradient while reducing variance. In this subsection, we explore the stochastic variance-reduced gradient (SVRG) method, first introduced in [122, 234] and later extended to a proximal version in [225]. SVRG achieves variance reduction by periodically computing a precise reference gradient, which is then used to adjust each stochastic gradient update, thereby enabling smoother and faster convergence. This method follows an epoch-based structure: At the beginning of each epoch, a reference point $\tilde{x}$ is set. During the inner iterations within each epoch, after sampling an index point, a stochastic gradient of the form (11.27) with $v_i^k = \nabla f_i(\tilde{x})$ is used for a variance-reduced update. A new reference point $\tilde{x}$ is chosen at the end of each epoch, and this process continues until convergence. The detailed steps of SVRG of the proximal version are provided below.

Algorithm 11.25 (Prox-SVRG) *Input an initial guess x^0, an integer $m \ge 1$ and a step-size $\gamma > 0$. Set $\tilde{x}^0 = x^0$. For $k = 0, 1, \cdots$, do the following:*

(i) *Calculate $\tilde{g}^k := \nabla f(\tilde{x}^k)$ and set $x^{k,0} = x^k$.*

(ii) *For* $s = 0, \cdots, m-1$ *do*

- *Pick* $i_{k,s} \in \{1, \cdots, N\}$ *randomly via a uniform distribution.*
- *Calculate* $g^{k,s} := \nabla f_{i_{k,s}}(x^{k,s}) - \nabla f_{i_{k,s}}(\tilde{x}^k) + \tilde{g}^k$.
- *Determine* $x^{k,s+1}$ *by* $x^{k,s+1} := \operatorname{prox}_{\gamma h}(x^{k,s} - \gamma g^{k,s})$.

(iii) *Update* $\tilde{x}^{k+1}$ *and* x^{k+1} *by*

$$\tilde{x}^{k+1} := \frac{1}{m}\sum_{s=1}^{m} x^{k,s} \quad \text{and} \quad x^{k+1} = x^{k,m}.$$

and repeat the above procedure.

Note that Prox-SVRG requires a periodic full pass over the data to compute the gradient, which can be computationally expensive in some applications. However, this cost is often offset by fewer required iterations to reach a given accuracy level, leading to overall computational savings, especially for problems where gradient evaluations are costly.

From the definition of $g^{k,s}$ in Prox-SVRG, it is easy to see that

$$\mathbb{E}_{i_{k,s}}[g^{k,s}] = \nabla f(x^{k,s}),$$

i.e., $g^{k,s}$ is an unbiased estimate of the full gradient $\nabla f(x^{k,s})$. For the corresponding variance of $g^{k,s}$, we have the following result.

Lemma 11.26 *Let Assumption 11.19 be fulfilled. There holds*

$$\mathbb{E}_{i_{k,s}}[\|g^{k,s} - \nabla f(x^{k,s})\|^2] \leq 4L\left(F(x^{k,s}) - F_* + F(\tilde{x}^k) - F_*\right).$$

Proof By the definition of $g^{k,s}$, we have

$$\|g^{k,s} - \nabla f(x^{k,s})\|^2 = \left\|\left(\nabla f_{i_{k,s}}(x^{k,s}) - \nabla f_{i_{k,s}}(\tilde{x}^k)\right) - \left(\nabla f(x^{k,s}) - \nabla f(\tilde{x}^k)\right)\right\|^2.$$

Since $\mathbb{E}_{i_{k,s}}[\nabla f_{i_{k,s}}(x^{k,s}) - \nabla f_{i_{k,s}}(\tilde{x}^k)] = \nabla f(x^{k,s}) - \nabla f(\tilde{x}^k)$, by the variance decomposition formula, we can obtain

$$\begin{aligned}\mathbb{E}_{i_{k,s}}\left[\|g^{k,s} - \nabla f(x^{k,s})\|^2\right] &= \mathbb{E}_{i_{k,s}}\left[\|\nabla f_{i_{k,s}}(x^{k,s}) - \nabla f_{i_{k,s}}(\tilde{x}^k)\|^2\right] \\ &\quad - \left\|\nabla f(x^{k,s}) - \nabla f(\tilde{x}^k)\right\|^2 \\ &\leq \mathbb{E}_{i_{k,s}}\left[\|\nabla f_{i_{k,s}}(x^{k,s}) - \nabla f_{i_{k,s}}(\tilde{x}^k)\|^2\right].\end{aligned}$$

Let $\bar{x}$ be any solution of (11.15). By using the inequality $\|a+b\|^2 \le 2(\|a\|^2+\|b\|^2)$, we can further obtain

$$
\begin{aligned}
&\mathbb{E}_{i_{k,s}}\left[\|g^{k,s} - \nabla f(x^{k,s})\|^2\right] \\
&\le 2\mathbb{E}_{i_{k,s}}\left[\|\nabla f_{i_{k,s}}(x^{k,s}) - \nabla f_{i_{k,s}}(\bar{x})\|^2 + \|\nabla f_{i_{k,s}}(\tilde{x}^k) - \nabla f_{i_{k,s}}(\bar{x})\|^2\right] \\
&= \frac{2}{N}\sum_{i=1}^{N}\|\nabla f_i(x^{k,s}) - \nabla f_i(\bar{x})\|^2 + \frac{2}{N}\sum_{i=1}^{N}\|\nabla f_i(\tilde{x}^k) - \nabla f_i(\bar{x})\|^2.
\end{aligned}
$$

By using the co-coercivity of smooth convex functions, see Proposition 2.34; we then have

$$
\begin{aligned}
\mathbb{E}_{i_{k,s}}\left[\|g^{k,s} - \nabla f(x^{k,s})\|^2\right] &\le \frac{4L}{N}\sum_{i=1}^{N}\left(f_i(x^{k,s}) - f_i(\bar{x}) - \langle \nabla f_i(\bar{x}), x^{k,s} - \bar{x}\rangle\right) \\
&\quad + \frac{4L}{N}\sum_{i=1}^{N}\left(f_i(\tilde{x}^k) - f_i(\bar{x}) - \langle \nabla f_i(\bar{x}), \tilde{x}^k - \bar{x}\rangle\right) \\
&= 4L\left(f(x^{k,s}) - f(\bar{x}) - \langle \nabla f(\bar{x}), x^{k,s} - \bar{x}\rangle\right) \\
&\quad + 4L\left(f(\tilde{x}^k) - f(\bar{x}) - \langle \nabla f(\bar{x}), \tilde{x}^k - \bar{x}\rangle\right).
\end{aligned}
$$

Since $\bar{x}$ is an optimal solution of (11.15), we have $-\nabla f(\bar{x}) \in \partial h(\bar{x})$. Thus

$$
h(x^{k,s}) - h(\bar{x}) \ge -\langle \nabla f(\bar{x}), x^{k,s} - \bar{x}\rangle, \quad h(\tilde{x}^k) - h(\bar{x}) \ge -\langle \nabla f(\bar{x}), \tilde{x}^k - \bar{x}\rangle.
$$

Consequently,

$$
\begin{aligned}
\mathbb{E}_{i_{k,s}}\left[\|g^{k,s} - \nabla f(x^{k,s})\|^2\right] &\le 4L\left(f(x^{k,s}) - f(\bar{x}) + h(x^{k,s}) - h(\bar{x})\right) \\
&\quad + 4L\left(f(\tilde{x}^k) - f(\bar{x}) + h(\tilde{x}^k) - h(\bar{x})\right) \\
&= 4L\left(F(x^{k,s}) - F(\bar{x}) + F(\tilde{x}^k) - F(\bar{x})\right).
\end{aligned}
$$

The proof is complete. □

According to the result given in Lemma 11.26, when both $x^{k,s}$ and $\tilde{x}^k$ approach an optimal solution of (11.15), the variance of $g^{k,s}$ converges to zero. As a result, the purpose of reducing variance is achieved, and therefore it is possible to use a constant step size to obtain faster convergence than the proximal SGD method studied in Sect. 11.2.2. To derive the convergence and convergence rates of Prox-SVRG under a constant step size, we first prove the following result.

Lemma 11.27 *Let Assumption 11.19 hold. Consider Algorithm 11.25 with* $0 < \gamma < 1/L$. *There holds*

$$\begin{aligned}\mathbb{E}_{i_{k,s}}[F(x^{k,s+1}) - F_*] \le \left(\frac{1}{2\gamma} - \mu_f\right)\|x^{k,s} - \bar{x}\|^2 \\ - \left(\frac{1}{2\gamma} + \mu_h\right)\mathbb{E}_{i_{k,s}}[\|x^{k,s+1} - \bar{x}\|^2] \\ + \frac{2L\gamma}{1 - L\gamma}\left(F(x^{k,s}) - F_* + F(\tilde{x}^k) - F_*\right),\end{aligned}$$

where $\bar{x}$ *denotes any solution of (11.15).*

Proof In view of (11.12) in Lemma 11.14 with $u = \bar{x}$, $v = x^{k.s}$, $w = x^{k,s} - \gamma g^{k,s}$, $\xi = \nabla f(x^{s,k})$, and $z = x^{k,s+1}$, we obtain

$$\begin{aligned}F(x^{k,s+1}) \le F_* + \left(\frac{1}{2\gamma} - \mu_f\right)\|x^{k,s} - \bar{x}\|^2 - \left(\frac{1}{2\gamma} + \mu_h\right)\|x^{k,s+1} - \bar{x}\|^2 \\ - \frac{1 - L\gamma}{2\gamma}\|x^{k,s+1} - x^{k,s}\|^2 + \langle \nabla f(x^{k,s}) - g^{k,s}, x^{k,s} - \bar{x}\rangle \\ + \langle \nabla f(x^{k,s}) - g^{k,s}, x^{k,s+1} - x^{k,s}\rangle.\end{aligned}$$

By the Young's inequality, we have

$$\begin{aligned}\langle \nabla f(x^{k,s}) - g^{k,s}, x^{k,s+1} - x^{k,s}\rangle &\le \|\nabla f(x^{k,s}) - g^{k,s}\|\|x^{k,s+1} - x^{k,s}\| \\ &\le \frac{\gamma}{2(1 - L\gamma)}\|\nabla f(x^{k,s}) - g^{k,s}\|^2 \\ &\quad + \frac{1 - L\gamma}{2\gamma}\|x^{k,s+1} - x^{k,s}\|^2.\end{aligned}$$

Therefore

$$\begin{aligned}F(x^{k,s+1}) \le F_* + \left(\frac{1}{2\gamma} - \mu_f\right)\|x^{k,s} - \bar{x}\|^2 - \left(\frac{1}{2\gamma} + \mu_h\right)\|x^{k,s+1} - \bar{x}\|^2 \\ + \langle \nabla f(x^{k,s}) - g^{k,s}, x^{k,s} - \bar{x}\rangle + \frac{\gamma}{2(1 - L\gamma)}\|g^{k,s} - \nabla f(x^{k,s})\|^2.\end{aligned}$$

Note that $\mathbb{E}_{i_{k,s}}[g^{k,s}] = \nabla f(x^{k,s})$. Taking the expectation with respect to $i_{k,s}$ then gives

$$\begin{aligned}\mathbb{E}_{i_{k,s}}[F(x^{k,s+1}) - F_*] \le & \left(\frac{1}{2\gamma} - \mu_f\right) \|x^{k,s} - \bar{x}\|^2 \\ & - \left(\frac{1}{2\gamma} + \mu_h\right) \mathbb{E}_{i_{k,s}}[\|x^{k,s+1} - \bar{x}\|^2] \\ & + \frac{\gamma}{2(1-L\gamma)} \mathbb{E}_{i_{k,s}}\left[\|g^{k,s} - \nabla f(x^{k,s})\|^2\right]\end{aligned}$$

which together with Lemma 11.26 completes the proof. □

For the Prox-SVRG method, we have the following convergence result under a constant step size.

Theorem 11.28 *Let Assumption 11.19 hold. Consider Algorithm 11.25 with $0 < \gamma < 1/(5L)$. Then, almost surely, $\{x^k\}$ converges to a solution of (11.15) and $\mathbb{E}[F(x^k) - F_*] \to 0$ as $k \to \infty$. If in addition $\mu_f + \mu_h > 0$, i.e., either f or h is strongly convex, then*

$$\mathbb{E}[\|x^k - x^*\|^2] \to 0, \quad \mathbb{E}[F(x^k) - F_*] \to 0 \quad \text{and} \quad \mathbb{E}[F(\tilde{x}^k) - F_*] \to 0$$

linearly as $k \to \infty$, where x^* denotes the unique solution of (11.15).

Proof Let $\bar{x}$ be any solution of (11.15). By multiplying the equation in Lemma 11.27 by 2γ, and rearranging the terms, we obtain

$$\begin{aligned}&(1 + 2\gamma\mu_h)\,\mathbb{E}_{i_{k,s}}\left[\|x^{k,s+1} - \bar{x}\|^2\right] \\ &\quad \le (1 - 2\gamma\mu_f)\,\|x^{k,s} - \bar{x}\|^2 - 2\gamma\mathbb{E}_{i_{k,s}}\left[F(x^{k,s+1}) - F_*\right] \\ &\qquad + \frac{4\gamma^2 L}{1-\gamma L}\left(F(x^{k,s}) - F_* + F(\tilde{x}^k) - F_*\right).\end{aligned}$$

Let $\mathbb{E}_k$ denote the expectation taking with respect to all $i_{k,s}$, $s = 0, \cdots, m-1$, we obtain from the above equation that

$$\begin{aligned}&(1 + 2\gamma\mu_h)\,\mathbb{E}_k\left[\|x^{k,s+1} - \bar{x}\|^2\right] \\ &\quad \le (1 - 2\gamma\mu_f)\,\mathbb{E}_k\left[\|x^{k,s} - \bar{x}\|^2\right] - 2\gamma\mathbb{E}_k\left[F(x^{k,s+1}) - F_*\right] \\ &\qquad + \frac{4\gamma^2 L}{1-\gamma L}\left(\mathbb{E}_k\left[F(x^{k,s}) - F_*\right] + F(\tilde{x}^k) - F_*\right).\end{aligned}$$

for all $s = 0, \cdots, m-1$. Summing over s from 0 to $m-1$ gives

$$
\begin{aligned}
&(1+2\gamma\mu_h)\sum_{s=1}^{m}\mathbb{E}_k\left[\|x^{k,s}-\bar{x}\|^2\right] \\
&\le \left(1-2\gamma\mu_f\right)\sum_{s=0}^{m-1}\mathbb{E}_k\left[\|x^{k,s}-\bar{x}\|^2\right]-2\gamma\sum_{s=1}^{m}\mathbb{E}_k\left[F(x^{k,s})-F_*\right] \\
&\quad+\frac{4\gamma^2L}{1-\gamma L}\sum_{s=0}^{m-1}\mathbb{E}_k\left[F(x^{k,s})-F_*\right]+\frac{4\gamma^2Lm}{1-\gamma L}\left(F(\tilde{x}^k)-F_*\right).
\end{aligned}
$$

Note that $\|x^{k,0}-\bar{x}\|^2$ and $F(x^{k,0})-F_*$ are independent of $i_{k,s}$ for $s = 0, \cdots, m-1$. Thus, after simplification, we obtain

$$
\begin{aligned}
&(1+2\gamma\mu_h)\mathbb{E}_k\left[\|x^{k,m}-\bar{x}\|^2\right]+2\gamma\mathbb{E}_k\left[F(x^{k,m})-F_*\right] \\
&\le -2\gamma(\mu_f+\mu_h)\sum_{s=1}^{m-1}\mathbb{E}_k[\|x^{k,s}-\bar{x}\|^2]+\left(1-2\gamma\mu_f\right)\|x^{k,0}-\bar{x}\|^2 \\
&\quad-\left(2\gamma-\frac{4\gamma^2L}{1-\gamma L}\right)\sum_{s=1}^{m-1}\mathbb{E}_k\left[F(x^{k,s})-F_*\right]+\frac{4\gamma^2L}{1-\gamma L}\left(F(x^{k,0})-F_*\right) \\
&\quad+\frac{4\gamma^2Lm}{1-\gamma L}\left(F(\tilde{x}^k)-F_*\right).
\end{aligned}
$$

Since $0 < \gamma < 1/(5L)$, we have $2\gamma - 4\gamma^2L/(1-\gamma L) \ge \gamma$. Thus, it follows from the above equation that

$$
\begin{aligned}
&(1+2\gamma\mu_h)\mathbb{E}_k\left[\|x^{k,m}-\bar{x}\|^2\right]+\gamma\mathbb{E}_k\left[F(x^{k,m})-F_*\right]+\gamma\sum_{s=1}^{m}\mathbb{E}_k\left[F(x^{k,s})-F_*\right] \\
&\le (1-2\gamma\mu_f)\|x^{k,0}-\bar{x}\|^2+\frac{4\gamma^2L}{1-\gamma L}\left(F(x^{k,0})-F_*\right)+\frac{4\gamma^2Lm}{1-\gamma L}\left(F(\tilde{x}^k)-F_*\right).
\end{aligned}
$$

Recall that $x^{k,0} = x^k$, $x^{k+1} = x^{k,m}$ and $\tilde{x}^{k+1} = \frac{1}{m}\sum_{s=1}^{m} x^{k,s}$, and thus, by the convexity of F, we have

$$
\frac{1}{m}\sum_{s=1}^{m}\mathbb{E}_k[F(x^{k,s})-F_*]=\mathbb{E}_k\left[\frac{1}{m}\sum_{s=1}^{m}F(x^{k,s})-F_*\right]\ge\mathbb{E}[F(\tilde{x}^{k+1})-F_*].
$$

Therefore

$$
\begin{aligned}
&(1+2\gamma\mu_h)\mathbb{E}_k\left[\|x^{k+1}-\bar{x}\|^2\right]+\gamma\mathbb{E}_k\left[F(x^{k+1})-F_*\right]+\gamma m\mathbb{E}_k\left[F(\tilde{x}^{k+1})-F_*\right] \\
&\le (1-2\gamma\mu_f)\|x^k-\bar{x}\|^2+\frac{4\gamma^2 L}{1-\gamma L}\left(F(x^k)-F_*\right) \\
&\quad+\frac{4\gamma^2 Lm}{1-\gamma L}\left(F(\tilde{x}^k)-F_*\right).
\end{aligned}
\tag{11.28}
$$

We first show the almost sure convergence when f and g are only convex functions. We may take $\mu_f=\mu_h=0$ in the above equation to obtain

$$
\mathbb{E}_k[X_{k+1}]\le X_k-\left(\gamma-\frac{4\gamma^2 L}{1-\gamma L}\right)\left\{\left(F(x^k)-F_*\right)+m\left(F(\tilde{x}^k)-F_*\right)\right\},
$$

where

$$
X_k:=\|x^k-\bar{x}\|^2+\gamma\left(F(x^k)-F_*\right)+\gamma m\left(F(\tilde{x}^k)-F_*\right).
$$

Therefore, we may use Lemma 11.10 to conclude that almost surely there hold

$$
\lim_{k\to\infty}X_k \text{ exists} \quad \text{and} \quad \sum_{k=0}^{\infty}\left\{\left(F(x^k)-F_*\right)+m\left(F(\tilde{x}^k)-F_*\right)\right\}<\infty.
$$

The second equation implies that

$$
\lim_{k\to\infty}\left\{\left(F(x^k)-F_*\right)+m\left(F(\tilde{x}^k)-F_*\right)\right\}=0 \text{ almost surely}
$$

which together with the first equation shows that

$$
\lim_{k\to\infty}\|x^k-\bar{x}\|^2 \text{ exists almost surely.}
$$

for any solution $\bar{x}$ of (11.15). Based on the above two equations, we may use the similar argument in the proof of Theorem 11.17 to conclude that almost surely $\{x^k\}$ converges to a solution of (11.15) as $k\to\infty$.

Next we derive the linear convergence result when $\mu_f + \mu_h > 0$. Taking the full expectation on Eq. (11.28) gives

$$
\begin{aligned}
&(1+2\gamma\mu_h)\mathbb{E}\left[\|x^{k+1}-\bar{x}\|^2\right]+\gamma\mathbb{E}\left[F(x^{k+1})-F_*\right]+\gamma m\mathbb{E}\left[F(\tilde{x}^{k+1})-F_*\right]\\
&\quad\le(1-2\gamma\mu_f)\mathbb{E}\left[\|x^k-\bar{x}\|^2\right]+\frac{4\gamma^2L}{1-\gamma L}\mathbb{E}\left[F(x^k)-F_*\right]\\
&\qquad+\frac{4\gamma^2Lm}{1-\gamma L}\mathbb{E}\left[F(\tilde{x}^k)-F_*\right].
\end{aligned}
$$

This implies that, for

$$
U_k := (1+2\gamma\mu_h)\mathbb{E}\left[\|x^k-\bar{x}\|^2\right]+\gamma\mathbb{E}\left[F(x^k)-F_*\right]+\gamma m\mathbb{E}\left[F(\tilde{x}^k)-F_*\right],
$$

there holds $U_{k+1} \le qU_k$ and hence $U_k \le q^k U_0$ for all $k \ge 0$, where

$$
q := \max\left\{\frac{1-2\gamma\mu_f}{1+2\gamma\mu_h}, \frac{4\gamma L}{1-\gamma L}\right\} < 1
$$

according to the given conditions. The desired conclusion on linear convergence thus follows. □

In the discussion above, we considered the Prox-SVRG method with a constant epoch size m, where each epoch consists of m stochastic gradient steps. In [2], a variant of the Prox-SVRG method was proposed that employs variable epoch sizes. Specifically, the epoch size doubles between consecutive epochs as the iterations progress. The algorithm presented in [2] can be formulated as follows.

Algorithm 11.29 (Prox-SVRG with Doubling Epoch [2]) *Input an initial guess $\tilde{x}^0$, an integer $m_0 \ge 1$ and a step size $\gamma > 0$. Set $x^{0,0} := \tilde{x}^0$. For $k = 0, 1, \cdots$, do the following:*

(i) *Calculate $\tilde{g}^k := \nabla f(\tilde{x}^k)$, and set $m_k := 2^k m_0$.*
(ii) *For $s = 0, \cdots, m_k - 1$, do*

- *Pick $i_{k,s} \in \{1, \cdots, N\}$ randomly via a uniform distribution.*
- *Calculate $g^{k,s} := \nabla f_{i_{k,s}}(x^{k,s}) - \nabla f_{i_{k,s}}(\tilde{x}^k) + \tilde{g}^k$.*
- *Determine $x^{k,s+1}$ by $x^{k,s+1} := \mathrm{prox}_{\gamma h}(x^{k,s} - \gamma g^{k,s})$.*

(iii) *Set $\tilde{x}^{k+1} := \frac{1}{m_k}\sum_{s=1}^{m_k} x^{k,s}$ and $x^{k+1,0} := x^{k,m_k}$. Repeat the above procedure.*

Without assuming the strong convexity of either f or h, the following convergence rate has been proved in [2].

Theorem 11.30 *Let Assumption 11.19 hold, and consider Algorithm 11.29. Let $\bar{x}$ denote any solution of (11.15). If $0 < \gamma < 1/(5L)$, then*

$$\mathbb{E}\left[F(\tilde{x}^k) - F_*\right] \leq q^{-k}\left(\frac{m_0+1}{m_0}(F(\tilde{x}^0) - F_*) + \frac{1-L\gamma}{4L\gamma^2 m_0}\|\tilde{x}^0 - \bar{x}\|^2\right)$$

for all integers $k \geq 0$, where $q := \min\{\frac{1-3L\gamma}{2L\gamma}, 2\} > 1$.

Proof By using Lemma 11.27 with $\mu_f = \mu_h = 0$, we can obtain

$$\begin{aligned}\mathbb{E}_{i_{k,s}}\left[F(x^{k,s+1}) - F_*\right] \leq \frac{2L\gamma}{1-L\gamma}\left(F(x^{k,s}) - F_* + F(\tilde{x}^k) - F_*\right)\\ + \frac{1}{2\gamma}\left(\|x^{k,s} - \bar{x}\|^2 - \mathbb{E}_{i_{k,s}}\left[\|x^{k,s+1} - \bar{x}\|^2\right]\right).\end{aligned}$$

Taking the full expectation, summing over $s = 0, \cdots, m_k - 1$, and then dividing by m_k, we have

$$\begin{aligned}&\mathbb{E}\left[\frac{1}{m_k}\sum_{s=0}^{m_k-1} F(x^{k,s+1}) - F_*\right]\\ &\leq \frac{2L\gamma}{1-L\gamma}\mathbb{E}\left[\frac{1}{m_k}\sum_{s=0}^{m_k-1} F(x^{k,s}) - F_* + F(\tilde{x}^k) - F_*\right]\\ &\quad + \frac{1}{2\gamma m_k}\left(\mathbb{E}\left[\|x^{k,0} - \bar{x}\|^2\right] - \mathbb{E}\left[\|x^{k,m_k} - \bar{x}\|^2\right]\right).\end{aligned}$$

After rearranging the terms, it gives

$$\begin{aligned}&\frac{1-3L\gamma}{1-L\gamma}\mathbb{E}\left[\frac{1}{m_k}\sum_{s=0}^{m_k-1} F(x^{k,s+1}) - F_*\right]\\ &\leq \frac{2L\gamma}{1-L\gamma}\mathbb{E}\left[\frac{1}{m_k}\left((F(x^{k,0}) - F_*) - (F(x^{k,m_k}) - F_*)\right) + F(\tilde{x}^k) - F_*\right]\\ &\quad + \frac{1}{2\gamma m_k}\left(\mathbb{E}\left[\|x^{k,0} - \bar{x}\|^2\right] - \mathbb{E}\left[\|x^{k,m_k} - \bar{x}\|^2\right]\right).\end{aligned}$$

By the convexity of F and the definition of $\tilde{x}^{k+1}$, we have

$$\frac{1}{m_k}\sum_{s=0}^{m_k-1} F(x^{k,s+1}) \geq F\left(\frac{1}{m_k}\sum_{s=0}^{m_k-1} x^{k,s+1}\right) = F(\tilde{x}^{k+1}).$$

Therefore, by recalling $x^{k+1,0} = x^{k,m_k}$, we have

$$\begin{aligned}
&\frac{1-3L\gamma}{1-L\gamma}\mathbb{E}\left[F(\tilde{x}^{k+1}) - F_*\right] \\
&\leq \frac{2L\gamma}{1-L\gamma}\mathbb{E}\left[\frac{F(x^{k,0}) - F_*}{m_k} - \frac{F(x^{k+1,0}) - F_*}{m_k} + F(\tilde{x}^k) - F_*\right] \\
&\quad + \frac{1}{2\gamma m_k}\left(\mathbb{E}\left[\|x^{k,0} - \bar{x}\|^2\right] - \mathbb{E}\left[\|x^{k+1,0} - \bar{x}\|^2\right]\right).
\end{aligned}$$

Consequently, by using $m_{k+1} = 2m_k$, we obtain

$$\begin{aligned}
&\mathbb{E}\left[\frac{1-3L\gamma}{1-L\gamma}(F(\tilde{x}^{k+1}) - F_*) + \frac{4L\gamma(F(x^{k+1,0}) - F_*)}{m_{k+1}(1-L\gamma)} + \frac{\|x^{k+1,0} - \bar{x}\|^2}{\gamma m_{k+1}}\right] \\
&\leq \mathbb{E}\left[\frac{2L\gamma}{1-L\gamma}(F(\tilde{x}^k) - F_*) + \frac{2L\gamma(F(x^{k,0}) - F_*)}{m_k(1-L\gamma)} + \frac{\|x^{k,0} - \bar{x}\|^2}{2\gamma m_k}\right].
\end{aligned}$$

Multiplying the both sides by $1-L\gamma$ and setting $q := \min\{\frac{1-3L\gamma}{2L\gamma}, 2\}$, it then follows from the above inequality that

$$\begin{aligned}
&q\mathbb{E}\left[2L\gamma(F(\tilde{x}^{k+1}) - F_*) + 2L\gamma\frac{F(x^{k+1,0}) - F_*}{m_{k+1}} + \frac{1-L\gamma}{2\gamma}\frac{\|x^{k+1,0} - \bar{x}\|^2}{m_{k+1}}\right] \\
&\leq \mathbb{E}\left[2L\gamma(F(\tilde{x}^k) - F_*) + 2L\gamma\frac{F(x^{k,0}) - F_*}{m_k} + \frac{1-L\gamma}{2\gamma}\frac{\|x^{k,0} - \bar{x}\|^2}{m_k}\right]
\end{aligned}$$

for all integers $k \geq 0$. Note that $0 < \gamma < 1/(5L)$ implies $q > 1$. Recursively using the above inequality shows in particular that

$$\mathbb{E}\left[F(\tilde{x}^k) - F_*\right] \leq q^{-k}\left(F(\tilde{x}^0) - F_* + \frac{1}{m_0}(F(\tilde{x}^0) - F_*) + \frac{1-L\gamma}{4L\gamma^2 m_0}\|\tilde{x}^0 - \bar{x}\|^2\right),$$

where we used $x^{0,0} = \tilde{x}^0$. The proof is thus complete. □

11.3.2 SAGA: Stochastic Average Gradient Accelerated Method

We next consider the SAGA method for solving (11.15). In SAGA, variance reduction is achieved by maintaining a table that stores the most recent gradient for each data point. At each iteration, SAGA samples a data point, updates its corresponding stored gradient, and uses this table to form a variance-reduced gradient estimate that is less noisy and closer to the true gradient. Specifically, at the k-th iteration, a data point i_k is sampled and a gradient estimate g_i^k of the form (11.27) is then constructed using the stored gradient estimates v_i^k, $i = 1, \cdots, N$. Meanwhile, the table $\{v_i^k\}$ is updated to $\{v_i^{k+1}\}$ by setting $v_{i_k}^{k+1} = \nabla f_{i_k}(x^k)$, while leaving the other gradients unchanged, i.e., $v_i^{k+1} = v_i^k$ for $i \neq i_k$. A detailed description of the SAGA method is as follows.

Algorithm 11.31 (SAGA [68]) *Input an initial guess x^0 and a step size $\gamma > 0$. Set $v_i^0 := \nabla f_i(x^0)$ for $i = 1, \cdots, N$. For $k = 0, 1, \cdots$, do the following:*

(i) *Pick $i_k \in \{1, \cdots, N\}$ randomly via a uniform distribution;*
(ii) *Define*

$$v_i^{k+1} = \begin{cases} \nabla f_{i_k}(x^k) & \text{if } i = i_k, \\ v_i^k & \text{otherwise;} \end{cases}$$

(iii) *Update x^{k+1} by $x^{k+1} = \mathrm{prox}_{\gamma h}(x^k - \gamma g^k)$, where*

$$g^k = v_{i_k}^{k+1} - v_{i_k}^k + \frac{1}{N}\sum_{i=1}^{N} v_i^k.$$

SAGA reduces variance by maintaining a table of previously computed gradients v_i^k and using this information in each update to adjust the current stochastic gradient. While SAGA requires $O(Nd)$ memory to store the gradients of all data points, which can lead to significant overhead for large datasets, it avoids periodic full gradient computations, unlike SVRG. Instead, SAGA leverages the stored gradients for variance reduction, making it preferable when full gradient evaluations are computationally expensive. SAGA can be seen as an improved version of the stochastic average gradient (SAG) method, proposed in [198], which was the first pioneering technique for variance reduction. In the k-th iteration of SAG, an estimate v_i^k is maintained for each data point i, and the average of the v_i^k values, i.e., $\bar{g}^k := \frac{1}{N}\sum_{i=1}^{N} v_i^k$, is used as an estimate of the full gradient $\nabla f(x^k)$. However, although v_i^k is updated similarly to SAGA, $\bar{g}^k$ remains a biased estimate of the full gradient, which complicates the analysis of SAG.

Remark 11.32 Note that, in the SAGA method, the update of g^k involves the term

$$\bar{v}^k := \frac{1}{N}\sum_{i=1}^{N} v_i^k.$$

Since N can be very large, computing this term directly from the formula can be very costly. Actually, this term can be computed in a cheap way. That is, at the starting point, we compute $\bar{v}^0$ by the given formula, and then for $k \geq 0$, we can compute $\bar{v}^{k+1}$ recursively from $\bar{v}^k$ because

$$\bar{v}^{k+1} = \frac{1}{N}\sum_{i\neq i_k} v_i^{k+1} + \frac{1}{N}v_{i_k}^{k+1} = \frac{1}{N}\sum_{i\neq i_k} v_i^k + \frac{1}{N}v_{i_k}^{k+1} = \bar{v}^k + \frac{1}{N}\left(v_{i_k}^{k+1} - v_{i_k}^k\right).$$

Therefore, from the previous stored $\bar{v}^k$, we can obtain $\bar{v}^{k+1}$ from $\bar{v}^k$ by simply adding $(v_{i_k}^{k+1} - v_{i_k}^k)/N$.

Remark 11.33 From the description of the algorithm, one can easily see that for every v_i^k, there is ϕ_i^k such that

$$v_i^k = \nabla f_i(\phi_i^k);$$

moreover, $\phi_{i_k}^{k+1} = x^k$ and $\phi_i^{k+1} = \phi_i^k$ if $i \neq i_k$. These ϕ_i^k will be used in the convergence analysis of SAGA. However, they are not needed in the implementation of the algorithm.

In the following, we will provide a convergence analysis on Algorithm 11.31. We first use Lemma 11.14 to obtain the following result.

Lemma 11.34 *Let Assumption 11.19 hold. Consider Algorithm 11.31 for solving (11.15). Let $\bar{x}$ be any solution of (11.15). If $0 < \gamma \leq 1/L$, then there holds*

$$\begin{aligned}\mathbb{E}_{i_k}[\|x^{k+1} - \bar{x}\|^2] \leq \|x^k - \bar{x}\|^2 - 2\gamma\mathbb{E}_{i_k}[F(x^{k+1}) - F_*] \\ + \frac{4\gamma^2}{N}\sum_{i=1}^N \|\nabla f_i(x^k) - \nabla f_i(\bar{x})\|^2 \\ + \frac{4\gamma^2}{N}\sum_{i=1}^N \|v_i^k - \nabla f_i(\bar{x})\|^2\end{aligned} \tag{11.29}$$

for all integers $k \geq 0$.

Proof By using $0 < \gamma \leq 1/L$ and (11.12) in Lemma 11.14 with $u = \bar{x}$, $v = x^k$, $w = x^k - \gamma g^k$, $\xi = \nabla f(x^k)$, and $z = x^{k+1}$, we have

$$F(x^{k+1}) \leq F_* + \frac{1}{2\gamma}\left(\|x^k - \bar{x}\|^2 - \|x^{k+1} - \bar{x}\|^2\right) - \langle g^k - \nabla f(x^k), x^{k+1} - \bar{x}\rangle.$$

Let $\hat{x}^{k+1} := \operatorname{prox}_{\gamma h}(x^k - \gamma\nabla f(x^k))$ which is independent of i_k and, note by the non-expansiveness of $\operatorname{prox}_{\gamma h}$ that

$$\begin{aligned}\|x^{k+1} - \hat{x}^{k+1}\| &= \|\operatorname{prox}_{\gamma h}(x^k - \gamma g^k) - \operatorname{prox}_{\gamma h}(x^k - \gamma\nabla f(x^k)\| \\ &\leq \gamma\|g^k - \nabla f(x^k)\|.\end{aligned}$$

Thus

$$
\begin{aligned}
F(x^{k+1}) \le F_* &+ \frac{1}{2\gamma}\left(\|x^k - \bar{x}\|^2 - \|x^{k+1} - \bar{x}\|^2\right) \\
&- \langle g^k - \nabla f(x^k), \hat{x}^{k+1} - \bar{x}\rangle \\
&+ \gamma\|g^k - \nabla f(x^k)\|^2.
\end{aligned}
$$

Note that $\mathbb{E}_{i_k}[g^k] = \nabla f(x^k)$. Taking the expectation with respect to i_k in the above equation gives

$$
\begin{aligned}
\mathbb{E}_{i_k}[F(x^{k+1})] \le F_* &+ \frac{1}{2\gamma}\left(\|x^k - \bar{x}\|^2 - \mathbb{E}_{i_k}[\|x^{k+1} - \bar{x}\|^2]\right) \\
&+ \gamma\mathbb{E}_{i_k}\left[\|g^k - \nabla f(x^k)\|^2\right].
\end{aligned}
$$

Therefore, by using the definition of g^k and the variance decomposition formula, we then obtain

$$
\begin{aligned}
\mathbb{E}_{i_k}[\|x^{k+1} - \bar{x}\|^2] &\le \|x^k - \bar{x}\|^2 - 2\gamma\mathbb{E}_{i_k}[F(x^{k+1}) - F_*] \\
&\quad + 2\gamma^2\mathbb{E}_{i_k}\left[\left\|\nabla f_{i_k}(x^k) - v_{i_k}^k + \frac{1}{n}\sum_{i=1}^n v_i^k - \nabla f(x^k)\right\|^2\right] \\
&\le \|x^k - \bar{x}\|^2 - 2\gamma\mathbb{E}_{i_k}[F(x^{k+1}) - F_*] \\
&\quad + 2\gamma^2\mathbb{E}_{i_k}\left[\|\nabla f_{i_k}(x^k) - v_{i_k}^k\|^2\right].
\end{aligned}
$$

With the help of the inequality $\|a + b\|^2 \le 2(\|a\|^2 + \|b\|^2)$, we thus have

$$
\begin{aligned}
\mathbb{E}_{i_k}\left[\|x^{k+1} - \bar{x}\|^2\right] \le \|x^k - \bar{x}\|^2 &- 2\gamma\mathbb{E}_{i_k}\left[F(x^{k+1}) - F_*\right] \\
&+ 4\gamma^2\mathbb{E}_{i_k}\left[\|\nabla f_{i_k}(x^k) - \nabla f_{i_k}(\bar{x})\|^2\right] \\
&+ 4\gamma^2\mathbb{E}_{i_k}\left[\|v_{i_k}^k - \nabla f_{i_k}(\bar{x})\|^2\right]
\end{aligned}
$$

which shows the conclusion. □

In order to use the estimate in Lemma 11.34, we need to develop ideas to deal with the last two terms on the right-hand side of (11.29). To this end, we will estimate $\mathbb{E}_{i_k}[\|x^{k+1} - \bar{x}\|^2]$ in a different way.

Since $\bar{x}$ is an optimal solution of (11.15), we have $\bar{x} = \mathrm{prox}_{\gamma h}(\bar{x} - \gamma \nabla f(\bar{x}))$. By the definition of x^{k+1}, the non-expansiveness of $\mathrm{prox}_{\gamma h}$, and the polarization identity we have

$$\begin{aligned}\|x^{k+1} - \bar{x}\|^2 &= \|\mathrm{prox}_{\gamma h}(x^k - \gamma g^k) - \mathrm{prox}_{\gamma h}(\bar{x} - \gamma \nabla f(\bar{x}))\|^2 \\ &\le \|(x^k - \bar{x}) - \gamma(g^k - \nabla f(\bar{x}))\|^2 \\ &= \|x^k - \bar{x}\|^2 + \gamma^2 \|g^k - \nabla f(\bar{x})\|^2 \\ &\quad - 2\gamma \langle g^k - \nabla f(\bar{x}), x^k - \bar{x}\rangle.\end{aligned}$$

Taking the expectation with respect to i_k gives

$$\begin{aligned}\mathbb{E}_{i_k}[\|x^{k+1} - \bar{x}\|^2] \le \|x^k - \bar{x}\|^2 &+ \gamma^2 \mathbb{E}_{i_k}[\|g^k - \nabla f(\bar{x})\|^2] \\ &- 2\gamma \left\langle \nabla f(x^k) - \nabla f(\bar{x}), x^k - \bar{x} \right\rangle,\end{aligned} \tag{11.30}$$

where we used the fact $\mathbb{E}_{i_k}[g^k] = \nabla f(x^k)$. Let us consider the term $\langle \nabla f(x^k), \bar{x} - x^k\rangle$. By the co-coercivity of ∇f_i for each i, we have

$$\frac{1}{2L}\|\nabla f_i(x^k) - \nabla f_i(\bar{x})\|^2 \le f_i(\bar{x}) - f_i(x^k) - \langle \nabla f_i(x^k), \bar{x} - x^k\rangle.$$

Summing it over i from 1 to N gives

$$\frac{1}{2NL}\sum_{i=1}^{N} \|\nabla f_i(x^k) - \nabla f_i(\bar{x})\|^2 \le f(\bar{x}) - f(x^k) - \langle \nabla f(x^k), \bar{x} - x^k\rangle$$

and hence

$$\langle \nabla f(x^k), \bar{x} - x^k\rangle \le f(\bar{x}) - f(x^k) - \frac{1}{2NL}\sum_{i=1}^{N} \|\nabla f_i(x^k) - \nabla f_i(\bar{x})\|^2.$$

Combining this with (11.30) gives

$$\begin{aligned}\mathbb{E}_{i_k}[\|x^{k+1} - \bar{x}\|^2] \le \|x^k - \bar{x}\|^2 &- 2\gamma \left(f(x^k) - f(\bar{x}) - \langle \nabla f(\bar{x}), x^k - \bar{x}\rangle \right) \\ &- \frac{\gamma}{NL}\sum_{i=1}^{N} \|\nabla f_i(x^k) - \nabla f_i(\bar{x})\|^2 \\ &+ \gamma^2 \mathbb{E}_{i_k}[\|g^k - \nabla f(\bar{x})\|^2].\end{aligned}$$

By using the definition of g^k, the inequality $\|a+b\|^2 \le 2(\|a\|^2+\|b\|^2)$, and the variance decomposition formula, we have

$$\begin{aligned}
\mathbb{E}_{i_k}[\|g^k-\nabla f(\bar{x})\|^2] &= \mathbb{E}_{i_k}[\|g^k-\nabla f(x^k)\|^2]+\|\nabla f(x^k)-\nabla f(\bar{x})\|^2 \\
&\le 2\mathbb{E}_{i_k}\left[\left\|\nabla f_{i_k}(x^k)-\nabla f_{i_k}(\bar{x})-(\nabla f(x^k)-\nabla f(\bar{x}))\right\|^2\right] \\
&\quad + 2\mathbb{E}_{i_k}\left[\left\|v_{i_k}^k-\nabla f_{i_k}(\bar{x})-\frac{1}{N}\sum_{i=1}^N(v_i^k-\nabla f_i(\bar{x}))\right\|^2\right] \\
&\quad + \|\nabla f(x^k)-\nabla f(\bar{x})\|^2 \\
&\le 2\left(\mathbb{E}_{i_k}[\|\nabla f_{i_k}(x^k)-\nabla f_{i_k}(\bar{x})\|^2]-\|\nabla f(x^k)-\nabla f(\bar{x})\|^2\right) \\
&\quad + 2\mathbb{E}_{i_k}[\|v_{i_k}^k-\nabla f_{i_k}(\bar{x})\|^2]+\|\nabla f(x^k)-\nabla f(\bar{x})\|^2 \\
&\le 2\mathbb{E}_{i_k}[\|\nabla f_{i_k}(x^k)-\nabla f_{i_k}(\bar{x})\|^2]+2\mathbb{E}_{i_k}[\|v_{i_k}^k-\nabla f_{i_k}(\bar{x})\|^2] \\
&= \frac{2}{N}\sum_{i=1}^N\|\nabla f_i(x^k)-\nabla f_i(\bar{x})\|^2+\frac{2}{N}\sum_{i=1}^N\|v_i^k-\nabla f_i(\bar{x})\|^2.
\end{aligned}$$

Therefore, we obtain

Lemma 11.35 *Let Assumption 11.19 hold. Consider Algorithm 11.31 with* $0<\gamma<1/(2L)$. *Let* $\bar{x}$ *be any solution of (11.15). There holds*

$$\begin{aligned}
\mathbb{E}_{i_k}[\|x^{k+1}-\bar{x}\|^2] \le \|x^k-\bar{x}\|^2 &- 2\gamma\left(f(x^k)-f(\bar{x})-\langle\nabla f(\bar{x}),x^k-\bar{x}\rangle\right) \\
&- \left(\frac{1}{L}-2\gamma\right)\frac{\gamma}{N}\sum_{i=1}^N\|\nabla f_i(x^k)-\nabla f_i(\bar{x})\|^2 \\
&+ \frac{2\gamma^2}{N}\sum_{i=1}^N\|v_i^k-\nabla f_i(\bar{x})\|^2
\end{aligned}\tag{11.31}$$

for all integers $k\ge 0$.

Note that, when $0<\gamma<1/(2L)$, by adding a suitable positive multiple of (11.31) in Lemma 11.35 to (11.29) in Lemma 11.34, we can successfully handle the term

$$\frac{4\gamma^2}{N}\sum_{i=1}^N\|\nabla f_i(x^k)-\nabla f_i(\bar{x})\|^2.$$

However, a positive multiple of the term

$$\frac{1}{N}\sum_{i=1}^{N}\|v_i^k-\nabla f_i(\bar{x})\|^2$$

remains there. To deal with such a term, we consider

$$B_k := \sum_{i=1}^{N}\Delta_i(\phi_i^k,\bar{x}), \tag{11.32}$$

where $\{\phi_i^k\}$ are elements given in Remark 11.33 and

$$\Delta_i(x,\bar{x}) := f_i(x)-f_i(\bar{x})-\langle\nabla f_i(\bar{x}),x-\bar{x}\rangle$$

is the Bregman distance induced by f_i at $\bar{x}$. Note that

$$B_{k+1}=\sum_{i=1}^{N}\Delta_i(\phi_i^{k+1},\bar{x})=\sum_{i\neq i_k}\Delta_i(\phi_i^k,\bar{x})+\Delta_{i_k}(x^k,\bar{x}).$$

Thus

$$\begin{aligned}
\mathbb{E}_{i_k}[B_{k+1}] &= \frac{1}{N}\sum_{j=1}^{N}\sum_{i\neq j}\Delta_i(\phi_i^k,\bar{x})+\frac{1}{N}\sum_{i=1}^{N}\Delta_i(x^k,\bar{x})\\
&= \frac{N-1}{N}\sum_{i=1}^{N}\Delta_i(\phi_i^k,\bar{x})+\Big(f(x^k)-f(\bar{x})-\langle\nabla f(x),x^k-\bar{x}\rangle\Big)\\
&= (1-\varepsilon)B_k-\left(\frac{1}{N}-\varepsilon\right)\sum_{i=1}^{N}\Delta_i(\phi_i^k,\bar{x})\\
&\quad+\Big(f(x^k)-f(\bar{x})-\langle\nabla f(\bar{x}),x^k-\bar{x}\rangle\Big),
\end{aligned}$$

where $0\le\varepsilon<1/N$ is any number. By the co-coercivity of ∇f_i for each i, we have

$$\Delta_i(\phi_i^k,\bar{x})\ge\frac{1}{2L}\|\nabla f_i(\phi_i^k)-\nabla f_i(\bar{x})\|^2=\frac{1}{2L}\|v_i^k-\nabla f_i(\bar{x})\|^2.$$

Combining this with the above equation, we therefore obtain the following result.

Lemma 11.36 *Let Assumption 11.19 hold. Consider Algorithm 11.31. Let $\bar{x}$ be any solution (11.15), and define B_k by (11.32). Then for any $0 \le \varepsilon < 1/N$, there holds*

$$\mathbb{E}_{i_k}[B_{k+1}] \le (1-\varepsilon)B_k - \left(\frac{1}{N} - \varepsilon\right)\frac{1}{2L}\sum_{i=1}^{N}\|v_i^k - \nabla f_i(\bar{x})\|^2$$
$$+ \left(f(x^k) - f(\bar{x}) - \langle \nabla f(\bar{x}), x^k - \bar{x}\rangle\right).$$

By multiplying the equation in Lemma 11.36 by 2γ and then adding to the equation in Lemma 11.35, we immediately obtain

Lemma 11.37 *Let Assumption 11.19 hold. Consider Algorithm 11.31 with $0 < \gamma < 1/(2L)$. Let $\bar{x}$ be any solution (11.15) and define B_k by (11.32). Then for any $0 \le \varepsilon < 1/N$, there holds*

$$\mathbb{E}_{i_k}\left[2\gamma B_{k+1} + \|x^{k+1} - \bar{x}\|^2\right]$$
$$\le 2\gamma(1-\varepsilon)B_k + \|x^k - \bar{x}\|^2 - \left(\frac{1}{L} - 2\gamma\right)\frac{\gamma}{N}\sum_{i=1}^{N}\|\nabla f_i(x^k) - \nabla f_i(\bar{x})\|^2$$
$$- \left(\frac{1-N\varepsilon}{L} - 2\gamma\right)\frac{\gamma}{N}\sum_{i=1}^{N}\|v_i^k - \nabla f_i(\bar{x})\|^2$$

for all integers $k \ge 0$.

Combining Lemmas 11.37 and 11.34, we can obtain the following result.

Proposition 11.38 *Let Assumption 11.19 hold. Consider Algorithm 11.31 with $0 < \gamma < 1/(2L)$. Let $\bar{x}$ be any solution of (11.15), and let B_k be defined by (11.32). Then for any $0 < \varepsilon < 1/N$ and $\delta > 0$, there holds*

$$\mathbb{E}_{i_k}\left[2\gamma B_{k+1} + (1+\delta)\|x^{k+1} - \bar{x}\|^2\right]$$
$$\le 2\gamma(1-\varepsilon)B_k + (1+\delta)\|x^k - \bar{x}\|^2 - 2\gamma\delta\mathbb{E}_{i_k}[F(x^{k+1}) - F_*]$$
$$- \left(\frac{1}{L} - 2\gamma - 4\gamma\delta\right)\frac{\gamma}{N}\sum_{i=1}^{N}\|\nabla f_i(x^k) - \nabla f_i(\bar{x})\|^2$$
$$- \left(\frac{1-N\varepsilon}{L} - 2\gamma - 4\gamma\delta\right)\frac{\gamma}{N}\sum_{i=1}^{N}\|v^{k,i} - \nabla f_i(\bar{x})\|^2$$

for all integers $k \ge 0$.

Proof The proof can be concluded by multiplying the equation in Lemma 11.34 by δ and then adding to the equation in Lemma 11.37. □

In the following, we will use Proposition 11.38 to establish convergence and derive convergence rates of Algorithm 11.31 when γ is chosen such that $0 < \gamma < 1/(2L)$.

Theorem 11.39 *Let Assumption 11.19 hold. Consider Algorithm 11.31 and assume $0 < \gamma < 1/(2L)$. Then almost surely $\{x^k\}$ converges to a solution x^* of (11.15) and $F(x^k) \to F_*$ as $k \to \infty$.*

Proof Let z be any solution of (11.15), and define B_k by (11.32) with $\bar{x}$ replaced by z. We may use Proposition 11.38 with $\delta = \frac{1-2\gamma L}{8\gamma L}$ and $\varepsilon = \frac{1-2\gamma L}{2N}$ to obtain

$$\mathbb{E}_{i_k}\left[2\gamma B_{k+1} + (1+\delta)\|x^{k+1} - z\|^2\right] \leq 2\gamma B_k + (1+\delta)\|x^k - z\|^2 - 2\gamma\varepsilon B_k - 2\gamma\delta\mathbb{E}_{i_k}\left[F(x^{k+1}) - F_*\right]. \tag{11.33}$$

This implies that $\{2\gamma B_k + (1+\delta)\|x^k - z\|^2\}$ is a supermartingale and hence

$$\lim_{k\to\infty}\left(2\gamma B_k + (1+\delta)\|x^k - z\|^2\right) \text{ exists almost surely.} \tag{11.34}$$

Furthermore, by taking the full expectation of (11.33), we can see that

$$\begin{aligned} &\mathbb{E}\left[2\gamma\varepsilon B_k + 2\gamma\delta(F(x^{k+1}) - F_*)\right] \\ &\leq \mathbb{E}\left[2\gamma B_k + (1+\delta)\|x^k - z\|^2\right] - \mathbb{E}\left[2\gamma B_{k+1} + (1+\delta)\|x^{k+1} - z\|^2\right] \end{aligned}$$

which implies that

$$\begin{aligned} \mathbb{E}&\left[\sum_{k=0}^{\infty}\left(2\gamma\varepsilon B_k + 2\gamma\delta(F(x^{k+1}) - F_*)\right)\right] \\ &= \sum_{k=0}^{\infty}\mathbb{E}\left[2\gamma\varepsilon B_k + 2\gamma\delta(F(x^{k+1}) - F_*)\right] \\ &\leq 2\gamma B_0 + (1+\delta)\|x^0 - z\|^2 < \infty. \end{aligned}$$

Hence

$$\sum_{k=0}^{\infty}\left(2\gamma\varepsilon B_k + 2\gamma\delta(F(x^{k+1}) - F_*)\right) < \infty \quad \text{almost surely.}$$

Consequently $B_k \to 0$ and $F(x^k) - F_* \to 0$ as $k \to \infty$ almost surely. This together with (11.34) then implies that $\lim_{k\to\infty} \|x^k - z\|^2$ exists almost surely. The above argument actually shows that for any solution z of (11.15), the event

$$\Omega_z := \left\{ \lim_{k\to\infty} \|x^k - z\| \text{ exists and } \lim_{k\to\infty} F(x^k) = F_* \right\}$$

has probability one, i.e., $\mathbb{P}(\Omega_z) = 1$. Now we may use the separability of Euclidean spaces and the similar argument in the proof of Theorem 11.17 to conclude that almost surely $\{x^k\}$ converges to a solution of (11.15) and $F(x^k) \to F_*$. □

We next show that Algorithm 11.31 enjoys a linear convergence if F is strongly convex.

Theorem 11.40 *Let Assumption 11.19 hold. Consider Algorithm 11.31 with* $0 < \gamma < 1/(2L)$. *If* F *is strongly convex, then* $\mathbb{E}[\|x^k - x^*\|^2] \to 0$ *linearly, where* x^* *denotes the unique solution of (11.15).*

Proof Since F is strongly convex, there is a constant $\mu > 0$ such that

$$F(x) - F_* \geq \mu \|x - x^*\|^2, \quad \forall x \in \mathbb{R}^d.$$

Thus, with the same choices of ε and δ in the proof of Theorem 11.39, it follows from (11.33) that

$$\begin{aligned} &\mathbb{E}_{i_k}[2\gamma B_{k+1} + (1+\delta)\|x^{k+1} - x^*\|^2] \\ &\leq 2\gamma(1-\varepsilon)B_k + (1+\delta)\|x^k - x^*\|^2 - 2\gamma\delta\mu\mathbb{E}_{i_k}[\|x^{k+1} - x^*\|^2] \end{aligned}$$

Therefore

$$\begin{aligned} &\mathbb{E}_{i_k}\left[2\gamma B_{k+1} + (1+\delta+2\gamma\delta\mu)\|x^{k+1} - x^*\|^2\right] \\ &\leq 2\gamma(1-\varepsilon)B_k + (1+\delta)\|x^k - x^*\|^2 \leq q\left(2\gamma B_k + (1+\delta+2\gamma\delta\mu)\|x^k - x^*\|^2\right), \end{aligned}$$

where

$$q := \max\left\{1-\varepsilon, \frac{1+\delta}{1+\delta+2\gamma\delta\mu}\right\} < 1.$$

By taking the full expectation and recursively using the result, we thus obtain

$$\begin{aligned} &\mathbb{E}\left[2\gamma B_k + (1+\delta+2\gamma\delta\mu)\|x^k - x^*\|^2\right] \\ &\leq q^k\left(2\gamma B_0 + (1+\delta+2\gamma\delta\mu)\|x^0 - x^*\|^2\right) \end{aligned}$$

which in particular implies that $\mathbb{E}[\|x^k - x^*\|^2] \le Cq^k$ for all integers $k \ge 0$, where C is a constant independent of k. □

The linear convergence of Algorithm 11.31 was first established in [68] under the condition that every f_i is strongly convex. Theorem 11.40 relaxes the condition by requiring only F to be strongly convex.

11.4 Randomized Block-Coordinate Descent Method

In this section, we consider the minimization problem of the form

$$\min_{x_1\in\mathbb{R}^{m_1},\cdots,x_N\in\mathbb{R}^{m_N}} \left\{ F(x) := f(x) + \sum_{i=1}^{N} \psi_i(x_i) \right\}, \tag{11.35}$$

where $x = (x_1, \cdots, x_N)$ and the following condition is assumed to be satisfied by f and ψ_i.

Assumption 11.41

(i) *f is convex and continuous differentiable; for each $i \in \{1, \cdots, N\}$, we use $\nabla_i f$ to denote the block gradient of f with respect to x_i.*
(ii) *For each $i \in \{1, \cdots, N\}$, there is a number L_i such that*

$$\|\nabla_i f(x_{<i}, x_i + h, x_{>i}) - \nabla_i f(x_{<i}, x_i, x_{>i})\| \le L_i \|h\|$$

for all $h \in \mathbb{R}^{m_i}$, where $x_{<i} = (x_1, \cdots, x_{i-1})$ and $x_{>i} = (x_{i+1}, \cdots, x_N)$.
(iii) *For each $i \in \{1, \cdots, N\}$, $\psi_i : \mathbb{R}^{m_i} \to (-\infty, \infty]$ is proper, lower semicontinuous, and convex. We set $\psi(x) := \sum_{i=1}^{N} \psi_i(x_i)$.*
(iv) *The problem (11.35) has an optimal solution; we use S to denote the collection of all optimal solutions and set F_* to be the minimum value of F.*

Note that in (11.35), the variable x is decomposed into N blocks, and the regularization term $\psi(x) := \sum_{i=1}^{N} \psi_i(x_i)$ has the corresponding separable structure. We will explore this structure to develop the randomized block-coordinate descent (RBCD) method; see [143, 149, 166, 187]. The RBCD method focuses on updating only one block of x at each iteration step, which simplifies the computations significantly, particularly for large-scale problems. The core idea is to randomly select one of the blocks, say block i, and apply the proximal gradient method to update the coordinates associated with x_i while keeping all other coordinates x_j (for $j \ne i$) fixed. This randomization reduces the computational complexity per iteration, since it avoids a full update of the entire vector x and a calculation of the full gradient of f. The detailed formulation of the algorithm is as follows.

Algorithm 11.42 (Randomized Block-Coordinate Descent Method) *Pick an initial guess $x^0 \in \mathbb{R}^{m_1} \times \cdots \times \mathbb{R}^{m_N}$ and step sizes t_i for $i = 1, \cdots, N$. For $k = 0, 1, \cdots$, do*

(i) *Pick $i_k \in \{1, \cdots, N\}$ randomly via the uniform distribution;*
(ii) *Update $x^{k+1} := (x_1^{k+1}, \cdots, x_N^{k+1})$ by setting $x_i^{k+1} = x_i^k$ for $i \neq i_k$ and*

$$x_{i_k}^{k+1} = \text{prox}_{t_{i_k}\psi_{i_k}}(x_{i_k}^k - t_{i_k}\nabla_{i_k} f(x^k)).$$

We present a convergence analysis of Algorithm 11.42 under Assumption 11.41. The following general sublinear convergence result for Algorithm 11.42 is established in [149].

Theorem 11.43 *Let Assumption 11.41 hold, and consider Algorithm 11.42. If $0 < t_i \leq 1/L_i$ for each $i = 1, \cdots, N$, then*

$$\mathbb{E}[F(x^k) - F_*] \leq \frac{N}{N+k}\left(F(x^0) - F_* + \sum_{i=1}^N \frac{1}{t_i}\|x^0 - \bar{x}\|^2\right)$$

for all integers $k \geq 0$, where $\bar{x}$ denotes any solution of (11.35).

Proof Based on x^k, we first introduce $\hat{x}^{k+1} := (\hat{x}_1^{k+1}, \cdots, \hat{x}_N^{k+1})$ by defining

$$\hat{x}_i^{k+1} = \text{prox}_{t_i\psi_i}(x_i^k - t_i\nabla_i f(x^k)), \quad i = 1, \cdots, N.$$

Note that $x_i^{k+1} = x_i^k$ for $i \neq i_k$ and $x_{i_k}^{k+1} = \hat{x}_{i_k}^{k+1}$. By the given conditions on f, we have

$$f(x^{k+1}) \leq f(x^k) + \langle \nabla_{i_k} f(x^k), \hat{x}_{i_k}^{k+1} - x_{i_k}^k\rangle + \frac{L_{i_k}}{2}\|\hat{x}_{i_k}^{k+1} - x_{i_k}^k\|^2$$

which implies that

$$\begin{aligned} F(x^{k+1}) - F(x^k) &\leq \langle \nabla_{i_k} f(x^k), \hat{x}_{i_k}^{k+1} - x_{i_k}^k\rangle + \frac{L_{i_k}}{2}\|\hat{x}_{i_k}^{k+1} - x_{i_k}^k\|^2 \\ &\quad + \psi_{i_k}(\hat{x}_{i_k}^{k+1}) - \psi_{i_k}(x_{i_k}^k). \end{aligned} \tag{11.36}$$

According to the definition of $\hat{x}_i^{k+1}$, we have

$$\hat{x}_i^{k+1} = \arg\min_{x_i \in \mathbb{R}^{m_i}} \left\{\psi_i(x_i) + \langle \nabla_i f(x^k), x_i - x_i^k\rangle + \frac{1}{2t_i}\|x_i - x_i^k\|^2\right\}$$

and hence, by using $0 < t_i \le 1/L_i$, the right-hand side of (11.36) is nonpositive. Consequently,

$$F(x^{k+1}) \le F(x^k), \quad \forall k \ge 0. \tag{11.37}$$

Next, by taking the expectation with respect to i_k, we can obtain from (11.36) that

$$\mathbb{E}_{i_k}[F(x^{k+1})] \le F(x^k) + \frac{1}{N}\langle \nabla f(x^k), \hat{x}^{k+1} - x^k\rangle + \frac{1}{2N}\sum_{i=1}^{N} L_i \|\hat{x}_i^{k+1} - x_i^k\|^2$$
$$+ \frac{1}{N}(\psi(\hat{x}^{k+1}) - \psi(x^k)). \tag{11.38}$$

We will use (11.38) to derive the estimate on $\mathbb{E}[F(x^k) - F_*]$. Let $\bar{x}$ be any solution of (11.35). By using again the definition of $\hat{x}_i^{k+1}$, we have

$$-\nabla_i f(x^k) - \frac{1}{t_i}(\hat{x}_i^{k+1} - x_i^k) \in \partial\psi_i(\hat{x}_i^{k+1})$$

and hence

$$-\nabla f(x^k) - \left(\frac{1}{t_1}(\hat{x}_1^{k+1} - x_1^k), \cdots, \frac{1}{t_N}(\hat{x}_N^{k+1} - x_N^k)\right)^T \in \partial\psi(\hat{x}^{k+1}).$$

Thus, by the convexity of ψ, we can obtain

$$\psi(\hat{x}^{k+1}) \le \psi(\bar{x}) + \langle \nabla f(x^k), \bar{x} - \hat{x}^{k+1}\rangle + \sum_{i=1}^{N}\frac{1}{t_i}\langle \hat{x}_i^{k+1} - x_i^k, \bar{x}_i - \hat{x}_i^{k+1}\rangle.$$

Combining this with (11.38) gives

$$\mathbb{E}_{i_k}[F(x^{k+1})] \le F(x^k) + \frac{1}{N}(\psi(\bar{x}) - \psi(x^k)) + \frac{1}{N}\langle \nabla f(x^k), \bar{x} - x^k\rangle$$
$$+ \frac{1}{N}\sum_{i=1}^{N}\frac{1}{t_i}\langle \hat{x}_i^{k+1} - x_i^k, \bar{x}_i - \hat{x}_i^{k+1}\rangle + \frac{1}{2N}\sum_{i=1}^{N} L_i\|\hat{x}_i^{k+1} - x_i^k\|^2.$$

By the convexity of f, we have

$$\langle \nabla f(x^k), \bar{x} - x^k\rangle \le f(\bar{x}) - f(x^k).$$

Consequently

$$\mathbb{E}_{i_k}[F(x^{k+1})] \le F(x^k) - \frac{1}{N}(F(x^k) - F_*) + \frac{1}{2N}\sum_{i=1}^{N} L_i \|\hat{x}_i^{k+1} - x_i^k\|^2$$
$$+ \frac{1}{N}\sum_{i=1}^{N} \frac{1}{t_i}\langle \hat{x}_i^{k+1} - x_i^k, \bar{x}_i - \hat{x}_i^{k+1}\rangle. \tag{11.39}$$

To proceed further, we consider the quantity

$$r_k := \sum_{i=1}^{N} \frac{1}{2t_i}\|x_i^k - \bar{x}_i\|^2.$$

Then by the definition of x^{k+1}, we have

$$r_{k+1} = r_k + \frac{1}{2t_{i_k}}\left(\|\hat{x}_{i_k}^{k+1} - \bar{x}_{i_k}\|^2 - \|x_{i_k}^k - \bar{x}_{i_k}\|^2\right).$$

Consequently

$$\mathbb{E}_{i_k}[r_{k+1}] = r_k + \frac{1}{2N}\sum_{i=1}^{N} \frac{1}{t_i}\left(\|\hat{x}_i^{k+1} - \bar{x}_i\|^2 - \|x_i^k - \bar{x}_i\|^2\right)$$
$$= r_k + \frac{1}{2N}\sum_{i=1}^{N} \frac{1}{t_i}\left(2\langle \hat{x}_i^{k+1} - x_i^k, \hat{x}_i^{k+1} - \bar{x}_i\rangle - \|\hat{x}_i^{k+1} - x_i^k\|^2\right).$$

Adding this equation with (11.39) gives

$$\mathbb{E}_{i_k}[F(x^{k+1}) - F_* + r_{k+1}] \le (F(x^k) - F_* + r_k) - \frac{1}{N}(F(x^k) - F_*)$$
$$- \frac{1}{2N}\sum_{i=1}^{N}\left(\frac{1}{t_i} - L_i\right)\|\hat{x}_i^{k+1} - x_i^k\|^2.$$

Since $0 < t_i \le 1/L_i$, we thus obtain

$$\mathbb{E}_{i_k}[F(x^{k+1}) - F_* + r_{k+1}] \le (F(x^k) - F_* + r_k) - \frac{1}{N}(F(x^k) - F_*).$$

Taking the full expectation gives

$$\mathbb{E}[F(x^{k+1}) - F_* + r_{k+1}] \le \mathbb{E}[F(x^k) - F_* + r_k] - \frac{1}{N}\mathbb{E}[F(x^k) - F_*].$$

Recursively using this inequality shows

$$\mathbb{E}[F(x^k) - F_*] \le \mathbb{E}[F(x^k) - F_* + r_k] \le \Delta_0 - \frac{1}{N}\sum_{l=0}^{k-1}\mathbb{E}[F(x^l) - F_*],$$

where $\Delta_0 := F(x^0) - F_* + r_0$. Since (11.37) implies that $\mathbb{E}[F(x^l) - F_*]$ is monotonically decreasing, we can further obtain

$$\mathbb{E}[F(x^k) - F_*] \le \Delta_0 - \frac{k}{N}\mathbb{E}[F(x^k) - F_*]$$

which implies the desired estimate. □

Next we provide a linear convergence result on Algorithm 11.42 under the bounded metric subregularity of ∂F; see [228].

Theorem 11.44 *Let Assumption 11.41 hold, and consider Algorithm 11.42 with* $0 < t_i \le 1/L_i$. *Assume that the sublevel set* $\{F \le F(x^0)\}$ *is bounded and* ∂F *is bounded metrically subregular at* $(\bar{x}, 0) \in Gr(\partial F)$ *for some* $\bar{x} \in S$. *Then* $\mathbb{E}[F(x^k) - F_*]$ *converges to* 0 *linearly.*

Proof According to (11.37), we have $F(x^k) \le F(x^0)$ for all $k \ge 0$. Since $\{F \le F(x^0)\}$ is bounded, there is a number R such that $\|x^k\| \le R$ for all k. Since ∂F is bounded metrically subregular at $(x^*, 0)$ for some $x^* \in S$, we can conclude from Theorems 8.8, 8.7, and Lemma 8.5 that there is a constant $\kappa \ge 0$ such that

$$d(x, S) \le \kappa \|x - \text{prox}_{t_m \psi}(x - t_m \nabla f(x))\| \tag{11.40}$$

for all x satisfying $\|x\| \le R$, where $t_m := \min\{t_i : 1 \le i \le N\}$.

Now we use (11.38) to derive the linear convergence of $\mathbb{E}[F(x^k) - F_*]$. According to (11.38) and $0 < t_i \le 1/L_i$, we have

$$\begin{aligned}
&\mathbb{E}_{i_k}[F(x^{k+1})] - F(x^k) \\
&\le \frac{1}{N}\sum_{i=1}^{N}\left(\psi_i(\hat{x}_i^{k+1}) + \langle \nabla_i f(x^k), \hat{x}_i^{k+1} - x_i^k\rangle + \frac{L_i}{2}\|\hat{x}_i^{k+1} - x_i^k\|^2 - \psi_i(x_i^k)\right) \\
&\le \frac{1}{N}\sum_{i=1}^{N}\left(\psi_i(\hat{x}_i^{k+1}) + \langle \nabla_i f(x^k), \hat{x}_i^{k+1} - x_i^k\rangle + \frac{1}{2t_i}\|\hat{x}_i^{k+1} - x_i^k\|^2 - \psi_i(x_i^k)\right).
\end{aligned}$$

By using the definition of $\hat{x}_i^{k+1}$ for $i = 1, \cdots, N$, we have for any $x = (x_1, \cdots, x_N) \in \mathbb{R}^{m_1} \times \cdots \times \mathbb{R}^{m_N}$ that

$$\begin{aligned}
&\mathbb{E}_{i_k}[F(x^{k+1})] - F(x^k) \\
&\leq \frac{1}{N}\sum_{i=1}^{N}\left(\psi_i(x_i) + \langle \nabla_i f(x^k), x_i - x_i^k\rangle + \frac{1}{2t_i}\|x_i - x_i^k\|^2 - \psi_i(x_i^k)\right) \\
&\leq \frac{1}{N}\left(\psi(x) + \langle \nabla f(x^k), x - x^k\rangle + \frac{1}{2t_m}\|x - x^k\|^2 - \psi(x^k)\right),
\end{aligned}$$

where for the second inequality we used $t_i \geq t_m$ for all i. Let

$$\Phi_k(x) := \psi(x) + \langle \nabla f(x^k), x - x^k\rangle + \frac{1}{2t_m}\|x - x^k\|^2 + f(x^k).$$

Then

$$\mathbb{E}_{i_k}[F(x^{k+1})] - F(x^k) \leq \frac{1}{N}\left(\Phi_k(x) - F(x^k)\right), \quad \forall x. \tag{11.41}$$

Note that Φ_k is $1/(2t_m)$-strongly convex and achieves its minimum at

$$\tilde{x}^k := \operatorname{prox}_{t_m \psi}(x^k - t_m \nabla f(x^k)).$$

Therefore

$$\Phi_k(x) - \Phi_k(\tilde{x}^k) \geq \frac{1}{2t_m}\|x - \tilde{x}^k\|^2, \quad \forall x.$$

Since $\Phi_k(x^k) = \psi(x^k) + f(x^k) = F(x^k)$, we thus have

$$F(x^k) - \Phi_k(\tilde{x}^k) \geq \frac{1}{2t_m}\|x^k - \tilde{x}^k\|^2. \tag{11.42}$$

Let $\bar{x} \in S$ be any element. Then by the minimality of $\tilde{x}^k$, we have

$$\Phi_k(\tilde{x}^k) \leq \psi(\bar{x}) + \langle \nabla f(x^k), \bar{x} - x^k\rangle + \frac{1}{2t_m}\|\bar{x} - x^k\|^2 + f(x^k).$$

By using the conditions on f, it is easy to derive that

$$\langle \nabla f(x^k), \bar{x} - x^k\rangle + f(x^k) - f(\bar{x}) \leq \frac{L}{2}\|x^k - \bar{x}\|^2.$$

Consequently

$$\begin{aligned}\Phi_k(\tilde{x}^k) &\le \psi(\bar{x}) + f(\bar{x}) + \left(\frac{L}{2} + \frac{1}{2t_m}\right)\|x^k - \bar{x}\|^2 \\ &= F_* + \left(\frac{L}{2} + \frac{1}{2t_m}\right)\|x^k - \bar{x}\|^2.\end{aligned}$$

Since $\bar{x} \in S$ is arbitrary, we thus obtain

$$\Phi_k(\tilde{x}^k) - F_* \le \left(\frac{L}{2} + \frac{1}{2t_m}\right) d(x^k, S)^2.$$

By using (11.40) and (11.42), we then have

$$\begin{aligned}\Phi_k(\tilde{x}^k) - F_* &\le \kappa^2\left(\frac{L}{2} + \frac{1}{2t_m}\right)\|x^k - \mathrm{prox}_{t_m\psi}(x^k - t_m\nabla f(x^k))\|^2 \\ &= \kappa^2\left(\frac{L}{2} + \frac{1}{2t_m}\right)\|x^k - \tilde{x}^k\|^2 \\ &\le \kappa^2(1 + Lt_m)(F(x^k) - \Phi_k(\tilde{x}^k))\end{aligned}$$

which implies that

$$F(x^k) - F_* \le \left(1 + \kappa^2(1 + Lt_m)\right)\left(F(x^k) - \Phi_k(\tilde{x}^k)\right).$$

Therefore, it follows from (11.41) with $x = \tilde{x}^k$ that

$$\begin{aligned}&\mathbb{E}_{i_k}[F(x^{k+1})] - F(x^k) \\ &\le \frac{1}{N}\left(\Phi_k(\tilde{x}^k) - F(x^k)\right) \le -\frac{1}{N(1 + \kappa^2(1 + Lt_m))}(F(x^k) - F_*).\end{aligned}$$

Consequently

$$\mathbb{E}_{i_k}[F(x^{k+1}) - F_*] \le \left(1 - \frac{1}{N(1 + \kappa^2(1 + Lt_m))}\right)(F(x^k) - F_*).$$

Taking the full expectation gives

$$\mathbb{E}[F(x^{k+1}) - F_*] \le \left(1 - \frac{1}{N(1 + \kappa^2(1 + Lt_m))}\right)\mathbb{E}[F(x^k) - F_*]$$

which shows the desired linear convergence. □

Example 11.45 Consider the linear system $Ax = b$, where $A \in \mathbb{R}^{m\times n}$ and $b \in \mathbb{R}^m$. Let $\mathcal{R} : \mathbb{R}^n \to (-\infty, \infty]$ be a proper, lower semicontinuous, strongly convex function. We assume $Ax = b$ has a solution lying in $\text{dom}(\mathcal{R})$. Our goal is to find a solution of $Ax = b$ with smallest $\mathcal{R}$ value, which is equivalent to solving the following minimization problem

$$\min\left\{\mathcal{R}(x) : x \in \mathbb{R}^n \text{ and } Ax = b\right\}. \tag{11.43}$$

By the given conditions on $\mathcal{R}$, it is easy to show that (11.43) has a unique optimal solution, which is denoted by $x^\dagger$. In the following, we will show that Algorithm 11.42 can be used to solve (11.43) by applying it to the dual problem. Let $\mathcal{R}^*$ denote the Legendre-Fenchel conjugate of $\mathcal{R}$ which, by Theorem 5.8, is continuous differentiable. Then the dual function is given by

$$d(\lambda) := \inf_{x\in\mathbb{R}^n} \{\mathcal{R}(x) - \langle\lambda, Ax - b\rangle\} = -\mathcal{R}^*(A^T\lambda) + \langle b, \lambda\rangle, \quad \forall\lambda \in \mathbb{R}^m.$$

Thus the dual problem of (11.43) is

$$\min\left\{\phi(\lambda) := -d(\lambda) = \mathcal{R}^*(A^T\lambda) - \langle b, \lambda\rangle\right\} \tag{11.44}$$

We can show that (11.44) has an optimal solution. To see this, we note by the KKT theory that, for the unique solution $x^\dagger$ of (11.43), there exists $\lambda^\dagger \in \mathbb{R}^m$ such that $A^T\lambda^\dagger \in \partial\mathcal{R}(x^\dagger)$. Thus $x^\dagger = \nabla\mathcal{R}^*(A^T\lambda^\dagger)$ by Theorem 4.9 and

$$\nabla\phi(\lambda^\dagger) = A\nabla\mathcal{R}^*(A^T\lambda^\dagger) - b = Ax^\dagger - b = 0.$$

Since ϕ is a convex function, $\lambda^\dagger$ is a minimizer of ϕ. Let A_i denote the i-th row of A, and let b_i denote the i-th component of b. Then the dual problem can be written as

$$\min\left\{\phi(\lambda) := \mathcal{R}^*\left(\sum_{i=1}^m \lambda_i A_i^T\right) - \sum_{i=1}^m b_i\lambda_i\right\}. \tag{11.45}$$

which takes the form (11.35) with $\psi_i = 0$ for all i. Thus, Algorithm 11.42 can be applied to solve the dual problem. Note that

$$\nabla_i\phi(\lambda) = A_i\nabla\mathcal{R}^*\left(\sum_{j=1}^m \lambda_j A_j^T\right) - b_i = A_i\nabla\mathcal{R}^*(A^T\lambda) - b_i$$

for $i = 1, \cdots, m$. Hence, applying Algorithm 11.42 to solve (11.44) gives the algorithm: Starting from $\lambda^0 = 0 \in \mathbb{R}^m$, once $\lambda^k \in \mathbb{R}^m$ is determined, we choose an index $i_k \in \{1, \cdots, m\}$ randomly via a uniform distribution and update

$\lambda^{k+1} = (\lambda_1^{k+1}, \cdots, \lambda_m^{k+1})$ by setting $\lambda_i^{k+1} = \lambda_i^k$ for $i \neq i_k$ and

$$\lambda_{i_k}^{k+1} = \lambda_{i_k}^k - t_{i_k}\left(A_{i_k}\nabla\mathcal{R}^*(A^T\lambda^k) - b_{i_k}\right).$$

By setting $\xi^k := A^T\lambda^k$ and $x^k := \nabla\mathcal{R}^*(A^T\lambda^k)$. It is straightforward to verify that

$$x^k = \nabla\mathcal{R}^*(\xi^k), \quad \xi^{k+1} = \xi^k - t_{i_k}A_{i_k}^T(A_{i_k}x^k - b_{i_k}) \tag{11.46}$$

for $k = 0, 1, \cdots$, where $\xi^0 = 0$. The update rule (11.46) is known as the stochastic mirror descent method for solving (11.43); see[121], which is a stochastic version of Algorithm 3.46.

Using Theorem 11.43, we will show that $\mathbb{E}[\|x^k - x^\dagger\|^2] \to 0$ as $k \to \infty$ with a sublinear rate. Indeed, a straightforward application of Theorem 11.43 gives

$$\mathbb{E}[\phi(\lambda^k) - \phi(\lambda^\dagger)] = O\left(\frac{1}{k+1}\right). \tag{11.47}$$

Furthermore, by using the argument in the proof of Theorem 5.45, we have for the Bregman distance

$$D_{\mathcal{R}}^{\xi^k}(x^\dagger, x^k) := \mathcal{R}(x^\dagger) - \mathcal{R}(x^k) - \langle\xi^k, x^\dagger - x^k\rangle$$

that

$$\sigma\|x^k - x^\dagger\| \le D_{\mathcal{R}}^{\xi^k}(x^\dagger, x^k) = \phi(\lambda^k) - \phi(\lambda^\dagger), \tag{11.48}$$

where $\sigma > 0$ is the modulus of convexity of $\mathcal{R}$. Thus, we can use (11.47) to conclude

$$\mathbb{E}[\|x^k - x^\dagger\|^2] = O\left(\frac{1}{k+1}\right).$$

Example 11.46 If we take $\mathcal{R}(x) = \frac{1}{2}\|x\|^2$ in Example 11.45, then the method (11.46) becomes

$$x^{k+1} = x^k - t_{i_k}A_{i_k}^T(A_{i_k}x^k - b_{i_k}) \tag{11.49}$$

with $i_k \in \{1, \cdots, m\}$ chosen uniformly at random, where $x^0 = 0$. It is interesting to note that (11.49) can be obtained by directly applying the stochastic gradient descent to minimize the function

$$\frac{1}{2}\|Ax - b\|^2 = \frac{1}{2}\sum_{i=1}^m \|A_i x - b_i\|^2.$$

Using Theorem 11.44, we will show that $\mathbb{E}[\|x^k - x^\dagger\|^2] \to 0$ linearly as $k \to \infty$. Since $\phi(\lambda) = \frac{1}{2}\|A^T\lambda\|^2 - \langle b, \lambda\rangle$, we have $\nabla\phi(\lambda) = AA^T\lambda - b$. Thus, we may use Theorem 8.23 to conclude that $\nabla\phi$ is bounded metrically subregular. While we cannot establish the boundedness of sublevel sets of ϕ, we can show that there is a constant C, independent of the sample paths, such that $\|\lambda^k\| \le C$ for all $k \ge 0$, which is sufficient for the proof of Theorem 11.44. Indeed, according to the proof of Theorem 11.43, the sequence $\{\phi(\lambda^k)\}$ is monotonically decreasing, implying $\phi(\lambda^k) \le \phi(\lambda^0) = 0$. Consequently, using $b = Ax^\dagger$, we have

$$\frac{1}{2}\|A^T\lambda^k\|^2 - \langle x^\dagger, A^T\lambda^k\rangle = \phi(\lambda^k) \le 0$$

which implies $\|A^T\lambda^k\| \le 2\|x^\dagger\|$ for all $k \ge 0$. Since, by the definition of λ^k, we have $\lambda^k \in \text{Ran}(A)$ for all k, we can apply Lemma 6.19 to conclude that $\|\lambda^k\| \le C_0\|A^T\lambda^k\| \le 2C_0\|x^\dagger\|$ for all $k \ge 0$, where C_0 is the constant from Lemma 6.19. Therefore, Theorem 11.44 guarantees that $\mathbb{E}[\phi(\lambda^k) - \phi(\lambda^\dagger)] \to 0$ linearly as $k \to \infty$. This together with (11.48) implies that $\mathbb{E}[\|x^k - x^\dagger\|^2] \to 0$ linearly as $k \to \infty$.

Example 11.47 If we take $\mathcal{R}(x) = \gamma\|x\|_1 + \frac{1}{2}\|x\|^2$ with a large number $\gamma > 0$, then the method (11.46) becomes

$$\begin{aligned}
\xi^{k+1} &= \xi^k - t_{i_k}A_{i_k}^T(A_{i_k}x_k - b_{i_k}), \\
x^{k+1} &= \text{sign}(\xi^k)\cdot\max\{|\xi^k| - \gamma, 0\}
\end{aligned}$$

with $x^0 = \xi^0 = 0$, where $i_k \in \{1, \cdots, m\}$ is chosen uniformly at random. This is known as the randomized sparse Kaczmarz method, studied in [200]. With appropriate choices of the step sizes t_{i_k}, the linear convergence of this method has been established in [200].

11.5 Stochastic Primal Dual Hybrid Gradient Method

We consider the minimization problem

$$\min_{x\in\mathbb{R}^d}\left\{\sum_{i=1}^N f_i(A_ix) + g(x)\right\}, \tag{11.50}$$

where, for each $i \in \{1, \cdots, N\}$, A_i is a matrix of size $m_i \times d$, and $f_i : \mathbb{R}^{m_i} \to (-\infty, \infty]$ and $g : \mathbb{R}^d \to (-\infty, \infty]$ are proper, lower semicontinuous, convex functions. Let $Y := \mathbb{R}^{m_1} \times \cdots \times \mathbb{R}^{m_N}$ with the natural inner product inherited from its component spaces Y_i. We define $f : Y \to (-\infty, \infty]$ by

$$f(y) := \sum_{i=1}^{N} f_i(y_i), \quad \forall y = (y_1; \cdots; y_N) \in Y$$

and introduce the matrix

$$A := \begin{pmatrix} A_1 \\ \vdots \\ A_N \end{pmatrix}$$

which has the size $\prod_{i=1}^{N} m_i \times d$. Clearly $A^T = (A_1^T, \cdots, A_N^T)$ and thus

$$A^T y = \sum_{i=1}^{N} A_i^T y_i, \quad \forall y = (y_1; \cdots; y_N) \in Y.$$

Using the Legendre-Fenchel conjugates f_i^* of f_i for $i = 1, \cdots, N$, we can rewrite (11.50) equivalently as the saddle point problem

$$\min_{x \in \mathbb{R}^d} \sup_{y \in Y} \left\{ \sum_{i=1}^{N} \left(\langle A_i x, y_i \rangle - f_i^*(y_i) \right) + g(x) \right\} \tag{11.51}$$

for which we may use the primal-dual splitting algorithm, introduced in Chap. 6, to solve. However, the primal-dual splitting algorithm requires to use A_i and A_i^* for all i at each iteration and thus can be time-consuming if N is large. In this section, we introduce a stochastic primal-dual hybrid gradient method which uses partial information of data at each iteration to reduce the computational cost. The detailed description of the algorithm is as follows [52].

Algorithm 11.48 (Stochastic Primal Dual Hybrid Gradient Method) *Pick initial guesses $x^0 \in \mathbb{R}^d$, $y^1 = \bar{y}^1 \in Y$ and step sizes $\tau > 0$, $\sigma_i > 0$ for $i = 1, \cdots, N$. Set up a probability distribution $\{p_1, \cdots, p_N\}$ with $p_i > 0$ for $i = 1, \cdots, N$, and let $P = \mathrm{diag}(p_1 I_1, \cdots, p_N I_N)$, where $I_i \in \mathbb{R}^{m_i \times m_i}$ is the identity matrix for each i. For $k = 1, 2, \cdots$, do the following*

(i) *Calculate $x^k = \mathrm{prox}_{\tau g}(x^{k-1} - \tau A^T \bar{y}^k)$;*
(ii) *Draw $i_k \in \{1, \cdots, N\}$ with $\mathbb{P}(i_k = i) = p_i$;*
(iii) *Calculate $y^{k+1} := (y_1^{k+1}, \cdots, y_N^{k+1}) \in Y$ by*

$$y_i^{k+1} = \begin{cases} \mathrm{prox}_{\sigma_{i_k} f_{i_k}^*}(y_{i_k}^k + \sigma_{i_k} A_{i_k} x^k) & \text{if } i = i_k, \\ y_i^k & \text{if } i \neq i_k; \end{cases}$$

(iv) *Calculate* $\bar{y}^{k+1} := y^{k+1} + P^{-1}(y^{k+1} - y^k)$, *i.e.,*

$$\bar{y}_i^{k+1} = \begin{cases} y_{i_k}^{k+1} + \frac{1}{p_{i_k}}(y_{i_k}^{k+1} - y_{i_k}^k) & \textit{if } i = i_k, \\ y_i^k, & \textit{if } i \neq i_k. \end{cases}$$

In the formulation of Algorithm 11.48, it appears that computing $A^T\bar{y}^k$ is necessary to obtain x^k. However, the direct computation of $A^T\bar{y}^k$ can actually be avoided. Indeed, by the definitions of $\bar{y}^{k+1}$ and y^{k+1}, we observe that

$$A^T\bar{y}^{k+1} = A^T y^{k+1} + \frac{1}{p_{i_k}} A_{i_k}^T(y_{i_k}^{k+1} - y_{i_k}^k) \tag{11.52}$$

and

$$A^T y^{k+1} = \sum_{i\neq i_k} A_i^T y_i^{k+1} + A_{i_k}^T y_{i_k}^{k+1} = A^T y^k + A_{i_k}^T(y_{i_k}^{k+1} - y_{i_k}^k).$$

Thus, $A^T y^k$ can be computed recursively. By storing the most recent $A^T y^{k+1}$, we can efficiently compute $A^T\bar{y}^{k+1}$ using (11.52).

In the following, we will provide a convergence analysis of Algorithm 11.48. For convenience, we introduce $\hat{y}^{k+1} := (\hat{y}_1^{k+1}; \cdots; \hat{y}_N^{k+1}) \in Y$, where

$$\hat{y}_i^{k+1} := \text{prox}_{\sigma_i f_i^*}(y_i^k + \sigma_i A_i x^k), \quad i = 1, \cdots, N.$$

By the definition of $\hat{y}_i^{k+1}$, we have $(y_i^k - \hat{y}_i^{k+1})/\sigma_i + A_i x^k \in \partial f_i^*(\hat{y}_i^{k+1})$, and hence by the convexity of f_i^*, we can obtain

$$\begin{aligned} f_i^*(y_i) &\geq f_i^*(\hat{y}_i^{k+1}) + \langle A_i x^k, y_i - \hat{y}_i^{k+1}\rangle + \frac{1}{\sigma_i}\langle y_i^k - \hat{y}_i^{k+1}, y_i - \hat{y}_i^{k+1}\rangle \\ &= f_i^*(\hat{y}_i^{k+1}) + \langle A_i x^k, y_i - \hat{y}_i^{k+1}\rangle + \frac{1}{2\sigma_i}\|y_i^k - \hat{y}_i^{k+1}\|^2 \\ &\quad + \frac{1}{2\sigma_i}\|\hat{y}_i^{k+1} - y_i\|^2 - \frac{1}{2\sigma_i}\|y_i^k - y_i\|^2 \end{aligned}$$

for any $y_i \in \mathbb{R}^{m_i}$. Let $\sigma := \text{diag}(\sigma_1 I_1, \cdots, \sigma_N I_N)$ and set

$$\|y\|_{\sigma^{-1}}^2 := \langle y, \sigma^{-1}y\rangle = \sum_{i=1}^N \frac{1}{\sigma_i}\|y_i\|^2$$

for any $y \in Y$. Then, summing the above inequality from $i = 1$ to N gives

$$\begin{aligned} f^*(y) \geq f^*(\hat{y}^{k+1}) + \langle Ax^k, y - \hat{y}^{k+1}\rangle + \frac{1}{2}\|y^k - \hat{y}^{k+1}\|^2_{\sigma^{-1}} \\ + \frac{1}{2}\|\hat{y}^{k+1} - y\|^2_{\sigma^{-1}} - \frac{1}{2}\|y^k - y\|^2_{\sigma^{-1}}. \end{aligned} \tag{11.53}$$

Similarly, the definition of x^k implies $(x^{k-1} - x^k)/\tau - A^T\bar{y}^k \in \partial g(x^k)$, we may use the convexity of g to obtain

$$\begin{aligned} g(x) &\geq g(x^k) - \langle A^T\bar{y}^k, x - x^k\rangle + \frac{1}{\tau}\langle x^{k-1} - x^k, x - x^k\rangle \\ &= g(x^k) - \langle A^T\bar{y}^k, x - x^k\rangle + \frac{1}{2\tau}\|x^k - x^{k-1}\|^2 \\ &\quad + \frac{1}{2\tau}\|x^k - x\|^2 - \frac{1}{2\tau}\|x^{k-1} - x\|^2 \end{aligned} \tag{11.54}$$

for all $x \in \mathbb{R}^d$. Adding (11.53) and (11.54) shows that

$$\begin{aligned} 0 \geq\ & g(x^k) - g(x) - \langle A^T\bar{y}^k, x - x^k\rangle + f^*(\hat{y}^{k+1}) - f^*(y) + \langle Ax^k, y - \hat{y}^{k+1}\rangle \\ & + \frac{1}{2\tau}\left(\|x^k - x^{k-1}\|^2 + \|x^k - x\|^2 - \|x^{k-1} - x\|^2\right) \\ & + \frac{1}{2}\left(\|y^k - \hat{y}^{k+1}\|^2_{\sigma^{-1}} + \|\hat{y}^{k+1} - y\|^2_{\sigma^{-1}} - \|y^k - y\|^2_{\sigma^{-1}}\right). \end{aligned}$$

Now we introduce

$$\begin{aligned} D_g(x^k; x, y) &:= g(x^k) - g(x) + \langle A^T y, x^k - x\rangle, \\ D_{f^*}(\hat{y}^{k+1}; x, y) &:= f^*(\hat{y}^{k+1}) - f^*(y) - \langle Ax, \hat{y}^{k+1} - y\rangle. \end{aligned}$$

Then we have

$$\begin{aligned} 0 \geq\ & D_g(x^k; x, y) + D_{f^*}(\hat{y}^{k+1}; x, y) + \langle A(x - x^k), \hat{y}^{k+1} - \bar{y}^k\rangle \\ & + \frac{1}{2\tau}\left(\|x^k - x^{k-1}\|^2 + \|x^k - x\|^2 - \|x^{k-1} - x\|^2\right) \\ & + \frac{1}{2}\left(\|y^k - \hat{y}^{k+1}\|^2_{\sigma^{-1}} + \|\hat{y}^{k+1} - y\|^2_{\sigma^{-1}} - \|y^k - y\|^2_{\sigma^{-1}}\right). \end{aligned}$$

This together with the formula $\bar{y}^k = y^k + P^{-1}(y^k - y^{k-1})$ shows that

$$\begin{aligned}
&D_g(x^k; x, y) + D_{f^*}(\hat{y}^{k+1}; x, y) \\
&\leq \langle A(x - x^k), y^k - \hat{y}^{k+1}\rangle + \langle A(x - x^k), P^{-1}(y^k - y^{k-1})\rangle \\
&\quad - \frac{1}{2\tau}\left(\|x^k - x^{k-1}\|^2 + \|x^k - x\|^2 - \|x^{k-1} - x\|^2\right) \\
&\quad - \frac{1}{2}\left(\|y^k - \hat{y}^{k+1}\|^2_{\sigma^{-1}} + \|\hat{y}^{k+1} - y\|^2_{\sigma^{-1}} - \|y^k - y\|^2_{\sigma^{-1}}\right) \qquad (11.55)
\end{aligned}$$

for any $x \in \mathbb{R}^d$ and $y \in Y$.

In order to proceed further, we need the following observations. Let $\mathbb{E}_{i_k}$ denote the expectation taken with respect to i_k. For any $y \in Y$, we have

$$\begin{aligned}
&\mathbb{E}_{i_k}\left[\|y^{k+1} - y\|^2_{\sigma^{-1}P^{-1}}\right] \\
&= \mathbb{E}_{i_k}\left[\frac{1}{\sigma_{i_k} p_{i_k}}\|\hat{y}^{k+1}_{i_k} - y_{i_k}\|^2 + \sum_{i \neq i_k} \frac{1}{\sigma_i p_i}\|y^k_i - y_i\|^2\right] \\
&= \mathbb{E}_{i_k}\left[\frac{1}{\sigma_{i_k} p_{i_k}}\|\hat{y}^{k+1}_{i_k} - y_{i_k}\|^2 - \frac{1}{\sigma_{i_k} p_{i_k}}\|y^k_{i_k} - y_{i_k}\|^2 + \|y^k - y\|^2_{\sigma^{-1}P^{-1}}\right] \\
&= \sum_{i=1}^N \frac{1}{\sigma_i}\|\hat{y}^{k+1}_i - y_i\|^2 - \sum_{i=1}^N \frac{1}{\sigma_i}\|y^k_i - y_i\|^2 + \|y^k - y\|^2_{\sigma^{-1}P^{-1}} \\
&= \|\hat{y}^{k+1} - y\|^2_{\sigma^{-1}} - \|y^k - y\|^2_{\sigma^{-1}} + \|y^k - y\|^2_{\sigma^{-1}P^{-1}}. \qquad (11.56)
\end{aligned}$$

In particular, by taking $y = y^k$, we can obtain

$$\mathbb{E}_{i_k}\left[\|y^{k+1} - y^k\|^2_{\sigma^{-1}P^{-1}}\right] = \|\hat{y}^{k+1} - y^k\|^2_{\sigma^{-1}}. \qquad (11.57)$$

Moreover, for any $x \in \mathbb{R}^d$, we have

$$\begin{aligned}
\mathbb{E}_{i_k}\left[\langle A(x^k - x), P^{-1}(y^{k+1} - y^k)\rangle\right] &= \mathbb{E}_{i_k}\left[\frac{1}{p_{i_k}}\langle A_{i_k}(x^k - x), \hat{y}^{k+1}_{i_k} - y^k_{i_k}\rangle\right] \\
&= \sum_{i=1}^N \langle A_i(x^k - x), \hat{y}^{k+1}_i - y^k_i\rangle \\
&= \langle A(x^k - x), \hat{y}^{k+1} - y^k\rangle. \qquad (11.58)
\end{aligned}$$

By using (11.56)–(11.58), we can obtain from (11.55) that

$$
\begin{aligned}
& D_g(x^k; x, y) + D_{f^*}(\hat{y}^{k+1}; x, y) \\
& \le \mathbb{E}_{i_k}[\langle A(x^k - x), P^{-1}(y^{k+1} - y^k)\rangle] - \langle A(x^k - x), P^{-1}(y^k - y^{k-1})\rangle \\
& \quad - \frac{1}{2\tau}\left(\|x^k - x^{k-1}\|^2 + \|x^k - x\|^2 - \|x^{k-1} - x\|^2\right) \\
& \quad - \frac{1}{2}\mathbb{E}_{i_k}[\|y^{k+1} - y\|^2_{\sigma^{-1}P^{-1}}] + \frac{1}{2}\|y^k - y\|^2_{\sigma^{-1}P^{-1}} \\
& \quad - \frac{1}{2}\mathbb{E}_{i_k}[\|y^{k+1} - y^k\|^2_{\sigma^{-1}P^{-1}}].
\end{aligned}
\tag{11.59}
$$

By introducing $z = (x, y)$ and

$$
\mathcal{V}(z) := \frac{1}{2\tau}\|x\|^2 + \frac{1}{2}\|y\|^2_{\sigma^{-1}P^{-1}} + \langle Ax, P^{-1}y\rangle, \tag{11.60}
$$

$$
\begin{aligned}
\mathcal{V}_k(x, y) := & \frac{1}{2\tau}\|x\|^2 - \langle Ax, P^{-1}(y^k - y^{k-1})\rangle + \frac{1}{2}\|y\|^2_{\sigma^{-1}P^{-1}} \\
& + \frac{1}{2}\|y^k - y^{k-1}\|^2_{\sigma^{-1}P^{-1}},
\end{aligned}
\tag{11.61}
$$

we can obtain from (11.59) immediately the following result.

Proposition 11.49 *For Algorithm 11.48, there holds*

$$
\begin{aligned}
& D_g(x^k; x, y) + D_{f^*}(\hat{y}^{k+1}; x, y) \\
& \le \mathcal{V}_k(x^{k-1} - x, y^k - y) - \mathbb{E}_{i_k}[\mathcal{V}_{k+1}(x^k - x, y^{k+1} - y)] - \mathcal{V}(z^k - z^{k-1})
\end{aligned}
$$

for any $x \in \mathbb{R}^d$, $y \in Y$ and all $k = 1, 2, \cdots$, where $z^k := (x^k, y^k)$.

For $\mathcal{V}(z^k - z^{k-1})$ and $\mathcal{V}_k(x, y)$, there hold the following estimates which in particular imply their nonnegativity.

Lemma 11.50 *If $p_i^{-1}\tau\sigma_i\|A_i\|^2 \le \gamma^2 < 1$ for all $i = 1, \cdots, N$, then there hold*

$$
\begin{aligned}
\mathcal{V}(z^k - z^{k-1}) & \ge c_1\left(\frac{1}{2\tau}\|x^k - x^{k-1}\|^2 + \frac{1}{2}\|y^k - y^{k-1}\|^2_{\sigma^{-1}P^{-1}}\right), \\
\mathcal{V}_k(x, y) & \ge c_1\left(\frac{1}{2\tau}\|x\|^2 + \frac{1}{2}\|y^k - y^{k-1}\|^2_{\sigma^{-1}P^{-1}}\right) + \frac{1}{2}\|y\|^2_{\sigma^{-1}P^{-1}},
\end{aligned}
$$

where $c_1 := 1 - \gamma$.

Proof By the definition of y^k, the Cauchy-Schwarz inequality and the condition $p_i^{-1}\tau\sigma_i\|A_i\|^2 \le \gamma^2 < 1$, we have

$$\begin{aligned}
|\langle Ax, P^{-1}(y^k - y^{k-1})\rangle| &= \left|\frac{1}{p_{i_{k-1}}}\langle A_{i_{k-1}}x, y^k_{i_{k-1}} - y^{k-1}_{i_{k-1}}\rangle\right| \\
&\le \frac{1}{p_{i_{k-1}}}\|A_{i_{k-1}}\|\|x\|\|y^k_{i_{k-1}} - y^{k-1}_{i_{k-1}}\| \\
&\le \frac{\gamma}{\sqrt{\tau p_{i_{k-1}}\sigma_{i_{k-1}}}}\|x\|\|y^k_{i_{k-1}} - y^{k-1}_{i_{k-1}}\| \\
&\le \gamma\left(\frac{1}{2\tau}\|x\|^2 + \frac{1}{2\sigma_{i_{k-1}}p_{i_{k-1}}}\|y^k_{i_{k-1}} - y^{k-1}_{i_{k-1}}\|^2\right) \\
&= \gamma\left(\frac{1}{2\tau}\|x\|^2 + \frac{1}{2}\|y^k - y^{k-1}\|^2_{\sigma^{-1}P^{-1}}\right).
\end{aligned}$$

According to the definition of $\mathcal{V}_k(x, y)$ and $\mathcal{V}(z^k - z^{k-1})$, we complete the proof. □

Now we are ready to show the following convergence result on Algorithm 11.48 (see [1]).

Theorem 11.51 *Let $\mathcal{Z}^*$ denote the set of all saddle points of (11.51) and assume $\mathcal{Z}^* \neq \emptyset$. Consider Algorithm 11.48 with*

$$p_i^{-1}\tau\sigma_i\|A_i\|^2 \le \gamma^2 < 1, \quad i = 1, \cdots, N.$$

Then, almost surely there exists $z^* := (x^*, y^*) \in \mathcal{Z}^*$ such that $(x^k, y^k) \to (x^*, y^*)$ as $k \to \infty$.

Proof Let $(\tilde{x}, \tilde{y}) \in \mathcal{Z}^*$ be any point. Then

$$-\sum_{i=1}^{N} A_i^T\tilde{y}_i \in \partial g(\tilde{x}) \quad \text{and} \quad A_i\tilde{x} \in \partial f_i^*(\tilde{y}_i), \quad i = 1, \cdots, N$$

which can be equivalently written as $-A^T\tilde{y} \in \partial g(\tilde{x})$ and $A\tilde{x} \in \partial f^*(\tilde{y})$. Thus, from the convexity of g and f^*, it follows that $D_g(x^k; \tilde{x}, \tilde{y}) \ge 0$ and $D_{f^*}(\hat{y}^{k+1}; \tilde{x}, \tilde{y}) \ge 0$. By taking $(x, y) = (\tilde{x}, \tilde{y})$ in Proposition 11.49, we then obtain

$$\mathbb{E}_{i_k}[\Delta_k] \le \Delta_{k-1} - \mathcal{V}(z^k - z^{k-1}), \tag{11.62}$$

where

$$\Delta_k := \mathcal{V}_{k+1}(x^k - \tilde{x}, y^{k+1} - \tilde{y}) \ge 0.$$

By Lemma 11.10, we can conclude that, almost surely, Δ_k converges to a finite valued random variable and $\mathcal{V}(z^k - z^{k-1}) \to 0$ as $k \to \infty$. By using Lemma 11.50, we then have $\|y^k - y^{k-1}\| \to 0$ and $\|x^k - x^{k-1}\| \to 0$ almost surely. Consequently, by the almost sure convergence of Δ_k, we can see that

$$\|z^k - \tilde{z}\|_M^2 := \frac{1}{\tau}\|x^k - \tilde{x}\|^2 + \|y^k - \tilde{y}\|^2_{\sigma^{-1}P^{-1}}$$

converges almost surely, where

$$M := \begin{pmatrix} \tau^{-1} I_d & 0 \\ 0 & \sigma^{-1} P^{-1} \end{pmatrix}.$$

The above argument actually shows that for any $\tilde{z} := (\tilde{x}, \tilde{y}) \in \mathcal{Z}^*$, the event

$$\Omega_{\tilde{z}} := \left\{ \lim_{k\to\infty} \|z^k - \tilde{z}\|_M \text{ exists} \right\}$$

has probability one. Next we will strengthen this result by showing that there is an event Ω_0 of probability one such that, for any $\tilde{z} \in \mathcal{Z}^*$ and any $\omega \in \Omega_0$, the limit $\lim_{k\to\infty} \|z^k(\omega) - \tilde{z}\|_M$ exists. We adapt the arguments in [27, 58]. By the separability of Euclidean space, we can find a countable set $C \subset \mathcal{Z}^*$ such that C is dense in $\mathcal{Z}^*$. Let

$$\Omega_0 := \bigcap_{z\in C} \Omega_z.$$

Since $\mathbb{P}(\Omega_z) = 1$ for each $z \in C$ and C is countable, we have $\mathbb{P}(\Omega_0) = 1$. Let $\tilde{z} \in \mathcal{Z}^*$ be any point. Then there is a sequence $\{z_l\} \subset C$ such that $z_l \to \tilde{z}$ as $l \to \infty$. For any $\omega \in \Omega_0$, we have by the triangle inequality that

$$-\|z_l - \tilde{z}\|_M \le \|z^k(\omega) - \tilde{z}\|_M - \|z^k(\omega) - z_l\|_M \le \|z_l - \tilde{z}\|_M.$$

Thus

$$\begin{aligned} -\|z_l - \tilde{z}\|_M &\le \liminf_{k\to\infty} \left\{ \|z^k(\omega) - \tilde{z}\|_M - \|z^k(\omega) - z_l\|_M \right\} \\ &\le \limsup_{k\to\infty} \left\{ \|z^k(\omega) - \tilde{z}\|_M - \|z^k(\omega) - z_l\|_M \right\} \\ &\le \|z_l - \tilde{z}\|_M. \end{aligned}$$

Since $\omega \in \Omega_0 \subset \Omega_{z_l}$, $\lim_{k\to\infty} \|z^k(\omega) - z_l\|_M$ exists. Thus, by the properties of lim inf and lim sup, we have

$$\begin{aligned} -\|z_l - \tilde{z}\|_M &\leq \liminf_{k\to\infty} \|z^k(\omega) - \tilde{z}\|_M - \lim_{k\to\infty} \|z^k(\omega) - z_l\|_M \\ &\leq \limsup_{k\to\infty} \|z^k(\omega) - \tilde{z}\|_M - \lim_{k\to\infty} \|z^k(\omega) - z_l\|_M \\ &\leq \|z_l - \tilde{z}\|_M. \end{aligned}$$

This implies that both $\liminf_{k\to\infty} \|z^k(\omega) - \tilde{z}\|_M$ and $\limsup_{k\to\infty} \|z^k(\omega) - \tilde{z}\|_M$ are finite with

$$0 \leq \limsup_{k\to\infty} \|z^k(\omega) - \tilde{z}\|_M - \liminf_{k\to\infty} \|z^k(\omega) - \tilde{z}\|_M \leq 2\|z_l - \tilde{z}\|_M.$$

Letting $l \to \infty$ shows that

$$\liminf_{k\to\infty} \|z^k(\omega) - \tilde{z}\|_M = \limsup_{k\to\infty} \|z^k(\omega) - \tilde{z}\|_M$$

and hence $\lim_{k\to\infty} \|z^k(\omega) - \tilde{z}\|_M$ exists for every $\omega \in \Omega_0$ and $\tilde{z} \in \mathcal{Z}^*$.

Next, taking the full expectation on (11.62) gives

$$\mathbb{E}[\Delta_k] \leq \mathbb{E}[\Delta_{k-1}] - \mathbb{E}[\mathcal{V}(z^k - z^{k-1})]$$

which implies that

$$\sum_{k=1}^{\infty} \mathbb{E}[\mathcal{V}(z^{k+1} - z^k)] \leq \mathbb{E}[\Delta_0] < \infty.$$

By using $\mathbb{E}[\mathcal{V}(z^{k+1} - z^k)] = \mathbb{E}[\mathbb{E}_{i_k}[\mathcal{V}(z^{k+1} - z^k)]]$, we can further obtain

$$\mathbb{E}\left[\sum_{k=1}^{\infty} \mathbb{E}_{i_k}[\mathcal{V}(z^{k+1} - z^k)]\right] \leq \mathbb{E}[\Delta_0] < \infty.$$

Thus $\sum_k \mathbb{E}_{i_k}[\mathcal{V}(z^{k+1} - z^k)] < \infty$ almost surely and hence $\mathbb{E}_{i_k}[\mathcal{V}(z^{k+1} - z^k)] \to 0$ almost surely. By using Lemma 11.50, we can conclude that there is an event Ω_1 with $\mathbb{P}(\Omega_1) = 1$ such that for every sample point in Ω_1 there holds

$$\lim_{k\to\infty} \mathbb{E}_{i_k}\left[\frac{1}{\tau}\|x^{k+1} - x^k\|^2 + \|y^{k+1} - y^k\|^2_{\sigma^{-1}P^{-1}}\right] = 0. \tag{11.63}$$

Let $\tilde{\Omega} := \Omega_0 \cap \Omega_1$. Then $\mathbb{P}(\tilde{\Omega}) = 1$ and, according to the above arguments, for every sample point in $\tilde{\Omega}$, (11.63) holds and $\lim_{k\to\infty} \|z^k - \tilde{z}\|_M$ exists for any

$\tilde{z} \in \mathcal{Z}^*$. The latter statement in particular implies that $\{z^k\}$ is bounded. Thus, there is a subsequence $\{z^{k_l}\}$ of $\{z^k\}$ such that $z^{k_l} \to z^* := (x^*, y^*)$ as $l \to \infty$. If $z^* \in \mathcal{Z}^*$, then we can conclude that $\lim_{k\to\infty} \|z^k - z^*\|_M$ exists. Since $\|z^{k_l} - z^*\|_M \to 0$, we must have $\|z^k - z^*\| \to 0$ as $k \to \infty$.

We now use (11.63) to show $z^* \in \mathcal{Z}^*$. To this end, we first use the definition of x^{k+1} to write

$$\begin{aligned} x^{k+1} &= \operatorname{prox}_{\tau g}(x^k - \tau A^T \bar{y}^{k+1}) \\ &= \operatorname{prox}_{\tau g}\left(x^k - \tau A^T (y^k + (I + P^{-1})(y^{k+1} - y^k))\right) \end{aligned}$$

In this expression, only the term $(I + P^{-1})(y^{k+1} - y^k)$ depends on i_k. For each $k \geq 0$ and $i \in \{1, \cdots, N\}$, we define $v_i^k := (v_{i,1}^k, \cdots, v_{i,N}^k) \in Y$ such that

$$v_{i,j}^k = \begin{cases} (1 + p_i^{-1})(\hat{y}_i^{k+1} - y_i^k) & \text{if } j = i \\ 0 & \text{if } j \neq i \end{cases}$$

for $j = 1, \cdots, N$. Then we can write

$$x^{k+1} = \operatorname{prox}_{\tau g}(x^k - \tau A^T (y^k + v_{i_k}^k)).$$

Consequently

$$\begin{aligned} \mathbb{E}_{i_k}[\|x^{k+1} - x^k\|^2] &= \mathbb{E}_{i_k}\left[\|\operatorname{prox}_{\tau g}(x^k - \tau A^T (y^k + v_{i_k}^k)) - x^k\|^2\right] \\ &= \sum_{i=1}^N p_i \|\operatorname{prox}_{\tau g}(x^k - \tau A^T (y^k + v_i^k) - x^k\|^2. \end{aligned}$$

Combining this and (11.57) with (11.63), we can obtain

$$\|\hat{y}^{k+1} - y^k\|_{\sigma^{-1}} \to 0 \tag{11.64}$$

and

$$\|\operatorname{prox}_{\tau g}(x^k - \tau A^T (y^k + v_i^k) - x^k\| \to 0, \quad i = 1, \cdots, N. \tag{11.65}$$

Recall the definition of $\hat{y}^{k+1}$; Eq. (11.64) is equivalent to

$$\|\operatorname{prox}_{\sigma_i f_i^*}(y_i^k + \sigma_i A_i x^k) - y_i^k\| \to 0, \quad i = 1, \cdots, N \tag{11.66}$$

By using (11.64) and the definition of v_i^k, we have $v_i^k \to 0$ as $k \to \infty$ for each $i = 1, \cdots, N$. Now replacing k by k_l in (11.65) and (11.66), letting $l \to \infty$, and using $(x^{k_l}, y^{k_l}) \to (x^*, y^*)$ and the continuity of proximal mapping, we can obtain

$$x^* = \text{prox}_{\tau g}(x^* - \tau A^T y^*), \quad y_i^* = \text{prox}_{\sigma_i f_i^*}(y_i^* + \sigma_i A_i x^*), \quad i = 1, \cdots, N$$

which is equivalent to

$$-A^T y^* \in \partial g(x^*), \quad A_i x^* \in \partial f_i^*(y_i^*), \quad i = 1, \cdots, N,$$

that is $z^* = (x^*, y^*) \in \mathcal{Z}^*$. □

Problems

11.1 Let $\{M_n\}$ be a martingale or a nonnegative submartingale with respect to a filtration $\{\mathcal{F}_n\}$. Let $p \geq 1$. Show that any $\epsilon > 0$, there holds

$$\mathbb{P}\left(\sup_{1 \leq k \leq n} M_k \geq \epsilon\right) \leq \frac{\mathbb{E}[|M_n|^p]}{\epsilon^p}.$$

This result is known as the Doob's maximal inequality.

11.2 Let $x_1 < x_2 < \cdots < x_n$ be given values, and let $p_1, p_2, \cdots, p_n$ be positive numbers satisfying $\sum_{i=1}^n p_i = 1$. We generate a random variable X as follows: Draw U from (0, 1) via the uniform distribution, and set $X = x_k$, where $k \in \{1, \cdots, n\}$ is the unique index such that

$$\sum_{i=1}^{k-1} p_i < U \leq \sum_{i=1}^{k} p_i.$$

Show that X is a discrete random variable with probability mass function $\mathbb{P}(X = x_k) = p_k$ for all $k = 1, \cdots, n$.

11.3 Let $\omega_1, \cdots, \omega_N$ be N given numbers, and consider the strong convex function

$$f(x) := \frac{1}{2N} \sum_{i=1}^{N} (x - \omega_i)^2$$

on $\mathbb{R}$ whose unique minimizer is given by $\bar{\omega} := \frac{1}{N}\sum_{i=1}^{N}\omega_i$. When the stochastic gradient descent is used to determine the minimizer of f, the corresponding iterative scheme takes the form

$$x^{k+1} = x^k - \gamma_k(x^k - \omega_{i_k}),$$

where γ_k is a step size and $i_k \in \{1, \cdots, N\}$ is sampled uniformly at random.

(i) If $\gamma_k = 1/(k+1)$ for all $k \geq 0$, show that

$$\mathbb{E}[(x^k - \bar{\omega})^2] = \frac{1}{(k+1)N}\sum_{i=1}^{N}(\omega_i - \bar{\omega})^2, \quad k = 0, 1, \cdots.$$

Thus $\{x^k\}$ converges to $\bar{\omega}$ sublinearly.

(ii) If $\gamma_k = \gamma > 0$ for all $k \geq 0$, show that

$$\mathbb{E}[(x^k - \bar{\omega})^2] = (1-\gamma)^{2k}(x^0 - \bar{\omega})^2 + \gamma^2\left(\sum_{l=1}^{k-1}(1-\gamma)^{2l}\right)\frac{1}{N}\sum_{i=1}^{N}(\omega_i - \bar{\omega})^2$$

for all $k \geq 0$. Thus $\{x^k\}$ does not converge to $\bar{\omega}$ unless $\omega_i = \bar{\omega}$ for all $i = 1, \cdots, N$.

11.4 Consider the problem of minimizing a smooth, convex function of the form

$$\min_{x\in\mathbb{R}^d}\left\{f(x) := \frac{1}{N}\sum_{i=1}^{N} f_i(x)\right\}, \tag{11.67}$$

where each $f_i : \mathbb{R}^d \to \mathbb{R}$ is a differentiable convex function with L-Lipschitz continuous gradient. The loopless SVRG method [126] is a variance reduction technique that updates iterates as follows: Input an initial guess $x^0 = \tilde{x}^0 \in \mathbb{R}^d$, a step size $\gamma > 0$, and a probability $p \in (0, 1]$. For $k = 0, 1, \cdots$, do

- Sample an index i_k from $\{1, \cdots, N\}$ uniformly at random;
- Compute a gradient estimate using

$$g^k = \nabla f_{i_k}(x^k) - \nabla f_{i_k}(\tilde{x}^k) + \nabla f(\tilde{x}^k);$$

- Update the iterate as $x^{k+1} = x^k - \gamma g^k$.
- Update $\tilde{x}^{k+1}$ by the probability scheme

$$\tilde{x}^{k+1} = \begin{cases} x^k & \text{with probability } p, \\ \tilde{x}^k & \text{with probability } 1-p. \end{cases}$$

The loopless SVRG is a variant of the classical SVRG method designed to improve its practical efficiency and ease of implementation. The classical SVRG relies on an outer loop, where a full gradient is computed periodically to construct variance-reduced updates. In contrast, loopless SVRG eliminates this explicit outer loop, replacing it with a probabilistic mechanism for updating the reference point.

Assume f is strongly convex with modulus of convexity $\mu > 0$. Perform a convergence analysis of the loopless SVRG by completing the following items:

(i) Let x^* denote the unique minimizer of f over $\mathbb{R}^d$. For all $k \geq 0$, show that

$$\begin{aligned}&\mathbb{E}_{i_k}[\|x^{k+1} - x^*\|^2]\\&\leq (1 - 2\gamma\mu)\|x^k - x^*\|^2 - 2\gamma(f(x^k) - f(x^*)) + \gamma^2\mathbb{E}_{i_k}[\|g^k\|^2].\end{aligned}$$

(ii) Show that

$$\mathbb{E}_{i_k}[\|g^k\|^2] \leq 4L(f(x^k) - f(x^*)) + 2D_k$$

for all $k \geq 0$, where $D_k := \frac{1}{N}\sum_{i=1}^N \|\nabla f_i(\tilde{x}^k) - \nabla f_i(x^*)\|^2$.

(iii) Let $\mathbb{E}_{\tilde{x}^{k+1}}$ denote the expectation taken with respect to $\tilde{x}^{k+1}$. Show that

$$\mathbb{E}_{\tilde{x}^{k+1}}[D_{k+1}] \leq (1 - p)D_k + 2pL(f(x^k) - f(x^*));$$

(iv) Let

$$\Phi_k := \|x^k - x^*\|^2 + \frac{4\gamma^2}{p}D_k.$$

If $0 < \gamma \leq 1/(6L)$, show that $\mathbb{E}[\Phi_k] \leq q^k\Phi_0$ for all $k \geq 0$, where $q := \max\{1 - 2\gamma\mu, 1 - p/2\} < 1$.

11.5 Consider the convex minimization problem (11.67), where each f_i is a continuously differentiable convex function. To determine an optimal solution, we consider the following stochastic recursive gradient algorithm: Input an initial guess x^0, an integer $m \geq 1$, and a step size $\gamma > 0$. For $k = 0, 1, \cdots$, do the following:

(i) Set $x^{k,0} = x^k$, and calculate $g^{k,0} := \nabla f(x^{k,0})$ and $x^{k,1} = x^{k,0} - \gamma g^{k,0}$.
(ii) For $s = 1, \cdots, m-1$, do

- pick $i_{k,s} \in \{1, \cdots, N\}$ uniformly at random.
- Calculate $g^{k,s} := \nabla f_{i_{k,s}}(x^{k,s}) - \nabla f_{i_{k,s}}(x^{k,s-1}) + g^{k,s-1}$.
- Determine $x^{k,s+1}$ by $x^{k,s+1} := x^{k,s} - \gamma g^{k,s}$.

(iii) Update x^{k+1} by $x^{k+1} = x^{k,m}$, and repeat the above procedure.

This algorithm was introduced in [168] in which it is termed as SARAH; it modifies the SVRG update by recursively refining gradient estimates. Here we adopt a slight modification on the updating formula for x^{k+1}; see [141, 145]. Assume that each f_i has a L-Lipschitz continuous gradient. Show the following results:

(a) For any $k \geq 0$ and $s = 1, \cdots, m$, there holds

$$\mathbb{E}\left[\|\nabla f(x^{k,s}) - g^{k,s}\|^2\right]$$
$$= \sum_{j=1}^{s} \mathbb{E}\left[\|g^{k,j} - g^{k,j-1}\|^2\right] - \sum_{j=1}^{s} \mathbb{E}\left[\|\nabla f(x^{k,j}) - \nabla f(x^{k,j-1})\|^2\right],$$

(b) If $0 < \gamma < 2/L$, then for any $k \geq 0$ and $s = 1, \cdots, m$, there holds

$$\sum_{j=1}^{s} \mathbb{E}\left[\|g^{k,j} - g^{k,j-1}\|^2\right] \leq \frac{\gamma L}{2 - \gamma L}\left(\mathbb{E}[\|g^{k,0}\|^2] - \mathbb{E}[\|g^{k,s}\|^2]\right).$$

(c) If $0 < \gamma < 2/L$, use (a) and (b) to derive for any $k \geq 0$ and $s - 1, \cdots, m$ that

$$\mathbb{E}\left[\|\nabla f(x^{k,s}) - g^{k,s}\|^2\right]$$
$$\leq \frac{\gamma L}{2 - \gamma L}\mathbb{E}\left[\|\nabla f(x^k)\|^2\right] - \sum_{j=1}^{s} \mathbb{E}\left[\|\nabla f(x^{k,j}) - \nabla f(x^{k,j-1})\|^2\right].$$

(d) If f is strongly convex with modulus of convexity $\mu > 0$ and $0 < \gamma < 2/L$, then for all $k \geq 0$ and $s = 1, \cdots, m$, there holds

$$\mathbb{E}\left[\|g^{k,s}\|^2\right] \leq \left(1 - 4\mu^2\gamma^2\left(\frac{2}{\gamma L} - 1\right)\right)^s \mathbb{E}\left[\|\nabla f(x^k)\|^2\right].$$

(c) Assume f is strongly convex with modulus of convexity $\mu > 0$. If $0 < \gamma < 2/(3L)$ and m are sufficiently large such that

$$q := \frac{2\gamma L}{2 - \gamma L} + (2 + 2\gamma L)\left(1 - 4\mu^2\gamma^2\left(\frac{2}{\gamma L} - 1\right)\right)^m < 1,$$

then

$$\mathbb{E}[\|\nabla f(x^k)\|^2] \leq q\mathbb{E}[\|\nabla f(x^{k-1})\|^2], \quad \forall k \geq 1$$

and consequently $\mathbb{E}[\|x^k - x^*\|^2] = O(q^k)$, where x^* denotes the unique solution of (11.67).

11.6 Under the same conditions in Theorem 11.43, show that the sequence $\{x^k\}$ defined by Algorithm 11.42 converges almost surely to a solution of (11.35).

References

1. A. Alacaoglu, O. Fercoq, V. Cevher, On the convergence of stochastic primal-dual hybrid gradient. SIAM J. Optim. **32**(2), 1288–1318 (2022)
2. Z. Allen-Zhu, Y. Yuan, Improved SVRG for non-strongly-convex or sum-of-non-convex objectives, in *International Conference on Machine Learning* (2016), pp. 1080–1089
3. F.J. Aragón Artacho, M.H. Geoffroy, Characterization of metric regularity of subdifferentials. J. Convex Anal. **15**, 365–380 (2008)
4. F.J. Aragón Artacho, M.H. Geoffroy, Metric subregularity of the convex subdifferential in Banach spaces. J. Nonlinear Convex Anal. **15**, 35–47 (2014)
5. K.J. Arrow, L. Hurwicz, H. Uzawa, *Studies in Linear and NonLinear Programming*, vol. II (Stanford University Press, Stanford, 1958)
6. H. Attouch, J. Bolte, On the convergence of the proximal algorithm for nonsmooth functions involving analytic features. Math. Program. Ser. B **116**, 5–16 (2009)
7. H. Attouch, J. Bolte, P. Redont, A. Soubeyran, Proximal alternating minimization and projection methods for nonconvex problems: an approach based on the Kurdyka-Lojasiewicz inequality. Math. Oper. Res. **35**(2), 438–457 (2010)
8. H. Attouch, J. Bolte, B.F. Svaiter, Convergence of descent methods for semi-algebraic and tame problems: proximal algorithms, forward-backward splitting, and regularized gauss-seidel methods. Math. Program. **137**, 91–129 (2013)
9. H. Attouch, J. Peypouquet, The rate of convergence of Nesterov accelerated forward-backward method is actually faster than $O(1/k^2)$. SIAM J. Optim. **26**, 1824–1834 (2016)
10. A. Auslender, Méthodes numériques pour la décomposition et la minimisation de fonctions non différentiables. Numer. Math. **18**, 213–223 (1971)
11. J.-B. Baillon, G. Haddad, Quelques propriétés des opérateurs angle-bornés et n-cycliquement monotones. Israel J. Math. **26**, 137–150 (1977)
12. S. Banert, M. Upadhyaya, P. Giselsson, The Chambolle–Pock method converges weakly with $\theta > 1/2$ and $\tau\sigma\|L\|^2 < 4/(1+2\theta)$, arXiv: 2309.03998v1 (2023)
13. H. Bauschke, J. Borwein, On projection algorithms for solving convex feasibility problems. SIAM Rev. **38**(3), 367–426 (1996)
14. H.H. Bauschke, P.L. Combettes, *Convex Analysis and Monotone Operator Theory in Hilbert Spaces* (Springer, Berlin, 2011)
15. H. Bauschke, J. Bolte, M. Teboulle, A descent lemma beyond Lipschitz gradient continuity: first-order methods revisited and applications. Math. Oper. Res. **42**, 330–348 (2017)
16. A. Beck, *First-Order Methods in Optimization*, MOS-SIAM Series on Optimization (Society for Industrial and Applied Mathematics, Philadelphia, 2017)

Q. Jin, *Lectures on Nonsmooth Optimization*, Texts in Applied Mathematics 82,
https://doi.org/10.1007/978-3-031-91417-1

17. A. Beck, M. Teboulle, Mirror descent and nonlinear projected subgradient methods for convex optimization. Oper. Res. Lett. **31**(3), 167–175 (2003)
18. A. Beck, M. Teboulle, A fast iterative shrinkage-thresholding algorithm for linear inverse problems. SIAM J. Imaging Sci. **2**, 183–202 (2009)
19. A. Beck, M. Teboulle, Fast gradient-based algorithms for constrained total variation image denoising and deblurring problems. IEEE Trans. Image Process **18**, 2419–2434 (2009)
20. A. Beck, M. Teboulle, Gradient-based algorithms with applications to signal-recovery problems, in *Convex Optimization in Signal Processing and Communications* (Cambridge University Press, Cambridge, 2010), pp. 42–88
21. A. Beck, M. Teboulle, A fast dual proximal gradient algorithm for convex minimization and applications. Oper. Res. Lett. **42**(1), 1–6 (2014)
22. Y. Bello-Cruz, T.T.A. Nghia, On the convergence of the proximal forward-backward splitting method with linesearches. Optim. Method Softw. **31**, 1209–1238 (2016)
23. Y. Bello-Cruz, G. Li, T.T.A. Nghia, On the linear convergence of forward-backward splitting method: part I - convergence analysis. J. Optim. Theory Appl. **188**, 378–401 (2021)
24. Y. Bello-Cruz, G. Li, T.T.A. Nghia, Quadratic Growth Conditions and Uniqueness of Optimal Solution to Lasso. J. Optim. Theory Appl. **194**, 167–190 (2022)
25. M. Bergounioux, K. Ito, K. Kunisch, Primal-dual strategy for constrained optimal control problems. SIAM J. Control Optim. **37**, 1176–1194 (1999)
26. A. Berman, R.J. Plemmons, *Nonnegative Matrices in the Mathematical Sciences*, Computer Science and Scientific Computing Series (Academic Press, New York, 1979)
27. D.P. Bertsekas, Incremental proximal methods for large scale convex optimization. Math. Program. **129**, 163–195 (2011)
28. R. Bhattacharya, E.C. Waymire, *A Basic Course in Probability Theory*, Universitext, 2nd edn. (Springer, Cham, 2016)
29. D. Blatt, A.O. Hero, H. Gauchman, A convergent incremental gradient method with a constant step size. SIAM J. Optim. **18**(1), 29–51 (2007)
30. J. Bochnak, M. Coste, M.-F. Roy, *Real Algebraic Geometry* (Springer, Berlin, 1998)
31. J. Bolte, A. Daniilidis, A. Lewis, The Lojasiewicz inequality for nonsmooth subanalytic functions with applications to subgradient dynamical systems. SIAM J. Optim. **17**, 1205–1223 (2006)
32. J. Bolte, A. Daniilidis, A. Lewis, M. Shiota, Clarke subgradients of stratifiable functions. SIAM J. Optim. **18**(2), 556–572 (2007)
33. J. Bolte, A. Daniilidis, A. Lewis, Tame functions are semismooth. Math. Program. **117**, 5–19 (2009)
34. J. Bolte, S. Sabach, M. Teboulle, Proximal alternating linearized minimization for nonconvex and nonsmooth problems. Math. Program. Ser. A **146**, 459–494 (2014)
35. J. Bolte, T.P. Nguyen, J. Peypouquet, B.W. Suter, From error bounds to the complexity of first-order descent methods for convex functions. Math. Program. Ser. A **165**, 471–507 (2017)
36. S. Bonettini, V. Ruggiero, On the convergence of primal-dual hybrid gradient algorithms for total variation image restoration. J. Math. Imaging Vis. **44**, 236–253 (2012)
37. J.F. Bonnans, A. Shapiro, *Perturbation Analysis of Optimization Problems*, Springer Series in Operations Research (Springer, Berlin Heidelberg, 2000)
38. W.M. Boothby, *An Introduction to Differentiable Manifolds and Riemannian Geometry*, Pure and Applied Mathematics, vol. 120, 2nd edn. (Academic Press, Orlando, 1986)
39. J.M. Borwein, Q.J. Zhu, *Techniques of Variational Analysis*, CMS Books in Mathematics/Ouvrages de Mathématiques de la SMC, vol. 20 (Springer, New York, 2005)
40. L. Bottou, F.E. Curtis, J. Nocedal, Optimization methods for large-scale machine learning. SIAM Rev. **60**(2), 223–311 (2018)
41. S. Boyd, N. Parikh, E. Chu, B. Peleato, J. Eckstein, Distributed optimization and statistical learning via the alternating direction method of multipliers. Found. Trends Mach. Learn. **3**(1), 1–122 (2011)

42. L.M. Bregman, The relaxation method for finding common points of convex sets and its application to the solution of problems in convex programming. USSR Comput. Math. Math. Phys. **7**, 200–217 (1967)
43. H. Brezis, P.L. Lions, Produits infinis de résolvantes. Israel J. Math. **29**, 329–345 (1978)
44. P. Brucker, An $O(n)$ algorithm for quadratic knapsack problems. Oper. Res. Lett. **3**(3), 163–166 (1984)
45. C. Byrne, Iterative oblique projection onto convex sets and the split feasibility problem. Inverse Probl. **18**, 441–453 (2002)
46. J.F. Cai, E.J. Candès, Z. Shen, A singular value thresholding algorithm for matrix completion. SIAM J. Optim. **20**(4), 1956–1982 (2010)
47. G. Carlier, Fenchel-Young inequality with a remainder and applications to convex duality and optimal transport. SIAM J. Optim. **33**(3), 1463–1472 (2023)
48. Y. Censor, T. Elfving, A multiprojection algorithm using Bregman projections in a product space. Numer. Algorithms **8**(2), 221–239 (1994)
49. A. Chambolle, C. Dossal, On the convergence of the iterates of the "Fast Iterative Shrinkage/Thresholding Algorithm". J. Optim. Theory Appl. **166**, 968–982 (2015)
50. A. Chambolle, T. Pock, A first-order primal-dual algorithm for convex problems with applications to imaging. J. Math. Imaging Vis. **40**, 120–145 (2011)
51. A. Chambolle, T. Pock, An introduction to continuous optimization for imaging. Acta Numer. **25**, 161–319 (2016)
52. A. Chambolle, M.J. Ehrhardt, P. Richtarik, C.B. Schonlieb, Stochastic primal-dual hybrid gradient algorithm with arbitrary sampling and imaging applications. SIAM J. Optim. **28**, 2783–2808 (2018)
53. Z.X. Chan, D. Sun, Constraint nondegeneracy, strong regularity, and nonsingularity in semidefinite programming. SIAM J. Optim. **19**, 370–396 (2008)
54. X. Chen, Z. Nashed, L. Qi, Smoothing methods and semismooth methods for nondifferentiable operator equations. SIAM J. Numer. Anal. **38**(4), 1200–1216 (2000)
55. C. Chen, R. Chan, S. Ma, J. Yang, Inertial proximal ADMM for linearly constrained separable convex optimization. SIAM J. Imaging Sci. **8**(4), 2239–2267 (2015)
56. W. Cheney, A. Goldstein, Proximity maps for convex sets. Proc. AMS **10**, 448–450 (1959)
57. F.H. Clarke, *Optimization and Nonsmooth Analysis* (Wiley, New York, 1983)
58. P.L. Combettes, J.-C. Pesquet, Stochastic quasi-Fejér block-coordinate fixed point iterations with random sweeping. SIAM J. Optim. **25**, 1221–1248 (2015)
59. P.L. Combettes, V.R. Wajs, Signal recovery by proximal forward-backward splitting. Multiscale Model. Simul. **4**(4), 1168–1200 (2005)
60. L. Condat, A primal-dual splitting method for convex optimization involving Lipschitzian, proximable and linear composite terms. J. Optim. Theory Appl. **158**(2), 460–479 (2013)
61. M. Coste, *An Introduction to O-Minimal Geometry*, RAAG Notes (Institut de Recherche Mathématiques de Rennes, Rennes, 1999), 81 pp.
62. M. Coste, An introduction to semialgebraic geometry, preprint (2002). Available online from http://perso.univ-rennes1.fr/michel.coste/polyens/SAG.pdf
63. R.W. Cottle, J.S, Pang, R.E. Stone, *The Linear Complementarity Problem* (Academic Press, Boston, 1992)
64. I. Csiszár, Why least squares and maximum entropy? An axiomatic approach to inference for linear inverse problems. Ann. Statist. **19**, 2032–2066 (1991)
65. I. Daubechies, M. Defrise, C. De Mol, An iterative thresholding algorithm for linear inverse problems with a sparsity constraint. Commun. Pure Appl. Math. **57**, 1413–1457 (2004)
66. D. Davis, W. Yin, Convergence rate analysis of several splitting schemes, in *Splitting Methods in Communication, Imaging, Science, and Engineering*, Sci. Comput. (Springer, Cham, 2016), pp. 115–163
67. A. De Cezaro, M. HaltmeieM, A. Leitão, O. Scherzer, On steepest-descent-Kaczmarz methods for regularizing systems of nonlinear ill-posed equations. Appl. Math. Comput. **202**, 596–607 (2008)

68. A. Defazio, F. Bach, S. Lacoste-Julien, SAGA: a fast incremental gradient method with support for non-strongly convex composite objectives, in *Proc. 27th Int. Conf. Adv. Neural Inf. Process. Syst.* (2014), pp. 1646–1654
69. W. Deng, W. Yin, On the global and linear convergence of the generalized alternating direction method of multipliers. J. Sci. Comput. **66**, 889–916 (2016)
70. A.L. Dontchev, R.T. Rockafellar, Regularity and conditioning of solution mappings in variational analysis. Set-Valued Anal. **12**(1–2), 79–109 (2004)
71. A.L. Dontchev, R.T. Rockafellar, *Implicit Functions and Solution Mappings, A View from Variational Analysis*, Springer Monographs in Mathematics (Springer, Dordrecht, 2009)
72. J. Douglas, H. Rachford, On the numerical solution of heat conduction problems in two or three space variables. Trans. Amer. Math. Soc. **82**, 421–439 (1956)
73. D. Driggs, M.J. Ehrhardt, C.-B. Schönlieb, Accelerating variance-reduced stochastic gradient methods. Math. Program. **191**, 671–715 (2022)
74. D. Drusvyatskiy, A.S. Lewis, Error bounds, quadratic growth, and linear convergence of proximal methods. Math. Oper. Res. **43**(3), 919–948 (2018)
75. J. Eckstein, D.P. Bertsekas, On the douglas-rachford splitting method and the proximal point algorithm for maximal monotone operators. Math. Program. **55**(1–3), 293–318 (1992)
76. I. Ekeland, On the variational principle. J. Math. Anal. Appl. **47**, 324–353 (1974)
77. H.W. Engl, M. Hanke, A. Neubauer, *Regularization of Inverse Problems*, vol. 375 (Springer Science & Business Media, New York, 1996)
78. E. Esser, X. Zhang, T. Chan, A general framework for a class of first order primal-dual algorithms for convex optimization in imaging science. SIAM J. Imaging Sci. **3**(4), 1015–1046 (2010)
79. L.C. Evans, R.F. Gariepy, *Measure Theory and Fine Propertiesof Functions*, Studies in Advanced Mathematics (CRC Press, Boca Raton, 1992)
80. F. Facchinei, Minimization of SC^1 functions and the Maratos effect. Oper. Res. Lett. **17**, 131–137 (1995)
81. M. Fiedler, V. Pták, On matrices with non-positive off-diagonal elements and positive principal minors. Czechoslov. Math. J. **12**, 382–400 (1962)
82. J.R. Fineup, Phase retrieval algorithms: a comparison. Appl. Opt. **21**(15), 2758–2769 (1982)
83. M. Fortin, R. Glowinski (eds.), *Augmented Lagrangian Methods: Applications to the Numerical Solution of Boundary-Value Problems* (North-Holland, Amsterdam, 1983)
84. K. Frick, M. Grasmair, Regularization of linear ill-posed problems by the augmented Lagrangian method and variational inequalities. Inverse Probl. **28**(10), 104005 (2012)
85. K. Frick, O. Scherzer, Regularization of ill-posed linear equations by the non-stationary augmented Lagrangian method. J. Integral Equ. Appl. **22**(2), 217–257 (2010)
86. K. Frick, D. Lorenz, E. Resmerita, Morozov's principle for the augmented Lagrangian method applied to linear inverse problems. Multiscale Model. Simul. **9**(4), 1528–1548 (2011)
87. D. Gabay, Applications of the method of multipliers to variational inequalities, in *Studies in Mathematics and Its Applications*, vol. 15 (Elsevier, Amsterdam, 1983), pp. 299–331
88. D. Gabay, B. Mercier, A dual algorithm for the solution of nonlinear variational problems via finite element approximation. Comput. Math. Appl. **2**, 17–40 (1976)
89. E.G. Gladyshev, On stochastic approximations. Theory Probab. Appl. **10**, 275–278 (1965)
90. R. Glowinski, A. Marrocco, Sur l'approximation, par éléments finis d'ordre un, et la résolution, par pénalisation-dualité, d'une classe de problémes de Dirichlet non linéaires. ESAIM: Math. Model. Numer. Anal. **9**(R2), 41–76 (1975)
91. T. Goldstein, B. O'Donoghue, S. Setzer, R. Baraniuk, Fast alternating direction optimization methods. SIAM J. Imaging Sci. **7**, 1588–1623 (2014)
92. R.M. Gower, M. Schmidt, F. Bach, P. Richtarik, Variance-reduced methods for machine learning. Proc. IEEE **108**(11), 1968–1983 (2020)
93. R. Griesse, D.A. Lorenz, A semismooth Newton method for Tikhonov functionals with sparsity constraints. Inverse Probl. **24**, 035007 (2008)
94. O. Güler, On the convergence if the proximal point algorithm for convex minimization. SIAM J. Control Optim. **29**(2), 403–419 (1991)

95. H.V. Ha, T.S. Pham, *Genericity in Polynomial Optimization*. Ser. Optim. Appl., vol. 3 (World Scientific, Singapore, 2017)
96. M. Haltmeier, A. Leitão, O. Scherzer, Kaczmarz methods for regularizing nonlinear illposed equations I. Convergence analysis. Inverse Probl. Imaging **1**, 289–298 (2007)
97. D. Han, A survey on some recent developments of alternating direction method of multipliers. J. Oper. Res. Soc. China **10**(1), 1–52 (2022)
98. M. Hanke, C.W. Groetsch, Nonstationary iterated Tikhonov regularization. J. Optim. Theory Appl. **98**(1), 37–53 (1998)
99. B. He, X. Yuan, On the $O(1/n)$ convergence rate of Douglas-Rachford alternating direction method. SIAM J. Numer. Anal. **50**, 700–709 (2012)
100. B. He, X. Yuan, On non-ergodic convergence rate of Douglas-Rachford alternating direction method of multipliers. Numer. Math. **130**, 567–577 (2015)
101. B. He, Y. You, X. Yuan, On the convergence of primal-dual hybrid gradient algorithm. SIAM J. Imaging Sci. **7**(4), 2526–2537 (2014)
102. B. He, F. Ma, X. Yuan, Convergence study on the symmetric version of ADMM with larger step sizes. SIAM J. Imaging Sci. **9**(3), 1467–1501 (2016)
103. G. Herman, *Fundamentals of Computerized Tomography: Image Reconstruction from Projection*, 2nd edn. (Springer, London, 2009)
104. M.R. Hestenes, Multiplier and gradient methods. J. Optim. Theory Appl. **4**, 303–320 (1969)
105. M. Hintermüller, K. Ito, K. Kunisch, The primal-dual active set strategy as a semismooth Newton method. SIAM J. Optim. **13**(3), 865–888 (2003)
106. A.J. Hoffman, On approximate solutions of systems of linear inequalities. J. Res. Nat. Bur. Stand. **49**, 263–265 (1952)
107. J. Hu, T. Tian, S. Pan, Z. Wen, On the local convergence of the semismooth Newton method for composite optimization, arXiv:2211.01127v2 (2022)
108. Y. Huang, Y. Dong, New properties of forward-backward splitting and a practical proximal-descent algorithm. Appl. Math. Comput. **237**, 60–68 (2014)
109. A. D. Ioffe, Necessary and sufficient conditions for a local minimum. I. A reduction theorem and first order conditions. SIAM J. Control Optim. **17**(2), 245–250 (1979)
110. A.D. Ioffe, An invitation to tame optimization. SIAM J. Optim. **19**, 1894–1917 (2009)
111. K. Ito, K. Kunisch, *Lagrange Multiplier Approach to Variational Problems and Applications*, vol. 15, Advances in Design and Control (Society for Industrial and Applied Mathematics, Philadelphia, 2008)
112. H. Jiang, L. Qi, Local uniqueness and convergence of iterative methods for nonsmooth variational inequalities. J. Math. Anal. Appl. **196**(1), 314–331 (1995)
113. Y. Jiao, Q. Jin, X. Lu, W. Wang, Alternating direction method of multipliers for linear inverse problems. SIAM J. Numer. Anal. **54**, 2114–2137 (2016)
114. Y. Jiao, Q. Jin, X. Lu, W. Wang, Preconditioned alternating direction method of multipliers for solving inverse problems with constraints. Inverse Probl. **33**, 025004 (34 pp.) (2017)
115. Q. Jin, Landweber-Kaczmarz method in Banach spaces with inexact inner solvers. Inverse Probl. **32**, 104005 (26 pp.) (2016)
116. Q. Jin, On a heuristic stopping rule for the regularization of inverse problems by the augmented Lagrangian method. Numer. Math. **136**(4), 973992 (2017)
117. Q. Jin, On convergence rates of proximal alternating direction method of multipliers. J. Sci. Comput. **97**(3), Paper No. 66 (2023)
118. Q. Jin, On growth error bound conditions with an application to heavy ball method, arXiv:2310.03947v2 (2023)
119. Q. Jin, L. Stals, Nonstationary iterated Tikhonov regularization for ill-posed problems in Banach spaces. Inverse Probl. **28**, 104011 (2012)
120. Q. Jin, W. Wang, Landweber iteration of Kaczmarz type with general non-smooth convex penalty functionals. Inverse Probl. **29**, 085011 (22 pp.) (2013)
121. Q. Jin, X. Lu, L. Zhang, Stochastic mirror descent method for linear ill-posed problems in Banach spaces. Inverse Probl. **39**(6), Paper No. 065010, 39 pp. (2023)

122. R. Johnson, T. Zhang, Accelerating stochastic gradient descent using predictive variance reduction, in *NIPS* (2013), pp. 315–323
123. J. Jost, *Riemannian geometry and geometric analysis*, Universitext, 7th edn. (Springer, Cham, 2017)
124. S. Kaczmarz, Angenäherte Auflösung von Systemen linearer Gleichungen. Bull. Int. Acad. Polon. Sci. Lett. A **35**, 355–357 (1937)
125. K.C. Kiwiel, Convergence of approximate and incremental subgradient methods for convex optimization. SIAM J. Optim. **14**(3), 807–840 (2003)
126. D. Kovalev, S. Horváth, P. Richtárik, Don't jump through hoops and remove those loops: SVRG and Katyusha are better without the outer loop. Proc. Mach. Learn. Res. **117**, 1–17 (2020)
127. A.Y. Kruger, Error bounds and metric subregularity. Optimization **64**(1), 49–79 (2015)
128. B. Kummer, Generalized Newton and NCP-methods: convergence, regularity, actions. Discuss. Math. Differential Inclusions Control Optim. **20**, 209–244 (2000)
129. K. Kurdyka, On gradients of functions definable in o-minimal structures. Ann. Inst. Fourier **48**(3), 769–783 (1998)
130. M.J. Lai, W.T. Yin, Augmented ℓ_1 and nuclear-norm models with a globally linearly convergent algorithm. SIAM J. Imaging Sci. **6**(2), 1059–1091 (2013)
131. L.J. Lardy, A series representation for the generalized inverse of a closed linear operator. Atti Accad. Naz. Lincei Rend. Classe Sci. Fisiche Mat. Nat. Ser. VIII **58**, 152–157 (1975)
132. T. Larsson, M. Patriksson, A. B. trömberg, On the convergence of conditional ε-subgradient methods for convex programs and convex–concave saddle-point problems. Eur. J. Oper. Res. **151**, 461–473 (2003)
133. P. Latafat, N.M. Freris, P. Patrinos, A new randomized block-coordinate primal-dual proximal algorithm for distributed optimization. IEEE Trans. Autom. Control **64**, 4050–4065 (2019)
134. J.H. Lee, T.S. Pham, Openness, Hölder metric regularity, and Hölder continuity properties of semialgebraic set-valued maps. SIAM J. Optim. **32**(1), 56–74 (2022)
135. H. Li, Z. Lin, Accelerated alternating direction method of multipliers: an optimal $O(1/k)$ nonergodic analysis. J. Sci. Comput. **79**, 671–699 (2019)
136. G. Li, T.K. Pong, Douglas-Rachford splitting for nonconvex optimization with application to nonconvex feasibility problems. Math. Program. Ser. A **159**, 371–401 (2016)
137. G. Li, T.K. Pong, Calculus of the exponent of Kurdyka-Lojasiewicz inequality and its applications to linear convergence of first-order methods. Found. Comput. Math. **18**, 1199–1232 (2018)
138. Y. Li, M. Yan, On the improved conditions for some primal–dual algorithms, arXiv: 2201.00139 (2022)
139. G. Li, T. Liu, T.K. Pong, Peaceman-Rachford splitting for a class of nonconvex optimization problems. Comput. Optim. Appl. **68**, 407–436 (2017)
140. X. Li, D. Sun, K. Toh, A highly efficient semismooth Newton augmented Lagrangian method for solving lasso problems. SIAM J. Optim. **28**(1), 433–458 (2018)
141. B. Li, M. Ma, G.B. Giannakis, On the convergence of SARAH and beyond, in *PMLR*, vol. 108, pp. 223–233 (2020)
142. B. Li, B. Shi, Y. Yuan, Linear convergence of forward-backward accelerated algorithms without knowledge of the modulus of strong convexity. SIAM J. Optim. **34**(2), 2150–2168 (2024)
143. Q. Lin, Z. Lu, L. Xiao, An accelerated randomized proximal coordinate gradient method and its application to regularized empirical risk minimization. SIAM J. Optim. **25**(4), 2244–2273 (2015)
144. P.L. Lions, B. Mercier, Splitting algorithms for the sum of two nonlinear operators. SIAM J. Numer. Anal. **16**(6), 964–979 (1979)
145. Y. Liu, C. Han, T. Guo, A class of stochastic variance reduced methods with an adaptive stepsize (2019), http://www.optimization-online.org/DB_FILE/2019/04/7170.pdf

146. S. Lojasiewicz, Une propriété topologique des sous-ensembles analytiques reéls, in *Les Equations aux Dériveés Partielles* (Editions du centre National ´ de la Recherche Scientifique, Paris, 1963)
147. S. Lojasiewicz, *Ensembles Semi-analytiques*, Lecture Notes (Institut des Hautes Etudes Scientifiques, Buressur-Yvette (Seine-et-Oise), 1965)
148. D.A. Lorenz, F. Schöpfer, S. Wenger, The linearized Bregman method via split feasibility problems: analysis and generalizations. SIAM J. Imaging Sci. **7**(2), 1237–1262 (2014)
149. Z. Lu, L. Xiao, On the complexity analysis of randomized block-coordinate descent methods. Math. Program. **152**, 615–642 (2015)
150. Z.Q. Luo, P. Tseng, Error bounds and convergence analysis of feasible descent methods: a general approach. Ann. Oper. Res. **46–47**(1), 157–178 (1993)
151. Y. Malitsky, T. Pock, A first-order primal-dual algorithm with linesearch. SIAM J. Optim. **28**(1), 411–432 (2018)
152. B. Martinet, *Régularisation d'inéquations variationnelles par approximations successives*, Revue Francaise de Informatique et Recherche Opérationelle (1970)
153. B. Martinet, Détermination approchée d'un point fixe d'une application pseudo-contractante. C. R. Acad. Sci. Paris **274A**, 163–165 (1972)
154. R. Mifflin, Semismooth and semiconvex functions in constrained optimization. SIAM J. Control Optim. **15**, 959–972 (1977)
155. A. Milzarek, F. Schaipp, M. Ulbrich, A semismooth Newton stochastic proximal point algorithm with variance reduction. SIAM J. Optim. **34**(1), 1157–1185 (2024)
156. B.S. Mordukhovich, *Variational Analysis and Generalized Differentiation I: Basic Theory* (Springer, Berlin, 2006)
157. J.J. Moreau, Fonctions convexes duales et points proximaux dans un espace hilbertien. C. R. Hebd. Séances Acad. Sci. Paris **255**, 2897–2899 (1962)
158. J.J. Moreau, Proximité et dualité dans un espace Hilbertien. Bull. Soc. Math. France **93**(2), 273–299 (1965)
159. F. Natterer, *The Mathematics of Computerized Tomography* (SIAM, Philadelphia, 2001)
160. A. Nedic, D. Bertsekas, Incremental subgradient methods for nondifferentiable optimization. SIAM J. Optim. **12**(1), 109–138 (2001)
161. A. Nedic, D. Bertsekas, Convergence rate of incremental subgradient algorithms, in *Stochastic Optimization: Algorithms and Applications* (Springer, 2001), pp. 223–264
162. A. Nedic, A. Ozdaglar, Distributed subgradient methods for multi-agent optimization. IEEE Trans. Autom. Control **54**(1), 48–61 (2009)
163. S. Nemirovski, D.B. Yudin, *Problem Complexity and Method Efficiency in Optimization* (Wiley, New York, 1983)
164. Y.E. Nesterov, A method for solving the convex programming problem with convergence rate $O(1/k^2)$. Dokl. Akad. Nauk SSSR **27**, 372–376 (1983)
165. Y.E. Nesterov, *Introductory Lectures on Convex Optimization*, Applied Optimization, vol. 87 (Kluwer Academic Publishers, Boston, 2004)
166. Y. Nesterov, Efficiency of coordinate descent methods on huge-scale optimization problems. SIAM J. Optim. **22**(2), 341–362 (2012)
167. Y.E. Nesterov, Gradient methods for minimizing composite functions. Math. Program. **140**, 125–161 (2013)
168. L.M. Nguyen, J. Liu, K. Scheinberg, M. Takác, SARAH: A novel method for machine learning problems using stochastic recursive gradient, in *Proceedings of the 34th International Conference on Machine Learning*, PMLR, vol. 70 (2017), pp. 2613–2621
169. J. Nocedal, S.J. Wright, *Numerical Optimization*, Springer Series in Operations Research (Springer, New York, 1999)
170. P. Ochs, Y. Chen, T. Brox, T. Pock, iPiano: inertial proximal algorithm for non-convex optimization. SIAM J. Imaging Sci. **7**, 1388–1419 (2014)
171. J.M. Ortega, W.C. Rheinboldt, *Iterative Solution of Nonlinear Equations in Several Variables* (SIAM, Philadelphia, 2000)

172. S. Osher, M. Burger, D. Goldfarb, J. Xu, W. Yin, An iterative regularization method for total variation-based image restoration. Multiscale Model. Simul. **4**(2), 460–489 (2005)
173. Y. Ouyang, Y. Chen, G. Lan, E. Pasiliao, An accelerated linearized alternating direction method of multipliers. SIAM J. Imaging Sci **8**(1), 644–681 (2015)
174. J.S. Pang, L. Qi, A globally convergent Newton method of convex SC^1 minimization problems. J. Optim. Theory Appl. **85**, 633–648 (1995)
175. N. Parikh, S. Boyd, Proximal algorithms. Found. Trends Optim. **1**(3), 127–239 (2014)
176. G.B. Passty, Ergodic convergence to a zero of the sum of monotone operators in Hilbert space. J. Math. Anal. Appl. **72**, 383–390 (1979)
177. P. Patrinos, L. Stella, A. Bemporad, Douglas–Rachford splitting: complexity estimates and accelerated variants, in *Proceedings of the 2014 IEEE 53rd Annual Conference on Decision and Control (CDC)* (2014), pp. 4234–4239
178. E.J.R. Pauwels, A. Beck, Y.C. Eldar, S. Sabach, On Fineup methods for sparse phase retrieval. IEEE Trans. Signal Processing **66**(4), 982–991 (2018)
179. D.W. Peaceman, H.H. Rachford, The numerical solution of parabolic and elliptic differential equations. J. Soc. Ind. Appl. Math. **3**, 28–41 (1955)
180. A. Pillay, C. Steinhorn, Definable sets in ordered structures. I. Trans. Am. Math. Soc. **295**(2), 565–592 (1986)
181. B.T. Polyak, Gradient methods for minimizing functionals (in Russian). Zh. Vychisl. Mat. Mat. Fiz. **3**, 643–653 (1963)
182. B.T. Polyak, *Introduction to optimization*, Translations Series in Mathematics and Engineering (Optimization Software, New York, 1987)
183. M.J.D. Powell, A method for nonlinear constraints in minimization problems, in *Optimization* ed. by R. Fletcher (Academic Press, New York, 1969), pp. 283–298
184. L. Qi, J. Sun, A nonsmooth version of Newton's method. Math. Program. Ser. A **58**(3), 353–367 (1993)
185. Y. Rahimi, C. Wang, H. Dong, Y. Lou, A scale invariant approach for sparse signal recovery. SIAM J. Sci. Comput. **41**, A3649–A3672 (2019)
186. S.S. Ram, A. Nedic, V.V. Veeravalli, Distributed stochastic subgradient projection algorithms for convex optimization. J. Optim. Theory Appl. **147**(3), 516–545 (2010)
187. P. Richtarik, M. Takacv, Iteration complexity of randomized block-coordinate descent methods for minimizing a composite function. Math. Program. **144**, 1–38 (2014)
188. H. Robbins, S. Monro, A stochastic approximation method. Ann. Math. Stat. **22**, 400–407 (1951)
189. H. Robbins, D. Siegmund, *A Convergence Theorem for Nonnegative Almost Supermartingales and Some Applications, in Optimizing Methods in Statistics* (Elsevier, New York, 1971), pp. 233–257
190. S.M. Robinson, Some continuity properties of polyhedral multifunctions. Math. Program. Study **14**, 206–214 (1981)
191. R.T. Rockafellar, *Convex Analysis* (Princeton University Press, Princeton, 1970)
192. R.T. Rockafellar, Monotone operators and the proximal point algorithm. SIAM J. Control Optim. **14**, 877–898 (1976)
193. R.T. Rockafellar, Augmented Lagrangians and applications of the proximal point algorithm in convex programming. Math. Oper. Res. **1**, 97–116 (1976)
194. R.T. Rockafellar, R.J.-B. Wets, *Variational Analysis* (Springer, Berlin, 2009)
195. L. Rudin, S. Osher, C. Fatemi, Nonlinear total variation based noise removal algorithm. Phys. D **60**, 259–268 (1992)
196. S. Salzo, The variable metric forward-backward splitting algorithm under mild differentiability assumptions. SIAM J. Optim. **27**, 2153–2181 (2017)
197. W. Schirotzek, *Nonsmooth Analysis*. Universitext (Springer, Berlin, 2007)
198. M. Schmidt, N. Le Roux, F. Bach, Minimizing finite sums with the stochastic average gradient. Math. Program. Ser. A **162**, 83–112 (2017)
199. F. Schöpfer, Linear convergence of descent methods for the unconstrained minimization of restricted strongly convex functions. SIAM J. Optim. **26**(3), 1883–1911 (2016)

200. F. Schöpfer, D.A. Lorenz, Linear convergence of the randomized sparse Kaczmarz method. Math. Program. **173**, 509–536 (2019)
201. A. Seidenberg, A new decision method for elementary algebra. Ann. Math. **60**(2), 365–374 (1954)
202. I. Selesnick, *Total Variation Denoising (an MM algorithm)*, NYU Polytechnic School of Engineering Lecture Notes (2012)
203. S. Shalev-Shwartz, Y. Singer, Efficient learning of label ranking by soft projections onto polyhedra. J. Mach. Learn. Res. **7**, 1567–1599 (2006)
204. A. Shapiro, On concepts of directional differentiability. J. Optim. Theory Appl. **66**, 477–487 (1990)
205. M. Shiota, *Geometry of Subanalytic and Semialgebraic Sets*. Progr. Math., vol. 150 (Birkhäuser, Boston, 1997)
206. T. Strohmer, R. Vershynin, A randomized Kaczmarz algorithm with exponential convergence. J. Fourier Anal. Appl. **15**(1), 262–278 (2009)
207. H. Sun, An investigation on semismooth Newton based augmented Lagrangian method for image restoration. J. Sci. Comput. **92**(3), Paper No. 82 (2022)
208. A. Tarski, *A Decision Method for Elementary Algebra and Geometry* (University of California Press, Berkeley1951)
209. F. Terkelsen, Some minimax theorems. Math. Scand. **31**, 405–413 (1972)
210. A. Themelis, P. Patrinos, Douglas-Rachford splitting and ADMM for nonconvex optimization: tight convergence results. SIAM J. Optim. **30**(1), 149–181 (2020)
211. R. Tibshirani, Regression shrinkage and selection via the lasso. J. R. Stat. Soc. Ser. B **58**(1), 267–288 (1996)
212. L.N. Trefethen, D. Bau III, *Numerical Linear Algebra* (SIAM, Philadelphia, 1997)
213. P. Tseng, Convergence of a block coordinate descent method for nondifferentiable minimization. J. Optim. Theory Appl. **109**(3), 475–494 (2001)
214. P. Tseng, S. Yun, A coordinate gradient descent method for nonsmooth separable minimization. Math. Program. **117**, 387–423 (2000)
215. L. van den Dries, Remarks on Tarski's problem concerning ($\mathbb{R}$, +, ·, exp), in *Studies in Logic and the Foundations of Mathematics*, vol. 112 (1984), pp. 97–121
216. L. van den Dries, *Tame Topology and o-minimal Structures*, London Mathematical Society, vol. 248, Lecture Note Series (Cambridge University Press, Cambridge, UK, 1998)
217. L. van den Dries, C. Miller, Geometric categories and o-minimal structures. Duke Math. J. **84**, 497–540 (1996)
218. Y. Vardi, L.A. Shepp, L. Kaufman, A statistical model for positron emission tomography. J. Amer. Stat. Assoc. **80**, 8–37 (1985)
219. B.C. Vũ, A splitting algorithm for dual monotone inclusions involving cocoercive operators. Adv. Comput. Math. **38**(3), 667–681 (2013)
220. C.T.C. Wall, *Singular Points of Plane Curves*, London Mathematical Society Student Texts, vol. 63 (Cambridge University Press, Cambridge, 2004)
221. W. Wang, M.Á. Carreira-Perpinán, Projection onto the probability simplex: an efficient algorithm with a simple proof, and an application, arXiv:1309.1541 (2013)
222. C. Wang, M. Yan, Y. Lou, Accelerated schemes for the L1/L2 minimization. IEEE Trans. Signal Process. **68**, 2660–2669 (2020)
223. Y.W. Wen, W.K. Ching, M. Ng, A semi-smooth Newton method for inverse problem with uniform noise. J. Sci. Comput. **75**(2), 713–732 (2018)
224. J.H. Wilkinson, *Rounding Errors in Algebraic Processes* (Prentice Hall, Englewood Cliffs, 1963)
225. L. Xiao, T. Zhang, A proximal stochastic gradient method with progressive variance reduction. SIAM J. Optim. **24**(4), 2057–2075 (2014)
226. W.H. Yang, D.R. Han, Linear convergence of alternating direction method of multipliers for a class of convex optimization problems. SIAM J. Numer. Anal. **54**(2), 625–640 (2016)
227. J. Yang, Y. Zhang, Alternating direction algorithms for l_1-problems in compressive sensing. SIAM J. Sci. Comput. **33**, 250–278 (2011)

228. J. Ye, X. Yuan, S. Zeng, J. Zhang, Variational analysis perspective on linear convergence of some first order methods for nonsmooth convex optimization problems. Set-Valued Var. Anal. **29**(4), 803–837 (2021)
229. P. Yin, E. Esser, J. Xin, Ratio and difference of ℓ_1 and ℓ_2 norms and sparse representation with coherent dictionaries. Commun. Inf. Syst. **14**, 87–109 (2014)
230. P. Yu, G. Li, T.K. Pong, Kurdyka-Łojasiewicz Exponent via Inf-projection. Found. Comput. Math. **22**, 1171–1217 (2022)
231. C. Zălinscu, *Convex Analysis in General Vector Spaces* (World Scientific Publishing, River Edge, 2002)
232. H. Zhang, L.Z. Cheng, Restricted strong convexity and its applications to convergence analysis of gradient-type methods. Optim. Lett. **9**(5), 961–979 (2015)
233. X. Zhang, M. Burger, S. Osher, A unified primal-dual algorithm framework based on Bregman iteration. J. Sci. Comput. **46**, 20–46 (2011)
234. L. Zhang, M. Mahdavi, R. Jin, Linear convergence with condition number independent access of full gradients, in *Advances in Neural Information Processing Systems* (2013), pp. 980–988
235. Y. Zhang, N. Zhang, D. Sun, K. Toh, An efficient Hessian based algorithm for solving large-scale sparse group Lasso problems. Math. Program. **179**, 223–263 (2020)
236. X.Y. Zhao, D.F. Sun, K.C. Toh, A Newton-CG augmented Lagrangian method for semidefinite programming. SIAM J. Optim. **20**, 1737–1765 (2010)
237. X.Y. Zheng, K.F. Ng, Metric subregularity of piecewise linear multifunctions and applications to piecewise linear multiobjective optimization. SIAM J. Optim. **24**, 154–174 (2014)
238. M. Zhu, T. F. Chan, An efficient primal-dual hybrid gradient algorithm for Total Variation image restoration, CAM Report 08-34, UCLA (2008)

Index

Q. Jin, *Lectures on Nonsmooth Optimization*, Texts in Applied Mathematics 82,
https://doi.org/10.1007/978-3-031-91417-1

www.ingramcontent.com/pod-product-compliance
Lightning Source LLC
Chambersburg PA
CBHW061012050826
49219CB00022B/107
* 9 7 8 3 0 3 1 9 1 4 1 6 4 *